Historical Geology
FIFTH EDITION

Evolution of Earth and Life Through Time

REED WICANDER
Central Michigan University

JAMES S. MONROE
Emeritus, Central Michigan University

THOMSON
BROOKS/COLE

Australia • Brazil • Canada • Mexico • Singapore • Spain • United Kingdom • United States

Historical Geology: Evolution of Earth and Life Through Time, **Fifth Edition**
Reed Wicander, James S. Monroe

Earth Sciences Editor: Peter Adams
Assistant Editors: Carol Benedict, Alex Brady
Editorial Assistant: Anna Jarzab
Technology Project Manager: Melinda Newfarmer
Marketing Manager: Joe Rogove
Marketing Assistant: Jennifer Liang
Marketing Communications Manager: Bryan Vann
Project Manager, Editorial Production: Hal Humphrey
Art Director: Vernon Boes
Print Buyer: Rebecca Cross
Permissions Editor: Bob Kauser

Production Service: Pre-Press Company, Inc.
Text Designer: Lisa Buckley
Photo Researcher: Kathleen Olson
Illustrator: Accurate Art, Precision Graphics, Rolin Graphics, Pre-Press Company, Inc.
Cover Designer: DiAnna VanEycke
Cover Image: *Archaeopteryx lithographica,* Michael W. Skrepnick, copyright 2001
Cover Printer: Courier Corporation/Kendallville
Copyeditor and Compositor: Pre-Press Company, Inc.
Printer: Courier Corporation/Kendallville

© 2007, 2004 Thomson Brooks/Cole, a part of The Thomson Corporation. Thomson, the Star logo, and Brooks/Cole are trademarks used herein under license.

ALL RIGHTS RESERVED. No part of this work covered by the copyright hereon may be reproduced or used in any form or by any means—graphic, electronic, or mechanical, including photocopying, recording, taping, web distribution, information storage and retrieval systems, or in any other manner—without the written permission of the publisher.

Printed in the United States of America
1 2 3 4 5 6 7 10 09 08 07 06

For more information about our products, contact us at:
Thomson Learning Academic Resource Center
1–800–423–0563

For permission to use material from this text or product, submit a request online at **http://www.thomsonrights.com**.
Any additional questions about permissions can be submitted by e-mail to **thomsonrights@thomson.com**.

ExamView® and *ExamView Pro*® are registered trademarks of FSCreations, Inc. Windows is a registered trademark of the Microsoft Corporation used herein under license. Macintosh and Power Macintosh are registered trademarks of Apple Computer, Inc. Used herein under license.

© 2007 Thomson Learning, Inc. All Rights Reserved. Thomson Learning WebTutor™ is a trademark of Thomson Learning, Inc.

Thomson Higher Education
10 Davis Drive
Belmont, CA 94002–3098
USA

Library of Congress Control Number: 2006933459

Student Edition: ISBN 0–495–01204–1

About the Authors

Reed Wicander

REED WICANDER is a geology professor at Central Michigan University where he teaches physical geology, historical geology, prehistoric life, and invertebrate paleontology. He has co-authored several geology textbooks with James S. Monroe. His main research interests involve various aspects of Paleozoic palynology, specifically the study of acritarchs, on which he has published many papers. He is a past president of the American Association of Stratigraphic Palynologists and a former councillor of the International Federation of Palynological Societies. He is the current chairman of the Acritarch Subcommission of the Commission Internationale de Microflore du Paléozoique.

James S. Monroe

JAMES S. MONROE is professor emeritus of geology at Central Michigan University where he taught physical geology, historical geology, prehistoric life, and stratigraphy and sedimentology since 1975. He has co-authored several textbooks with Reed Wicander and has interests in Cenozoic geology and geologic education.

Brief Contents

Chapter 1	The Dynamic and Evolving Earth	1
Chapter 2	Earth Materials—Minerals and Rocks	16
Chapter 3	Plate Tectonics: A Unifying Theory	35
Chapter 4	Geologic Time: Concepts and Principles	60
Chapter 5	Rocks, Fossils, and Time—Making Sense of the Geologic Record	77
Chapter 6	Sedimentary Rocks—The Archives of Earth History	101
Chapter 7	Evolution—The Theory and Its Supporting Evidence	122
Chapter 8	Precambrian Earth and Life History—The Archean Eon	147
Chapter 9	Precambrian Earth and Life History—The Proterozoic Eon	166
Chapter 10	Early Paleozoic Earth History	189
Chapter 11	Late Paleozoic Earth History	209
Chapter 12	Paleozoic Life History: Invertebrates	231
Chapter 13	Paleozoic Life History: Vertebrates and Plants	249
Chapter 14	Mesozoic Earth History	272
Chapter 15	Life of the Mesozoic Era	295
Chapter 16	Cenozoic Geologic History: The Paleogene and Neogene	321
Chapter 17	Cenozoic Geologic History: The Pleistocene and Holocene Epochs	347
Chapter 18	Life of the Cenozoic Era	371
Chapter 19	Primate and Human Evolution	397
	Epilogue	412
Appendix A	Metric Conversion Chart	414
Appendix B	Classification of Organisms	415
Appendix C	Mineral Identification	420
	Glossary	423
	Answers to Multiple-Choice Review Questions	430
	Index	431

Contents

Chapter 1
The Dynamic and Evolving Earth 1
Introduction 2
What Is Geology? 4
Perspective 1.1 Interpreting Earth History 5
Historical Geology and the Formulation
  of Theories 5
Origin of the Universe and Solar System,
  and Earth's Place in Them 6
  Origin of the Universe—Did It Begin with a Big Bang? 6
  Our Solar System—Its Origin and Evolution 6
  Earth—Its Place in Our Solar System 8
  Forming the Earth—Moon System 8
Why Is Earth a Dynamic and Evolving Planet? 9
Organic Evolution and the History of Life 11
Geologic Time and Uniformitarianism 11
How Does the Study
  of Historical Geology Benefit Us? 13
Summary 13

Chapter 2
Earth Materials—
Minerals and Rocks 16
Introduction 17
Matter and Its Composition 18
  Elements and Atoms 18
  Bonding and Compounds 18
Minerals—The Building Blocks of Rocks 20
How Many Minerals Are There? 21
  Silicate Minerals 21
  Other Mineral Groups 21
Rock-Forming Minerals and the Rock Cycle 22
Igneous Rocks 23
  Texture and Composition of Igneous Rocks 23
  Classifying Igneous Rocks 23
Sedimentary Rocks 26
  Sediment Transport, Deposition, and Lithification 26
  Types of Sedimentary Rocks 26
Metamorphic Rocks 29
  The Agents of Metamorphism 29
  Types of Metamorphism 30
  Classifying Metamorphic Rock 30
Plate Tectonics and the Rock Cycle 31
Earth Materials and Historical Geology 33
Summary 33

Chapter 3
Plate Tectonics: A Unifying Theory 35
Introduction 36
Early Ideas About Continental Drift 36
Perspective 3.1 Oil, Plate Tectonics, and Politics 37
  Alfred Wegener and the Continental Drift Hypothesis 38
  Additional Support for Continental Drift 38
Paleomagnetism and Polar Wandering 41
How Do Magnetic Reversals Relate to Seafloor
  Spreading? 42
Plate Tectonics and Plate Boundaries 45
  Divergent Boundaries 46
  An Example of Ancient Rifting 47
  Convergent Boundaries 48
  Recognizing Ancient Convergent Boundaries 51
  Transform Boundaries 51
What Are Hot Spots and Mantle Plumes? 52
How Are Plate Movement and Motion Determined? 52
What Is The Driving Mechanism
  of Plate Tectonics? 53
How Are Plate Tectonics
  and Mountain Building Related? 55
  Terrane Tectonics 55
How Does Plate Tectonics Affect
  the Distribution of Life? 55
How Does Plate Tectonics Affect
  the Distribution of Natural Resources? 57
Summary 58

Chapter 4
Geologic Time: Concepts and Principles 60

Introduction 61
How Is Geologic Time Measured? 61
How Has the Concept of Geologic Time and Earth's Age Changed Throughout Human History? 62

Perspective 4.1 Geologic Time and Climate Change 63

What Are Relative Dating Methods, and Why Are They Important? 64
   Fundamental Principles of Relative Dating 64
Establishment of Geology as a Science—The Triumph of Uniformitarianism over Neptunism and Catastrophism 66
   Neptunism and Catastrophism 66
   Uniformitarianism 66
   Modern View of Uniformitarianism 67
Lord Kelvin and a Crisis in Geology 68
What Are Absolute Dating Methods, and Why Are They Important? 68
   Atoms and Isotopes 68
   Radioactive Decay and Half-Lives 69
   Long-Lived Radioactive Isotope Pairs 72
   Fission Track Dating 72
   Radiocarbon and Tree Ring Dating 73
Summary 74

Chapter 5
Rocks, Fossils, and Time—Making Sense of the Geologic Record 77

Introduction 78
Stratigraphy 79
   Vertical Stratigraphic Relationships 79
   Lateral Relationships—Facies 81
   Marine Transgressions and Regressions 81
   Extent, Rates, and Causes of Marine Transgressions and Regressions 84
Fossilization and Fossils 85
   How Do Fossils Form? 86

Perspective 5.1 Fossils and Uniformitarianism 87

   Fossils and Telling Time 88
The Relative Geologic Time Scale 91
Stratigraphic Terminology 93
Correlation 94
Absolute Dates and the Relative Geologic Time Scale Summary 97
Summary 98

Chapter 6
Sedimentary Rocks—The Archives of Earth History 101

Introduction 102
Sedimentary Rock Properties 102
   Composition and Texture 103
   Sedimentary Structures 104

Perspective 6.1 Determining the Relative Ages of Deformed Sedimentary Rocks 107

   Geometry of Sedimentary Rocks 108
   Fossils—The Biologic Content of Sedimentary Rocks 109
Depositional Environments 109
   Continental Environments 109
   Transitional Environments 111
   Marine Environments 114
Interpreting Depositional Environments 117
Paleogeography 119
Summary 120

Chapter 7
Evolution—The Theory and Its Supporting Evidence 122

Introduction 123
Evolution: What Does It Mean? 124
   Jean-Baptiste de Lamarck and His Ideas on Evolution 125
   The Contributions of Charles Darwin and Alfred Wallace 126

Perspective 7.1 The Tragic Lysenko Affair 127

   Natural Selection—What Is Its Significance? 128
Mendel and the Birth of Genetics 128
   Mendel's Experiments 128
   Genes and Chromosomes 129
The Modern View of Evolution 130
   What Brings About Variation? 130
   Speciation and the Rate of Evolution 131
   Divergent, Convergent, and Parallel Evolution 133

Microevolution and Macroevolution 134
Cladistics and Cladograms 135
Evolutionary Trends and Mosaic Evolution 135
Extinctions 137

What Kinds of Evidence Support Evolutionary Theory? 137
Classification—A Nested Pattern of Similarities 138
How Does Biological Evidence Support Evolution? 140
Fossils: What Do We Learn from Them? 141

Perspective 7.2 The Fossil Record and Missing Links 143

The Evidence—A Summary 144

Summary 144

Chapter 8
Precambrian Earth and Life History—The Archean Eon 147

Introduction 148

What Happened During the Eoarchean? 149

Continental Foundations—Shields, Platforms, and Craton 150
Archean Rocks 152
Greenstone Belts 152

Perspective 8.1 Geology of Grand Teton National Park 153
Evolution of Greenstone Belts 154

Archean Plate Tecontics and the Origin of Cratons 156

The Atmosphere and Hydrosphere 157
How Did the Atmosphere Form and Evolve? 157
Earth's Surface Waters—The Hydrosphere 158

The Origin of Life 159
Experimental Evidence and the Origin of Life 159
Submarine Hydrothermal Vents and the Origin of Life 161
The Oldest Known Organisms 162

Archean Mineral Resources 163

Summary 164

Chapter 9
Precambrian Earth and Life History—The Proterozoic Eon 166

Introduction 167

Evolution of Proterozoic Continents 168
Paleoproterozoic History of Laurentia 168
Paleo- and Mesoproterozoic Igneous Activity 170
Mesoproterozoic Orogeny and Rifting 171
Meso- and Neoproterozoic Sedimentation 172

Proterozoic Supercontinents 172

Ancient Glaciers and Their Deposits 175
Paleoproterozoic Glaciers 175
Glaciers of the Neoproterozoic 176

The Evolving Atmosphere 176
Banded Iron Formations (BIFs) 176
Continental Red Beds 177

Important Events in Life History 178
Eukaryotic Cells Evolve 178
Endosymbiosis and the Origin of Eukaryotic Cells 179
The Dawn of Multicelled Organisms 180
Neoproterozoic Animals 182

Proterozoic Mineral Resources 184

Perspective 9.1 Bif: From Mine to Steel Mill 185

Summary 186

Chapter 10
Early Paleozoic Earth History 189

Introduction 190

Continental Architecture: Cratons and Mobile Belts 190

Paleozoic Paleogeography 191
Early Paleozoic Global History 193

Early Paleozoic Evolution of North America 194

The Sauk Sequence 195

Perspective 10.1 Pictured Rocks National Lakeshore 196
The Cambrian of the Grand Canyon Region: A Transgressive Facies Model 197

The Tippecanoe Sequence 198
Tippecanoe Reefs and Evaporites 199
The End of The Tippecanoe Sequence 203

The Appalachian Mobile Belt and the Taconic Orogeny 204

Early Paleozoic Mineral Resources 206

Summary 206

Chapter 11
Late Paleozoic Earth History 209

Introduction 210
Late Paleozoic Paleogeography 210
   The Devonian Period 211
   The Carboniferous Period 211
   The Permian Period 211
Late Paleozoic Evolution of North America 214
The Kaskaskia Sequence 214
   Reef Development in Western Canada 214

Perspective 11.1 The Canning Basin, Australia—A Devonian Great Barrier Reef 216
   Black Shales 216
   The Late Kaskaskia—A Return to Extensive Carbonate Deposition 217
The Absaroka Sequence 218
   What Are Cyclothems, and Why Are They Important? 218
   Cratonic Uplift—The Ancestral Rockies 221
The Middle Absaroka—More Evaporite Deposits and Reefs 222
History of the Late Paleozoic Mobile Belts 224
   Cordilleran Mobile Belt 224
   Ouachita Mobile Belt 224
   Appalachian Mobile Belt 224
What Role Did Microplates and Terranes Play in the Formation of Pangaea? 226
Late Paleozoic Mineral Resources 227
Summary 228

Chapter 12
Paleozoic Life History: Invertebrates 231

Introduction 232
What Was the Cambrian Explosion? 232
The Emergence of a Shelly Fauna 233
Paleozoic Invertebrate Marine Life 234
   The Present Marine Ecosystem 235
   Cambrian Marine Community 237
   The Burgess Shale Biota 238
   Ordovician Marine Community 239
   Silurian and Devonian Marine Communities 241

Perspective 12.1 Mass Extinctions and Their Possible Causes 242
   Carboniferous and Permian Marine Communities 243
   The Permian Mass Extinction 244
Summary 245

Chapter 13
Paleozoic Life History: Vertebrates and Plants 249

Introduction 250
Vertebrate Evolution 251
Fish 251
Amphibians—Vertebrates Invade the Land 257
Evolution of the Reptiles—The Land Is Conquered 258
Plant Evolution 260

Perspective 13.1 Palynology: A Link between Geology and Biology 262
   Silurian and Devonian Floras 263
   Late Carboniferous and Permian Floras 266
Summary 268

Chapter 14
Mesozoic Earth History 272

Introduction 273
The Breakup of Pangaea 273
   The Effects of the Breakup of Pangaea on Global Climates and Ocean Circulation Patterns 276
Mesozoic History of North America 277
Continental Interior 278
Eastern Coastal Region 279
Gulf Coastal Region 280
Western Region 282
   Mesozoic Tectonics 282
   Mesozoic Sedimentation 284

Perspective 14.1 Petrified Forest National Park 287

What Role Did Accretion of Terranes Play in the Growth of Western North America? 290
Mesozoic Mineral Resources 291
Summary 292

Chapter 15
Life of the Mesozoic Era 295

Introduction 296
Marine Invertebrates and Phytoplankton 297
Aquatic and Semiaquatic Vertebrates—Fish and Amphibians 299
Plants—Primary Producers on Land 300
The Diversification of Reptiles 300
    Archosaurs and the Origin of Dinosaurs 301
    Dinosaurs 301
    Warm-Blooded Dinosaurs? 305
    Flying Reptiles 307
    Mesozoic Marine Reptiles 307
    Crocodiles, Turtles, Lizards, and Snakes 308

Perspective 15.1 Mary Anning and Her Contributions To Paleontology 309

From Reptiles to Birds 309
Origin and Early Evolution of Mammals 310
    Cynodonts and the Origin of Mammals 310
    Mesozoic Mammals 312
Mesozoic Paleobiogeography 314
Mass Extinctions—A Crisis in the History of Life 316
Summary 317

Chapter 16
Cenozoic Geologic History: The Paleogene and Neogene 321

Introduction 322
Cenozoic Plate Tectonics—An Overview 323
Cenozoic Orogenic Belts 325
    The Alpine–Himalayan Orogenic Belt 326
    The Circum-Pacific Orogenic Belt 328
The North American Cordillera 328
    The Laramide Orogeny 329
    Cordillera Igneous Activity 332
    Basin and Range Province 333
    Colorado Plateau 334
    The Rio Grande Rift 335
    Pacific Coast 335
The Continental Interior 337

Perspective 16.1 Shiprock, New Mexico 338

Cenozoic History of The Appalachian Mountains 339
North America's Southern and Eastern Continental Margins 339
    The Gulf Coastal Plain 340
    The Atlantic Continental Margin 341
Paleogene and Neogene Mineral Resources 343
Summary 344

Chapter 17
Cenozoic Geologic History: The Pleistocene and Holocene Epochs 347

Introduction 348
Pleistocene and Holocene Tectonism and Volcanism 349

Perspective 17.1 Supervolcanoes and the Origin of the Yellowstone Caldera 350

Pleistocene Stratigraphy 352
    Terrestrial Stratigraphy 354
    Deep-Sea Stratigraphy 354
Onset of the Ice Age 355
    Climate of the Pleistocene 355
    Glaciers—What Are They and How Do They Form? 356
Glaciation and Its Effects 357
    Glacial Landforms 357
    Changes in Sea Level 358
    Glaciers and Isostasy 358
    Pluvial and Proglacial Lakes 359
What Caused Pleistocene Glaciation? 362
    The Milankovitch Theory 363
    Short-Term Climatic Changes 366
Glaciers Today 366
Pleistocene Mineral Resources 368
Summary 368

Chapter 18
Life of the Cenozoic Era   371

Introduction   372
Marine Invertebrates and Phytoplankton   373
Cenozoic Vegetation and Climate   373
Cenozoic Birds   377
The Age of Mammals Begins   377
Diversification of Placental Mammals   378
Paleogene and Neogene Mammals   381
    Small Mammals—Insectivores, Rodents, Rabbits, and Bats   381

Perspective 18.1 A Miocene Catastrophe in Nebraska   382
    A Brief History of the Primates   383
    The Meat Eaters—Carnivorous Mammals   383
    The Ungulates Or Hoofed Mammals   385
    Giant Land-Dwelling Mammals—Elephants   387
    Giant Aquatic Mammals—Whales   388
Pleistocene Faunas   390
    Ice Age Mammals   390
    Pleistocene Extinctions   391
Intercontinental Migrations   393
Summary   394

Chapter 19
Primate and Human Evolution   397

Introduction   398
What are Primates?   398
Prosimians   399
Anthropoids   400
Hominids   401
    Australopithecines   404

Perspective 19.1 Footprints At Laetoli   406
    The Human Lineage   407
    Neanderthals   407
    Cro-Magnons   408
Summary   410

Epilogue   412

Appendix A
Metric Conversion Chart   414

Appendix B
Classification of Organisms   415

Appendix C
Mineral Identification   420

Glossary   423

Answers to Multiple-Choice Review Questions   430

Index   431

Preface

Earth is a dynamic planet that has changed continuously during its 4.6 billion years of existence. The size, shape, and geographic distribution of the continents and ocean basins have changed through time, as have the atmosphere and biota. As scientists and concerned citizens, we have become increasingly aware of how fragile our planet is and, more importantly, how interdependent all of its various systems and subsystems are. Furthermore, we are coming to realize how central geology is to our everyday lives. For these and other reasons, geology is one of the most important college or university courses a student can take.

Historical geologists are concerned with all aspects of Earth and its life history. They seek to determine what events occurred during the past, place those events into an orderly chronological sequence, and provide conceptual frameworks for explaining such events. Equally important is using the lessons learned from the geologic past to understand and place in context some of the global issues facing the world today, such as depletion of natural resources, global warming, and decreasing biodiversity. Thus, what makes historical geology both fascinating and relevant is that, like the dynamic Earth it seeks to understand, it is an exciting and ever-changing science in which new discoveries are continually being made.

Historical Geology: Evolution of Earth and Life Through Time, fifth edition, is designed for a one-semester geology course and is written with students in mind. One of the problems with any science course is that students are overwhelmed by the amount of material that must be learned. Furthermore, most of the material does not seem to be linked by any unifying theme and does not always appear to be relevant to their lives.

The goal of this book is to provide students with an understanding of the principles of historical geology and how these principles are applied in unraveling Earth's history. It is our intent to present the geologic and biologic history of Earth, not as a set of encyclopedic facts to memorize, but rather as a continuum of interrelated events reflecting the underlying geologic and biologic principles and processes that have shaped our planet and life upon it. Instead of emphasizing individual, and seemingly unrelated events, we seek to understand the underlying causes of why things happened the way they did and how all of Earth's systems and subsystems are interrelated. Using this approach, students will gain a better understanding of how everything fits together, and why events occurred in a particular sequence.

Because of the nature of the science, all historical geology textbooks share some broad similarities. Most begin with several chapters on concepts and principles, followed by a chronological discussion of Earth history. In this respect we have not departed from convention. We have, however, attempted to place greater emphasis on basic concepts and principles, their historical development, and their importance in deciphering Earth history; in other words, how do we know what we know. By approaching Earth history in this manner, students come to understand Earth's history as part of a dynamic and complex integrated system, and not as a series of isolated and unrelated events.

New and Retained Features in the Fifth Edition

The fifth edition has undergone considerable rewriting and updating to produce a book that is easier to read with a high level of current information, many new photographs and figures, as well as several new perspectives, all of which are designed to help students maximize their learning and understanding of their planet's history. Drawing on the comments and suggestions of reviewers and users of the fourth edition, we have incorporated several new innovations into this edition, as well as keeping the features that were successful in the previous edition.

Some of the features retained include Chapter 2 (Earth Materials), which provides the necessary background for those students who are unfamiliar with minerals, rocks, and the rock cycle; incorporating the history of the universe, solar system, and the planets in the first chapter so as to give students a complete view of Earth's earliest development;

and covering plate tectonics in Chapter 3. This early discussion of plate tectonics provides students with a firm grounding in geology's unifying theory and how it affects Earth's geologic and biologic history.

We have also retained the *Chapter Objectives* outline at the beginning of each chapter to alert students to the key points that the chapter will address. In addition, there are several new *Introductions*. These start each chapter with a story and also address the question of why each chapter's material is relevant and important to the student's overall understanding of the topic.

The popular *What Would You Do?* boxes in each chapter continue to encourage students to think critically about what they're learning. These brief sections incorporate material from each chapter and ask open-ended questions of the student to get them to think critically about incorporating what they've learned into hypothetical or real life situations.

In addition to the standard multiple choice and short answer questions, we have added *Apply Your Knowledge* and *Field Questions* to the end-of-chapter material. The *Apply Your Knowledge* questions include both thought-provoking and quantitative questions that require the student to apply what he or she has learned to solving geologic problems. The *Field Questions* are photographs or drawings in which the student is asked to describe a geologic feature or apply the principles learned in the chapter to answering the question about the feature illustrated.

Just as in previous editions, the fifth edition has many new photos and figures to help students understand the principles, concepts, and material being presented. The highly acclaimed paleogeographic maps illustrating the geography during the various geological periods that were introduced in the last edition have been retained in this edition. Students will find two global views of Earth for each time period.

The extensive rewriting and updating done in the text, as well as the new photographs and illustrations have greatly improved the fifth edition by making it easier to read and comprehend, as well as engaging students in the learning process, and thereby fostering better understanding of the material.

Text Organization

As in the previous editions, we develop three major themes in this textbook that are essential to the interpretation and appreciation of historical geology, introduce these themes early, and reinforce them throughout the book. These themes are *plate tectonics* (Chapter 3), a unifying theory for interpreting much of Earth's physical history and, to a large extent, its biological history; *time* (Chapter 4), the dimension that sets historical geology apart from most of the other sciences; and *evolutionary theory* (Chapter 7), the explanation for inferred relationships among living and fossil organisms. Additionally, we have emphasized the intimate interrelationship existing between physical and biological events, and the fact that Earth is a complex, dynamic, and evolving planet whose history is best studied by using a systems approach.

This book was written for a one-semester course in historical geology to serve both majors and nonmajors in geology and in the Earth sciences. Many students taking a historical geology course will have had an introductory physical geology course. In this case, Chapter 1 can be used as a review of the principles and concepts of geology and as an introduction to the science of historical geology and the three themes this book emphasizes. The text is written at an appropriate level for those students taking historical geology with no prerequisites, but the instructor may have to spend more time expanding some of the concepts and terminology discussed in Chapter 1. Chapter 2 "Earth Materials — Minerals and Rocks," can be used to introduce those students who have not had an introductory geology course to minerals and rocks, or as a review for those students that have had such a course.

Chapter 3 explores plate tectonics, which is the first major theme of this book. Particular emphasis is placed on the evidence substantiating plate tectonic theory, why this theory is one of the cornerstones of geology, and why plate tectonic theory serves as a unifying paradigm in explaining many apparently unrelated geologic phenomena.

The second major theme of this book, the concepts and principles of geologic time, is examined in Chapter 4. Chapter 5 expands on that theme by integrating geologic time with rocks and fossils. Depositional environments are sometimes covered rather superficially (perhaps little more than a summary table) in some historical geology textbooks. Chapter 6, "Sedimentary Rocks – The Archives of Earth History," is completely devoted to this topic; it contains sufficient detail to be meaningful, but avoids an overly detailed discussion more appropriate for advanced courses.

The third major theme of this book, organic evolution, is examined in Chapter 7. In this chapter, the theory of evolution is covered as well as its supporting evidence. A short discussion on micro- and macroevolution has been added in this edition as well as a more detailed treatment and new figures regarding allopatric speciation.

Precambrian time—fully 88% of all geologic time— is sometimes considered in a single chapter in other historical geology textbooks. However, in this book, Chapters 8 and 9 are devoted to the geologic and biologic histories of the Archean and Proterozoic eons, respectively, with much updating on this early period of Earth history, as well as a discussion on the Precambrian time subdivisions.

Chapters 10 through 19 constitute our chronological treatment of the Phanerozoic geologic and biologic history of Earth. These chapters are arranged so that the geologic history of an era is followed by a discussion of the biologic history of that era. We think that this format allows easier integration of life history with geologic history.

In these chapters, there is an integration of the three themes of this book as well as an emphasis on the underlying principles of geology and how they helped decipher Earth's history. An Epilogue summarizes the major topics and themes of this book.

Chapter Organization

All chapters have the same organizational format. Each chapter opens with a photograph relating to the chapter material, a detailed *Outline* that engages students by having many of the headings as questions, and a *Chapter Objectives* outline that serves to alert the student to the learning outcome objectives of the chapter. The chapter text begins with an *Introduction* that is intended to stimulate interest in the chapter by discussing some aspect of the chapter material and showing students how that material fits into the larger geologic context.

The text is written in a clear informal style, making it easy for students to comprehend. Numerous color diagrams and photographs complement the text, providing a visual representation of the concepts and information presented.

Most chapters contain at least one *Perspective* that presents a brief discussion of an interesting aspect of historical geology or geological research pertinent to that chapter. Many of these have been revised or replaced from the fourth edition. *What Would You Do?* boxes, usually two per chapter, are designed to encourage critical thinking by students as they attempt to solve hypothetical problems or issues relating to the chapter material. Each of the chapters on geologic history in the second half of this book contains a final section on mineral resources characteristic of that time period. These sections provide applied economic material of interest to students.

The end-of-chapter materials begin with a concise review of important concepts and ideas in the *Summary*. The *Important Terms*, which are printed in boldface type in the chapter text, are listed at the end of each chapter for easy review, along with the page number where that term is first defined. A full glossary of important terms appears at the end of the text. The *Review Questions* are another important feature of this book. They include multiple-choice questions with answers as well as short essay questions. Many new multiple-choice questions as well as short answer and essay questions have been added in each chapter for this edition.

Two new features to the end-of-chapter material are the thought-provoking and quantitative questions found in the *Apply Your Knowledge* section, and interpretations of actual outcrops in the *Field Questions* section. The photographs or drawings in the *Field Questions* section ask the student to describe the geologic feature illustrated or answer questions relating to the photograph or drawing.

The global paleogeographic maps that illustrate in stunning relief the geography of the world during various time periods have been retained in this edition. These maps enable students to visualize what the world looked like during the time period being studied and add a visualization dimension to the text material.

As in the previous editions, end-of-chapter summary tables are provided for the chapters on geologic and biologic history. These tables are designed to give an overall perspective of the geologic and biologic events that occurred during a particular time interval and to show how these events are interrelated. The emphasis in these tables is on the geologic evolution of North America. Global tectonic events and sea-level changes are also incorporated into these tables to provide global insights.

Ancillary Materials

We are pleased to offer a full suite of text and multimedia products to accompany the fifth edition of *Historical Geology: Evolution of Earth and Life Through Time*.

For Instructors

Online Instructor's Manual with Test Bank This comprehensive manual is designed to help instructors prepare for lectures. It contains lecture outlines, teaching suggestions and tips, and an expanded list of references, plus a bank of test questions. (ISBN: 0–495–01208–4)

ExamView® Computerized Testing Create, deliver, and customize tests and study guides (both print and online) in minutes with this easy-to-use assessment and tutorial system. ExamView offers both a *Quick Test Wizard* and an *Online Test Wizard* that guide you step-by-step through the process of creating tests, while its unique capabilities allow you to see the test you are creating on the screen exactly as it will print or display online. You can build tests of up to 250 questions using up to 12 question types. Using *ExamView's* complete word processing capabilities, you can enter an unlimited number of new questions or edit existing questions. (ISBN: 0-495-01210-6)

Multimedia Manager with Living Lecture™ Tools This easy-to-use multimedia lecture tool allows you to quickly assemble art and database files with notes to create fluid lectures. Available on CD, the *Multimedia Manager* includes a complete set of prepared lectures using Microsoft® PowerPoint® plus database of animations from Brooks/Cole earth science titles and nearly all the images from *Historical*

Geology, fifth edition. This simple interface makes it easy for you to incorporate graphics, digital video, animations, and audio clips into your lectures. (ISBN: 0–495–01211–4)

Transparency Acetates A collection of 133 acetates featuring important line drawings, maps, charts, and graphs from the text. (ISBN: 0–495–01212–2)

For Students

ThomsonNow A web-based learning resource seamlessly tied to the text through interactive and animated active figures.

Study Guide An essential tool for any student, the *Study Guide* enhances students' understanding of historical geology. The *Study Guide* provides a broad overview of each chapter's content and helps students identify the most important facts, concepts, and underlying explanations with learning objectives, important terms, interesting facts and learning activities, links to authoritative websites, and study questions and answers for each chapter. (ISBN: 0–495–01209–2)

Acknowledgments

As authors, we are of course responsible for the organization, style, and accuracy of the text, and any mistakes, omissions, or errors are our responsibility. The finished product is the culmination of many years of work during which we received numerous comments and advice from many geologists who reviewed parts of the text. We wish to express our sincere appreciation to the reviewers who reviewed the fourth edition and made many helpful and useful comments that led to the improvements seen in this fifth edition.

Rex E. Crick, *University of Texas at Arlington*
Chris Dewey, *Mississippi State University*
Joseph C. Gould, *University of South Florida*
Glen K. Merrill, *University of Houston-Downtown*
N. S. Parate, *HCC-NE College, Pinemont Center*
Scott Ritter, *Brigham Young University*
Gary D. Rosenberg, *Indiana University-Purdue University, Indianapolis*
Matthew S. Tomaso, *Montclair State University*

We would also like to thank the reviewers of the third edition. Their comments and suggestions resulted in many improvements and pedagogical innovations in the fourth edition.

Paul Belasky, *Ohlone College*
Mark J. Camp, *University of Toledo*
Dean A. Dunn, *The University of Southern Mississippi*
William W. Korth, *Buffalo State College*
George F. Maxey, *University of North Texas*
William A. Smith, *Charleston Southern University*
Jane L. Teranes, *Scripps Institution of Oceanography*
Robert A. Vargo, *California University of Pennsylvania*

We also thank the reviewers of the first and second editions whose comments and suggestions helped make the book a success for students and instructors alike.

Thomas W. Broadhead, *University of Tennessee at Knoxville*
Mark J. Camp, *University of Toledo*
James F. Coble, *Tidewater Community College*
William C. Cornell, *University of Texas at El Paso*
Rex E. Crick, *University of Texas at Arlington*
Richard Fluegeman, Jr., *Ball State University*
Annabelle Foos, *University of Akron*
Susan Goldstein, *University of Georgia*
Bryan Gregor, *Wright State University*
Thor A. Hansen, *Western Washington University*
Paul D. Howell, *University of Kentucky*
Jonathan D. Karr, *North Carolina State University*
R. L. Langenheim, Jr., *University of Illinois at Urbana-Champaign*
Michael McKinney, *University of Tennessee at Knoxville*
Arthur Mirsky, *Indiana University, Purdue University at Indianapolis*
William C. Parker, *Florida State University*
Anne Raymond, *Texas A & M University*
G. J. Retallack, *University of Oregon*
Mark Rich, *University of Georgia*
Barbara L. Ruff, *University of Georgia*
W. Bruce Saunder, *Bryn Mawr College*
Ronald D. Stieglitz, *University of Wisconsin, Green Bay*
Carol M. Tang, *Arizona State University*
Michael J. Tevesz, *Cleveland State University*
Art Troell, *San Antonio College*

We also wish to thank Kathy Benison, Richard V. Dietrich (Professor Emeritus), David J. Matty, Jane M. Matty, Wayne E. Moore (Professor Emeritus), and Sven Morgan of the Geology Department, and Bruce M. C. Pape (Emeritus) of the Geography Department of Central Michigan University, as well as Eric Johnson (Hartwick College, New York) and Stephen D. Stahl (St. Bonaventure, New York) for providing us with photographs and answering our questions concerning various topics. We are also grateful for the generosity of the various agencies and individuals from many countries who provided photographs.

Special thanks must go to Keith Dodson, former earth sciences editor at Thomson Brooks/Cole, who initiated this fifth edition and encouraged us throughout its revision. We are equally indebted to our project manager editorial production Hal Humphrey for all his help, and to Jan Turner and Alyssa Morin of Pre-Press Company for their help and guidance as our production managers. We would also like to thank Pre-Press Company for their copyediting skills. We appreciate their help in improving our manuscript. We thank Kathleen Olsen for her invaluable help in locating appropriate photos and checking on photo permissions and Bob Kauser, permissions editor.

Because historical geology is largely a visual science, we extend thanks to Carlyn Iverson, who rendered the reflective art, and to the artists at Precision Graphics, Graphic World, and Pre-Press Company who were responsible for much of the rest of the art program. We also thank the artists at Magellan Geographix, who rendered many of the maps, and Dr. Ron Blakey, who allowed us to use his global paleogeographic maps.

As always, our families were very patient and encouraging when most of our spare time and energy were devoted to this book. We thank them for their support and understanding.

Reed Wicander
James S. Monroe

NASA Goddard Space Flight Center

CHAPTER 1

THE DYNAMIC AND EVOLVING EARTH

Satellite-based image of Earth. North America can clearly be seen in the center of this view, as well as Central America and the northern part of South America. The present locations of continents and ocean basins are the result of plate movements. The interaction of plates through time has affected the physical and biological history of Earth.

[OUTLINE]

INTRODUCTION

WHAT IS GEOLOGY?

PERSPECTIVE 1.1 *Interpreting Earth History*

HISTORICAL GEOLOGY AND THE FORMULATION OF THEORIES

ORIGIN OF THE UNIVERSE AND SOLAR SYSTEM, AND EARTH'S PLACE IN THEM

Origin of the Universe—Did It Begin with a Big Bang?

Our Solar System—Its Origin and Evolution

Earth—Its Place in Our Solar System

Forming the Earth–Moon System

WHY IS EARTH A DYNAMIC AND EVOLVING PLANET?

ORGANIC EVOLUTION AND THE HISTORY OF LIFE

GEOLOGIC TIME AND UNIFORMITARIANISM

HOW DOES THE STUDY OF HISTORICAL GEOLOGY BENEFIT US?

SUMMARY

ThomsonNOW™ Explore interactive tutorials, animations, or practice problems available on the ThomsonNow website at **www.thomsonedu.com/login**.

CHAPTER OBJECTIVES

At the end of this chapter, you will have learned that

- Earth is a complex, dynamic planet that has continually evolved since its origin some 4.6 billion years ago.
- To help understand Earth's complexity and history, it can be viewed as an integrated system of interconnected components that interact and affect each other in various ways.
- Theories are based on the scientific method and can be tested by observation or experiment.
- The universe is thought to have originated about 15 billion years ago with a Big Bang, and the solar system and planets evolved from a turbulent, rotating cloud of material surrounding the embryonic Sun.
- Earth consists of three concentric layers—core, mantle, and crust—and this orderly division resulted during Earth's early history.
- Plate tectonics is the unifying theory of geology.
- An appreciation of geologic time and the principle of uniformitarianism is central to understanding the evolutionary history of Earth and its biota.
- Geology is an integral part of our lives.

Introduction

What kind of movie would we have if it were possible to travel back in time and film Earth's history from its beginning 4.6 billion years ago? It would certainly be a story of epic proportions, with incredible special effects, a cast of trillions, a plot with twists and turns—and an ending that is still a mystery!

Unfortunately, we can't travel back in time, but we *can* tell the story of Earth and its inhabitants, because that history is preserved in its geologic record. In this book you will learn how to decipher that history from the clues Earth provides. Before we learn about the underlying principles of historical geology and how to use those principles and the clues preserved in Earth's geologic record, let's take a sneak preview of the full-length feature film *The History of Earth*.

In this movie we would see a planet undergoing remarkable change as continents moved about its surface. As a result of these movements, ocean basins would open and close, and mountain ranges would form along continental margins or where continents collided with each other. Oceanic and atmospheric circulation patterns would shift in response to the moving continents, sometimes causing massive ice sheets to form, grow, and then melt away. At other times, extensive swamps or vast interior deserts would appear.

We would also witness the first living cells evolving from a primordial organic soup, sometime between 4.6 and 3.6 billion years ago. Somewhere around 1.5 billion years later, cells with a nucleus would evolve, and not long thereafter multicelled soft-bodied animals would make their appearance in the world's oceans, followed in relatively short order by animals with skeletons and then animals with backbones.

Up until about 450 million years ago, Earth's landscape was essentially barren. At that time, however, the landscape comes to life as plants and animals move from their home in the water to take up residency on land. Viewed from above, Earth's landmasses would take on new hues and colors as different life-forms began inhabiting the terrestrial environment. From that moment on, Earth would never be the same, as plants, insects, amphibians, reptiles, birds, and mammals made the land their home. Near the end of our film, humans evolve, and we see how their activities greatly impact the global ecosystem. It seems only fitting that the movie's final image is of Earth, a shimmering blue-green oasis in the black void of space and a voice-over saying, "To be continued."

Every good movie has a theme, and *The History of Earth* is no exception. Because of its epic nature, not one but three interrelated themes run throughout it. The first is that Earth's outermost part is composed of a series of moving plates *(plate tectonics)* whose interactions have affected the planet's physical and biological history. The second is that Earth's biota has evolved or changed throughout its history *(organic evolution)*. The third is that the physical and biological changes that occurred did so over long periods of time *(geologic* or *deep time)*. These three interrelated themes are central to our understanding and appreciation of our planet's history. As you read this book, keep in mind that the different topics you study are parts of a system of interconnected components and not isolated pieces of information.

By viewing Earth as a whole—that is, thinking of it as a system—we not only see how its various components are interconnected, but we can also better appreciate its complex and dynamic nature. The system concept makes it easier for us to study a complex subject, such as Earth, because it divides the whole into smaller components we can easily understand, without losing sight of how the separate components fit together as a whole.

A **system** is a combination of related parts that interact in an organized fashion. We can thus consider Earth as a system of interconnected components that interact and affect each other in many different ways. The principal subsystems of Earth are the *atmosphere, biosphere, hydrosphere, lithosphere, mantle,* and *core* (• Figure 1.1). The complex interactions among these subsystems result in a dynamically changing planet in which matter and energy are continuously recycled into different forms (Table 1.1). For example, the movement of plates has profoundly affected the formation of landscapes, the distribution of mineral resources, and atmospheric and oceanic circulation patterns, which in turn have affected global climate changes. Examined in this manner, the continuous evolution of Earth and its life is not a series of isolated and

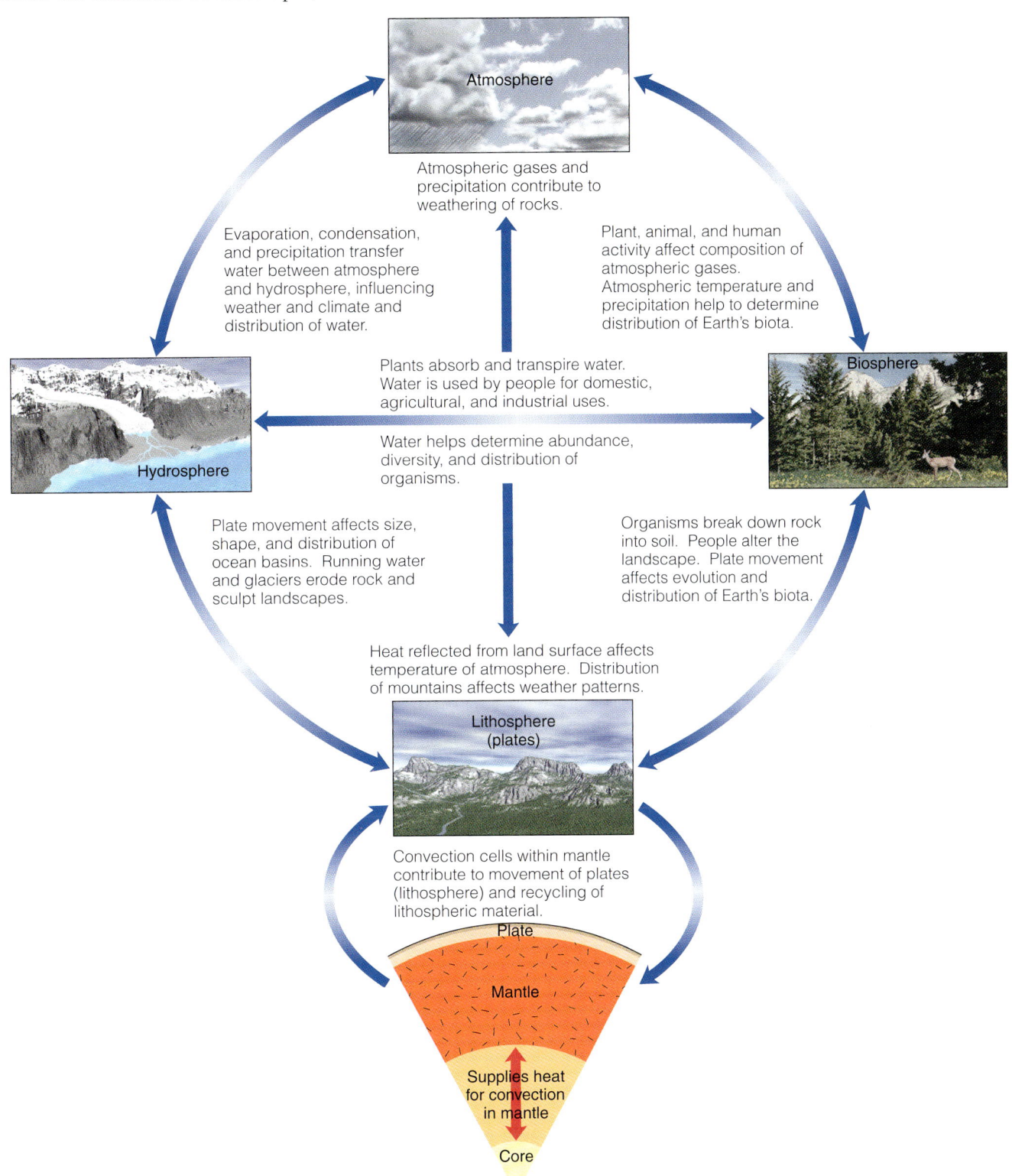

• **Figure 1.1 Subsystems of Earth** The atmosphere, hydrosphere, biosphere, lithosphere, mantle, and core are all subsystems of Earth. This simplified diagram shows how these subsystems interact, with some examples of how materials and energy are cycled throughout the Earth system. The interactions of these subsystems make Earth a dynamic planet that has evolved and changed since its origin 4.6 billion years ago.

TABLE 1.1 Interactions Between Major Earth Systems

	Earth–Universe System	Atmosphere	Hydrosphere	Biosphere	Solid Earth
Earth–Universe System	Gravitational interaction Recycling of stellar materials	Solid energy input Creation of ozone layer	Heating by solar energy Tides	Solar energy for photosynthesis Daily/seasonal rhythms	Heating by solar energy Meteor impacts
Atmosphere	Escape of heat radiation to space	Interaction of air masses	Winds drive surface currents Evaporation	Gases for respiration Transport of seeds, spores	Wind erosion Transport of water vapor for precipitation
Hydrosphere	Tidal friction affects Moon's orbit	Input of water vapor and stored solar heat	Mixing of oceans Deep ocean circulation Hydrologic cycle	Water for cell fluids Medium for aquatic organisms Transport of organisms	Precipitation Water and glacial erosion Solution of minerals
Biosphere	Biogenic gases affect escape of heat radiation to space	Gases from respiration	Removal of dissolved materials by organisms	Predator–prey interactions Food cycles	Modification of weathering and erosion processes Formation of soil
Solid Earth	Gravity of Earth affects other bodies	Input of stored solar heat Mountains divert air movements	Source of sediment and dissolved materials	Source of mineral nutrients Modification of habitats by crustal movements	Plate tectonics Crustal movements

Source: Adapted by permission from Stephen Dutch, James S. Monroe, and Joseph Moran, *Earth Science* (Minneapolis/St. Paul: West Publishing Co.).

unrelated events but a dynamic interaction among its various subsystems.

What Is Geology?

Geology, from the Greek *geo* and *logos,* is defined as the study of Earth, but now it must also include the study of the planets and moons in our solar system. Geology is generally divided into two broad areas: physical geology and historical geology. *Physical geology* studies Earth materials, such as minerals and rocks, as well as the processes operating within the Earth and on its surface. *Historical geology* examines the origin and evolution of Earth, its continents, atmosphere, oceans, and life.

Historical geology is, however, more than just a recitation of past events. It is the study of a dynamic planet that has changed continuously during the past 4.6 billion years.

In addition to determining what occurred in the past, geologists are also concerned with explaining how and why past events happened (see Perspective 1.1). It is one thing to observe in the fossil record that dinosaurs went extinct but quite another to ask how and why they became extinct, and, perhaps more important, what implications that holds for today's global ecosystem.

Not only do the basic principles of historical geology aid in interpreting Earth's history, but they also have practical applications. For example, William Smith, an English surveyor and engineer, recognized that by studying the sequences of rocks and the fossils they contained, he could predict the kinds and thicknesses of rocks that would have to be excavated in the construction of canals. The same principles Smith used in the late 18th and early 19th centuries are still used today in mineral and oil exploration and also in interpreting the geologic history of the planets and moons of our solar system.

Perspective 1.1

Interpreting Earth History

Historical geology is the study of the origin and evolution of Earth. Geologists are interested not only in placing events in a chronological sequence but, more important, in explaining how and why past events took place. Recently, historical geology has taken on even greater importance because scientists in many disciplines are looking to the past to help explain current events (such as short- and long-term climatic changes) and using this information to try and predict future trends.

We look at Earth as a system consisting of a collection of various subsystems or related parts that interact with each other in complex ways. By using this systems approach, we can see that the evolution of Earth, far from being a series of isolated events, is a continuum in which the different components both affect and are affected by each other. An example of this is the early history of Earth in which the evolution of the atmosphere, hydrosphere, lithosphere, and biosphere are intimately related (see Chapters 8 and 9). Today, scientists are examining the effect humans are having on short-term climate changes and the environment, as well as what a decrease in global biodiversity means for both humans and the planet in general.

Geologists seek to know not only *what* happened in the past but also *why* something happened and what the implications are for Earth today and in the future. Thus it is important to understand present-day processes and to have an accurate means of measuring geologic time so as to appreciate the duration of past events and how these events might affect Earth and its inhabitants today.

An important component of historical geology is understanding how we know what we know. How do we know dinosaurs became extinct around 66 million years ago or that glacial conditions prevailed over what is now the Sahara Desert during the Carboniferous Period? How can we be so sure that the early atmosphere was devoid of oxygen and evolved over millions of years to one that today has oxygen? These are all questions that historical geology addresses, and it does so by seeking answers in rocks and fossils. As more information becomes available from new observations or scientific techniques, geologists become more confident in their interpretations of past events.

One of the many exciting aspects of geology, and of science in general, is that there are still so many unanswered questions. For example, heated debate continues on what caused the Permian mass extinctions (see Chapter 12). Another exciting area of research is the determination of past environments. New studies indicate that changes in the chemistry of the oceans may have significantly affected the carbon cycle and have important implications in terms of present-day reef ecology and evolution.

What is important to remember is that rocks and fossils provide the clues to Earth's evolution. By applying the principles of historical geology, we can interpret Earth's history. It is also equally important to remember that historical geology is not a static science but one that, like the dynamic Earth it seeks to understand, is constantly evolving as new information becomes available.

Historical Geology and the Formulation of Theories

The term **theory** has various meanings and is frequently misunderstood and consequently misused. In colloquial usage, it means a speculative or conjectural view of something—hence the widespread belief that scientific theories are little more than unsubstantiated wild guesses. In scientific usage, however, a theory is a coherent explanation for one or several related natural phenomena supported by a large body of objective evidence. From a theory, scientists derive predictive statements that can be tested by observations and/or experiments so that their validity can be assessed.

For example, one prediction of plate tectonic theory is that oceanic crust is young near spreading ridges but progressively older with increasing distance from ridges. This prediction has been verified by observations (see Chapter 3). Likewise, according to the theory of evolution, fish should appear in the fossil record before amphibians, followed by reptiles, mammals, and birds—and that is indeed the case (see Chapter 7).

Theories are formulated through the process known as the **scientific method.** This method is an orderly, logical approach that involves gathering and analyzing the facts or data about the problem under consideration. Tentative explanations, or **hypotheses,** are then formulated to explain the observed phenomena. Next, the hypotheses are tested to see if what they predicted actually occurs in a given situation. Finally, if, after repeated tests, one of the hypotheses is found to explain the phenomena, then that hypothesis is proposed as a theory. Remember, however, that in science even a theory is still subject to further testing and refinement as new data become available.

The fact that a scientific theory can be tested and is subject to such testing separates science from other forms of human inquiry. Because scientific theories can be tested, they have the potential of being supported or even proved wrong. Accordingly, science must proceed without any appeal to beliefs or supernatural explanations, not because such beliefs or explanations are necessarily untrue but because we have no way to investigate them. For this reason, science makes no claim about the existence or nonexistence of a supernatural or spiritual realm.

Each scientific discipline has certain theories that are of particular importance. For example, the theory of organic evolution revolutionized biology when it was proposed in the 19th century. In geology, plate tectonic theory has changed the way geologists view Earth. Geologists now look at Earth from a global perspective in which all its subsystems and cycles are interconnected, and Earth history is seen as a continuum of interrelated events that are part of a global pattern of change.

Origin of the Universe and Solar System, and Earth's Place in Them

How did the universe begin? What has been its history? Is it infinite? What is its eventual fate? These are just some of the basic questions people have asked and wondered about since they first looked into the nighttime sky and saw the vastness of the universe beyond Earth.

Origin of the Universe—Did It Begin with a Big Bang?

Most scientists think that the universe originated about 15 billion years ago in what is popularly called the **Big Bang**. The Big Bang is a model for the evolution of the universe in which a dense, hot state was followed by expansion, cooling, and a less dense state. In a region infinitely smaller than an atom, both time and space were set at zero. Therefore, there is no "before the Big Bang" but only what occurred after it. The reason is that space and time are unalterably linked to form a space–time continuum demonstrated by Einstein's theory of relativity. Without space, there can be no time.

How do we know the Big Bang took place approximately 15 billion years ago? Why couldn't the universe have always existed as we know it today? Two fundamental phenomena indicate that the Big Bang occurred. First, the universe is expanding. When astronomers look beyond our own solar system, they observe that everywhere in the universe galaxies are moving away from each other at tremendous speeds. By measuring this expansion rate, astronomers can calculate how long ago the galaxies were all together at a single point. Second, everywhere in the universe there is a pervasive background radiation of 2.7 Kelvin (K) above absolute zero (absolute zero equals –273°C; 2.7 K = –270.3°C). This background radiation is thought to be the faint afterglow of the Big Bang.

According to the currently accepted theory, matter as we know it did not exist at the moment of the Big Bang, and the universe consisted of pure energy. During the first second following the Big Bang, the four basic forces—*gravity* (the attraction of one body toward another), *electromagnetic force* (combines electricity and magnetism into one force and binds atoms into molecules), *strong nuclear force* (binds protons and neutrons together), and *weak nuclear force* (responsible for the breakdown of an atom's nucleus, producing radioactive decay)—separated, and the universe experienced enormous expansion.

About 300,000 years later, the universe was cool enough for complete atoms of hydrogen and helium to form, and photons (the energetic particles of light) separated from matter and light burst forth for the first time.

During the next 200 million years, as the universe continued expanding and cooling, stars and galaxies began to form and the chemical makeup of the universe changed. Initially, the universe was 100% hydrogen and helium, whereas today it is 98% hydrogen and helium and 2% all other elements by weight. How did such a change in the universe's composition occur? Throughout their life cycle, stars undergo many nuclear reactions whereby lighter elements are converted into heavier elements by nuclear fusion. When a star dies, often explosively, the heavier elements that were formed in its core are returned to interstellar space and are available for inclusion in new stars. In this way, the composition of the universe is gradually enhanced in heavier elements.

Our Solar System—Its Origin and Evolution

Our solar system, which is part of the Milky Way Galaxy, consists of a Sun, eight planets, 153 known moons or satellites (although this number keeps changing with the discovery of new moons and satellites surrounding the Jovian planets), a tremendous number of asteroids—most of which orbit the Sun in a zone between Mars and Jupiter—and millions of comets and meteorites, as well as interplanetary dust and gases (• Figure 1.2). Any theory formulated to explain the origin and evolution of our solar system must therefore take into account its various features and characteristics.

Many scientific theories for the origin of the solar system have been proposed, modified, and discarded since the French scientist and philosopher René Descartes first proposed, in 1644, that the solar system formed from a gigantic whirlpool within a universal fluid. Today, the **solar nebula theory** for the origin of our solar system involves the condensation and collapse of interstellar material in a spiral arm of the Milky Way Galaxy.

The collapse of this cloud of gases and small grains into a counterclockwise-rotating disk concentrated about 90% of the material in the central part of the disk and formed an embryonic Sun around which swirled a rotating cloud of material called a *solar nebula*. Within this solar nebula were localized eddies in which gases and solid particles condensed. During the condensation process, gaseous, liquid, and solid particles began accreting into ever-larger

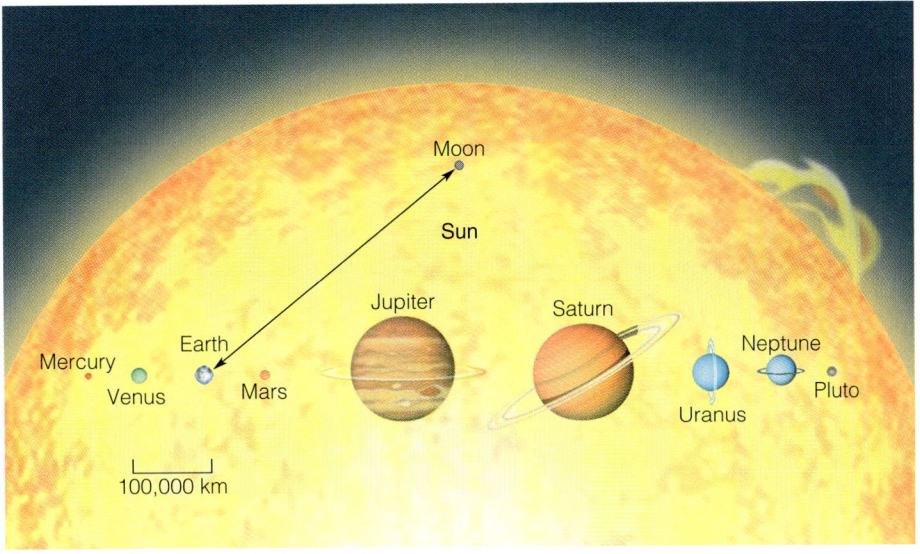

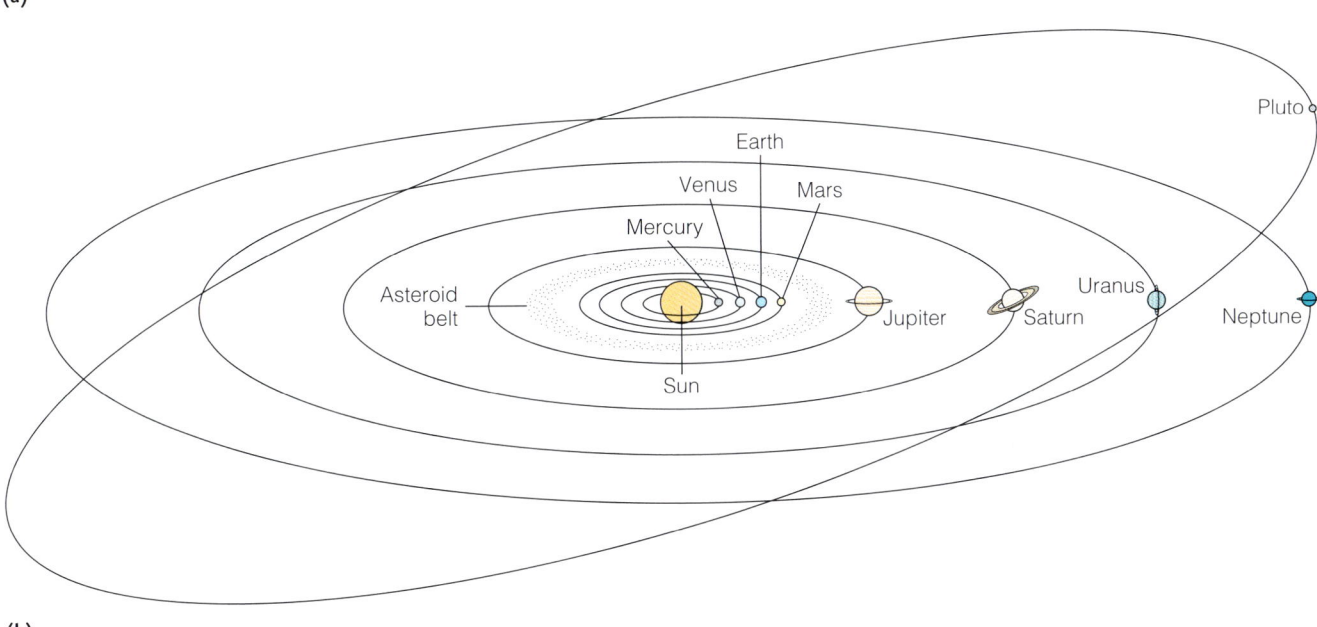

• **Figure 1.2 Diagrammatic Representation of the Solar System** This representation of the solar system shows **(a)** the relative sizes of the eight planets and Pluto **(b)** their orbits around the Sun. (On August 24, 2006, the International Astronomical Union (IAU) reclassified Pluto from planet to a "dwarf planet".)

masses called *planetesimals* (• Figure 1.3) which collided and grew in size and mass until they eventually became planets.

The composition and evolutionary history of the planets are a consequence, in part, of their distance from the Sun. The **terrestrial planets**—Mercury, Venus, Earth, and Mars—so named because they are similar to *terra*, which is Latin for "earth," are all small and composed of rock and metallic elements that condensed at the high temperatures of the inner nebula. The **Jovian planets**—Jupiter, Saturn, Uranus, and Neptune—so named because they resemble Jupiter (the Roman god was also named Jove) all have small central rocky cores compared to their overall size and are composed mostly of hydrogen, helium, ammonia, and methane, which condense at low temperatures.

While the planets were accreting, material that had been pulled into the center of the nebula also condensed, collapsed, and was heated to several million degrees by gravitational compression. The result was the birth of a star: our Sun.

During the early accretionary phase of the solar system's history, collisions between various bodies were common, as indicated by the craters on many planets and moons. Asteroids probably formed as planetesimals in a localized eddy between what eventually became Mars and Jupiter in much the same way as other planetesimals formed the terrestrial planets. The tremendous gravitational field of Jupiter, however, prevented this material from ever accreting into a planet. Comets, which are interplanetary bodies composed

• **Figure 1.3 Planetesimals** At the stage of development shown here, planetesimals have formed in the inner solar system, and large eddies of gas and dust remain at great distances from the embryonic Sun.

of loosely bound rocky and icy material, are thought to have condensed near the orbits of Uranus and Neptune.

The solar nebula theory of the formation of the solar system thus accounts for most of the characteristics of the planets and their moons, the differences in composition between the terrestrial and Jovian planets, and the presence of the asteroid belt. Based on the available data, the solar nebula theory best explains the features of the solar system and provides a logical explanation for its evolutionary history.

Earth—Its Place in Our Solar System

Some 4.6 billion years ago, various planetesimals in our solar system gathered enough material together to form Earth and seven other planets. Scientists think that this early Earth was probably cool, of generally uniform composition and density throughout, and composed mostly of silicates—compounds consisting of silicon and oxygen, iron and magnesium oxides, and smaller amounts of all the other chemical elements (• Figure 1.4a). Subsequently, when the combination of meteorite impacts, gravitational compression, and heat from radioactive decay increased the temperature of Earth enough to melt iron and nickel, this homogeneous composition disappeared (Figure 1.4b) and was replaced by a series of concentric layers of differing composition and density, resulting in a differentiated planet (Figure 1.4c).

This differentiation into a layered planet is probably the most significant event in Earth history. Not only did it lead to the formation of a crust and eventually continents, but it also was probably responsible for the emission of gases from the interior that eventually led to the formation of the oceans and atmosphere (see Chapter 8).

Forming the Earth–Moon System

Although we probably know more about our Moon (• Figure 1.5) than any other celestial object except Earth, scientists do not completely agree how the Moon originated. However, the model that seems to account best for the Moon's particular composition and structure involves an impact by a large planetesimal with the young Earth.

In this model, a giant planetesimal, the size of Mars or larger, crashed into Earth about 4.4–4.6 billion years ago, causing the ejection of a large quantity of hot material that formed the Moon. The material that was ejected was mostly in the liquid and vapor phase and came primarily from the mantle of the colliding planetesimal. As the material cooled, the various lunar layers crystallized, forming a zoned body (Figure 1.5b).

The Moon's light-colored highlands, which are its oldest features, are heavily cratered (Figure 1.5a), providing striking evidence of the massive meteorite bombardment

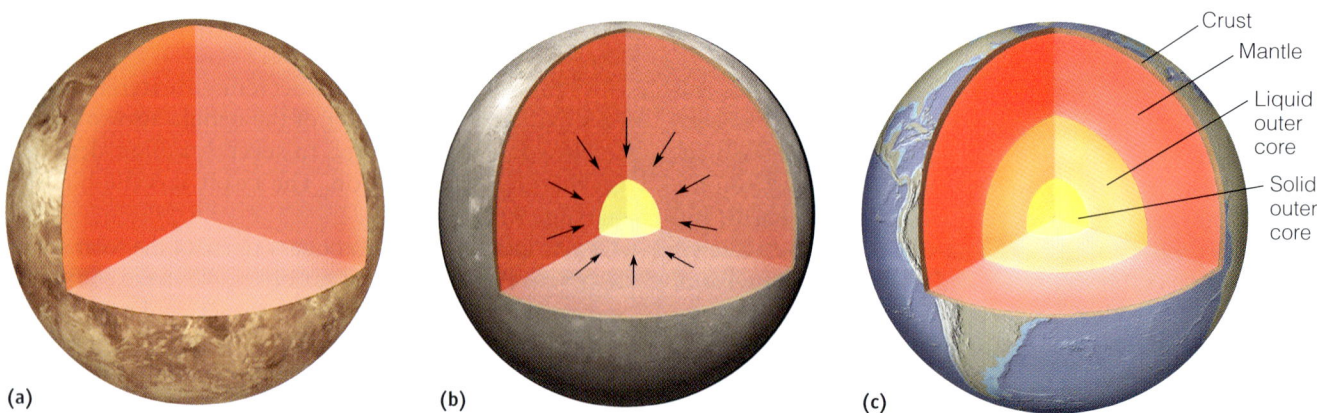

• **Figure 1.4 Homogeneous Accretion Theory for the Formation of a Differentiated Earth** (a) Early Earth probably had a uniform composition and density throughout. (b) The temperature of early Earth reached the melting point of iron and nickel, which, being denser than silicate minerals, settled to Earth's center. At the same time, the lighter silicates flowed upward to form the mantle and the crust. (c) In this way, a differentiated Earth formed, consisting of a dense iron-nickel core, an iron-rich silicate mantle, and a silicate crust with continents and ocean basins.

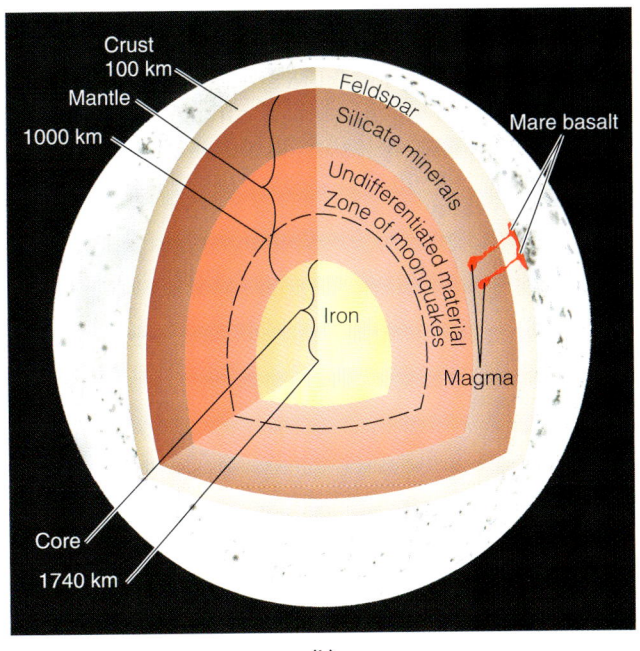

• **Figure 1.5 Earth's Moon and Its Internal Structure** (a) High-quality image of the Moon taken through a telescope on Earth. The light-colored lunar highlands are heavily cratered. The dark-colored areas are maria, which formed when lava flowed onto the surface. (b) The Moon's internal structure. Because seismic waves are not transmitted below 1000 km, it is likely that the innermost mantle is liquid.

that occurred in the solar system more than 4 billion years ago. And why isn't Earth's present surface also densely cratered if it was subjected to the same meteorite barrage that pockmarked the Moon? You can find the answer to that question in the next section.

Why Is Earth a Dynamic and Evolving Planet?

The reason Earth's surface isn't heavily cratered is because it is a dynamic planet that has continuously changed during its 4.6-billion-year existence. The size, shape, and geographic distribution of continents and ocean basins have changed through time, the composition of the atmosphere has evolved, and life-forms existing today differ from those that lived during the past. Mountains, hills, and craters have been worn away by erosion, and the forces of wind, water, and ice have sculpted a diversity of landscapes. Volcanic eruptions and earthquakes reveal an active interior, and folded and fractured rocks indicate the tremendous power of Earth's internal forces.

Earth consists of three concentric layers: the core, the mantle, and the crust (• Figure 1.6). This orderly division results from density differences between the layers as a function of variations in composition, temperature, and pressure.

The **core** has a calculated density of 10–13 grams per cubic centimeter (g/cm^3) and occupies about 16% of Earth's total volume. Seismic (earthquake) data indicate that the core consists of a small, solid, inner region and a larger, apparently liquid, outer portion. Both are thought to consist largely of iron and a small amount of nickel.

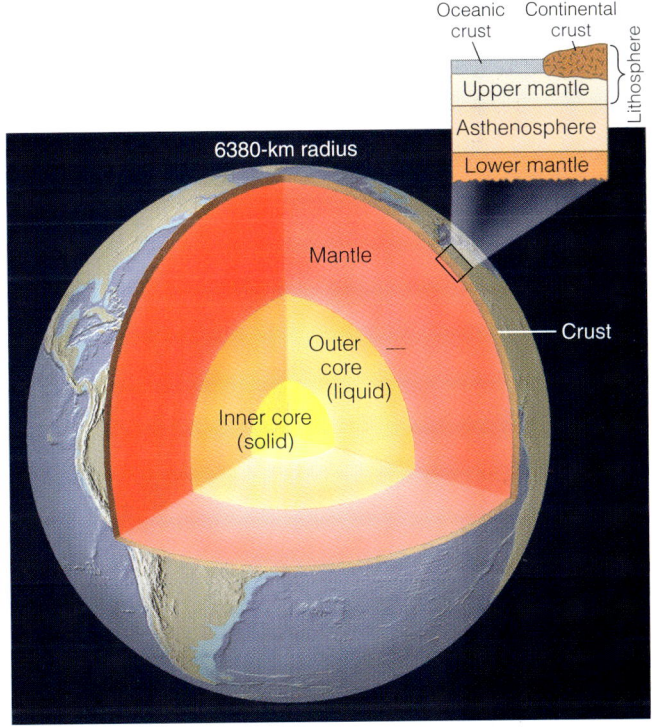

• **Figure 1.6 Cross Section of Earth Illustrating the Core, Mantle, and Crust** The enlarged portion shows the relationship between the lithosphere (composed of the continental crust, oceanic crust, and upper mantle) and the underlying asthenosphere and lower mantle.

WHY IS EARTH A DYNAMIC AND EVOLVING PLANET? **9**

The **mantle** surrounds the core and comprises about 83% of Earth's volume. It is less dense than the core (3.3–5.7 g/cm^3) and is thought to be composed largely of *peridotite*, a dark, dense igneous rock containing abundant iron and magnesium (see Figure 2.10). The mantle can be divided into three distinct zones based on physical characteristics. The lower mantle is solid and forms most of the volume of Earth's interior. The **asthenosphere** surrounds the lower mantle. It has the same composition as the lower mantle but behaves plastically and slowly flows. Partial melting within the asthenosphere generates *magma* (molten material), some of which rises to the surface because it is less dense than the rock from which it was derived. The upper mantle surrounds the asthenosphere. The solid upper mantle and the overlying crust constitute the **lithosphere,** which is broken into numerous individual pieces called **plates** that move over the asthenosphere as a result of underlying *convection cells* (• Figure 1.7).

The **crust,** Earth's outermost layer, consists of two types. *Continental crust* is thick (20–90 km), has an average density of 2.7 g/cm^3, and contains considerable silicon and aluminum. *Oceanic crust* is thin (5–10 km), denser than continental crust (3.0 g/cm^3), and is composed of the dark igneous rock *basalt* and *gabbro* (see Figures 2.11a and b).

The recognition that the lithosphere is divided into rigid plates that move over the asthenosphere forms the foundation of **plate tectonic theory.** Zones of volcanic activity, earthquakes, or both mark most plate boundaries. Along these boundaries, plates diverge, converge, or slide sideways past each other (• Figure 1.8).

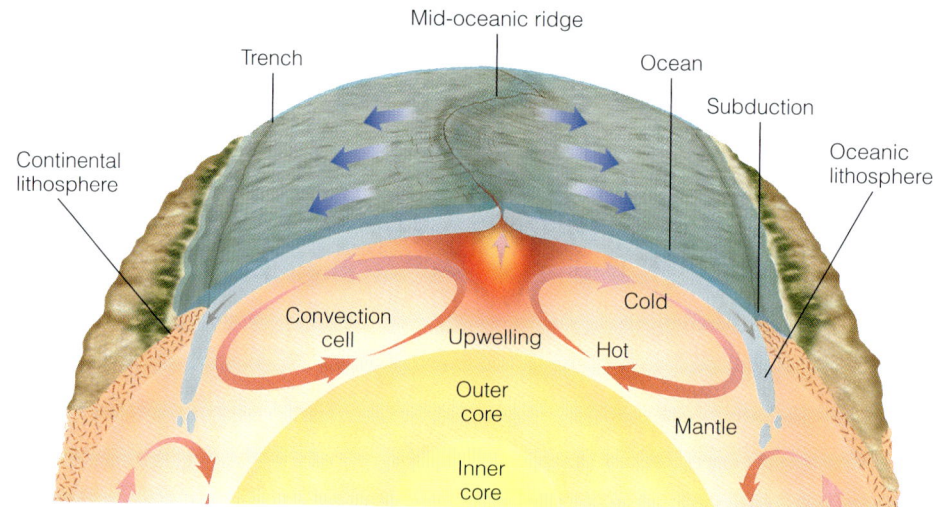

• **Figure 1.7 Movement of Earth's Plates** Earth's plates are thought to move partially as a result of underlying mantle convection cells in which warm material from deep within Earth rises toward the surface, cools, and then, upon losing heat, descends back into the interior, as shown in this diagrammatic cross section.

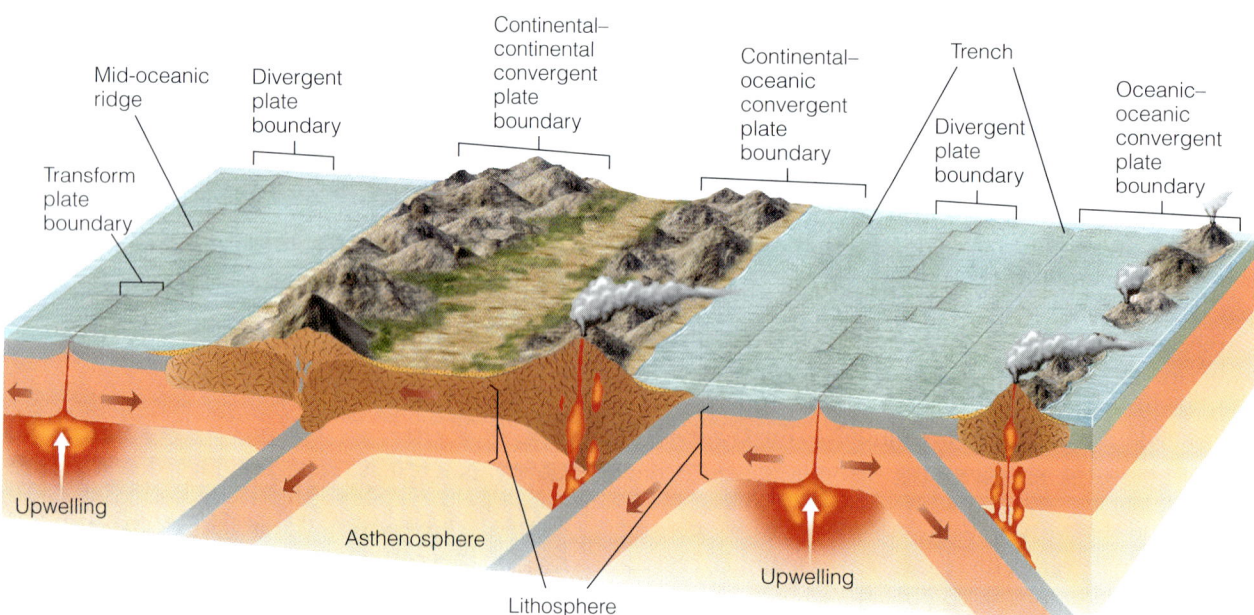

• **Figure 1.8 Relationship Between Lithosphere, Asthenosphere, and Plate Boundaries** An idealized block diagram illustrating the relationship between the lithosphere and the underlying asthenosphere and the three principal types of plate boundaries: divergent, convergent, and transform.

The acceptance of plate tectonic theory is recognized as a major milestone in the geologic sciences, comparable to the revolution caused by Darwin's theory of evolution in biology. Plate tectonics has provided a framework for interpreting the composition, structure, and internal processes of Earth on a global scale. It has led to the realization that the continents and ocean basins are part of a lithosphere-atmosphere-hydrosphere (water portion of the planet) system that evolved together with Earth's interior (Table 1.2).

A revolutionary concept when it was proposed in the 1960s, plate tectonic theory has had significant and far-reaching consequences in all fields of geology because it provides the basis for relating many seemingly unrelated phenomena. Besides being responsible for the major features of Earth's crust, plate movements also affect the formation and occurrence of Earth's natural resources, as well as influencing the distribution and evolution of the world's biota.

The impact of plate tectonic theory has been particularly notable in the interpretation of Earth's history. For example, the Appalachian Mountains in eastern North America and the mountain ranges of Greenland, Scotland, Norway, and Sweden are not the result of unrelated mountain-building episodes but, rather, are part of a larger mountain-building event that involved the closing of an ancient "Atlantic Ocean" and the formation of the supercontinent Pangaea about 251 million years ago (see Chapter 11).

TABLE 1.2 Plate Tectonics and Earth Systems

Solid Earth

Plate tectonics is driven by convection in the mantle and in turn drives mountain-building and associated igneous and metamorphic activity.

Atmosphere

Arrangement of continents affects solar heating and cooling, and thus winds and weather systems. Rapid plate spreading and hot-spot activity may release volcanic carbon dioxide and affect global climate.

Hydrosphere

Continental arrangement affects ocean currents. Rate of spreading affects volume of mid-oceanic ridges and hence sea level. Placement of continents may contribute to onset of ice ages.

Biosphere

Movement of continents creates corridors or barriers to migration, the creation of ecological niches, and transport of habitats into more or less favorable climates.

Extraterrestrial

Arrangement of continents affects free circulation of ocean tides and influences tidal slowing of Earth's rotation.

Source: Adapted by permission from Stephen Dutch, James S. Monroe, and Joseph Moran, *Earth Science* (Minneapolis/St. Paul: West Publishing Co.).

Organic Evolution and the History of Life

Plate tectonic theory provides us with a model for understanding the internal workings of Earth and its effect on Earth's surface. The theory of **organic evolution** provides the conceptual framework for understanding the history of life. Together, the theories of plate tectonics and organic evolution have changed the way we view our planet, and we should not be surprised at the intimate association between them. Although the relationship between plate tectonic processes and the evolution of life is incredibly complex, paleontologic data provide indisputable evidence of the influence of plate movement on the distribution of organisms.

The publication in 1859 of Darwin's *On the Origin of Species by Means of Natural Selection* revolutionized biology and marked the beginning of modern evolutionary biology. With its publication, most naturalists recognized that evolution provided a unifying theory that explained an otherwise encyclopedic collection of biologic facts.

The central thesis of organic evolution is that all present-day organisms are related and that they have descended with modifications from organisms that lived during the past. When Darwin proposed his theory of organic evolution, he cited a wealth of supporting evidence, including the way organisms are classified, embryology, comparative anatomy, the geographic distribution of organisms, and, to a limited extent, the fossil record. Furthermore, Darwin proposed that *natural selection*, which results in the survival to reproductive age of those organisms best adapted to their environment, is the mechanism that accounts for evolution.

Perhaps the most compelling evidence in favor of evolution can be found in the fossil record. Just as the geologic record allows geologists to interpret physical events and conditions in the geologic past, **fossils,** which are the remains or traces of once-living organisms, not only provide evidence that evolution has occurred but also demonstrate that Earth has a history extending beyond that recorded by humans. The succession of fossils in the rock record provides geologists with a means for dating rocks and allowed for a relative geologic time scale to be constructed in the 1800s.

Geologic Time and Uniformitarianism

An appreciation of the immensity of geologic time is central to understanding the evolution of Earth and its biota. Indeed, time is one of the main aspects that sets geology apart from the other sciences except astronomy. Most people have difficulty comprehending geologic time because they tend to think in terms of the human perspective—seconds, hours, days, and years. Ancient history is what occurred hundreds or even thousands of years ago. When geologists talk of ancient geologic history, however, they are referring to events that happened hundreds of millions, or even billions, of years ago. To a geologist, recent geologic events are those that occurred within the last million years or so.

It is also important to remember that Earth goes through cycles of much longer duration than the human perspective of time. Although they may have disastrous effects on the human species, global warming and cooling are part of a larger cycle that has resulted in numerous glacial advances and retreats during the past 1.8 million years. Because of their geologic perspective on time and how the various Earth subsystems and cycles are interrelated, geologists can make valuable contributions to many of the current environmental debates, such as those involving global warming and sea-level changes.

The **geologic time scale** subdivides geologic time into a hierarchy of increasingly shorter time intervals; each time subdivision has a specific name. The geologic time scale resulted from the work of many 19th-century geologists who pieced together information from numerous rock exposures and constructed a chronology based on changes in Earth's biota through time. Subsequently, with the discovery of radioactivity in 1895 and the development of various radiometric dating techniques, geologists have been able to assign ages (also known as absolute ages) in years to the subdivisions of the geologic time scale (• Figure 1.9).

One of the cornerstones of geology is the **principle of uniformitarianism,** which is based on the premise that present-day processes have operated throughout geologic time. Therefore, in order to understand and interpret geologic events from evidence preserved in rocks, we must first understand present-day processes and their results. In fact, uniformitarianism fits in completely with the system approach we are following for the study of Earth.

Uniformitarianism is a powerful principle that allows us to use present-day processes as the basis for interpreting the past and for predicting potential future events. We should keep in mind, however, that uniformitarianism does not exclude sudden or catastrophic events such as volcanic eruptions, earthquakes, landslides, or flooding. These are processes that shape our modern world, and some geologists view Earth history as a series of such short-term or punctuated events. This view is certainly in keeping with the modern principle of uniformitarianism.

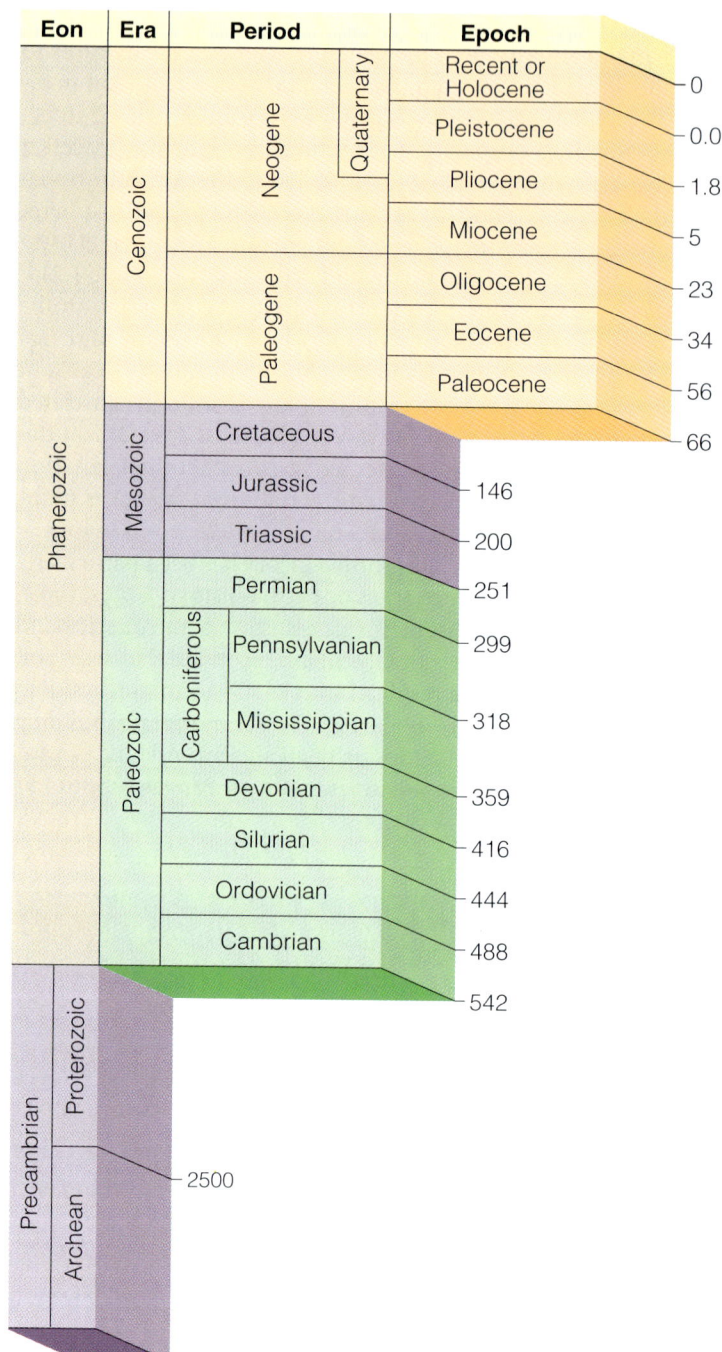

• **Figure 1.9 The Geologic Time Scale** The numbers to the right of the columns are the ages in millions of years before the present. Dates are from Gradstein, F., Ogg, J., and Smith, A., *A Geologic Time Scale 2004* (Cambridge, UK: Cambridge University Press, 2005), Figure 1.2.

Furthermore, uniformitarianism does not require that the rates and intensities of geologic processes be constant through time. We know that volcanic activity was more intense in North America 5 to 10 million years ago than it is today and that glaciation has been more prevalent during the last several million years than in the previous 300 million years.

What Would You Do?

An important environmental issue facing the world today is global warming. How can this problem be approached from a global systems perspective? What are the possible consequences of global warming, and can we really do anything about it? Are there ways to tell if global warming occurred in the geologic past?

What uniformitarianism means is that even though the rates and intensities of geologic processes have varied during the past, the physical and chemical laws of nature have remained the same. Although Earth is in a dynamic state of change and has been ever since it was formed, the processes that have shaped it during the past are the same ones operating today.

How Does the Study of Historical Geology Benefit Us?

The most meaningful lesson to learn from the study of historical geology is that Earth is an extremely complex planet in which interactions are taking place between its various subsystems and have been for the past 4.6 billion years. If we want to ensure the survival of the human species, we must understand how the various subsystems work and interact with each other. We can do this, in part, by studying what has happened in the past, particularly on the global scale, and use that information to try to determine how our actions might affect the delicate balance between Earth's various subsystems in the future.

The study of geology goes beyond learning numerous facts about Earth. In fact, we don't just study geology—we "live" it. Geology is an integral part of our lives. Our standard of living depends directly on our consumption of natural resources, resources that formed millions and billions of years ago. However, the way we consume natural resources and interact with the environment, as individuals and as a society, also determines our ability to pass on this standard of living to the next generation.

As you study the three themes of this book, keep in mind how, like a system, plate tectonics, geologic time, and organic evolution are interconnected and responsible for the 4.6-billion-year history of Earth. View each chapter's topic in the context of its place in the entire Earth system, as well as remembering that Earth history is a continuum and the result of interaction between Earth's various subsystems.

Historical geology is not a dry history of Earth but a vibrant, dynamic science in which what we see today is based on what went on before. As Stephen Jay Gould states in his book *Wonderful Life: The Burgess Shale and the Nature of History* (1989), "Play the tape of life again starting with the Burgess Shale, and a different set of survivors—not including vertebrates this time—would grace our planet today."

What Would You Do?

Because of budget problems, someone at your local school board meeting proposed eliminating geology as one of the science courses taught in high school. Why would this be a bad idea, and what arguments would you use to dissuade the school board from dropping geology from the high school curriculum?

SUMMARY

- Earth can be viewed as a system of interconnected components that interact and affect each other. The principal subsystems of Earth are the atmosphere, hydrosphere, biosphere, lithosphere, mantle, and core. Earth is considered a dynamic planet that continually changes because of the interaction among its various subsystems and cycles. Geology, the study of Earth, is divided into two broad areas: Physical geology is the study of Earth materials as well as the processes that operate within and on Earth's surface; historical geology examines the origin and evolution of Earth, its continents, oceans, atmosphere, and life.
- The scientific method is an orderly, logical approach that involves gathering and analyzing facts about a particular phenomenon, formulating hypotheses to explain the phenomenon, testing the hypotheses, and finally proposing a theory. A theory is a testable explanation for some natural phenomenon that has a large body of supporting evidence. Both the theory of organic evolution and plate tectonic theory are theories that revolutionized biology and geology, respectively.
- The universe began with a Big Bang approximately 15 billion years ago. Astronomers have deduced this age by observing that celestial objects are moving away from each other in an ever-expanding universe. Furthermore, the universe has a pervasive background radiation of 2.7 K above absolute zero (2.7 K = –270.3°C), which is thought to be the faint afterglow of the Big Bang.
- About 4.6 billion years ago, our solar system formed from a rotating cloud of interstellar matter. As this cloud condensed, it eventually collapsed under the influence of gravity and flattened into a counterclockwise rotating disk. Within this rotating disk, the Sun, planets, and moons formed from the turbulent eddies of nebular gases and solids.
- Earth formed from a swirling eddy of nebular material 4.6 billion years ago, accreting as a solid body and soon thereafter differentiated into a layered planet during a period of internal heating.
- The Moon probably formed as a result of a Mars-sized planetesimal crashing into Earth 4.4–4.6 billion years ago, causing it to eject a large quantity of hot material.

- As the material cooled, the various lunar layers crystallized, forming a zoned body.
- Earth is differentiated into layers. The outermost layer is the crust, which is divided into continental and oceanic portions. The crust and underlying solid part of the upper mantle, also known as the lithosphere, overlie the asthenosphere, a zone that behaves plastically and flows slowly. The asthenosphere is underlain by the solid lower mantle. Earth's core consists of an outer liquid region and an inner solid portion.
- The lithosphere is broken into a series of plates that diverge, converge, and slide sideways past one another.
- Plate tectonic theory provides a unifying explanation for many geological features and events. The interaction between plates is responsible for volcanic eruptions, earthquakes, the formation of mountain ranges and ocean basins, and the recycling of rock materials.
- The central thesis of the theory of organic evolution is that all living organisms evolved (descended with modifications) from organisms that existed in the past.
- Time sets geology apart from the other sciences except astronomy, and an appreciation of the immensity of geologic time is central to understanding Earth's evolution. The geologic time scale is the calendar geologists use to date past events.
- The principle of uniformitarianism is basic to the interpretation of Earth history. This principle holds that the laws of nature have been constant through time and that the same processes operating today have operated in the past, although at different rates.
- Geology is an integral part of our lives. Our standard of living depends directly on our consumption of natural resources, resources that formed millions and billions of years ago.

IMPORTANT TERMS

asthenosphere, p. 10
Big Bang, p. 6
core, p. 9
crust, p. 10
fossil, p. 11
geologic time scale, p. 12
geology, p. 4
hypothesis, p. 5
Jovian planets, p. 7
lithosphere, p. 10
mantle, p. 10
organic evolution, p. 11
plate, p. 10
plate tectonic theory, p. 10
principle of uniformitarianism, p. 12
scientific method, p. 5
solar nebula theory, p. 6
system, p. 2
terrestrial planets, p. 7
theory, p. 5

REVIEW QUESTIONS

1. The premise that present-day processes have operated throughout geologic time is the principle of
   a. _____ organic evolution; b. _____ plate tectonics; c. _____ uniformitarianism; d. _____ geologic time; e. _____ scientific deduction.
2. The currently accepted theory for the origin of the solar system is
   a. _____ a huge nebula collapsed under its own gravitational attraction; b. _____ the nebula formed a disk with the Sun in the center; c. _____ planetesimals accreted from gaseous, liquid, and solid particles; d. _____ all of the previous answers; e. _____ none of the previous answers.
3. The study of the origin and evolution of Earth is
   a. _____ astronomy; b. _____ historical geology; c. _____ astrobiology; d. _____ physical geology; e. _____ paleontology.
4. The concentric layer that comprises most of Earth's volume is the
   a. _____ inner core; b. _____ outer core; c. _____ mantle; d. _____ asthenosphere; e. _____ crust.
5. Plates are composed of
   a. _____ the crust and upper mantle; b. _____ the asthenosphere and upper mantle; c. _____ the crust and asthenosphere; d. _____ continental and oceanic crust only; e. _____ the core and mantle.
6. Which of the following is *not* a terrestrial planet?
   a. _____ Mercury; b. _____ Jupiter; c. _____ Mars; d. _____ Earth; e. _____ Venus.
7. The movement of plates is thought to result from
   a. _____ density differences between the inner and outer core; b. _____ rotation of the mantle around the core; c. _____ gravitational forces; d. _____ the Coriolis effect; e. _____ convection cells.
8. Which of the following statements about a scientific theory is *not* true?
   a. _____ It is an explanation for some natural phenomenon. b. _____ It is a conjecture or guess. c. _____ It has a large body of supporting evidence. d. _____ It is testable. e. _____ Predictive statements can be derived from it.
9. What two observations led scientists to conclude that the Big Bang occurred approximately 15 billion years ago?
   a. _____ A steady-state universe and background radiation of 2.7 K above absolute zero. b. _____ A steady-state universe and opaque background radiation. c. _____ An expanding universe and opaque background radiation. d. _____ An expanding universe and background radiation of 2.7 K above absolute zero. e. _____ A shrinking universe and opaque background radiation.

10. That all living organisms are the descendants of different life-forms that existed in the past is the central claim of
    a. _____ the principle of fossil succession; b. _____ the principle of uniformitarianism; c. _____ plate tectonics; d. _____ organic evolution; e. _____ none of the previous answers.
11. The model that best accounts for the formation of Earth's Moon is
    a. _____ capture from an independent orbit; b. _____ an independent origin from Earth; c. _____ breaking off from Earth during Earth's accretion; d. _____ a collision between Earth and a large planetesimal; e. _____ none of the previous answers.
12. Which layer has the same composition as the mantle but behaves plastically?
    a. _____ continental crust; b. _____ oceanic crust; c. _____ outer core; d. _____ inner core; e. _____ asthenosphere.
13. Why is viewing Earth as a system a good way to study it?
14. Discuss how the three major layers of Earth differ from each other and why the differentiation into a layered planet is probably the most significant event in Earth history.
15. Explain how the principle of uniformitarianism allows for catastrophic events.
16. Discuss why plate tectonic theory is a unifying theory of geology.
17. How does plate tectonic theory fit into a systems approach to the study of Earth?
18. How does the solar nebula theory account for the formation of our solar system, its features, and evolutionary history?
19. How was the age of the universe determined?
20. Why is it important to have a basic knowledge and understanding of historical geology?

APPLY YOUR KNOWLEDGE

1. Describe how you would use the scientific method to formulate a hypothesis explaining the similarity of mountain ranges on the east coast of North America and those in England, Scotland, and the Scandinavian countries. How would you test your hypothesis?
2. Discuss why an accurate geologic time scale is particularly important for geologists in examining global temperature changes during the past and how an understanding of geologic time is crucial to the current debate on global warming and its consequences.

CHAPTER 2
EARTH MATERIALS—MINERALS AND ROCKS

Sue Monroe

The mountains around Tenaya Lake in Yosemite National Park in California are made up of granite which in turn is composed of the minerals quartz, potassium feldspars, plagioclase feldspars, and small amounts of one or two other minerals. The valley that is now the site of Tenaya Lake was occupied by a large glacier during the Pleistocene Epoch.

[OUTLINE]

INTRODUCTION

MATTER AND ITS COMPOSITION

Elements and Atoms

Bonding and Compounds

MINERALS—THE BUILDING BLOCKS OF ROCKS

HOW MANY MINERALS ARE THERE?

Silicate Minerals

Other Mineral Groups

ROCK-FORMING MINERALS AND THE ROCK CYCLE

IGNEOUS ROCKS

Texture and Composition of Igneous Rocks

Classifying Igneous Rocks

SEDIMENTARY ROCKS

Sediment Transport, Deposition, and Lithification

Types of Sedimentary Rocks

METAMORPHIC ROCKS

The Agents of Metamorphism

Types of Metamorphism

Classifying Metamorphic Rocks

PLATE TECTONICS AND THE ROCK CYCLE

EARTH MATERIALS AND HISTORICAL GEOLOGY

SUMMARY

ThomsonNOW Explore interactive tutorials, animations, or practice problems available on the ThomsonNow website at www.thomsonedu.com/login.

[CHAPTER OBJECTIVES]

At the end of this chapter, you will have learned that

- A mineral is a naturally occurring, inorganic, crystalline solid, with characteristic physical properties, and a narrowly defined chemical composition.

- Chemical elements are made up of atoms, all of the same kind, whereas compounds form when different atoms bond together. Most minerals are compounds.

- Only a very few of the 3500 or so minerals known are common in rocks, but many others are important as natural resources.

- The rock cycle illustrates how Earth's internal and surface processes yield the three major families of rocks, any one of which can be derived from the others.

- Igneous rocks result from cooling and crystallization of magma or lava or the consolidation of pyroclastic materials. Geologists use composition and texture to classify intrusive (plutonic) and extrusive (volcanic) igneous rocks.

- Weathering yields sediment that is transported, deposited, and then lithified to form detrital sedimentary rocks and chemical sedimentary rocks.

- Texture and composition are the criteria geologists use to classify sedimentary rocks.

- Heat, pressure, and fluids convert rocks below the surface to metamorphic rocks.

- Many metamorphic rocks are foliated—that is, they have a platy aspect developed during metamorphism, but some do not possess this feature.

- Interactions among Earth's systems, especially at divergent and convergent plate boundaries, account for much of the recycling of Earth materials in the rock cycle.

Introduction

When you hear the word *mineral*, you most likely think of calcium, iron, and potassium that we need for good nutrition, but these are actually chemical elements and not minerals, at least in the geologic sense. Ice probably does not come to mind either, and yet ice is a mineral because it is a naturally occurring, inorganic, crystalline solid, meaning that its atoms of hydrogen and oxygen are arranged in a specific three-dimensional pattern, as opposed to liquids and gases, which have no such orderly arrangement of atoms. In addition, ice has a specific chemical composition (H_2O) and, like all other minerals, it has characteristic physical properties such as hardness and density. To summarize, then, a **mineral** is a naturally occurring, crystalline solid with distinctive physical properties and a specified chemical composition (• Figure 2.1a). So how is a mineral different from a rock? By definition, a **rock** is made up of one or more minerals (Figure 2.1b), although masses of mineral-like matter, such as partly altered vegetation in coal and natural glass, are rocks, too.

Some minerals, such as gold, diamond, and emerald, are very beautiful and have been a source of fascination for thousands of years. Indeed, several minerals and some rocks have served as religious symbols or talismans, or have been worn, carried, applied externally, or ingested for their presumed mystical or curative powers. Diamond, a well-known gemstone, is supposed to ward off evil spirits, sickness, and floods, and relating gemstones to birth month gives them even more appeal to many people.

Gemstones, as well as the precious metals gold, silver, and platinum, certainly warrant our attention, but be aware that many other minerals and rocks are essential to industrialized societies. No nation is totally self-sufficient and must import some resources from elsewhere; thus various governments form complex economic and political ties. The United States has no domestic production of manganese, a necessary element for the production of steel, nor does it produce any cobalt that is used in gas turbine engines, magnets, and corrosion- and wear-resistant alloys. In addition to manganese and cobalt, the United States imports all of the aluminum ore it uses, as well as all or some of many other important resources.

Even some very common minerals are important but otherwise would attract little attention. Quartz, the mineral that makes up most of the world's sand, is used to make glass, optical instruments, and sandpaper, whereas the phosphate used for fertilizers comes from phosphorus-rich rocks. The vast iron ore deposits in the Lake Superior region and eastern Canada account for most iron ore mined in North America.

Rocks, too, find many uses. Some are simply crushed for aggregate in cement or for roadbeds, but several varieties are sawed and polished for tombstones, monuments, mantelpieces, and countertops. Indeed, some granite, a common rock, is attractive, and natural exposures of it are quite impressive (see the chapter opening photo). In fact, some rocks adjacent to granite were altered by heat and fluids and now contain important copper deposits. Even the soils we depend on for most of our food formed as a result of alteration of minerals and rocks by several physical and chemical processes, thus providing a good example of the interactions among Earth's systems.

• **Figure 2.1 Some Minerals and Rocks Are Interesting and Attractive** (a) These minerals are on display at the California Academy of Sciences in San Francisco, whereas the metamorphic rock in (b) lies on the Lake Superior shoreline at Marquette, Michigan.

So why should you study minerals and rocks? To understand rocks, you must first be familiar with the building blocks of rocks—that is, with minerals. In addition, rocks of various kinds provide our only record of prehistoric events, including ancient life. In short, to understand Earth's physical and biological history, you must have some knowledge of Earth materials.

Matter and Its Composition

Anything that has mass and occupies space is *matter;* it exists as solids, liquids, gases, and plasma. The last is an ionized gas, as in neon lights and the Sun. Here we are concerned with solids, because by definition minerals are solids.

Elements and Atoms

Matter consists of chemical **elements,** each of which is composed of tiny particles called **atoms,** the smallest units of matter that retain the characteristics of an element. Atoms have a compact nucleus made up of one or more *protons*—particles with a positive electrical charge—and electrically neutral *neutrons.* Negatively charged *electrons* rapidly orbit the nucleus at specific distances in one or more *electron shells* (• Figure 2.2).

The number of protons in an atom's nucleus determines its **atomic number;** carbon has 6 protons, whereas potassium has 19 (Figure 2.2). An atom's **atomic mass number,** in contrast, is found by adding the number of protons and neutrons in the nucleus (electrons have negligible mass). However, the number of neutrons in the nucleus of an element might vary. Carbon atoms (with 6 protons), for instance, have 6, 7, or 8 neutrons, making three *isotopes,* or different forms, of carbon (• Figure 2.3). Some elements have only one isotope, but many have several.

Bonding and Compounds

The process whereby atoms join to other atoms is known as **bonding.** Should atoms of two or more elements bond,

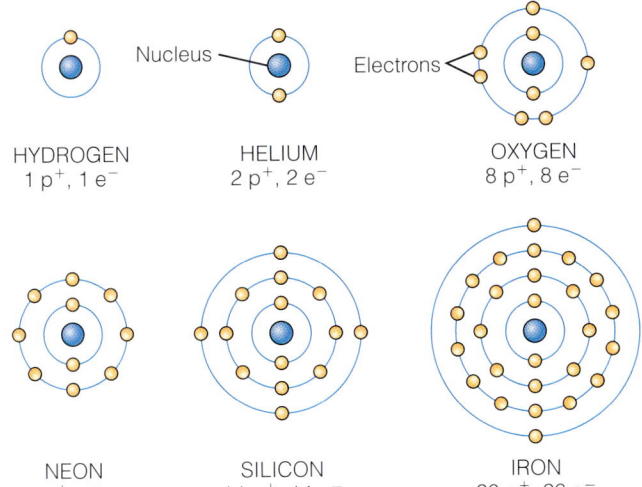

• **Figure 2.2 Shell Model for Atoms** The shell model for several atoms and their electron configurations. A circle represents the nucleus of each atom, but remember that atomic nuclei are made up of protons and neutrons as shown in Figure 2.3.

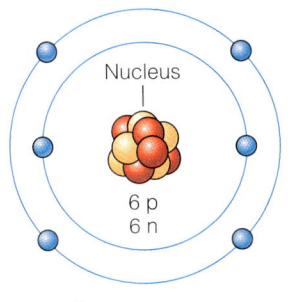
^{12}C (Carbon 12)

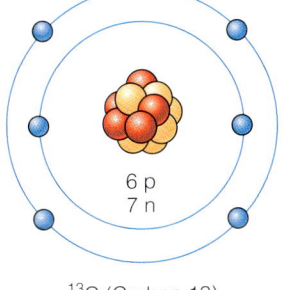
^{13}C (Carbon 13)

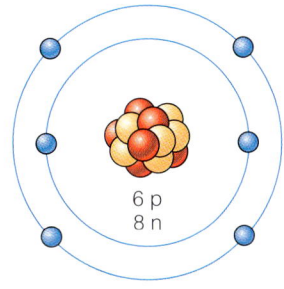
^{14}C (Carbon 14)

• **Figure 2.3 Isotopes of Carbon**
Schematic representation of the isotopes of carbon. Carbon has an atomic number of 6 and an atomic mass number of 12, 13, or 14 depending on the number of neutrons (n) in its nucleus.

the resulting substance is a **compound.** Thus gaseous oxygen is an element, whereas ice, made up of hydrogen and oxygen (H_2O), is a compound. Most minerals are compounds, but there are a few important exceptions.

With the exception of hydrogen with one proton and one electron, the innermost electron shell of an atom has no more than two electrons, and the outermost shell contains no more than eight; these outer ones are those involved in chemical bonding. *Ionic* and *covalent bonding* are the most important types in minerals, although some useful properties of certain minerals result from metallic bonding and van der Waals bonds or forces. A few elements, known as *noble gases,* have complete outer shells with eight electrons (Figure 2.2), so they rarely react with other elements to form compounds.

One way for the noble gas configuration of eight outer electrons to be attained is by transfer of one or more electrons from one atom to another. A good example is sodium (Na) and chlorine (Cl); sodium has only one electron in its outer shell, whereas chlorine has seven. Sodium loses its outer electron, leaving the next shell with eight (• Figure 2.4a). But now sodium has one fewer electron (negative charge) than it has protons (positive charge), so it is a positively charged *ion,* symbolized Na^{+1}.

The electron lost by sodium goes into chlorine's outer shell, which had seven to begin with, so now it has eight (Figure 2.4a). In this case, though, chlorine has one too many negative charges and is thus an ion symbolized Cl^{-1}. An attractive force exists between the Na^{+1} and Cl^{-1} ions, so an *ionic bond* forms between them, yielding the mineral halite (NaCl).

Covalent bonds result when the electron shells of adjacent atoms overlap and they share electrons. A carbon atom in diamond shares all four of its outer electrons with a neighbor to produce the stable noble gas configuration (Figure 2.4b). Among the most common minerals, the silicates, silicon forms partly covalent and partly ionic bonds with oxygen.

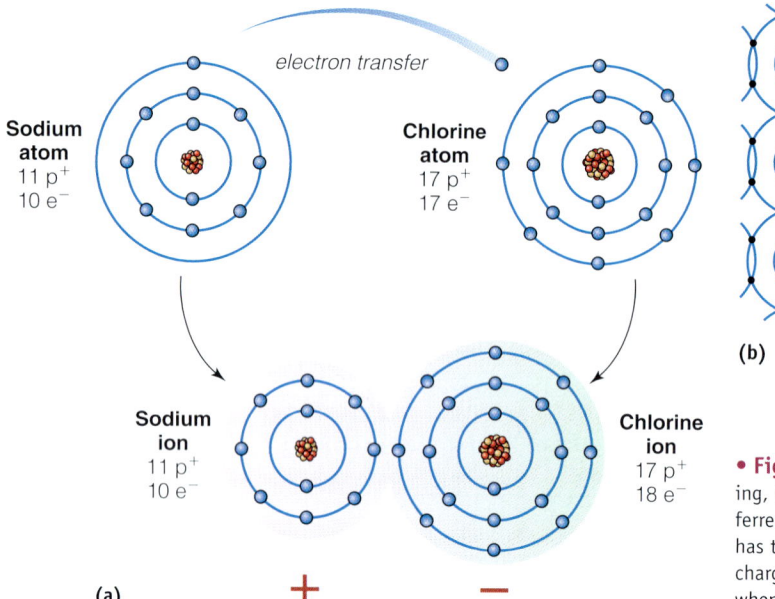
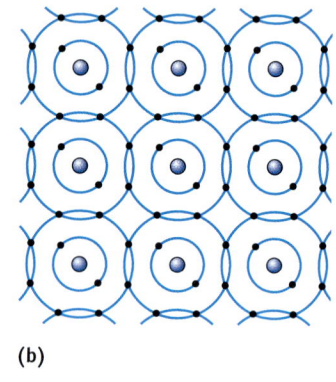

• **Figure 2.4 Ionic and Covalent Bonding** **(a)** In ionic bonding, the electron in the outermost electron shell of sodium is transferred to the outermost electron shell of chlorine. Once the transfer has taken place, the positively charged sodium ion and negatively charged chlorine ion attract one another. **(b)** Covalent bonds form when adjacent atoms share electrons, as in these carbon atoms.

Minerals—The Building Blocks of Rocks

A mineral's composition is shown by a chemical formula, a shorthand way of indicating how many atoms of different kinds it contains. For example, quartz (SiO_2) is made up of one silicon atom for every two oxygen atoms, whereas the formula for orthoclase is $KAlSi_3O_8$. A few minerals known as *native elements,* such as gold (Au) and diamond (C), consist of only one element and accordingly are not compounds.

Recall our formal definition of minerals. The adjective *inorganic* reminds us that animal and vegetable matter are not minerals. Nevertheless, corals and clams and some other organisms build shells of calcite ($CaCO_3$) or silica (SiO_2). By definition minerals are **crystalline solids** in which their atoms are arranged in a specific three-dimensional framework. Ideally, minerals grow and form perfect crystals with planar surfaces (crystal faces), sharp corners, and straight edges (• Figure 2.5). In many cases numerous minerals grow in proximity, as in a cooling lava flow, and thus do not develop well-formed crystals.

Some minerals have very specific chemical compositions, but others have a range of compositions because one element can substitute for another if the atoms of the two elements have the same electrical charge and are about the same size. Iron and magnesium meet these criteria and thus substitute for one another as in olivine {$(Fe,Mg)_2SiO_3$}, which may have magnesium, iron, or a combination of the two. Calcium (Ca) and sodium (Na) substitute for one another in the plagioclase feldspars, which vary from calcium-rich ($CaAl_2Si_2O_8$) to sodium-rich ($NaAlSi_3O_8$) varieties.

Composition and structure control the characteristic physical properties of minerals. These properties are particularly useful for mineral identification (see Appendix C).

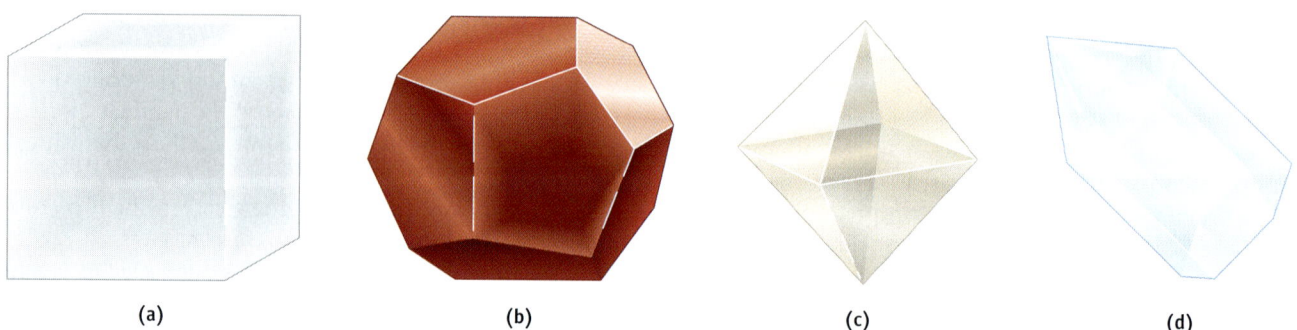

• **Figure 2.5 Mineral Crystals Are Found in a Variety of Shapes** **(a)** Cubic crystals typically develop in the minerals halite and galena. **(b)** Pyritohedron crystals such as those of pyrite are 12-sided. **(c)** Diamond has octahedral or 8-sided crystals. **(d)** A prism terminated by pyramids is found in quartz.

How Many Minerals Are There?

Geologists recognize several mineral groups, each of which is composed of minerals sharing the same negatively charged ion or ion group (Table 2.1). More than 3500 minerals are known, but only about two dozen are particularly common. Many others are important resources, though.

TABLE 2.1 Some of the Mineral Groups Recognized by Geologists

Mineral Group	Negatively Charged Ion or Ion Group	Examples	Composition
Carbonate	$(CO_3)^{-2}$	Calcite	$CaCO_3$
		Dolomite	$CaMg(CO_3)_2$
Halide	Cl^{-1}, F^{-1}	Halite	$NaCl$
		Fluorite	CaF_2
Native element	—	Gold	Au
		Silver	Ag
		Diamond	C
		Graphite	C
Oxide	O^{-2}	Hematite	Fe_2O_3
		Magnetite	Fe_3O_4
Silicate	$(SiO_4)^{-4}$	Quartz	SiO_2
		Potassium feldspar	$KAlSi_3O_8$
		Olivine	$(Mg,Fe)_2SiO_4$
Sulfate	$(SO_4)^{-2}$	Anhydrite	$CaSO_4$
		Gypsum	$CaSO_4 \cdot 2H_2O$
Sulfide	S^{-2}	Galena	PbS
		Pyrite	FeS_2

Silicate Minerals

Given that oxygen (62.6%) and silicon (21.2%) account for nearly 84% of all atoms in Earth's crust, you might expect these elements to be common in minerals. And, indeed, they are. In fact, a combination of silicon and oxygen is called *silica*, and the minerals made up of silica are **silicates**. Silicates account for about one-third of all known minerals and make up perhaps 95% of Earth's crust.

All silicates are composed of a basic building block called the *silica tetrahedron*, consisting of one silicon atom surrounded by four oxygen atoms. These tetrahedra exist in minerals as isolated units bonded to other elements, or they may be arranged in single chains, double chains, sheets, or complex three-dimensional networks, thus accounting for the incredible diversity of these minerals.

Among the silicates, geologists recognize *ferromagnesian silicates*, which contain iron (Fe), magnesium (Mg), or both (• Figure 2.6a). They tend to be dark colored and denser than the *nonferromagnesian silicates*, which, of course, lack these elements (Figure 2.6b).

Other Mineral Groups

Carbonate minerals include those with the carbonate ion $(CO_3)^{-2}$ as in the mineral calcite $(CaCO_3)$. Many other carbonate minerals are known, but of these only dolomite $[CaMg(CO_3)_2]$ is very common. These two minerals are important as the constituents of the sedimentary rocks limestone and dolostone, respectively. Several other mineral groups are also important but more as resources than as constituents of rocks. Their characteristics are summarized in Table 2.1.

(a) Ferromagnesian silicates

(b) Nonferromagnesian silicates

• **Figure 2.6 Some of the Common Silicate Minerals** (a) The ferromagnesian silicates tend to be darker colored and denser than the nonferromagnesian silicates (b).

What Would You Do?

Imagine you are a science teacher. You know that rocks are made up of minerals and that minerals are composed of chemical elements, but despite your best efforts to define them, your students commonly mistake one for the other. Can you think of any analogies that might help them distinguish between minerals and rocks?

Rock-Forming Minerals and the Rock Cycle

We already mentioned that rocks are made up of one or more minerals, but there are a few exceptions. For instance, coal and volcanic glass are both classified as rocks because they are composed of mineral-like matter and they are found associated with other rocks. Geologists have identified hundreds of minerals in rocks, but only a few are common enough to be designated as **rock-forming minerals**—that is, minerals that are essential for the identification and classification of rocks. As you might expect from our previous discussion, most rock-forming minerals are silicates (Figure 2.6), but the carbonate minerals calcite and dolomite are also important. In most cases, minerals present in small amounts, called *accessory minerals*, can be ignored.

The **rock cycle** is a pictorial representation of events leading to the origin, destruction and/or changes, and reformation of rocks as a consequence of Earth's internal and surface processes (• Figure 2.7). Furthermore it shows that the three major rock groups—igneous, sedimentary, and metamorphic—are interrelated; that is, any rock type can be derived from the others. Notice in Figure 2.7 that the ideal cycle involves those events depicted on the circle

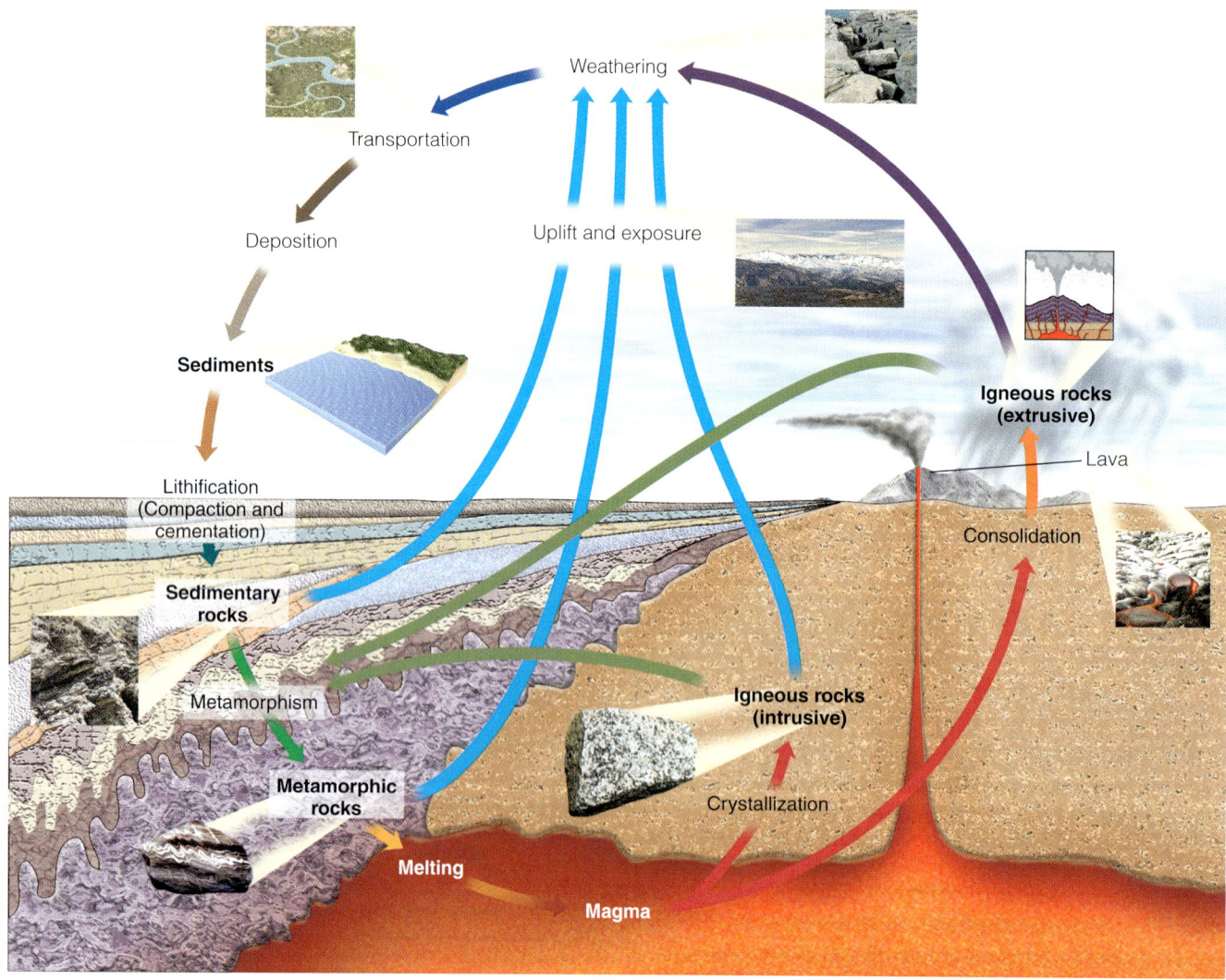

• **Figure 2.7 The Rock Cycle** This cycle shows the interrelationships among Earth's internal and surface processes and how the three major rock groups are related. An ideal cycle includes the events on the outer margin of the cycle, but interruptions indicated by internal arrows are common.

leading from magma to igneous rocks and so on. Notice also that the circle has several internal arrows indicating interruptions in the cycle.

Igneous Rocks

Igneous rocks cool and crystallize from **magma,** molten material below the surface, or from **lava,** molten material on the surface, or they form by the consolidation of particles known as **pyroclastic materials** such as volcanic ash. In any case, geologists categorize igneous rocks as (1) **extrusive** or **volcanic,** meaning they formed at the surface from lava or pyroclastic materials, and (2) **intrusive** or **plutonic,** meaning they formed from magma injected into or formed in place in the crust. The various *plutons* (intrusive bodies) shown in • Figure 2.8 account for the intrusive/plutonic rocks.

Texture and Composition of Igneous Rocks

Several igneous rock textures tell us something about how the rocks formed in the first place. For instance, rapid cooling in a lava flow results in a fine-grained or *aphanitic* texture, in which individual minerals are too small to see without magnification (• Figure 2.9a). In contrast, a coarse-grained or *phaneritic texture* is the outcome of comparatively slow cooling that takes place in plutons (Figure 2.9c). Thus, these two textures are usually sufficient to determine whether an igneous rock is volcanic or plutonic.

Some igneous rocks, though, have a combination of markedly different-sized minerals—a so-called *porphyritic texture;* the large minerals are *phenocrysts,* whereas the smaller ones constitute the rock's *groundmass* (Figure 2.9e). A porphyritic texture might indicate a two-stage cooling history in which magma began cooling below the surface and then was expelled onto the surface where cooling continued. The resulting igneous rocks are characterized as *porphyry*—basalt porphyry, for example.

A *glassy texture* results from cooling so rapidly that the atoms in lava have too little time to form the three-dimensional framework of minerals. As a result, the natural glass *obsidian* forms (Figure 2.9g). Cooling lava might have a large content of trapped water vapor and other gases that form small holes or cavities called *vesicles;* rocks with numerous vesicles are *vesicular* (Figure 2.9h). And finally, a *pyroclastic* or *fragmental texture* characterizes igneous rocks composed of pyroclastic materials (Figure 2.9i).

With few exceptions the primary constituent of magma is silica, but the silica content varies enough for us to recognize *felsic* (>65% silica), *intermediate* (53–65% silica), *mafic* (45–52% silica), and *ultramatic* (<45% silica) *magmas*. Felsic magma also contains considerable sodium, potassium, and aluminum, but little calcium, iron, and magnesium. In contrast, mafic magma has proportionately more calcium, iron, and magnesium.

Intermediate magma, of course, has a composition between those of felsic and mafic magmas.

Classifying Igneous Rocks

Texture and composition are the criteria geologists use to classify all but a few igneous rocks. Notice in • Figure 2.10 that all the rocks shown, except peridotite, are pairs; each member of a pair has the same composition but a different texture. Thus basalt (aphanitic) and gabbro (phaneritic) have the same minerals, and because they contain a large proportion of ferromagnesian silicates they tend to be dark colored (• Figure 2.11a and b). Rhyolite (aphanitic) and granite (phaneritic) also have the

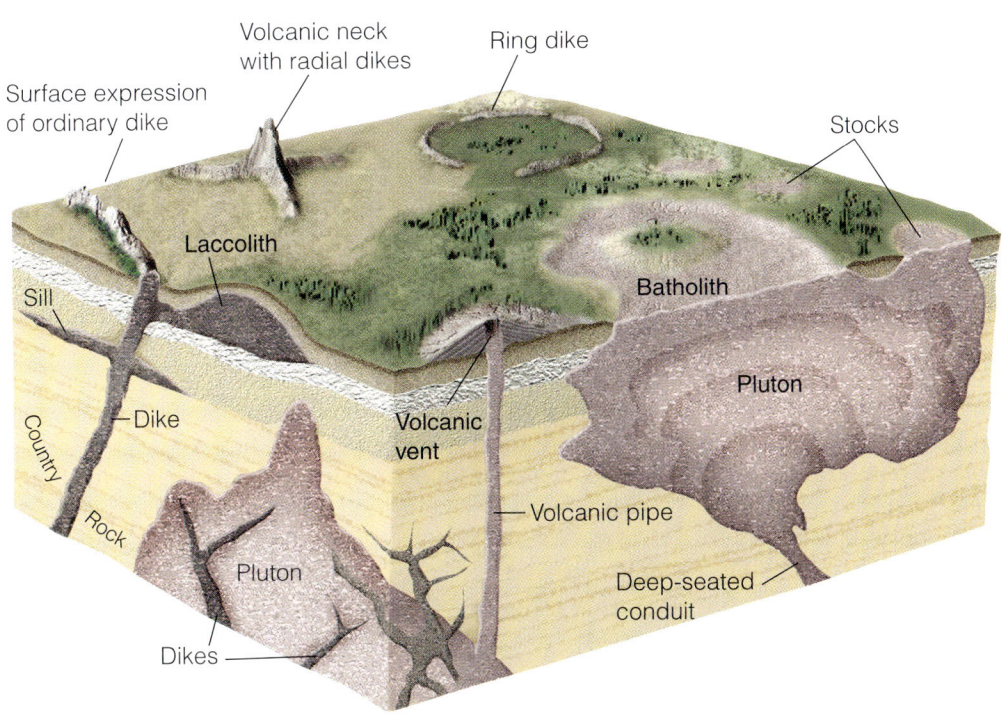

• **Figure 2.8 Block Diagram Showing Plutons and Other Intrusive Igneous Bodies**
Plutons are irregularly shaped bodies of magma that cool to form plutonic rocks, also called intrusive igneous rocks. Other intrusive bodies are sheetlike dikes and sills, and cylindrical volcanic pipes.

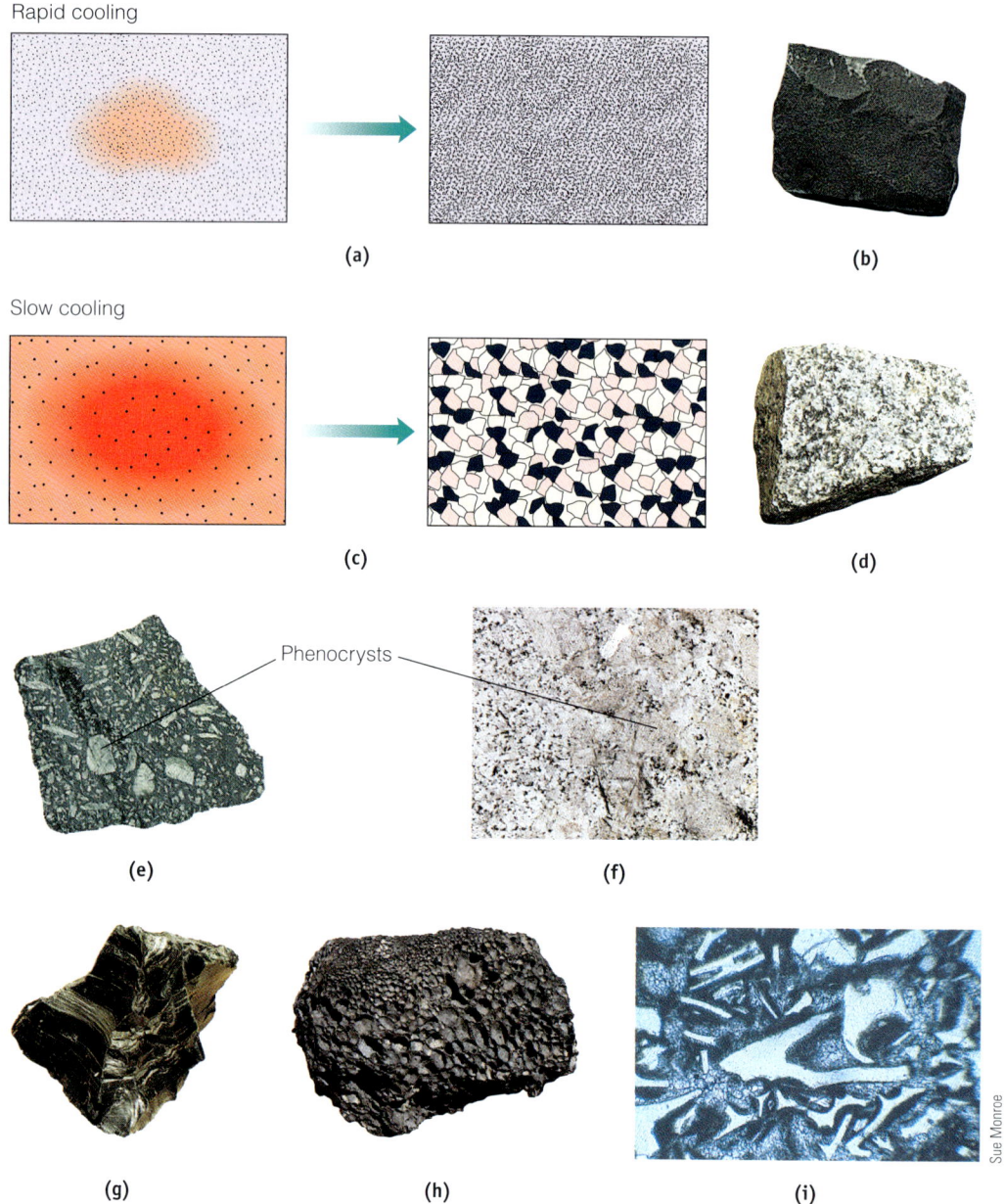

• **Figure 2.9 The Textures of Igneous Rocks** Texture is one criterion used to classify igneous rocks. **(a, b)** Rapid cooling as in lava flows results in many small minerals and an aphanitic (fine-grained) texture. **(c, d)** Slower cooling in plutons yields a phaneritic texture. **(e, f)** These porphyritic textures indicate a complex cooling history. **(g)** Obsidian has a glassy texture because magma cooled too quickly for mineral crystals to form. **(h)** Gases expand in lava and yield a vesicular texture. **(i)** Microscopic view of an igneous rock with a fragmental texture. The colorless, angular objects are pieces of volcanic glass measuring up to 2 mm.

same composition but differ in texture, and because they are made up largely of nonferromagnesian silicates, they are light colored (Figure 2.11e and f). We can infer from their textures that basalt, andesite, and rhyolite are extrusive rocks, whereas gabbro, diorite, and granite are intrusive. Notice also from Figure 2.10 that composition is related to the type of magma from which the rocks cooled.

For a few igneous rocks, texture is the only criterion for classification (• Figure 2.12a). Tuff is composed of volcanic ash, a designation for pyroclastic materials measuring less than 2 mm (Figure 2.12b). Some ash deposits are so hot when they form that the particles fuse together, forming *welded tuff*. Consolidation of larger pyroclastic materials such as *lapilli* (2–64 mm) and blocks and bombs* (>64 mm)

Blocks are angular, whereas bombs are smooth and commonly shaped like a tear drop.

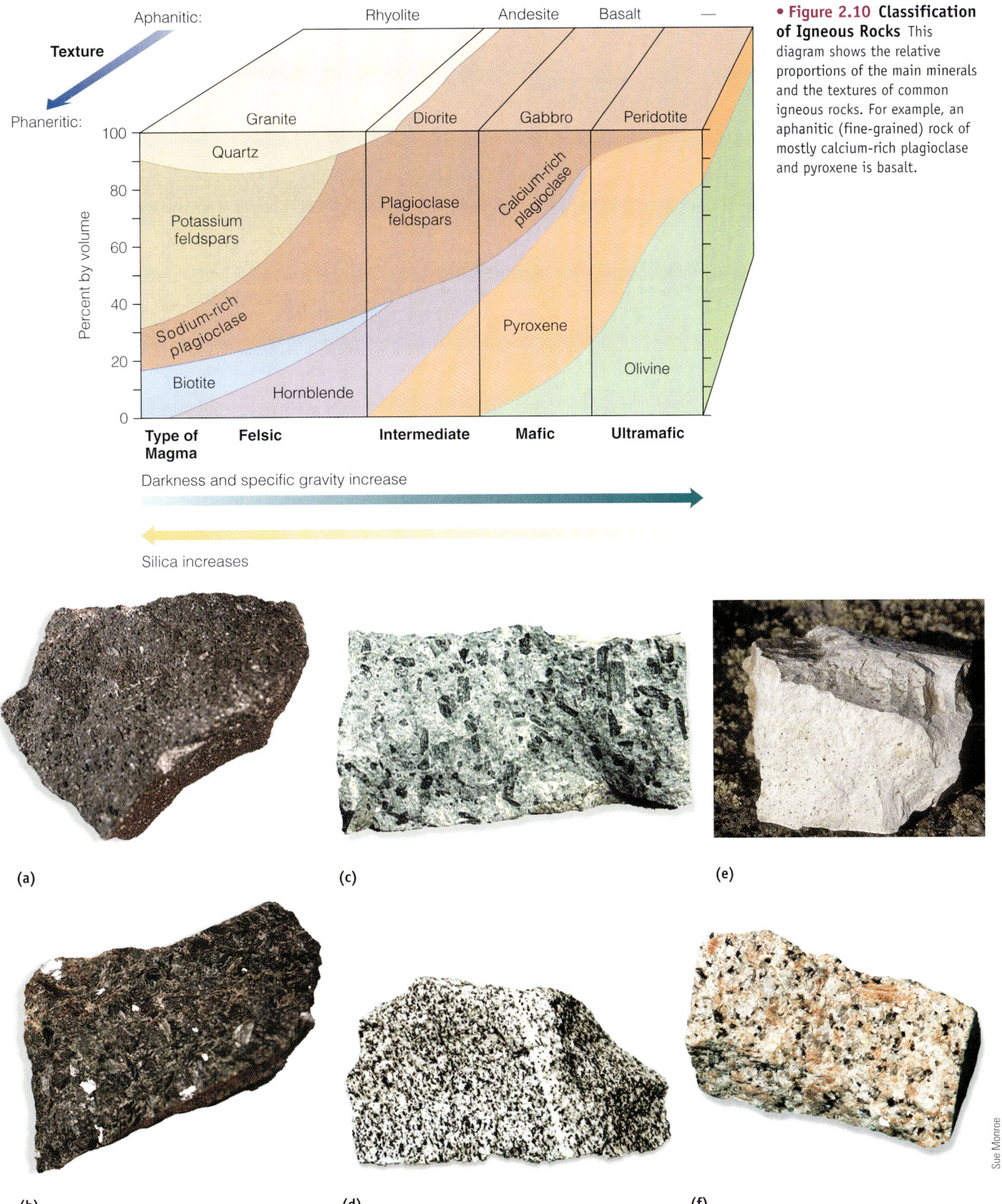

• **Figure 2.10 Classification of Igneous Rocks** This diagram shows the relative proportions of the main minerals and the textures of common igneous rocks. For example, an aphanitic (fine-grained) rock of mostly calcium-rich plagioclase and pyroxene is basalt.

• **Figure 2.11 The Common Igneous Rocks** (a) Basalt, (b) gabbro, (c) andesite porphyry, (d) diorite, (e) rhyolite, and (f) granite. Can you tell from the images which are volcanic and which are plutonic?

IGNEOUS ROCKS 25

(a)

(b)

(c)

• **Figure 2.12 Classification of Igneous Rocks for Which Texture Is the Main Criterion** (a) Vesicular, glassy, and pyroclastic are textures of some igneous rocks. (b) This rock exposure (outcrop) in Colorado shows a basalt lava flow at the top that is underlain by tuff (composed of volcanic ash) and volcanic breccia which is made up of angular volcanic fragments. (c) Pumice is glassy and vesicular. Its density is so low that it floats.

yields *volcanic breccia*. *Obsidian* and *pumice* are varieties of volcanic glass. The former looks like red, brown, or black glass (Figure 2.9g), whereas the latter has a frothy appearance, because it has numerous vesicles (Figure 2.12c).

Sedimentary Rocks

Notice in the rock cycle (Figure 2.7) that weathering processes tend to disintegrate or decompose any rock at the surface. Accordingly, rocks are broken into smaller particles of gravel (>2 mm), sand (1/16–2 mm), silt (1/256–1/16 mm), and clay (<1/256 mm), and some minerals may be dissolved—that is, taken into solution. In either case, these materials are commonly transported elsewhere and deposited as *sediment*, perhaps as sand on a beach or as minerals extracted from solution by inorganic chemical processes or by the activities of organisms. Any sedimentary deposit may be lithified—that is, transformed into **sedimentary rock**, which by definition is any rock made up of sediment.

Sediment Transport, Deposition, and Lithification

Sediment transport involves removing sediment from its source area and carrying it elsewhere. Running water is the most effective transport agent, but glaciers, wind, and waves are also important, at least in some areas. In any case, transported sediment must eventually be deposited as in a stream channel, on a beach, or on the seafloor. Sediment also accumulates as minerals that precipitate from solution or that organisms extract from solution. In all cases, though, sediment is an unconsolidated aggregate of solid particles, such as sand or mud.

Lithification is the geologic term for the phenomenon of converting sediment into sedimentary rock. As sediment accumulates, *compaction,* resulting from pressure exerted by the weight of overlying sediments, reduces the amount of pore space (the void space between particles) and thus the volume of the deposit (• Figure 2.13). Compaction alone is usually sufficient for lithification of mud (particles measuring less than 1/16 mm). But for gravel and sand, *cementation* is necessary, which involves the precipitation of minerals within pores that effectively binds the sediment together (Figure 2.13). The most common cements are calcium carbonate ($CaCO_3$), silica (SiO_2), and, in a distant third place, iron oxides and hydroxides such as hematite (Fe_2O_3) and limonite [$FeO(OH)$], respectively.

Types of Sedimentary Rocks

There are two broad categories of sedimentary rocks. **Detrital sedimentary rocks** are made up of detritus—that is, the solid particles such as gravel and sand derived from preexisting rocks. In contrast, **chemical sedimentary rocks** are made up of minerals derived from materials in solution and extracted by either inorganic chemical processes or

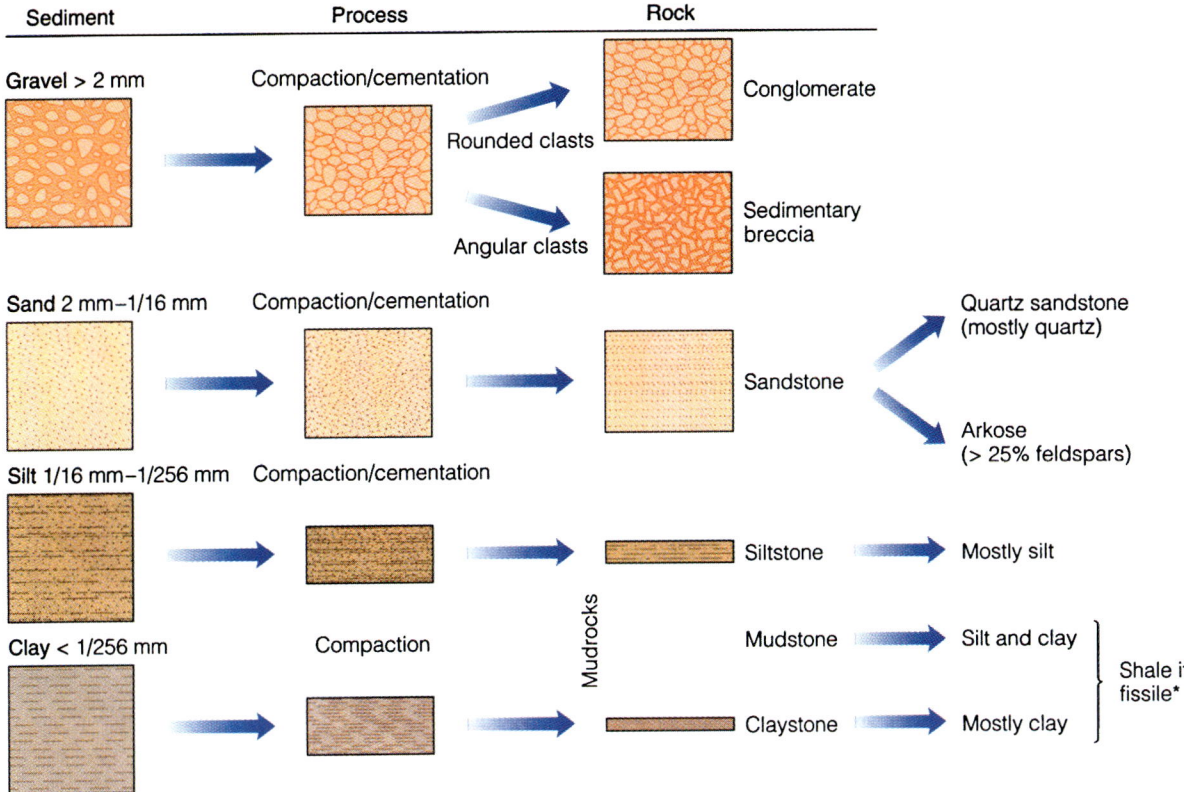

• **Figure 2.13 Lithification and Classification of Detrital Sedimentary Rocks** Notice that little compaction takes place in gravel and sand, that two types of sandstone are shown, and that mudrock is a collective term for detrital sedimentary rocks made up of silt and clay.

by the activities of organisms. The subcategory called *biochemical sedimentary rocks* is so named because of the importance of organisms.

Detrital Sedimentary Rocks All detrital sedimentary rocks are composed of fragments or particles, known as *clasts*, derived from any preexisting rock, and they are defined primarily by the size of their constituent clasts (Figure 2.13). *Conglomerate* and *sedimentary breccia* are both made up of gravel, detrital particles measuring more than 2 mm. The only difference between them is that conglomerate (• Figure 2.14a) has rounded clasts, as opposed to the angular ones in sedimentary breccia (Figure 2.14b). Particle rounding results from wear and abrasion during transport.

The term *sandstone* applies to any detrital sedimentary rock composed of sand (Figure 2.14d). *Quartz sandstone* (Figure 2.14c) is the most common, and, as its name implies, is made up mostly of quartz. Another fairly common type of sandstone is *arkose,* which in addition to quartz has at least 25% feldspar minerals (Figure 2.13).

All detrital sedimentary rocks composed of particles measuring less than 1/16 mm are collectively called *mudrocks* (Figure 2.13). We distinguish between *siltstone* composed of silt-sized particles, *mudstone* with a mixture of silt- and clay-sized particles, and *claystone,* which is composed mostly of particles measuring less than 1/256 mm. However, some mudstones and claystones are further designated as *shale* if they are fissile—meaning that they break along closely spaced parallel planes (Figures 2.13 and 2.14e).

Chemical Sedimentary Rocks You already know that chemical sedimentary rocks result when minerals are extracted from solution by inorganic chemical processes or by organisms. Some of the resulting rocks have a *crystalline texture*—that is, they are made up of an interlocking mosaic of mineral crystals. But others have a clastic texture—an accumulation of broken shells on a beach if lithified becomes a type of limestone.

Limestone and *dolostone* are the most common rocks in this category, and geologists refer to both as **carbonate rocks** because they are composed of carbonate minerals: calcite and dolomite, respectively (Table 2.2, • Figure 2.15a). Actually, most dolomite was originally limestone that was altered when magnesium replaced some of the calcium in calcite. Furthermore, most limestone is also biochemical, because organisms are so important in its origin.

• **Figure 2.14 Detrital Sedimentary Rocks** (a) Exposure (outcrop) of conglomerate at San Gregorio State Beach in California. (b) Sedimentary breccia. (c) Quartz sandstone. (d) Microscopic view of sandstone showing quartz grains that measure mostly from 0.5 to 1.0 mm. (e) Outcrop of the mudrock shale in Tennessee. Notice that the rock splits along closely spaced planes, so it is fissile.

TABLE 2.2 Classification of Chemical and Biochemical Sedimentary Rocks

Chemical Sedimentary Rocks

Texture	Composition	Rock Name	
Crystalline	Calcite ($CaCO_3$)	Limestone	⎤ Carbonates
Crystalline	Dolomite [$CaMg(CO_3)_2$]	Dolostone	⎦
Crystalline	Gypsum ($CaSO_4 \cdot 2H_2O$)	Rock gypsum	⎤ Evaporites
Crystalline	Halite (NaCl)	Rock salt	⎦

Biochemical Sedimentary Rocks

Texture	Composition	Rock Name
Clastic	Calcium carbonate ($CaCO_3$) shells	Limestone (various types such as chalk and coquina)
Usually crystalline	Altered microscopic shells of silicon dioxide (SiO_2)	Chert
—	Mostly carbon from altered plant remains	Coal

The **evaporites** such as *rock salt* and *rock gypsum* form by inorganic chemical precipitation of minerals from solutions made concentrated by evaporation (Table 2.2, Figures 2.15b and c). If you were to evaporate seawater, for instance, you would end up with a small amount of rock gypsum and considerably more rock salt. Several other evaporites would also form but only in very small amounts.

Chert, a dense, hard rock, consists of silica, some of which forms as oval to spherical masses within other rocks by inorganic chemical precipitation. Other chert consists of layers of microscopic shells of silica-secreting organisms and is thus a biochemical sedimentary rock (Table 2.2, Figure 2.15d).

A biochemical sedimentary rock of great economic importance is *coal,* which consists of partially altered, compressed remains of land plants. Coal forms in oxygen-deficient swamps and bogs where the decomposition process is interrupted and the accumulating vegetation alters first to *peat* and if buried and compressed, it changes to coal (Table 2.2, Figure 2.15e).

Metamorphic Rocks

All sedimentary and igneous rocks form under their own specific conditions of temperature, pressure, and fluid activity. However, should they be subjected to different conditions, they may be altered in the solid state—that is, they become **metamorphic rocks.** Metamorphic changes may be compositional (new minerals form) or textural (minerals become aligned) or both (• Figure 2.16). Some of these changes are minor, so the parent rock is easily recognized, but changes can be so great that it can be very difficult to identify the parent rock.

The Agents of Metamorphism

Heat, pressure, and chemical fluids—the agents responsible for metamorphism—may act singly or in any combination, and the time during which they are effective varies considerably. For example, a lava flow may bake the underlying rocks for a short time but otherwise has little effect, whereas the rocks adjacent to a batholith (Figure 2.8) may be altered over a long period and for a great distance from the batholith.

(a) Limestone
(b) Rock salt 
(c) Rock gypsum 
(d) Chert 
(e) Coal 

• **Figure 2.15 Chemical Sedimentary Rocks** (a) Limestone with numerous fossil shells. (b) Rock salt. This specimen was retrieved from an oil well core thus accounting for its cylindrical shape. (c) Rock gypsum. (d) Chert. (e) Coal. Which of these rock specimens are also biochemical?

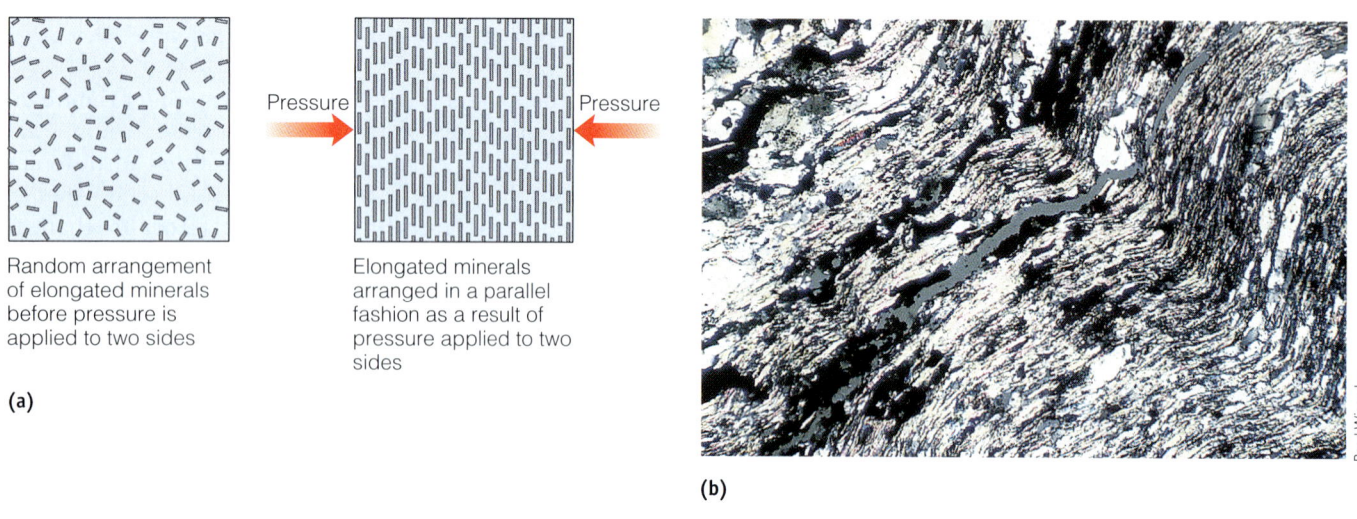

• **Figure 2.16 Textural Change During Metamorphism** (a) When rocks are subjected to differential pressure, the minerals are typically arranged in a parallel fashion, producing a foliated texture. (b) Microscopic view of a metamorphic rock with foliation showing the parallel arrangement of minerals.

Heat is important because it increases the rates of chemical reactions that may yield minerals different from those in the parent rock. Sources of heat include plutons, especially the larger ones such as stocks and batholiths, and deep burial, because Earth's temperature increases with depth. Deep burial also subjects rocks to *lithostatic pressure* resulting from the weight of the overlying rocks. Under these conditions minerals in a rock become more closely packed and may *recrystallize*—that is, form smaller and denser minerals. Lithostatic pressure operates with the same intensity in all directions, but *differential pressure* exerts force more intensely from one direction, as occurs at convergent plate boundaries.

In any region where metamorphism takes place, water is present in varying amounts and may contain ions in solution that enhance metamorphism by increasing the rate of chemical reactions. Accordingly, *fluid activity* is also an important metamorphic agent.

Types of Metamorphism

Contact metamorphism takes place when heat and chemical fluids from an igneous body alter adjacent rocks. The rocks in contact with a batholith may be heated to nearly 1000°C, and of course fluids from the magma also bring about changes. The degree of metamorphism decreases with increasing distance from the body of magma until the surrounding rocks are unaffected.

Most metamorphic rocks result from **regional metamorphism,** which takes place over large but elongated areas as the result of tremendous pressure, elevated temperatures, and fluid activity. This kind of metamorphism is most obvious along convergent plate boundaries where the rocks are intensely deformed during convergence and subduction. It also takes place at divergent plate boundaries, though usually at shallower depths, and here only high temperature and fluid activity are important.

Classifying Metamorphic Rocks

Many metamorphic rocks, especially those subjected to intense differential pressure, have their platy and elongate minerals aligned in a parallel fashion, giving them a *foliated texture* (Figure 2.16). In contrast, some metamorphic rocks do not develop any discernable orientation of their minerals. Instead, they consist of a mosaic of roughly equidimensional minerals and have a *nonfoliated texture.*

Foliated Metamorphic Rocks *Slate,* a very fine-grained rock, results from low-grade metamorphism of mudrocks or, more rarely, volcanic ash (Table 2.3, • Figure 2.17a). *Phyllite* is coarser-grained than slate, but the minerals are still too small to see without magnification. Actually, the change from some kind of mudrock (shale perhaps) to slate and then to phyllite is part of a continuum, and so is the origin of *schist.* In schist, though, the elongate and platy minerals are clearly visible and they impart a *schistosity* or *schistose foliation* to the rock (Table 2.3, Figure 2.17b).

Gneiss, with its alternating dark and light bands of minerals, is one of the more attractive metamorphic rocks (Table 2.3, Figure 2.17c). Quartz and feldspars are most common in the light-colored bands, and biotite and hornblende are found in the dark-colored bands. Most gneiss probably forms from regional metamorphism of clay-rich sedimentary rocks, but metamorphism of granite and other metamorphic rocks also yields gneiss.

In some areas of regional metamorphism, exposures of "mixed rocks" having both igneous and metamorphic

TABLE 2.3 Classification of Common Metamorphic Rocks

Texture	Metamorphic Rock	Typical Minerals	Metamorphic Grade	Characteristics of Rocks	Parent Rock
Foliated	Slate	Clays, micas, chlorite	Low	Fine-grained, splits easily into flat pieces	Mudrocks, volcanic ash
	Phyllite	Fine-grained quartz, micas, chlorite	Low to medium	Fine-grained, glossy or lustrous sheen	Mudrocks
	Schist	Micas, chlorite, quartz, talc, hornblende, garnet, staurolite, graphite	Low to high	Distinct foliation, minerals visible	Mudrocks, carbonates, mafic igneous rocks
	Gneiss	Quartz, feldspars, hornblende, micas	High	Segregated light and dark bands visible	Mudrocks, sandstones, felsic igneous rocks
	Amphibolite	Hornblende, plagioclase	Medium to high	Dark-colored, weakly foliated	Mafic igneous rocks
	Migmatite	Quartz, feldspars, hornblende, micas	High	Streaks or lenses of granite intermixed with gneiss	Felsic igneous rocks mixed with metamorphic rocks
Nonfoliated	Marble	Calcite, dolomite	Low to high	Interlocking grains of calcite or dolomite, reacts with HCl	Limestone or dolostone
	Quartzite	Quartz	Medium to high	Interlocking quartz grains, hard, dense	Quartz sandstone
	Greenstone	Chlorite, epidote, hornblende	Low to high	Fine-grained, green	Mafic igneous rocks
	Hornfels	Micas, garnet, andalusite, cordierite, quartz	Low to medium	Fine-grained, equidimensional grains, hard, dense	Mudrocks
	Anthracite	Carbon	High	Black, lustrous, subconcoidal fracture	Coal

characteristics are present. These rocks, called *migmatites*, consist of streaks or lenses of granite intermixed with ferromagnesian mineral-rich metamorphic rocks (Figure 2.17d). Where do you think these rocks would be in the rock cycle?

Nonfoliated Metamorphic Rocks Nonfoliated metamorphic rocks may form by either contact or regional metamorphism, but in all cases they lack the platy and elongate minerals found in foliated rocks. *Marble*, a well-known metamorphic rock composed of calcite or dolomite, results from metamorphism of limestone or dolostone (Table 2.3, • Figure 2.18a). Some marble is attractive and has been a favorite with sculptors and for building material for centuries.

Metamorphism of quartz sandstone yields *quartzite*, a hard, compact rock (Table 2.3, Figure 2.18b).

The name *greenstone* applies to any compact, dark green, altered mafic igneous rock that formed under metamorphic conditions. Minerals such as chlorite and epidote give it its green color. Several varieties of *hornfels* are known, most of which formed when clay-rich sedimentary rocks were altered during contact metamorphism.

Anthracite is a black, lustrous, hard coal with mostly carbon and a low amount of volatile matter. It forms from other types of coal.

Plate Tectonics and the Rock Cycle

Notice in the rock cycle (Figure 2.7) that interactions among various Earth systems are responsible for the origin and alternation of rocks. The atmosphere, hydrosphere, and biosphere acting on Earth materials account for weathering, erosion, and deposition, whereas Earth's internal heat is responsible for melting and contributes to metamorphism. Plate tectonics also plays an important role in recycling Earth materials.

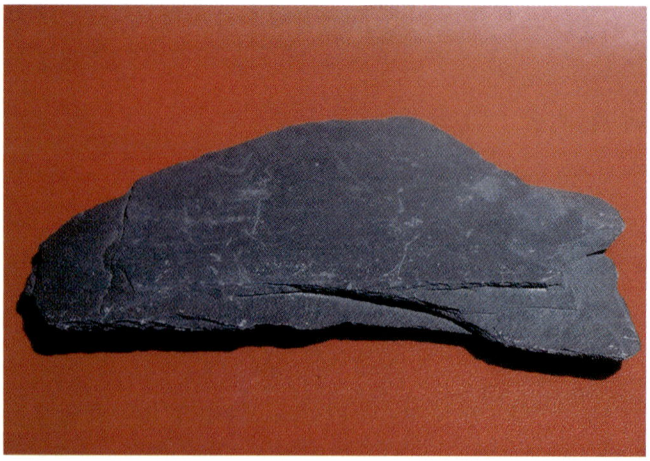

(a) Slate
(c) Gneiss 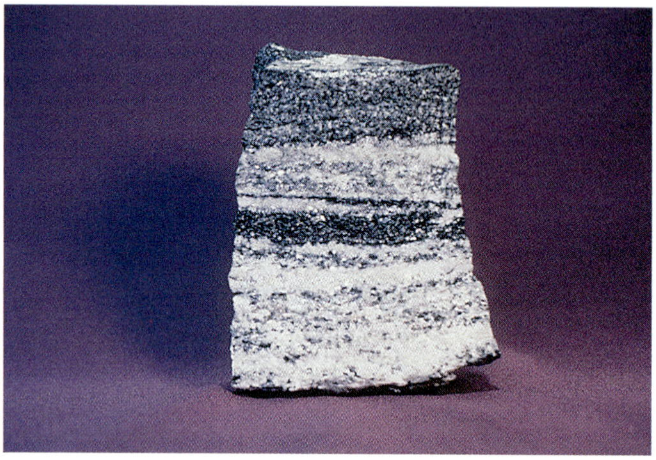
(b) Schist 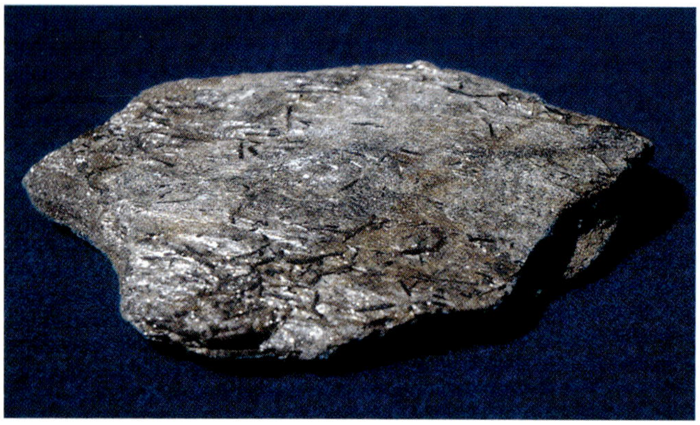
(d) Migmatite

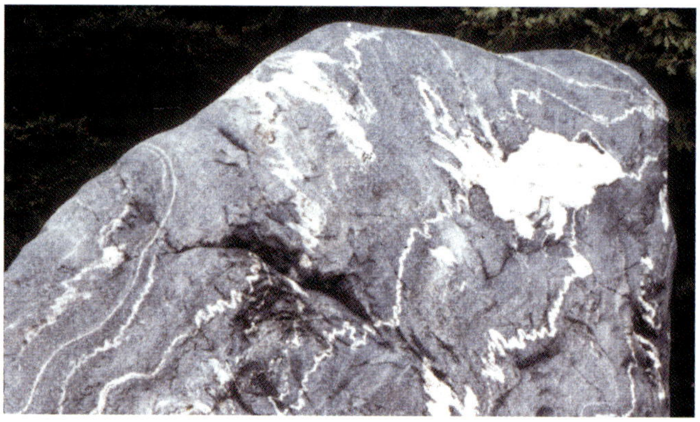

• Figure 2.17 **Foliated Metamorphic Rocks** (a) Slate, (b) schist, (c) gneiss. (d) This rock, known as migmatite, shows features of both metamorphic and igneous rocks. Where would it fit in the rock cycle (see Figure 2.7)?

(a) Marble
(b) Quartzite

• Figure 2.18 **Nonfoliated Metamorphic Rocks** (a) Marble and (b) quartzite. What were these rocks before they were metamorphosed?

Sediment along continental margins may become lithified and be incorporated into a moving plate along with underlying oceanic crust. Where plates collide at convergent plate boundaries, heat and pressure commonly lead to metamorphism and igneous activity and the origin of new rocks. In addition, some of the rocks in a subducted plate are deformed and incorporated into an evolving mountain system that in turn is weathered and eroded, yielding sediment to begin yet another cycle.

Earth Materials and Historical Geology

No human observers were present to record the events that took place during most of geologic time, so our only record of these events is preserved in rocks. Sedimentary rocks have a special place in deciphering Earth and life history, so we will have much more to say about them in Chapters 5 and 6. However, the origin and distribution of igneous rocks reveal much about ancient and continuing plate tectonic activity. Studies of metamorphic rocks provide information about processes operating deep within the crust, such as the pressures and temperatures that prevailed during metamorphism.

> ### What Would You Do?
> Let's say some reputable businesspeople tell you of opportunities to invest in the minerals industry. Two ventures look promising: a gold mine and a sand/gravel pit. Gold sells for about $690 per ounce, and sand and gravel are worth $4 or $5 per ton. In which venture would you likely invest? Explain not only how market price would influence your decision but also what other factors you would consider.

SUMMARY

- Elements are composed of atoms, which have a nucleus of protons and neutrons around which electrons orbit in electron shells.
- The number of protons in a nucleus determines the atomic number of an element; the atomic mass number is the number of protons plus neutrons in the nucleus.
- Atoms join together or bond by transferring electrons from one atom to another (ionic bond) or by sharing electrons (covalent bond). Most minerals are compounds of two or more different elements bonded together.
- By far the most common minerals are the silicates (composed of at least silicon and oxygen), but carbonate minerals (containing the CO_3 ion) are prevalent in some rocks.
- The two broad groups of igneous rocks, intrusive (or plutonic) and extrusive (or volcanic) are classified by composition and texture. However, for a few extrusive rocks texture is the main consideration.
- Sedimentary rocks are also grouped into two broad categories: detrital (made up of solid particles of preexisting rocks) and chemical/biochemical (composed of minerals derived by inorganic chemical processes or the activities of organisms).
- Lithification involving compaction and cementation is the process whereby sediment is transformed into sedimentary rock.
- Metamorphic rocks result from compositional and/or textural transformation of other rocks by heat, pressure, and fluid activity. Most metamorphism is regional, occurring deep within the crust over large areas, but some, called contact metamorphism, takes place adjacent to hot igneous rocks.
- Metamorphism imparts a foliated texture to many rocks (parallel alignment of minerals), but some rocks have a mosaic of equidimensional minerals and are nonfoliated.
- The several varieties of metamorphic rocks are classified largely by their textures, but composition is a consideration for some of these rocks.
- Plate tectonics, which is driven by Earth's internal heat coupled with surface processes such as weathering, erosion, and deposition, accounts for the recycling of Earth materials in the rock cycle.
- Our only record of prehistoric physical and biological events is preserved in rocks, especially sedimentary rocks.

IMPORTANT TERMS

atom, p. 18
atomic mass number, p. 18
atomic number, p. 18
bonding, p. 18
carbonate mineral, p. 21
carbonate rock, p. 27
chemical sedimentary rock, p. 26
compound, p. 19
contact metamorphism, p. 30
crystalline solid, p. 20

detrital sedimentary rock, p. 26
element, p. 18
evaporite, p. 28
extrusive igneous (volcanic) rock, p. 23
igneous rock, p. 23
intrusive igneous (plutonic) rock, p. 23
lava, p. 23
lithification, p. 26
magma, p. 23
metamorphic rock, p. 29

mineral, p. 17
pyroclastic material p. 23
regional metamorphism, p. 30
rock, p. 17
rock cycle, p. 22
rock-forming mineral, p. 22
sedimentary rock, p. 26
silicate, p. 21

REVIEW QUESTIONS

1. An atom with 6 protons and 8 neutrons has an atomic mass number of
   a. ____ 14; b. ____ 2; c. ____ 8; d ____ 6; e. ____ 48.
2. In which type of bonding do atoms transfer electrons to other atoms?
   a. ____ covalent; b. ____ tetrahedral; c. ____ ionic; d. ____ carbon; e. ____ silicate.
3. One process involved in the lithification of sand is
   a. ____ weathering; b. ____ erosion; c. ____ transport; d. ____ cementation; e. ____ metamorphism.
4. If you were to encounter an igneous rock in which you could see no minerals, you would be justified in concluding that the rock is
   a. ____ plutonic; b. ____ detrital; c. ____ foliated; d. ____ metamorphic; e. ____ volcanic.
5. Crystalline means
   a. ____ a mineral has an orderly arrangement of atoms; b. ____ igneous rocks are derived from metamorphic rocks; c. ____ ferromagnesian silicates tend to be light colored; d. ____ compounds are made up of two or more elements bonded together; e. ____ metamorphic rocks have minerals too small to see without magnification.
6. Marble is
   a. ____ a sedimentary rock made up of partly altered vegetation; b. ____ a metamorphic rock formed from limestone; c. ____ an igneous rock that solidified from mafic magma; d. ____ the most common nonferromagnesian silicate mineral; e. ____ a type of natural glass.
7. Which of the following pairs of igneous rocks have the same mineral composition?
   a. ____ andesite-basalt; b. ____ gabbro-shale; c. ____ gneiss-slate; d. ____ rhyolite-granite; e. ____ obsidian-diorite.
8. Most limestones have a large component of calcite that was extracted from seawater by
   a. ____ organisms; b. ____ lithificiation; c. ____ chemical weathering; d. ____ melting; e. ____ evaporation.
9. The alteration of rocks adjacent to a batholith is called
   a. ____ chemical bonding; b. ____ erosion; c. ____ contact metamorphism; d. ____ crystal differentiation; e. ____ mechanical weathering.
10. Which one of the following is a carbonate mineral?
    a. ____ basalt; b. ____ quartz; c. ____ potassium feldspar; d. ____ calcite; e. ____ olivine.
11. How do evaporites form, and which ones are common?
12. How does contact metamorphism compare with regional metamorphism?
13. What kinds of igneous rock textures would you expect to find in a batholith and a lava flow? Why is there a difference in textures?
14. What are the two basic categories of sedimentary rock, and how do they differ?
15. Compare covalent and ionic bonding.
16. Why is ice a mineral, but water vapor and liquid water are not minerals?
17. What are the two groups of silicate minerals, and how do they differ from one another?

APPLY YOUR KNOWLEDGE

1. Cubic crystals of pyrite, galena, and halite are common, so how could you tell them apart? See Appendix C.

FIELD QUESTION

1. Examine the two igneous rock specimens shown here. Can you tell which is volcanic and which is plutonic? If so, how? Specimen 1 is made up of 38% calcium-rich plagioclase, 10% hornblende, 40% pyroxene, and 12% olivine. Specimen 2 has 15% quartz, 55% potassium feldspar, 20% sodium-rich plagioclase, and 10% biotite and hornblende. Classify both rock specimens.

Dita Alangkara/ AP/ Wide World Photo

CHAPTER 3
PLATE TECTONICS: A UNIFYING THEORY

An earthquake survivor walks among the ruins of houses in Bantul, central Indonesia, where a 6.3-magnitude earthquake on May 27, 2006, left more than 6,200 dead and over 200,000 people homeless. Devastating earthquakes such as this are the result of movement along plate boundaries. Such earthquakes are part of the interaction between plates, and, unfortunately for humans, will continue to result in tremendous loss of life and property damage in seismically active areas.

[OUTLINE]

INTRODUCTION

EARLY IDEAS ABOUT CONTINENTAL DRIFT

PERSPECTIVE 3.1 *Oil, Plate Tectonics, and Politics*

Alfred Wegener and the Continental Drift Hypothesis

Additional Support for Continental Drift

PALEOMAGNETISM AND POLAR WANDERING

HOW DO MAGNETIC REVERSALS RELATE TO SEAFLOOR SPREADING?

PLATE TECTONICS AND PLATE BOUNDARIES

Divergent Boundaries

An Example of Ancient Rifting

Convergent Boundaries

Recognizing Ancient Convergent Boundaries

Transform Boundaries

WHAT ARE HOT SPOTS AND MANTLE PLUMES?

HOW ARE PLATE MOVEMENT AND MOTION DETERMINED?

WHAT IS THE DRIVING MECHANISM OF PLATE TECTONICS?

HOW ARE PLATE TECTONICS AND MOUNTAIN BUILDING RELATED?

Terrane Tectonics

HOW DOES PLATE TECTONICS AFFECT THE DISTRIBUTION OF LIFE?

HOW DOES PLATE TECTONICS AFFECT THE DISTRIBUTION OF NATURAL RESOURCES?

SUMMARY

ThomsonNOW   Explore interactive tutorials, animations, or practice problems available on the ThomsonNow website at www.thomsonedu.com/login.

CHAPTER OBJECTIVES

At the end of this chapter, you will have learned that

- Plate tectonics is a unifying theory of geology that has revolutionized geology.
- The hypothesis of continental drift was based on considerable geologic, paleontologic, and climatologic evidence.
- The hypothesis of seafloor spreading accounts for continental movement, and thermal convection cells provide a mechanism for plate movement.
- The three types of plate boundaries are divergent, convergent, and transform, and along these boundaries new plates are formed, consumed, or slide past one another.
- Interaction along plate boundaries accounts for most of Earth's earthquake and volcanic activity.
- The rate of movement and motion of plates can be calculated in several ways.
- Plate movements have affected the distribution of natural resources, as well as the evolution and distribution of the world's biota.

Introduction

At 5:27 A.M. on December 26, 2003, violent shaking from an earthquake awakened hundreds of thousands of people in the Bam area of southeastern Iran. Soon after the earthquake was over, an estimated 43,000 people were dead, at least 30,000 were injured, and approximately 75,000 survivors were left homeless. The amount of destruction this 6.6-magnitude earthquake caused is staggering. At least 85% of the structures in the Bam area were destroyed or damaged. Collapsed buildings were everywhere, streets were strewn with rubble, and all communications were knocked out. All in all, this was a disaster of epic proportions. Yet it was not the first, nor will it be the last, major devastating earthquake in this region or other parts of the world.

Now go back another 12½ years to June 15, 1991, when Mount Pinatubo in the Philippines erupted violently, discharging huge quantities of ash and gases into the atmosphere. Fortunately, in this case, warnings of an impending eruption were broadcast and heeded, resulting in the evacuation of 200,000 people from areas around the volcano. Unfortunately, the eruption still caused at least 364 deaths, not only from the eruption but also from ensuing mudflows.

What do these two tragic events and other equally destructive volcanic eruptions and earthquakes have in common? They are part of the dynamic interactions involving Earth's plates. When two plates come together, one plate is pushed or pulled under the other plate, triggering large earthquakes such as the one that shook India in 2001, Iran in 2003, Pakistan in 2005, and Indonesia in 2006.

As the descending plate moves downward and is assimilated into Earth's interior, magma is generated. Being less dense than the surrounding material, the magma rises toward the surface, where it may erupt as a volcano, as Mount Pinatubo did in 1991, and others have before and since. It therefore should not be surprising that the distribution of volcanoes and earthquakes closely follows plate boundaries.

As we stated in Chapter 1, plate tectonic theory has had significant and far-reaching consequences in all fields of geology, because it provides the basis for relating many seemingly unrelated phenomena. The interactions between moving plates determine the location of continents, ocean basins, and mountain systems, which in turn affect atmospheric and oceanic circulation patterns that ultimately determine global climates. Plate movements have also profoundly influenced the geographic distribution, evolution, and extinction of plants and animals. Furthermore, the formation and distribution of many geologic resources, such as metal ores, are related to plate tectonic processes, so geologists incorporate plate tectonic theory in their prospecting efforts.

If you're like most people, you probably have no idea or only a vague notion of what plate tectonic theory is. Yet plate tectonics affects us all, whether in terms of the destruction caused by volcanoes and earthquakes or politically and economically (see Perspective 3.1). It is therefore important to understand this unifying theory, not only because it impacts us as individuals and citizens of nation-states but also because it ties together all aspects of the geology you will be studying.

Early Ideas About Continental Drift

The idea that Earth's past geography was different from today is not new. The earliest maps showing the east coast of South America and the west coast of Africa probably provided people with the first evidence that continents may have once been joined together, then broken apart and moved to their present locations.

Perspective 3.1

Oil, Plate Tectonics, and Politics

It is certainly not surprising that oil and politics are closely linked. The Iran–Iraq War of 1980–1989 and the Gulf War of 1990–1991 were both fought over oil (Figure 1). Indeed, many of the conflicts in the Middle East are a result of desires to control the vast deposits of petroleum in the region. Most people, however, are not aware of why there is so much oil in this part of the world.

Although large concentrations of petroleum occur in many areas of the world, more than 50% of all proven reserves are in the Persian Gulf region. It is interesting, however, that this region did not become a significant petroleum-producing area until the economic recovery following World War II (1939–1945). After the war, Western Europe and Japan in particular became dependent on Persian Gulf oil, and they still rely heavily on this region for most of their supply. The United States is also dependent on imports from the Persian Gulf but receives significant quantities of petroleum from other sources, such as Mexico and Venezuela.

Why is there so much oil in the Persian Gulf region? The answer lies in the paleogeography and plate movements of this region during the Mesozoic and Cenozoic eras. During the Mesozoic Era, and particularly the Cretaceous Period when most of the petroleum formed, the Persian Gulf area was a broad, stable marine shelf extending eastward from Africa. This passive continental margin lay near the equator, where countless microorganisms lived in the surface waters. The remains of these organisms accumulated with the bottom sediments and were buried, beginning the complex process of petroleum generation and the formation of source beds in which petroleum forms.

As a consequence of rifting in the Red Sea and Gulf of Aden during the Cenozoic Era, the Arabian plate is moving northeast away from Africa and subducting beneath Iran. As the sediments of the passive continental margin were initially subducted, during the early stages of collision between Arabia and Iran, the heating broke down the organic molecules and led to the formation of petroleum. The tilting of the Arabian block to the northeast allowed the newly formed petroleum to migrate upward into the interior of the Arabian plate. The continued subduction and collision with Iran folded the rocks, creating traps for petroleum to accumulate, such that the vast area south of the collision zone (known as the Zagros suture) is a major oil-producing region.

Figure 1 The Kuwaiti night skies were illuminated by 700 blazing oil wells set afire by Iraqi troops during the 1991 Gulf War. The fires continued for nine months.

During the late 19th century, Austrian geologist Edward Suess noted the similarities between the Late Paleozoic plant fossils of India, Australia, South Africa, and South America, as well as evidence of glaciation in the rock sequences of these southern continents. The plant fossils comprise a unique flora that occurs in the coal layers just above the glacial deposits of these southern continents. This flora is very different from the contemporaneous coal swamp flora of the northern continents and is collectively known as the *Glossopteris* **flora,** after its most conspicuous genus (• Figure 3.1).

In his book, *The Face of the Earth*, published in 1885, Suess proposed the name *Gondwanaland* (or **Gondwana,** as we will use here) for a supercontinent composed of the aforementioned southern continents. Abundant fossils of the *Glossopteris* flora are found in coal seams in Gondwana, a province in India. Suess thought these southern continents were at one time connected by land bridges over which plants and animals migrated. Thus, in his view, the similarities of fossils on these continents were due to the appearance and disappearance of the connecting land bridges.

The American geologist Frank Taylor published a pamphlet in 1910 presenting his own theory of continental drift. He explained the formation of mountain ranges as a result of the lateral movement of continents. He also envisioned the present-day continents as parts of larger polar continents that eventually broke apart and migrated toward the equator after Earth's rotation was supposedly slowed by gigantic tidal forces. According to Taylor, these tidal forces were generated when Earth captured the Moon about 100 million years ago.

Although we now know that Taylor's mechanism is incorrect, one of his most significant contributions was his suggestion that the Mid-Atlantic Ridge, discovered by the 1872–1876 British HMS *Challenger* expeditions, might mark the site along which an ancient continent broke apart, forming the present-day Atlantic Ocean.

Alfred Wegener and the Continental Drift Hypothesis

Alfred Wegener, a German meteorologist (• Figure 3.2), is generally credited with developing the hypothesis of **continental drift.** In his monumental book, *The Origin of Continents and Oceans* (first published in 1915), Wegener proposed that all landmasses were originally united into a single supercontinent that he named **Pangaea,** from the Greek meaning "all land." Wegener portrayed his grand concept of continental movement in a series of maps showing the breakup of Pangaea and the movement of the various continents to their present-day locations. Wegener amassed a tremendous amount of geologic, paleontologic, and climatologic evidence in support of continental drift, but the initial reaction of scientists to his then-heretical ideas can best be described as mixed.

What evidence did Wegener use to support his hypothesis of continental drift? First, Wegener noted that the shorelines of continents fit together, forming a large supercontinent (• Figure 3.3), and that marine, nonmarine, and glacial rock sequences of Pennsylvanian to Jurassic age are almost identical for all five Gondwana continents, strongly indicating that they were joined together at one time (• Figure 3.4). Furthermore, mountain ranges and glacial deposits match up when continents are united into a single landmass (• Figure 3.5). And last, many of the same extinct plant and animal groups are found today on widely separated continents, indicating that the continents must have been close to each other at one time (• Figure 3.6). Wegener argued that this vast amount of evidence from a variety of sources surely indicated the continents must have been in close proximity to each other at some time in the past.

Additional Support for Continental Drift

With the publication of *The Origin of Continents and Oceans* and its subsequent four editions, some scientists began to take Wegener's unorthodox views seriously.

Alexander du Toit, a South African geologist, was one of his more ardent supporters. He further developed Wegener's arguments and introduced more geologic evidence in support of continental drift. In 1937 du Toit published *Our Wandering Continents,* in which he contrasted the glacial deposits of Gondwana with coal deposits of the same age found in the continents of the Northern Hemisphere. To resolve this apparent climatologic paradox, du Toit placed the southern continents of Gondwana at or near the South Pole and arranged the northern continents together such that the coal deposits were located at the equator. This northern landmass, consisting of present-day North America, Greenland, Europe, and Asia (except for India), he named **Laurasia.**

• Figure 3.1 Fossil *Glossopteris* Leaves Plant fossils, such as these *Glossopteris* leaves from the Upper Permian Dunedoo Formation, Australia, are found on all five of the Gondwana continents. The presence of these fossil plants on continents with widely varying climates today is evidence that the continents were at one time connected. The distribution of the plants at that time was in the same climatic latitudinal belt.

• **Figure 3.2 Alfred Wegener** Alfred Wegener, a German meteorologist, proposed the continental drift hypothesis in 1912 based on a tremendous amount of geologic, paleontologic, and climatologic evidence. He is shown here waiting out the Arctic winter in an expedition hut in Greenland.

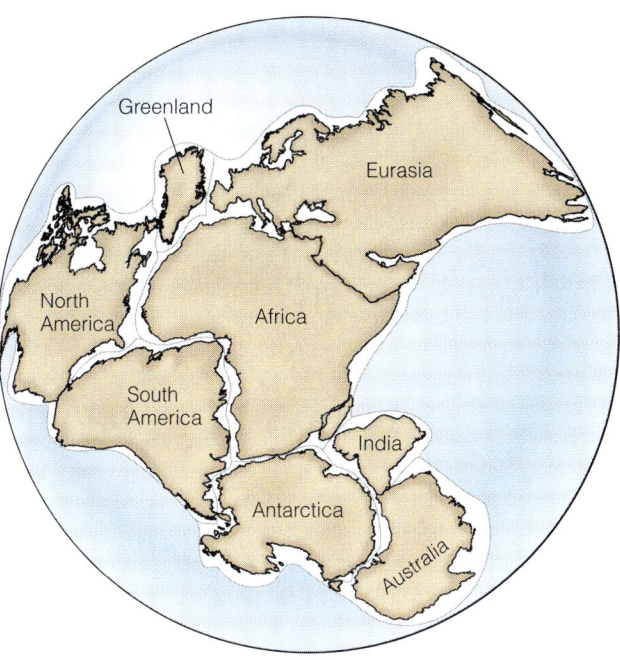

• **Figure 3.3 Continental Fit** When continents are brought together based on their outlines, the best fit isn't along their present-day coastlines but rather along the continental slope at a depth of about 2000 m.

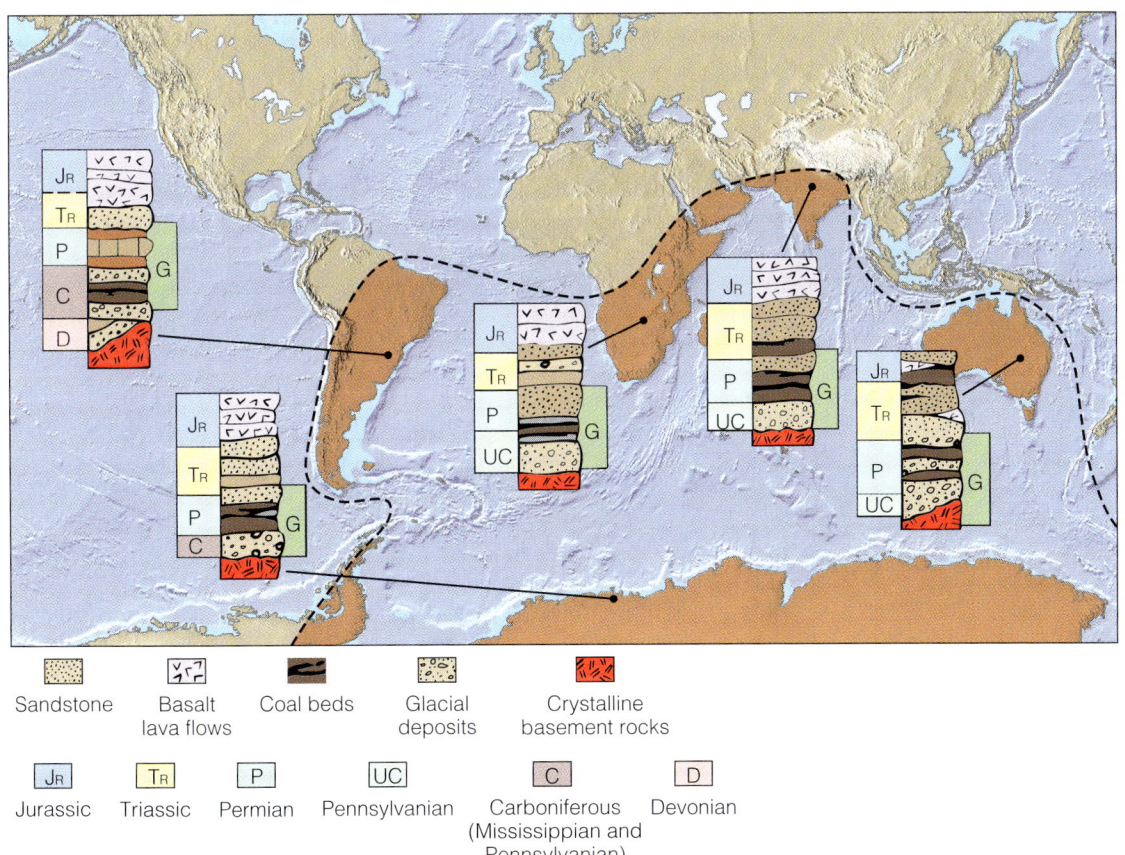

• **Figure 3.4 Similarity of Rock Sequences on the Gondwana Continents** Sequences of marine, nonmarine, and glacial rocks of Pennsylvanian (UC) to Jurassic (JR) age are nearly the same on all five Gondwana continents (South America, Africa, India, Australia, and Antarctica). These continents are widely separated today and have different environments and climates ranging from tropical to polar. Thus, the rocks forming on each continent are very different. When the continents were all joined together in the past, however, the environments of adjacent continents were similar and the rocks forming in those areas were similar. The range indicated by G in each column is the age range (Carboniferous-Permian) of the *Glossopteris* flora.

EARLY IDEAS ABOUT CONTINENTAL DRIFT

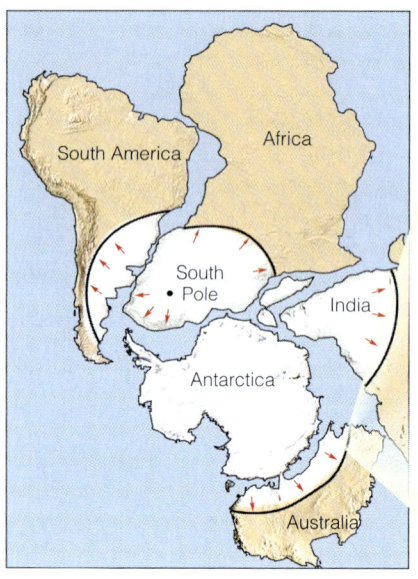

• **Figure 3.5 Glacial Evidence Indicating Continental Drift** (a) When the continents are placed together so that South Africa is located at the South Pole, the glacial movements indicated by striations (red arrows) found on rock outcrops on each continent make sense. In this situation, the glacier (white area), is located in a polar climate and has moved radially outward from its thick central area toward its periphery. (b) Glacial striations (scratch marks) on an outcrop of Permian-aged bedrock exposed at Hallet's Cove, Australia, indicate the general direction of glacial movement more than 200 million years ago. As a glacier moves over a continent's surface, it grinds and scratches the underlying rock. The scratch marks that are preserved on a rock's surface (glacial striations) thus provide evidence of the direction (red arrows) the glacier moved at that time.

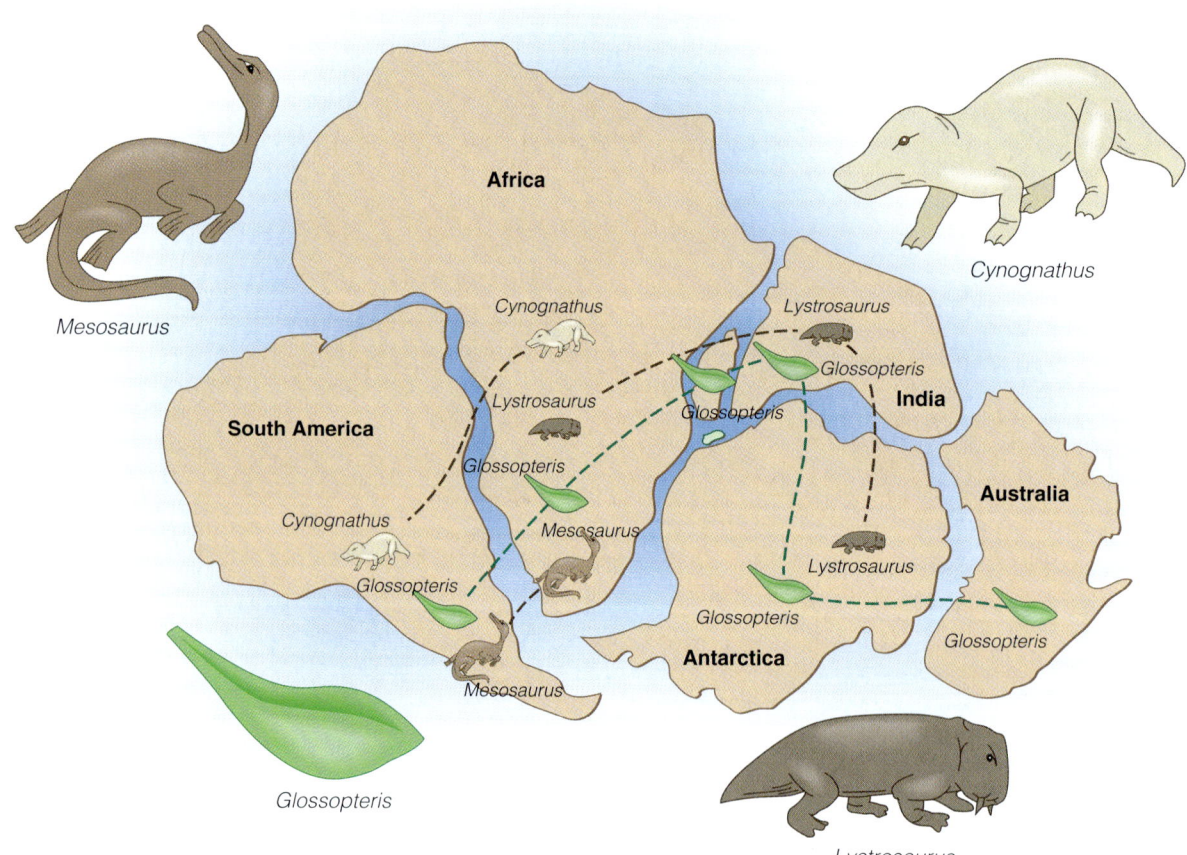

• **Figure 3.6 Fossil Evidence Supporting Continental Drift** Some of the animals and plants whose fossils are found today on the widely separated continents of South America, Africa, India, Australia, and Antarctica. During the Late Paleozoic Era, these continents were joined together to form Gondwana, the southern landmass of Pangaea. Plants of the *Glossopteris* flora are found on all five continents, which today have widely different climates, but during the Pennsylvanian and Permian periods, they were all located in the same general climatic belt. *Mesosaurus* is a freshwater reptile whose fossils are found only in similar nonmarine Permian-aged rocks in Brazil and South Africa. *Cynognathus* and *Lystrosaurus* are land reptiles that lived during the Early Triassic Period. Fossils of *Cynognathus* are found in South America and Africa, whereas fossils of *Lystrosaurus* have been recovered from Africa, India, and Antarctica. It is hard to imagine how freshwater and land-dwelling reptiles could have swum across the wide oceans that presently separate these continents. It is more logical to assume that the continents were at one time connected. Modified from E.H. Colbert, *Wandering Lands and Animals,* 1973, 72, Figure 31.

Du Toit also provided additional paleontologic support for continental drift by noting that fossils of the Permian freshwater reptile *Mesosaurus* occurred in rocks of the same age in both Brazil and South Africa (Figure 3.6). Because the physiologies of freshwater and marine animals are completely different, it is hard to imagine how a freshwater reptile could have swum across the Atlantic Ocean and then found a freshwater environment nearly identical to its former habitat. Moreover, if *Mesosaurus* could have swum across the ocean, its fossil remains should be widely dispersed and not just limited to Brazil and South Africa. It is more logical to assume that *Mesosaurus* lived in lakes in what are now adjacent areas of South America and Africa but were then united into a single continent.

Despite all of the empirical evidence presented by Wegener and later by du Toit and others, most geologists simply refused to entertain the idea that continents might have moved in the past. The geologists were not necessarily being obstinate about accepting new ideas; rather, they found the proposed mechanisms for continental drift inadequate and unconvincing. In part, this was because no one could provide a suitable mechanism to explain how continents could move over Earth's surface. Not until new evidence from studies of Earth's magnetic field and oceanographic research showed that the present-day ocean basins were geologically young features did interest in continental drift theory revive.

Paleomagnetism and Polar Wandering

Interest in continental drift revived during the 1950s as a result of new evidence from paleomagnetic studies. **Paleomagnetism** is the remanent magnetism in ancient rocks recording the direction and intensity of Earth's magnetic poles at the time of the rock's formation. Earth can be thought of as a giant dipole magnet in which the magnetic poles essentially coincide with the geographic poles (• Figure 3.7). This arrangement means that the strength of the magnetic field is not constant but varies, being weakest at the equator and strongest at the poles. Earth's magnetic field is thought to result from the different rotation speeds of the outer core and mantle.

When magma cools, the magnetic iron-bearing minerals align themselves with Earth's magnetic field, recording both its direction and strength. The temperature at which iron-bearing minerals gain their magnetization is called the **Curie point.** As long as the rock is not subsequently heated above the Curie point, it will preserve that remanent magnetism. Thus an ancient lava flow provides a record of the orientation and strength of Earth's magnetic field at the time the lava flow cooled.

As paleomagnetic research progressed in the 1950s, some unexpected results emerged. When geologists measured the paleomagnetism of geologically recent rocks,

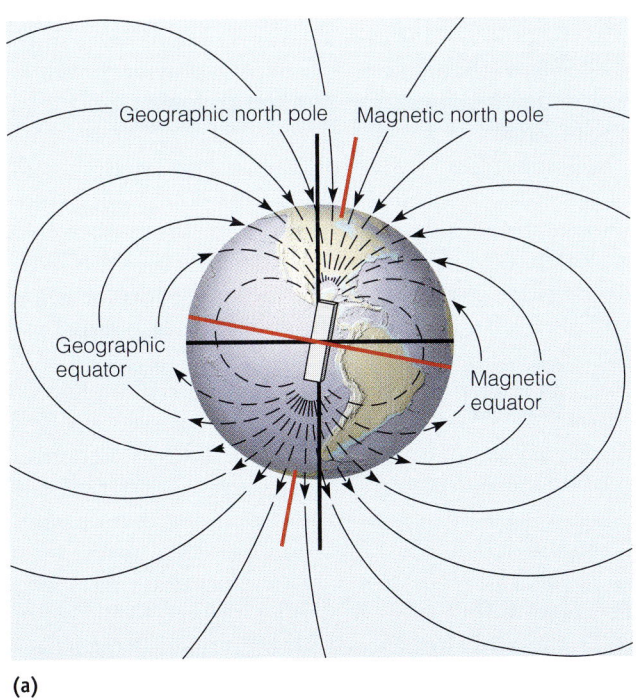

(a)

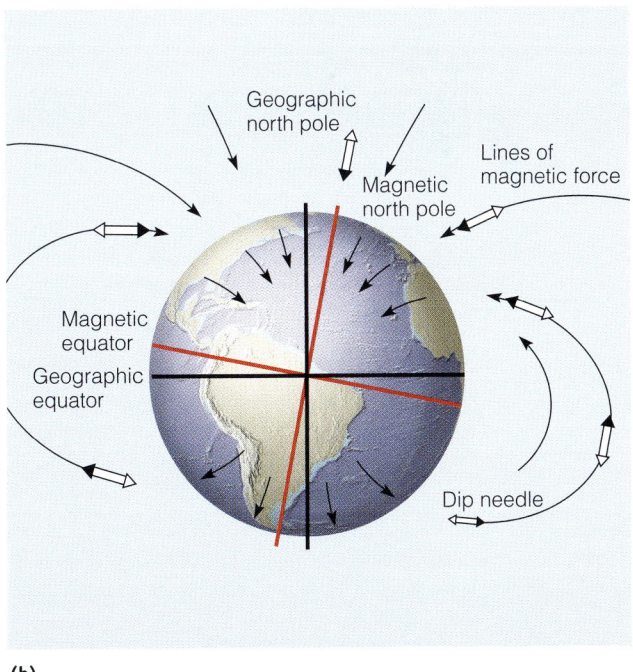

(b)

• **Figure 3.7 Earth's Magnetic Field (a)** Earth's magnetic field has lines of force just like those of a bar magnet. **(b)** The strength of the magnetic field changes from the magnetic equator to the magnetic poles. This change in strength causes a dip needle (a magnetic needle that is balanced on the tip of a support so that it can freely move vertically) to be parallel to Earth's surface only at the magnetic equator, where the strength of the magnetic north and south poles are equally balanced. Its inclination or dip with respect to Earth's surface increases as it moves toward the magnetic poles, until it is at 90 degrees or perpendicular to Earth's surface at the magnetic poles.

they found it was generally consistent with Earth's current magnetic field. The paleomagnetism of ancient rocks, though, showed different orientations. For example, paleomagnetic studies of Silurian lava flows in North America indicated that the north magnetic pole was located in the western Pacific Ocean at that time, whereas the paleomagnetic evidence from Permian lava flows pointed to yet another location in Asia. When plotted on a map, the paleomagnetic readings of numerous lava flows from all ages in North America trace the apparent movement of the magnetic pole (called *polar wandering*) through time (• Figure 3.8). This paleomagnetic evidence from a single continent could be interpreted in three ways: (1) the continent remained fixed, and the north magnetic pole moved, (2) the north magnetic pole stood still and the continent moved, or (3) both the continent and the north magnetic pole moved.

Upon additional analysis, magnetic minerals from European Silurian and Permian lava flows pointed to a different magnetic pole location than those of the same age in North America (Figure 3.8.) Furthermore, analysis of lava flows from all continents indicated that each continent seemingly had its own series of magnetic poles. Does this really mean there were different north magnetic poles for each continent? That would be highly unlikely and difficult to reconcile with the theory accounting for Earth's magnetic field.

The best explanation for such data is that the magnetic poles have remained near their present locations at the geographic north and south poles and the continents have moved. When the continental margins are fitted together so that the paleomagnetic data point to only one magnetic pole, we find, just as Wegener did, that the rock sequences and glacial deposits match up, and the fossil evidence is consistent with the reconstructed paleogeography.

How Do Magnetic Reversals Relate to Seafloor Spreading?

Geologists refer to Earth's present magnetic field as being normal—that is, with the north and south magnetic poles located approximately at the north and south geographic poles. At various times in the geologic past, however, Earth's magnetic field has completely reversed. The existence of such **magnetic reversals** was discovered by dating and determining the orientation of the remanent magnetism in lava flows on land (• Figure 3.9).

Once magnetic reversals were well established for continental lava flows, magnetic reversals were also discovered in igneous rocks in the oceanic crust as part of the large-scale mapping of the ocean basins that took place during the 1960s. Although the cause of magnetic reversals is still uncertain, their occurrence in the geologic record is well documented.

Besides the discovery of magnetic reversals, mapping of the ocean basins also revealed an oceanic ridge system more than 65,000 km long, constituting the most extensive mountain range in the world. Perhaps the best-known part of the ridge system is the Mid-Atlantic Ridge, which divides the Atlantic Ocean basin into two nearly equal parts (• Figure 3.10).

As a result of the oceanographic research conducted in the 1950s, Harry Hess of Princeton University proposed the theory of **seafloor spreading** in 1962 to account for continental movement. He suggested that continents do not move across oceanic crust, but rather that the continents and oceanic crust move together. Hess further postulated that the seafloor separates at oceanic ridges where new crust is formed by upwelling magma. As the magma cools, the newly formed oceanic crust moves laterally away from the ridge.

As a mechanism to drive this system, Hess revived the idea of **thermal convection cells** in the mantle; that is, hot magma rises from the mantle, intrudes along fractures defining oceanic ridges, and thus forms new crust. Cold crust is subducted back into the mantle at oceanic

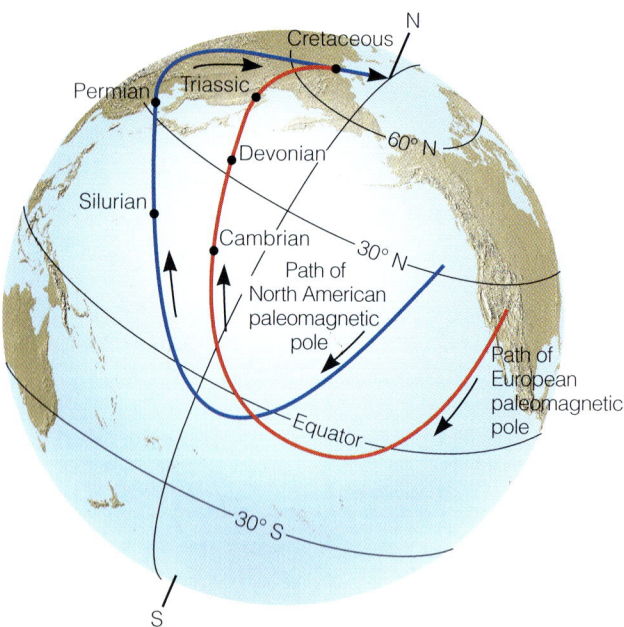

• **Figure 3.8 Polar Wandering** The apparent paths of polar wandering for North America and Europe. The apparent location of the north magnetic pole is shown for different periods on each continent's polar wandering path. If the continents have not moved through time, and because Earth has only one magnetic pole, the paleomagnetic readings for the same time in the past taken on different continents should all point to the same location. However, the north magnetic pole has different locations for the same time in the past when measured on different continents, indicating multiple north magnetic poles. The logical explanation for this dilemma is that the magnetic north pole has remained at the same approximate geographic location during the past, and the continents have moved.

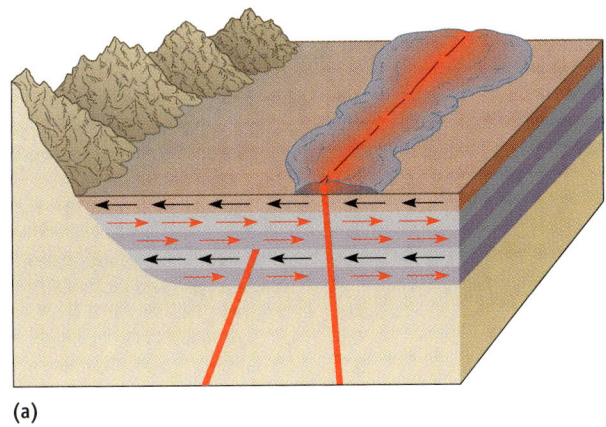

trenches, where it is heated and recycled, thus completing a thermal convection cell (see Figure 1.7).

How could Hess's hypothesis be confirmed? Magnetic surveys of the oceanic crust revealed striped **magnetic anomalies** (deviations from the average strength of Earth's magnetic field) in the rocks that are both parallel to and symmetric around the oceanic ridges (• Figure 3.11). Furthermore, the pattern of oceanic magnetic anomalies matches the pattern of magnetic reversals already known from studies of continental lava flows (Figure 3.9). When magma wells up and cools along a ridge summit, it records Earth's magnetic field at that time as either normal or reversed. As new crust forms at the summit, the previously formed crust moves laterally away from the ridge. These magnetic stripes represent times of normal and reversed polarity at oceanic ridges (where upwelling magma forms new oceanic crust), thus conclusively confirming Hess's theory of seafloor spreading.

• **Figure 3.9 Magnetic Reversals**
(a) Magnetic reversals recorded in a succession of lava flows are shown diagrammatically by red arrows, and the record of normal polarity events is shown by black arrows. The lava flows containing a record of such magnetic-polarity events can be radiometrically dated so that a magnetic time scale as in (b) can be constructed.
(b) Magnetic reversals for the last 4.5 million years as determined from lava flows on land. Black bands represent normal magnetism, and purple bands represent reverse magnetism.
Source: (b) Reprinted with permission from A. Cox, *Science,* Vol 163, pp. 237–245, January 17, 1969. Copyright © 1969 American Association for the Advancement of Science.

One of the consequences of the seafloor spreading theory is its confirmation that ocean basins are geologically young features whose openings and closings are partially responsible for continental movement (• Figure 3.12). Radiometric dating reveals that the oldest oceanic crust is somewhat less than 180 million years old, whereas the oldest continental crust is 3.96 billion years old. Although geologists do not universally accept the idea of thermal convection cells as a driving mechanism for plate movement, most accept that plates are created at oceanic ridges and destroyed at deep-sea trenches, regardless of the driving mechanism involved.

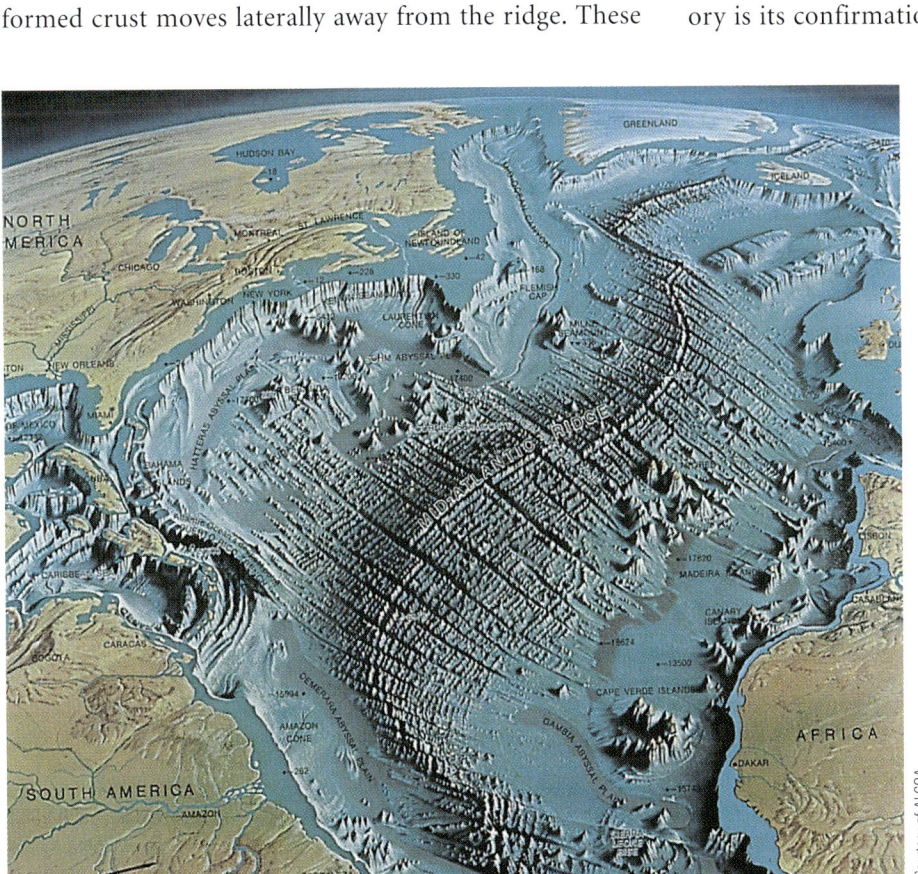

• **Figure 3.10 Topography of the Atlantic Ocean Basin** Artistic view of what the Atlantic Ocean basin would look like without water. The major feature is the Mid-Atlantic Ridge, an oceanic ridge system that is longer than 65,000 km and divides the Atlantic Ocean basin in half. It is along such oceanic ridges that the seafloor is separating and new oceanic crust is forming from upwelling magma in Earth's interior.

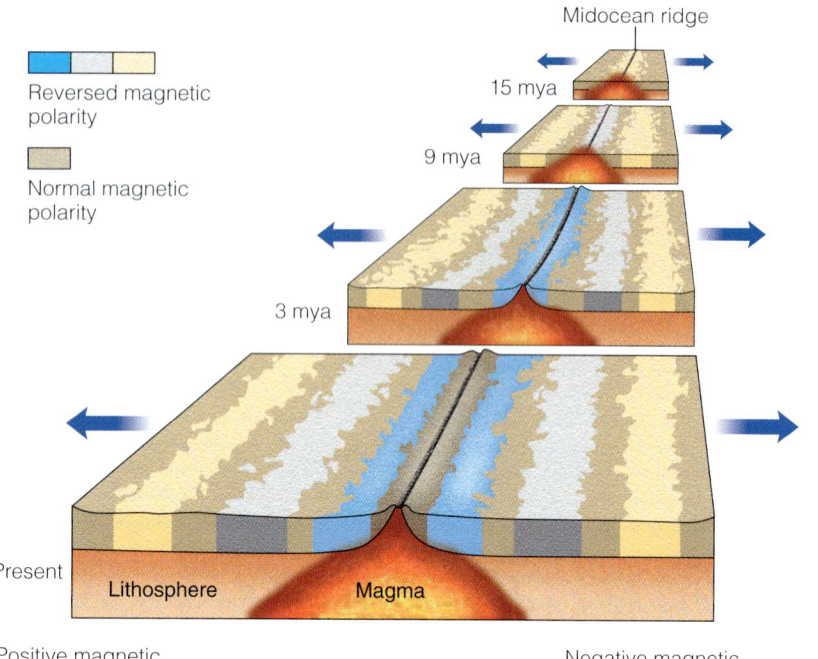

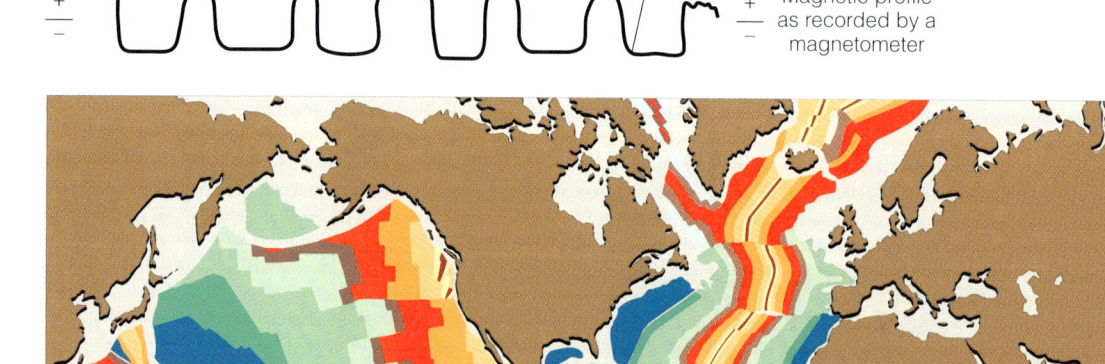

- **Figure 3.11 Magnetic Anomalies and Seafloor Spreading** The sequence of magnetic anomalies preserved within the oceanic crust is both parallel to and symmetric around oceanic ridges. Basaltic lava intruding into an oceanic ridge today and spreading laterally away from the ridge records Earth's current magnetic field or polarity (considered by convention to be normal). Basaltic intrusions 3, 9, and 15 million years ago record Earth's reversed magnetic field at that time. This schematic diagram shows how the solidified basalt moves away from the oceanic ridge (or spreading center), carrying with it the magnetic anomalies that are preserved in the oceanic crust. Magnetic anomalies are magnetic readings that are either higher (positive magnetic anomalies) or lower (negative magnetic anomalies) than Earth's current magnetic field strength. The magnetic anomalies are recorded by a magnetometer, which measures the strength of the magnetic field.

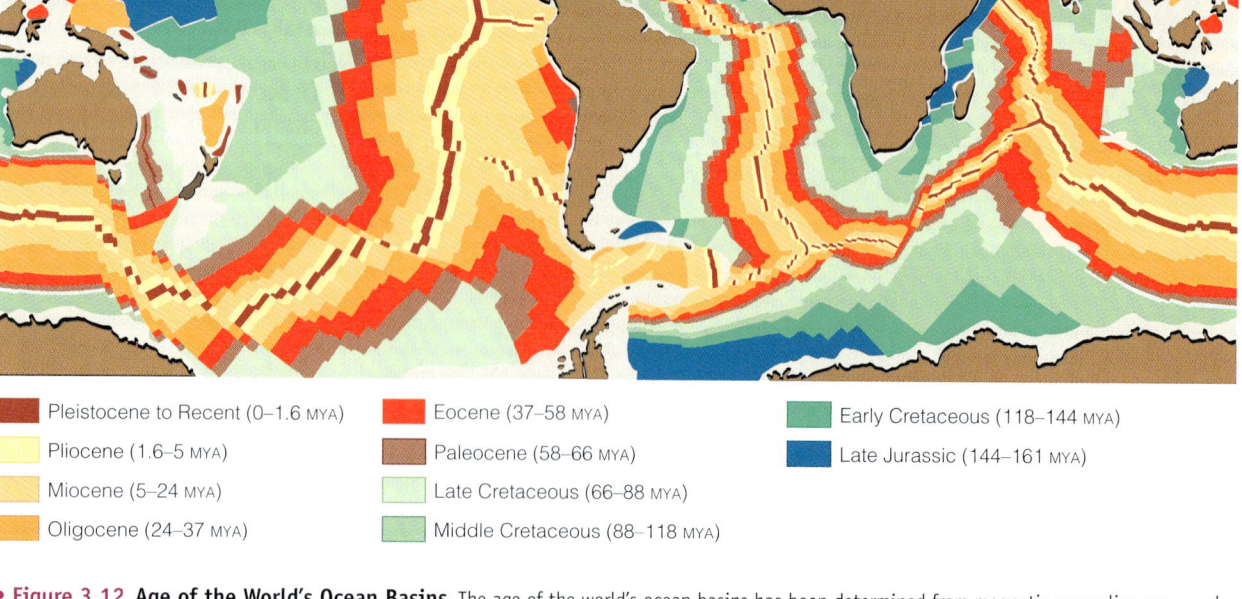

Legend:
- Pleistocene to Recent (0–1.6 MYA)
- Pliocene (1.6–5 MYA)
- Miocene (5–24 MYA)
- Oligocene (24–37 MYA)
- Eocene (37–58 MYA)
- Paleocene (58–66 MYA)
- Late Cretaceous (66–88 MYA)
- Middle Cretaceous (88–118 MYA)
- Early Cretaceous (118–144 MYA)
- Late Jurassic (144–161 MYA)

- **Figure 3.12 Age of the World's Ocean Basins** The age of the world's ocean basins has been determined from magnetic anomalies preserved in oceanic crust. Magnetic anomalies demonstrate that the youngest oceanic crust is adjacent to the spreading ridges and that its age increases away from the ridge axis. Based on *The Bedrock Geology of the World*, by R. L. Larson and W. C. Pitman, 1985 by W. H. Freeman and Company.

Plate Tectonics and Plate Boundaries

Plate tectonic theory is based on a simple model of Earth. The rigid lithosphere, composed of both oceanic and continental crust as well as the underlying upper mantle, consists of numerous variable-sized pieces called **plates** (• Figure 3.13). The plates vary in thickness; those composed of upper mantle and continental crust are as much as 250 km thick, whereas those of upper mantle and oceanic crust are up to 100 km thick.

The lithosphere overlies the hotter and weaker semiplastic asthenosphere. It is thought that movement resulting from some type of heat-transfer system within the asthenosphere causes the overlying plates to move. As plates move over the asthenosphere, they separate, mostly at oceanic ridges; in other areas such as at oceanic trenches, they collide and are subducted back into the mantle.

An easy way to visualize plate movement is to think of a conveyor belt moving luggage from an airplane's cargo hold to a baggage cart. The conveyor belt represents convection currents within the mantle, and the luggage represents Earth's lithospheric plates. The luggage is moved along by the conveyor belt until it is dumped into the baggage cart in the same way plates are moved by convection cells until they are subducted into Earth's interior.

What Would You Do?

You've been selected to be part of the first astronaut team to go to Mars. While your two fellow crew members descend to the Martian surface, you'll be staying in the command module and circling the Red Planet. As part of the geologic investigation of Mars, one of the crew members will be mapping the geology around the landing site and deciphering the geologic history of the area. Your job will be to observe and photograph the planet's surface and try to determine whether Mars had an active plate tectonic regime in the past and whether there is current plate movement. What features would you look for, and what evidence might reveal current or previous plate activity?

Although this analogy allows you to visualize how the mechanism of plate movement takes place, remember that this analogy is limited. The major limitation is that, unlike the luggage, plates consist of continental and oceanic crust, which have different densities; only oceanic crust, because it is denser than continental crust, is subducted into Earth's interior. Nonetheless, this analogy does provide an easy way to visualize plate movement.

Most geologists accept plate tectonic theory, in part, because the evidence for it is overwhelming, and it ties together many seemingly unrelated geologic features and

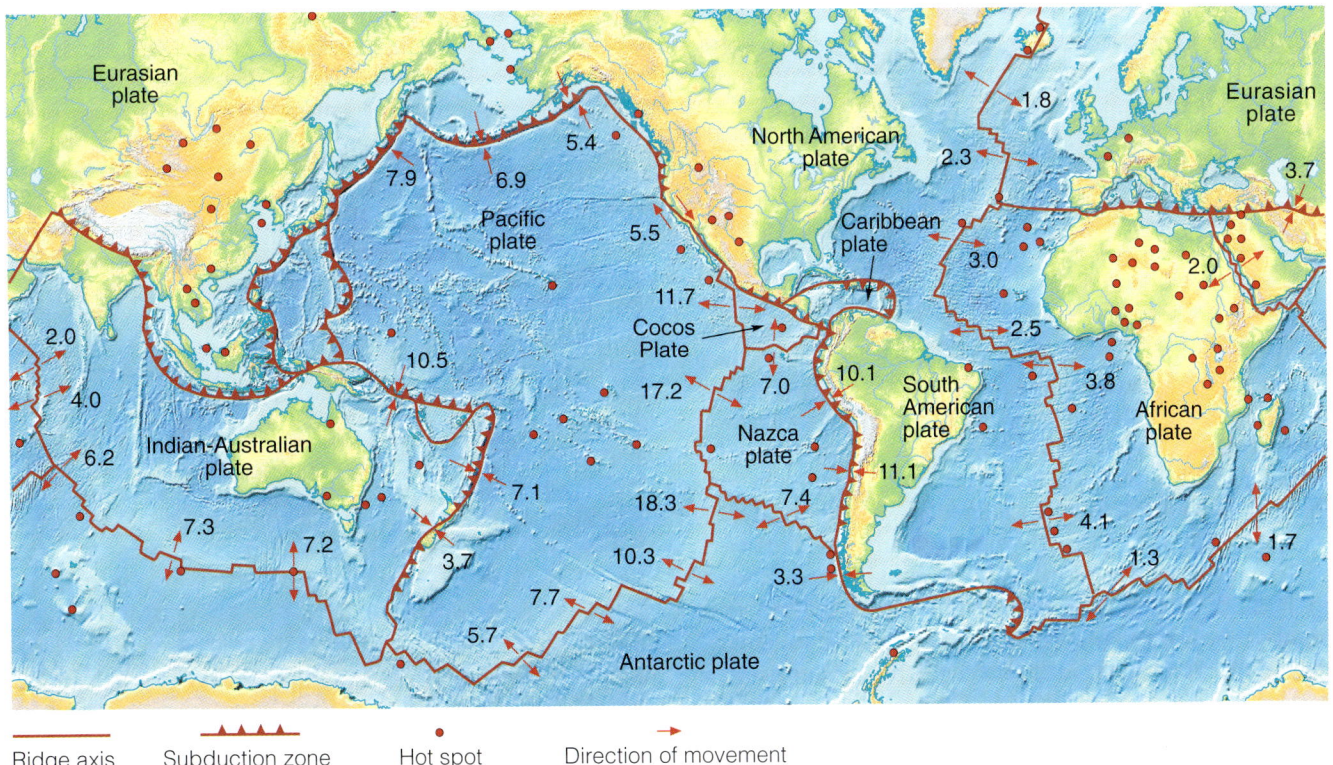

• **Figure 3.13 Earth's Plates** A world map showing Earth's plates, their boundaries, their relative motion and average rates of movement in centimeters per year, and hot spots. Data from J. B. Minster and T. H. Jordan "Present Day Plate Motions," *Journal of Geophysical Research, 83* (1978) 5331–51. Copyright © 1978 American Geophysical Union. Modified by permission of American Geophysical Union.

TABLE 3.1 Types of Plate Boundaries

Type	Example	Landforms	Volcanism
Divergent			
Oceanic	Mid-Atlantic Ridge	Mid-oceanic ridge with axial rift valley	Basalt
Continental	East African Rift Valley	Rift valley	Basalt and rhyolite, no andesite
Convergent			
Oceanic–oceanic	Aleutian Islands	Volcanic island arc, offshore oceanic trench	Andesite
Oceanic–continental	Andes	Offshore oceanic trench, volcanic mountain chain, mountain belt	Andesite
Continental–continental	Himalayas	Mountain belt	Minor
Transform	San Andreas fault	Fault valley	Minor

events and shows how they are interrelated. Consequently, geologists now view such geologic processes as mountain building, earthquake activity, and volcanism from the perspective of plate tectonics. Furthermore, because all the inner planets have had a similar origin and early history, geologists are interested in determining whether plate tectonics is unique to Earth or whether it operates in the same way on other terrestrial planets.

Because it appears plate tectonics has operated since at least the Proterozoic Eon (see Chapter 9), it is important to understand how plates move and interact with each other and how ancient plate boundaries are recognized. After all, the movement of plates has profoundly affected the geologic and biologic history of this planet.

Geologists recognize three major types of plate boundaries: *divergent, convergent,* and *transform* (Table 3.1). Along these boundaries new plates are formed, consumed, or slide laterally past one another. To understand the implications of plate interactions as they have affected Earth history, geologists must study present plate boundaries.

Divergent Boundaries

Divergent plate boundaries, or *spreading ridges,* occur where plates are separating and new oceanic lithosphere is forming. Divergent boundaries are places where the crust is extended, thinned, and fractured as magma—derived from the partial melting of the mantle—rises to the surface. The magma is almost entirely basaltic and intrudes into vertical fractures to form dikes and pillow lava flows (• Figure 3.14). As successive injections of magma cool and solidify, they form new oceanic crust and record the intensity and orientation of Earth's magnetic field (Figure 3.11). Divergent boundaries most commonly occur along the crests of oceanic ridges—for example, the Mid-Atlantic Ridge. Oceanic ridges are thus characterized by rugged topography with high relief resulting from displacement of rocks along large fractures, shallow-depth earthquakes, high heat flow, and basaltic flows or pillow lavas.

Divergent boundaries are also present under continents during the early stages of continental breakup. When magma wells up beneath a continent, the crust is initially elevated, stretched, and thinned, producing fractures, faults, rift valleys, and volcanic activity (• Figure 3.15a). As magma intrudes into faults and fractures, it solidifies or flows out onto the surface as lava flows; the latter often covering the rift valley floor (Figure 3.15b). The East African Rift Valley is an excellent example of continental breakup at this stage (• Figure 3.16).

As spreading proceeds, some rift valleys continue to lengthen and deepen until the continental crust eventually breaks and a narrow linear sea is formed that separates two continental blocks (Figure 3.15c). The Red Sea, which separates the Arabian Peninsula from Africa (Figure 3.16a), and the Gulf of California, which separates Baja California from mainland Mexico, are good examples of this more advanced stage of rifting.

• **Figure 3.14 Pillow Lavas** Pillow lavas forming along the Mid-Atlantic Ridge. Their distinctive bulbous shape is the result of underwater eruption.

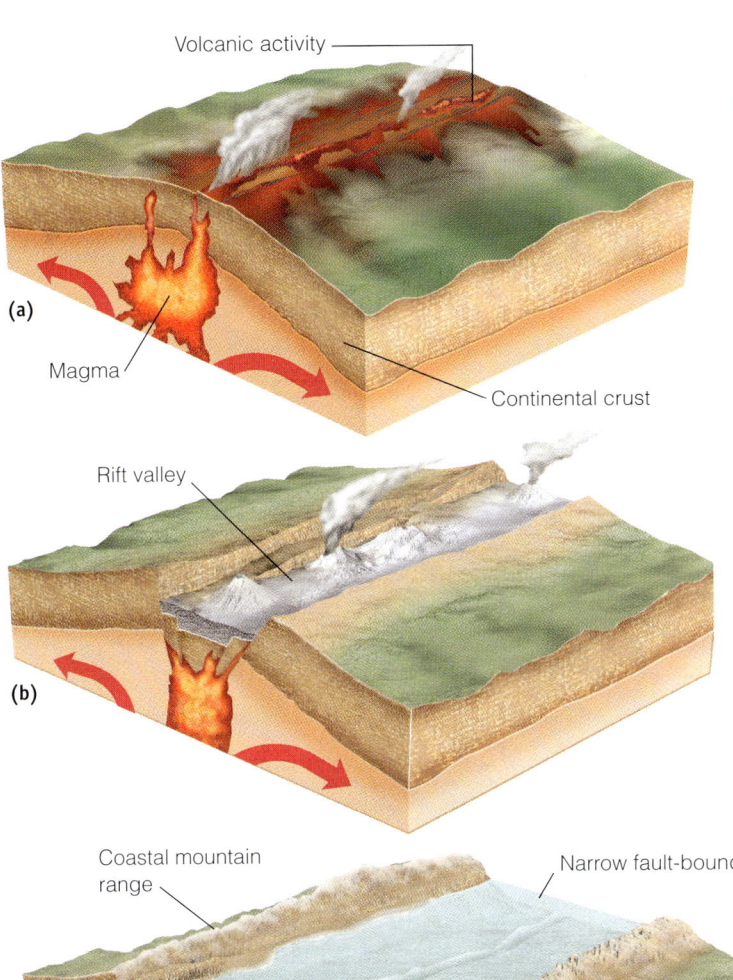

As a newly created narrow sea continues to enlarge, it may eventually become an expansive ocean basin such as the Atlantic Ocean basin is today, separating North and South America from Europe and Africa by thousands of kilometers (Figure 3.15d). The Mid-Atlantic Ridge is the boundary between these diverging plates (Figure 3.10); the American plates are moving westward, and the Eurasian and African plates are moving eastward.

An Example of Ancient Rifting

What features in the geologic record can geologists use to recognize ancient rifting? Associated with regions of continental rifting are faults, dikes, sills, lava flows, and thick sedimentary sequences within rift valleys, all features that are preserved in the geologic record. The Triassic fault-block basins of the eastern United States are a good example of ancient continental rifting (• Figure 3.17a). These fault-block basins mark the zone of rifting that occurred when North America split apart from Africa. They contain thousands of meters of continental sediment and are riddled with dikes and sills (see Chapter 14).

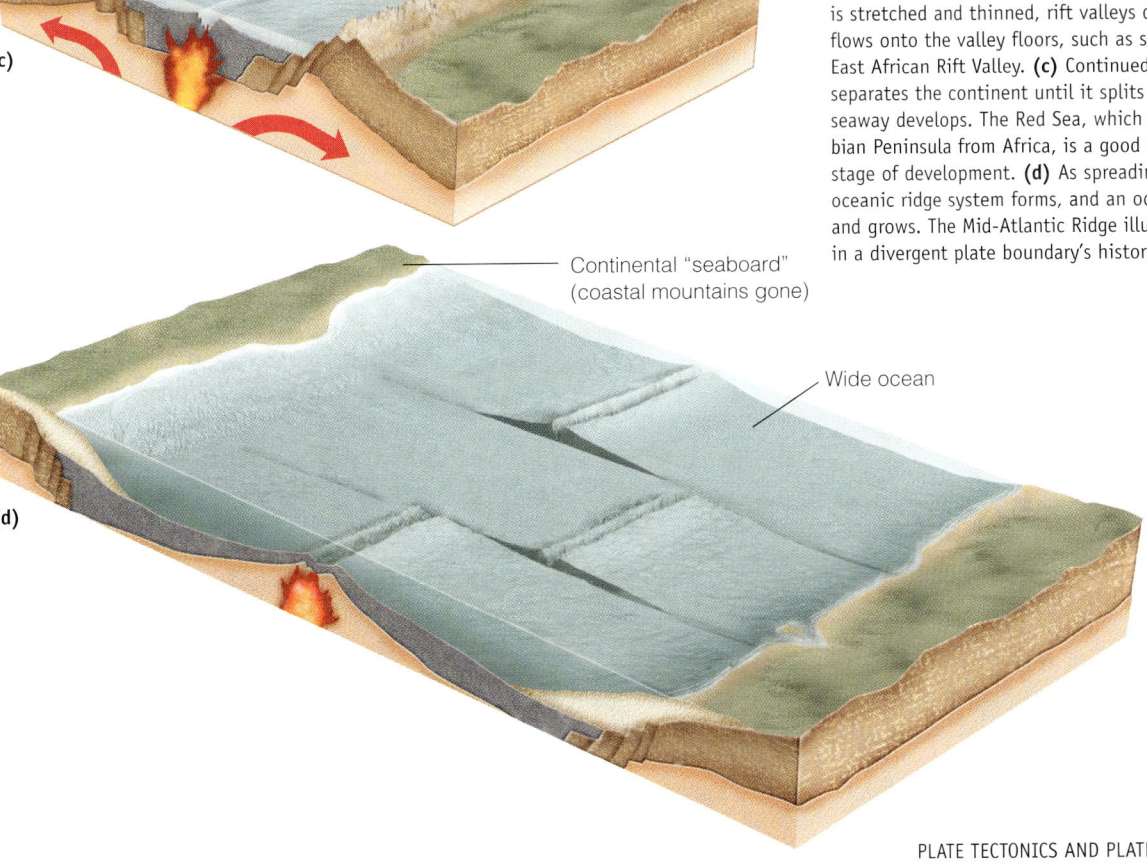

• **Figure 3.15 History of a Divergent Plate Boundary** (a) Rising magma beneath a continent pushes the crust up, producing numerous fractures, faults, rift valleys, and volcanic activity. (b) As the crust is stretched and thinned, rift valleys develop, and lava flows onto the valley floors, such as seen today in the East African Rift Valley. (c) Continued spreading further separates the continent until it splits apart and a narrow seaway develops. The Red Sea, which separates the Arabian Peninsula from Africa, is a good example of this stage of development. (d) As spreading continues, an oceanic ridge system forms, and an ocean basin develops and grows. The Mid-Atlantic Ridge illustrates this stage in a divergent plate boundary's history.

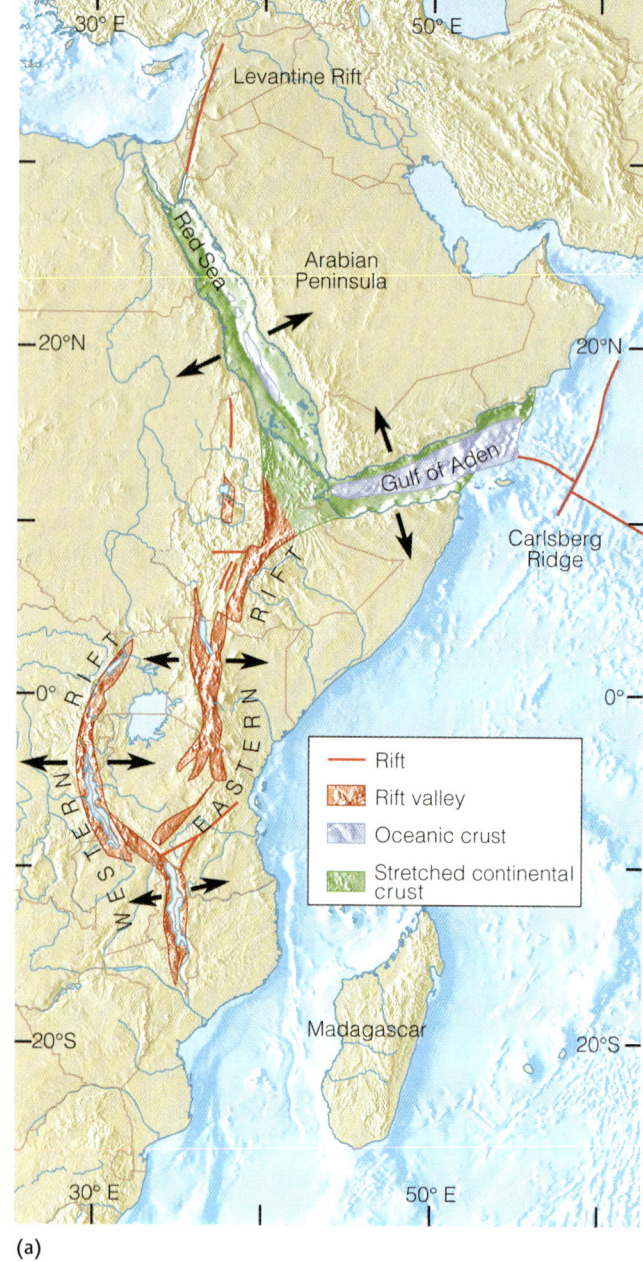

(a)

(b)

• Figure 3.16 **East African Rift Valley** (a) The East African Rift Valley is being formed by the separation of East Africa from the rest of the continent along a divergent plate boundary. The Red Sea represents a more advanced stage of rifting, in which two continental blocks (Africa and the Arabian Peninsula) are separated by a narrow sea. (b) View looking down the Great Rift Valley of Africa. Little Magadi, seen in the background, is one of numerous soda lakes forming in the valley. Because of high evaporation rates and lack of any drainage outlets, these lakes are very saline. The Great Rift Valley is part of the system of rift valleys resulting from stretching of the crust as plates move away from each other in Eastern Africa.

Convergent Boundaries

Whereas new crust forms at divergent plate boundaries, older crust must be destroyed and recycled in order for the entire surface area of Earth to remain the same. Otherwise, we would have an expanding Earth. Such plate destruction takes place at **convergent plate boundaries** where two plates collide and the leading edge of one plate descends beneath the margin of the other by a process known as *subduction*. As the subducting plate moves down into the asthenosphere, it is heated and eventually incorporated into the mantle.

Convergent boundaries are characterized by deformation, volcanism, mountain building, metamorphism, earthquake activity, and valuable mineral deposits. Three types of convergent plate boundaries are recognized: *oceanic–oceanic*, *oceanic–continental*, and *continental–continental*.

Oceanic–Oceanic Boundaries When two oceanic plates converge, one is subducted beneath the other along an **oceanic–oceanic plate boundary** (• Figure 3.18). The subducting plate bends downward to form the outer wall of an oceanic trench. A *subduction complex*, composed of wedge-shaped slices of highly folded and faulted marine sediments and oceanic lithosphere scraped off the descending plate, forms along the inner wall of the oceanic trench. As the subducting plate descends into the mantle, it is heated and partially melted, generating magma commonly of andesitic composition. This magma is less dense than the

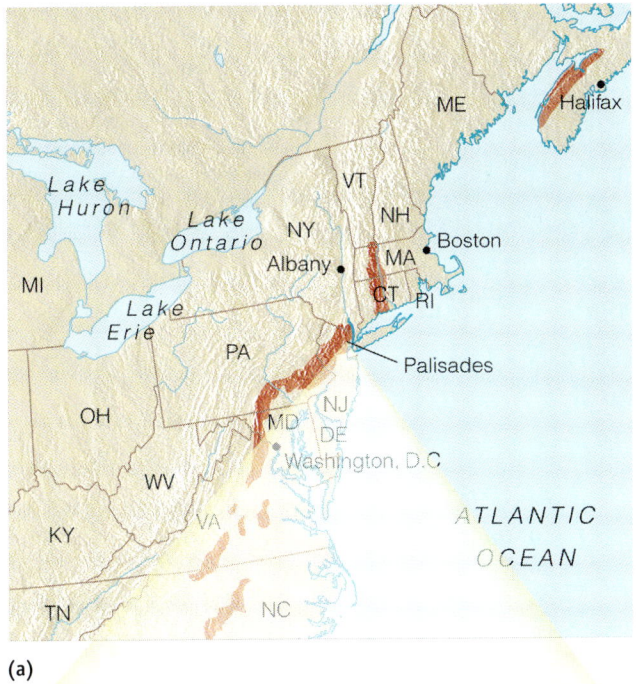

(a)

(b)

• **Figure 3.17 Triassic Fault-Block Basins of Eastern North America—An Example of Ancient Rifting** (a) Triassic fault-block basin deposits appear in numerous locations throughout eastern North America. These fault-block basins are good examples of ancient continental rifting, and during the Triassic Period they looked like today's fault-block basins (rift valleys) of the East African Rift Valley. (b) Palisades of the Hudson River. This sill (tabular-shaped horizontal igneous intrusion) was one of many that were intruded into the fault-block basin sediments during the Late Triassic rifting that marked the separation of North America from Africa.

surrounding mantle rocks and rises to the surface of the nonsubducted plate, forming a curved chain of volcanic islands called a **volcanic island arc** (any plane intersecting a sphere makes an arc). This arc is nearly parallel to the oceanic trench and is separated from it by a distance of up to several hundred kilometers—the distance depending on the angle of dip of the subducting plate (Figure 3.18).

In those areas where the rate of subduction is faster than the forward movement of the overriding plate, the lithosphere on the landward side of the volcanic island arc may be subjected to tensional stress and stretched and thinned, resulting in the formation of a *back-arc basin*. This back-arc basin may grow by spreading if magma breaks through the thin crust and forms new oceanic crust (Figure 3.18). A good example of a back-arc basin associated with an oceanic–oceanic plate boundary is the Sea of Japan between the Asian continent and the islands of Japan.

Most present-day active volcanic island arcs are in the Pacific Ocean basin and include the Aleutian Islands, the Kermadec–Tonga arc, and the Japanese (Figure 3.18) and Philippine Islands. The Scotia and Antillean (Caribbean) island arcs are present in the Atlantic Ocean basin.

Oceanic–Continental Boundaries An **oceanic–continental plate boundary** occurs when a denser oceanic plate is subducted under a continental plate (• Figure 3.19). The magma generated by subduction rises beneath the continent and either crystallizes as large igneous bodies before reaching the surface or erupts at the surface to produce a chain of andesitic volcanoes (also called a *volcanic arc*).

An excellent example of an oceanic–continental plate boundary is the Pacific coast of South America where the oceanic Nazca plate is currently being subducted under

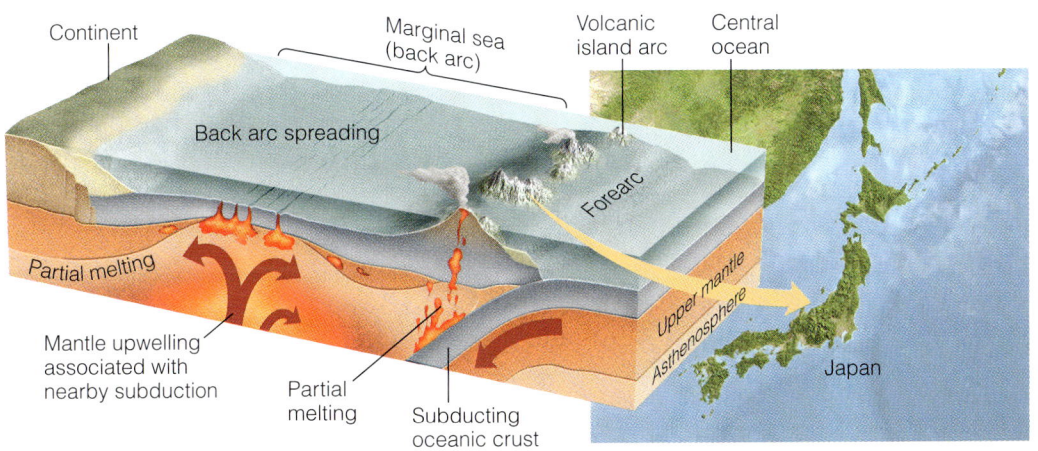

• **Figure 3.18 Oceanic–Oceanic Convergent Plate Boundary** An oceanic trench forms where one oceanic plate is subducted beneath another. On the nonsubducted plate, a volcanic island arc forms from the rising magma generated from the subducting plate. The Japanese Islands are a volcanic island arc resulting from the subduction of one oceanic plate beneath another oceanic plate.

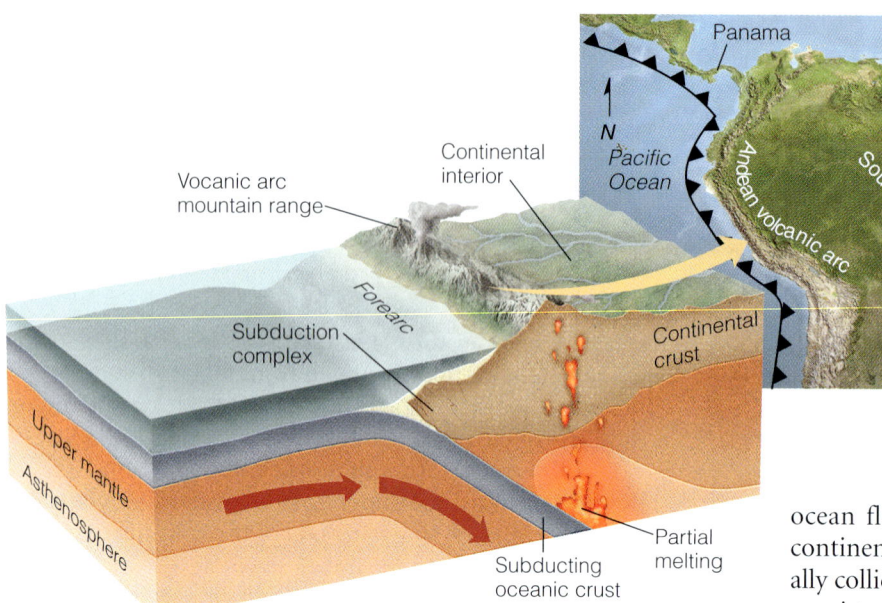

• **Figure 3.19 Oceanic–Continental Convergent Plate Boundary** When an oceanic plate is subducted beneath a continental plate, an andesitic volcanic mountain range is formed on the continental plate as a result of rising magma. The Andes Mountains are one of the best examples of continuing mountain building at an oceanic–continental convergent plate boundary.

South America (Figure 3.19). The Peru-Chile Trench marks the site of subduction, and the Andes Mountains are the resulting volcanic mountain chain on the nonsubducting plate. This particular example demonstrates the effect plate tectonics has on our lives. For instance, earthquakes are commonly associated with subduction zones, and the western side of South America is the site of frequent and devastating earthquakes.

Continental–Continental Boundaries Two continents approaching each other are initially separated by an ocean floor that is being subducted under one continent.

The edge of that continent will display the features characteristic of oceanic–continental convergence. As the ocean floor continues to be subducted, the two continents come closer together until they eventually collide. Because continental lithosphere, which consists of continental crust and the upper mantle, is less dense than oceanic lithosphere (oceanic crust and upper mantle), it cannot sink into the asthenosphere. Although one continent may partly slide under the other, it cannot be pulled or pushed down into a subduction zone (• Figure 3.20).

When two continents collide, they are welded together along a zone marking the former site of subduction. At this **continental–continental plate boundary,** an interior mountain belt is formed that consists of deformed sediments and sedimentary rocks, igneous intrusions, metamorphic rocks, and fragments of oceanic crust. In addition, the entire region is subjected to numerous earthquakes. The Himalayas in central Asia, the world's youngest and highest mountain system, resulted from the collision between India and Asia that began about 40 to 50 million years ago and is still continuing (Figure 3.20).

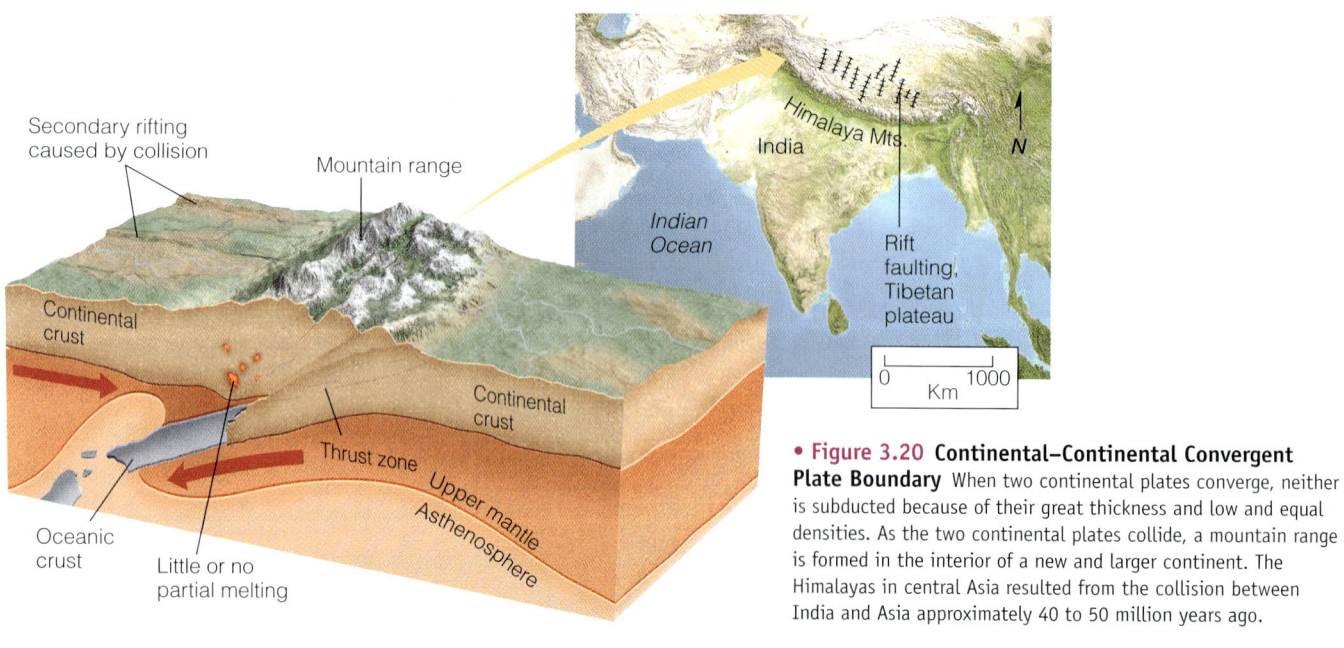

• **Figure 3.20 Continental–Continental Convergent Plate Boundary** When two continental plates converge, neither is subducted because of their great thickness and low and equal densities. As the two continental plates collide, a mountain range is formed in the interior of a new and larger continent. The Himalayas in central Asia resulted from the collision between India and Asia approximately 40 to 50 million years ago.

Recognizing Ancient Convergent Plate Boundaries

How can former subduction zones be recognized in the geologic record? Igneous rocks provide one clue. The magma erupted at the surface, forming island arc volcanoes and continental volcanoes, is of andesitic composition. Another clue is the zone of intensely deformed rocks between the deep-sea trench where subduction is taking place and the area of igneous activity. Here, sediments and submarine rocks are folded, faulted, and metamorphosed into a chaotic mixture of rocks termed a *mélange*.

During subduction, pieces of oceanic lithosphere are sometimes incorporated into the mélange and accreted onto the edge of the continent. Such slices of oceanic crust and upper mantle are called **ophiolites** (• Figure 3.21). They consist of a layer of deep-sea sediments that include graywackes (poorly sorted sandstones containing abundant feldspars and rock fragments, usually in a clay-rich matrix), black shales, and cherts. These deep-sea sediments are underlain by pillow lavas, a sheeted dike complex, massive gabbro, and layered gabbro, all of which form the oceanic crust. Beneath the gabbro is peridotite, which probably represents the upper mantle. The presence of ophiolites in an outcrop or drilling core is a key feature in recognizing plate convergence along a subduction zone.

Elongate belts of folded and faulted marine sedimentary rocks, andesites, and ophiolites are found in the Appalachians, Alps, Himalayas, and Andes mountains. The combination of such features is good evidence that these mountain ranges resulted from deformation along convergent plate boundaries.

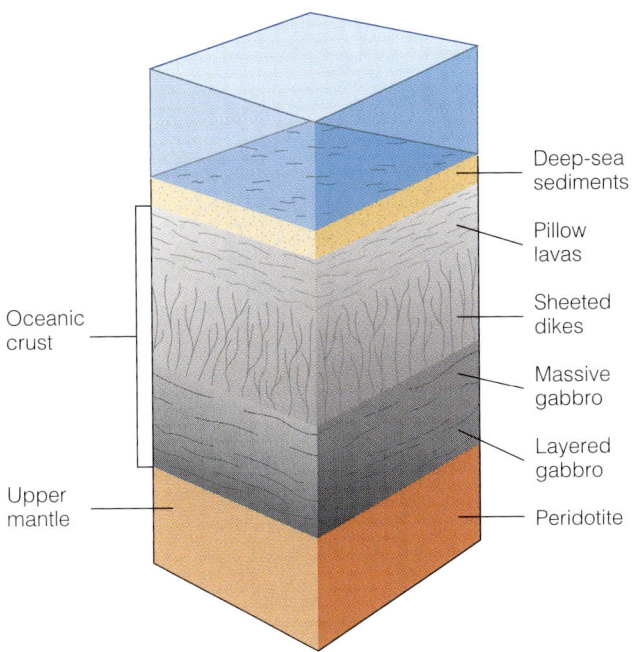

• **Figure 3.21 Ophiolites** Ophiolites are sequences of rock on land that consist of deep-sea sediments, oceanic crust, and upper mantle. Ophiolites are one feature used to recognize ancient convergent plate boundaries.

Transform Boundaries

The third type of plate boundary is a **transform plate boundary.** These mostly occur along fractures in the seafloor, known as *transform faults,* where plates slide laterally past each other, roughly parallel to the direction of plate movement. Although lithosphere is neither created nor destroyed along a transform boundary, the movement between plates results in a zone of intensely shattered rock and numerous shallow-depth earthquakes.

Transform faults are particular types of faults that "transform" or change one type of motion between plates into another type of motion. Most commonly, transform faults connect two oceanic ridge segments (• Figure 3.22), but they can also connect ridges to trenches, and trenches to trenches. Although the majority of transform faults are in oceanic crust and are marked by distinct fracture zones, they may also extend into continents.

One of the best-known transform faults is the San Andreas fault in California. It separates the Pacific plate from the North American plate and connects spreading ridges in the Gulf of California with the Juan de Fuca and Pacific plates off the coast of northern California (• Figure 3.23). Many of the earthquakes affecting California are the result of movement along this fault (see Chapter 16).

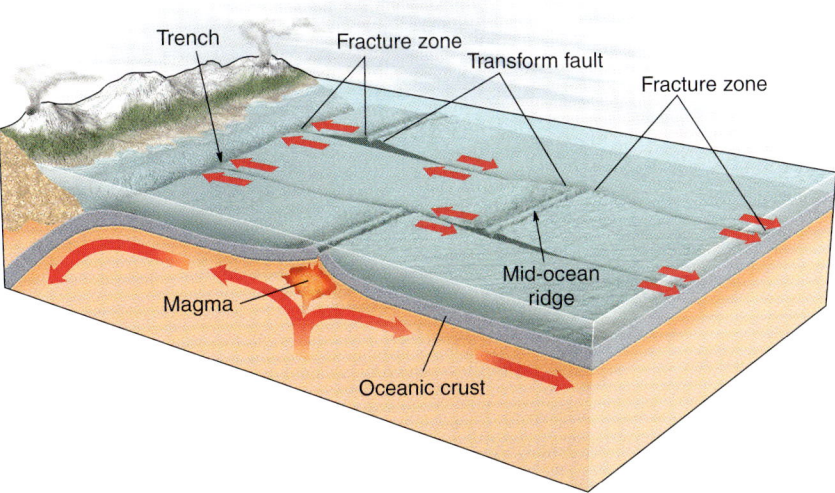

• **Figure 3.22 Transform Plate Boundaries** Horizontal movement between plates occurs along transform faults. Extensions of transform faults on the seafloor form fracture zones. The majority of transform faults connect two oceanic ridge segments. Note that relative motion between the plates only occurs between the two ridges.

• **Figure 3.23 The San Andreas Fault—A Transform Plate Boundary** The San Andreas fault is a transform fault separating the Pacific plate from the North American plate. It connects the spreading ridges in the Gulf of California with the Juan de Fuca and Pacific plates off the coast of northern California. Movement along the San Andreas fault has caused numerous earthquakes. The insert photograph shows a segment of the San Andreas fault as it cuts through the Carrizo Plain, California.

Unfortunately, transform faults generally do not leave any characteristic or diagnostic features except for the obvious displacement of the rocks with which they are associated. This displacement is usually large, on the order of tens to hundreds of kilometers. Such large displacements in ancient rocks can sometimes be related to transform fault systems.

What Are Hot Spots and Mantle Plumes?

Before leaving the topic of plate boundaries, we should briefly mention an intraplate feature found beneath both oceanic and continental plates. A **hot spot** is the location on Earth's surface where a stationary column of magma, originating deep within the mantle (*mantle plume*), has slowly risen to the surface and formed a volcano (• Figure 3.24). Because the mantle plumes apparently remain stationary (although some evidence suggests that they might not) within the mantle while plates move over them, the resulting hot spots leave a trail of extinct and progressively older volcanoes called *aseismic ridges* that record the movement of the plate. Some examples of aseismic ridges and hot spots are the Emperor Seamount–Hawaiian Island chain and Yellowstone National Park in Wyoming.

How Are Plate Movement and Motion Determined?

How fast and in what direction are Earth's various plates moving? Do they all move at the same rate? Rates of plate movement can be calculated in several ways. The least accurate method is to determine the age of the sediments immediately above any portion of the oceanic crust and then divide the distance from the spreading ridge by that age. Such calculations give an average rate of movement.

A more accurate method of determining both the average rate of movement and the relative motion is by dating the magnetic anomalies in the crust of the seafloor (Figure 3.12). The distance from an oceanic ridge axis to any magnetic anomaly indicates the width of new seafloor that formed during that time interval. Thus, for a given interval of time, the wider the strip of seafloor, the faster the plate has moved. In this way, not only can the present average rate of movement and relative motion be determined (Figure 3.13), but the average rate of movement during the past can also be calculated by dividing the distance between anomalies by the amount of time elapsed between anomalies.

Geologists use magnetic anomalies not only to calculate the average rate of plate movement but also to determine plate positions at various times in the past. Because magnetic anomalies are parallel and symmetric with

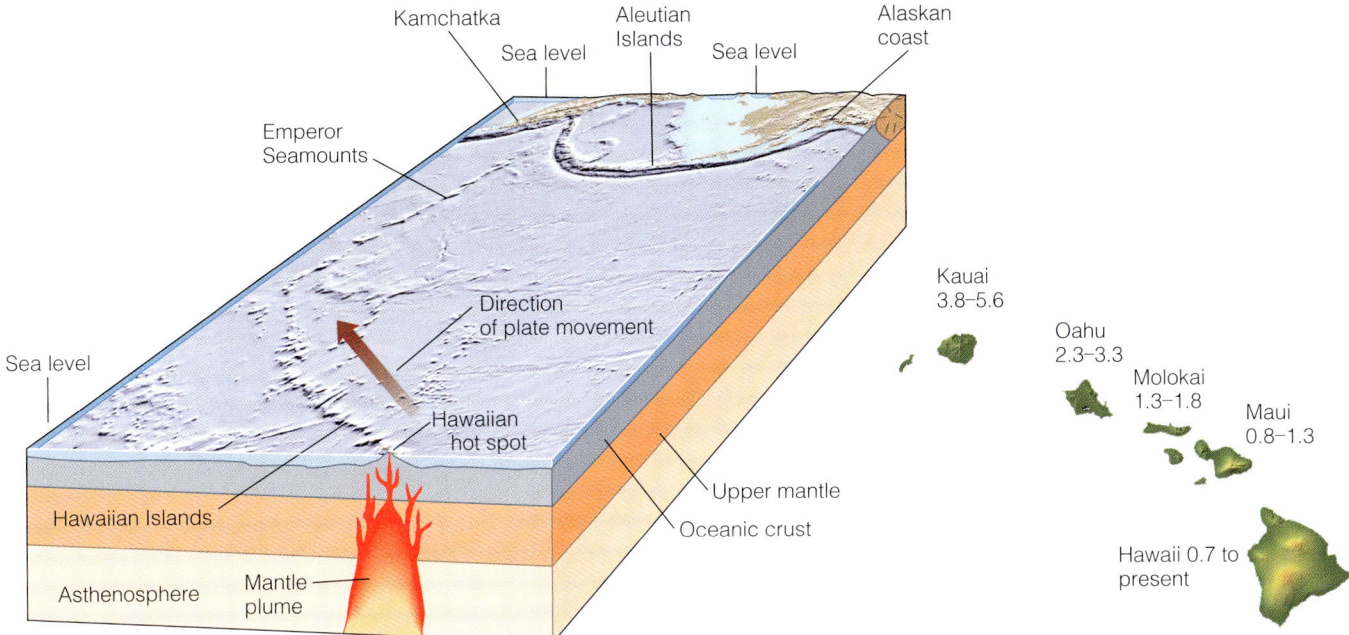

• **Figure 3.24 Hot Spots** A hot spot is the location where a stationary mantle plume has risen to the surface and formed a volcano. The Emperor Seamount–Hawaiian Island chain formed as a result of the Pacific plate moving over a mantle plume, and the line of volcanic islands in this chain traces the direction of plate movement. The Hawaiian hot spot currently underlies the southern half of the island of Hawaii and adjoining offshore area. The numbers indicate the age of the islands in millions of years.

respect to spreading ridges, all one must do to determine the position of continents when particular anomalies formed is to move the anomalies back to the spreading ridge, which will also move the continents with them (• Figure 3.25). Unfortunately, subduction destroys oceanic crust and the magnetic record it carries. Thus we have an excellent record of plate movements since the breakup of Pangaea but not as good an understanding of plate movement before that time.

The average rate of movement as well as the relative motion between any two plates can also be determined by satellite-laser ranging techniques. Laser beams from a station on one plate are bounced off a satellite (in geosynchronous orbit) and returned to a station on a different plate. As the plates move away from each other, the laser beam takes more time to go from the sending station to the stationary satellite and back to the receiving station. This difference in elapsed time is used to calculate the rate of movement and the relative motion between plates.

Plate motions derived from magnetic reversals and satellite-laser ranging techniques give only the relative motion of one plate with respect to another. Hot spots enable geologists to determine absolute motion because they provide an apparently fixed reference point from which the rate and direction of plate movement can be measured. The previously mentioned Emperor Seamount-Hawaiian Island chain (Figure 3.24) formed as a result of movement over a hot spot. Thus the line of the volcanic islands traces the direction of plate movement, and dating the volcanoes enables geologists to determine the rate of movement.

What Is the Driving Mechanism of Plate Tectonics?

A major obstacle to the acceptance of continental drift was the lack of a driving mechanism to explain continental movement. When it was shown that continents and ocean floors moved together, not separately, and that new crust formed at spreading ridges by rising magma, most geologists accepted some type of convective heat system as the basic process responsible for plate motion. The question, however, remains: What exactly drives the plates?

Two models that both involve thermal convection cells have been proposed to explain plate movement (• Figure 3.26). In one model, thermal convection cells are restricted to the asthenosphere; in the second model, the entire mantle is involved. In both models, spreading ridges mark the ascending limbs of adjacent convection cells, and trenches are present where convection cells descend back into Earth's interior. The convection cells therefore determine the location of spreading ridges and trenches, with the lithosphere lying above the thermal convection cells. Each plate thus corresponds to a single convection cell and moves as a result of the convective movement of the cell itself.

Although most geologists agree that Earth's internal heat plays an important role in plate movement, there are problems with both models. The major problem associated with the first model is the difficulty in explaining the

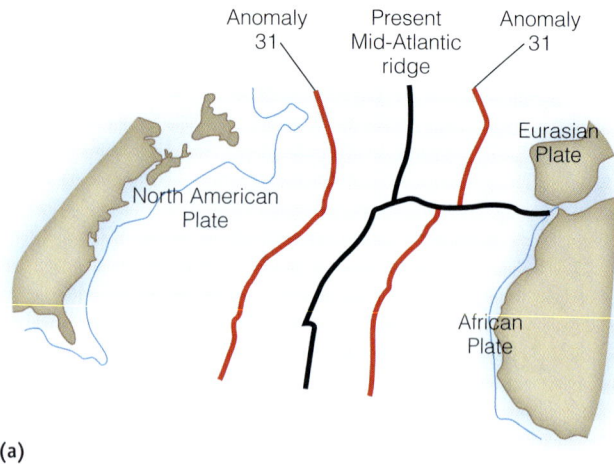

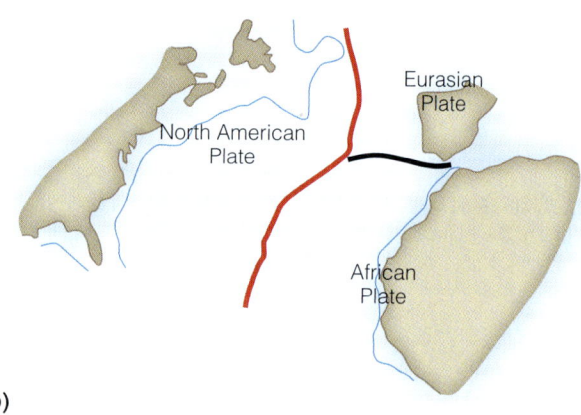

• **Figure 3.25 Reconstructing Plate Positions Using Magnetic Anomalies** (a) The present North Atlantic, showing the Mid-Atlantic Ridge and magnetic anomaly 31, which formed 67 million years ago. (b) The Atlantic Ocean 67 million years ago. Anomaly 31 marks the plate boundary 67 million years ago. By moving the anomalies back together, along with the plates they are on, we can reconstruct the former positions of the continents.

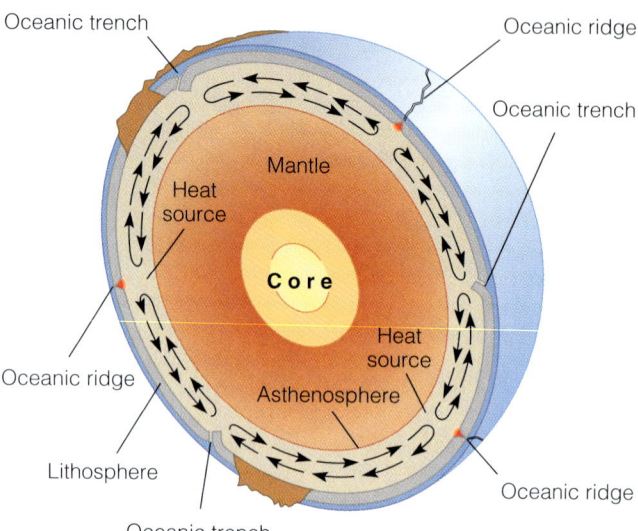

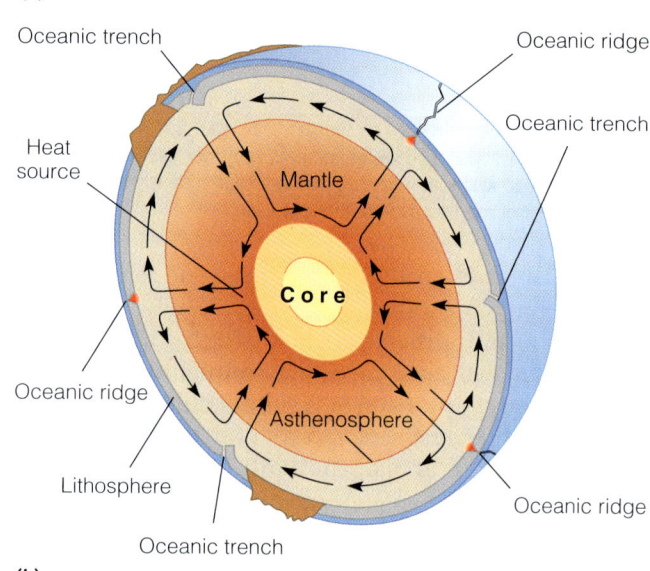

• **Figure 3.26 Thermal Convection Cells as the Driving Force of Plate Movement** Two models involving thermal convection cells have been proposed to explain plate movement. (a) In one model, thermal convection cells are restricted to the asthenosphere. (b) In the other model, thermal convection cells involve the entire mantle.

source of heat for the convection cells and why they are restricted to the asthenosphere. In the second model, the heat comes from the outer core, but it is still not known how heat is transferred from the outer core to the mantle. Nor is it clear how convection can involve both the lower mantle and the asthenosphere.

In addition to some type of thermal convection system driving plate movement, some geologists think plate movement occurs because of a mechanism involving "slab-pull" or "ridge-push," both of which are gravity driven but still dependent on thermal differences within Earth (• Figure 3.27). In slab-pull, the subducting cold slab of lithosphere, being denser than the surrounding warmer asthenosphere, pulls the rest of the plate along as it descends into the asthenosphere. As the lithosphere moves downward, there is a corresponding upward flow back into the spreading ridge.

Operating in conjunction with slab-pull is the ridge-push mechanism. As a result of rising magma, the oceanic ridges are higher than the surrounding oceanic crust. It is thought that gravity pushes the oceanic lithosphere away from the higher spreading ridges and toward the trenches.

Currently, geologists are fairly certain that some type of convective system is involved in plate movement, but the extent to which other mechanisms such as slab-pull and ridge-push are involved is still unresolved. However, the fact that plates have moved in the past and are still moving today has been proven beyond a doubt. And although a comprehensive theory of plate movement has not yet been developed, more and more of the pieces are falling into place as geologists learn more about Earth's interior.

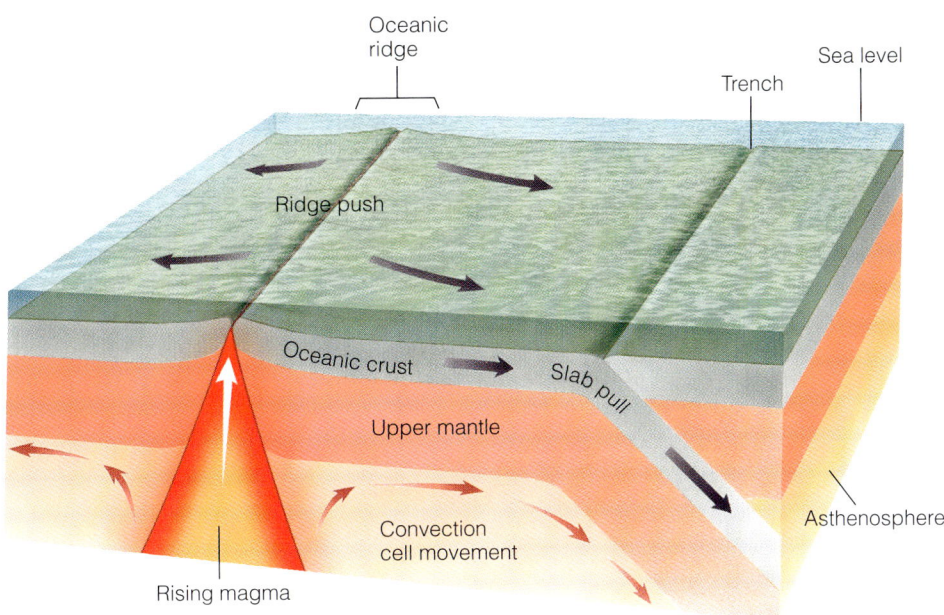

• **Figure 3.27 Plate Movement Resulting from Gravity-Driven Mechanisms** Plate movement is also thought to occur because of gravity-driven "slab-pull" or "ridge-push" mechanisms. In slab-pull, the edge of the subducting plate descends into the interior, and the rest of the plate is pulled downward. In ridge-push, rising magma pushes the oceanic ridges higher than the rest of the oceanic crust. Gravity thus pushes the oceanic lithosphere away from the ridges and toward the trenches.

How Are Plate Tectonics and Mountain Building Related?

What role does plate tectonics play in mountain building? An **orogeny** is an episode of intense rock deformation or mountain building. Orogenies are the consequence of compressive forces related to plate movement. As one plate is subducted under another, sedimentary and volcanic rocks are folded and faulted along the plate margin, while the more deeply buried rocks are subjected to regional metamorphism. Magma generated within the mantle either rises to the surface to erupt as andesitic volcanoes or cools and crystallizes beneath the surface, forming intrusive igneous bodies. Typically, most orogenies occur along either oceanic–continental or continental–continental plate boundaries.

As we discussed earlier in this chapter, ophiolites are evidence of ancient convergent plate boundaries. The slivers of ophiolites found in the interiors of such mountain ranges as the Alps, Himalayas, and Urals mark the sites of former subduction zones. The relationship between mountain building and the opening and closing of ocean basins is called the **Wilson cycle** in honor of the Canadian geologist J. Tuzo Wilson, who first suggested that an ancient ocean had closed to form the Appalachian Mountains and then reopened and widened to form the present Atlantic Ocean.

According to some geologists, much of the geology of continents can be described in terms of a succession of Wilson cycles. We will see in Chapter 14, however, that new evidence concerning the movement of microplates and their accretion at the margin of continents must also be considered when dealing with the tectonic history of a continent.

Terrane Tectonics

During the 1970s and 1980s, geologists discovered that parts of many mountain systems are composed of small, accreted lithospheric blocks that clearly originated elsewhere. These **terranes**, as they are called, differ completely in their fossil content, stratigraphy, structural trends, and paleomagnetic properties from the rocks of the surrounding mountain system. In fact, these terranes are so different from adjacent rocks that most geologists think that they formed elsewhere and were carried great distances as parts of other plates until they collided with other terranes or continents.

To date, most terranes have been identified in mountains of the North American Pacific coast region, but a number of such terranes are suspected to be present in other mountain systems, such as the Appalachians. Terrane tectonics is, however, providing a new way of viewing Earth and gaining a better understanding of the geologic history of the continents.

How Does Plate Tectonics Affect the Distribution of Life?

Plate tectonic theory is as revolutionary and far-reaching in its implications for geology as the theory of evolution was for biology when it was proposed. Interestingly, it was the fossil evidence that convinced Wegener, Suess, and du Toit, as well as many other geologists, of the correctness of continental drift. Together, the theories of plate tectonics and evolution have changed the way we view our planet, and we should not be surprised at the intimate association between them. Although the relationship between plate tectonic processes and the evolution of life is incredibly complex, paleontologic data provide convincing evidence of the influence of plate movement on the distribution of organisms.

The present distribution of plants and animals is not random but is controlled largely by climate and geographic barriers. The world's biotas occupy *biotic provinces,* which are regions characterized by a distinctive assemblage of plants and animals. Organisms within a province have similar ecologic requirements, and the boundaries separating provinces are therefore natural ecologic breaks. Climatic or geographic barriers are the most common province boundaries, and they are largely controlled by plate movement.

Because adjacent provinces usually have less than 20% of their species in common, global diversity is a direct reflection of the number of provinces; the more provinces there are, the greater the global diversity. When continents break up, for example, the opportunity for new provinces to form increases, with a resultant increase in diversity. Just the opposite occurs when continents come together. Plate tectonics thus plays an important role in the distribution of organisms and their evolutionary history.

The complex interaction between wind and ocean currents has a strong influence on the world's climates. Wind and ocean currents are thus strongly influenced by the number, distribution, topography, and orientation of continents. For example, the southern Andes Mountains act as an effective barrier to moist, easterly blowing Pacific winds, resulting in a desert east of the southern Andes that is virtually uninhabitable. Temperature is one of the major limiting factors for organisms, and province boundaries often reflect temperature barriers. Because atmospheric and oceanic temperatures decrease from the equator to the poles, most species exhibit a strong climatic zonation. This biotic zonation parallels the world's latitudinal atmospheric and oceanic circulation patterns. Changes in climate thus have a profound effect on the distribution and evolution of organisms.

The distribution of continents and ocean basins not only influences wind and ocean currents but also affects provinciality by creating physical barriers to, or pathways for, the migration of organisms. Intraplate volcanoes, island arcs, mid-oceanic ridges, mountain ranges, and subduction zones all result from the interaction of plates, and their orientation and distribution strongly influence the number of provinces and hence total global diversity. Provinciality and diversity will thus be highest when there are numerous small continents spread across many zones of latitude.

When a geographic barrier separates a once-uniform fauna, species may undergo divergence. If conditions on opposite sides of the barrier are sufficiently different, then species must adapt to the new conditions, migrate, or become extinct. Adaptation to the new environment by various species may involve enough change that new species eventually evolve. The marine invertebrates found on opposite sides of the Isthmus of Panama provide an excellent example of divergence caused by the formation of a geographic barrier. Prior to the rise of this land connection between North and South America, a homogeneous population of bottom-dwelling invertebrates inhabited the shallow seas of the area. After the formation of the Isthmus of Panama by subduction of the Pacific plate about 5 million years ago, the original population was divided into two populations—one in the Caribbean Sea, and the other in the Pacific Ocean—by the Isthmus of Panama. In response to the changing environment, new species evolved on opposite sides of the isthmus (• Figure 3.28).

The formation of the Isthmus of Panama also influenced the evolution of the North and South American mammalian faunas (see Chapter 18). During most of the Cenozoic Era, South America was an island continent, and its mammalian fauna evolved in isolation from the rest of the world's faunas. When North and South America were connected by the Isthmus of Panama, most of the indigenous South American mammals were replaced by migrants from North America. Surprisingly, only a few South American mammal groups migrated northward.

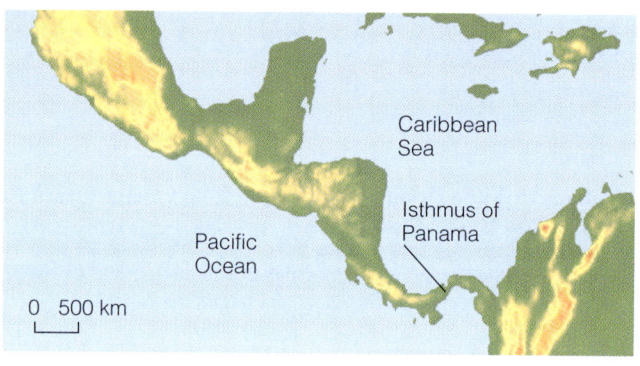

• **Figure 3.28 Plate Tectonics and the Distribution of Organisms**
(a) The Isthmus of Panama forms a barrier that divides a once-uniform fauna of molluscs. **(b)** Divergence of gastropod and bivalve species after the formation of the Isthmus of Panama. Each pair belongs to the same genus but is a different species.

How Does Plate Tectonics Affect the Distribution of Natural Resources?

Besides being responsible for the major features of Earth's crust and influencing the distribution and evolution of the world's biota, plate movements also affect the formation and distribution of some natural resources. Consequently, geologists are using plate tectonic theory in their search for petroleum, natural gas, and mineral deposits and in explaining the occurrence of these natural resources.

Although large concentrations of petroleum occur in many areas of the world, more than 50% of all proven reserves are in the Persian Gulf region (see Perspective 3.1). The reason for this is paleogeography and plate movement. Elsewhere in the world, plate tectonics is also responsible for concentrations of petroleum. The formation of the Appalachians, for example, resulted from the compressive forces generated along a convergent plate boundary and provided the structural traps necessary for petroleum to accumulate (see Chapters 10 and 11).

Many metallic mineral deposits such as copper, gold, lead, silver, tin, and zinc are related to igneous and associated hydrothermal activity. So it is not surprising that a close relationship exists between plate boundaries and the occurrence of these valuable deposits.

The magma generated by partial melting of a subducting plate rises toward the surface, and as it cools, it precipitates and concentrates various metallic ores. Many of the world's major metallic ore deposits, such as the porphyry copper deposits of western North and South America (• Figure 3.29), are associated with convergent plate boundaries.

Divergent plate boundaries also yield valuable ore deposits. Hydrothermal vents are the sites of considerable metallic mineral precipitation. The island of Cyprus in the Mediterranean is rich in copper and has been supplying all or part of the world's needs for the last 3000 years. The concentration of copper on Cyprus formed as a result of precipitation adjacent to hydrothermal vents along a divergent plate boundary.

Studies indicate that minerals containing such metals as copper, gold, iron, lead, silver, and zinc are currently forming as sulfides in the Red Sea. The Red Sea is opening as a result of plate divergence and represents the earliest stage in the growth of an ocean basin (Figures 3.15c and 3.16a).

> ## What Would You Do?
>
> You are part of a mining exploration team that is exploring a promising, remote area of central Asia. You know that former convergent and divergent plate boundaries are frequently sites of ore deposits. What evidence would you look for to determine if the area you're exploring might be an ancient convergent or divergent plate boundary? Is there anything you can do before visiting the area that might help you determine what the geology of the area is?

(a) (b)

• **Figure 3.29 Copper Deposits and Convergent Plate Boundaries** (a) Valuable copper deposits are located along the west coasts of North and South America in association with convergent plate boundaries. The rising magma and associated hydrothermal activity resulting from subduction carried small amounts of copper, which became trapped and concentrated in the surrounding rocks through time. (b) Bingham Mine in Utah is a huge open-pit copper mine with reserves estimated at 1.7 billion tons. More than 400,000 tons of rock are removed for processing each day.

It is becoming increasingly clear that if we are to keep up with the continuing demands of a global industrialized society, the application of plate tectonic theory to the origin and distribution of mineral resources is essential. We will discuss natural resources and their formation as part of the geologic history of Earth in later chapters in this book.

SUMMARY

- The concept of continental movement is not new. The earliest maps showing the similarity between the east coast of South America and the west coast of Africa provided people with the first evidence that continents might once have been united and subsequently separated from each other.
- Alfred Wegener is generally credited with developing the hypothesis of continental drift. He provided abundant geologic and paleontologic evidence to show that the continents were once united into one supercontinent he named Pangaea. Unfortunately, Wegener could not explain how the continents moved, and most geologists ignored his ideas.
- The hypothesis of continental drift was revived during the 1950s when paleomagnetic studies of rocks indicated the presence of multiple magnetic north poles instead of just one as there is today. This paradox was resolved by constructing a map in which the continents could be moved into different positions such that the paleomagnetic data would then be consistent with a single magnetic north pole.
- Magnetic surveys of the oceanic crust revealed magnetic anomalies in the rocks, indicating that Earth's magnetic field has reversed itself numerous times during the past. Because the anomalies are parallel to and form symmetric belts adjacent to the oceanic ridges, new oceanic crust must have formed as the seafloor was spreading.
- Radiometric dating reveals the oldest oceanic crust is less than 180 million years old, whereas the oldest continental crust is 3.96 billion years old. Clearly, the ocean basins are recent geologic features.
- Plate tectonic theory became widely accepted by the 1970s because the evidence overwhelmingly supports it and because it provides geologists with a powerful theory for explaining such phenomena as volcanism, earthquake activity, mountain building, global climatic changes, the distribution of the world's biota, and the distribution of some mineral resources.
- Three types of plate boundaries are recognized: divergent boundaries, where plates move away from each other; convergent boundaries, where two plates collide; and transform boundaries, where two plates slide past each other.
- Ancient plate boundaries can be recognized by their associated rock assemblages and geologic structures. For divergent boundaries, these may include rift valleys with thick sedimentary sequences and numerous dikes and sills. For convergent boundaries, ophiolites and andesitic rocks are two characteristic features. Transform faults generally do not leave any characteristic or diagnostic features in the rock record.
- Although a comprehensive theory of plate movement has yet to be developed, geologists think that some type of convective heat system is the major driving force.
- The average rate of movement and relative motion of plates can be calculated in several ways. The results of these different methods all agree and indicate that the plates move at different average velocities.
- Absolute motion of plates can be determined by the movement of plates over mantle plumes. A mantle plume is an apparently stationary column of magma that rises to the surface where it becomes a hot spot and forms a volcano.
- Geologists now realize that plates can grow when terranes collide with the margins of continents.
- A close relationship exists between the formation of some mineral deposits and petroleum, and plate boundaries. Furthermore, the formation and distribution of some natural resources are related to plate movements.

IMPORTANT TERMS

continental–continental plate boundary, p. 50
continental drift, p. 38
convergent plate boundary, p. 48
Curie point, p. 41
divergent plate boundary, p. 46
Glossopteris flora, p. 38
Gondwana, p. 38
hot spot, p. 52
Laurasia, p. 38
magnetic anomaly, p. 43
magnetic reversal, p. 42
oceanic–continental plate boundary, p. 49
oceanic–oceanic plate boundary, p. 48
ophiolite, p. 51
orogeny, p. 55
paleomagnetism, p. 41
Pangaea, p. 38
plate, p. 45
plate tectonic theory, p. 45
seafloor spreading, p. 42
terrane, p. 55
thermal convection cell, p. 42
transform fault, p. 51
transform plate boundary, p. 51
volcanic island arc, p. 49
Wilson cycle, p. 55

REVIEW QUESTIONS

1. The man credited with developing the continental drift hypothesis is
   a. _____ Wilson; b. _____ Hess; c. _____ Vine; d. _____ Wegener; e. _____ du Toit.
2. The southern part of Pangaea, consisting of South America, Africa, India, Australia, and Antarctica, is called
   a. _____ Gondwana; b. _____ Laurentia; c. _____ Atlantis; d. _____ Laurasia; e. _____ Pacifica.
3. The Hawaiian Island chain and Yellowstone National Park, Wyoming, are examples of
   a. _____ oceanic–oceanic plate boundaries; b. _____ hot spots; c. _____ divergent plate boundaries; d. _____ transform plate boundaries; e. _____ oceanic–continental plate boundaries.
4. Along what type of plate boundary does subduction occur?
   a. _____ divergent; b. _____ transform; c. _____ convergent; d. _____ answers a and b; e. _____ answers a and c.
5. Magnetic surveys of the ocean basins indicate that
   a. _____ the oceanic crust is youngest adjacent to mid-oceanic ridges; b. _____ the oceanic crust is oldest adjacent to mid-oceanic ridges; c. _____ the oceanic crust is youngest adjacent to the continents; d. _____ the oceanic crust is the same age everywhere; e. _____ answers b and c.
6. The driving mechanism of plate movement is thought to be
   a. _____ isostasy; b. _____ Earth's rotation; c. _____ thermal convection cells; d. _____ magnetism; e. _____ polar wandering.
7. Divergent plate boundaries are areas where
   a. _____ new continental lithosphere is forming; b. _____ new oceanic lithosphere is forming; c. _____ two plates come together; d. _____ two plates slide past each other; e. _____ answers b and c.
8. The formation and distribution of copper deposits are associated with what type(s) of plate boundaries?
   a. _____ convergent; b. _____ hot spots; c. _____ divergent; d. _____ transform; e. _____ answers a and c.
9. The most common biotic province boundaries are
   a. _____ geographic barriers; b. _____ biologic barriers; c. _____ climatic barriers; d. _____ answers a and b; e. _____ answers a and c.
10. The San Andreas fault is an example of what type of plate boundary?
    a. _____ divergent; b. _____ convergent; c. _____ transform; d. _____ oceanic–continental; e. _____ continental–continental.
11. Which of the following allows geologists to determine absolute plate motion?
    a. _____ hot spots; b. _____ the age of the sediment directly above any portion of the oceanic crust; c. _____ magnetic reversals in the oceanic crust; d. _____ satellite-laser ranging techniques; e. _____ all of the previous answers.
12. The Himalayas are a good example of what type of plate boundary?
    a. _____ continental–continental; b. _____ oceanic–oceanic; c. _____ oceanic–continental; d. _____ divergent; e. _____ transform.
13. Iron-bearing minerals in magma gain their magnetism and align themselves with the magnetic field when they cool through the
    a. _____ Curie point; b. _____ magnetic anomaly point; c. _____ thermal convection point; d. _____ hot spot point; e. _____ isostatic point.
14. Explain why global diversity increases with an increase in biotic provinces. How does plate movement affect the number of biotic provinces?
15. What evidence convinced Wegener and others that continents must have moved in the past and at one time formed a supercontinent?
16. Why is some type of thermal convection system thought to be the major force driving plate movement? How have "slab-pull" and "ridge-push," both mainly gravity driven, modified a purely thermal convection model for plate movement?
17. Explain how magnetic anomalies recorded in oceanic crust help confirm the seafloor spreading theory.
18. How have plate tectonic processes affected the formation and distribution of natural resources?
19. Explain why such natural disasters as volcanic eruptions and earthquakes are associated with divergent and convergent plate boundaries.
20. Plate tectonic theory builds on the continental drift hypothesis and the theory of seafloor spreading. As such, it is a unifying theory of geology. Explain why it is a unifying theory.

APPLY YOUR KNOWLEDGE

1. Using the age for each of the Hawaiian Islands in Figure 3.24 and an atlas in which you can measure the distance between islands, calculate the average rate of movement per year for the Pacific plate since each island formed. Is the average rate of movement the same for each island? Would you expect it to be? Explain why it may not be.
2. If the movement along the San Andreas fault, which separates the Pacific plate from the North American plate, averages 5.5 cm per year, how long will it take before Los Angeles is opposite San Francisco?
3. Based on your knowledge of biology and the distribution of organisms throughout the world, how do you think plate tectonics has affected this distribution both on land and in the oceans?

CHAPTER 4

GEOLOGIC TIME: CONCEPTS AND PRINCIPLES

Reed Wicander

The Grand Canyon, Arizona. Major John Wesley Powell led two expeditions down the Colorado River and through the canyon in 1869 and 1871. He was struck by the seemingly limitless time represented by the rocks exposed in the canyon walls and by the recognition that these rock layers, like the pages in a book, contain the geologic history of this region.

{ OUTLINE }

INTRODUCTION

HOW IS GEOLOGIC TIME MEASURED?

HOW HAS THE CONCEPT OF GEOLOGIC TIME AND EARTH'S AGE CHANGED THROUGHOUT HUMAN HISTORY?

PERSPECTIVE 4.1 *Geologic Time and Climate Change*

WHAT ARE RELATIVE DATING METHODS, AND WHY ARE THEY IMPORTANT?

Fundamental Principles of Relative Dating

ESTABLISHMENT OF GEOLOGY AS A SCIENCE—THE TRIUMPH OF UNIFORMITARIANISM OVER NEPTUNISM AND CATASTROPHISM

Neptunism and Catastrophism

Uniformitarianism

Modern View of Uniformitarianism

LORD KELVIN AND A CRISIS IN GEOLOGY

WHAT ARE ABSOLUTE DATING METHODS, AND WHY ARE THEY IMPORTANT?

Atoms and Isotopes

Radioactive Decay and Half-Lives

Long-Lived Radioactive Isotope Pairs

Fission-Track-Dating

Radiocarbon and Tree-Ring Dating

SUMMARY

ThomsonNOW™ Explore interactive tutorials, animations, or practice problems available on the ThomsonNow website at **www.thomsonedu.com/login**.

CHAPTER OBJECTIVES

At the end of this chapter, you will have learned that

- The concept of geologic time and its measurement have changed through human history.
- The fundamental principles of relative dating provide a means to interpret geologic history.
- The principle of uniformitarianism is fundamental to geology and prevailed over the concepts of neptunism and catastrophism because it provides a better explanation for observed geologic phenomena.
- The discovery of radioactivity provided geologists with a clock that could measure Earth's age and validate that Earth was very old.
- Different absolute dating methods are used to date geologic events in terms of years before present.
- The most accurate radiometric dates are obtained from igneous rocks.

Introduction

In 1869, Major John Wesley Powell, a Civil War veteran who had lost his right arm in the battle of Shiloh, led a group of hardy explorers down the uncharted Colorado River through the Grand Canyon. With no maps or other information, Powell and his group ran the many rapids of the Colorado River in fragile wooden boats, hastily recording what they saw. Powell wrote in his diary that "all about me are interesting geologic records. The book is open and I read as I run."

From this initial reconnaissance, Powell led a second expedition down the Colorado River in 1871. This second trip included a photographer, a surveyor, and three topographers. Members of the expedition made detailed topographic and geologic maps of the Grand Canyon area as well as the first photographic record of the region.

Probably no one has contributed as much to the understanding of the Grand Canyon as Major Powell. In recognition of his contributions, the Powell Memorial was erected on the South Rim of the Grand Canyon in 1969 to commemorate the 100th anniversary of this history-making first expedition.

Most tourists today, like Powell and his fellow explorers in 1869, are astonished by the seemingly limitless time represented by the rocks exposed in the walls of the Grand Canyon. For most visitors, viewing a 1.5-kilometer-deep cut into Earth's crust is the only encounter they'll ever have with the vastness of geologic time. When we stand on the rim and look down into the Grand Canyon, we are really looking far back in time, all the way back to the early history of our planet. In fact, more than 1 billion years of history are preserved in the rocks of the Grand Canyon, and reading what is preserved in those rocks is, just as Powell noted in his diary more than 100 years ago, like reading the pages in a history book.

In reading this "book"—the rock layers of the Grand Canyon—we learn that this area underwent episodes of mountain building as well as periods of advancing and retreating shallow seas. How do we know this? The answer lies in applying the principles of relative dating to the rocks we see exposed and also in recognizing that present-day processes have operated throughout Earth history. In this chapter you will learn what those principles are and how they can be used to determine how long ago geologic events took place, as well as how to decipher the geologic history of an area by reading the clues provided by the rocks.

We begin this chapter by asking the question "What is time?" We seem obsessed with time, and we organize our lives around it. Yet most of us feel we don't have enough of it—we are always running "behind" or "out of time." Whereas physicists deal with extremely short intervals of time, and geologists deal with incredibly long periods of time, most of us tend to view time from the perspective of our own existence; that is, we partition our lives into seconds, hours, days, weeks, months, and years. Ancient history is what occurred hundreds or even thousands of years ago. Yet when geologists talk of ancient geologic history, they are referring to events that happened millions or even billions of years ago!

Vast periods of time set geology apart from most of the other sciences, and an appreciation of the immensity of geologic time is fundamental to understanding the physical and biological history of our planet. In fact, understanding and accepting the magnitude of geologic time are major contributions geology has made to the sciences.

How Is Geologic Time Measured?

In some respects, time is defined by the methods used to measure it. Geologists use two different frames of reference when discussing geologic time. **Relative dating** is placing geologic events in a sequential order as determined from their positions in the geologic record. Relative dating will not tell us how long ago a particular event occurred, only that one event preceded another. A useful analogy for relative dating is a television guide that does not list the times programs are shown. In this example, you cannot tell what time

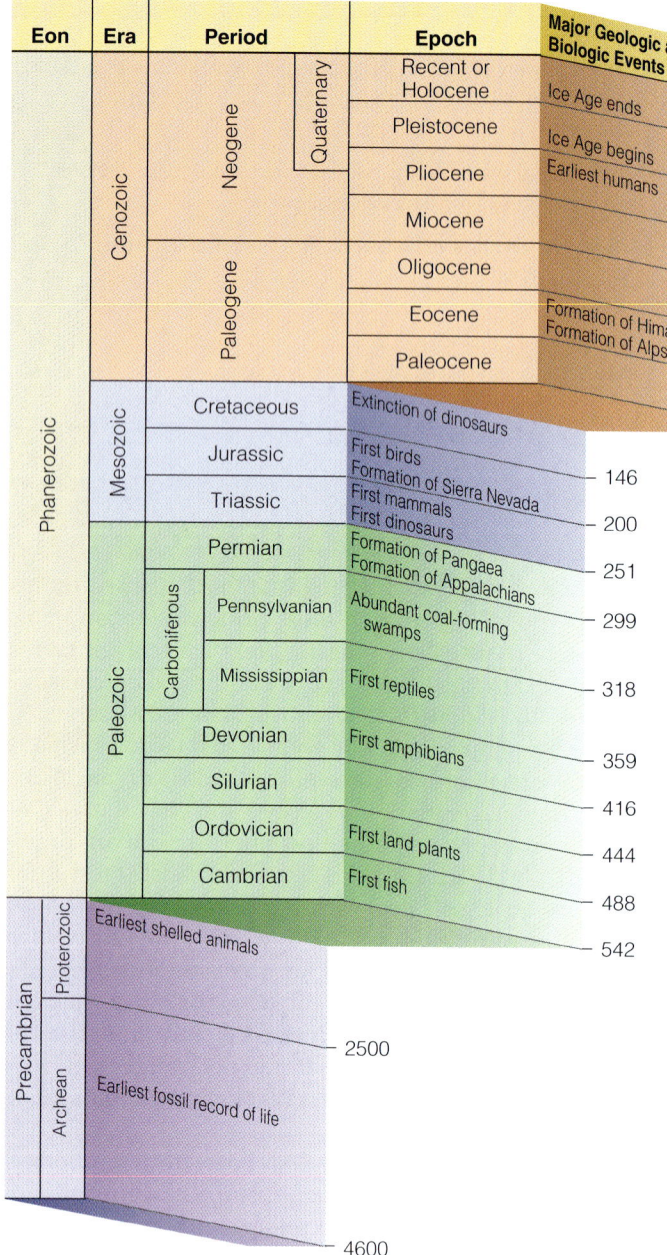

• **Figure 4.1 The Geologic Time Scale**
Some of the major geologic and biologic events are indicated along the right-hand margin. Dates are from Gradstein, F., Ogg, J. and Smith, A., *A Geologic Timescale 2004* (Cambridge, UK: Cambridge University Press, 2005), Figure 1.2.

Radiometric dating is the most common method of obtaining absolute ages. Dates are calculated from the natural rates of decay of various radioactive elements present in trace amounts in some rocks. It was not until the discovery of radioactivity near the end of the 19th century that absolute ages could be accurately applied to the relative geologic time scale. Today, the geologic time scale is really a dual scale—a relative scale based on rock sequences with radiometric dates expressed as years before the present (Figure 4.1).

Besides providing an appreciation for the immensity of geologic time, why is the study of geologic time important? One of the most valuable lessons you will learn in this chapter is how to reason and apply the fundamental geologic principles in solving geologic problems. The logic used in applying the principles of relative dating to interpret the geologic history of an area involves basic reasoning skills that can be transferred to, and used in almost any profession or discipline.

Advances and refinements in absolute dating techniques during the 20th century have changed the way we view Earth in terms of when events occurred in the past and the rates of geologic change through time. The ability to accurately determine past climatic changes and their causes has important implications for the current debate on global warming and its effects on humans (see Perspective 4.1).

a particular program will be broadcast, but by watching a few shows and checking the guide, you can determine whether you have missed the show or how many shows are scheduled before the one you want to see.

The various principles used to determine relative dating were discovered hundreds of years ago, and since then they have been used to construct the *relative geologic time scale* (• Figure 4.1). Furthermore, these principles are still widely used by geologists today.

Absolute dating provides specific dates for rock units or events expressed in years before the present. In our analogy of the television guide, the time when the programs are actually shown would be the absolute dates. In this way, you can not only see if you have missed a program (relative dating), but can also find out when the show you want to see will be aired (absolute dating).

How Has the Concept of Geologic Time and Earth's Age Changed Throughout Human History?

The concept of geologic time and its measurement have changed throughout human history. Early Christian theologians were largely responsible for formulating the idea that time is linear rather than circular. When St. Augustine of Hippo (A.D. 354–430) stated that the Crucifixion was a unique event from which all other events could be measured, he helped establish the idea of the B.C. and A.D. time scale. This prompted many religious scholars and clerics to try to establish the date of creation by analyzing historical records and the genealogies found in Scripture.

Perspective 4.1

Geologic Time and Climate Change

Given the debate concerning global warming and its possible implications, it is extremely important to be able to reconstruct past climatic regimes as accurately as possible. To model how Earth's climate system has responded to changes in the past and use that information for simulations of future climatic scenarios, geologists must have a geologic calendar that is as precise as possible.

New dating techniques with greater precision are providing geologists with more accurate dates for when and how long past climate changes occurred. The ability to accurately determine when past climate changes took place helps geologists correlate these changes with regional and global geologic events to see whether there are any possible connections.

One interesting method that is becoming more common in reconstructing past climates is to analyze stalagmites from caves. Stalagmites are icicle-shaped structures rising from a cave floor and formed of calcium carbonate precipitated from evaporating water (Figure 1a). A stalagmite therefore records a layered history, because each newly precipitated layer of calcium carbonate is younger than the previously precipitated layer. Thus a stalagmite's layers are oldest in the center at its base and are progressively younger as they move outward (principle of superposition) (Figure 1b).

Using techniques based on ratios of uranium 234 to thorium 230, geologists can achieve very precise radiometric dates on individual layers of a stalagmite (Figure 1c). This technique enables geologists to determine the age of materials much older than they can date by the carbon-14 method, and it is reliable back to about 500,000 years.

A study of stalagmites from Crevice Cave in Missouri revealed a history of climatic and vegetation change in the midcontinent region of the United States during the interval between 75,000 and 25,000 years ago. Dates obtained from the Crevice Cave stalagmites were correlated with major changes in vegetation and average temperature fluctuations, obtained from carbon 13 and oxygen 18 isotope profiles, to reconstruct a detailed picture of climate changes during this time period.

It was determined that during the interval between 75,000 and 55,000 years ago, the climate oscillated between warm and cold, and vegetation alternated among forest, savannah, and prairie. Fifty-five thousand years ago the climate cooled, and there was a sudden change from grasslands to forest, which persisted until 25,000 years ago. This corresponds to the time when global ice sheets began building and advancing.

Thus precise dating techniques in stalagmite studies using uranium 234 and thorium 230 provide an accurate chronology that allows geologists to model climate systems of the past and perhaps to determine what causes global climatic changes and their duration. Without these sophisticated dating techniques and others like them, geologists would not be able to make precise correlations and accurately reconstruct past environments and climates. By analyzing past environmental and climate changes and their duration, geologists hope they can use these data, sometime in the near future, to predict and possibly modify regional climatic changes.

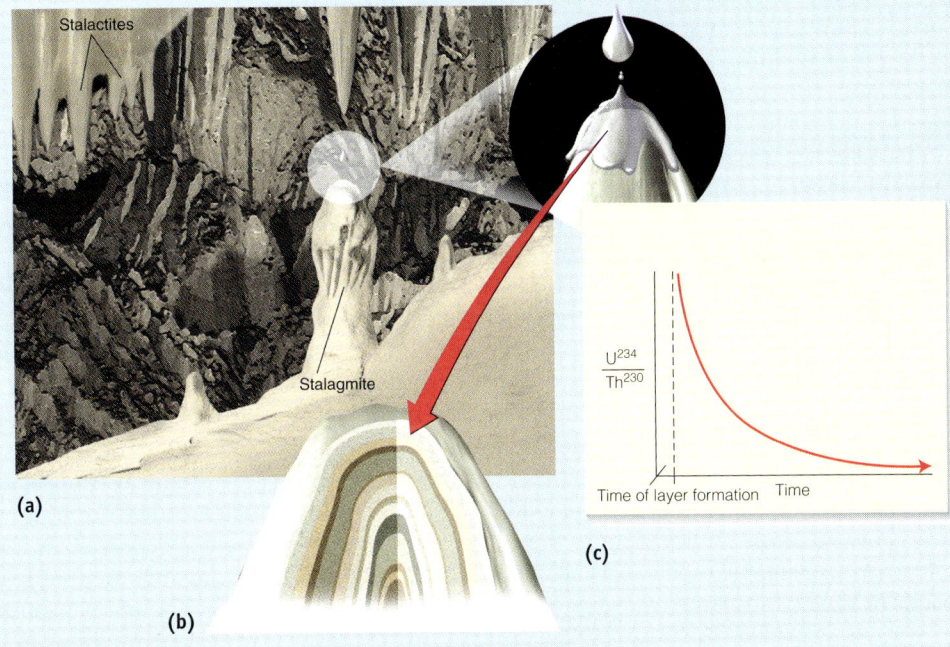

Figure 1 (a) Stalagmites are icicle-shaped structures rising from the floor of a cave, and formed by the precipitation of calcium carbonate from evaporating water. (b) A stalagmite is thus layered, with the oldest layer in the center and the youngest layers on the outside. (c) Uranium 234 frequently substitutes for the calcium ion in the calcium carbonate of the stalagmite. Uranium 234 decays to thorium 230 at a predictable and measurable rate. Therefore, the age of each layer of the stalagmite can be dated by measuring the ratio of uranium 234 to thorium 230.

One of the most influential and famous Christian scholars was James Ussher (1581–1665), archbishop of Armagh, Ireland, who, based on Old Testament genealogy, asserted that God created Earth on Sunday, October 23, 4004 B.C. In 1701, an authorized version of the Bible made this date accepted Church doctrine. For nearly a century thereafter, it was considered heresy to assume that Earth and all its features were more than about 6000 years old. Thus, the idea of a very young Earth provided the basis for most Western chronologies of Earth history prior to the 18th century.

During the 18th and 19th centuries, several attempts were made to determine Earth's age on the basis of scientific evidence rather than revelation. The French zoologist Georges Louis de Buffon (1707–1788) assumed Earth gradually cooled to its present condition from a molten beginning. To simulate this history, he melted iron balls of various diameters and allowed them to cool to the surrounding temperature. By extrapolating their cooling rate to a ball the size of Earth, he determined that Earth was at least 75,000 years old. Although this age was much older than that derived from Scripture, it was still vastly younger than we now know the planet to be.

Other scholars were equally ingenious in attempting to calculate Earth's age. For example, if deposition rates could be determined for various sediments, geologists reasoned that they could calculate how long it would take to deposit any rock layer. They could then extrapolate how old Earth was from the total thickness of sedimentary rock in its crust. Rates of deposition vary, however, even for the same type of rock. Furthermore, it is impossible to estimate how much rock has been removed by erosion or how much a rock sequence has been reduced by compaction. As a result of these variables, estimates of Earth's age ranged from younger than 1 million years to older than 2 billion years.

Another attempt to determine Earth's age involved ocean salinity. Scholars assumed that Earth's ocean waters were originally fresh and that their present salinity was the result of dissolved salt being carried into the ocean basins by streams. Knowing the volume of ocean water and its salinity, John Joly, a 19th-century Irish geologist, measured the amount of salt currently in the world's streams. He then calculated that it would have taken at least 90 million years for the oceans to reach their present salinity level. This was still much younger than the now accepted age of 4.6 billion years for Earth, mainly because Joly had no way to calculate either how much salt had been recycled or the amount of salt stored in continental salt deposits and seafloor clay deposits.

Besides trying to determine Earth's age, the naturalists of the 18th and 19th centuries were also formulating some of the fundamental geologic principles that are used in deciphering Earth history. From the evidence preserved in the geologic record, it was clear to them that Earth is very old and that geologic processes have operated over long periods of time.

What Are Relative Dating Methods, and Why Are They Important?

Before the development of radiometric dating techniques, geologists had no reliable means of absolute dating and therefore depended solely on relative dating methods. Relative dating places events in sequential order but does not tell us how long ago an event took place. Although the

What Would You Do?

You have been chosen to be part of the first astronaut crew to land on Mars. You were selected because you are a geologist, and therefore your primary responsibility is to map the geology of the landing site area. An important goal of the mission is to determine the geologic history of the area. How will you go about this? Will you be able to use the principles of relative dating? How will you correlate the various rock units? Will you be able to determine absolute ages? How will you do this?

principles of relative dating may now seem self-evident, their discovery was an important scientific achievement because they provided geologists with a means to interpret geologic history and develop a relative geologic time scale.

Six fundamental geologic principles are used in relative dating: superposition, original horizontality, lateral continuity, cross-cutting relationships (all discussed in this chapter), and inclusions and fossil succession (discussed in Chapter 5).

Fundamental Principles of Relative Dating

The 17th century was an important time in the development of geology as a science because of the widely circulated writings of the Danish anatomist Nicolas Steno (1638–1686). Steno observed that when streams flood, they spread out across their floodplains and deposit layers of sediment that bury organisms dwelling on the floodplain. Subsequent floods produce new layers of sediments that are deposited or superposed over previous deposits.

When lithified, these layers of sediment become sedimentary rock. Thus, in an undisturbed succession of sedimentary rock layers, the oldest layer is at the bottom and the youngest layer is at the top. This **principle of superposition** is the basis for relative-age determinations of strata and their contained fossils (• Figure 4.2a).

Steno also observed that, because sedimentary particles settle from water under the influence of gravity, sediment is deposited in essentially horizontal layers, thus illustrating the **principle of original horizontality** (Figure 4.2a). Therefore, a sequence of sedimentary rock layers that is steeply inclined from the horizontal must have been tilted after deposition and lithification (Figure 4.2b).

Steno's third principle, the **principle of lateral continuity,** states that sediment extends laterally in all directions until it thins and pinches out or terminates against the edge of the depositional basin (Figure 4.2a).

The Scottish geologist James Hutton (1726–1797) is considered by many to be the founder of modern geology and is credited with discovering the **principle of cross-cutting relationships.** Based on his detailed studies and observations of rock exposures in Scotland, Hutton recognized that an igneous intrusion or a fault must be younger than the rocks it intrudes or displaces (• Figure 4.3).

• **Figure 4.2 The Principles of Original Horizontality, Superposition, and Lateral Continuity** (a) Bryce Canyon National Park, Utah, illustrates three of the six fundamental principles of relative dating. The sedimentary rocks of Bryce Canyon were originally deposited horizontally in a variety of continental environments (principle of original horizontality). The oldest rocks are at the bottom of this highly dissected landscape, and the youngest rocks are at the top, forming the rims (principle of superposition). The exposed rock layers extend laterally in all directions for some distance (principle of lateral continuity). (b) These shales and limestones of the Postolonnec Formation, at Postolonnec Beach, Crozon Peninsula, France, were originally deposited horizontally but have been significantly tilted since their formation.

• **Figure 4.3 The Principle of Cross-Cutting Relationships** (a) A dark gabbro dike cuts across granite in Acadia National Park, Maine. The dike is younger than the granite it intrudes. (b) A small fault (arrows show direction of movement) cuts across, and thus displaces, tilted sedimentary beds along Templin Highway in Castaic, California. The fault is therefore younger than the youngest beds that are displaced.

WHAT ARE RELATIVE DATING METHODS, AND WHY ARE THEY IMPORTANT? **65**

Establishment of Geology as a Science—The Triumph of Uniformitarianism over Neptunism and Catastrophism

Steno's principles were significant contributions to early geologic thought, but the prevailing concepts of Earth history continued to be those that could be easily reconciled with a literal interpretation of Scripture. Two of these ideas, *neptunism* and *catastrophism*, were particularly appealing and accepted by many naturalists. In the final analysis, however, another concept, *uniformitarianism*, became the underlying philosophy of geology because it provided a better explanation for observed geologic phenomena than either neptunism or catastrophism.

Neptunism and Catastrophism

The concept of **neptunism** was proposed in 1787 by a German professor of mineralogy, Abraham Gottlob Werner (1749–1817). Although Werner was an excellent mineralogist, he is best remembered for his incorrect interpretation of Earth history. He thought that all rocks, including granite and basalt, were precipitated in an orderly sequence from a primeval, worldwide ocean (Table 4.1).

Werner's subdivision of Earth's crust by supposed relative age attracted a large following in the late 1700s and became almost universally accepted as the standard geologic column. Two factors account for this. First, Werner's charismatic personality, enthusiasm for geology, and captivating lectures popularized the concept. Second, neptunism included a worldwide ocean that easily conformed to the biblical deluge.

Despite Werner's personality and arguments, his neptunian theory failed to explain what happened to the tremendous amount of water that once covered Earth. An even greater problem was Werner's insistence that all igneous rocks were precipitated from seawater. It was Werner's failure to recognize the igneous origin of basalt that finally led to the downfall of neptunism.

From the late 18th century to the mid-19th century, the concept of **catastrophism,** proposed by the French zoologist Baron Georges Cuvier (1769–1832), dominated European geologic thinking. Cuvier explained the physical and biologic history of Earth as resulting from a series of sudden widespread catastrophes. Each catastrophe accounted for significant and rapid changes in Earth, including exterminating existing life in the affected area. Following a catastrophe, new organisms were either created or migrated in from elsewhere.

According to Cuvier, six major catastrophes had occurred in the past. These conveniently corresponded to the six days of biblical creation. Furthermore, the last catastrophe was taken to be the biblical deluge, so catastrophism had wide appeal, especially among theologians.

Eventually, both neptunism and catastrophism were abandoned as untenable hypotheses because their basic assumptions could not be supported by field evidence. The simplistic sequence of rocks predicted by neptunism (Table 4.1) was contradicted by field observations from many different areas of the world. Moreover, basalt was shown to be of igneous origin, and subsequent discoveries of volcanic rocks interbedded with secondary and primitive deposits proved that volcanic activity had occurred throughout Earth history. As more field observations from widely separated areas were made, naturalists realized that far more than six catastrophes were needed to account for Earth history. With the demise of neptunism and catastrophism, the principle of uniformitarianism, advocated by James Hutton (1726–1797) and Charles Lyell, became the guiding philosophy of geology.

Uniformitarianism

Hutton observed the processes of wave action, erosion by running water, and sediment transport, and concluded that, given enough time, these processes could account for the geologic features in his native Scotland. He thought that "the past history of our globe must be explained by what can be seen to be happening now." This assumption that present-day processes have operated throughout geologic time was the basis for the **principle of uniformitarianism.**

Although Hutton developed a comprehensive theory of uniformitarian geology, it was Charles Lyell (1797–1875) who became the principal advocate and interpreter of uniformitarianism. The term itself was coined by William Whewell in 1832.

Hutton viewed Earth history as cyclical—that is, continents are worn down by erosion, the eroded sediment is deposited in the sea, and uplift of the seafloor creates new continents, thus completing a cycle. He thought the

TABLE 4.1 Werner's Subdivision of the Rocks of Earth's Crust

Alluvial rocks	Unconsolidated sediments
Secondary rocks	Various fossiliferous, detrital, and chemical rocks such as sandstones, limestones, and coal, as well as basalts
Transition rocks	Chemical and detrital rocks including fossiliferous rocks
Primitive rocks	Oldest rocks of Earth (igneous and metamorphic)

mechanism for uplift was thermal expansion from Earth's hot interior. Hutton's field observations, and experiments performed by his contemporaries involving the melting of basalt samples, convinced him that igneous rocks were the result of cooling magmas. This interpretation of the origin of igneous rocks, called *plutonism,* eventually displaced the neptunian view that igneous rocks precipitated from seawater.

Hutton also recognized the importance of unconformities in his cyclical view of Earth history. At Siccar Point, Scotland, he observed steeply inclined metamorphic rocks that had been eroded and covered by flat-lying younger rocks (• Figure 4.4). It was clear to him that severe upheavals had tilted the lower rocks and formed mountains. These were then worn away and covered by younger, flat-lying rocks. The erosion surface meant there was a gap in the geologic record, and the rocks above and below this surface provided evidence that both mountain building and erosion had occurred. Although Hutton did not use the word *unconformity,* he was the first to understand and explain the significance of such gaps in the geologic record.

Hutton was also instrumental in establishing the concept that geologic processes had vast amounts of time in which to operate. Because Hutton relied on known processes to account for Earth history, he concluded that Earth must be very old. However, he estimated neither how old Earth was nor how long it took to complete a cycle of erosion, deposition, and uplift. He merely allowed that "we find no vestige of a beginning, and no prospect of an end," which was in keeping with a cyclical view of Earth history.

Unfortunately, Hutton was not a good writer, and thus his ideas were not widely disseminated or accepted. In fact, neptunism and catastrophism continued to be the dominant geologic concepts well into the 1800s. In 1830, however, Charles Lyell published a landmark book, *Principles of Geology,* in which he championed Hutton's concept of uniformitarianism.

• **Figure 4.4 Angular Unconformity at Siccar Point, Scotland** It was at this location in 1788 that James Hutton first realized the significance of unconformities in interpreting Earth history.

Instead of relying on catastrophic events to explain various Earth features, Lyell recognized that small, imperceptible changes brought about by present-day processes could, over long periods of time, have tremendous cumulative effects. Not only did Lyell effectively reintroduce and establish the concept of unlimited geologic time, but he also discredited catastrophism as a viable explanation of geologic phenomena. Thus, through his writings, Lyell firmly established uniformitarianism as the guiding principle of geology. Furthermore, the recognition of virtually limitless amounts of time was also necessary for, and instrumental in, the acceptance of Darwin's 1859 theory of evolution (see Chapter 7).

Perhaps because uniformitarianism is such a general concept, scientists have interpreted it in different ways. Lyell's concept of uniformitarianism embodied the idea of a steady-state Earth in which present-day processes have operated at the same rate in the past as they do today. For example, the frequency of earthquakes and volcanic eruptions for any given period of time in the past has been the same as it is today. If the climate in one part of the world became warmer, another area would have to become cooler so that overall the climate remains the same. By such reasoning, Lyell claimed that conditions for Earth as a whole had been constant and unchanging through time.

Modern View of Uniformitarianism

Geologists today assume that principles, or laws, of nature are constant but that rates and intensities of change have varied through time. For example, volcanic activity was more intense in North America during the Miocene Epoch than today, whereas glaciation has been more prevalent during the last 1.8 million years than in the previous 300 million years. Because rates and intensities of geologic processes have varied through time, some geologists prefer to use the term *actualism* rather than *uniformitarianism* to remove the idea of "uniformity" from the concept. Most geologists, though, still use the term *uniformitarianism* because it indicates that, even though rates and intensities of change have varied in the past, laws of nature have remained the same.

Uniformitarianism is a powerful concept that allows us through analogy and inductive reasoning to use present-day processes as the basis for interpreting the past and for predicting potential future events. It does not eliminate occasional, sudden, short-term events such as volcanic eruptions, earthquakes, floods, or even meteorite impacts as forces that shape our modern world. In fact, some geologists view Earth history as a series of such short-term, or punctuated, events, and this view is certainly in keeping with the modern principle of uniformitarianism.

Earth is in a state of dynamic change and has been since it formed. Although rates of change may have varied in the past, natural laws governing the processes have not.

Lord Kelvin and a Crisis in Geology

Lord Kelvin (1824–1907), a highly respected English physicist, claimed in a paper written in 1866 to have destroyed the uniformitarian foundation on which Huttonian–Lyellian geology was based. Kelvin did not accept Lyell's strict uniformitarianism, in which chemical reactions in Earth's interior were supposed to continually produce heat, allowing for a steady-state Earth. Kelvin rejected this idea as perpetual motion—impossible according to the known laws of physics—and accepted instead the assumption that Earth was originally molten.

Kelvin knew from deep mines in Europe that Earth's temperature increases with depth, and he reasoned that Earth is losing heat from its interior. By knowing the size of Earth, the melting temperature of rocks, and the rate of heat loss, Kelvin calculated the age at which Earth was entirely molten. From these calculations, he concluded that Earth could not be older than 400 million years or younger than 20 million years. This wide discrepancy in age reflected uncertainties in average temperature increases with depth and the various melting points of Earth's constituent materials.

After establishing that Earth was very old and that present-day processes operating over long periods of time account for geologic features, geologists were in a quandary. If they accepted Kelvin's dates, they would have to abandon the concept of seemingly limitless time that was the underpinning of uniformitarian geology and one of the foundations of Darwinian evolution and squeeze events into a shorter time frame. Some geologists objected to such a young age for Earth, but their objections were seemingly based more on faith than on hard facts. Kelvin's quantitative measurements and arguments seemed flawless and unassailable to scientists at the time.

Kelvin's reasoning and calculations were sound, but his basic premises were false, thereby invalidating his conclusions. Kelvin was unaware that Earth has an internal heat source—radioactivity—that has allowed it to maintain a fairly constant temperature through time.* His 40-year campaign for a young Earth ended with the discovery of radioactivity near the end of the 19th century and the insight in 1905 that natural radioactive decay can be used to date the age of formation of rocks. His "unassailable calculations" were no longer valid, and his proof for a geologically young Earth collapsed. Kelvin's theory, like neptunism, catastrophism, and a worldwide flood, thus became another interesting footnote in the history of geology.

Although the discovery of radioactivity destroyed Kelvin's arguments, it provided geologists with a clock that could measure Earth's age and validate what geologists had been saying all along—namely, that Earth was indeed very old!

What Are Absolute Dating Methods, and Why Are They Important?

Although most of the isotopes of the 92 naturally occurring elements are stable, some are radioactive and spontaneously decay to other, more stable isotopes of elements, releasing energy in the process. The discovery in 1903 by Pierre and Marie Curie that radioactive decay produces heat meant that geologists finally had a mechanism for explaining Earth's internal heat that did not rely on residual cooling from a molten origin. Furthermore, geologists now had a powerful tool to date geologic events accurately and to verify the long time periods postulated by Hutton and Lyell.

Atoms and Isotopes

As we discussed in Chapter 2, all matter is made up of chemical elements, each composed of extremely small particles called *atoms*. The *nucleus* of an atom is composed of *protons* (positively charged particles) and *neutrons* (neutral particles) with *electrons* (negatively charged particles) encircling it (see Figure 2.2). The number of protons defines an element's *atomic number* and helps determine its properties and characteristics.

What Would You Do?

You are a member of a regional planning commission that is considering a plan for constructing what is said to be a much-needed river dam that will create a recreational lake. Opponents of the dam project have come to you with a geologic report and map showing that a fault underlies the area of the proposed dam, and the fault trace can be clearly seen at the surface. The opponents say the fault may be active, so someday it will move suddenly, bursting the dam and sending a wall of water downstream. You seek the advice of a local geologist who has worked in the dam area. She tells you that she found a lava flow covering the fault less than a mile from the proposed dam project site. Can you use this information along with a radiometric date from the lava flow to help convince the opponents that the fault has not moved in any direction (vertically or laterally) anytime in the recent past? How would you do this, and what type of reasoning would you use?

---

*Actually, Earth's temperature has decreased through time because the original amount of radioactive materials has been decreasing and therefore is not supplying as much heat. However, the temperature is decreasing at a rate considerably slower than would be required to lend any credence to Kelvin's calculations (see p. 159, Chapter 8).

The combined number of protons and neutrons in an atom is its *atomic mass number*. However, not all atoms of the same element have the same number of neutrons in their nuclei. These variable forms of the same element are called *isotopes* (see Figure 2.3). Most isotopes are stable, but some are unstable and spontaneously decay to a more stable form. It is the decay rate of unstable isotopes that geologists measure to determine the absolute age of rocks.

Radioactive Decay and Half-Lives

Radioactive decay is the process whereby an unstable atomic nucleus is spontaneously transformed into an atomic nucleus of a different element. Scientists recognize three types of radioactive decay, all of which result in a change of atomic structure (• Figure 4.5). In **alpha decay**, two protons and two neutrons are emitted from the nucleus, resulting in a loss of two atomic numbers and four atomic mass numbers. In **beta decay**, a fast-moving electron is emitted from a neutron in the nucleus, changing that neutron to a proton and consequently increasing the atomic number by 1, with no resultant atomic mass number change. **Electron capture decay** is when a proton captures an electron from an electron shell and thereby converts to a neutron, resulting in a loss of 1 atomic number but not changing the atomic mass number.

Some elements undergo only one decay step in the conversion from an unstable form to a stable form. For example, rubidium 87 decays to strontium 87 by a single beta emission, and potassium 40 decays to argon 40 by a single electron capture. Other radioactive elements undergo several decay steps. Uranium 235 decays to lead 207 by 7 alpha steps and 6 beta steps, whereas uranium 238 decays to lead 206 by 8 alpha and 6 beta steps (• Figure 4.6).

When discussing decay rates, it is convenient to refer to them in terms of half-lives. The **half-life** of a radioactive element is the time it takes for one-half of the atoms of the original unstable **parent element** to decay to atoms of a new, more stable **daughter element.** The half-life of a given radioactive element is constant and can be precisely measured. Half-lives of various radioactive elements range from less than a billionth of a second to 49 billion years.

Radioactive decay occurs at a geometric rate rather than a linear rate. Therefore, a graph of the decay rate produces a curve rather than a straight line (• Figure 4.7). For example, an element with *1,000,000* parent atoms will have *500,000* parent atoms and *500,000* daughter atoms after one half-life. After two half-lives, it will have *250,000* parent atoms (one-half of the previous parent atoms, which is equivalent to one-fourth of the original parent atoms) and *750,000* daughter atoms. After three half-lives, it will have *125,000* parent atoms (one-half of the previous parent atoms, or one-eighth of the original parent atoms) and *875,000* daughter atoms, and so on, until the number of parent atoms remaining is so few that they cannot be accurately measured by present-day instruments.

By measuring the parent–daughter ratio and knowing the half-life of the parent (which has been determined in the laboratory), geologists can calculate the age of a

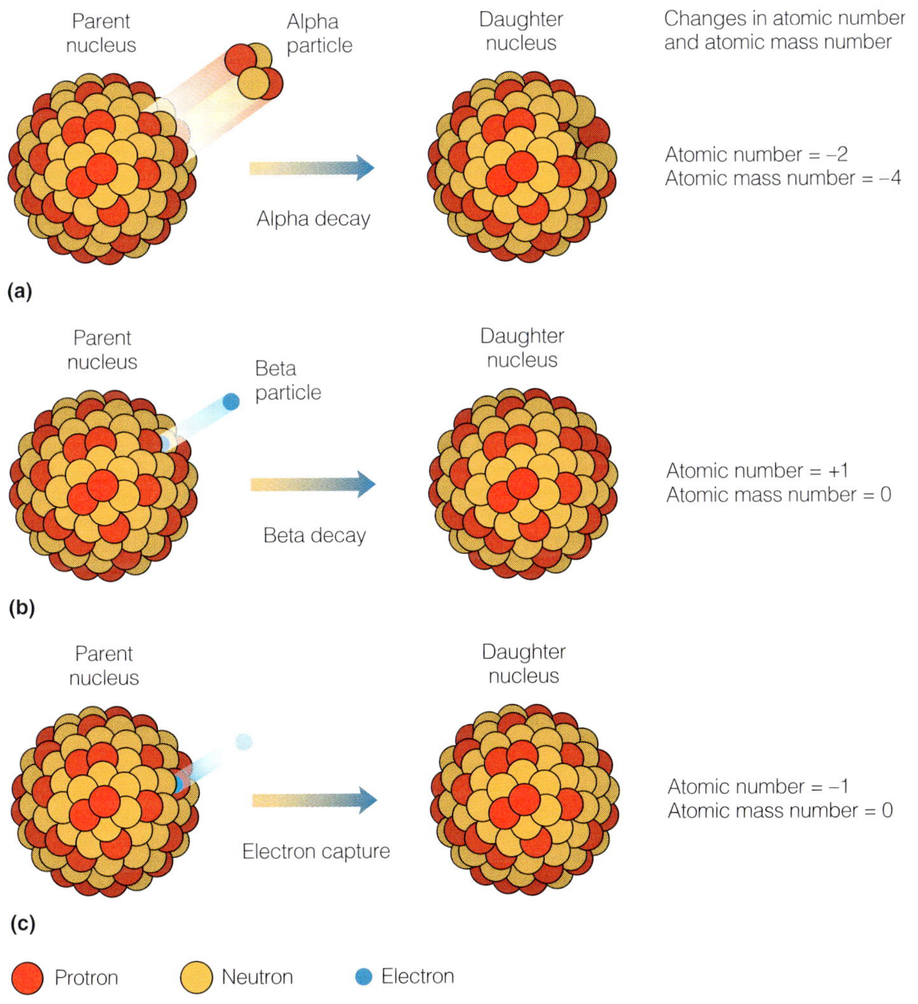

• **Figure 4.5 Three Types of Radioactive Decay** (a) Alpha decay, in which an unstable parent nucleus emits 2 protons and 2 neutrons. (b) Beta decay, in which an electron is emitted from a neutron in the nucleus. (c) Electron capture, in which a proton captures an electron and is thereby converted to a neutron.

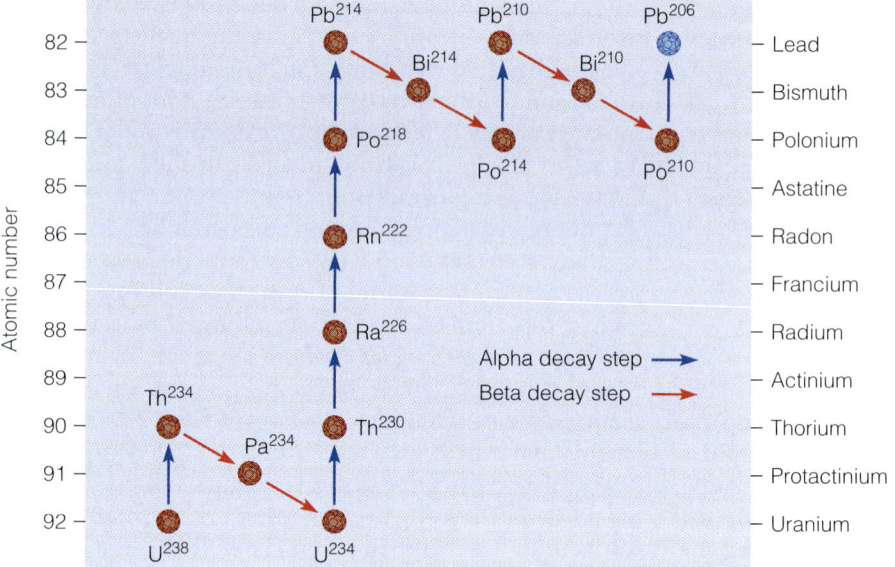

• **Figure 4.6 Radioactive Decay Series for Uranium 238 and Lead 206** Radioactive uranium 238 decays to its stable daughter product, lead 206, by 8 alpha and 6 beta decay steps. A number of different isotopes are produced as intermediate steps in the decay series.

sample that contains the radioactive element. The parent–daughter ratio is usually determined by a *mass spectrometer,* an instrument that measures the proportions of elements of different masses.

Sources of Uncertainty The most accurate radiometric dates are obtained from igneous rocks. As magma cools and begins to crystallize, radioactive parent atoms are separated from previously formed daughter atoms. Because they are the right size, some radioactive parent atoms are incorporated into the crystal structure of certain minerals. The stable daughter atoms, though, are a different size from the radioactive parent atoms and consequently cannot fit into the crystal structure of the same mineral as the parent atoms. Therefore, a mineral crystallizing in a cooling magma will contain radioactive parent atoms but no stable daughter atoms (• Figure 4.8). Thus the time that is being measured is the time of crystallization of the mineral containing the radioactive atoms and not the time of formation of the radioactive atoms.

Except in unusual circumstances, sedimentary rocks cannot be radiometrically dated because one would be measuring the age of a particular mineral rather than the time that it was deposited as a sedimentary particle. One of the few instances in which radiometric dates can be obtained on sedimentary rocks is when the mineral glauconite is present. Glauconite is a greenish mineral containing radioactive potassium 40, which decays to argon 40 (Table 4.2). It forms in certain marine environments as a result of chemical reactions with clay minerals during the conversion of sediments to sedimentary rock. Thus glauconite forms when the sedimentary rock forms, and a radiometric date indicates the time of the sedimentary rock's origin. Being a gas, however, the daughter product argon can easily escape from a mineral. Therefore, any date obtained from glauconite, or any

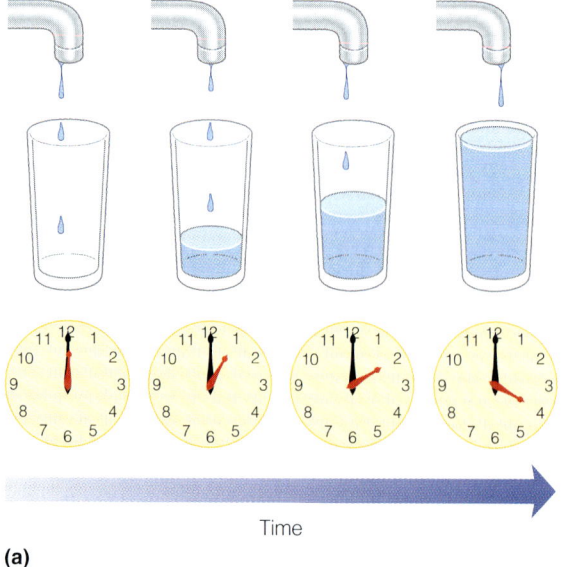

(a)

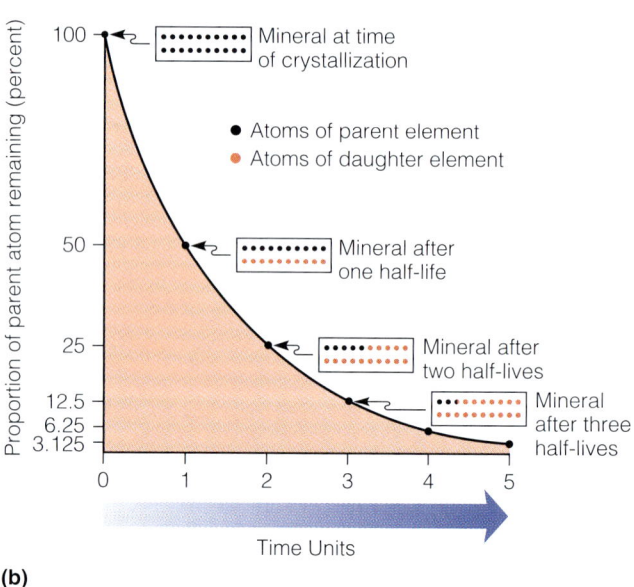

(b)

• **Figure 4.7 Uniform, Linear Change Compared to Geometric Radioactive Decay** (a) Uniform, linear change is characteristic of many familiar processes. In this example, water is being added to a glass at a constant rate. (b) Geometric radioactive decay curve, in which each time unit represents one half-life, and each half-life is the time it takes for half of the parent element to decay to the daughter element.

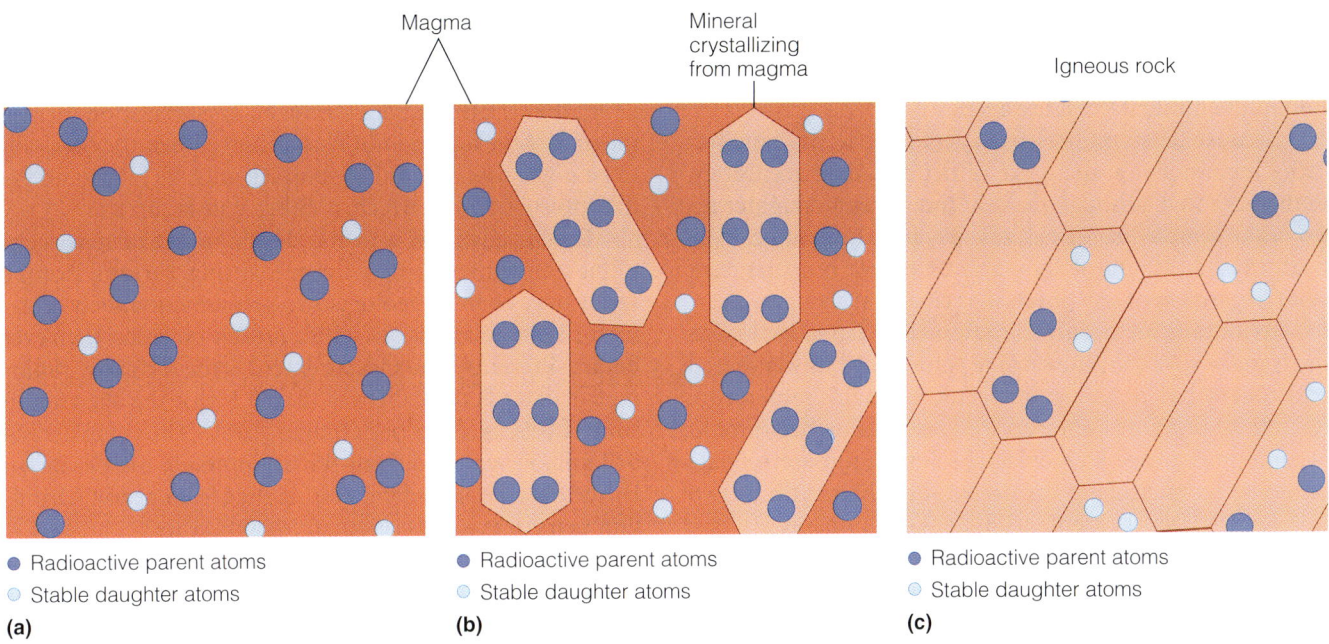

• **Figure 4.8 Crystallization of Magma Containing Radioactive Parent and Stable Daughter Atoms** (a) Magma contains both radioactive parent atoms and stable daughter atoms. The radioactive parent atoms are larger than the stable daughter atoms. (b) As magma cools and begins to crystallize, some of the radioactive parent atoms are incorporated into certain minerals because they are the right size and can fit into the crystal structure. In this example, only the larger radioactive parent atoms fit into the crystal structure. Therefore, at the time of crystallization, minerals in which the radioactive parent atoms can fit into the crystal structure will contain 100% radioactive parent atoms and 0% stable daughter atoms. (c) After one half-life, 50% of the radioactive parent atoms will have decayed to stable daughter atoms, such that those minerals that had radioactive parent atoms in their crystal structure will now have 50% radioactive parent atoms and 50% stable daughter atoms.

TABLE 4.2	Five of the Principal Long-Lived Radioactive Isotope Pairs Used in Radiometric Dating			
Isotopes		**Half-Life of Parent (Years)**	**Effective Dating Range (Years)**	**Minerals and Rocks That Can Be Dated**
Parent	**Daughter**			
Uranium 238	Lead 206	4.5 billion	10 million to 4.6 billion	Zircon Uraninite
Uranium 235	Lead 207	704 million		
Thorium 232	Lead 208	14 billion		
Rubidium 87	Strontium 87	48.8 billion	10 million to 4.6 billion	Muscovite Biotite Potassium feldspar Whole metamorphic or igneous rock
Potassium 40	Argon 40	1.3 billion	100,000 to 4.6 billion	Glauconite Hornblende Muscovite Whole volcanic rock Biotite

other mineral containing the potassium 40 and argon 40 pair, must be considered a minimum age.

To obtain accurate radiometric dates, geologists must be sure that they are dealing with a *closed system,* meaning that neither parent nor daughter atoms have been added or removed from the system since crystallization and that the ratio between them results only from radioactive decay. Otherwise, an inaccurate date will result. If daughter atoms have leaked out of the mineral being analyzed, the calculated age will be too young; if parent atoms have been removed, the calculated age will be too old.

Leakage may take place if the rock is heated or subjected to intense pressure as can sometimes occur during metamorphism. If this happens, some of the parent or daughter atoms may be driven from the mineral being analyzed, resulting in an inaccurate age determination. If the daughter product was completely removed, then one would be measuring the time since metamorphism (a useful measurement itself) and not the time since crystallization of the mineral (• Figure 4.9).

Because heat and pressure affect the parent–daughter ratio, metamorphic rocks are difficult to date accurately. Remember that although the resulting parent–daughter ratio of the sample being analyzed may have been affected by heat, the decay rate of the parent element remains constant, regardless of any physical or chemical changes.

To obtain an accurate radiometric date, geologists must make sure that the sample is fresh and unweathered and that it has not been subjected to high temperature or intense pressures after crystallization. Furthermore, it is sometimes possible to cross-check the radiometric date obtained by measuring the parent–daughter ratio of two different radioactive elements in the same mineral.

For example, naturally occurring uranium consists of both uranium 235 and uranium 238 isotopes. Through various decay steps, uranium 235 decays to lead 207, whereas uranium 238 decays to lead 206 (Figure 4.6). If the minerals that contain both uranium isotopes have remained closed systems, the ages obtained from each parent–daughter ratio should agree closely. If they do, they are said to be *concordant*, thus reflecting the time of crystallization of the magma. If the ages do not closely agree, then they are said to be *discordant*, and other samples must be taken and ratios measured to see which, if either, date is correct.

Recent advances and the development of new techniques and instruments for measuring various isotope ratios have enabled geologists to analyze not only increasingly smaller samples but with a greater precision than ever before. Presently, the measurement error for many radiometric dates is typically less than 0.5% of the age and, in some cases, is even better than 0.1%. Thus, for a rock 540 million years old (near the beginning of the Cambrian Period), the possible error could range from nearly 2.7 million years to as low as less than 540,000 years.

Long-Lived Radioactive Isotope Pairs

Table 4.2 shows the five common, long-lived parent–daughter isotope pairs used in radiometric dating. Long-lived pairs have half-lives of millions or billions of years. All of these pairs were present when Earth formed and are still present in measurable quantities. Other shorter-lived radioactive isotope pairs have decayed to the point that only small quantities near the limit of detection remain.

The most commonly used isotope pairs are the uranium–lead and thorium–lead series, which are used principally to date ancient igneous intrusives, lunar samples, and some meteorites. The rubidium–strontium pair is also used for very old samples and has been effective in dating the oldest rocks on Earth as well as meteorites.

The potassium–argon method is typically used for dating fine-grained volcanic rocks from which individual crystals cannot be separated; hence, the whole rock is analyzed. Because argon is a gas, great care must be taken to ensure that the sample has not been subjected to heat, which would allow argon to escape; such a sample would yield an age that is too young. Other long-lived radioactive isotope pairs exist, but they are rather rare and used only in special situations.

Fission-Track Dating

The emission of atomic particles resulting from the spontaneous decay of uranium within a mineral damages its crystal structure. The damage appears as microscopic linear tracks that are visible only after the mineral has been etched with hydrofluoric acid, an acid so powerful that its vapors can destroy one's sense of smell without careful handling. The age of the sample is determined from the number of fission tracks present and the amount of uranium the sample contains. The older the sample, the greater the number of tracks (• Figure 4.10).

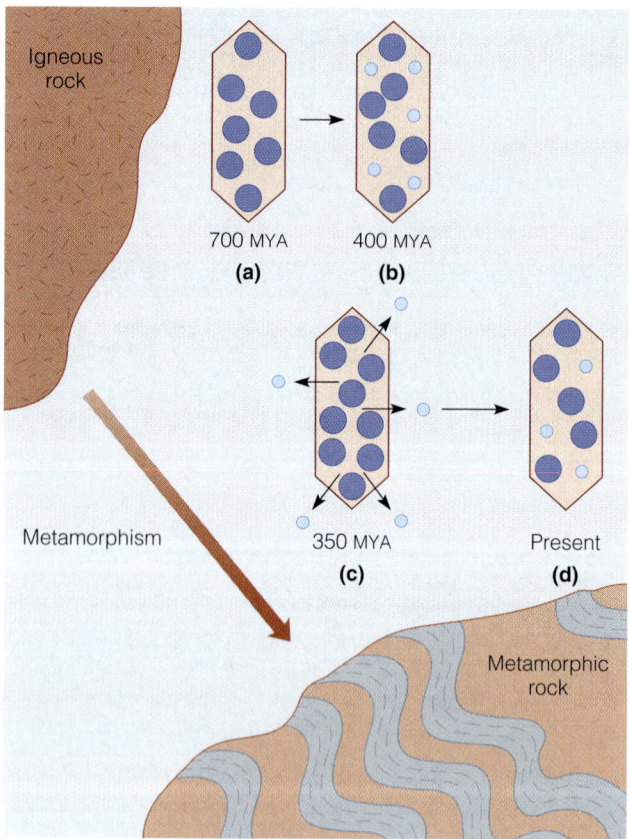

• **Figure 4.9 Effects of Metamorphism on Radiometric Dating** The effect of metamorphism in driving out daughter atoms from a mineral that crystallized 700 million years ago (MYA). The mineral is shown immediately after crystallization **(a)**, then at 400 MYA **(b)**, when some of the parent atoms had decayed to daughter atoms. Metamorphism at 350 MYA **(c)** drives the daughter atoms out of the mineral into the surrounding rock. **(d)** If the rock has remained a closed chemical system throughout its history, dating the mineral today yields the time of metamorphism, whereas dating the whole rock provides the time of its crystallization, 700 MYA.

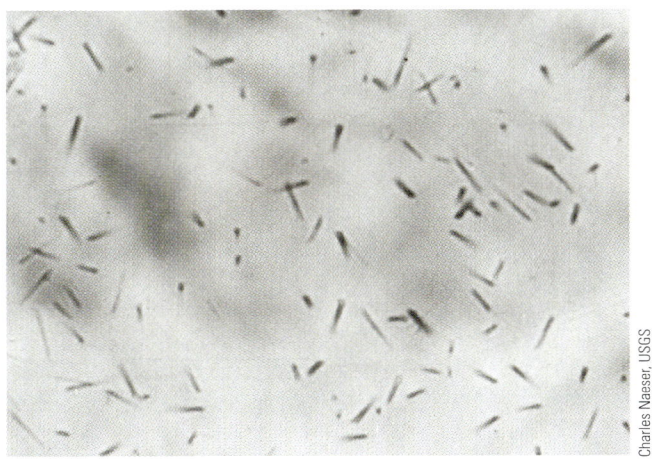

• Figure 4.10 **Fission-Track Dating** Each fission track (about 16 microns long) in this apatite crystal is the result of the radioactive decay of a uranium atom. The apatite crystal, which has been etched with hydrofluoric acid to make the fission tracks visible, comes from one of the dikes at Shiprock, New Mexico, and has a calculated age of 27 million years.

Fission-track dating is of particular interest to archaeologists and geologists because the technique can be used to date samples ranging from only a few hundred to hundreds of millions of years old. It is most useful for dating samples between about 40,000 and 1.5 million years ago, a period for which other dating techniques are not always particularly suitable. One of the problems in fission-track dating occurs when the rocks have later been subjected to high temperatures. If this happens, the damaged crystal structures are repaired by annealing, and, consequently, the tracks disappear. In such instances, the calculated age will be younger than the actual age.

Radiocarbon and Tree-Ring Dating

Carbon is an important element in nature and is one of the basic elements found in all forms of life. It has three isotopes; two of these, carbon 12 and 13, are stable, whereas carbon 14 is radioactive (see Figure 2.3). Carbon 14 has a half-life of 5730 years plus or minus 30 years. The **carbon-14 dating** technique is based on the ratio of carbon 14 to carbon 12 and is generally used to date once-living material.

The short half-life of carbon 14 makes this dating technique practical only for specimens younger than about 70,000 years. Consequently, the carbon-14 dating method is especially useful in archaeology and has greatly helped unravel the events of the latter portion of the Pleistocene Epoch. For example, carbon-14 dates of maize from the Tehuacan Valley of Mexico have forced archaeologists to rethink their ideas of where the first center for maize domestication in Mesoamerica arose. Carbon-14 dating is also helping to answer the question of when humans began populating North America.

Carbon 14 is constantly formed in the upper atmosphere when cosmic rays, which are high-energy particles (mostly protons), strike the atoms of upper-atmospheric gases, splitting their nuclei into protons and neutrons. When a neutron strikes the nucleus of a nitrogen atom (atomic number 7, atomic mass number 14), it may be absorbed into the nucleus and a proton emitted. Thus the atomic number of the atom decreases by 1, whereas the atomic mass number stays the same. Because the atomic number has changed, a new element, carbon 14 (atomic number 6, atomic mass number 14), is formed. The newly formed carbon 14 is rapidly assimilated into the carbon cycle and, along with carbon 12 and 13, is absorbed in a nearly constant ratio by all living organisms (• Figure 4.11). When an organism dies, however, carbon 14 is not replenished, and the ratio of carbon 14 to carbon 12 decreases as carbon 14 decays back to nitrogen by a single beta-decay step (Figure 4.5).

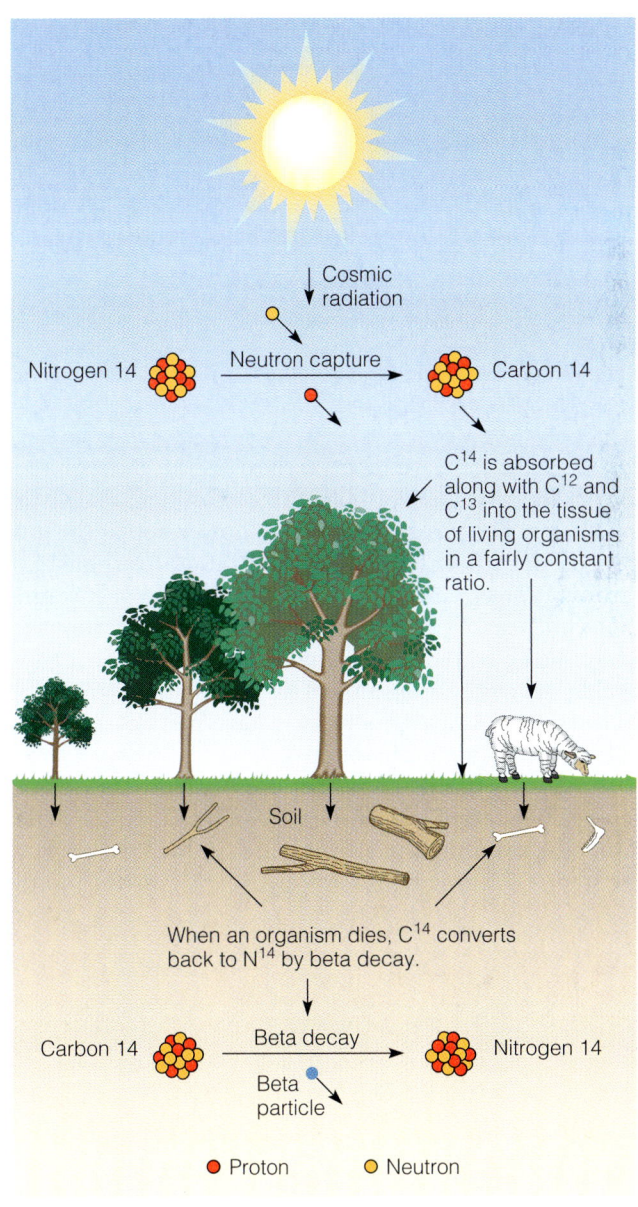

• Figure 4.11 **Carbon-14 Dating Method** The carbon cycle involves the formation of carbon 14 in the upper atmosphere, its dispersal and incorporation into the tissue of all living organisms, and its decay back to nitrogen 14 by beta decay.

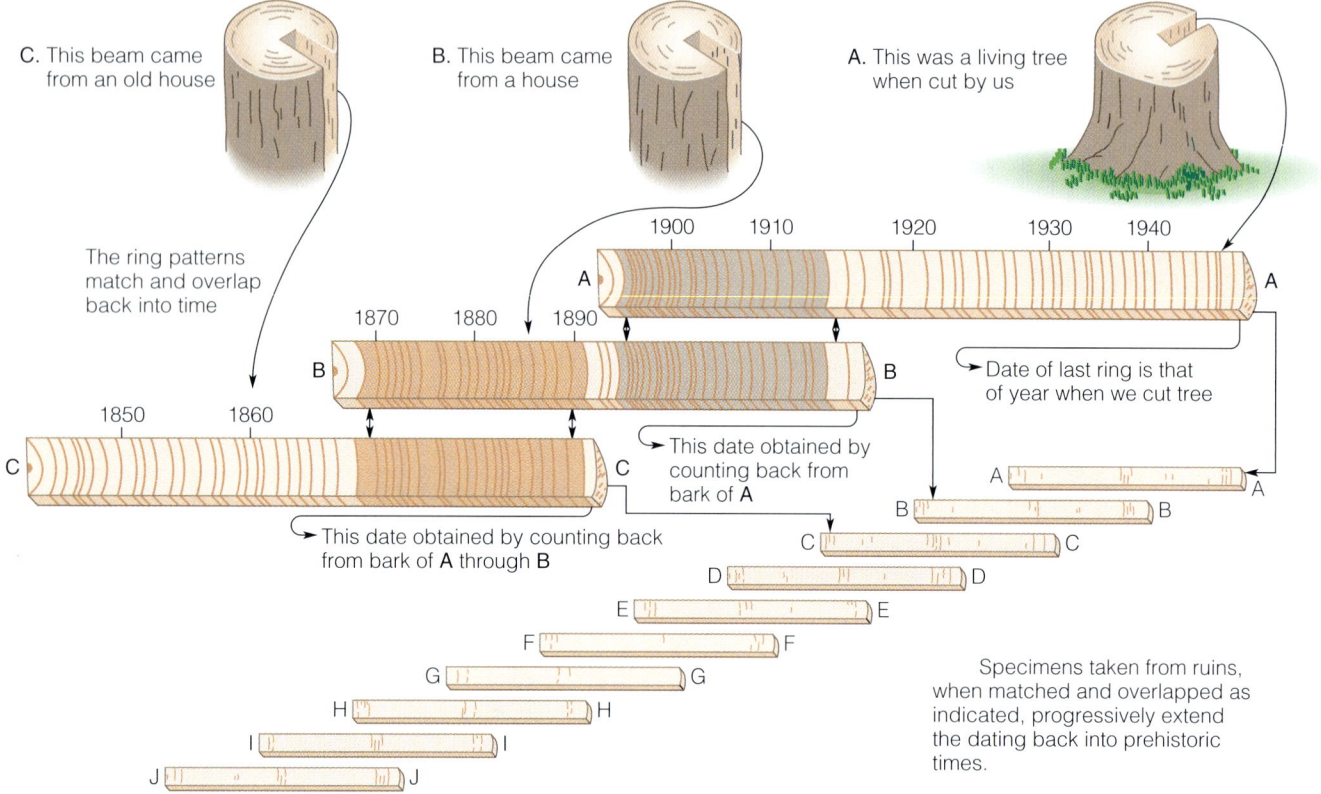

• **Figure 4.12 Tree-Ring Dating Method** In the cross-dating method, tree-ring patterns from different woods are compared to establish a ring-width chronology backward in time.

Currently, the ratio of carbon 14 to carbon 12 is remarkably constant in both the atmosphere and living organisms. There is good evidence, however, that the production of carbon 14, and thus the ratio of carbon 14 to carbon 12, has varied somewhat during the past several thousand years. This was determined by comparing ages established by carbon-14 dating of wood samples with ages established by counting annual tree rings in the same samples. As a result, carbon-14 ages have been corrected to reflect such variations in the past.

Tree-ring dating is another useful method for dating geologically recent events. The age of a tree can be determined by counting the growth rings in the lower part of the stem. Each ring represents one year's growth, and the pattern of wide and narrow rings can be compared among trees to establish the exact year in which the rings were formed. The procedure of matching ring patterns from numerous trees and wood fragments in a given area is called *cross-dating*.

By correlating distinctive tree-ring sequences from living to nearby dead trees, a time scale can be constructed that extends back to about 14,000 years ago (• Figure 4.12). By matching ring patterns to the composite ring scale, wood samples whose ages are not known can be accurately dated.

The applicability of tree-ring dating is somewhat limited because it can be used only where continuous tree records are found. It is therefore most useful in arid regions, particularly the southwestern United States.

SUMMARY

- Early Christian theologians were responsible for formulating the idea that time is linear and that Earth was very young. Archbishop Ussher calculated Earth's age at about 6000 years based on his interpretation of the scriptures.

- During the 18th and 19th centuries, attempts were made to determine Earth's age based on scientific evidence rather than revelation. Although some attempts were quite ingenious, they yielded a variety of ages that are now known to be much too young.

- Considering the religious, political, and social climate of the 17th, 18th, and early 19th centuries, it is easy to see why concepts such as neptunism, catastrophism, and a very young Earth were eagerly embraced. As geologic data accumulated, it became apparent that these concepts were not supported by evidence and that Earth must be much older than 6000 years.

- James Hutton thought present-day processes operating over long periods of time could explain all the geologic features of Earth. He also viewed Earth history as cyclical and thought Earth to be very old. Hutton's observations were instrumental in establishing the principle of uniformitarianism.

- Uniformitarianism, as articulated by Charles Lyell, soon became the guiding principle of geology. According to the principle of uniformitarianism, the laws of nature have been constant through time, and the same processes operating today have also operated in the past, although not necessarily at the same rates.

- Besides uniformitarianism, the principles of superposition, original horizontality, lateral continuity, and cross-cutting relationships are basic for determining relative geologic ages and for interpreting Earth history.

- Radioactivity was discovered during the late 19th century, and soon thereafter radiometric dating techniques allowed geologists to determine absolute ages for geologic events.

- Absolute ages for rock samples are usually obtained by determining how many half-lives of a radioactive parent element have elapsed since the sample originally crystallized. A half-life is the time it takes for one-half of the radioactive parent element to decay to a stable daughter element.

- The most accurate radiometric dates are obtained from long-lived radioactive isotope pairs in igneous rocks. The five common long-lived radioactive isotope pairs are uranium 238–lead 206, uranium 235–lead 207, thorium 232–lead 208, rubidium 87–strontium 87, and potassium 40–argon 40. The most reliable dates are those obtained by using at least two different radioactive-decay series in the same rock.

- Carbon-14 dating is effective back to about 70,000 years ago and can only be used on organic matter such as wood, bones, and shells. Unlike the long-lived isotopic pairs, carbon-14 dating determines age by the ratio of radioactive carbon 14 to stable carbon 12.

IMPORTANT TERMS

absolute dating, p. 62
alpha decay, p. 69
beta decay, p. 69
carbon-14 dating, p. 73
catastrophism, p. 66
daughter element, p. 69
electron capture decay, p. 69

fission-track dating, p. 73
half-life, p. 69
neptunism, p. 66
parent element, p. 69
principle of cross-cutting relationships, p. 64
principle of lateral continuity, p. 64

principle of original horizontality, p. 64
principle of superposition, p. 64
principle of uniformitarianism, p. 66
radioactive decay, p. 69
relative dating, p. 61
tree-ring dating, p. 74

REVIEW QUESTIONS

1. Who is generally considered the founder of modern geology?
   a. _____ Werner; b. _____ Lyell; c. _____ Steno; d. _____ Cuvier; e. _____ Hutton.

2. In which type of radioactive decay are two protons and two neutrons emitted from the nucleus?
   a. _____ alpha decay; b. _____ beta decay; c. _____ electron capture; d. _____ fission track; e. _____ none of the previous answers.

3. Considering that the half-life of uranium 235 is 704 million years, what fraction of the original uranium 235 will remain after 2,816,000,000 years?
   a. _____ 1/2; b. _____ 1/4; c. _____ 1/8; d. _____ 1/16; e. _____ 1/32.

4. Because of the heat and pressure exerted during metamorphism, daughter atoms were driven out of a mineral being analyzed for a radiometric date. The date obtained from this mineral will therefore be _____ its actual age of formation.
   a. _____ younger than; b. _____ older than; c. _____ the same as; d. _____ it can't be determined; e. _____ none of the previous answers.

5. Which of the following is *not* a long-lived radioactive isotope pair?
   a. _____ potassium–argon; b. _____ thorium–lead; c. _____ carbon 14–carbon 12; d. _____ rubidium–strontium; e. _____ uranium–lead.

6. Placing geologic events in sequential or chronologic order as determined by their position in the geologic record is
   a. _____ absolute dating; b. _____ correlation; c. _____ historical dating; d. _____ relative dating; e. _____ uniformitarianism.

7. If a flake of biotite within a sedimentary rock (such as a sandstone) is radiometrically dated, the date obtained indicates when
   a. _____ the biotite crystal formed; b. _____ the sedimentary rock formed; c. _____ the parent radioactive isotope formed; d. _____ the daughter radioactive isotope(s) formed; e. _____ none of the previous answers.

8. The author of *Principles of Geology* and the principal advocate and interpreter of uniformitarianism was
   a. _____ Werner; b. _____ Kelvin; c. _____ Steno; d. _____ Hutton; e. _____ Lyell.

9. The atomic number of an element is determined by the number of _____ in its nucleus.
   a. _____ protons; b. _____ neutrons; c. _____ electrons; d. _____ protons and neutrons; e. _____ protons and electrons.
10. What is being measured in radiometric dating is
    a. _____ the time when a radioactive isotope formed; b. _____ the time of crystallization of a mineral containing an isotope; c. _____ the amount of the parent isotope only; d. _____ when the dated mineral became part of a sedimentary rock; e. _____ when the stable daughter isotope was formed.
11. As carbon 14 decays back to nitrogen in radiocarbon dating, what isotopic ratio decreases?
    a. _____ nitrogen 14 to carbon 12; b. _____ carbon 14 to carbon 12; c. _____ carbon 13 to carbon 12; d. _____ nitrogen 14 to carbon 14; e. _____ nitrogen 14 to carbon 13.
12. What is the difference between relative dating and absolute dating?
13. How does metamorphism affect the potential for accurate radiometric dating using any and all techniques discussed in this chapter? How would such radiometric dates be affected by metamorphism, and why?
14. Describe the principle of uniformitarianism according to Hutton and Lyell. What is the significance of this principle?
15. If you wanted to calculate the absolute age of an intrusive body, what information would you need?
16. A volcanic ash fall was radiometrically dated using the potassium 40 to argon 40 and rubidium 87 to strontium 87 isotope pairs. The isotope pairs yielded distinctly different ages. What possible explanation could be offered as to why these two isotope pairs yielded different ages? What would you do to rectify the discrepancy in ages?
17. Describe the uncertainties associated with radiometrically dating any sedimentary rock.
18. Why were Lord Kelvin's arguments and calculations so compelling, and what was the basic flaw in his assumption?
19. What is the major difference between the carbon-14 dating technique and the techniques used for the five common, long-lived radioactive isotope pairs?
20. Why are the fundamental principles used in relative dating so important in geology?

APPLY YOUR KNOWLEDGE

1. If a radioactive element has a half-life of 64 million years, what fraction of the original amount of parent material will remain after 448 million years? What percentage will that be?
2. How many half-lives are required in which there are presently 1,250,000,000 atoms of rubidium 87 and 38,750,000,000 strontium 87 in its crystal structure? What is the percentage of rubidium 87 remaining after this many half-lives?
3. Given the current debate concerning global warming and the many possible short-term consequences for humans, can you visualize how the world might look in 10,000 years or even 1 million years from now? Use what you have learned about plate tectonics and the direction and rate of movement of plates, as well as how plate movement and global warming will affect ocean currents, weather patterns, weathering rates, and other factors, to make your prediction. Do you think such short-term changes can be extrapolated to long-term trends in trying to predict what Earth will be like in the future?

FIELD QUESTION

1. Based on what is shown in this photograph and your knowledge of relative dating principles, provide a geologic history of this area.

CHAPTER 5
ROCKS, FOSSILS, AND TIME—MAKING SENSE OF THE GEOLOGIC RECORD

Sue Monroe

Mesozoic-aged sedimentary rocks, mostly sandstone, in the Valley of the Gods in Utah. Notice the layering or stratification in the rocks. The rocks contain a small smount of iron oxide that gives them their red color. Differential weathering and erosion have yielded a spectacular landscape of spires, pillars, monuments, and buttes. According to Najavo legend, the spires and pillars are ancient warriors that have turned to stone.

[OUTLINE]

INTRODUCTION
STRATIGRAPHY
Vertical Stratigraphic Relationships
Lateral Relationships—Facies
Marine Transgressions and Regressions
Extent, Rates, and Causes of Marine Transgressions and Regressions
FOSSILIZATION AND FOSSILS
How Do Fossils Form?

PERSPECTIVE 5.1 *Fossils and Uniformitarianism*
Fossils and Telling Time
THE RELATIVE GEOLOGIC TIME SCALE
STRATIGRAPHIC TERMINOLOGY
CORRELATION
ABSOLUTE DATES AND THE RELATIVE GEOLOGIC TIME SCALE
SUMMARY

ThomsonNOW Explore interactive tutorials, animations, or practice problems available on the ThomsonNow website at www.thomsonedu.com/login.

[CHAPTER OBJECTIVES]

At the end of this chapter, you will have learned that

- To analyze the geologic record, you must first determine the correct vertical sequence of rocks—that is, from oldest to youngest—even if they have been deformed.

- Even though rocks provide our only evidence of prehistoric events, the record is incomplete at any one locality because discontinuities are common.

- Several marine transgressions and regressions occurred during Earth history, at times covering much of the continents and at other times leaving the land above sea level.

- Fossils, the remains or traces of prehistoric organisms, are preserved in several ways, and some types of fossils are much more common than most people realize.

- Distinctive groups of fossils found in sedimentary rocks are useful for determining the relative ages of rocks in widely separated areas.

- Superposition and the principle of fossil succession were used to piece together a composite geologic column, which is the basis for the relative geologic time scale.

- Geologists have developed terminology to refer to rocks and to time.

- Several criteria are used to match up (correlate) similar rocks over large regions or to demonstrate that rocks in different areas are the same age.

- Absolute ages of sedimentary rocks are most often determined by radiometric dating of associated igneous or metamorphic rocks.

Introduction

The fact that Earth has changed throughout time is apparent from evidence in the **geologic record**—that is, the record of events preserved in rocks (see chapter opening photo). In short, Earth has been, and continues to be, a dynamic planet, with sufficient internal heat to account for ongoing volcanism, seismic activity, and plate movements. In addition, its atmosphere, hydrosphere, and biosphere continually modify the planet's surface. The other terrestrial planets and Earth's Moon, with the possible exception of Venus, are essentially dead planets. That is, they show no signs of the kinds of internal and surface processes that we find on Earth.

In our effort to decipher Earth's geologic history, geologists pay special attention to sedimentary rocks, although all types of rocks are important. So how do we analyze the geologic record? After all, no human witnesses existed to record these events.

Perhaps an analogy will help. Suppose you are walking in the woods and encounter a shattered, charred tree. Maybe the tree always existed in its present condition, or an exploding bomb may have caused the damage, or maybe it was hit by lightning. Given your knowledge of trees, the first option seems quite unlikely, and we can check the second one easily. Assuming no evidence exists for the explosion hypothesis, you would no doubt choose the last proposal because it coincides with your knowledge about the effects of a lightning strike. So even though you did not observe the event, you were able to determine what happened because evidence exists for the event having occurred. Likewise, the geologic record provides evidence of past events having taken place.

By the late 1700s scientists had made some progress in deciphering Earth history, but in 1857 Philip Henry Gosse, a British naturalist, proposed that Earth had been created only a few thousands of years ago, with the appearance of great age. That is, rocks that looked like they formed from lava flows or that appeared to have been sand that was later converted to sandstone were created with that appearance, and fossils were created with the appearance of clams, corals, bones, and so on.

Most people reasoned that if Gosse were correct, a deceitful creator must have fabricated the geologic record, a thesis they could not accept. In addition, there is no way to check Gosse's idea, because the geologic record would look exactly the same if he were right or if the record was produced by natural processes operating over many millions of years. The scientists of the day opted for the latter—the uniformitarian approach to Earth history—because they had every reason to think that the features preserved in rocks were the products of processes they could observe operating at present.

As for Earth's age, scientists rejected Gosse's idea that Earth was only a few thousand years old, reasoning that it must be much older than Gosse allowed. Even Gosse had to admit that nothing in the geologic record supported his age determination, and his critics were quick to point out that if his idea was correct, Earth could have been created at any time during the past, including only moments ago, along with all records and memories.

Philip Henry Gosse's ideas on Earth history now serve only as a failed attempt to reconcile Earth history with a particular interpretation of Judeo-Christian scripture. The uniformitarian method offers a much more fruitful approach, but keep in mind that the geologic record is complex, incomplete at any one location, and it requires interpretation. In this chapter we endeavor to bring some organization to this record.

Stratigraphy

In this chapter our main concern is sedimentary rocks, most of which show some kind of layering or stratification. The branch of geology called **stratigraphy** is concerned with the composition, origin, age relationships, and geographic extent of sedimentary rocks, but the principles of stratigraphy apply to any sequence of stratified rocks. For example, a succession of lava flows, or lava flows and ash beds, as well as many metamorphic rocks are stratified (• Figure 5.1).

Where sedimentary rocks are well exposed, as in the walls of deep canyons, you can easily determine the relationships among individual layers (*strata;* singular *stratum*). Lateral relationships are equally important in analyzing the geologic record, but they usually must be determined from a number of separate rock exposures, or what geologists call *outcrops*.

Vertical Stratigraphic Relationships

In vertical successions of sedimentary rocks, surfaces known as *bedding planes* separate individual strata from one another (Figure 5.1b), or the strata grade vertically from one rock type into another. The rocks below and above a bedding plane differ in composition, texture, color, or a combination of these features. These differences indicate a rapid change in sedimentation or perhaps a period of nondeposition and/or erosion followed by renewed deposition. In contrast, gradually changing conditions of sedimentation account for those rocks that show a vertical gradation. Regardless of the nature of the vertical relationships among strata, the correct order in which they were deposited must be determined.

Superposition   In Chapter 4 we discussed the *principle of superposition* that resulted from the work of Nicolas Steno during the 17th century. Recall that according to this principle, you can determine the correct relative ages of underformed strata by their position in a sequence. Accordingly, the oldest layer is at the bottom of the sequence with successively younger layers upward in the sequence (see Figure 4.2). Of course, if strata are deformed by faulting, folding, or both, the task is more difficult, but several sedimentary structures and some fossils allow geologists to resolve these kinds of problems.

The **principle of inclusions** is yet another way to figure out relative ages. According to this principle, inclusions, or fragments in a rock, are older than the rock itself. So granite with basalt inclusions is older than the basalt, and sandstone with granite particles is younger than the granite (• Figure 5.2). How can you determine whether a layer of basalt in a sequence of sedimentary rocks is a buried lava flow or a sill (• Figure 5.3)? If it is a lava flow, it formed after the layer below but before the one above, whereas a sill—a sheetlike, intrusive body—is younger than the underlying

(a)

(b)

(c)

• **Figure 5.1 Stratification in Igneous, Sedimentary, and Metamorphic Rocks** Stratification in **(a)** several layers of pyroclastic materials in Oregon. Note that the layers are cut by two small faults. **(b)** These alternating layers of sandstone and shale in California show stratification even though they have been tilted from their original horizontal or nearly horizontal position. **(c)** The Siamo Slate in Michigan also shows stratification.

• **Figure 5.2 Principle of Inclusions** Outcrop in northern Wisconsin showing basalt inclusions (dark gray) in light-colored granite. Which rock is older—the basalt or the granite?

• **Figure 5.3 How to Determine the Relative Ages of Lava Flows, Sills, and Associated Sedimentary Rocks** **(a)** A buried lava flow has baked underlying bed 2, and clasts of the lava were deposited along with other sediment in bed 4. The lava flow is older than beds 4, 5, and 6. **(b)** The layers above and below the sill have been baked, so the sill is younger than layers 2 and 4. Can you determine its age relative to beds 5, 6, and 7? **(c)** This lava flow in Yellowstone National Park in Wyoming shows well-developed columnar jointing. There is a baked zone below the lava but none above, indicating that it is a buried lava flow rather than a sill.

rock and younger than the layer immediately above it. Study Figure 5.3 closely and note that a lava flow bakes the underlying rock, but the sill bakes the underlying and the overlying rocks (Figure 5.3).

Unconformities So far, we have discussed vertical relationships among **conformable** strata—that is, sequences of rocks in which deposition was more or less continuous. A bedding plane between strata may represent a depositional break of anywhere from minutes to tens of years but is inconsequential in the context of geologic time. However, in many sequences of strata, surfaces known as **unconformities** that represent times of nondeposition and/or erosion are present. These unconformities encompass long periods of geologic time, perhaps millions or tens of millions of years. Accordingly, the geologic record is incomplete at that particular location, just as a book with missing pages is incomplete, and the interval of geologic time not represented by strata is a *hiatus* (• Figure 5.4).

Unconformity is a general term that encompasses three distinct types of surfaces called *disconformity*, *nonconformity*, and *angular unconformity*. Furthermore, unconformities of regional extent may change from one type to another, and they do not necessarily encompass equivalent amounts of geologic time everywhere (• Figure 5.5a). We should also note that unconformities are common, so the geologic record is incomplete at any location where an unconformity is present. Nevertheless, the geologic time not recorded by rocks in one area is represented by rocks elsewhere.

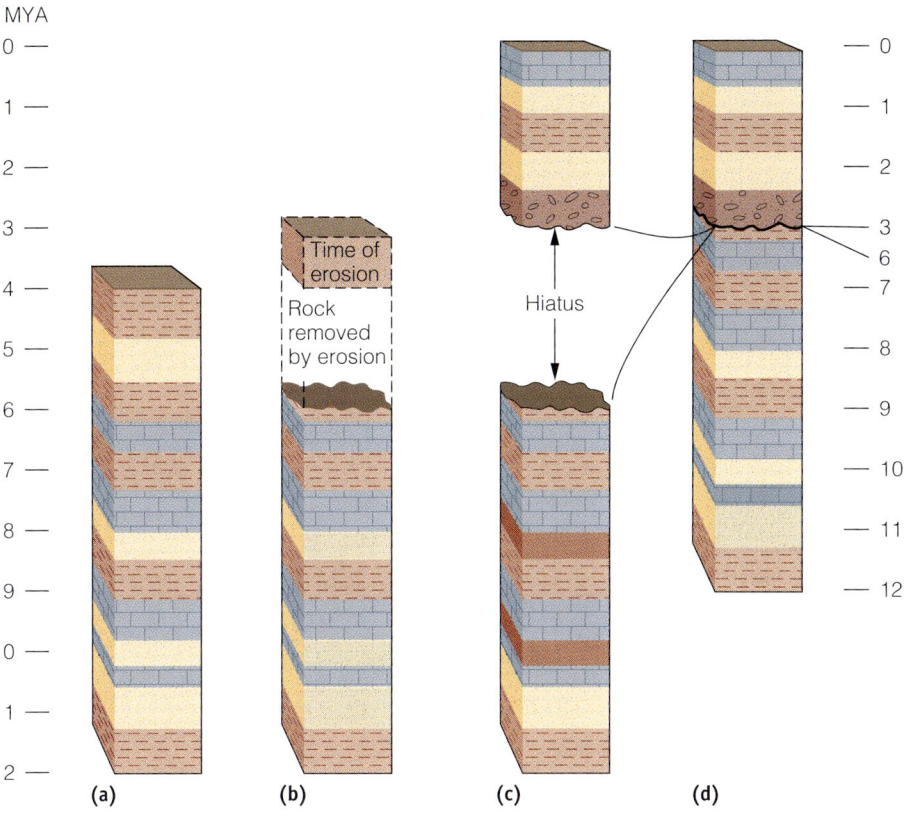

• **Figure 5.4 The Origin of an Unconformity and a Hiatus** (a) Deposition began 12 million years ago (MYA) and continued until 4 MYA. (b) A 1-million-year episode of erosion took place during which rocks representing 2 million years of geologic time were eroded. (c) A 3-million-year hiatus exists between the older rocks and those that formed during renewed deposition beginning 3 MYA. (d) The actual stratigraphic record showing the unconformity, which is a discontinuity in the record of geologic time.

A **disconformity** is an erosion surface in sedimentary rocks that separates younger rocks from older rocks, both of which are parallel to each other (Figure 5.5b). However, an erosion surface cut into plutonic rocks or metamorphic rocks that is overlain by sedimentary rocks is a **nonconformity** (Figure 5.5c). And finally, an **angular unconformity** is present if the strata below an erosion surface are inclined at some angle to the strata above (Figure 5.5d). In this case we can infer that the sequence of events included deposition, lithification, deformation, erosion, and, finally, renewed deposition.

Lateral Relationships—Facies

In 1669, Nicolas Steno proposed his principle of lateral continuity, meaning that layers of sediment extend outward in all directions until they terminate (see Chapter 4). Rock layers may terminate abruptly where they abut the edge of a depositional basin, where they are eroded, or where they are truncated by faults (• Figure 5.6). Termination may also take place when a rock unit becomes progressively thinner until it *pinches out,* or a rock unit splits laterally into thinner units each of which pinches out—a phenomenon known as *intertonguing*. And finally, a rock unit might change by *lateral gradation* as its composition and/or texture become increasingly different (Figure 5.6).

Both intertonguing and lateral gradation indicate the simultaneous operation of different depositional processes in adjacent environments. For example, on the continental shelf sand may accumulate in the high-energy nearshore environment, while at the same time mud and carbonate deposition takes place in offshore low-energy environments (• Figure 5.7). Deposition in each of these laterally adjacent environments yields a **sedimentary facies,** a body of sediment with distinctive physical, chemical, and biological attributes.

Armanz Gressly, in 1838, was the first to use the term *facies* when he carefully traced sedimentary rocks in the Jura Mountains of Switzerland. He noticed lateral changes such as sandstone grading into shale and reasoned that these changes indicated deposition in different environments that lie next to one another. Any attribute of sedimentary rocks that makes them recognizably different from laterally adjacent rocks of about the same age is sufficient to establish a sedimentary facies. In Figure 5.7 we recognize three sedimentary facies based on rock type—a sandstone facies, a shale facies, and a limestone facies.

Marine Transgressions and Regressions

In the Grand Canyon of Arizona three rock units exposed in the canyon's walls consist of sandstone followed upward by shale and finally limestone (• Figure 5.8). Thus we have three facies, all with fossils of marine-dwelling trilobites and brachiopods, that were deposited on one another. The question is, What accounts for their presence in Arizona far from the sea, and how were these facies deposited in the order observed? These deposits formed during a time when sea level rose with respect to the land, giving rise to a **marine transgression.** During a marine transgression the shoreline migrates landward, as

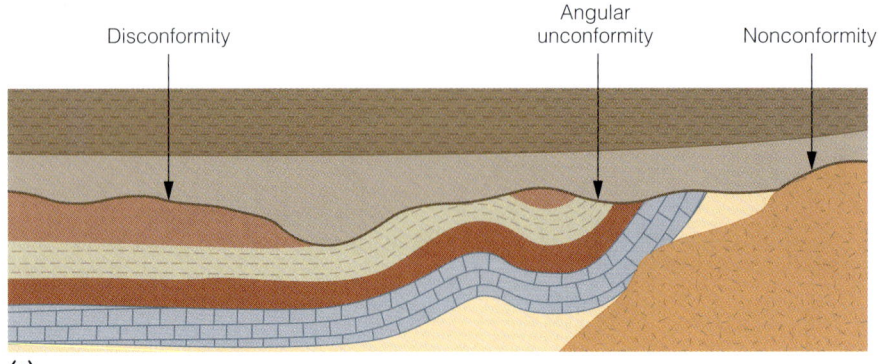

• **Figure 5.5 Types of Unconformities**
(a) Cross section showing regional variation in an unconformity. (b) The geologist at the upper left is sitting on Jurassic rocks, and his foot is resting on Mississippian strata. The surface between these rocks is a disconformity that represents a 165-million-year break in the geologic record at this location in Montana. (c) Nonconformity between Precambrian metamorphic rocks and overlying Paleozoic sedimentary rocks in the Grand Canyon of Arizona. (d) This angular unconformity is between the Ordovician Austin Glen Formation and the overlying, but steeply dipping, Upper Silurian Rondout Formation.

do the environments that parallel the shoreline as the sea progressively covers more and more of a continent (• Figure 5.9). Remember that each laterally adjacent depositional environment is an area where a sedimentary facies develops. So during a marine transgression the facies that formed in offshore environments become superposed on facies deposited in nearshore environments.

Another aspect of a marine transgression is that the rocks making up each facies become younger in a landward direction. The sandstone, shale, and limestone facies in Figure 5.9 were deposited continuously as the shoreline moved landward, but none of the facies was deposited simultaneously over its entire geographic extent. In other words, the facies are *time transgressive*, meaning that their ages vary from place to place.

Obviously the sea is no longer present in the Grand Canyon area, so the transgression responsible for the rocks in

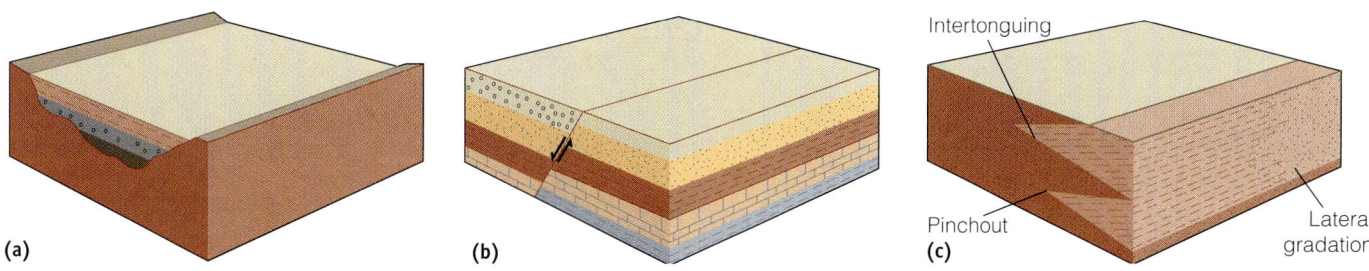

• **Figure 5.6 Lateral Termination of Rock Layers** (a) Lateral termination at the edge of a depositional basin. (b) Faulting followed by erosion and lateral termination. Note that the layer of conglomerate on the left of the fault is missing on the right side of the fault. (c) Lateral termination by intertonguing, pinchout, and lateral gradation.

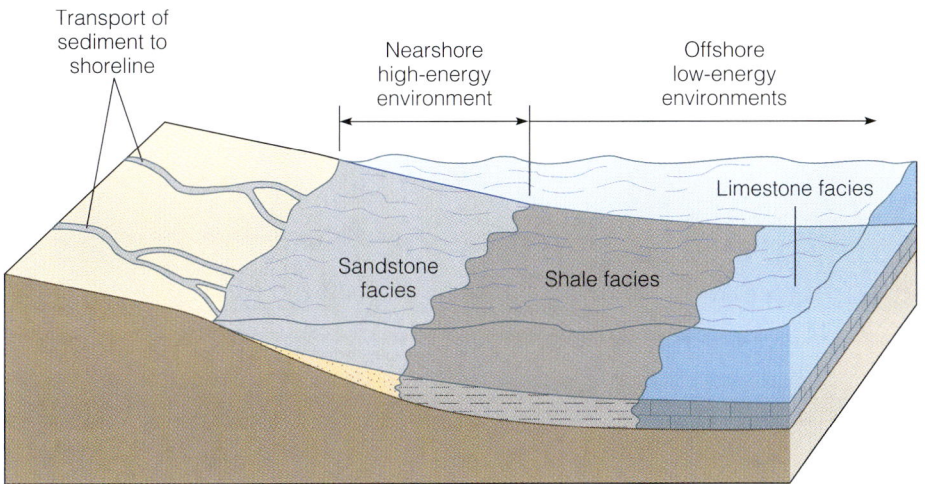

• **Figure 5.7 Sedimentary Facies** Simultaneous deposition of sediments in adjacent environments results in distinctive rock associations called sedimentary facies.

• **Figure 5.8 Rocks in the Grand Canyon That Were Deposited During a Marine Transgression** The vertical sequence of facies compares closely to the sequence shown in Figure 5.9d. Some of the features in these rocks that indicate deposition in a marine environment include wave-formed ripple marks in the Tapeats Sandstone and fossils of marine animals such as trilobites and brachiopods in all three of the formations shown.

Figure 5.8 must have ended. In fact, it did—during a **marine regression** when sea level fell with respect to the continent and the environments that paralleled the shoreline migrated seaward. In other words, a marine regression is the opposite of a transgression and it yields a vertical sequence with nearshore facies overlying offshore facies (Figure 5.9). In this case individual rock units become younger in a seaward direction.

In our discussion of marine transgressions and regressions, we considered both vertical and lateral facies relationships, the significance of which was first recognized by Johannes Walther (1860–1937). When Walther traced rock units laterally, he reasoned, as Gressly had, that each sedimentary facies he encountered was deposited in laterally adjacent environments. In addition, Walther noticed that the same facies he found laterally were also present in a vertical sequence. His observations have since been formulated into **Walther's law,** which holds that the facies seen in a conformable vertical sequence will also replace one another laterally.

The application of Walther's law is well illustrated by the marine transgression and regression shown in Figure 5.9. In practice, it is usually difficult to follow rock units far enough laterally to demonstrate facies changes. It is much easier to observe vertical facies relationships and use Walther's law to work out the lateral relationships. Remember, though, that Walther's law applies only to conformable sequences of rocks; rocks above and below an unconformity are unrelated, so Walther's law does not apply.

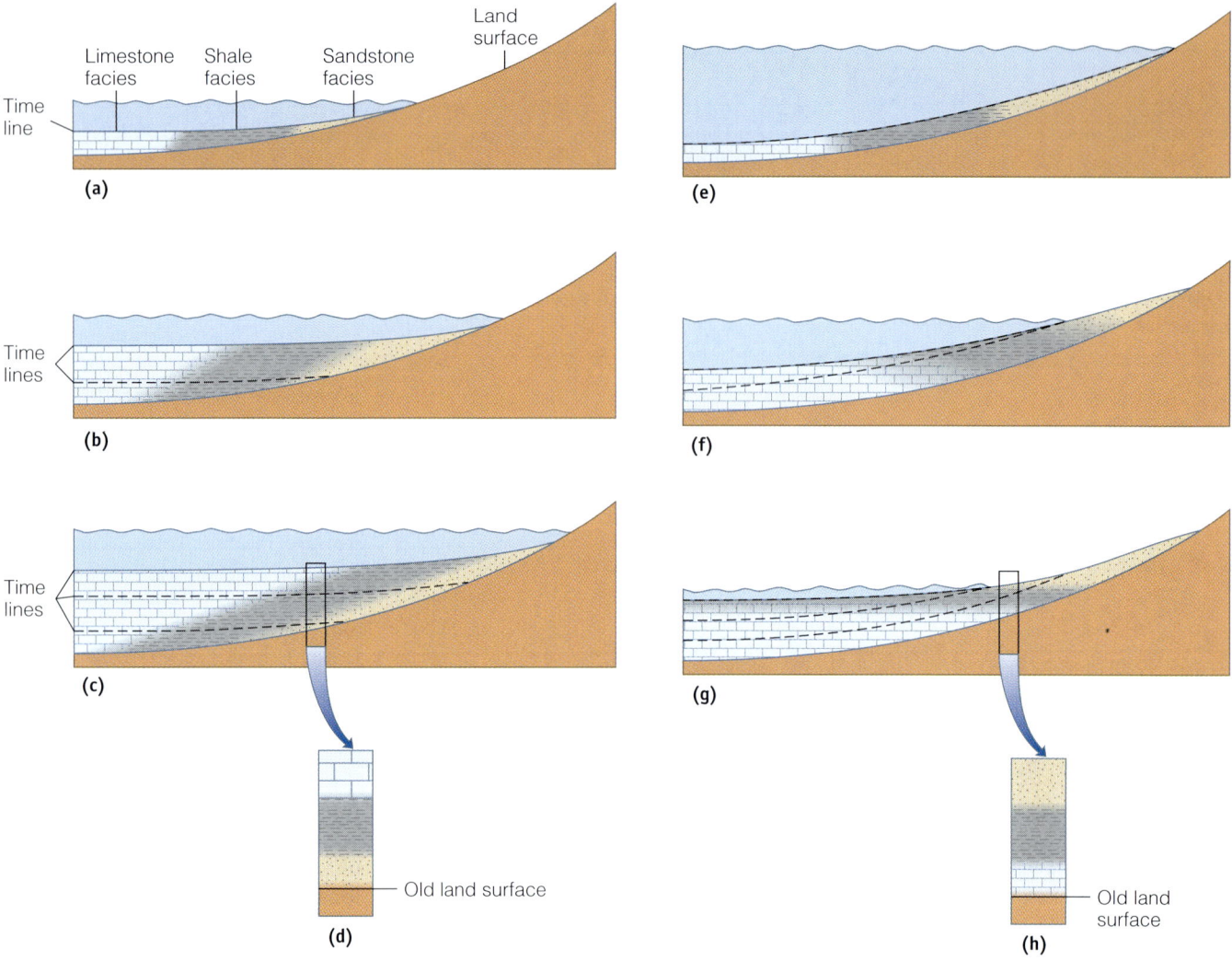

- **Figure 5.9 Marine Transgression and Regression** (a), (b), and (c) show three stages during a marine transgression. (d) Diagrammatic column showing the vertical succession of facies resulting from this transgression. (e), (f), and (g) show three stages in a marine regression. (h) The vertical sequence of facies resulting from deposition during a regression.

Extent, Rates, and Causes of Marine Transgressions and Regressions

Careful analysis of sedimentary rocks enables geologists to determine the maximum extent of marine transgressions and the greatest withdrawal of the sea during regressions. Six major marine transgressions followed by regressions have taken place in North America since the Late Precambrian, yielding unconformity-bounded rock sequences that provide the stratigraphic framework for our discussions of Paleozoic and Mesozoic geologic history. The paleogeographic maps in Chapters 10, 11, 14, and 16 show the extent to which North America was covered by the sea or exposed during these events.

Shoreline movements during transgressions and regressions probably amount to no more than a few centimeters per year. Suppose that a shoreline moves landward 1000 km in 20 million years, giving 5 cm/yr as the average rate of transgression. Our average is reasonable, but we must point out that large-scale transgressions are not simply events during which the shoreline steadily moves landward. In fact, they are characterized by a number of reversals in the overall transgressive trend, thus accounting for the intertonguing we see among some sedimentary rock units (see Figure 5.6c).

Geologists agree that uplift and subsidence (downward movement) of the continents, the amount of water frozen in glaciers, and rates of seafloor spreading are responsible for marine transgressions and regressions. During uplift of a continent, the shoreline moves seaward, and just the

What Would You Do?

You live in the continental interior where flat-lying sedimentary rocks are well exposed. At one location the rocks consist of mudstone with dinosaur fossils overlain first by seashell-bearing sandstone, then upward by shale, and finally by limestone with fossil clams, oysters, and corals. How would you explain the presence of marine fossils so far from the sea and how this vertical sequence of rocks came to be deposited?

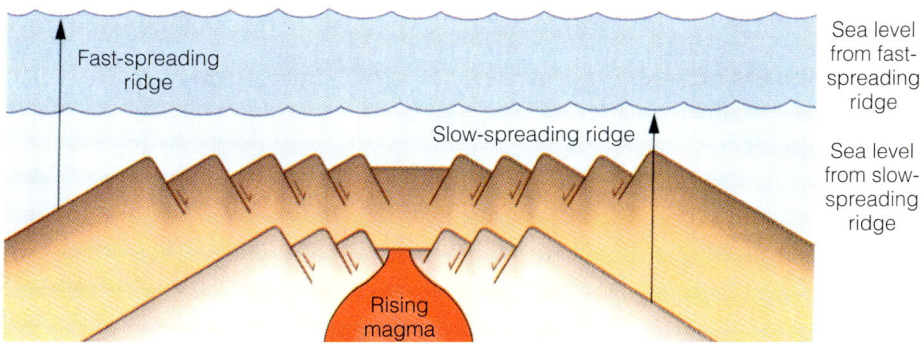

• **Figure 5.10 Mid-Oceanic Ridges and Marine Transgressions and Regressions** The cross-sectional area of a mid-oceanic ridge depends on its spreading rate. The ridge forms due to expansion by heating. A fast-spreading ridge has a larger cross section and consequently displaces more seawater, resulting in a rise in sea level and marine transgressions onto the continents.

Today it is apparent that bones, teeth, and shells in rocks are the remains of once-living creatures, yet this view is rather recent. Indeed, during most of historic time in the West, people variously believed fossils to be inorganic objects formed within rocks by some kind of molding force, or even objects placed in rocks by the Creator to test our faith or by Satan to sow seeds of doubt. Some perceptive observers— such as Leonardo da Vinci in 1508, Robert Hook in 1665, and Nicolas Steno in 1667—recognized the true nature of fossils, but their views were largely ignored. By the 18th and 19th centuries, though, it was apparent that fossils were truly the remains of organisms, and it was also clear that many fossils were of organisms now extinct.

opposite takes place during subsidence. Widespread glaciers expanded and contracted during the Pennsylvanian Period (see Chapter 11), which caused a number of sea level changes and resulted in transgressions and regressions. Indeed, if all of Earth's present-day glacial ice were to melt, sea level would rise by about 70 m.

Geologic evidence indicates that sea level may have been as much as 250 m higher during the Cretaceous Period, and as a result, there was a widespread marine transgression during which much of the continents was invaded by the sea (see Figure 14.6). The probable cause of this event was comparatively rapid seafloor spreading during which heat beneath the mid-oceanic ridges caused them to expand and displace water onto the continents (• Figure 5.10). When seafloor spreading slows, the mid-oceanic ridges subside, increasing the volume of the ocean basins, and the seas retreat from the continents.

Fossilization and Fossils

In Chapter 4 and earlier sections of this chapter we discussed cross-cutting relationships, superposition, and original horizontality, all of which are essential for interpreting geologic history. • Figure 5.11 is a graphic representation of two columns of rock, or what geologists call *columnar sections*, in which you can easily determine the relative ages of rocks at each location. But with the data provided, you cannot determine the ages of rocks in one section relative to those in the other section. The solution to such problems involves using fossils, if present, and some physical events of short duration.

Fossils, the remains or traces of prehistoric organisms, are most common in sedimentary rocks and in some accumulations of pyroclastic materials such as ash and volcanic mudflows. They are extremely useful for determining relative ages of strata, but geologists also use them to ascertain environments of deposition (see Chapter 6), and fossils provide some of the evidence for the theory of evolution (see Chapter 7).

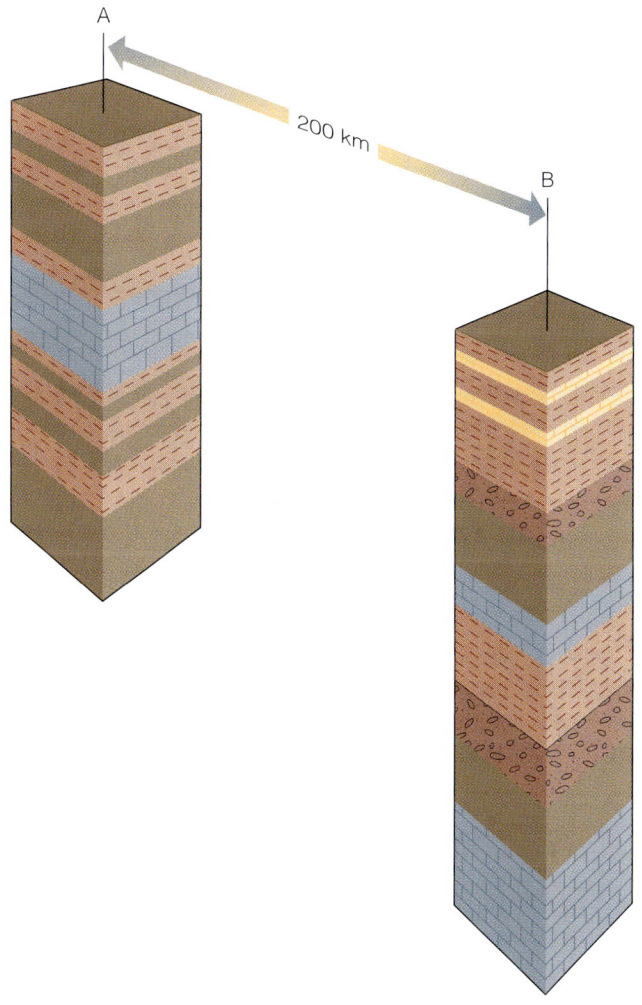

• **Figure 5.11 Relative Ages of Rocks** You can determine the relative ages of the rocks in either column A or B, but you cannot tell the relative ages of the rocks in A compared to B. The rocks in A may be younger than those in B, the same age as some in B, or older than all the rocks in B.

How Do Fossils Form?

In our definition of *fossil* we used the phrase "remains or traces." Remains, usually called **body fossils** (• Figure 5.12), consist mostly of durable skeletal parts such as bones, teeth, and shells, but in some exceptional cases we might find entire animals preserved by freezing or mummification. In contrast, indications of organic activity, including tracks, trails, burrows, and nests, are called **trace fossils** (• Figure 5.13). One type of trace fossil known as a *coprolite* consists of fossilized feces that may provide information about the size and diet of the animal that produced it.

The most favorable conditions for preserving body fossils are that the organism in question has some kind of durable skeleton and lives in an area where burial is likely. Consequently, the fossil record for brachiopods, clams, and corals is quite good because they meet these criteria. In contrast, jellyfish with no skeletons are poorly represented in the fossil record, and bats live where burial is unlikely. Moreover, bat skeletons are quite delicate. Even given these qualifications and the fact that bacterial decay, physical processes, scavenging, and metamorphism destroy organic remains, fossils are quite common. This is especially true for shelled marine invertebrate animals, including various microorganisms, but even many land-dwelling animals are more common in the fossil record than most people realize (see Perspective 5.1).

• **Figure 5.12 Body Fossils** (a) Skeleton of a mammoth, an extinct relative of today's elephants, on display in the Museum of Geology and Paleontology in Florence, Italy. (b) Shells of Mesozoic-aged invertebrate animals known as ammonoids and nautiloids in the Comstock Rock Shop in Virginia City, Nevada.

• **Figure 5.13 Trace Fossils** (a) Bird tracks. This is the slab of rock that formed over the actual tracks, so it is a cast of the tracks. (b) Fossilized feces (coprolite) of a carnivorous mammal. The specimen measures about 5 cm long and it contains small fragments of bone.

5.1 Perspective

Fossils and Uniformitarianism

As you might expect, organisms that have durable skeletons and live where burial is likely are the ones best represented in the fossil record. So we have a good record for organisms that meet these criteria but a poor record of those that do not—jellyfish and bats, for instance. Furthermore, even if organic remains are buried, they may be destroyed by dissolution or metamorphism, or they may remain deeply buried and undiscovered. Nevertheless, fossils of many organisms are far more common than most people realize.

The shells of marine invertebrate animals are the most easily collected, but the remains of some vertebrate animals—fish, amphibians, reptiles, birds, and mammals—are found in huge concentrations in some areas. Single surfaces within the Eocene-aged Green River Formation in Wyoming have thousands of fossil fish (Figure 1a), and an area measuring about 0.4 km by 2 km in Montana has an estimated 10,000 duck-billed dinosaurs that were overcome by volcanic gases and buried in ash. A deposit of Jurassic rocks in Wyoming has yielded 4000 bones of about 20 large dinosaurs that became mired in mud and perished. Hundreds of rhinoceroses, three-toed horses, saber-toothed deer, and other animals have been recovered from Miocene-aged ash in Nebraska (Figure 1b).

Were there some kinds of unknown processes operating during the past that account for such remarkable concentrations of fossils? In nearly all cases, we can find present-day examples that help us understand the conditions for fossilization. During the 1830s in Uruguay, flooding rivers buried the remains of thousands of horses and cattle that had died during a previous drought. Likewise, many of the fossil mammals found in the White River Badlands in South Dakota were buried in river deposits. Although a rare occurrence, African elephants today become mired in mud and die just as Jurassic dinosaurs did in what is now Wyoming. In 1984, thousands of caribou died in the Caniapiscau River of Quebec, Canada, and in Africa wildebeests die by the hundreds during river crossings. A bone bed in Canada with the remains of hundreds of horned dinosaurs probably formed in the same way.

We mentioned that rapid burial in sediment or pyroclastic materials is essential for fossilization. But the term rapid implies only that the burial takes place before the remains disintegrate or decompose, which might take years in arid environments or in the oxygen-deficient bottom waters of some lakes or parts of the oceans. Scientists in a submersible in the Gulf of California saw the remains of a whale that was slowly being covered by sediment. In fact, the sediment is quite similar to deposits in southern California from which fossil whales have been recovered.

When Mount St. Helens erupted in 1980, mudflows buried trees in their original growth position, and erosion exposed trees buried in growth position in 1885. In 1924 a debris flow on Mount Shasta in northern California buried trees several meters high in their original position (Figure 2); the trees are rooted in the deposits beneath the

Figure 1 (a) Thousands of fossil fish are found on bedding planes in the Eocene Green River Formation in Wyoming, indicating mass mortality. (b) Paleontologists excavating horses (background) and rhinoceroses (foreground) from Miocene volcanic ash near Orchard, Nebraska.

5.1 Perspective (continued)

debris flow. Fossil forests in what is now Yellowstone National Park in Wyoming and Florissant Fossil Beds National Monument in Colorado were buried in a similar manner.

The famous La Brea Tar Pits in downtown Los Angeles, California, with their numerous fossil saber-toothed cats, dire wolves, horses, camels, vultures, and armored mammals called *glyptodonts*, also have present-day parallels. The tar in the pits is actually asphalt, the sticky residue that forms when the easily vaporized constituents of oil are lost (see Chapter 17). Of course, the area no longer supports the animals just noted, but small animals are still trapped exactly as were their ancient counterparts.

It is important to note from these examples that the first step in fossilization, the burial of organisms, requires only processes with which we are familiar. The burial conditions may be very rapid as when mudflows engulf trees or animals, but they may just as well be very slow, as in the case of the whale in the Gulf of California. In all cases, though, we have every reason to think that organisms in the fossil record were buried under conditions like those of the present.

Figure 2 In 1924, a debris flow spilled over the banks of Mud Creek near Mount Shasta in California and buried these trees, which are rooted in the deposits below. Subsequent erosion has exposed some of the trees, but others remain buried and are now potential fossils.

Body fossils may be preserved as *unaltered remains* (• Figure 5.14), meaning they retain their original composition and structure, or as *altered remains* (• Figure 5.15), in which case some change has taken place in their composition and/or structure. Table 5.1 summarizes the several subcategories of each type of preservation. In addition, **molds** and **casts** are common in the fossil record (• Figure 5.16). A mold forms when buried organic remains dissolve, leaving a cavity shaped like a clam or bone, for example. If minerals or sediment should later fill the cavity, it forms a cast—that is, a replica of the original.

The *fossil record*, the record of ancient life, just as the geologic record of which it is a part, must be analyzed and interpreted. As a repository of prehistoric organisms, this record provides our only knowledge of such extinct animals as trilobites and dinosaurs. Furthermore, the study of fossils has several practical applications that we will discuss in the following sections and in Chapter 6.

Fossils and Telling Time

We began this section on fossils by referring to the columnar sections in Figure 5.11 and asking how you might determine the relative ages of the strata in these geographically separate areas. Our answer was "fossils"—but exactly how do fossils resolve this problem? To fully understand the usefulness of fossils, we must examine the historical development of an important geologic principle.

• **Figure 5.14 Unaltered Remains** (a) Insects preserved in amber. (b) This Harlan's ground sloth skeleton was preserved in tar in the La Brea Tar Pits in Los Angeles. (c) Frozen baby mammoth found in Russia in 1989.

The use of fossils in relative dating and geologic mapping was clearly demonstrated during the early 1800s in England and France. William Smith (1769–1839), an English civil engineer involved in surveying and building canals in southern England, independently discovered Steno's principle of superposition. He reasoned that in a sequence of strata, the oldest is at the bottom and the youngest is at the top, but he also extended this reasoning to include any fossils the rocks might contain. Smith made numerous observations at outcrops, mines, and quarries and discovered that the sequence of fossils, and especially groups of fossils, is consistent from area to area. In short, he discovered a method whereby the relative ages of sedimentary rocks at different locations could be determined by their fossil content (• Figure 5.17).

By recognizing the relationship between strata and fossils, Smith was able to predict the order in which fossils would appear in rocks at some locality he had not previously visited. In addition, his knowledge of rocks and fossils allowed him to predict the best route for a canal and the best areas for bridge foundations. As a result, his services were in great demand.

Smith gets much of the credit for developing the idea of using fossils to determine relative ages, but other geologists, such as Alexander Brongniart in France, also recognized this relationship. In any case, their observations served as the basis for what we now call the **principle of fossil succession** (also known as the *principle* or *law of faunal succession*). This important principle holds that fossil assemblages (groups of fossils) succeed one another through time in a regular and determinable order.

But why not simply match up similar rock types in Figure 5.11 and conclude they are the same age? Rock type will not work because the same kind of rock—sandstone, for instance—has formed repeatedly through time. Fossils have also formed continuously through time, but because different organisms existed at different times, fossil assemblages are unique. In short, an assemblage of fossils has a distinctive aspect compared with younger or older fossil assemblages. Accordingly, if we match up rocks containing similar fossils, we can assume they are of the same relative age.

• **Figure 5.15 Altered Remains** (a) Petrified redwood tree at the Petrified Forest in California. Numerous redwood trees were blown down by a volcanic eruption about 3 million years ago and covered with volcanic ash. (b) Eocene-aged carbonized palm frond on display at the Natural History Museum in Vienna, Austria. (c) This carbonized insect is from Oligocene-aged deposits in Montana.

• **Figure 5.16 Mold and Cast** (a) Burial of a shell in sediment followed by dissolution (b), leaving a cavity—a mold. (c) The mold is filled by sediment, thereby forming a cast. (d) Cast of a clam shell that formed by the processes outlined in a–c. (e) This fossil turtle shows some of the original shell material (body fossil) and a cast. The hollow interior of the shell was a mold that was filled with mud, thus forming a cast of the shell's interior.

90 CHAPTER 5 ROCKS, FOSSILS, AND TIME—MAKING SENSE OF THE GEOLOGIC RECORD

TABLE 5.1	Types of Fossil Preservation
Body fossils—unaltered remains	Original composition and structure preserved
Freezing	Large Ice Age mammals frozen in sediment
Mummification	Air drying and shriveling of soft tissues
Preservation in amber	Leaves, insects, small reptiles trapped and preserved in hardened tree resin
Preservation in tar	Bones, insects preserved in asphaltlike substance at oil seeps
Body fossils—altered remains	Change in composition and/or structure of original
Permineralization	Addition of minerals to pores and cavities in shells and bones
Recrystallization	Change in the crystal structure—for example, aragonite in shells recrystallizes as calcite
Replacement	One chemical compound replaces another—for example, pyrite (FeS_2) replaces calcium carbonate ($CaCO_3$) of shells
Carbonization	Volatile elements lost from organic matter leaving a carbon film; most common for leaves and insects
Trace fossils	Any indication of organic activity such as tracks, trails, burrows, droppings (coprolites), and nests
Molds and casts	Mold—a cavity with the shape of a bone or shell; cast—a mold filled by minerals or sediment

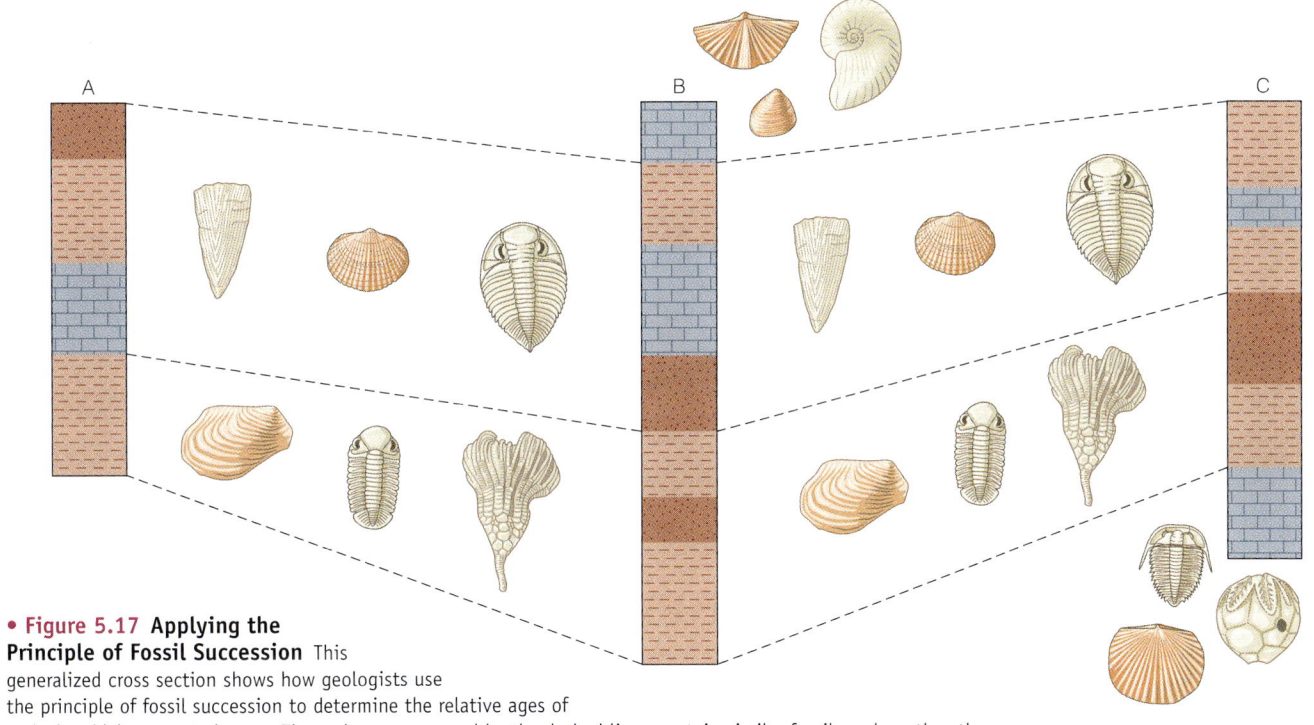

• Figure 5.17 **Applying the Principle of Fossil Succession** This generalized cross section shows how geologists use the principle of fossil succession to determine the relative ages of rocks in widely separated areas. The rocks encompassed by the dashed lines contain similar fossils and are thus the same age. Note that the youngest rocks are in column B, whereas the oldest ones are in column C.

The Relative Geologic Time Scale

In Chapter 4 we noted that the 18th-century Neptunists thought they could use composition to determine the relative ages of rocks (see Table 4.1). During the 19th century, however, it became apparent that composition is insufficient to make relative age determinations, whereas superposition and fossil content work quite well. The investigations of rocks by naturalists between 1830 and 1842 resulted in the recognition of rock bodies called *systems* and the construction of a composite geologic column that is the basis for the relative geologic time scale (• Figure 5.18). A short discussion about the methods used to recognize and define the systems will help you understand how the geologic column and relative geologic time scale were established.

During the 1830s, Adam Sedgwick studied rocks in northern Wales and described the Cambrian System, and Sir Roderick Impey Murchison named the Silurian System in southern Wales. Sedgwick paid little attention to the few fossils present, and his Cambrian System could not be recognized beyond the area where it was first described. Murchison carefully described fossils typical of the Silurian, so his system could be identified elsewhere.

• **Figure 5.18 The Geologic Column and the Relative Geologic Time Scale**
(a) Geologists in Great Britain and mainland Europe defined the geologic systems at the locations shown. **(b)** A composite geologic column was constructed by placing the systems in their correct relative order, so the column is a relative geologic time scale. The absolute ages were added much later. Notice that the Cenozoic consists of the Tertiary and Quaternary periods, which was the terminology used then.

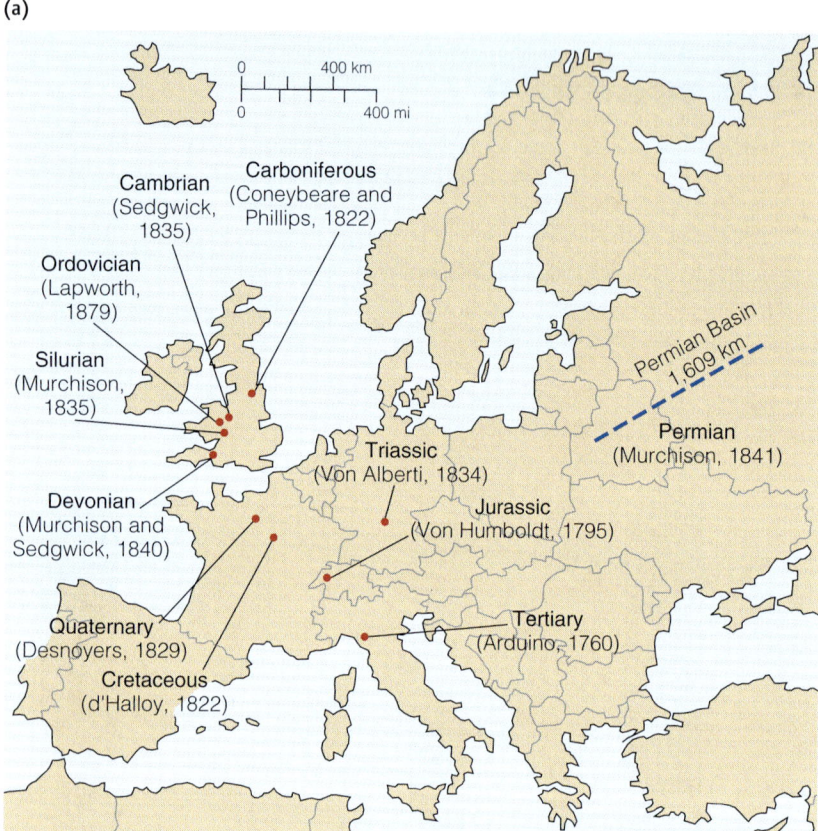

When both men published the results of their studies in 1835, it was apparent that Silurian rocks were younger than Cambrian strata, but it was also apparent that their two systems partially overlapped (• Figure 5.19), resulting in a boundary dispute. This dispute was not resolved until 1879 when Charles Lapworth suggested that the strata in the area of overlap be assigned to a new system, the Ordovician. So finally three systems were named, their stratigraphic positions were known, and distinctive fossils of each system were described.

Before Sedgwick's and Murchison's dispute, they had named the Devonian System, which contained fossils distinctly different from those in the underlying Silurian System rocks and the overlying rocks of the Carboniferous System.* Then in 1841, Murchison visited western Russia, where he identified strata as

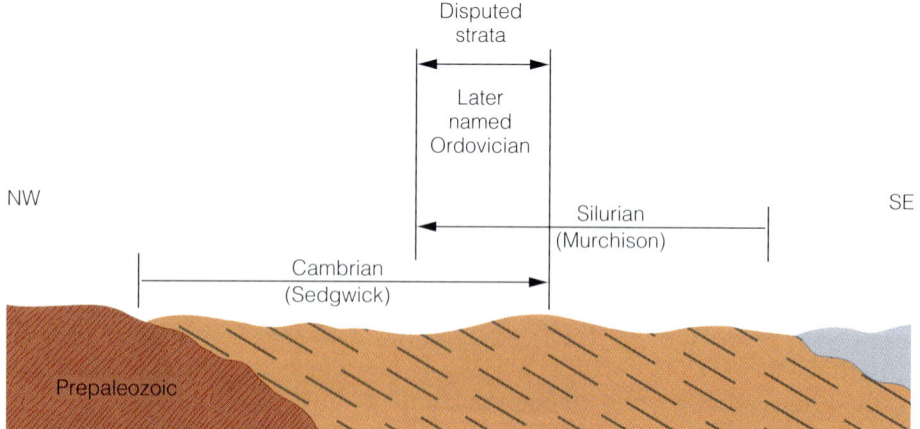

• **Figure 5.19 Cambrian, Ordovician, and Silurian Systems** Simplified cross section showing the disputed interval of strata that Sedgwick and Murchison each claimed belonged to their respective systems. The dispute was settled in 1879 when Charles Lapworth proposed assigning the area of overlap to the Ordovician System.

*In North America, two systems, the Mississippian and Pennsylvanian, correspond to the Lower and Upper Carboniferous, respectively.

TABLE 5.2	Classification of Stratigraphic Units		
Units Defined by Content		Units Expressing or Related to Geologic Time	
Lithostratigraphic Units	Biostratigraphic Units	Time-Stratigraphic Units	Time Units
Supergroup	Biozones	Eonothem	Eon
Group		Erathem	Era
Formation		System	Period
Member		Series	Epoch
Bed		Stage	Age

Silurian, Devonian, and Carboniferous by their fossil content. And overlying the Carboniferous strata were fossil-bearing rocks assigned to a Permian System (Figure 5.18).

We need not discuss the specifics of where and when the other systems were established, except to say that superposition and fossil content were the criteria used to define them. The important point is that geologists were piecing together a composite **geologic column,** which is in effect a **relative geologic time scale,** because the systems are arranged in their correct chronologic order. You should be aware of another aspect of the relative geologic time scale: All rocks beneath Cambrian strata are called Precambrian (Figure 5.18b). Long ago, geologists realized that these rocks contain few fossils and that they could not effectively be subdivided into systems. In Chapters 8 and 9, we will discuss the terminology for Precambrian rocks and time.

It is important to realize that although scientists today accept that organic evolution accounts for the differences in organisms through time, evolution is not the basis for the geologic time scale. Indeed, the geologic column and time scale were pieced together by the 1840s by scientists who fully accepted the biblical account of creation and had little or no idea about evolution—a concept that would not be formalized until 1859. The only requirement for fossils to be useful, as we have outlined, is that they differ through time, regardless of why they differ.

Stratigraphic Terminology

The recognition of systems and a relative geologic time scale brought some order to stratigraphy, but problems remained. Because sedimentary rock units are time transgressive (see Figure 5.9), they may belong to one system in one area and to another system elsewhere. And at some localities a rock unit simply straddles the boundary between systems. Accordingly, geologists developed terminology to deal both with rocks and with time. The terminology now used consists of two fundamentally different kinds of units: those defined by their content, and those expressing or related to geologic time (Table 5.2).

Rock type with no consideration of time of origin is the only parameter for defining a **lithostratigraphic unit*** (Table 5.2). The basic lithostratigraphic unit is the **formation,** which is a mappable body of rock with distinctive upper and lower boundaries. A formation may consist of a single rock type (the Redwall Limestone) or a variety of related rock types (the Morrison Formation). Formations are commonly subdivided into *members* and *beds,* and they may be parts of larger units known as *groups* and *supergroups* (• Figure 5.20, Table 5.2).

A body of strata recognized only on the basis of its fossil content is a **biostratigraphic unit** (Table 5.2), the boundaries of which do not necessarily correspond to those of lithostratigraphic units. The fundamental biostratigraphic unit is the **biozone,** which is discussed more fully in the following section on correlation.

The category of units expressing or related to geologic time includes time-stratigraphic units (also called chronostratigraphic units) and time units (Table 5.2). **Time-stratigraphic units** consist of rocks deposited during a particular interval of geologic time. The **system,** the basic time-stratigraphic unit, is based on a *stratotype* consisting of rocks in an area where the system was first described. Systems are recognized beyond their stratotype areas by their fossil content. Recall our discussion of how the systems were defined and recognized and served as the basis for the relative geologic time scale (Figure 5.18).

Time units are simply designations for certain parts of geologic time. **Period** is the most commonly used time designation, but two or more periods may be designated as an *era,* and two or more eras constitute an *eon.* Periods also consist of shorter time designations such as *epoch* and *age.* All these time units have corresponding time-stratigraphic units, although the terms *erathem* and *eonothem* are not often used (Table 5.2).

Time-stratigraphic units, time units, and their relationships are particularly confusing to beginning students. Remember, though, that time-stratigraphic units are defined by their position in a sequence of strata, whereas time

---
**Lith-* and *litho-* are prefixes meaning "stone" or "stonelike."

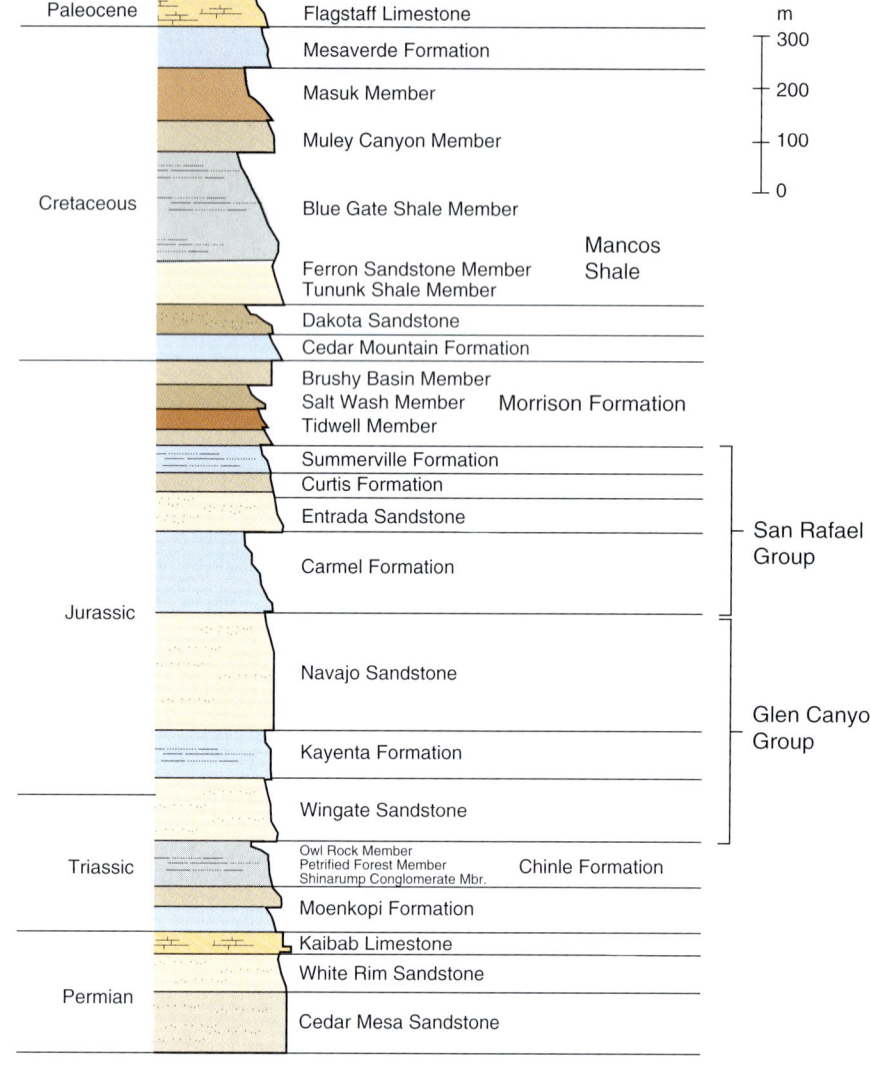

• **Figure 5.20 Graphic Representation of the Lithostratigraphic Units in Capital Reef National Park in Utah** Notice that some of the formations are further divided into members, and some are parts of more inclusive groups.

Correlation

In a previous section we noted that the systems were recognized beyond their stratotype areas by applying the principle of fossil succession. In other words, geologists matched up rocks of the same relative age in different areas, a process known as **correlation.** In this example *time-stratigraphic correlation* demonstrated time equivalence of events. *Lithostratigraphic correlation,* in contrast, is simply a matter of matching up the same rock units over a large area with no regard for time.

Correlation of lithostratigraphic units such as formations involves demonstrating that the same rock unit extends over the area in question. If surface exposures are adequate, rock units may be traced laterally even if occasional gaps are present (• Figure 5.21a). We can also correlate rock units based on composition, their position in a sequence, and the presence of a distinctive key bed (Figure 5.21b and c). Rock cores and well cuttings from drilling operations as well as geophysical data are useful for correlating rock units below the surface.

Because most rock units of regional extent are time transgressive, we cannot rely on lithostratigraphic correlation to demonstrate time equivalence. For example, geologists have accurately correlated sandstone in Arizona with similar rocks in Colorado and South Dakota, but the age of these rocks varies from Early Cambrian in the west to Middle Cambrian further east. Time-stratigraphic correlation using various biozones is the most effective way to demonstrate that sedimentary rocks in different areas are the same age, but several other methods are useful as well.

units refer to time only. Thus we may refer to the Devonian System as a material body of rock made up of lower, medial, and upper parts. In contrast, the Devonian Period is a designation for the time during which the Devonian System originated, and we can refer to the early, middle, and late Devonian Period.

Maybe an analogy will help you understand the distinction between these units. Think of a three-story building that was built one story at a time, the lowest story first and so on. The building—let us call it Brooks Hall—represents the Brooks Hall System, a material body that was built during a specific interval of time, the Brooks Hall Period. Lower Brooks Hall was built during the Early Brooks Hall Period and so on. Remember that system refers to position in a sequence, so we would refer to the third floor as Upper Brooks Hall but not Late Brooks Hall.

For all organisms now extinct, their time of existence marks two points in time—time of origin and time of extinction. One type of biozone, the **range zone,** is defined by the geologic range (the total time of existence) of a particular fossil group (a species or a group of related species called a *genus,* for example). Particularly useful fossils are those that are easily identified, that were geographically widespread, and that had a rather short geologic range. The brachiopod *Lingula* meets the first two of these criteria, but its geologic range makes it of little use, whereas the brachiopod *Atrypa* and the trilobite *Paradoxides* are well

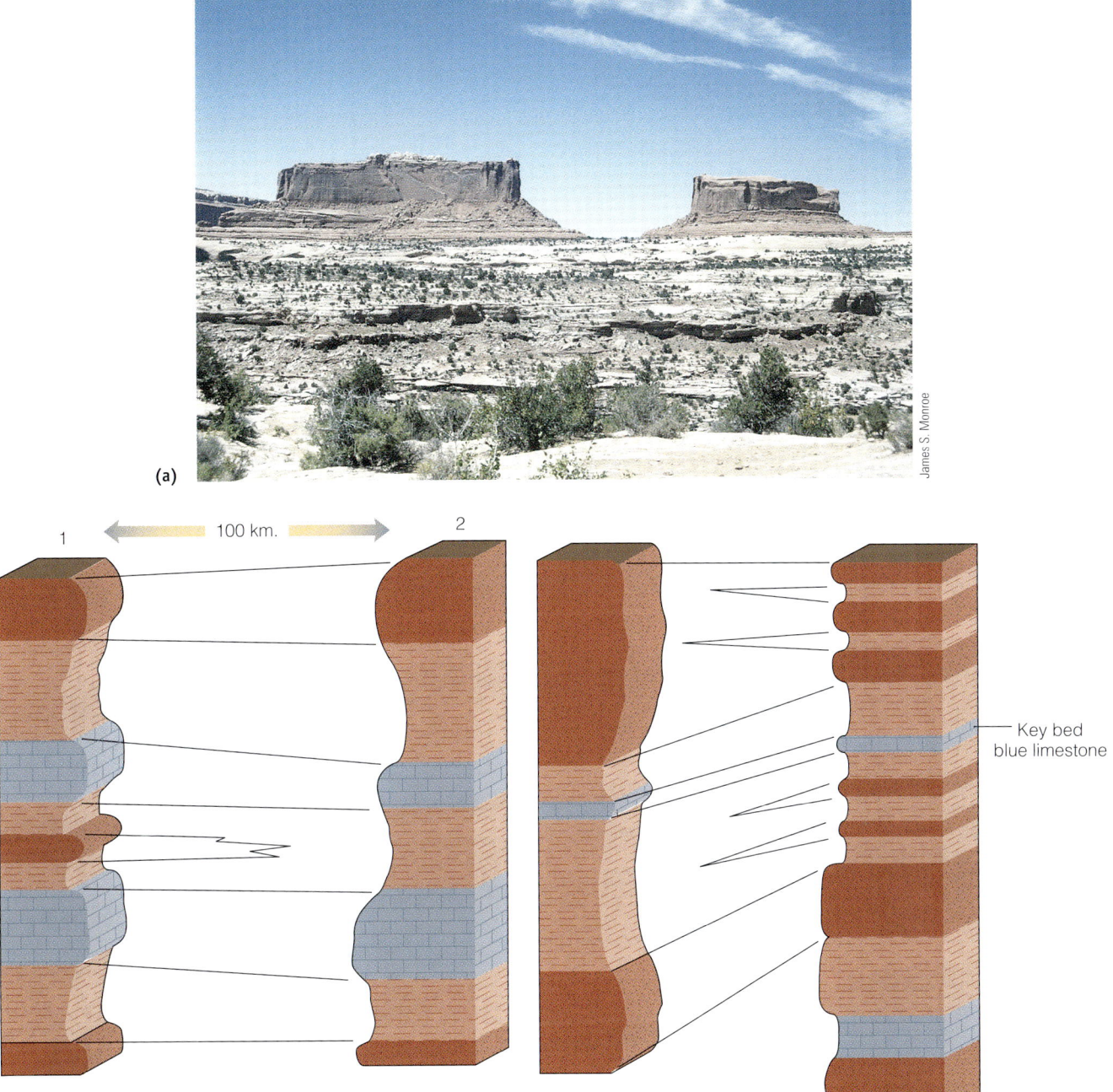

• **Figure 5.21 Lithostratigraphic Correlation** (a) In areas of adequate exposures, geologists trace rocks laterally and match them up even if occasional gaps exist. These rocks on the skyline in Utah were once part of a continuous unit that covered a much larger area. (b) Lithostratigraphic correlation by similarities in rock type and position in a sequence. We assume the middle sandstone in column 1 intertongues or grades laterally into the shale in column 2. (c) Correlation using a key bed, a distinctive blue limestone in this case.

suited for time-stratigraphic correlation and are therefore called **guide fossils** (• Figure 5.22).

Several types of biozones known as *interval zones* are used, but the most important one is the **concurrent range zone,** which is established by plotting the overlapping ranges of two or more fossils with different geologic ranges (• Figure 5.23). The first and last occurrences of fossils are used to determine zone boundaries. Correlating concurrent range zones is probably the most accurate method of determining time equivalence.

Some physical events of short duration are also used to demonstrate time equivalence. Correlating outcrops with a distinctive lava flow, although based on rock type, is time significant. Ash falls take place in a matter of hours or days and may cover large areas (• Figure 5.24). Furthermore, ash falls are not restricted to a specific environment,

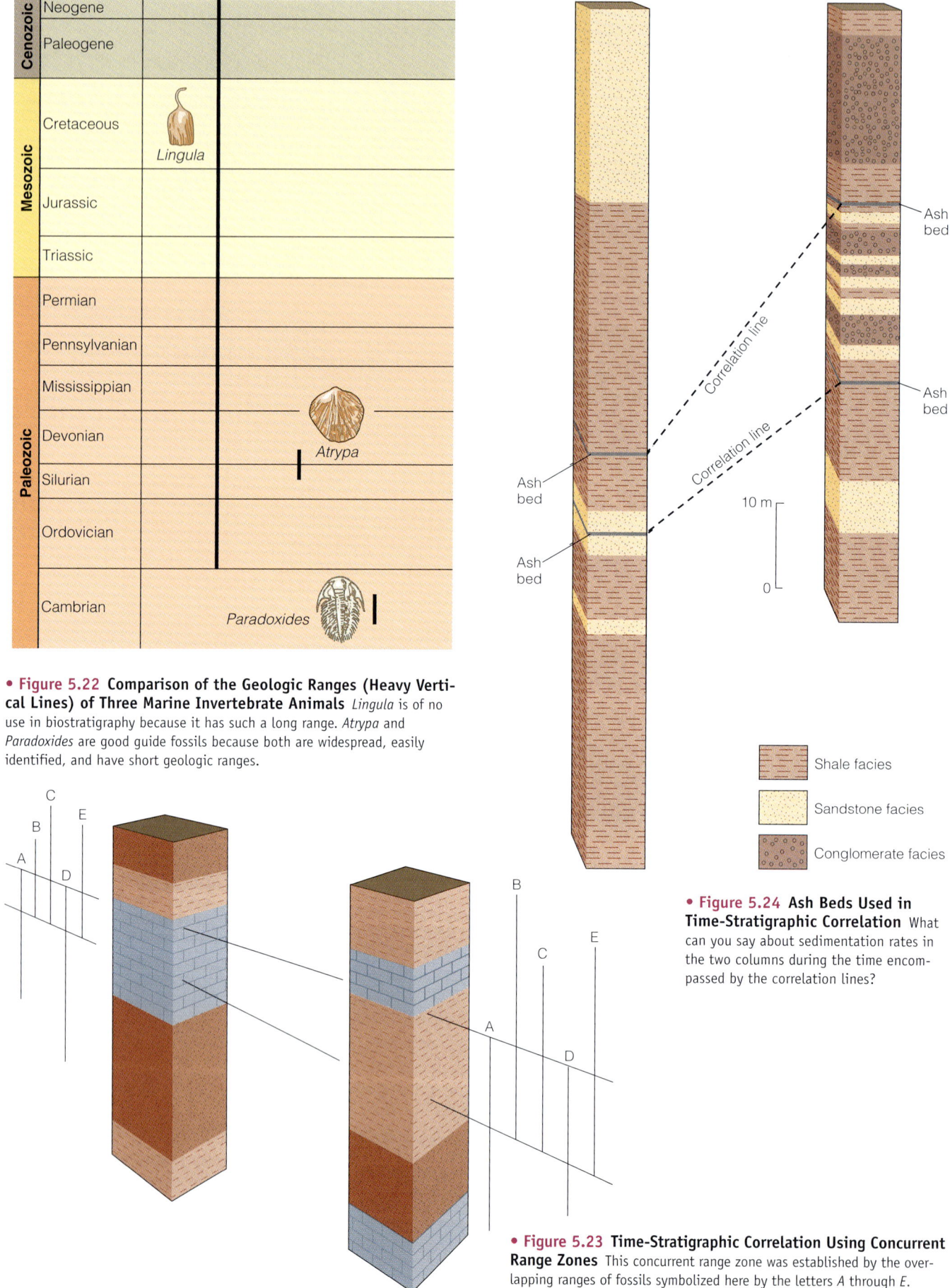

• **Figure 5.22 Comparison of the Geologic Ranges (Heavy Vertical Lines) of Three Marine Invertebrate Animals** *Lingula* is of no use in biostratigraphy because it has such a long range. *Atrypa* and *Paradoxides* are good guide fossils because both are widespread, easily identified, and have short geologic ranges.

• **Figure 5.24 Ash Beds Used in Time-Stratigraphic Correlation** What can you say about sedimentation rates in the two columns during the time encompassed by the correlation lines?

• **Figure 5.23 Time-Stratigraphic Correlation Using Concurrent Range Zones** This concurrent range zone was established by the overlapping ranges of fossils symbolized here by the letters *A* through *E*.

96 CHAPTER 5 ROCKS, FOSSILS, AND TIME—MAKING SENSE OF THE GEOLOGIC RECORD

What Would You Do?

You are one of the first astronauts to visit an Earth-like planet that has widespread fossil-bearing sedimentary rocks. How would you go about establishing a geologic column and relative geologic time scale? Also, how could you determine the absolute ages of the fossil-bearing rocks?

so they may extend from continental to marine environments. Finally, radiometric ages might be useful for some igneous and metamorphic rocks, particularly for those of Precambrian age.

Absolute Dates and the Relative Geologic Time Scale

Since the 1840s geologists have known that Ordovician rocks are younger than those of the Cambrian and older than Silurian rocks. But how old are they? That is, when did the Ordovician begin and end? Beginners might think it simple to obtain radiometric ages for fossil-bearing Ordovician rocks and thus resolve the problem. Unfortunately, absolute ages determined for minerals in sedimentary rocks give only the age of the source that supplied the minerals, not the age of the rock itself. For instance, sandstone that formed 10 million years ago might contain minerals 3 billion years old.

One mineral, however, forms in some sedimentary rocks shortly after deposition, and it can be dated by the potassium–argon method (see Chapter 4). This greenish mineral known as *glauconite* easily loses argon (the daughter product), so absolute ages determined are generally too young, giving only a minimum age for the host sedimentary rock. In most cases, absolute ages for sedimentary rocks must be determined indirectly by dating associated igneous and metamorphic rocks.

According to the principle of cross-cutting relationships, a dike must be younger than the rock it cuts, so an absolute age for a dike gives a minimum age for the host rock and a maximum age for any overlying rocks (• Figure 5.25). Absolute dates obtained from regionally metamorphosed rocks also give a maximum age for overlying sedimentary rocks.

Lava flows and ash falls interbedded with sedimentary rocks are the most useful for determining absolute ages (Figure 5.25). Both are time-equivalent, providing a maximum age for any rocks above and a minimum age for rocks below. Multiple lava flows or ash fall deposits in a stratigraphic sequence are particularly useful in this effort.

In the nonscientific literature one sometimes hears that radiometric dates are unreliable because geologists determined that a 200-year-old Hawaiian lava flow was 22 million years old. Actually, what they were dating were inclusions within the lava flow that were indeed about 22 million years old. For the lava flow itself, the scientists reported a date of essentially zero because the margin of error in potassium–argon dating with a half-life of 1.3 billion years is greater than the age of the lava.

In any case, sufficiently accurate radiometric dates are now available for many lava flows, ash falls, plutons, and metamorphic rock with associated fossil-bearing sedimentary rocks. Geologists have added these absolute ages to the geologic time scale, so we now can answer questions such as when the Ordovician Period began and ended (Figure 5.18b). In addition, we now know when a particular organism lived. For example, the fossil *Baculites reesidei* biostratigraphic zone in the Bearpaw Formation in Saskatchewan, Canada, is about 72 to 73 million years old because absolute ages have been determined for associated volcanic ash layers (• Figure 5.26).

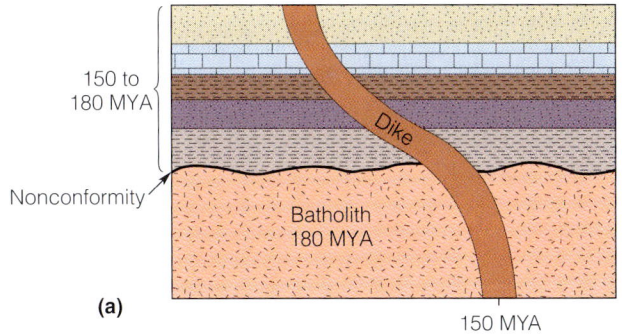

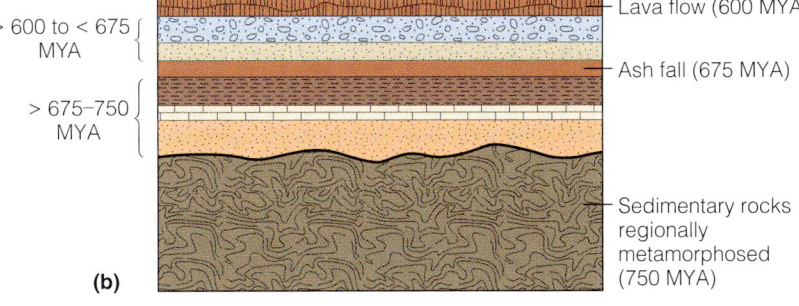

• **Figure 5.25 Determining the Absolute Ages of Sedimentary Rocks** Absolute ages for sedimentary rocks are most often found by radiometric dating of associated igneous and metamorphic rocks. In both **a** and **b**, the absolute ages of sedimentary rocks are bracketed by rocks for which absolute ages have been determined.

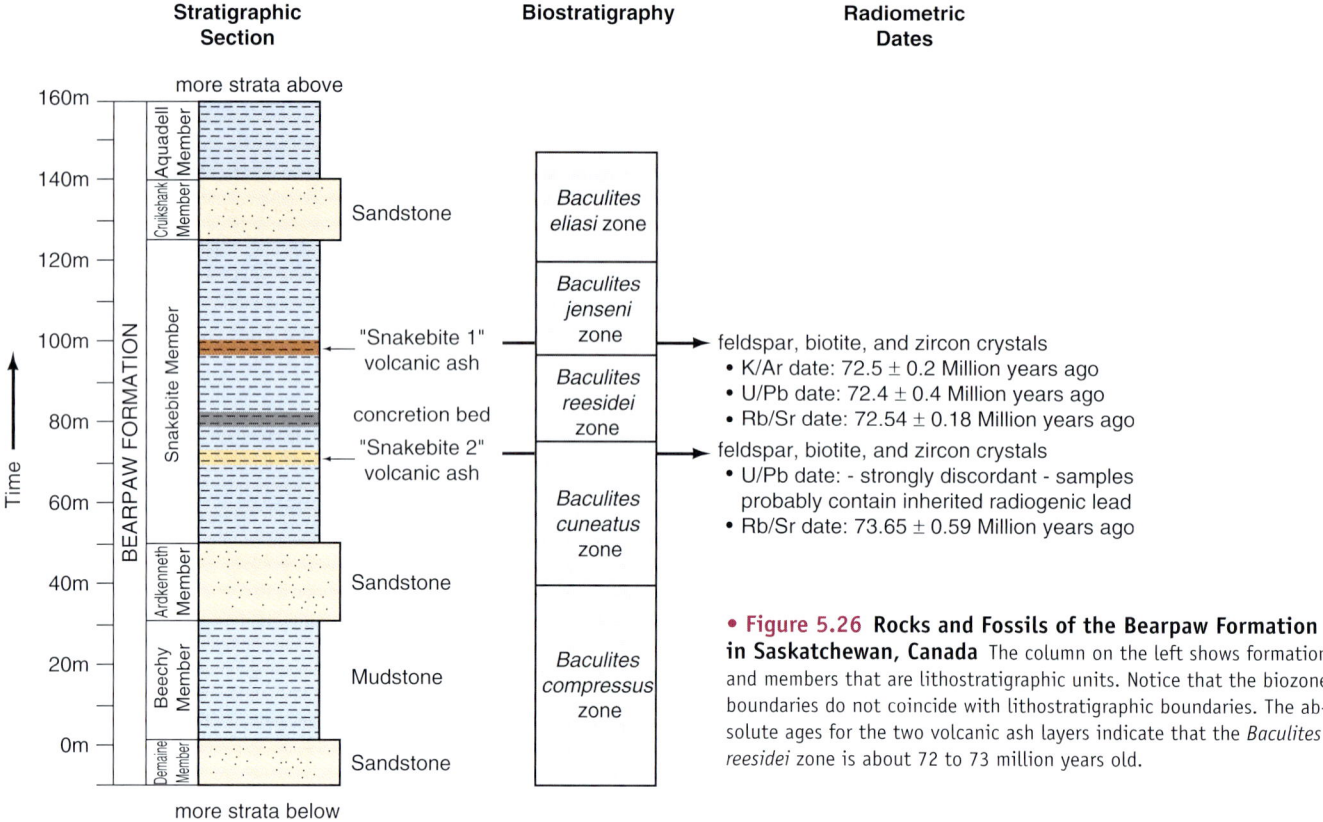

• **Figure 5.26 Rocks and Fossils of the Bearpaw Formation in Saskatchewan, Canada** The column on the left shows formations and members that are lithostratigraphic units. Notice that the biozone boundaries do not coincide with lithostratigraphic boundaries. The absolute ages for the two volcanic ash layers indicate that the *Baculites reesidei* zone is about 72 to 73 million years old.

SUMMARY

- The first step in deciphering the geologic history of a region involves determining the correct relative ages of the rocks present. Thus the vertical relationships among rock layers must be ascertained even if they have been complexly deformed.
- The geologic record is an accurate chronicle of ancient events, but it has many discontinuities or unconformities representing times of nondeposition, erosion, or both.
- Simultaneous deposition in adjacent but different environments yields sedimentary facies, which are bodies of sediment, or sedimentary rock, with distinctive lithologic and biologic attributes.
- According to Walther's law, the facies in a conformable vertical sequence replace one another laterally.
- During a marine transgression a vertical sequence of facies results with offshore facies superposed over nearshore facies. Just the opposite facies sequence results from a marine regression.
- Uplift and subsidence of continents, the amount of water frozen in glaciers, and the rate of seafloor spreading are responsible for marine transgressions and regressions.
- Most fossils are found in sedimentary rocks, although they might also be in volcanic ash and volcanic mudflows, but rarely in other rocks.
- Fossils are actually quite common, but the fossil record is strongly biased toward those organisms that have durable skeletons and that lived where burial was likely.
- The work of William Smith, among others, is the basis for the principle of fossil succession that holds that fossil assemblages succeed one another through time in a predictable order.
- Superposition and fossil succession were used to piece together a composite geologic column, which was the basis for the relative time scale.
- To bring order to stratigraphic terminology, geologists recognize units based entirely on content (lithostratigraphic and biostratigraphic units) and those related to time (time-stratigraphic and time units).
- Lithostratigraphic correlation involves demonstrating the original continuity of a rock unit over a given area even though it may not now be continuous over this area.
- Correlation of biostratigraphic zones, especially concurrent range zones, demonstrates that rocks in different areas, even though they may differ in composition, are of the same relative age.
- The best way to determine absolute ages of sedimentary rocks and their contained fossils is to obtain absolute ages for associated igneous rocks and metamorphic rocks.

IMPORTANT TERMS

angular unconformity, p. 81
biostratigraphic unit, p. 93
biozone, p. 93
body fossil, p. 86
cast, p. 88
concurrent range zone, p. 95
conformable, p. 80
correlation, p. 94
disconformity, p. 81
formation, p. 93
fossil, p. 85

geologic column, p. 93
geologic record, p. 78
guide fossil, p. 95
lithostratigraphic unit, p. 93
marine regression, p. 83
marine transgression, p. 81
mold, p. 88
nonconformity, p. 81
period, p. 93
principle of fossil succession, p. 89
principle of inclusions, p. 79

range zone, p. 94
relative geologic time scale, p. 93
sedimentary facies, p. 81
stratigraphy, p. 79
system, p. 93
time-stratigraphic unit, p. 93
time unit, p. 93
trace fossil, p. 86
unconformity, p. 80
Walther's law, p. 83

REVIEW QUESTIONS

1. A mappable rock unit with distinctive upper and lower boundaries is a(n)
   a. ____ formation; b. ____ concurrent range zone; c. ____ unconformity; d. ____ regression; e. ____ geologic column.

2. The process whereby minerals are added to the pores and cavities in bones, teeth, or shells is
   a. ____ compaction; b. ____ sorting; c. ____ permineralization; d. ____ rounding; e. ____ lithification.

3. The information about prehistoric events that is preserved in rocks is known as the
   a. ____ stratigraphic column; b. ____ geologic record; c. ____ principle of inclusions; d. ____ carbonization; e. ____ Hutton's dictum.

4. Which one of the following found in ancient rocks is a trace fossil?
   a. ____ horse tooth; b. ____ clam shell; c. ____ petrified tree; d. ____ frozen mammoth; e. ____ dinosaur footprint.

5. Which one of the following statements is incorrect?
   a. ____ Among other things, a good guide fossil must be geographically widespread; b. ____ An era consists of two or more periods; c. ____ Biostratigraphic zone boundaries do not necessarily coincide with lithostratigraphic boundaries. d. ____ Most fossils are found in igneous and metamorphic rocks; e. ____ During a marine transgression, offshore facies are superimposed on nearshore facies.

6. An erosion surface cut into plutonic or metamorphic rocks that is overlain by sedimentary rocks is a
   a. ____ system; b. ____ nonconformity; c. ____ lithostratigraphic unit; d. ____ sedimentary facies; e. ____ mold.

7. According to the principle of fossil succession,
   a. ____ a dike must be older than the sedimentary rocks it cuts through; b. ____ time-stratigraphic units are defined by rock type; c. ____ a marine regression takes place when the sea rises and invades a continent; d. ____ fossil assemblages succeed one another in a regular and predictable order; e. ____ the geologic column was pieced together by correlating rocks of similar types in different areas.

8. A lithostratigraphic unit made up of two or more formations is a
   a. ____ group; b. ____ facies; c. ____ range zone; d. ____ member; e. ____ disconformity.

9. According to _____, a vertical sequence of conformable facies replace one another laterally.
   a. ____ Steno's rule; b. ____ Hutton's uniformitarianism; c. ____ Darwin's theory; d. ____ Smith's corollary; e. ____ Walther's law.

10. Any aspect of sedimentary rocks that makes them recognizably different from adjacent rocks of the same age or approximately the same age is the definition of a sedimentary
    a. ____ biozone; b. ____ system; c. ____ facies; d. ____ epoch; e. ____ conformity.

11. Describe an angular unconformity and explain how one develops.

12. How were the principles of superposition and fossil succession used to establish the relative geologic time scale?

13. Give an example of how some types of rocks can be used in time-stratigraphic correlation.

14. What are the causes of marine transgressions and regressions?

15. How do geologists use the principle of uniformitarianism to interpret depositional environments of ancient rocks and to determine how fossils were preserved?

16. Refer to Figure 5.11. If you correlate the limestone in the left column with the lower limestone in the right column, can you assume they are of the same relative age? Explain.

17. What is a concurrent range zone, and how can it be used in time-stratigraphic correlation?

18. While visiting one of our national parks, you observe the following sequence of rocks: marine-fossil-bearing

sandstone, shale, and limestone, all inclined at 40 degrees. Lying over these rocks is a horizontal layer of mudstone baked by basalt overlying it. Finally, there is a sandstone layer with inclusions of basalt. Decipher the geologic history of these rocks.

19. Unconformities are common in the geologic record, and some are found over large areas, but none of them are worldwide. Why?
20. How does the vertical sequence of facies of a marine transgression differ from that of a marine regression?

FIELD QUESTIONS

1. In the photo below, which type of rock is oldest, and how do you know?
2. Refer to Figure 5.24 and consider this. Suppose that the upper ash bed formed 1.7 million years ago and the lower one is 1.9 million years old. What was the average sedimentation rate in cm/yr for the interval between the ash beds in the column on the right? Do you think this average is accurate for the actual rate of sedimentation? Explain.

CHAPTER 6
SEDIMENTARY ROCKS—THE ARCHIVES OF EARTH HISTORY

David Carriere/Index Stock Imagery

These sedimentary rocks in the 600-m-deep canyon at Dead Horse State Park near Moab, Utah, are made up of shale, sandstone, and conglomerate that were deposited during the Late Paleozoic and Mesozoic eras. Detailed studies of the rocks indicate that they were deposited in stream channels and on their adjacent flood-plains and as desert dunes.

[OUTLINE]

INTRODUCTION

SEDIMENTARY ROCK PROPERTIES

Composition and Texture

Sedimentary Structures

PERSPECTIVE 6.1 *Determining the Relative Ages of Deformed Sedimentary Rocks*

Geometry of Sedimentary Rocks

Fossils—The Biologic Content of Sedimentary Rocks

DEPOSITIONAL ENVIRONMENTS

Continental Environments

Transitional Environments

Marine Environments

INTERPRETING DEPOSITIONAL ENVIRONMENTS

PALEOGEOGRAPHY

SUMMARY

ThomsonNOW™ Explore interactive tutorials, animations, or practice problems available on the ThomsonNow website at **www.thomsonedu.com/login**.

CHAPTER OBJECTIVES

At the end of this chapter, you will have learned that

- Sedimentary rocks have a special place in deciphering Earth history because they preserve evidence of surface processes responsible for deposition, and many contain fossils.

- Geologists recognize three broad areas of deposition—continental, transitional, and marine—each with several specific depositional environments.

- The distinctive attributes of sedimentary rocks result from processes operating in specific depositional environments.

- Textures such as sorting and rounding as well as sedimentary structures provide evidence about depositional processes.

- Three-dimensional geometry of sedimentary rock bodies and fossils are also useful for determining depositional environments.

- The interpretations of sedimentary rocks in the chapters on geologic history are based on the considerations reviewed in this chapter.

- Interpretation of how and where rocks formed, especially sedimentary rocks, is the basis for determining Earth's ancient geographic features.

Introduction

The Jurassic-aged Navajo Sandstone of the southwestern United States was deposited as a vast blanket of sand dunes (• Figure 6.1), whereas the Cambrian-aged Redwall Limestone in Arizona was deposited in a shallow, warm sea. As we noted in the last chapter, no human observers were present to record these events, so how do geologists make such determinations? Our focus in this chapter is on specific areas or environments where sediment accumulates—that is, *depositional environments*—as well as on the criteria for recognizing the deposits of specific depositional environments.

We mentioned earlier that sedimentary rocks have a special place in historical geology because (1) they preserve evidence of the surface processes responsible for deposition, and (2) many contain fossils, our only record of prehistoric life. Our task here is to understand the physical, biological, and chemical processes accounting for the distinctive attributes of sedimentary rocks. Accordingly, geologists examine the features of sedimentary rocks, while keeping in mind that the processes responsible for these features are the same as those operating now.

Another important aspect of sedimentary rocks is that some are resources in their own right—sand and gravel, for example—or they contain resources such as liquid petroleum and natural gas. Certainly coal is an important energy resource, and phosphorus-rich sedimentary rock is mined for use in fertilizers, animal feed supplements, metallurgy, matches, ceramics, and preserved foods. A type of chemical sedimentary rock known as banded iron formation is the source of most of the world's iron ore (see Chapter 9).

• **Figure 6.1 Jurassic-aged Navajo Sandstone in Zion National Park, Utah** Geologists are convinced that the Navajo Sandstone was deposited as a vast blanket of wind-blown sand dunes. They came to this conclusion based on several features of these rocks as well as comparisons with present-day deposits. Much of this chapter covers the criteria geologists use to make such interpretations.

Sedimentary Rock Properties

The first step in investigating sedimentary rocks—or any other rock type, for that matter—is observation and data gathering by visiting rock exposures (outcrops) and carefully examining textures, composition, fossils (if present), thickness, and relationships to other rocks. During these initial field studies, geologists may make some preliminary interpretations. For example, color is a useful feature of some sedimentary rocks: Red rocks likely were deposited on land (see the chapter opening photo), whereas greenish ones more likely were deposited in a marine environment. However, exceptions are numerous, so color must be used with caution.

After completing fieldwork, geologists study the rocks more carefully by microscopic examination, chemical analyses, fossil identification, and construction of diagrams that show vertical and lateral facies relationships. A particularly important aspect of these studies is to compare the features of sedimentary rocks with those in present-day sediment accumulations that formed in known depositional environments. That is, geologists apply the principle of uniformitarianism. When all the data have been analyzed, geologists make an environmental interpretation.

Composition and Texture

More than 100 minerals are found in detrital sedimentary rocks, but only quartz, feldspars, and clay minerals are very common. Composition depends mostly on the composition of the rocks in the source area—that is, the area from which the detrital sediment was derived—but it tells little of how the sediment was deposited. Quartz sand, for example, may have been deposited in a stream channel, in desert dunes, or on a beach. So, taken alone, composition is of little help except for identifying the source of the sediment.

Among the chemical sedimentary rocks, including the biochemical ones, limestone (composed of calcite [$CaCO_3$], and dolostone (composed of dolomite [$CaMg(CO_3)_2$]) are by far the most common. Most limestone forms in warm, shallow seas, but a small amount is deposited in lakes. Evaporates such as rock salt (made up of halite [$NaCl$]) and rock gypsum (composed of gypsum [$CaSO_4 \cdot 2H_2O$]) invariably indicate arid environments where evaporation rates were high, as in some desert lakes and in marginal parts of the seas in arid regions (see Figure 2.15b and c). Of course, coal originates in swamps and bogs on land.

Grain size in detrital sedimentary rocks gives some clues about the energy conditions during transport and deposition. Conglomerate is made up of gravel, so we can be sure that energetic processes such as swiftly flowing streams, waves, or glaciers were responsible for transport and deposition. Sand also requires high-energy transport and tends to accumulate in desert dunes, in stream channels, and on beaches, but silt and clay are transported by weak currents and are deposited only under low-energy conditions, as in lakes and lagoons.

The term *texture* refers to the size, size distribution, shape, and arrangement of particles (clasts) in detrital sedimentary rocks. You already know that gravel, sand, silt, and clay are size designations for detrital particles (see Figure 2.13). Other textural features of detrital sedimentary rocks are rounding and sorting.

The degree to which detrital particles have had their sharp corners and edges smoothed off by abrasion is called **rounding** (• Figure 6.2a). Be aware that rounding does not mean spherical or ball-shaped; it only refers to the smoothness of particles. Rounding is less useful than sorting for determining the environment of deposition. Nevertheless, gravel in transport is rounded very quickly as the particles collide with one another, and with considerably more transport even sand grains become rounded.

(a)

(b)

(c)

• **Figure 6.2 Rounding and Sorting in Sediments** (a) Beginning students often mistake rounding to mean spherical or ball-shaped. All three of these stones are rounded—that is, the sharp corners and edges have been worn smooth—but only the one at the upper left is spherical. (b) A deposit of moderately sorted, well-rounded gravel. The particles average about 5 cm across. (c) Angular, poorly sorted gravel. Notice the quarter for scale.

Sorting refers to the size variation of particles in sediment or sedimentary rock. If most of the particles are of about the same size, the sediment or rock is well sorted, but if a wide range of sizes is present, the material is poorly sorted (Figure 6.2b and c). Wind has a limited capability to transport and deposit sand, so sand dunes tend to be well sorted, but glaciers can transport anything supplied to them, so their deposits are poorly sorted.

Sedimentary Structures

Sedimentary rocks contain a variety of features called **sedimentary structures** that formed during deposition or shortly thereafter.* All are manifestations of physical and biological processes that operated in the depositional environment. Analysis of sedimentary structures is the single most useful aspect of sedimentary rocks for determining how they were deposited in the first place. The origin of most sedimentary structures is well known because geologists have seen them forming in present-day environments, and many have been produced experimentally (Table 6.1).

Most sedimentary rocks have a layered aspect known as **stratification** or **bedding** (see the chapter opening photo and • Figure 6.3). Layers less than 1 cm thick are called *laminations*, whereas *beds* are thicker. Laminations are common in mudrocks, but coarser-grained sedimentary rocks are more commonly bedded. Individual beds are separated from one

TABLE 6.1 Summary of Sedimentary Structures

Physical Sedimentary Structures (produced by processes such as currents)	Characteristics	Origin
Laminations (or laminae)	Layers less than 1 cm thick	Form mostly as particles settle from suspension
Beds	Layers more than 1 cm thick	Form as particles settle from suspension and from moving sediment as sand in a stream channel
Graded bedding	Individual layers with an upward decrease in grain size	Deposition by turbidity currents or during the waning stages of floods
Cross-bedding	Layers deposited at an angle to the surface on which they accumulated	Deposition on a sloping surface as the downwind side of a sand dune
Ripple marks	Small (<3 cm high) ridges and troughs on bedding planes	
Current ripple marks	Asymmetric ripple marks	Result from deposition by water or air currents flowing in one direction
Wave-formed ripple marks	Symmetric ripple marks; generally with a sharp crest and broad trough	Formed by oscillating currents (waves)
Mud cracks	Intersecting cracks in clay-rich sediments	Drying and shrinkage of mud along a lakeshore or on tidal flats
Biogenic Sedimentary Structures (produced by organisms)		
Trace fossils	Tracks, trails, tubes, and burrows	Indications of organic activity. Intense activity results in *bioturbation* involving disruption of sediment

(a)

(b)

• **Figure 6.3 Stratification** (a) Alternating beds of sandstone and shale. The sandstone beds are 10 to 15 cm thick. (b) Laminations, layers less than 1 cm thick, in the chemical sedimentary rock called banded iron formation. The pencil measures about 14 cm long.

*There are also structures that form long after deposition, mostly by chemical processes, which are not useful for determining depositional environments.

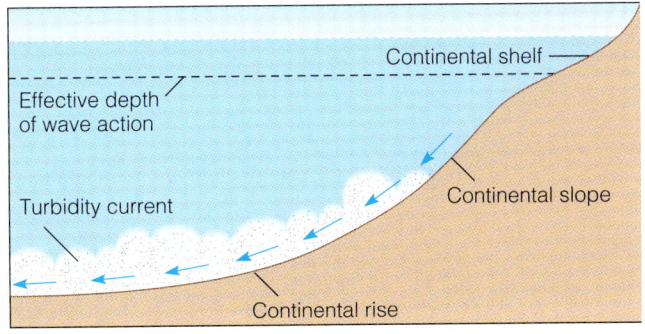

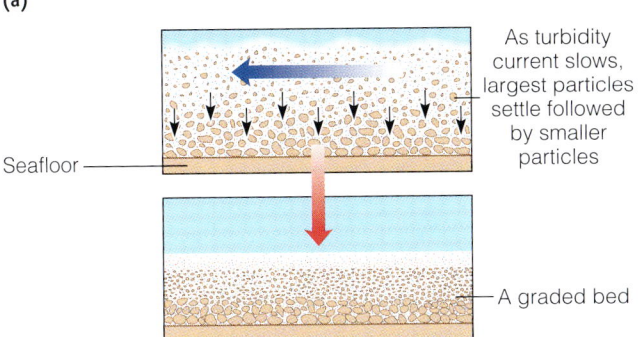

• **Figure 6.4 Graded Bedding** (a) Turbidity currents flow downslope along the seafloor (or lake bottom) because they are denser than sediment-free water. (b) Graded bedding forms as deposition of progressively smaller particles takes place. (c) The propeller of a submarine caused this turbidity current that flowed down a slope near Jamaica.

another by surfaces called *bedding planes*, above and below which the rocks differ in composition, grain size, color, or a combination of features. Such transitions indicate an abrupt change in the type of sediment deposited, or perhaps a time of nondeposition or erosion followed by renewed deposition. In contrast, some layers grade upward from one rock type into another, indicating a gradual change in deposition.

Graded bedding is found in layers that have an upward decrease in particle size—from gravel to sand, for example. This type of bedding is particularly common in turbidity current deposits that form when sediment-water mixtures flow along the seafloor (or a lake floor) and deposit the largest particles followed by smaller ones as flow velocity diminishes (• Figure 6.4). Graded bedding is also found in some stream channels where flow velocity decreases rapidly.

A very common sedimentary structure is **cross-bedding**, in which layers are deposited at an angle to the surface on which they accumulate, as on the downwind, sloping side of a sand dune (Table 6.1, • Figure 6.5). Cross-bedding results from transport and deposition by wind or water, and the cross-beds are inclined downward or dip in the direction of the current responsible for them. Thus they indicate ancient current directions, or *paleocurrents*.

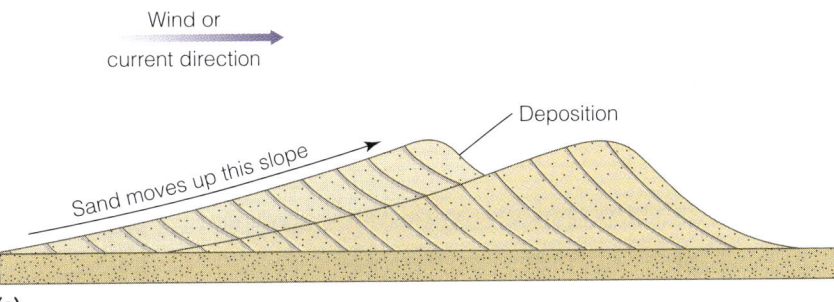

• **Figure 6.5 Cross-Bedding** Cross-bedding takes place when individual beds or strata are deposited at an angle to the surface upon which they accumulate. (a) Origin of cross-bedding by deposition on the sloping surface of a desert dune. Cross-bedding is also common in dune-like structures in stream and river channels. (b) Cross-bedding in sandstone at Natural Bridges National Monument in Utah. Can you determine the current direction from the cross-beds?

SEDIMENTARY ROCK PROPERTIES **105**

• **Figure 6.6 Current Ripple Marks** Current ripple marks are small (<3 cm high) sedimentary structures that have an asymmetrical profile. **(a)** Current ripple marks form where water flows in one direction over sand. The enlargement shows the internal structure of one ripple mark. **(b)** Current ripple marks in a stream channel. Which way did the current flow? **(c)** Ripple marks formed by wind.

• **Figure 6.7 Wave-Formed Ripple Marks** Wave-formed ripple marks are usually symmetrical and, like current ripple marks, are small-scale sedimentary structures found on sand beds. **(a)** Wave-formed ripple marks form where waves move to and fro. **(b)** Wave-formed ripple marks in shallow seawater. **(c)** These are also wave-formed ripple marks, but after one set of ripples formed, the waves approached from a different direction, thereby superimposing another set of ripples on the original ones.

Bedding planes, especially in layers of sand, frequently have small-scale alternating ridges and troughs known as **ripple marks** (Table 6.1). *Current ripple marks* form in response to water or wind currents that flow in one direction, and their asymmetrical profiles allow geologists to determine paleocurrent directions (• Figure 6.6). *Wave-formed ripple marks* result from the to-and-fro motion of waves, and they tend to be symmetrical in profile (• Figure 6.7). Cross-beds and ripple marks, as well as some other sedimentary structures, are useful for figuring out the correct relative ages of deformed sedimentary rocks (see Perspective 6.1).

6.1 Perspective

Determining the Age of Deformed Sedimentary Rocks

In Chapter 5 we emphasized that to accurately decipher the geologic history of an area, geologists must first determine the correct relative ages of the rocks present. Where sedimentary rocks are in their original horizontal position or have been only slightly deformed, it is easy to determine their relative ages. In some areas, though, the rocks have so intensely deformed that they are now overturned. That is, they have been rotated more than 90 degrees from their original position, so their order is the opposite of what we expect from the principle of superposition. In fact, in the nonscientific literature one sometimes hears that these "out-of-order" layers invalidate the geologic column and time scale.

Although intensely deformed sedimentary rocks pose a problem, it is usually one that is easily solved. Geologists use several sedimentary structures to work out these problems, and some fossils are useful as well. Look carefully at Figure 1a, which shows sandstone layers in Lucerne, Switzerland, that are inclined or dip about 50 degrees toward the left. We can infer that the youngest layer shown is the one toward the upper left or the lower right, the latter only if the beds have been overturned. Now look more closely at the layer with the cross-bedding (Figure 1b). This kind of cross-bedding has a sharp angular contact with the overlying beds and a nearly parallel contact with the beds below (see Figure 6.5b). Accordingly, we conclude that the youngest layer present is the one toward the upper left; the rock layers have not been overturned.

Many sedimentary structures work well for determining relative ages in deformed strata, but how is it possible to use fossils for this purpose? After all, many fossils are simply shells or bones that lay on the surface and were subsequently buried. Even so, shells of clams may be deposited consistently with their convex sides up, and oyster

(a)

(b)

Cross-bedding

Figure 1 (a) These sandstone layers in Lucerne, Switzerland, dip about 50 degrees toward the left. Can you tell which is the younger layer—the one toward the upper left or toward the lower right? The cross-bedding indicates that the layer toward the upper left is younger. Notice in (b) that the cross-beds form a sharp, angular contact with the younger rocks above them but they are nearly parallel with the older rocks below. Compare with the side view of the cross-beds in Figure 6.5b.

∽∽ Wave-formed ripples

⋁⋁⋁ Mud cracks

∴∴ Graded bedding

Figure 2 This cross section shows a recumbent fold in which some of the rock layers show various sedimentary structures. Your task is to figure out which rock layer is youngest and to determine whether this is a recumbent anticline or recumbent syncline.

SEDIMENTARY ROCK PROPERTIES **107**

Perspective (continued)

beds may be buried as they existed when they formed a living complex. And even some bulbous structures produced by the activities of blue-green algae tell geologists which layers are younger versus older (see Figure 9.4c).

Finally, let us solve a hypothetical problem in which we not only determine the relative ages of the rock layers, but also figure out whether folded rocks are part of a recumbent syncline or recumbent anticline. Recall that a recumbent fold is one lying on its side, so that one fold limb has been overturned. Study Figure 2 carefully and determine (1) which rock layer is youngest and (2) whether the fold is a recumbent syncline or recumbent anticline. The answers appear in the Review Questions section at the end of the chapter.

We must remember that the principle of superposition itself is quite straightforward, but applying it in actual geologic situations is not always easy, especially when sedimentary rocks have been complexly deformed during episodes of mountain building.

When clay-rich sediments dry, they shrink and crack into polygonal patterns bounded by fractures called **mud cracks** (Table 6.1, • Figure 6.8). The presence of mud cracks indicates that the original sediment was deposited where alternate wetting and drying took place, as along a lakeshore, on a river floodplain, or where mud is exposed at low tide along a seashore.

Biologic processes yield **biogenic sedimentary structures** and include tracks, burrows, and trails (Table 6.1). Recall from Chapter 5 that these biologically produced features are also called *trace fossils* (see Figure 5.13). Extensive burrowing by organisms, a process called **bioturbation,** may alter sediments so thoroughly that other structures are disrupted or even destroyed (• Figure 6.9).

As important as sedimentary structures are in environmental analyses, it is important to know that no single structure is unique to a specific environment. Current ripples, for example, are found in stream channels, tidal channels, and even on the seafloor. Associations of sedimentary structures taken along with other sedimentary rock properties, however, usually allow geologists to make environmental determinations with a high degree of confidence.

Geometry of Sedimentary Rocks

The three-dimensional shape, or geometry, of a sedimentary rock body may be helpful in environmental analyses, but it must be used with caution because the same geometry may be found in more than one environment. Moreover, geometry can be modified by sediment compaction during lithification, and by erosion and deformation. Nevertheless, it is useful when considered in conjunction with other features.

Some of the most extensive sedimentary rocks in the geologic record resulted from marine transgressions and regressions (see Figure 5.9). These rocks commonly cover

(a)

(b)

• **Figure 6.8 Mud Cracks** Mud cracks form in clay-rich sediments when they dry and contract. **(a)** Mud cracks in a present-day environment. **(b)** Ancient mud cracks in Glacier National Park in Montana. Notice that the cracks have been filled in by sediment.

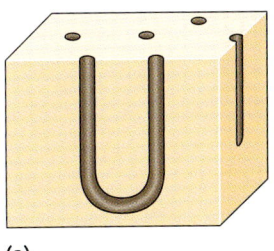

(a)

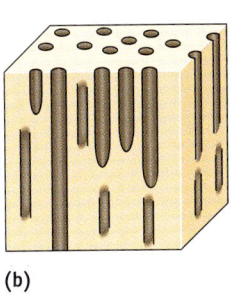

(b)

(c)

• Figure 6.9 **Bioturbation** Bioturbation results from the activities of organisms, especially those that burrow through sediment. **(a)** U-shaped burrows and **(b)** vertical burrows. **(c)** The vertical, dark-colored areas in this rock are sediment-filled burrows. Could you use burrows such as these to determine which layers are youngest in deformed sedimentary rocks? How?

hundreds or thousands of square kilometers but are perhaps only a few tens to hundreds of meters thick. That is, they are not very thick compared to their dimensions of length and width. Thus they are characterized as having a *blanket* or *sheet geometry*.

Some sand deposits have an *elongate* or *shoestring geometry*, especially those deposited in stream channels or barrier islands. Delta deposits tend to be lens shaped when viewed in cross profile or long profile, but lobate when observed from above. Buried reefs are irregular, but many are long and narrow, although rather circular ones are known, too.

Fossils—The Biologic Content of Sedimentary Rocks

We already defined *fossils* as the remains or traces of prehistoric organisms, and we discussed how geologists use fossils in some aspects of stratigraphy—to establish biostratigraphic units, for example (see Chapter 5). However, fossils are also important constituents of some rocks, especially limestones that may be composed largely of shells of marine-dwelling animals such as brachiopods, clams, and corals (see Figure 2.15a), or even the droppings (pellets) of these organisms. Furthermore, much of the sediment on the deep seafloor is made up of the microscopic shells of organisms. Fossils are not present in all sedimentary rocks, but if they are, they are important for determining depositional environments.

We must consider two factors when using fossils in environmental analyses. First, did the organisms in question live where they were buried, or were their remains transported there? Fossil dinosaurs, for example, usually indicate deposition in some land environment such as a river floodplain, but if their bones are found in rocks with clams, corals, and sea lilies, we must assume a carcass was washed out to sea. Second, what kind of habitat did the organisms originally occupy? Studies of a fossil's structure and its living relatives, if any, are helpful. Clams with heavy, thick shells typically live in shallow, turbulent water, whereas those with thin shells are found in low-energy environments. Most corals live in warm, clear, shallow marine environments where symbiotic bacteria can carry out photosynthesis.

Microfossils are particularly useful because geologists can recover many individuals from small rock samples. In oil-drilling operations, small rock chips called *well cuttings* are brought to the surface. These cuttings rarely contain complete fossils of large organisms, but they may have thousands of microfossils that aid in relative age determinations and environmental analyses. Even some trace fossils, which of course are not transported from their place of origin, are also characteristic of particular environments.

Depositional Environments

We defined a **depositional environment** as any area where sediment accumulates, but more specifically it entails a particular area where physical, chemical, and biological processes operate to yield a distinctive kind of deposit. Geologists recognize three broad areas of deposition—*continental*, *transitional*, and *marine*—each of which has several specific environments (• Figure 6.10, Table 6.2).

Continental Environments

Deposition on the continents—that is, on land—may take place in fluvial systems (rivers and streams), lakes, deserts, and areas covered by or adjacent to glaciers. The sedimentary deposits in each of these environments show combinations of features that allow us to determine that they were in fact deposited on land and to distinguish one from the other.

The term **fluvial** refers to river and stream activity and to their deposits. These fluvial deposits accumulate in either **braided stream** systems or **meandering stream** systems. Braided streams have multiple broad, shallow channels in which mostly sheets of gravel and cross-bedded sand are

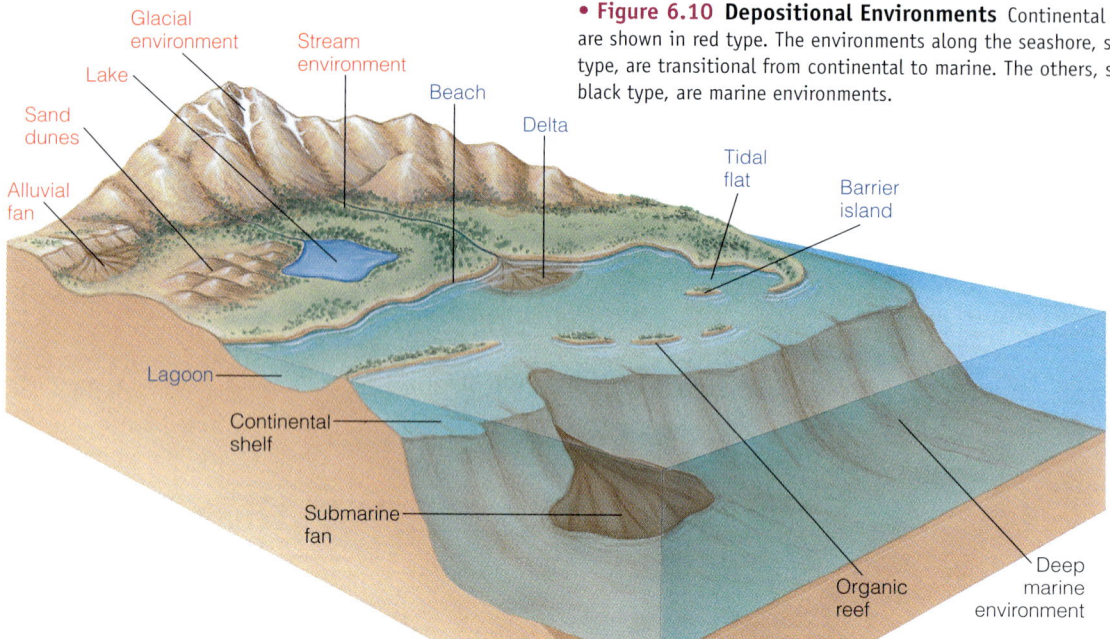

• **Figure 6.10 Depositional Environments** Continental environments are shown in red type. The environments along the seashore, shown in blue type, are transitional from continental to marine. The others, shown in black type, are marine environments.

TABLE 6.2	Depositional Environments Discussed in the Text	
Continental Environments	Fluvial	Braided stream
		Meandering stream
	Desert	Sand dunes
		Alluvial fans
		Playa lakes
	Glacial	Ice deposition (moraines)
		Fluvial deposition (outwash)
		Glacial lakes
Transitional Environments	Delta	Stream dominated
		Wave dominated
		Tide dominated
	Beach	Beach
	Barrier island	Sand dune
		Lagoon
Marine Environments	Continental shelf	Tidal flats
		Detrital deposition
	Carbonate platform	Carbonate deposition
	Continental slope and rise	Submarine fans (turbidities)
	Deep-ocean basin	Oozes
		Pelagic clay
	Evaporite environments*	

*Evaporites may be deposited in a variety of environments including playa lakes, saline lakes, lagoons, and tidal flats in arid regions, and in marine environments.

deposited, and mud is conspicuous by its near absence (• Figure 6.11a and b).

In contrast, meandering streams deposit mostly fine-grained sediments on floodplains but do have cross-bedded sand bodies, each with a shoestring geometry (Figure 6.11c and d). One of the most distinctive features of meandering streams is *point bar* deposits consisting of a sand body overlying an erosion surface that developed on the convex side of a meander loop (Figure 6.11c).

Desert environments are commonly inferred from an association of features found in **sand dune, alluvial fan,** and **playa lake** deposits (• Figure 6.12). Wind-blown dunes are typically composed of well-sorted, well-rounded sand with cross-beds meters to tens of meters high (Figure 6.12b and c). And, of course, any fossils are those of land-dwelling plants and animals. Alluvial fans form best along the margins of desert basins where streams and debris flows discharge from mountains onto a valley floor and form a triangular deposit of sand and gravel (Figure 6.12a). The more central part of a desert basin might be the site of a temporary lake (playa lake), in which laminated mud and evaporites accumulate.

What Would You Do?

No one was present millions of years ago to record data about the climate, geography, and geologic processes. So how is it possible to decipher unobserved past events? In other words, what features in rocks, and especially in sedimentary rocks, would you look for to determine what happened in the far distant past? Can you think of any economic reasons to decipher Earth history from the record preserved in rocks?

110 CHAPTER 6 SEDIMENTARY ROCKS—THE ARCHIVES OF EARTH HISTORY

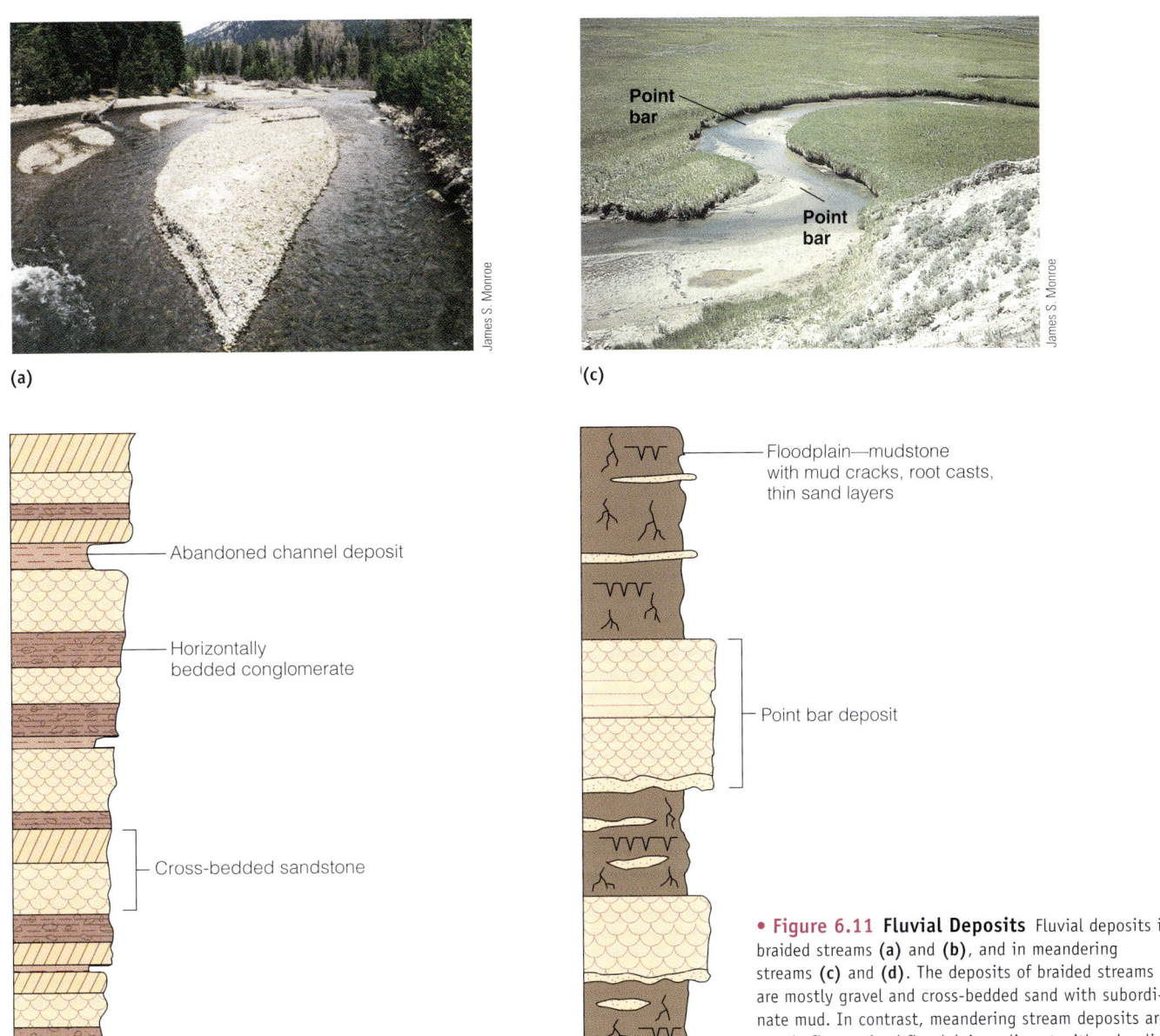

• **Figure 6.11 Fluvial Deposits** Fluvial deposits in braided streams **(a)** and **(b)**, and in meandering streams **(c)** and **(d)**. The deposits of braided streams are mostly gravel and cross-bedded sand with subordinate mud. In contrast, meandering stream deposits are mostly fine-grained floodplain sediment with subordinate sand bodies.

All sediments deposited in glacial environments are collectively called **drift,** but we must distinguish between two kinds of drift. **Till** is poorly sorted, nonstratified drift deposited directly by glacial ice, mostly in ridgelike deposits called *moraines* (• Figure 6.13). A second type of drift, called **outwash,** is sand and gravel deposited by braided streams issuing from melting glaciers. The association of these deposits along with scratched (striated) and polished bedrock generally indicates that glaciers were responsible for deposition. Another feature that may be helpful is glacial lake deposits showing alternating dark and light laminations; each dark–light couplet is a **varve** (Figure 6.13d). Varves represent yearly accumulations of sediment—the light layers form in spring and summer and consist of silt and clay, whereas the dark layers consist of clay and organic matter that settled when the lake froze over. Dropstones liberated from icebergs might also be present in glacial lake deposits (Figure 6.13d).

Transitional Environments

Transitional environments include those in which both marine processes and processes typical of continental environments operate (Figure 6.10). For instance, deposition where a river or stream (fluvial system) enters the sea yields a body of sediment called a **delta,** but the deposit is modified by marine processes, especially waves and tides. Small deltas in

• Figure 6.12 **Deposits in Deserts** (a) This view in Death Valley, California, shows huge alluvial fans along the base of the Panamint Mountains. (b) Sand dunes also in Death Valley. An association of the deposits in (a) and (b) is found in desert depositional environments. (c) Large-scale cross-beds in a Permian-aged wind-blown dune deposit in Arizona.

lakes are also common, but the ones along seashores are much larger and far more complex. Furthermore, many deltas along seashores are important economically because they form the reservoir rocks for petroleum and natural gas.

The simplest deltas, those in lakes, have a threefold subdivision of bottomset, foreset, and topset beds much like the one shown in • Figure 6.14. As delta sedimentation takes place, bottomset beds are deposited some distance offshore, over which foreset beds come to rest. And finally deposition also takes place in stream channels forming the topset beds. As a result, the delta builds seaward (or lakeward), or **progrades,** and forms a vertical sequence of rocks that become coarser-grained from bottom to top. In addition, the bottomset beds may contain marine fossils (or fossils of lake-dwelling animals), whereas the topset beds usually have fossils of land plants or animals.

Marine deltas rarely conform precisely to this simple threefold division, because they are strongly influenced by fluvial processes, as well as by waves and tides. Thus a stream/river–dominated delta has long distributary channels extending far seaward, as in the Mississippi River delta (• Figure 6.15a). Wave-dominated deltas also have distributary channels, but as their entire seaward margin progrades, it is simultaneously modified by wave action (Figure 6.15b). Tidal sand bodies that parallel the direction of tidal flow are characteristic of tide-dominated deltas (Figure 6.15c).

On broad continental margins with abundant sand, long **barrier islands** lie offshore separated from the mainland by a lagoon. These islands are common along the Gulf Coast, and much of the Atlantic Coast of the United States and many ancient deposits formed in this environment. Notice from • Figure 6.16a that a barrier island complex has subenvironments in which beach sands grade offshore into finer deposits of the shoreface, dune sands that are differentiated from desert dune sands by the presence of shell fragments, and fine-grained lagoon deposits with marine fossils and bioturbation.

Tidal flats are present along many coastlines where part of the shoreline environment is periodically covered by seawater at high tide and then exposed at low tide. Many

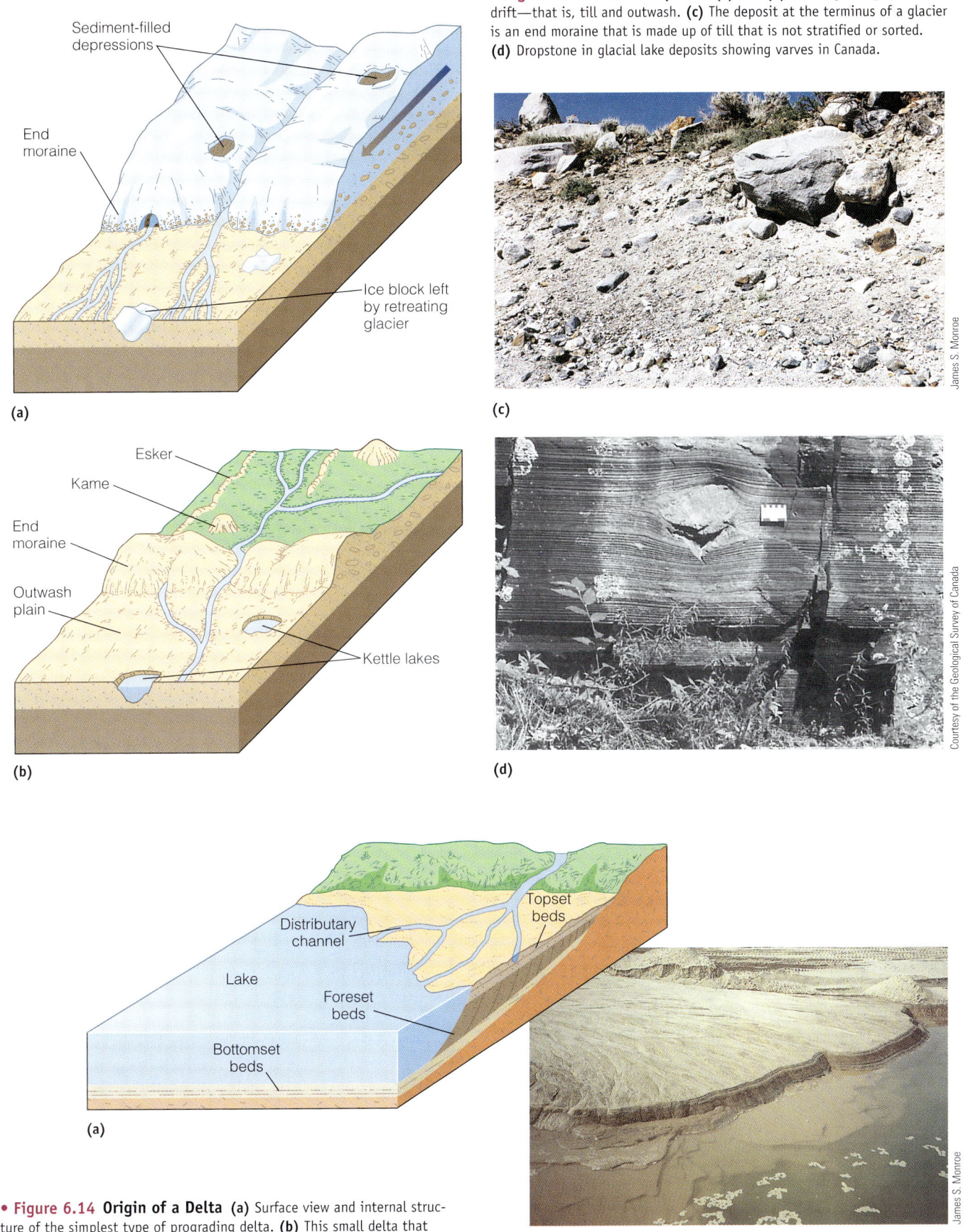

• **Figure 6.13 Glacial Deposits** (a) and (b) The origin of glacial drift—that is, till and outwash. (c) The deposit at the terminus of a glacier is an end moraine that is made up of till that is not stratified or sorted. (d) Dropstone in glacial lake deposits showing varves in Canada.

• **Figure 6.14 Origin of a Delta** (a) Surface view and internal structure of the simplest type of prograding delta. (b) This small delta that measures about 20 m across is prograding into a small lake.

DEPOSITIONAL ENVIRONMENTS 113

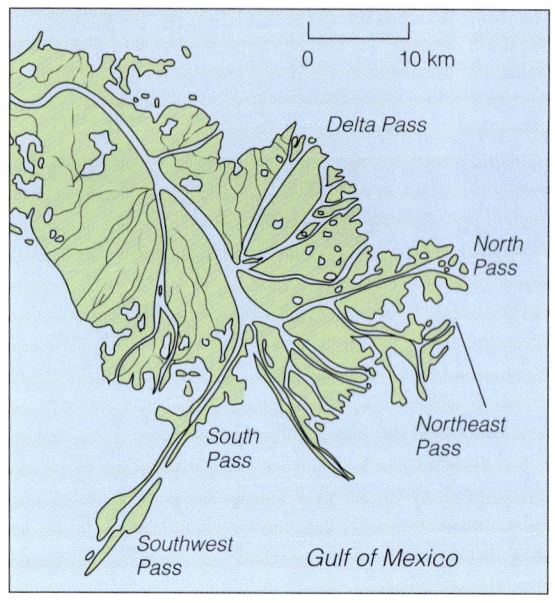
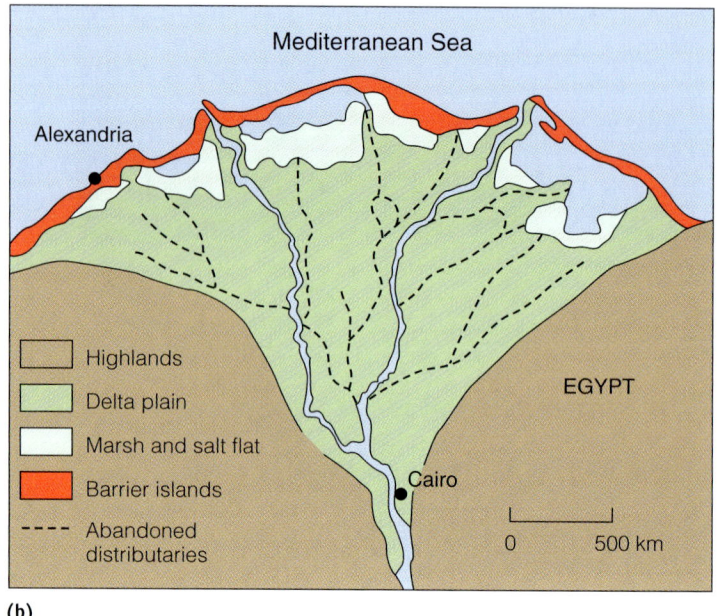
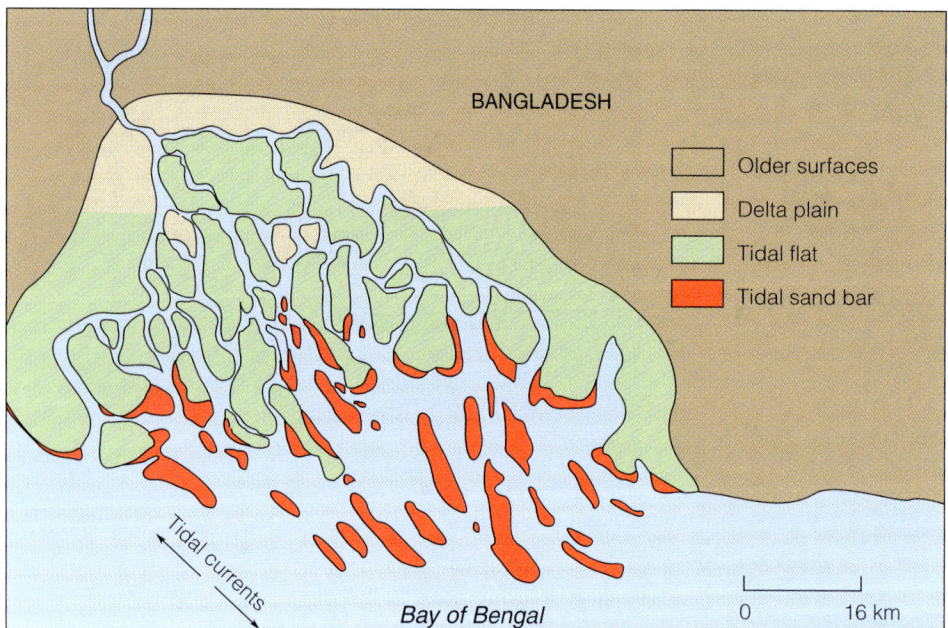

• **Figure 6.15 Types of Deltas**
(a) The Mississippi River delta of the U.S. Gulf Coast is stream dominated, and (b) the Nile delta of Egypt is wave dominated. (c) The Ganges-Brahmaputra delta of Bangladesh is tide dominated.

tidal flats build or prograde seaward and yield a sequence of rocks grading upward from sand to mud. One of their most distinctive features is herringbone cross-bedding—sets of cross-beds that dip in opposite directions (Figure 6.16b and c).

Marine Environments

Marine environments include depositional settings on the continental shelf, slope, and rise, and the deep seafloor (• Figure 6.17). Much of the detritus eroded from continents is eventually deposited in marine environments, but other types of sediments are found here, too. Much of the limestone in the geologic record, as well as many evaporites, were deposited in shallow marine environments.

Detrital Marine Environments

The gently sloping area adjacent to a continent, the **continental shelf**, consists of two parts. The high-energy inner part is periodically stirred up by waves and tidal currents. Its sediment is mostly sand shaped into large cross-bedded dunes, and bedding planes are commonly marked by wave-formed ripple marks. Of course, marine fossils are typical, and so is bioturbation. The low-energy outer part of the shelf has mostly mud with marine fossils, but at the transition between the inner and outer shelf, layers of sand and mud intertongue.

Much of the sediment derived from continents crosses the continental shelf and is funneled into deeper water through submarine canyons. It eventually comes to rest on the **continental slope** and **continental rise** as a series of overlapping submarine fans (Figure 6.17). Once sediment

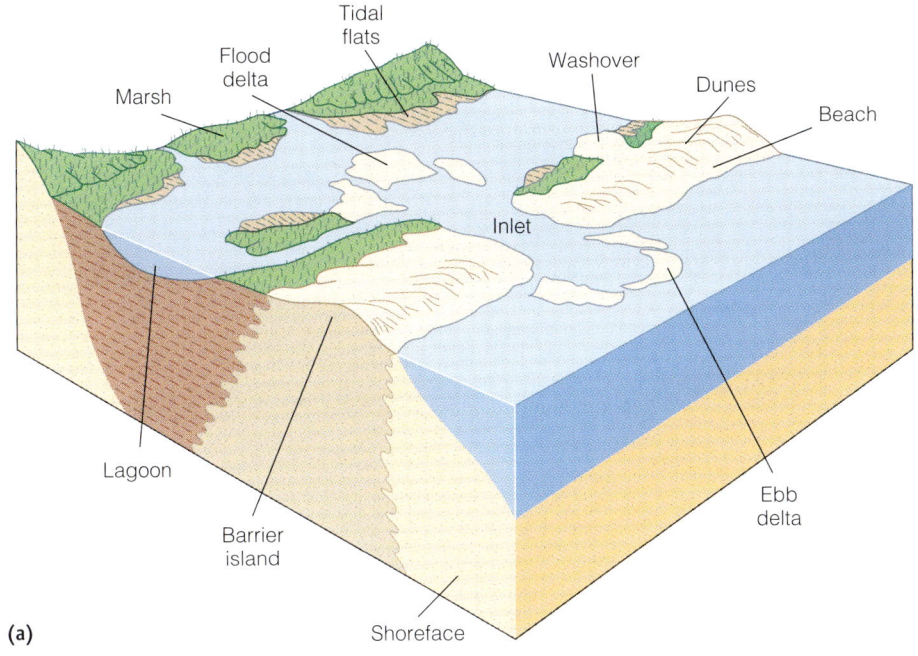

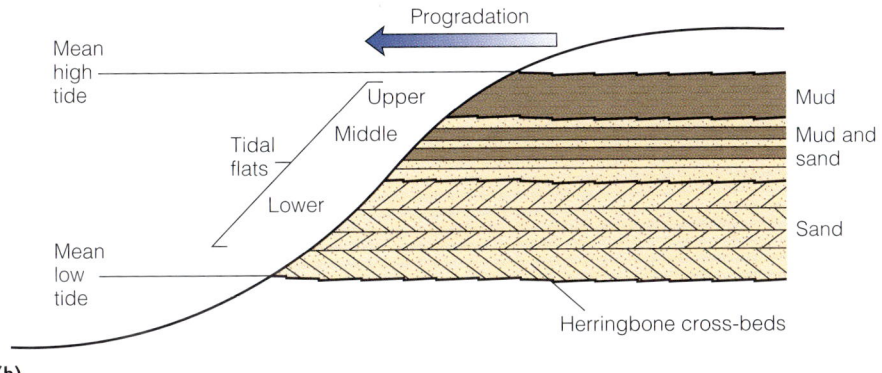

• **Figure 6.16 Barrier Island Complex** (a) A barrier island is made up mostly of sand, but it has subenvironments where mud or mixtures of mud and sand are deposited. (b) These tidal flat deposits formed where a shoreline prograded into the sea. Notice the distinctive herringbone cross-beds that dip in opposite directions because of reversing tidal currents. (c) Herringbone cross-beds in the Loyalhanna Formation in Pennsylvania.

passes the outer margin of the shelf, turbidity currents transport it, so sands with graded bedding are common—but so is mud that settled from seawater (Figure 6.4).

Beyond the continental shelf, the seafloor is nearly covered by fine-grained deposits known as *pelagic clay* and *ooze* (• Figure 6.18). The notable exceptions are the mid-oceanic ridges, which are areas where new oceanic crust forms. In fact, one prediction of plate tectonic theory is that the thickness of seafloor sediments should increase with increasing distance from spreading ridges (see Chapter 3). Sand and gravel are notable by their absence on the deep seafloor, because there are few mechanisms that can effectively transport coarse-grained sediment far from land. However, ice rafting accounts for a band of sand and gravel adjacent to Greenland and Antarctica.

The main sources of deep-sea sediments are dust blown from continents or oceanic islands, volcanic ash, and shells of microorganisms that dwelled in the ocean's surface waters. One type of sediment called pelagic clay consists of clay covering most of the deeper parts of the seafloor. Calcareous ($CaCO_3$) and siliceous (SiO_2) oozes, in contrast, are made up of microscopic shells.

Carbonate Depositional Environments Various types of limestone (composed of calcite) and dolostone (composed of dolomite) are the only widespread carbonate rocks, and we already know that most dolostone is altered limestone. Thus our discussion focuses on the deposition of sediment that when lithified becomes limestone.

In some respects limestone is similar to detrital sedimentary rocks—many limestones are made up of gravel- and sand-sized grains, and microcrystalline carbonate mud called **micrite**. However, these materials form in the environment of deposition rather than being transported there. Nevertheless, local currents and waves sort them and account for various sedimentary structures such as cross-bedding and ripple marks, and micrite along shorelines commonly shows mud cracks.

Some limestone forms in lakes, but by far most of it was deposited in warm, shallow seas on carbonate shelves

DEPOSITIONAL ENVIRONMENTS **115**

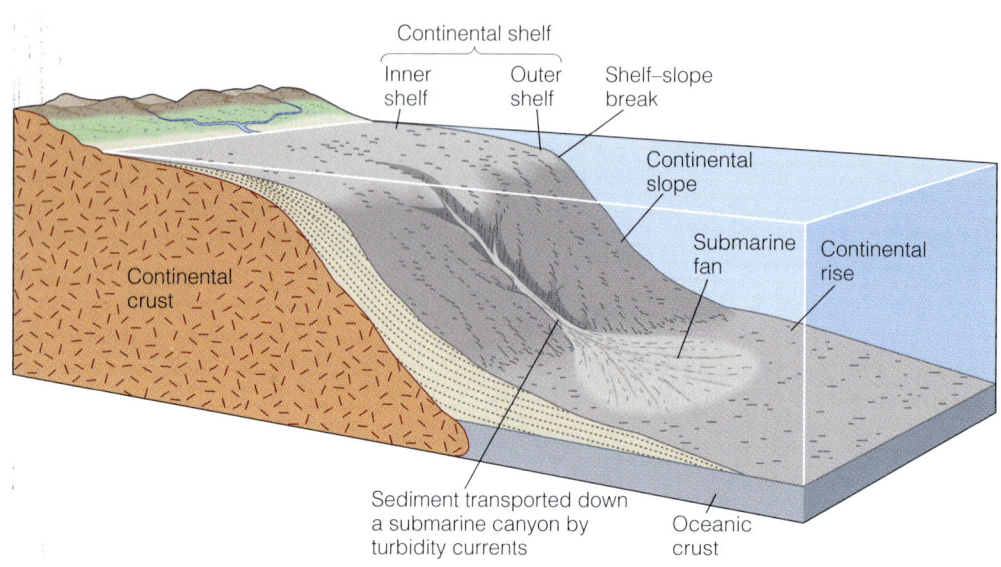

• Figure 6.17 **Depositional Environments on the Continental Shelf, Slope, and Rise** Submarine canyons are the main avenues of sediment transport across the shelf. Turbidity currents carry sediment beyond the shelf, where they accumulate as submarine fans.

or on carbonate platforms rising from oceanic depths (• Figure 6.19). In either case, deposition occurs where little detrital sediment is present, especially mud. Carbonate barriers form in high-energy areas and may be reefs, banks of skeletal particles, or accumulations of spherical carbonate grains known as *ooids*—limestone composed mostly of ooids is called *oolitic limestone*. Reef rock tends to be structureless, composed of skeletons of corals, mollusks, sponges, and many other organisms, whereas carbonate banks are made up of layers with horizontal beds, cross-beds, and wave-formed ripple marks. The lagoons, however, tend to have micrite with marine fossils and bioturbation.

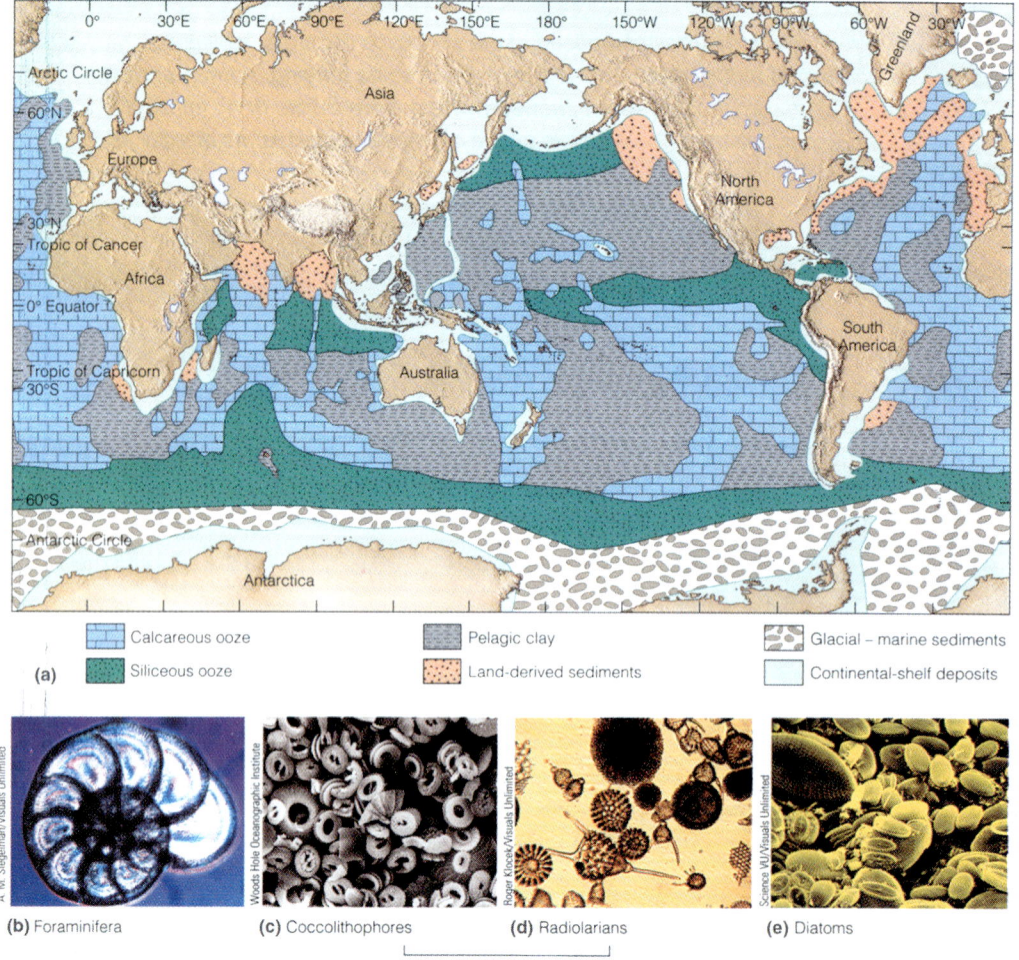

• Figure 6.18 **Deep-Seafloor Sediments** (a) Most of the seafloor is covered by pelagic clay and calcareous and siliceous ooze. Calcareous ooze is made up of skeletons of (b) foraminifera (floating single-celled animals) and (c) coccolithophores (floating single-celled plants), whereas siliceous ooze is composed of (d) radiolarians (single-celled floating animals) and (e) diatoms (single-celled floating plants).

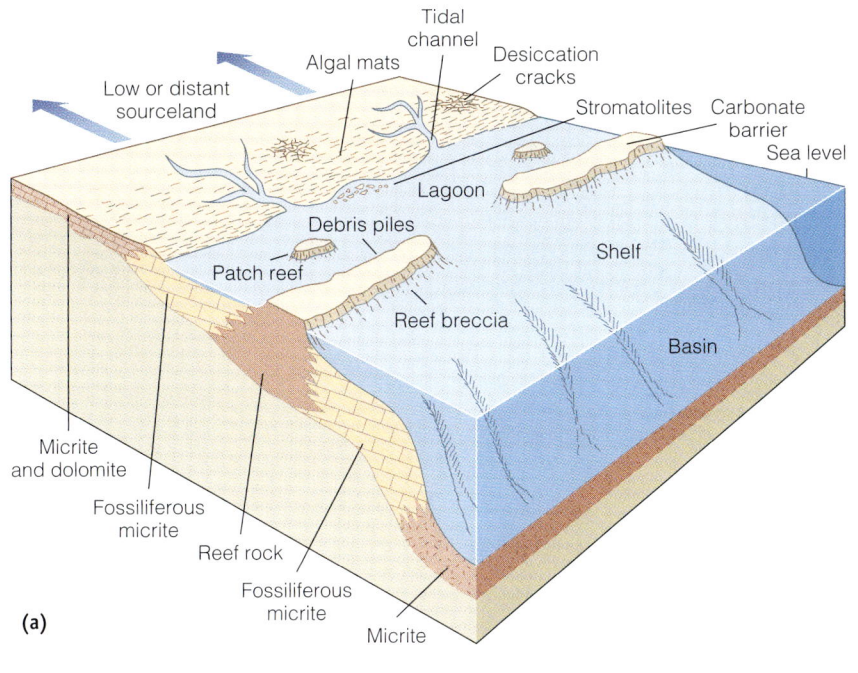

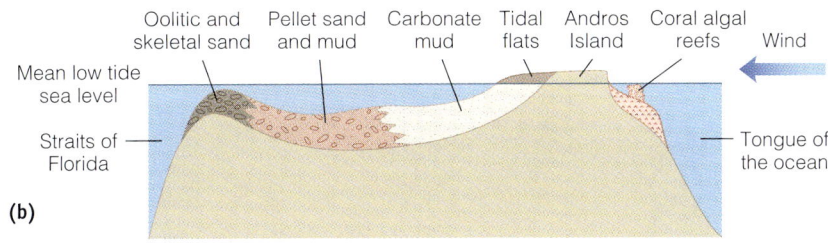

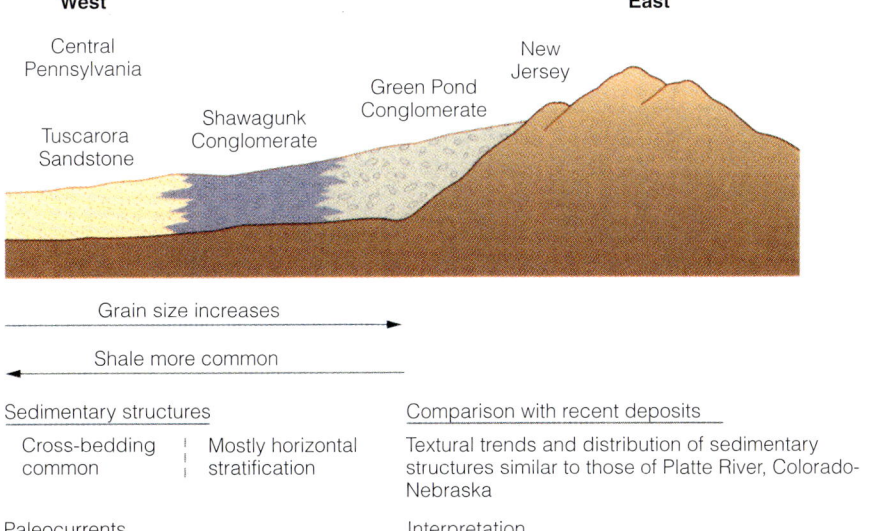

• **Figure 6.20 Lower Silurian Strata in the Eastern United States** Simplified cross section showing the probable lateral relationships for the Green Pond Conglomerate, Shawangunk Conglomerate, and Tuscarora Sandstone and the criteria used to interpret these rocks as braided stream deposits.

• **Figure 6.19 Carbonate Depositional Environments** (a) Deposition on a shelf attached to a continent is now taking place in southern Florida and the Persian Gulf. (b) The Great Bahama Bank in the Atlantic Ocean is an isolated carbonate shelf that rises from oceanic depths.

Evaporite Environments Evaporites consisting mostly of rock salt and rock gypsum* are found in a variety of environments, including playa lakes and saline lakes, but most of the extensive deposits formed in the seas. Although locally abundant, evaporites are not nearly as common as sandstone, mudrocks, and limestone. Deposits more than 2 km thick lie beneath the floor of the Mediterranean Sea, and large deposits are also known in Michigan, Ohio, and New York as well as in several Gulf Coast states and western Canada.

Exactly how some of these deposits originated is controversial, but all agree they formed where evaporation rates were high enough for seawater to become sufficiently concentrated for minerals to precipitate from solution. These conditions are best met in coastal environments in arid regions such as the present-day Persian Gulf. If the area shown in Figure 6.16a were an arid region, inflow of normal seawater into the lagoon would be restricted, and evaporation might lead to increased salinity and perhaps evaporite deposition.

Interpreting Depositional Environments

We began this chapter by stating that the Navajo Sandstone of the southwestern United States was deposited as a vast blanket of sand dunes (Figure 6.1),

---

*Gypsum ($CaSO_4 \cdot 2H_2O$) is the common sulfate mineral precipitated from seawater, but when deeply buried, gypsum loses its water and is converted to anhydrite ($CaSO_4$).

INTERPRETING DEPOSITIONAL ENVIRONMENTS   **117**

but couldn't it just as well have been deposited along the seashore or in river channels? Studies show that the sandstone is made up mostly of well-sorted, well-rounded quartz grains measuring 0.2 to 0.5 mm in diameter. In addition, the rocks have tracks of land-dwelling animals, including dinosaurs, excluding the possibility of a marine origin, and the sandstone has cross-beds up to 30 m high and current ripple marks like those produced by wind. All the evidence taken together justifies our interpretation of a wind-blown dune deposit. In fact, the cross-beds dip generally southwest, so we can confidently say that the wind blew mostly from the northeast.

Another good example of using the combined features of sedimentary rocks and comparisons with present-day deposits is Lower Silurian strata exposed in New Jersey and Pennsylvania. The features of these rocks are summarized in • Figure 6.20, all of which lead to the conclusion that they were deposited in braided streams that flowed generally from east to west. The preceding examples were taken from continental depositional environments, but the same reasoning is used for marine deposits. For example, geologists have evaluated vertical facies relationships, rock types, sedimentary structures, and fossils in Ordovician rocks in Arkansas and conclude that they formed as transgressive shelf carbonate deposits (• Figure 6.21).

In several of the following chapters we refer to fluvial deposits, ancient deltas, carbonate shelf deposits, and others as we examine the geologic record, especially for North America. We cannot include the supporting evidence for all of these interpretations, but we can say that they are based on the kinds of criteria discussed in this chapter.

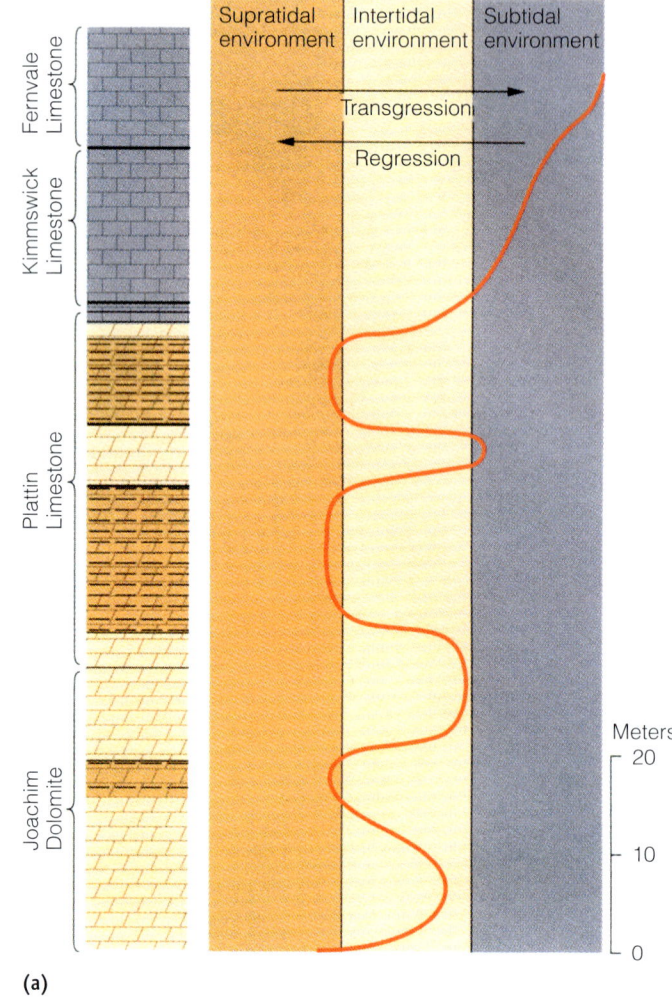

(a)

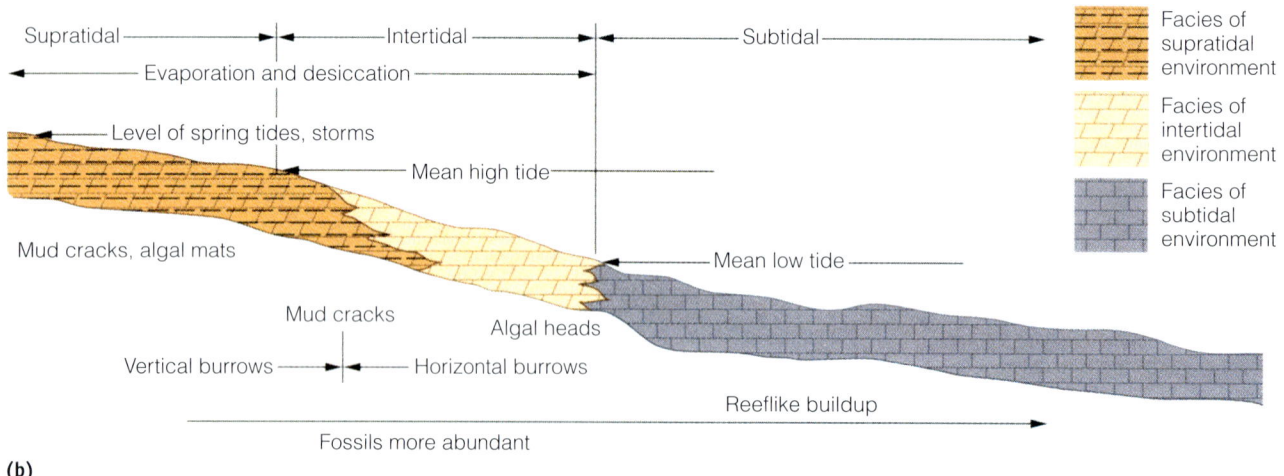

(b)

• **Figure 6.21 Middle and Upper Ordovician Strata in Northern Arkansas** (a) Vertical stratigraphic relationships and inferred environments of deposition. Notice that the trend was transgressive even though several minor regressions also occurred. (b) Simplified cross section showing lateral facies relationships.

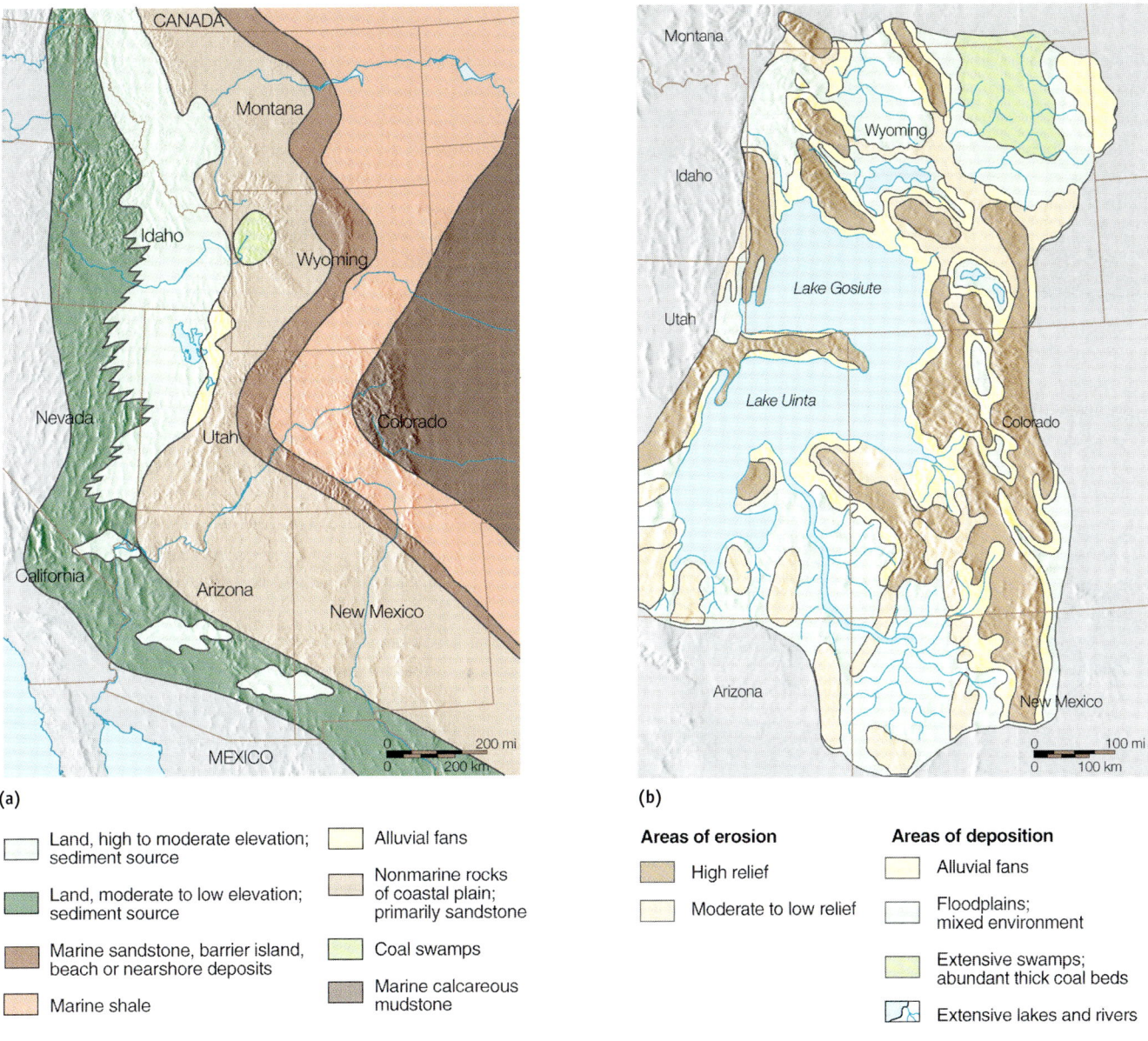

• **Figure 6.22 Paleogeographic Maps** Maps showing the **(a)** Late Cretaceous and **(b)** Eocene paleogeography of several western states. Notice that marine deposition took place over much of the area during the Late Cretaceous, but by Eocene time all deposition was in continental environments.

Paleogeography

The Navajo Sandstone is a wind-blown desert dune deposit, so we know that during the Jurassic Period a vast desert was present in what is now the southwestern United States (Figure 6.1). We also know that a marine transgression took place from the Late Precambrian to Middle Cambrian in North America as the shoreline migrated inland from east and west (see Figure 5.8). In short, Earth's geography—that is, the distribution of its various surface features—has varied through time.

Paleogeography deals with Earth's geography of the past. In the chapters on geologic history, paleogeographic maps show the distribution of continents at various times; they have not remained fixed because of plate movements. The distribution of land and sea constitutes Earth's first-order features, but paleogeography also applies on a more local scale. For example, detailed studies of rocks in several western states allow us to determine with some accuracy how the area appeared during the Late Cretaceous (• Figure 6.22a). According to this paleogeographic map, a broad coastal plain sloped gently eastward from a mountainous region to the sea. The rocks also tell us that many animals lived here, including dinosaurs and mammals. Later, however, vast lakes, river floodplains, and alluvial fans covered much of this area, and the sea had withdrawn from the continent (Figure 6.22b).

SUMMARY

- The physical and biologic features of sedimentary rocks provide information about the depositional processes responsible for them.
- Sedimentary structures and fossils are the most useful attributes of sedimentary rocks for environmental analyses, but textures and rock body geometry are also helpful.
- Geologists recognize three primary depositional areas—continental, transitional, and marine—each of which has several specific environments.
- Fluvial systems may be braided or meandering. Braided streams deposit mostly sand and gravel, whereas deposits of meandering streams are mostly mud and subordinate sand bodies with shoestring geometry.
- An association of alluvial fan, sand dune, and playa lake deposits is typical of desert depositional environments. Glacial deposits consist mostly of till in moraines and outwash.
- The simplest deltas, those in lakes, consist of a three-part sequence of rocks, grading from finest at the base upward to coarser-grained rocks. Marine deltas dominated by fluvial processes, waves, or tides are much larger and more complex.
- A barrier island system includes beach, dune, and lagoon subenvironments, each characterized by a unique association of rocks, sedimentary structures, and fossils.
- Inner shelf deposits are mostly sand, whereas those of the outer shelf are mostly mud; both have marine fossils and bioturbation. Much of the sediment from land crosses the shelves and is deposited on the continental slope and rise as submarine fans.
- Either pelagic clay or oozes derived from the shells of microscopic floating organisms cover most of the deep seafloor.
- Most limestone originates in shallow, warm seas where little detrital mud is present. Just like detrital rocks, carbonate rocks may possess cross-beds, ripple marks, mud cracks, and fossils that provide information about depositional processes.
- Evaporites form in several environments, but the most extensive ones were deposited in marine environments. In all cases, they formed in arid regions with high evaporation rates.
- With information from sedimentary rocks, as well as other rocks, geologists determine the past distribution of Earth's surface features.

IMPORTANT TERMS

alluvial fan, p. 110
barrier island, p. 112
bedding (stratification), p. 104
biogenic sedimentary structure, p. 108
bioturbation, p. 108
braided stream, p. 109
continental rise, p. 114
continental shelf, p. 114
continental slope, p. 114
cross-bedding, p. 105

delta, p. 111
depositioned environment, p. 109
drift, p. 111
fluvial, p. 109
graded bedding, p. 105
meandering stream, p. 109
micrite, p. 115
mud crack, p. 108
outwash, p. 111
paleogeography, p. 119

playa lake, p. 110
progradation, p. 112
ripple mark, p. 106
rounding, p. 103
sand dune, p. 110
sedimentary structure, p. 104
sorting, p. 103
tidal flat, p. 112
till, p. 111
varve, p. 111

REVIEW QUESTIONS

1. Braided stream deposits are made up mostly of
   a. ____ mud and point bars; b ____ sheets of gravel and cross-bedded sand; c. ____ silt, clay, and evaporates; d. ____ poorly sorted, nonstratified ridges of gravel; e. ____ coal with interbedded limestone.

2. An upward decrease in grain size in a single bed is known as
   a. ____ bioturbation; b. ____ inverted bedding; c. ____ cross-bedding; d. ____ graded bedding; e. ____ laminations.

3. The phenomenon of a delta building out into the sea is called
   a. ____ progradation; b. ____ assimilation; c. ____ recapitulation; d. ____ sorting; e. ____ bioturbation.

4. A deposit made up of laminated mud and evaporates probably formed
   a. ____ on the deep seafloor; b. ____ on a river floodplain; c. ____ in a playa lake; d. ____ where a river discharged into the sea; e. ____ on the convex side of a meander loop.

5. Which one of the following statements is correct?
   a. ____ Rounding refers to how nearly spherical sedimentary particles are; b. ____ The sand in desert dunes is typically poorly sorted; c. ____ The deep seafloor is covered by vast sheets of sand and gravel; d. ____ Cross-beds are good indicators of paleocurrent directions; e. ____ Micrite is a term for limestone made up of broken shells.

6. One of the most distinctive features of meandering stream deposits is
   a. ____ pelagic clay; b. ____ dark and light layers of mud with droptones; c. ____ bottomset, foreset, and topset beds; d. ____ ooids and micrite; e. ____ point bar deposits.
7. Cross-beds dipping in opposite directions are found in _____ environments.
   a. ____ braided stream; b. ____ submarine fan; c. ____ tidal flat; d. ____ glacial outwash; e. ____ alluvial fan.
8. Spherical to oval carbonate grains that form around a nucleus are called
   a. ____ ooids; b. ____ ripple marks; c. ____ ooze; d. ____ fluvial; e. ____ varves.
9. A well-sorted detrital sedimentary rock is one in which
   a. ____ several different minerals are present; b. ____ all the grains are about the same size; c. ____ micrite and shell fragments are present in equal amounts; d. ____ at least 50% of the grains are more than 2 mm across; e. ____ magnesium replaced calcium.
10. Turbidity currents account for much of the deposition on
    a. ____ glacial moraines; b. ____ river floodplains; c. ____ carbonate shelves; d. ____ submarine fans; e. ____ abandoned channels.
11. What are the most common deposits on the deep seafloor and why are they not found at mid-oceanic ridges?
12. Describe the vertical sequence of sedimentary rocks that form as a result of deposition on the simplest type of prograding delta.
13. Why are sets of cross-beds dipping in opposite directions found on tidal flats?
14. How and why do the detrital deposits on outer and inner continental shelves differ?
15. What sedimentary rocks and sedimentary structures can you use to differentiate between braided and meandering stream deposits?
16. Diagram and explain how wave-formed ripple marks and current ripple marks form. Can both be used to determine paleocurrents?
17. What distinctive features would you expect to find in sedimentary rocks that were deposited in desert dunes?
18. How do textures and sedimentary structures in detrital sedimentary rocks compare with those in carbonate rocks?
19. Under what conditions are evaporites deposited, and what are the common evaporite rocks?

Answer to question in Perspectives 6.1:
The sedimentary structures such as graded bedding indicate that the sandstone layer in the center of the fold is the youngest. Accordingly, this is a recumbent syncline.

APPLY YOUR KNOWLEDGE

1. How can you account for a vertical sequence of rocks made up of oolitic limestone overlain by micrite with marine fossils and bioturbation and finally by laminated micrite with mud cracks?

FIELD QUESTIONS

1. In what depositional environment was the sandstone in the diagram below deposited? How do you know? Also, refer to Appendix B, and see if you can figure out when the sandstone was deposited.

2. The image below shows deformed sedimentary rocks in Switzerland. What features would you look for in these rocks to determine which layer is oldest and which is youngest?

CHAPTER 7

EVOLUTION—THE THEORY AND ITS SUPPORTING EVIDENCE

Sue Monroe

Some of the evidence for evolution comes from fossils such as this example of a 2.3-meter-long specimen of *Ceresiosaurus* which belonged to a group of Triassic marine reptiles known as nothosaurs. The smaller skeletons on the slab are representatives of the nothosaur genus *Pachypleurosaurus* which was hunted by *Ceresiosaurus*. These fossils are on display at the Glacier Garden in Lucerne, Switzerland.

[OUTLINE]

INTRODUCTION

EVOLUTION: WHAT DOES IT MEAN?

Jean-Baptiste de Lamarck and His Ideas on Evolution

The Contributions of Charles Darwin and Alfred Wallace

PERSPECTIVE 7.1 *The Tragic Lysenko Affair*

Natural Selection—What Is Its Significance?

MENDEL AND THE BIRTH OF GENETICS

Mendel's Experiments

Genes and Chromosomes

THE MODERN VIEW OF EVOLUTION

What Brings About Variation?

Speciation and the Rate of Evolution

Divergent, Convergent, and Parallel Evolution

Microevolution and Macroevolution

Cladistics and Cladograms

Evolutionary Trends and Mosaic Evolution

Extinctions

WHAT KINDS OF EVIDENCE SUPPORT EVOLUTIONARY THEORY?

Classification—A Nested Pattern of Similarities

How Does Biological Evidence Support Evolution?

Fossils: What Do We Learn from Them?

PERSPECTIVE 7.2 *The Fossil Record and Missing Links*

The Evidence—A Summary

SUMMARY

ThomsonNOW™ Explore interactive tutorials, animations, or practice problems available on the ThomsonNow website at **www.thomsonedu.com/login**.

{ CHAPTER OBJECTIVES }

At the end of this chapter, you will have learned that

- The central claim of the theory of evolution is that today's organisms descended, with modification, from ancestors that lived in the past.

- Jean-Baptiste de Lamarck in 1809 proposed the first widely accepted mechanism—inheritance of acquired characteristics—to account for evolution.

- In 1859, Charles Darwin and Alfred Wallace simultaneously published their views on evolution and proposed natural selection as a mechanism to bring about change.

- Experiments carried out by Gregor Mendel during the 1860s demonstrated that variations in populations are maintained rather than blending during inheritance, as previously thought.

- In the modern view of evolution, variation is accounted for mostly by sexual reproduction and by mutations in sex cells.

- The fossil record provides many examples of macroevolution—that is, changes that result in the origin of new species, genera, and so on—but these changes are simply the cumulative effect of microevolution, which involves changes within a species.

- A number of evolutionary trends, such as size increase or changing configuration of shells, teeth, or limbs, are well known for organisms for which sufficient fossil material is available.

- The theory of evolution is truly scientific, because we can think of observations or experiments that would support it or render it incorrect.

- Fossils are important as evidence for evolution, but additional evidence comes from classification, biochemistry, molecular biology, genetics, and geographic distribution.

Introduction

A rugged group of 13 large islands, 8 smaller ones, and 40 islets, all belonging to Ecuador, lies in the Pacific Ocean about 1000 km west of South America. Officially called the Archipelago de Colon after Christopher Columbus, the group is better known as the Galápagos Islands (• Figure 7.1). All are composed of basalt, much of it from rather recent volcanic eruptions, giving them their distinctive gray color.

In 1831, Charles Robert Darwin (1809–1882) (• Figure 7.2) made a five-year voyage as an unpaid naturalist aboard the British research vessel HMS *Beagle*. In the Galápagos Islands, he made several observations that changed his ideas about the then widely held concept called *fixity of species*. According to this idea, all present-day species had been created in their present form and had changed little or not at all.

Darwin began his voyage not long after graduating from Christ's College of Cambridge University with a degree in theology, and although he was rather indifferent to religion, he fully accepted the biblical account of creation. During the voyage, though, his ideas began to change. For example, some of the fossil mammals he collected in South America were similar to present-day llamas, sloths, and armadillos, but they had interesting differences. He realized that these animals had descended with modification from ancestral species, and so he began to question the idea of fixity of species.

The finches and giant tortoises living on the Galápagos Islands particularly fascinated Darwin. He postulated that the 13 species of finches had evolved from a common ancestor species that somehow reached the islands as an accidental immigrant from South America. Indeed, their ancestor was very likely a single species resembling the blue-back grassquit finch now living along South America's Pacific Coast. The islands' scarcity of food accounts for the ancestral species evolving different physical characteristics, especially beak shape, to make a living (• Figure 7.3). And likewise, the tortoises, which vary from island to island, still resemble the tortoises that live along South America's Pacific coast, even though they differ in subtle ways.

Charles Darwin became convinced that organisms descended with modification from ancestors that lived during the past, which is the central claim of the **theory of evolution**. So why should you study evolution? For one thing, evolution that involves inheritable changes in organisms through time is fundamental to biology and **paleontology**, the study of life history as revealed by fossils. Furthermore, like plate tectonic theory, it is a unifying theory that explains an otherwise encyclopedic collection of facts. And finally, it serves as the framework for discussions of life history in the following chapters.

Unfortunately, many people have a poor understanding of the theory of evolution and hold a number of misconceptions. For example, many think that (1) evolution proceeds strictly by chance; (2) nothing less than structures in their present form, such as eyes, are of any use; (3) no transitional fossils (so-called missing links) connect ancestors and descendants; and (4) humans evolved from monkeys, so monkeys should no longer exist. We address these and

• **Figure 7.1 The Galápagos Islands** The Galápagos Islands are specks of land composed of basalt in the Pacific Ocean west of Ecuador. The red line in the map shows the route followed by Charles Darwin when he was aboard HMS *Beagle* from 1831 to 1836.

• **Figure 7.2 Charles Robert Darwin in 1840** During his five year voyage aboard the *Beagle* Darwin collected fossils and made observations that would later be instrumental in his ideas about evolution.

other misconceptions in this and in some of the later chapters on life history.

Evolution: What Does It Mean?

Evolution is the process whereby organisms have changed since life originated—that is, they have descended with modification from ancestors that lived during the past. It is important to note that the theory of evolution does not address how life originated, only how it has changed and diversified through time. There are indeed theories that attempt to explain life's origin (see Chapter 8), but they must be evaluated on their own merits.

The idea that organisms have changed—that is, evolved—is commonly attributed solely to Charles Darwin, but it was seriously considered long before he was born, even by some ancient Greeks and by philosophers and theologians during the Middle Ages (A.D. 476–1453). Nevertheless, the prevailing belief among Europeans well into the 1700s was that the works of Aristotle (384–322 B.C.) and the first two chapters of the book of Genesis contained all-important knowledge. Literally

124 CHAPTER 7 EVOLUTION—THE THEORY AND ITS SUPPORTING EVIDENCE

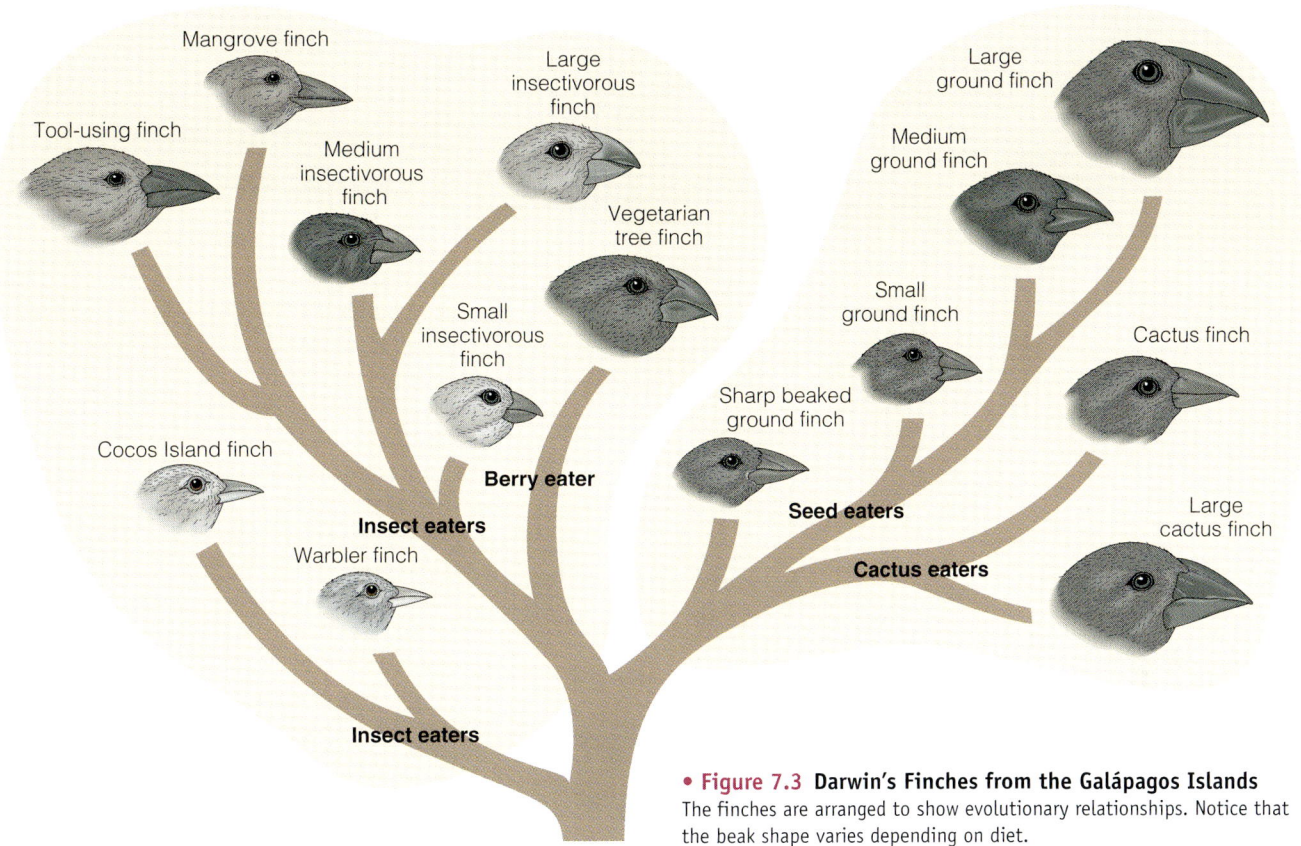

• **Figure 7.3 Darwin's Finches from the Galápagos Islands**
The finches are arranged to show evolutionary relationships. Notice that the beak shape varies depending on diet.

interpreted, Genesis was taken as the final word on the origin and variety of organisms, as well as much of Earth history. To question any aspect of this interpretation was heresy, which was usually dealt with harshly.

The social and intellectual climate changed in 18th century Europe, and the absolute authority of the Christian church in all matters declined. And ironically, the very naturalists trying to find physical evidence supporting Genesis found more and more evidence that could not be reconciled with a literal reading of Judeo-Christian scripture. For example, observations of sedimentary rocks previously attributed to a single worldwide flood led naturalists to conclude that they were deposited during a vast amount of time, in environments like those existing now. They reasoned that this evidence truly indicated how the rocks originated, for to infer otherwise implied a deception on the part of the Creator (see the Chapter 5 Introduction).

In this changing intellectual atmosphere, scientists gradually accepted the principle of uniformitarianism and Earth's great age, and the French zoologist Georges Cuvier demonstrated that many types of plants and animals had become extinct. In view of the accumulating evidence, particularly from studies of living organisms, scientists became convinced that change from one species to another actually took place. However, they lacked a theoretical framework to explain evolution.

Jean-Baptiste de Lamarck and His Ideas on Evolution

Jean-Baptiste de Lamarck (1744–1829) was not the first to propose a mechanism to account for evolution, but in 1809 he was the first to be taken seriously. Lamarck contributed greatly to our understanding of the natural world, but unfortunately he is best remembered for his theory of **inheritance of acquired characteristics.** According to this idea, new traits arise in organisms because of their needs, and somehow these characteristics are passed on to their descendants. In an ancestral population of short-necked giraffes, for instance, neck stretching to browse in trees gave them the capacity to have offspring with longer necks (• Figure 7.4a). In short, Lamarck thought that characteristics acquired during an individual's lifetime were inheritable.

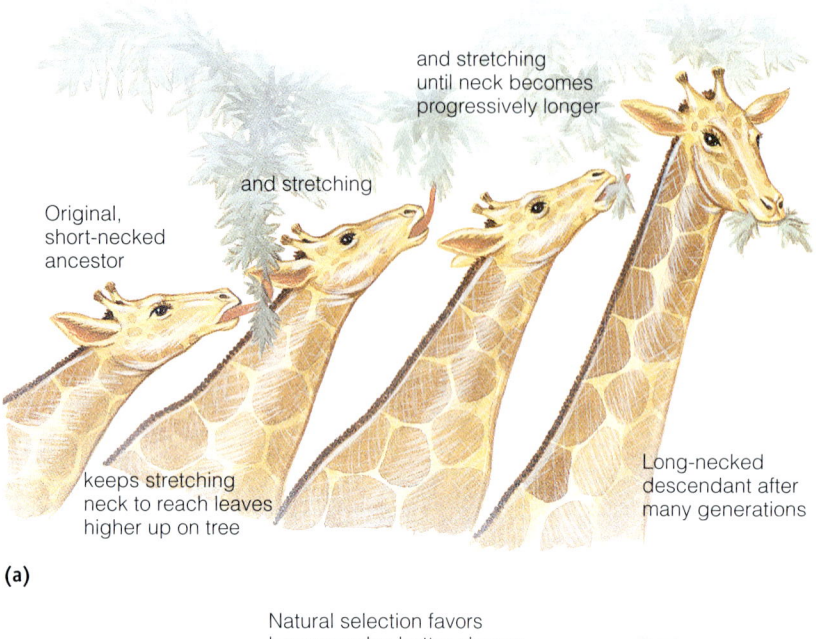

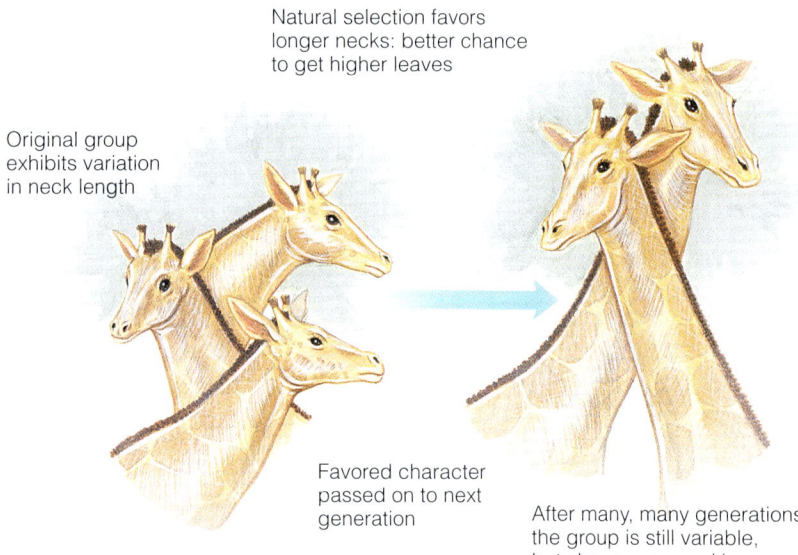

• **Figure 7.4 Inheritance of Acquired Characteristics and Natural Selection**
(a) According to Lamarck's inheritance of acquired characteristics, ancestral short-necked giraffes stretched their necks, and their offspring were born with longer necks. (b) Darwin and Wallace proposed natural selection to account for the same data. That is, giraffes with long necks evolved because more animals with favorable variations survived to reproduce compared to those lacking these variations.

If this were true, it would follow that girls in Asian countries where foot binding is practiced would have children with small feet!

Considering the information available at the time, Lamarck's theory seemed quite logical and was accepted by some as a viable mechanism for evolution. Indeed, it was not totally refuted until decades later, when scientists discovered that the units of heredity known as *genes* cannot be altered by any effort by an organism during its lifetime.

In fact, all attempts to demonstrate inheritance of acquired characteristics have been unsuccessful (see Perspective 7.1).

The Contributions of Charles Darwin and Alfred Wallace

In 1859, Charles Robert Darwin (1809-1882) (Figure 7.2) published *On the Origin of Species,* in which he detailed his ideas on evolution and proposed a mechanism whereby evolution could take place. Although 1859 marks the beginning of modern evolutionary thought, Darwin actually formulated his ideas more than 20 years earlier but, being aware of the furor they would cause, was reluctant to publish them.

Darwin had concluded during his 1831–1836 voyage aboard the *Beagle* that species are not immutable and fixed (see the Introduction), but he had no idea what might bring about change in organisms through time. However, his observations of selection practiced by plant and animal breeders and a chance reading of Thomas Malthus's essay on population gave him the elements necessary to formulate his theory.

Plant and animal breeders practice **artificial selection** by selecting those traits they deem desirable and then breed plants and animals with those traits, thereby bringing about a great amount of change (• Figure 7.5). The fantastic variety of plants and animals so produced made Darwin wonder if a process selecting among variant types in nature could also bring about change. He came to fully appreciate the power of selection when he read in Malthus's essay that far more animals are born than reach maturity, yet the adult populations remain rather constant. Malthus reasoned that competition for resources resulted in a high infant mortality rate, thus limiting population size.

In 1858, Darwin received a letter from Alfred Russel Wallace (1823–1913), a naturalist working in southern Asia, who had also read Malthus's essay and had come to the same conclusion: A natural process was selecting only a few individuals for survival. Darwin's and Wallace's idea, called *natural selection,* was presented simultaneously in 1859 to the Linnaean Society in London.

Perspective 7.1

The Tragic Lysenko Affair

When Jean-Baptiste de Lamarck (1744–1829) proposed his *theory of inheritance of acquired characteristics* some scientists accepted it as a viable explanation to account for evolution. In fact, it was not fully refuted for several decades, and even Charles Darwin, at least for a time, accepted some kind of Lamarckian inheritance. Nevertheless, the notion that acquired characteristics could be inherited is now no more than an interesting footnote in the history of science. There was, however, one instance in recent times when the idea enjoyed some popularity—but with tragic consequences.

One of the most notable examples of adherence to a disproved scientific theory involved Trofim Denisovich Lysenko (1898–1976), who became president of the Soviet Academy of Agricultural Sciences in 1938. He lost this position in 1953 but in the same year was appointed director of the Institute of Genetics, a post he held until 1965. Lysenko endorsed Lamarck's theory of inheritance of acquired characteristics because he thought plants and animals could be changed in desirable ways by exposing them to a new environment. Lysenko reasoned that seeds exposed to dry conditions or to cold would acquire a resistance to drought or cold weather, and these traits would be inherited by future generations.

Lysenko accepted inheritance of acquired characteristics because he thought it was compatible with Maxist-Leninist philosophy. Accordingly, beginning in 1929, Soviet scientists were encouraged to develop concepts consistent with this philosophy.

As president of the Academy of Agricultural Sciences and with the endorsement of the Central Committee of the Soviet Union, Lysenko did not allow any other research concerning inheritance. Those who publicly disagreed with him lost their jobs or were sent to labor camps.

Unfortunately for the Soviet people, inheritance of acquired characteristics had been discredited more than 50 years earlier. The results of Lysenko's belief in the political correctness of this theory and its implementation were widespread crop failures and famine and the thorough dismantling of Soviet genetic research in agriculture. In fact, it took decades for the Soviet genetic research programs to fully recover from this setback, after scientists finally became free to experiment with other theories of inheritance.

Lysenko's ideas on inheritance were not mandated in the Soviet Union because of their scientific merit but rather because they were deemed compatible with a belief system. The fact that Lysenko retained power and only his type of genetic research was permitted for more than 25 years is a testament to the absurdity of basing scientific theories on philosophic or political beliefs. In other words, a government mandate does not validate a scientific theory, nor does a popular vote or a decree by some ecclesiastic body. Legislating or decreeing that acquired traits are inheritable does not make it so.

Wild mustard

Broccoli

Cauliflower

Kale

Cabbage

• **Figure 7.5 Artificial Selection** Humans have practiced artificial selection for thousands of years, thereby giving rise to the dozens of varieties of domestic dogs, pigeons, and vegetables. Wild mustard was selectively bred to yield broccoli, kale, cauliflower, cabbage, and two other vegetables not shown here.

> **What Would You Do?**
>
> Suppose that a powerful group in Congress were to mandate that all future genetic research had to conform to strict guidelines—specifically, that plants and animals should be exposed to particular environments so that they would acquire characteristics that would allow them to live in otherwise inhospitable areas. Furthermore, this group enacted legislation that prohibited any other type of genetic research. Why would it be unwise to implement this research program?

Natural Selection—What Is Its Significance?

We can summarize the salient points on **natural selection,** a mechanism that accounts for evolution, as follows:

1. Organisms in all populations possess heritable variations—size, speed, agility, visual acuity, digestive enzymes, color, and so forth.
2. Some variations are more favorable than others; that is, some variant types have a competitive edge in acquiring resources and/or avoiding predators.
3. Not all young survive to reproductive maturity.
4. Those with favorable variations are *more likely* to survive and pass on their favorable variations.

In colloquial usage natural selection is sometimes expressed as "survival of the fittest," which is misleading because it reduces to "the fittest are those that survive and are thus the fittest." But it actually involves differential rates of survival and reproduction. Therefore, it is largely a matter of reproductive success. Having favorable variations does not guarantee survival for an individual, but in a population of perhaps thousands, those with favorable variations are more likely to survive and reproduce.

Natural selection works on the existing variation in a population, thus giving a competitive edge to some individuals, as in a population of giraffes (Figure 7.4b). So evolution by natural selection and evolution by inheritance of acquired characteristics are both testable, but evidence supports the former, whereas attempts to verify the latter have failed (see Perspective 7.1).

Darwin was not unaware of potential problems for his newly proposed theory of natural selection. In fact, in *On the Origin of Species,* he said the following:

> If it could be demonstrated that any complex organ existed which could not possibly have been formed by numerous, successive, slight modifications, my theory would absolutely break down. (p. 171)

Critics of natural selection have cited vertebrate eyes, birds' wings, and many other structures that they claim could not have evolved by natural selection, because, according to them, anything less than the structure as it exists now would be useless. Eyes of vertebrate animals are a favorite example. Of course, this argument presupposes that existing structures were always used exactly as they are now, which is not necessarily true. All eyes, no matter if they are light-sensitive spots, crude image makers, or image makers such as those of vertebrate animals, are all fully developed and useful. Their uses simply differ.

One misconception about natural selection, and especially the phrase "survival of the fittest," is that among animals only the biggest, strongest, and fastest are likely to survive. Indeed, size and strength are important when male bighorn sheep compete for mates, but remember that females pass along their genes, too. Speed is certainly an advantage to some predators, but weasels and skunks are not very fast and survive quite nicely. In fact, natural selection may favor the smallest where resources are limited, as on small islands, or the most easily concealed, or those that adapt most readily to a new food source, or those having the ability to detoxify some natural or human-made substance.

Mendel and the Birth of Genetics

Critics of natural selection were quick to point out that Darwin and Wallace could not account for the origin of variations or explain how variations were maintained in populations. These critics reasoned that should a variant trait arise, it would blend with other traits and would be lost. Actually, the answers to these criticisms existed even then, but they remained in obscurity until 1900.

Mendel's Experiments

During the 1860s, Gregor Mendel, an Austrian monk, performed a series of controlled experiments with true-breeding strains of garden peas (strains that when self-fertilized always display the same trait, such as flower color). He concluded from these experiments that traits such as flower color are controlled by a pair of factors, or what we now call **genes,** and that these factors (genes) that control the same trait occur in alternate forms, now called **alleles.** He further realized that one allele may be dominant over another and that offspring receive one allele of each pair from each parent (• Figure 7.6).

We can summarize the most important aspects of Mendel's work as follows: The factors (genes) that control traits do not blend during inheritance; and even though traits may not be expressed in each generation, they are not lost (Figure 7.6). Therefore, some variation in populations is accounted for by alternate expressions of genes (alleles), because traits do not blend, as previously thought. Mendelian

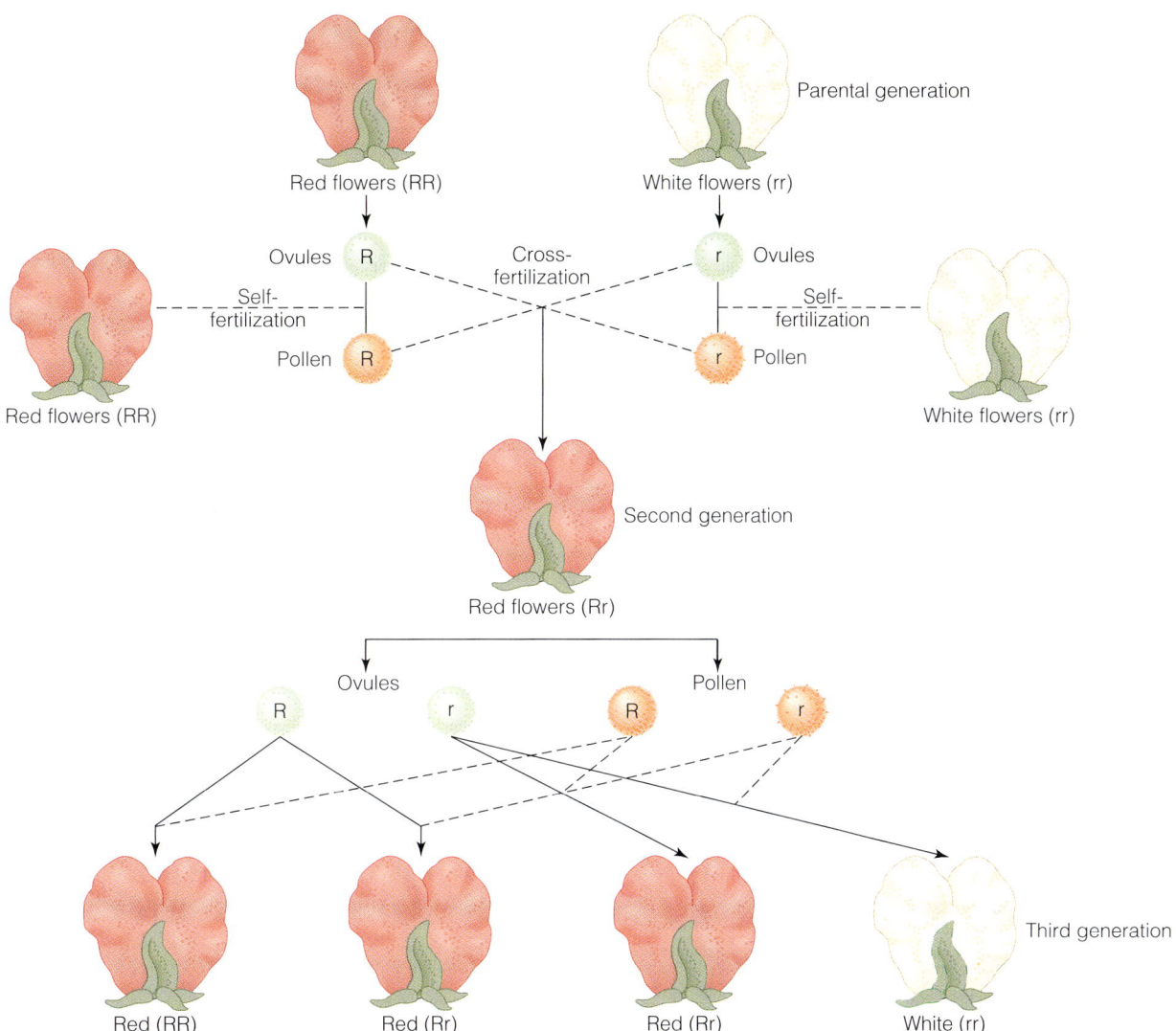

• **Figure 7.6 Mendel's Experiments with Flower Color** In his experiments with flower color in garden peas, Mendel used true-breeding strains. These plants, shown as the parental generation, when self-fertilized always yield offspring with the same trait as the parent. If the parental generation is cross-fertilized, however, all plants in the second generation receive the Rr combination of alleles; these plants will have red flowers because R is dominant over r. The second generation of plants produces ovules and pollen with the alleles shown and, when left to self-fertilize, produces a third generation with a ratio of three red-flowered plants to one white-flowered plant.

genetics explains much about heredity, but we now know the situation is much more complex than he realized. Our discussion focused on a single gene controlling a trait (flower color in Figure 7.6), but in fact many genes control most traits, and some genes show incomplete dominance. Nevertheless, Mendel provided the answers Darwin and Wallace needed, although they were published in an obscure journal and went unnoticed until three independent researchers rediscovered them in 1900.

Genes and Chromosomes

Complex, double-stranded, helical molecules of **deoxyribonucleic acid** (**DNA**), called **chromosomes,** are found in the cells of all organisms (• Figure 7.7). Specific segments or regions of the DNA molecule are the basic hereditary units: the genes. The number of chromosomes is specific for a single species but varies among species. For instance, fruit flies have 8 chromosomes (4 pairs), humans have 46, and domestic horses have 64; they are always found in pairs carrying genes controlling the same traits.

In sexually reproducing organisms, the production of *sex cells* (pollen and ovules in plants and sperm and eggs in animals) results when cells undergo a type of cell division known as **meiosis.** This process yields cells with only one chromosome of each pair, so all sex cells have only half the chromosome number of the parent cell (• Figure 7.8a). During reproduction, a sperm fertilizes an egg (or pollen fertilizes an ovule), yielding an egg (or ovule) with

The Modern View of Evolution

During the 1930s and 1940s, the ideas developed by paleontologists, geneticists, population biologists, and others were merged to form a **modern synthesis** or neo-Darwinian view of evolution. The chromosome theory of inheritance was incorporated into evolutionary thinking, changes in genes (mutations) were seen as one source of variation in populations, Lamarck's idea of inheritance of acquired characteristics was completely rejected, and the importance of natural selection was reaffirmed. The modern synthesis also emphasized that evolution is gradual, a point that has been challenged.

What Brings About Variation?

Natural selection works on variations in populations. Most variation results from reshuffling of genes from generation to generation by sexual reproduction in all organisms except bacteria, and even bacteria under some circumstances exchange genetic material. Given that each of thousands of genes might have several alleles, and that offspring receive half of their genes from each parent, the potential for variation is enormous. However, this variation was already present, so new variations arise by **mutations**—that is, changes in the chromosomes or genes. In other words, a mutation is a change in the hereditary information carried by an organism.

Whether a *chromosomal mutation* (affecting a large segment of a chromosome) or a *point mutation* (a change in a particular gene), as long as it takes place in a sex cell, it is inheritable.

To fully understand mutations, we must explore them further. For one thing, they are random with respect to fitness, meaning they may be beneficial, neutral, or harmful. If a species is well adapted to its environment, most mutations would not be particularly useful and perhaps be harmful. Some mutations are absolutely harmful, but some that were harmful can become useful if the environment changes. For instance, some plants have developed a resistance for contaminated soils around mines. Plants of the same species from the normal environment do poorly or die in contaminated soils, whereas contaminant-resistant plants do very poorly in the normal environment. Mutations for contaminant resistance probably occurred repeatedly in the population, but they were not beneficial until contaminated soils were present.

How can a mutation be neutral? In cells information carried on chromosomes directs the formation of proteins by selecting the appropriate amino acids and arranging them into a specific sequence. However, some mutations have no effect on the type of protein synthesized. In other words, the same protein is synthesized

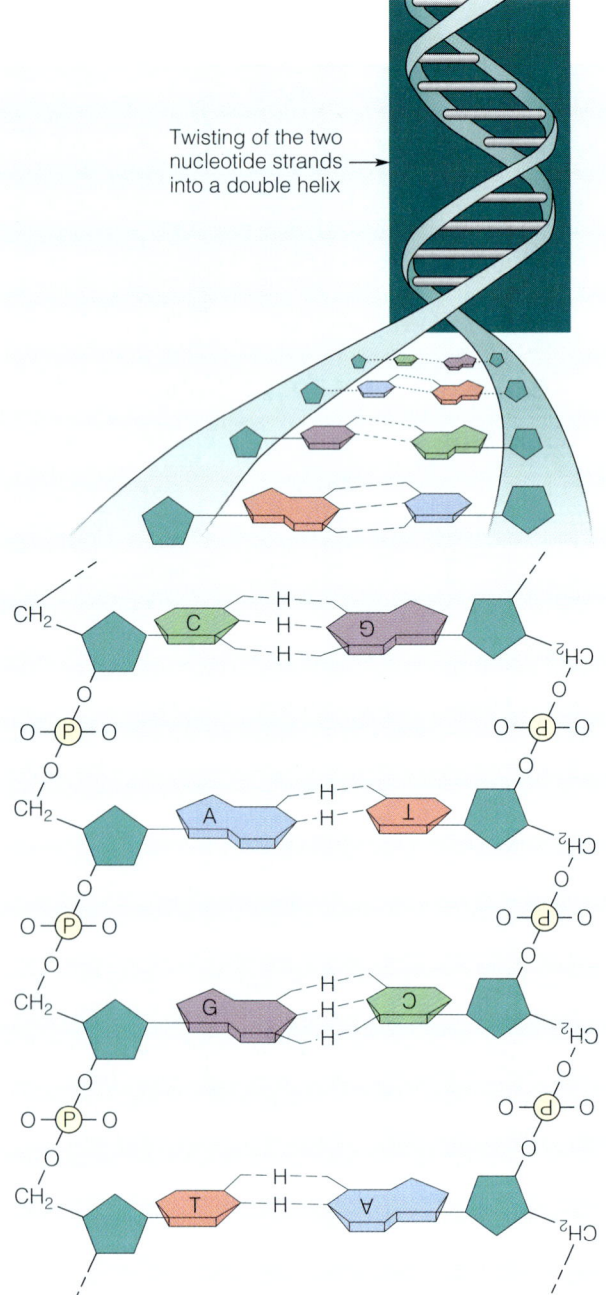

• Figure 7.7 **The Double-Stranded, Helical DNA Molecule** The molecule's two strands are joined by hydrogen bonds (H). Notice the bases A, G, C, and T, combinations of which code for amino acids during protein synthesis in cells.

the full set of chromosomes typical for that species (Figure 7.8b).

As Mendel deduced from his experiments, half of the genetic makeup of a fertilized egg comes from each parent. The fertilized egg, however, develops and grows by a cell division process called **mitosis** during which cells are simply duplicated—that is, there is no reduction in the chromosome number as in meiosis (Figure 7.8c).

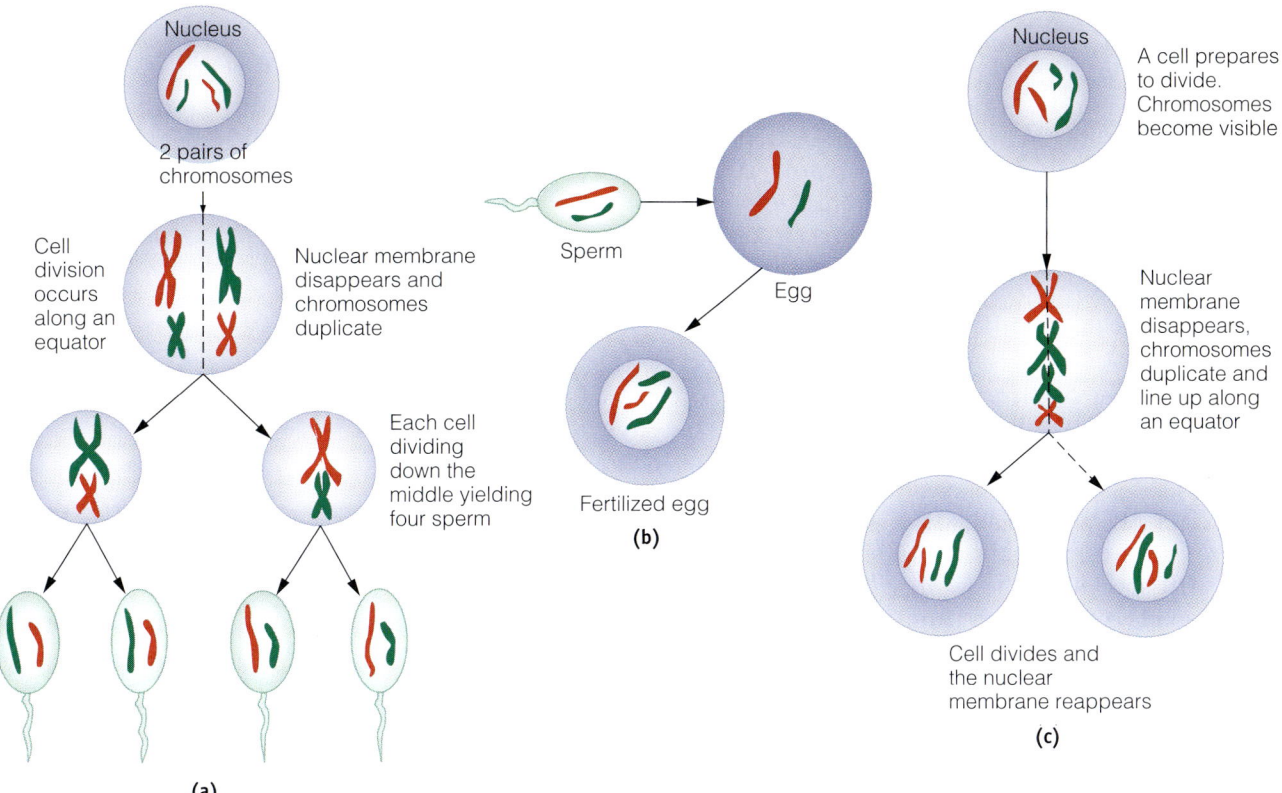

• **Figure 7.8 Meiosis and Mitosis (a)** During meiosis, sex cells form that contain one member of each chromosome pair. The formation of sperm is shown here; eggs form the same way, but only one of the four final eggs is functional. **(b)** The full number of chromosomes is restored when a sperm fertilizes an egg. **(c)** Mitosis results in the complete duplication of a cell. In this example, a cell with four chromosomes (two pairs) produces two cells, each with four chromosomes. Mitosis takes place in all cells except sex cells. Once an egg has been fertilized, the developing embryo grows by mitosis.

before and after the mutation, and thus the mutation is neutral (• Figure 7.9).

What causes mutations? Some are induced by *mutagens*, agents that bring about higher mutation rates. Exposure to some chemicals, ultraviolet radiation, X-rays, and extreme temperature changes might cause mutations. But some mutations are spontaneous, taking place in the absence of any known mutagen.

Sexual reproduction and mutations account for much of the variation in populations, but another factor is *genetic drift*, a random change in the genetic makeup of a population due to chance. Genetic drift is significant in small populations but probably of little or no importance in large ones. Good examples of genetic drift include evolution in small populations that reach remote areas where only a few individuals with limited genetic diversity rebuilt a larger population.

Speciation and the Rate of Evolution

Speciation, the phenomenon of a new species arising from an ancestral species, is well documented, but the rate and ways in which it takes place vary. First, though, let us be clear on what we mean by **species,** a biologic term for a population of similar individuals that in nature interbreed and produce fertile offspring. Thus a species is reproductively isolated from other species. This definition does not apply to organisms such as bacteria that reproduce

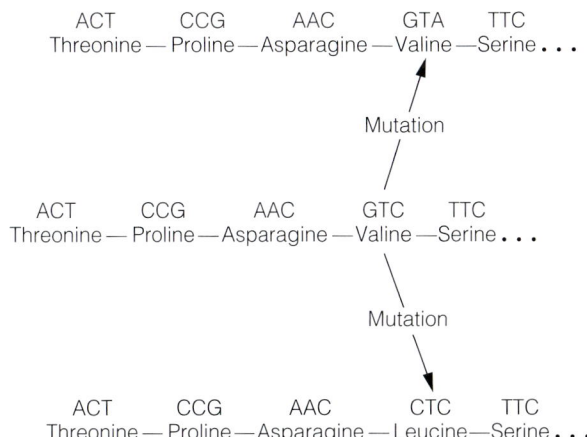

• **Figure 7.9 Hypothetical Protein Showing Its Sequence of Amino Acids and Two Possible Mutations** Notice that both GTC and GTA code for the amino acid valine, so the mutation shown on the top is neutral. However, a change from GTC to CTC results in the insertion of leucine for valine. This mutation may be harmful or beneficial.

asexually, but it is nevertheless useful for our discussion of plants, animals, fungi, and single-cell organisms called protistans.

Goats and sheep are distinguished by physical characteristics, and they do not interbreed in nature, so they are separate species. Yet in captivity they can produce fertile offspring. Lions and tigers can also interbreed in captivity, although they do not interbreed in nature and their offspring are sterile, so they too are separate species. Domestic horses that have gone wild can interbreed with zebras to yield a *zebroid*, which is sterile; yet horses and zebras are separate species. It should be obvious from these examples that reproductive barriers are not complete in some species, indicating varying degrees of change from a common ancestral species.

The process of speciation involves a change in the genetic makeup of a population, which also may bring about changes in form and structure. According to the concept of **allopatric speciation,** species arise when a small part of a population becomes isolated from its parent population (• Figure 7.10). Isolation may result from a marine transgression that effectively separates a once-interbreeding species, or a few individuals may somehow get to a remote area and no longer exchange genes with the parent population (Figure 7.3). Another example involves at least 15 iguanas that in 1995 were rafted 280 km on floating vegetation from Guadeloupe to Anguilla in the Caribbean Sea when the trees they were on were blown out to sea by hurricane winds. Given these conditions and the fact that different selective pressures are likely, they may eventually give rise to a reproductively isolated species.

Although widespread agreement exists on allopatric speciation, scientists disagree on how rapidly a new species may evolve. According to Darwin and reaffirmed by the modern synthesis, the gradual accumulation of minor changes eventually brings about the origin of a new species, a phenomenon called **phyletic gradualism.** Another view, known as **punctuated equilibrium,** holds that little or no change takes place in a species during most of its existence, and then

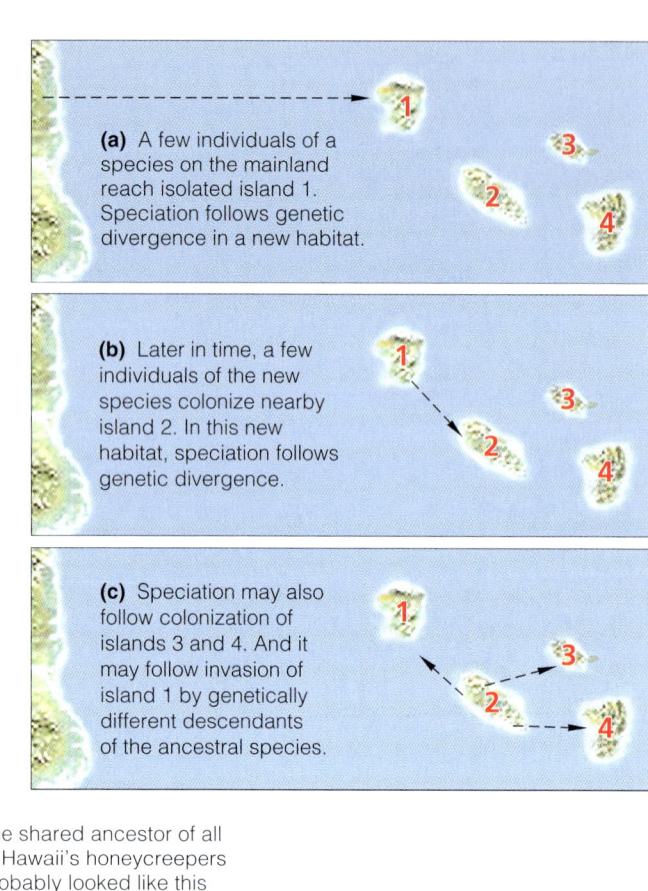

(a) A few individuals of a species on the mainland reach isolated island 1. Speciation follows genetic divergence in a new habitat.

(b) Later in time, a few individuals of the new species colonize nearby island 2. In this new habitat, speciation follows genetic divergence.

(c) Speciation may also follow colonization of islands 3 and 4. And it may follow invasion of island 1 by genetically different descendants of the ancestral species.

The shared ancestor of all of Hawaii's honeycreepers probably looked like this housefinch *(Carpodacus)*

Akepa (*Loxous coccineus*) Akekee (*L. caeruleirostris*) Nihoa finch (*Telespyza ultima*) Palila (*Loxioides bailleui*)

Maiu parrotbill (*Pseudonestor xanthrophrys*) Alauahio (*Paroreomyza montana*) Kauai Amakihi (*Hemignathus kauaiensis*) Akiapolaau (*H. munroi*)

(d)   Akohekohe (*Palmeria doli*)   Apap (*Himatione sanguinea*)   Iiwi (*Vestiaria coccinea*)

• **Figure 7.10 Allopatric Speciation (a–c)** An example of allopatric speciation on some remote islands. **(d)** More than 20 species of Hawaiian honeycreepers have evolved from a common ancestor as they adapted to diverse food sources on the islands.

evolution occurs rapidly, giving rise to a new species in perhaps as little as a few thousands of years.

Proponents of punctuated equilibrium argue that few examples of gradual transitions from one species to another are found in the fossil record. Critics, however, point out that neither Darwin nor those who formulated the modern synthesis insisted that all evolutionary change was gradual and continuous, a view shared by many present-day biologists and paleontologists. Indeed, they allowed for times during which evolutionary change in small populations could be quite rapid. Furthermore, deposition of sediments in most environments is not continuous; thus the lack of gradual transitions in many cases is simply an artifact of the fossil record. And finally, despite the incomplete nature of the fossil record, a number of examples of gradual transitions from ancestral to descendant species are well known.

If speciation proceeds as we have described, evidence of it taking place should be available from present-day organisms, and indeed it is. This is especially true in some new plant species that have arisen by *polyploidy* when their chromosome count doubled. Speciation, or at least incipient speciation, has also taken place in populations of mosquitoes, bees, mice, salamanders, fish, and birds that are isolated or partially isolated from one another (• Figure 7.11).

Divergent, Convergent, and Parallel Evolution

The phenomenon of an ancestral species giving rise to diverse descendants adapted to various aspects of the environment is referred to as **divergent evolution.** An impressive example involves the mammals whose diversification from a common ancestor during the Late Mesozoic gave rise to such varied animals as platypuses, armadillos, rodents, bats, primates, whales, and rhinoceroses (• Figure 7.12).

Divergent evolution leads to descendants that differ markedly from their ancestors. *Convergent evolution* and *parallel evolution* are processes whereby similar adaptations arise in different groups. Unfortunately, they differ in degree and are not always easy to distinguish. Nevertheless, both are common phenomena with **convergent evolution** involving the development of similar characteristics in distantly related organisms, whereas when similar characteristics arise in closely related organisms, it is considered **parallel evolution.** In both cases, similar characteristics develop independently, because the organisms in question adapt to comparable environments. Perhaps the following examples will clarify the distinction between these two concepts.

During much of the Cenozoic, South America was an island continent with a unique mammalian fauna that evolved

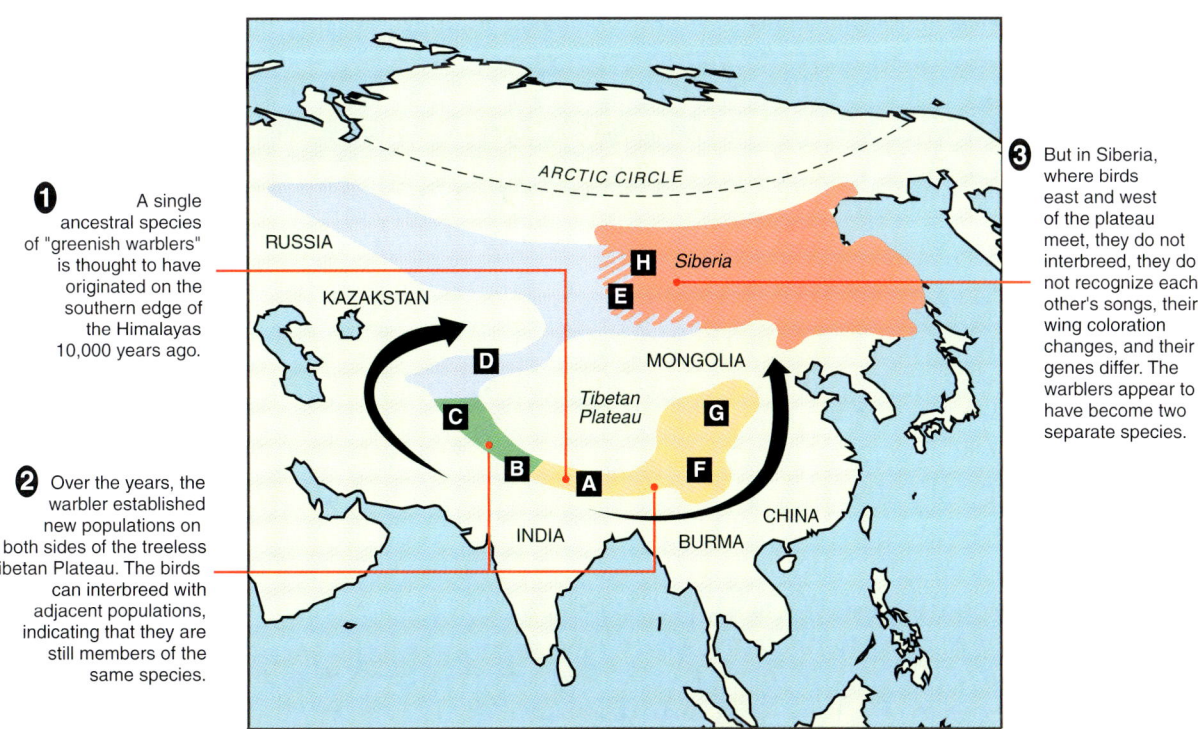

• **Figure 7.11 Speciation in Songbirds** Two species of Eurasian songbirds appear to have evolved from an ancestral species. Adjacent populations of these birds, A and B, or F and G, for instance, can interbreed even though they differ slightly. However, where populations E and H overlap in Siberia, they cannot interbreed.

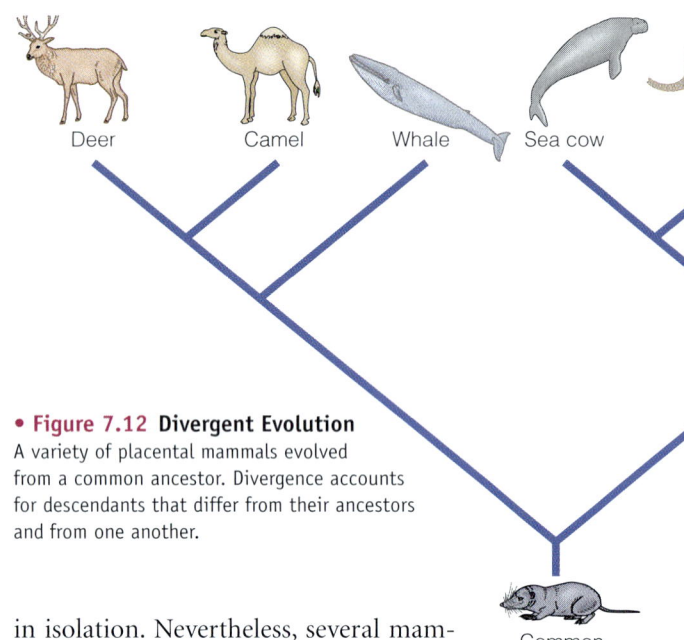

• **Figure 7.12 Divergent Evolution** A variety of placental mammals evolved from a common ancestor. Divergence accounts for descendants that differ from their ancestors and from one another.

in isolation. Nevertheless, several mammals in South and North America adapted in similar ways so that they superficially resembled one another (• Figure 7.13a)—a good example of convergent evolution. Likewise, convergence also accounts for the superficial similarities between Australian marsupial (pouched) mammals and placental mammals elsewhere—for example, catlike marsupial carnivores and true cats. Parallel evolution, in contrast, involves closely related organisms, such as jerboas and kangaroo rats, that independently evolved comparable features (Figure 7.13b).

Microevolution and Macroevolution

Micro and *macro* at the beginning of words mean "small" and "large," respectively. Actually **microevolution** is any change in the genetic makeup of a species, so microevolution involves changes within a species. For instance, house sparrows were introduced into North America in 1852 and have evolved so that members of northern populations are larger than those of southern populations, probably a response to climate. Likewise, organisms that develop resistance to insecticides and pesticides and plants that adapt to contaminated soils are examples of microevolution.

In contrast, **macroevolution** involves changes such as the origin of a new species or changes at even higher levels in the classification of organisms, such as the origin of new genera, families, orders, and classes. Good examples abound in the fossil record—the origin of birds from reptiles, the origin of mammals from mammal-like reptiles, the evolution of whales from land-dwelling ancestors, and many others. Although macroevolution encompasses greater

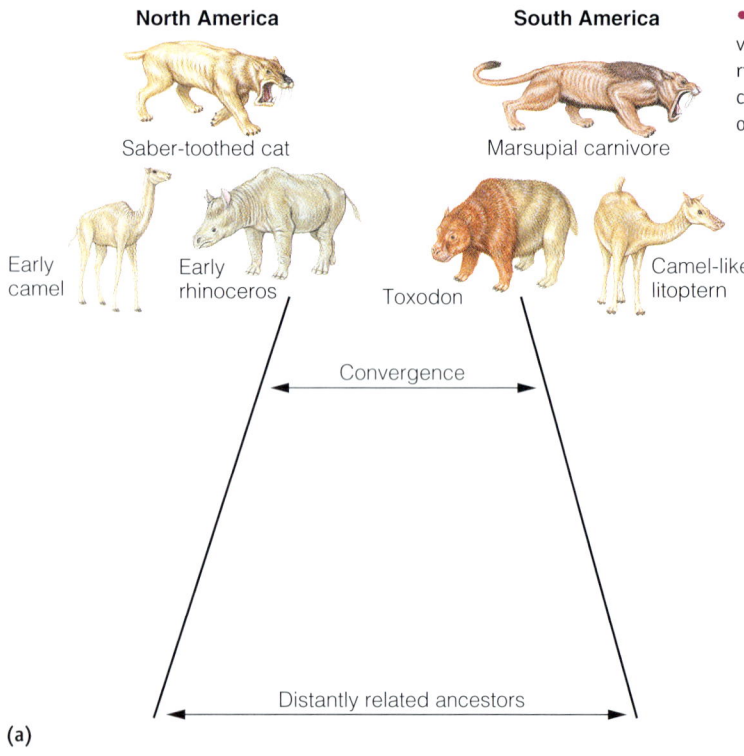

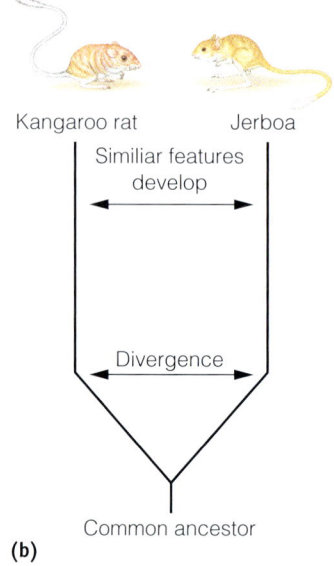

• **Figure 7.13 Convergent and Parallel Evolution** (a) Convergent evolution takes place when distantly related organisms give rise to species that resemble one another because they adapt in comparable ways. (b) Parallel evolution involves the independent origin of similar features in closely related organisms.

134 CHAPTER 7 EVOLUTION—THE THEORY AND ITS SUPPORTING EVIDENCE

changes than microevolution, the cumulative effects of microevolution are responsible for macroevolution. They differ only in the degree of change.

Cladistics and Cladograms

Traditionally, scientists have depicted evolutionary relationships with *phylogenetic trees,* in which the horizontal axis represents anatomic differences and the vertical axis denotes time (• Figure 7.14a). The patterns of ancestor–descendant relationships shown are based on a variety of characteristics, although the ones used are rarely specified. In contrast, a **cladogram** shows the relationships among members of a *clade,* a group of organisms including its most recent common ancestor (Figure 7.14b).

Cladistics is a type of biological analysis in which organisms are grouped together based on derived as opposed to primitive characteristics. For instance, all land-dwelling vertebrates have bone and paired limbs, so these characteristics are primitive and of little use in establishing relationships among them. However, hair and three middle ear bones are derived characteristics, sometimes called *evolutionary novelties,* because only one subclade, the mammals, has them. If you consider only mammals, hair and middle ear bones are primitive characteristics, but live birth is a derived characteristic that serves to distinguish most mammals from the egg-laying mammals.

Any number of organisms can be depicted in a cladogram, but the more shown, the more complex and difficult it is to construct. • Figure 7.15 shows three cladograms, each with a different interpretation of the relationships among bats, birds, and dogs. Bats and birds fly, so we might conclude that they are more closely related to one another than to dogs (Figure 7.15a). On the other hand, perhaps birds and dogs are more closely related than either is to bats (Figure 7.15b). However, if we concentrate on evolutionary novelties, such as hair and giving birth to live young, we conclude that the cladogram in Figure 7.15c shows the most probable relationships—that is, bats and dogs are more closely related than they are to birds.

Cladistics and cladograms work well for living organisms, but when applied to fossils, care must be taken in determining what are primitive versus derived characteristics, especially in groups with poor fossil records. Furthermore, cladistic analysis depends solely on characteristics inherited from a common ancestor, so paleontologists must be especially careful of characteristics that result from convergent evolution. Nevertheless, cladistics is a powerful tool that has more clearly elucidated the relationships among many fossil lineages and is now used extensively by paleontologists.

Evolutionary Trends and Mosaic Evolution

Evolutionary changes do not involve all aspects of an organism simultaneously, because selection pressure is greater on some features than on others. As a result, a key feature we associate with a descendant group may appear before other features typical of that group. For example, the oldest known bird had feathers and the typical fused clavicles (the furcula or wishbone) of birds, but it also retained many reptile characteristics (see Chapter 15). Accordingly, it represents the concept of **mosaic evolution,**

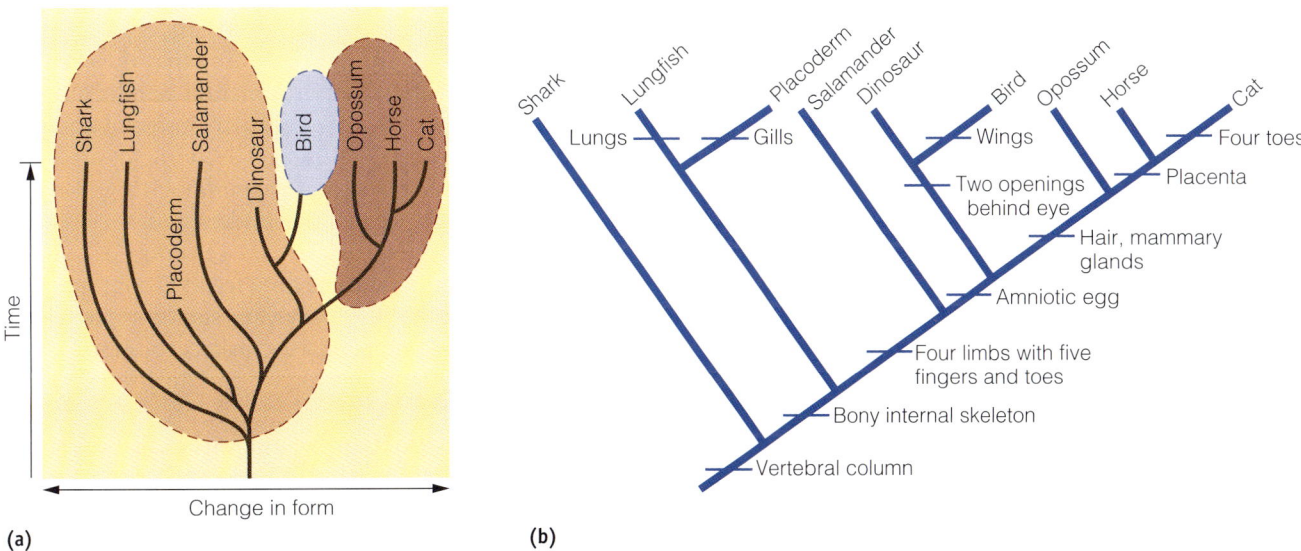

• **Figure 7.14 Phylogenetic Tree and Cladogram** A phylogenetic tree **(a)** and a cladogram **(b)** showing inferred relationships. Since a clade is a group of organisms along with its most recent common ancestor, opossums, horses, and cats form a clade, but birds and opossums do not. However, birds are in the same clade with opossums, horses, and cats if we also include dinosaurs and the most recent common ancestor for all of these animals.

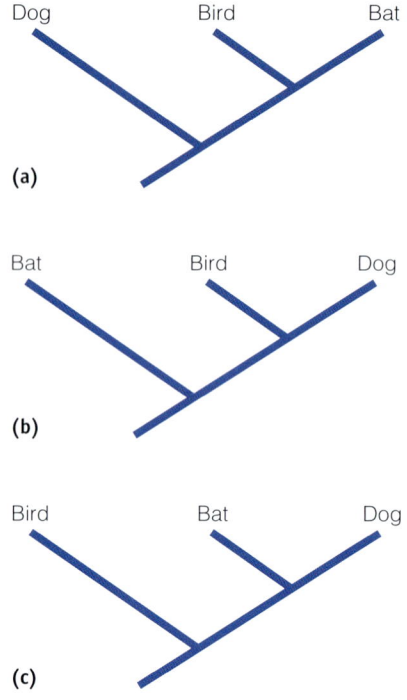

• **Figure 7.15 Cladograms** Three hypotheses for the relationships among birds, bats, and dogs. Derived characteristics such as hair and giving birth to live young indicate that dogs and bats are most closely related, as shown in **(c)**.

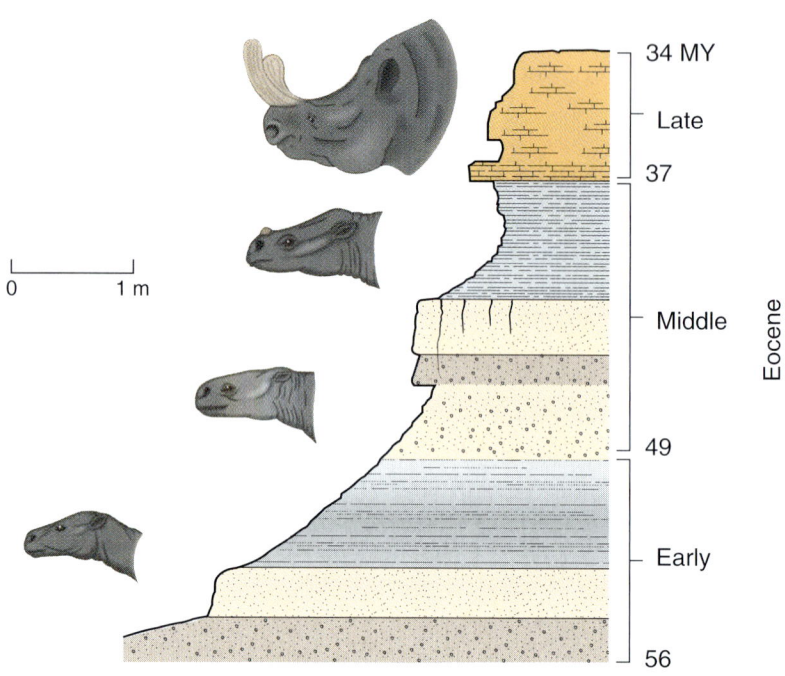

• **Figure 7.16 Evolutionary Trends in Titanotheres** These relatives of horses and rhinoceroses existed for about 20 million years during the Eocene Epoch. During that time they evolved from small ancestors to giants standing 2.4 m at the shoulder. In addition, they developed large horns, and the shape of the skull changed. Only 4 of the 16 known genera of titanotheres are shown here.

meaning that organisms have recently evolved characteristics as well as some features of their ancestral group.

Paleontologists determine in some detail the *phylogeny,* or evolutionary history, and various *evolutionary trends* for groups of organisms if sufficient fossil material is available. For instance, one trend in ammonoids, extinct relatives of squid and octopus, was the evolution of an increasingly complex shell. Abundant fossils show that the Eocene mammals called *titanotheres* not only increased in size but also developed large nasal horns; moreover, the shape of their skull changed (• Figure 7.16).

Size increase is one of the most common evolutionary trends, but trends are extremely complex, they may be reversed, and several trends taking place may not all proceed at the same rate. One evolutionary trend in horses was larger size, but some now-extinct horse species actually showed a size decrease. Other trends in horses include changes in their teeth and skull, as well as lengthening of their legs and reduction in the number of toes, but these trends did not all take place at the same rate (see Chapter 18).

We can think of evolutionary trends as a series of adaptations to a changing environment or adaptations that occur in response to exploitation of new habitats. Some organisms, however, show little evidence of any evolutionary trends for long periods. A good example is the brachiopod *Lingula,* whose shell at least has not changed significantly since the Ordovician Period (see Figure 5.22). Several other animals including horseshoe crabs and the coelacanth, as well as plants such as ginkgos, have shown little or no change for long periods (• Figure 7.17).

Of course, we have no way to evaluate changes that are not obvious in shells, teeth, and bones. For instance, the immune system or digestive enzymes of horseshoe crabs may have change significantly. In addition, we do know that some of these so-called **living fossils,** such as opossums, are generalized animals, meaning that they can live under a wide variety of environmental conditions and therefore are not so sensitive to environmental changes. Nevertheless, the remarkable stasis of living fossils is one aspect of the organic world that paleontologists and evolutionary biologists do not yet fully understand.

But isn't evolution by natural selection a random process? If so, how is it possible for a trend to continue long enough to account, just by chance, for such complex structures as eyes, wings, and hands? Actually, evolution by natural selection is a two-step process, and only the first step involves chance. First, variation must be present or arise in some population. Whether or not variations arising by mutations are favorable is indeed a matter of chance, but the second step involving natural selection is not, because only individuals with favorable variations are most likely to survive and reproduce. In one sense, then, natural selection is a process of elimination, weeding out those individuals not as well adapted to a particular set of environmental

• **Figure 7.17 Two Examples of So-Called Living Fossils** (a) *Latimaria* belongs to a group of fish thought to have gone extinct at the end of the Mesozoic Era. One was caught off the east coast of Africa in 1938, and since then many more have been captured. (b) Ginkgo trees with their fan-shaped leaves (inset) look much like their ancient ancestors.

circumstances. Of course, such individuals may survive and reproduce just by luck, but in the long run their genes will be eliminated as selection favors individuals with favorable variations—better visual acuity, appropriate digestive enzymes, more effective camouflage, and so on.

Extinctions

Judging from the fossil record, most organisms that ever existed are now extinct—perhaps as many as 99% of all species. Now, if species actually evolve as natural selection favors certain traits, shouldn't organisms be evolving toward some kind of higher order of perfection or greater complexity? Certainly vertebrates are more complex, at least in overall organization, than are bacteria, but complexity does not necessarily mean they are superior in some survival sense—after all, bacteria have persisted for at least 3.5 billion years. Actually, natural selection does not yield some kind of perfect organism, but rather those adapted to a specific set of circumstances at a particular time. Thus a clam or lizard existing now is not somehow superior to those that lived millions of years ago.

The continual extinction of species is referred to as *background extinction,* to clearly differentiate it from a **mass extinction,** during which accelerated extinction rates sharply reduce Earth's biotic diversity. Extinction is a continuous occurrence, but so is the evolution of new species that usually, but not always, quickly exploit the opportunities created by another species' extinction. When the dinosaurs and their relatives died out, the mammals began a remarkable diversification as they began occupying the niches left temporarily vacant.

Everyone is familiar with the mass extinction of dinosaurs and other animals at the end of the Mesozoic Era (see Chapter 15). The greatest extinction, though, during which perhaps more than 90% of all species died out, was at the end of the Paleozoic Era. In some of the following chapters, we discuss these extinctions and their possible causes as well as other extinctions of lesser magnitude.

What Kinds of Evidence Support Evolutionary Theory?

When Charles Darwin proposed his theory of evolution, he cited supporting evidence such as classification, embryology, comparative anatomy, geographic distribution, and, to a limited extent, the fossil record. He had little knowledge of the mechanism of inheritance, and both biochemistry and molecular biology were unknown during his time. Studies in these areas, coupled with a more complete and much better understood fossil record, have convinced scientists that the theory is as well supported by evidence as any other major theory.

But is the theory of evolution truly scientific? That is, can testable predictive statements be made from it? First we must be clear on what a theory is and what we mean by "predictive." Scientists propose hypotheses for natural phenomena, test them, and in some cases raise them to the status of a theory—an explanation of some natural phenomenon well supported by evidence from experiments and/or observations.

Almost everything in the sciences has some kind of theoretical underpinning—optics, heredity, the nature of matter, the present distribution of continents, diversity in the organic world, and so on. Of course, no theory is ever proven in some final sense, although it might be supported by substantial evidence; all are always open to question, revision, and occasionally to replacement by a more comprehensive theory. In geology, plate tectonic theory has replaced geosyncline theory to explain the origin of mountains, but the new theory contains many elements of its predecessor.

Prediction is commonly taken to mean to foresee some event that has not yet occurred, as in predicting the next solar eclipse. However, not all predictions are about future events. For instance, one prediction of the Big Bang theory is that if the universe was extremely hot when it formed, traces of that heat should still be present. Investigators detected the faint afterglow of the Big Bang as the pervasive background radiation 2.7 K above absolute zero (see Chapter 1). Accordingly, the Big Bang theory is verified but not proven in some final sense. Perhaps it will not account for data that become available in the future, in which case it will be modified or replaced.

Evolutionary theory cannot make predictions about the far distant future. No one knows which existing species might become extinct or what the descendants of clams or horses, if any, will look like millions of years from now. Nevertheless, we can make many predictions about the present-day natural world as well as the fossil record that should be consistent with evolutionary theory if it is correct. It is just as important that we be able to think of observations or experiments that would be negative evidence for the theory.

If the theory of evolution is correct, closely related species such as wolves and coyotes should be similar not only in their anatomy but also in terms of biochemistry, genetics, and embryonic development (number 4 in Table 7.1). Suppose that they differed in their biochemical mechanisms as well as their embryology. Obviously our prediction would fail, and we would at least have to modify our theory. If other predictions also failed—say, mammals appear in the fossil record before fishes—we would have to abandon our theory and find a better explanation for our observations. Accordingly, the theory of evolution is truly scientific because it can at least in principle "be falsified"—that is, proven wrong. In contrast, an explanation that can account for all conceivable data or observations is not a scientific theory because it cannot be tested. A good example is Philip Henry Gosse's idea that Earth was created a few thousand years ago with the appearance of great age (see Chapter 5).

Classification—A Nested Pattern of Similarities

Carolus Linnaeus (1707–1778) proposed a classification system in which organisms are given a two-part genus and species name; the coyote, for instance, is *Canis latrans*. Table 7.2 shows Linnaeus's classification scheme, which is a hierarchy of categories that becomes more inclusive as one

TABLE 7.1 Some Predictions from the Theory of Evolution

1. If evolution has taken place, the oldest fossil-bearing rocks should have remains of organisms very different from those existing now, and more recent rocks should have fossils more similar to today's organisms.
2. If evolution by natural selection actually occurred, Earth must be very old, perhaps many millions of years old.
3. If today's organisms descended with modification from ones in the past, there should be fossils showing characteristics connecting orders, classes, and so on.
4. If evolution is true, closely related species should be similar not only in details of their anatomy, but also in their biochemistry, genetics, and embryonic development, whereas distantly related species should show fewer similarities.
5. If the theory of evolution is correct—that is, living organisms descended from a common ancestor—classification of organisms should show a nested pattern or similarities.
6. If evolution actually took place, we would expect cave-dwelling plants and animals to most closely resemble those immediately outside their respective caves rather than being most similar to those in caves elsewhere.
7. If evolution actually took place, we would expect land-dwelling organisms on oceanic islands to most closely resemble those of nearby continents rather than those on other distant islands.
8. If evolution has taken place, a mechanism should exist that accounts for the evolution of one species to another.
9. If evolution occurred, we would expect mammals to appear in the fossil record long after the appearance of the first fish. Likewise, we would expect reptiles to appear before the first mammals or birds.
10. If we examine the fossil record of related organisms such as horses and rhinoceroses, we should find that they were quite similar when they diverged from a common ancestor but became increasingly different as their divergence continued.

TABLE 7.2		**Expanded Linnaen Classification Scheme (the animal classified in this example is the coyote, *Canis latrans*)**
Kingdom		**Animalia**—Multicelled organisms; cells with nucleus (see Chapter 9); reproduce sexually; ingest preformed organic molecules for nutrients
Phylum		**Chordata**—Possess notochord, gill slits, dorsal hollow nerve cord at some time during life cycle (see Chapter 13)
Subphylum		**Vertebrata**—Those chordates with a segmented vertebral column (see Chapter 13)
Class		**Mammalia**—Warm-blooded vertebrates with hair or fur and mammary glands (see Chapter 15)
Order		**Carnivora**—Mammals with teeth specialized for a diet of meat (see Chapter 18)
Family		**Canidae**—The doglike carnivores (excludes hyenas, which are more closely related to cats)
Genus		***Canis***—Made up only of closely related species—coyotes and wolves (also includes domestic dogs)
Species		***latrans***—Consists of similar individuals that in nature can interbreed and produce fertile offspring

proceeds up the list. The coyote (*Canis latrans*) and the wolf (*Canis lupus*) share numerous characteristics, so they are members of the same genus, whereas both share some but fewer characteristics with the red fox (*Vulpes fulva*), and all three are members of the family Canidae (• Figure 7.18). All canids share some characteristics with cats, bears, and weasels and are grouped together in the order Carnivora, which is one of 18 living orders of the class

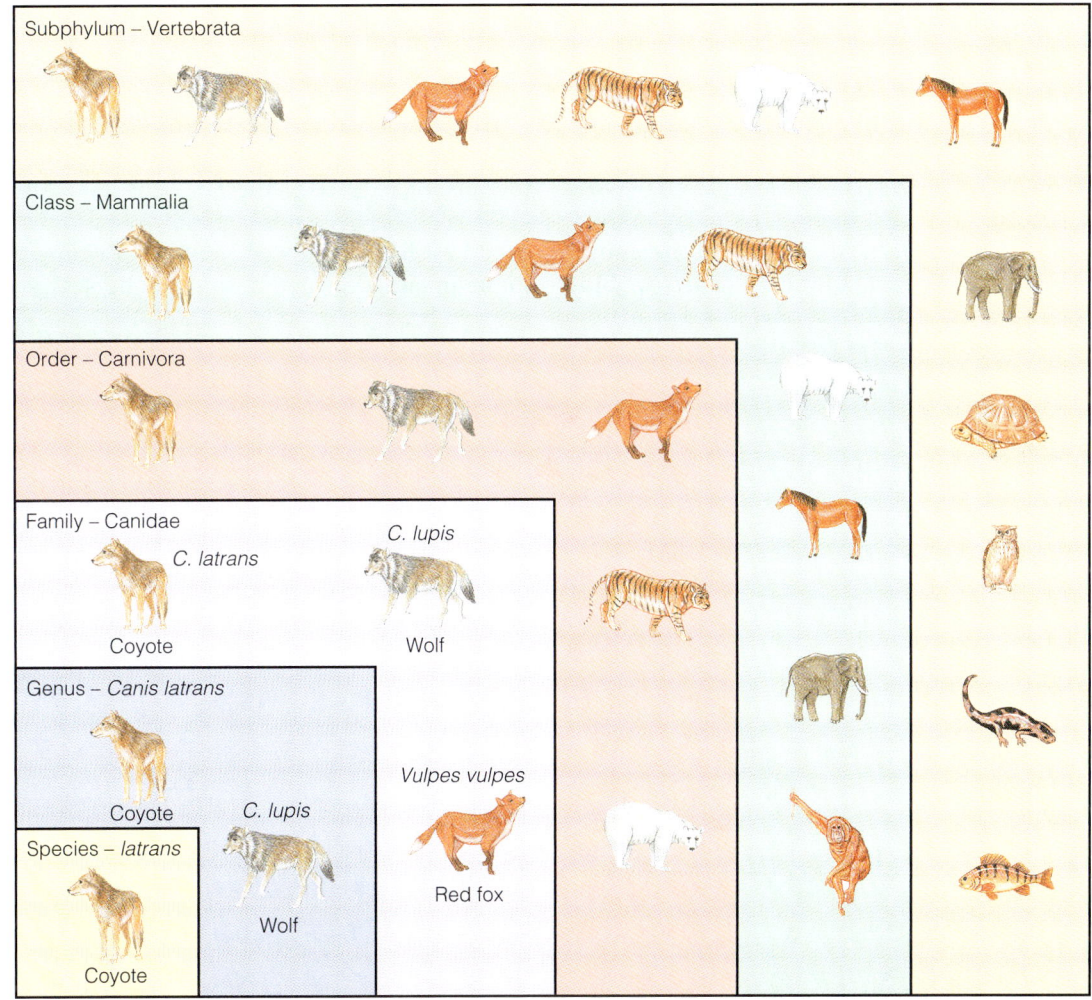

• **Figure 7.18 Classification of Organisms Based on Shared Characteristics** All members of the subphylum Vertebrata—fishes, amphibians, reptiles, birds, and mammals—have a segmented vertebral column. Subphyla belong to more inclusive phyla (singular, *phylum*), and these in turn belong to kingdoms (see Table 7.2). Among vertebrates only warm-blooded animals with hair or fur and mammary glands are mammals. Eighteen orders of mammals exist, including the order Carnivora shown here. The family Canidae includes only doglike carnivores, and the genus *Canis* includes only closely related species. The coyote, *Canis latrans*, stands alone as a species.

Mammalia, all of whom are warm-blooded, possess fur or hair, and have mammary glands.

Linnaeus certainly recognized shared characteristics among organisms, but his intent was simply to categorize species he thought were specially created and immutable. Following the publication of Darwin's *On the Origin of Species* in 1859, however, scientists quickly realized that shared characteristics constituted a strong argument for evolution. After all, if present-day organisms actually descended from ancient species, we should expect a pattern of similarities between closely related species and fewer between more distantly related ones.

How Does Biological Evidence Support Evolution?

If all existing organisms actually descended with modification from ancestors that lived during the past, fundamental similarities should exist among all life-forms. As a matter of fact, all living things, be they bacteria, redwood trees, or whales, are composed mostly of carbon, nitrogen, hydrogen, and oxygen. Furthermore, their chromosomes consist of DNA, and all cells synthesize proteins in essentially the same way.

Studies in biochemistry also provide evidence for evolutionary relationships. Blood proteins are similar among all mammals but also indicate that humans are most closely related to great apes, followed in order by Old World monkeys, New World monkeys, and lower primates such as lemurs. Biochemical tests support the idea that birds descended from reptiles, a conclusion supported by evidence in the fossil record.

The forelimbs of humans, whales, dogs, and birds are superficially dissimilar (• Figure 7.19). Yet all are made up of the same bones; have basically the same arrangement of muscles, nerves, and blood vessels; are similarly arranged with respect to other structures; and have a similar pattern of embryonic development. These **homologous structures,** as they are called, are simply basic vertebrate forelimbs modified for different functions; that is, they indicate derivation from a common ancestor. However, some similarities are unrelated to evolutionary relationships. For instance, wings of insects and birds serve the same function but are quite dissimilar in both structure and development and are thus termed **analogous structures** (• Figure 7.20).

Why do dogs have tiny remnants of toes, called dewclaws, on their forefeet or hind feet or, in some cases, on all four feet (• Figure 7.21a)? These dewclaws are **vestigial structures**—that is, remnants of structures that were fully functional in their ancestors. Ancestral dogs had five toes on each foot, each of which contacted the ground. As they evolved, though, dogs became toe walkers with only four digits on the ground, and the thumbs and big toes were either lost or reduced to their present state. Although vestigial structures need not be totally functionless, many are, or at best have a reduced function. Remnants of toes are found in pigs, deer, and horses, and horses are occasionally born with extra toes. Whales and some snakes have a pelvis

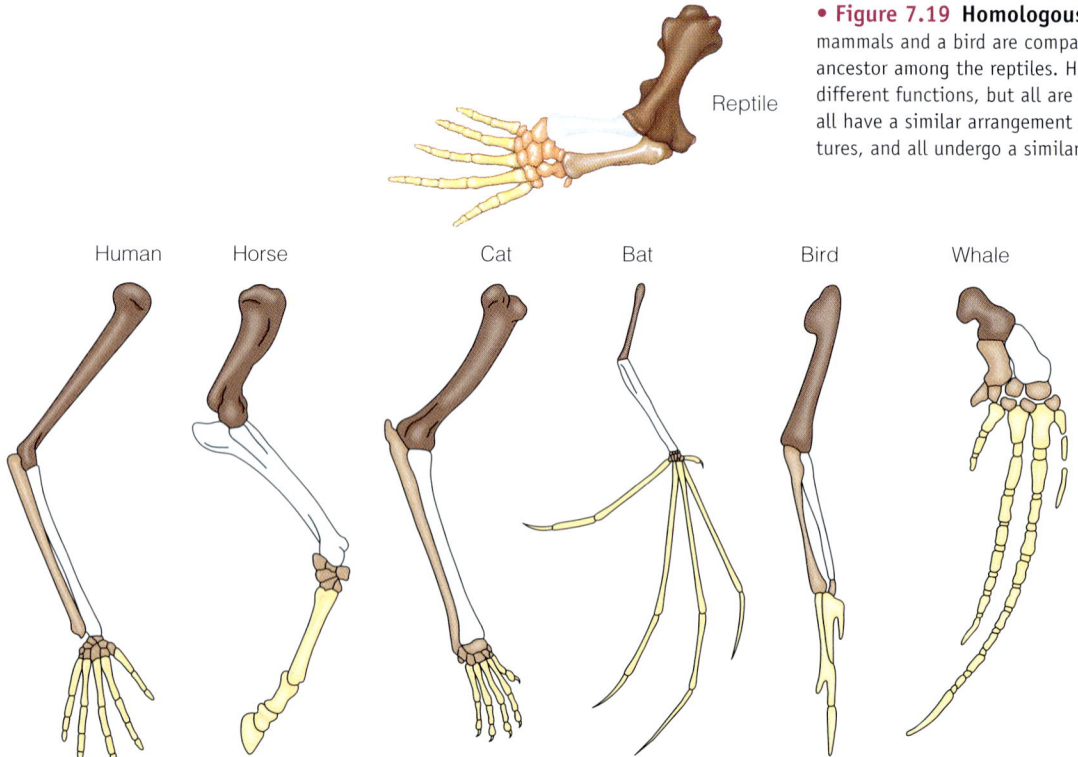

• **Figure 7.19 Homologous Organs** The forelimbs of mammals and a bird are compared with the forelimb of their ancestor among the reptiles. Homologous organs may serve different functions, but all are composed of the same bones, all have a similar arrangement with respect to other structures, and all undergo a similar embryonic development.

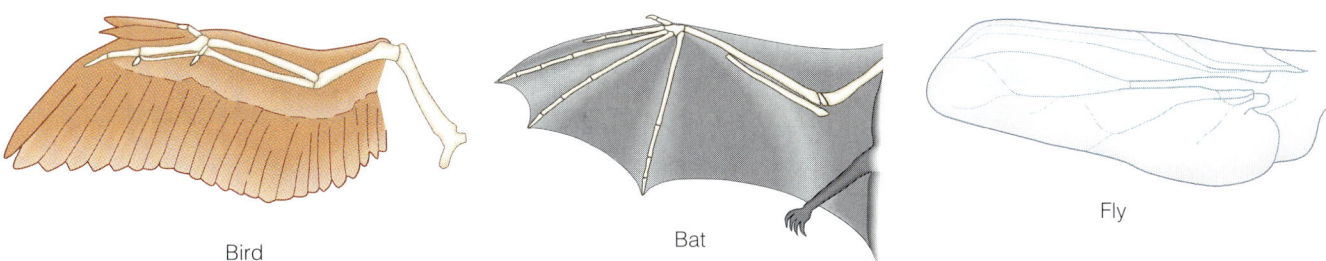

• **Figure 7.20 Analogous Organs** The fly's wings serve the same function as wings of birds and bats. Thus they are analogous, because they have a different structure and different embryologic development. Are any of these wings both analogous and homologous?

but no rear limbs (Figure 7.21b), except for a rare whale with hind limbs and two Cretaceous-aged fossil snakes with back legs.

No one would seriously argue that a dog's dewclaws are truly necessary, and it would be difficult to convince most people that wisdom teeth in humans are essential. After all, some people never have wisdom teeth, and in those that do, they commonly never erupt from the gums, or they come in impacted and cause significant pain. The reason? This is because the human jaw is too short to accommodate the ancestral number of teeth. Cave-dwelling salamanders and fish have vestigial eyes, whereas flightless birds have only remnants of wings, and flightless beetles have useless but fully developed wings beneath fused wing covers. In *On the Origin of Species*, Darwin said of vestigial structures, "They have been partially retained by the power of inheritance, and relate to a former state of things" (pp. 419–420).

Another type of evidence for evolution is observations of small-scale evolution—that is, microevolution—in living organisms. We have already mentioned one example: the adaptations of some plants to contaminated soils. As a matter of fact, small-scale changes take place rapidly enough that new insecticides and pesticides must be developed continually because insects and rodents develop resistance to existing ones. And development of antibiotic-resistant strains of bacteria is a continuing problem in medicine. Whether the variation in these populations previously existed or was established by mutations is irrelevant. In either case, some variant types lived and reproduced, bringing about a genetic change.

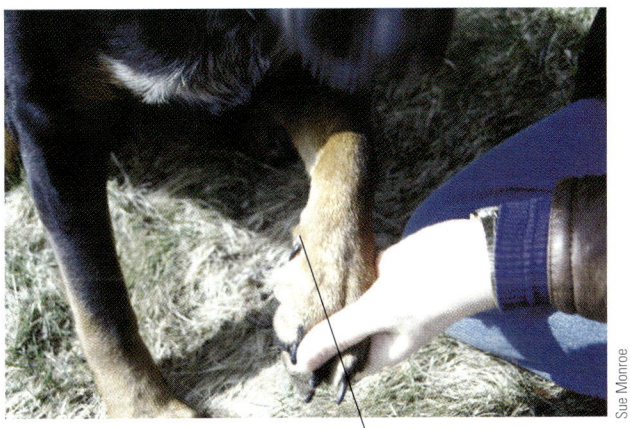

(a)

(b)

• **Figure 7.21 Vestigial Structures** (a) Notice the dewclaw, a vestige of the thumb, on the forefoot of this dog. (b) The Eocene-aged whale *Basilosaurus* had tiny vestigial back limbs, but it did not use limbs to support its body weight. Even today, whales have a vestige of a pelvis, and on a few occasions, whales with rear limbs have been caught.

Fossils: What Do We Learn from Them?

Fossil marine invertebrates found far from the sea, and even high in mountains, led early naturalists to conclude that the fossil-bearing rocks were deposited during a worldwide flood. In 1508, Leonardo da Vinci realized that the fossil distribution was not what one would expect from a rising flood, but the flood explanation persisted, and John Woodward (1665–1728) proposed a testable hypothesis. According to him, the weight of organic remains determined the order in which they settled from floodwaters, so, logically, it would seem that fossils in the oldest rocks should be heavier than those in younger ones. Woodward's hypothesis was quickly rejected because observations did not support it; fossils of various sizes, shapes, and weights are found throughout the fossil record.

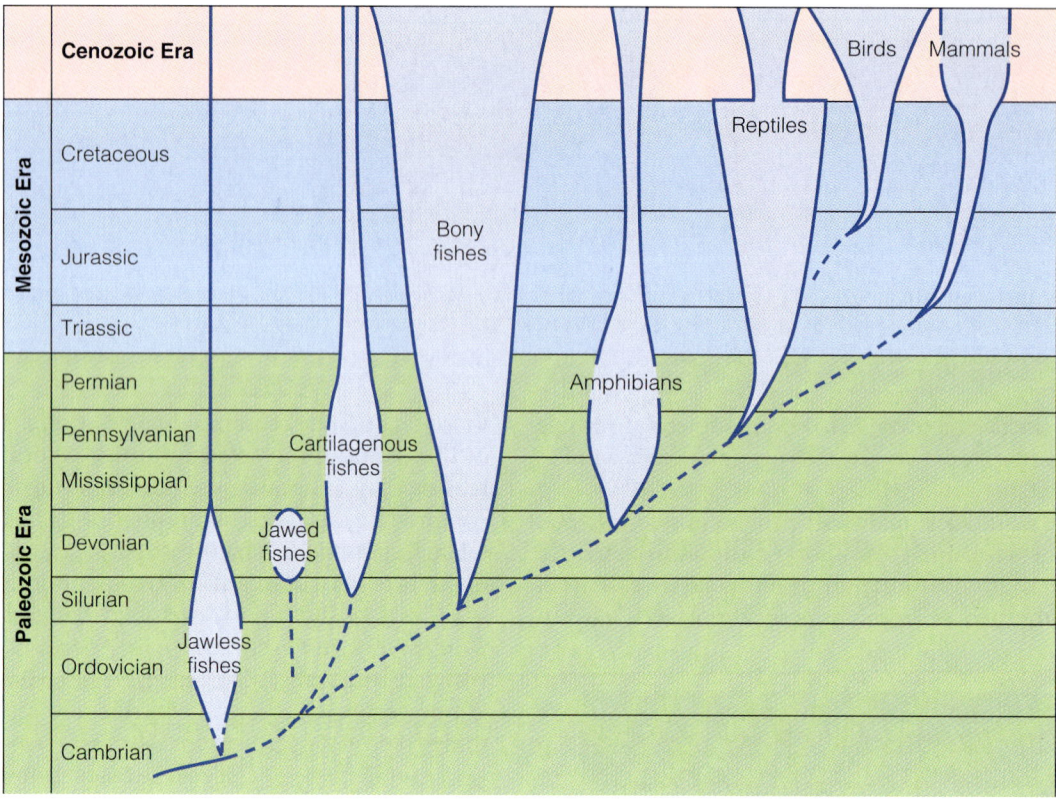

• **Figure 7.22 The Geologic Ranges of the Major Groups of Vertebrate Animals** The spindles indicate the relative abundance of members of each group, so there are far more existing species of bony fishes than there are amphibians, reptiles, birds, and mammals. The mass extinction at the end of the Mesozoic Era is shown by the marked restriction of the reptile spindle.

The fossil record does show a sequence of different organisms but not one based on density, size, shape, or habitat. Rather, the sequence consists of first appearances of various organisms through time (• Figure 7.22). One-celled organisms appear before multicelled ones, plants before animals, and invertebrates before vertebrates. Among vertebrates, fish appear first, followed in succession by amphibians, reptiles, mammals, and birds.

Fossils are much more common than many people realize, so we might ask, "Where are the fossils showing the diversification of horses, rhinoceroses, and tapirs from a common ancestor; or the origin of birds from reptiles; or the evolution of whales from a land-dwelling ancestor?" It is true that the origin and initial diversification of a group is the most poorly represented in the fossil record, but in these cases, as well as many others, fossils of the kind we would expect are known (see Perspective 7.2).

Horses, rhinoceroses, and tapirs may seem an odd assortment of animals, but fossils and studies of living animals indicate that they, along with the extinct titanotheres and chalicotheres, share a common ancestor (• Figure 7.23). If this statement is correct, then we can predict that as we trace these animals back in the fossil record, differentiating one from the other should become increasingly difficult (Table 7.1). And, in fact, the earliest members of each are remarkably similar, differing mostly in size and details of their teeth. As their diversification proceeded, though, differences became more apparent.

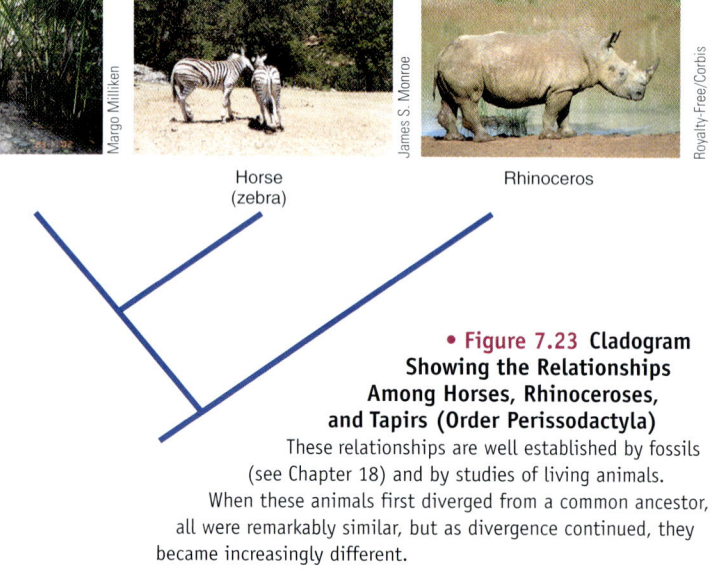

• **Figure 7.23 Cladogram Showing the Relationships Among Horses, Rhinoceroses, and Tapirs (Order Perissodactyla)** These relationships are well established by fossils (see Chapter 18) and by studies of living animals. When these animals first diverged from a common ancestor, all were remarkably similar, but as divergence continued, they became increasingly different.

7.2 Perspective

The Fossil Record and Missing Links

One of the most profound misconceptions about the fossil record is that so-called missing links invalidate the theory of evolution. In other words, paleontologists supposedly have no fossils that bridge the gaps between presumed ancestor and descendant groups of organisms. Indeed, there are many missing links so defined, but fossils do exist that are about as intermediate between groups as we could ever hope to find. Paleontologists call these *transitional fossils* to emphasize the fact that they possess characteristics that show relationships between ancestors and descendants. It is true that finding "missing links" between species is difficult, simply because species in an evolutionary lineage differ little from one another. However, when you consider higher levels in the classification scheme such as genera, families, orders, and classes, transitional forms become much more common. For example, some Mesozoic mammal-like reptiles so closely resemble mammals that it is difficult to distinguish them from the true mammals, and yet they also clearly possessed reptile characteristics (see Chapter 15).

We have already mentioned the fossil record of whales and noted that it has improved in the last two decades. The transitional forms between their land-dwelling ancestors and fully aquatic whales are now known, and their overall evolutionary history is firmly established (see Chapter 18). There are, however, still gaps in our knowledge, some of which will no doubt be filled as paleontologists find more fossils.

Among mammals whales are thoroughly adapted to an aquatic lifestyle, but they are not the only ones that have returned to an aqueous environment.

Members of the mammalian order Sirenia include four living species of dugong and manatee (Figure 1a), and the recently extinct Steller's sea cow. The latter, a large creature up to 8.5 m long and weighing several metric tons, was discovered in 1741 and hunted to extinction by 1768.

Dugongs and manatees are endangered species with perhaps as few as 10,000 existing worldwide. Here, however, our main interest is their evolutionary history.

Dugongs and manatees, collectively called sea cows, are highly specialized aquatic herbivorous mammals living in the temperate and tropical coastal waters of Africa, Asia, and North America. Some of these large, bulky, slow-moving animals (the largest measures 4.5 m long and weighs 1360 kg) inhabit freshwater environments, as in the inland waterways of Florida and Georgia. Paleontologists long assumed that sea cows evolved from land-dwelling mammal ancestors but had no definitive fossil evidence to support this opinion—until 1994, that is.

Today, sea cows swim with up-and-down movements of a muscular tail, much as whales do. They have paddle-like forelimbs, and the back legs have been lost; only a rodlike remnant of the pelvis remains as a vestigial structure. In addition, they have heavy, rounded ribs that act as ballast, and nostrils that open high on the skull. Prior to 1994, the oldest known fossil sea cows also possessed these features, although they differed in detail from those existing now. In 1994, however, fossils found in Jamaica revealed that a Middle Eocene manatee, now called *Pezosiren portelli*, had four well-developed limbs and a pelvis capable of supporting the animal on land (Figure 1b). In short, it is a sea cow with legs—a truly transitional fossil.

As opposed to sea cows today, *Pezosiren* probably swam more like an otter by flexing its back and kicking with its hind legs.

What features would we expect the ancestor of *Pezosiren* to possess? Certainly, it should have fewer aquatic characteristics than *Pezosiren* and more closely resemble the common ancestor of sea cows and the closely related elephants. Perhaps such fossils will be found in the future. For the present, though, we must be satisfied with the fact that this recently discovered transitional fossil allows us to more fully understand the evolution of sea cows.

(a)

(b)

Figure 1 (a) The Florida manatee, or sea cow, is a large, fully aquatic mammal living in the coastal waters of Florida and Georgia. (b) Skeleton of *Pezosiren portelli* found in Jamaica, an Eocene sea cow with four well-developed legs.

Some Jurassic-aged fossils from Germany have anatomic characteristics much like those of small, carnivorous dinosaurs, including dinosaur-like teeth, a long tail, braincase, and hind limbs, but they also have feathers. These creatures, known as *Archaeopteryx,* have characteristics we would expect in an animal transition between reptiles and birds. Until several years ago, the origin of whales was poorly understood. Fossil whales were common enough, but those linking fully aquatic animals with land-dwelling ancestors were largely absent. It turns out that this transition took place in southern Asia, a part of the world where the fossil record was poorly known. Now, however, a number of fossils are available that show how and when whales evolved (see Chapter 18).

Of course, we will never have enough fossils to document the evolutionary history of all living creatures because the remains of some organisms are more likely to be preserved than those of others, and the accumulation of sediments varies in both time and space. Nevertheless, more and more fossils are found that clearly illustrate ancestor–descendant relationships among many organisms.

The Evidence—A Summary

As already noted, scientists think that evolutionary theory is as well supported by evidence as any major theory. Indeed, several lines of investigation from studies as diverse as molecular biology to paleontology all confirm the basic concept of descent with modification. Disagreement exists on specific issues such as rates of evolution, the significance of some fossils, and precise relationships among some organisms, but overall the scientific community is overwhelmingly united in its support of the theory. And despite stories holding that scientists are unyieldingly committed to this idea and unwilling to investigate other explanations for the same data, just the opposite is true. If another scientific (testable) hypothesis offered a better explanation evolutionary scientists would be scrambling to investigate it. There is probably no better way in the sciences to gain respect and lasting recognition than to modify or replace an existing widely accepted theory.

What Would You Do?

Suppose someone were to tell you that evolution is "only a theory that has never been proven" and that "the fossil record shows a sequence of organisms in older to younger rocks that was determined by their density and habitat." Why is the first statement irrelevant to theories in general, and what kinds of evidence could you cite to refute the second statement?

SUMMARY

- The central claim of the theory of evolution is that all organisms have descended with modification from ancestors that lived during the past.
- The idea of evolution is not new, but the first widely accepted mechanism to account for evolution—inheritance of acquired characteristics—was proposed in 1809 by Jean-Baptiste de Lamarck.
- Charles Darwin's observations of variation in populations and artificial selection, as well as his reading of Thomas Malthus's essay on population, helped him formulate his idea of natural selection as a mechanism for evolution.
- In 1859 Charles Darwin and Alfred Wallace published their ideas of natural selection, which hold that in populations of organisms, some have favorable traits that make it more likely that they will survive and reproduce and pass on these variations.
- Gregor Mendel's breeding experiments with garden peas provided some of the answers regarding variation and how it is maintained in populations. Mendel's work is the basis for present-day genetics.
- Genes that are specific segments of chromosomes are the hereditary units in all organisms. Only the genes in sex cells are inheritable.
- Sexual reproduction and mutations, changes in chromosomes or genes, account for most variation in populations.
- Evolution by natural selection is a two-step process. First, variation must exist or arise and be maintained in interbreeding populations, and second, favorable variants must be selected for survival so that they reproduce.
- Many species evolve by allopatric speciation, which involves isolation of a small population from its parent population that is then subjected to different selection pressures.
- When diverse species arise from a common ancestor, it is called divergent evolution. The development of similar adaptive features in different groups of organisms results from convergent evolution and parallel evolution.
- Microevolution involves changes within a species, whereas macroevolution encompasses all changes above the species level. Macroevolution is simply the outcome of microevolution over time.
- Scientists have traditionally used phylogenetic trees to depict evolutionary relationships, but now they more commonly use cladistic analyses and cladograms to show these relationships.

- Background extinction occurs continually, but several mass extinctions have also taken place, during which Earth's biotic diversity has been decreased markedly.
- The theory of evolution is truly scientific because we can think of observations that would falsify it, that is, prove it wrong.
- Much of the evidence supporting evolutionary theory comes from classification, comparative anatomy, embryology, genetics, biochemistry, molecular biology, and present-day examples of microevolution.
- The fossil record also provides evidence for evolution in that it shows a sequence of different groups appearing through time, and some fossils show the features we would expect in the ancestors of birds, mammals, horses, whales, and so on.

IMPORTANT TERMS

allele, p. 128
allopatric speciation, p. 132
analogous structure, p. 140
artificial selection, p. 126
chromosome, p. 129
cladistics, p. 135
cladogram, p. 135
convergent evolution, p. 133
deoxyribonucleic acid (DNA), p. 129
divergent evolution, p. 133
gene, p. 128
homologous structure, p. 140
inheritance of acquired characteristics, p. 125
living fossil, p. 136
macroevolution, p. 134
mass extinction, p. 137
meiosis, p. 129
microevolution, p. 134
mitosis, p. 130
modern synthesis, p. 130
mosaic evolution, p. 135
mutation, p. 130
natural selection, p. 128
paleontology, p. 123
parallel evolution, p. 133
phyletic gradualism, p. 132
punctuated equilibrium, p. 132
species, p. 131
theory of evolution, p. 123
vestigial structure, p. 140

REVIEW QUESTIONS

1. Large-scale evolutionary changes resulting in the origin of new species or genera is an example of
   a. _____ phyletic gradualism; b. _____ binomial nomenclature; c. _____ macroevolution; d. _____ vestigial mutation; e. _____ Wallace's law.
2. Complex, helical molecules of deoxyribonucleic acid (DNA) are known as
   a. _____ paleofossils; b. _____ chromosomes; c. _____ allopatric cells; d. _____ homologous structures; e. _____ embryos.
3. One prediction of evolutionary theory is
   a. _____ species are fixed and immutable; b. _____ mammals should appear in the fossil record after fish; c. _____ acquired characteristics are inheritable; d. _____ sex cells are produced by mitosis; e. _____ humans evolved from monkeys.
4. A living animal or plant that is much like its ancient ancestors is a
   a. _____ structural dead end; b. _____ phyletic gradualist; c. _____ allopatric species; d. _____ chromosomal aberration; e. _____ living fossil.
5. Divergent evolution is a phenomenon that results in
   a. _____ the extinction of poorly adapted organisms; b. _____ interbreeding with other species to produce hybrids; c. _____ diversification of a species into many descendant species; d. _____ development of similar features in distantly related plants; e. _____ inheritance of acquired characteristics.
6. Studies of fossils and living animals indicate that horses are most closely related to which one of the following?
   a. _____ rhinoceroses; b. _____ pigs; c. _____ deer; d. _____ elephants; e. _____ whales.
7. The study of life as revealed by the fossil record is
   a. _____ archaeology; b. _____ physiology; c. _____ herpetology; d. _____ paleontology; e. _____ mineralogy.
8. Charles Darwin and _____ proposed _____ to account for evolution.
   a. _____ Gregor Mendel/chromosome theory of inheritance; b. _____ Alfred Wallace/natural selection; c. _____ Jean-Baptiste de Lamarck/inheritance of acquired characteristics; d. _____ Charles Lyell/uniformitarianism; e. _____ James Hutton/punctuated equilibrium.
9. According to the concept of phyletic gradualism,
   a. _____ species remain unchanged during most of their existence; b. _____ natural selection favors the largest for survival; c. _____ organisms evolve slowly but continuously; d. _____ organisms from different classes interbreed and give rise to new species; e. _____ the origin of a new species takes about 1000 years.
10. The type of cell division that yields sex cells is known as
    a. _____ symbiosis; b. _____ mutation; c. _____ meiosis; d. _____ cladistics; e. _____ artificial selection.
11. Does natural selection really mean that only the biggest, strongest, and fastest will survive? Explain.
12. Discuss the concept of allopatric speciation, and give two examples of how it might take place.

13. What kinds of evidence should we find in the fossil record if the theory of evolution is correct?
14. What is the significance of analogous and homologous structures? Give examples of each.
15. Inheritance of acquired characteristics seems so logical, and yet it does not work. Why not?
16. When we speak of classification, what do we mean by a *nested pattern of similarities*?
17. How do phyletic gradualism and punctuated equilibrium differ? Is there any evidence for one or both?

APPLY YOUR KNOWLEDGE

1. Draw three cladograms showing the possible relationships among sharks, whales, and bears. Which one shows the relationships best? Explain how you came to your conclusion.
2. Suppose that someone told you that Earth was visited during the Archean Eon by technologically advanced aliens who designed all organisms and then simply left them to evolve. Does this idea meet the criteria of a scientific theory? Explain.

CHAPTER 8

PRECAMBRIAN EARTH AND LIFE HISTORY—THE ARCHEAN EON

View of one of the Twin Lakes in the Beartooth Plateau on the border between Montana and Wyoming. The Beartooth Scenic Byway is advertised as one of the most beautiful in the United States. Most of the rocks of the plateau are Archean-aged gneisses.

John Boyd III

[OUTLINE]

INTRODUCTION

WHAT HAPPENED DURING THE EOARCHEAN?

CONTINENTAL FOUNDATIONS – SHIELDS, PLATFORMS, AND CRATONS

Archean Rocks

Greenstone Belts

PERSPECTIVE 8.1: *Geology of Grand Teton National Park*

Evolution of Greenstone Belts

ARCHEAN PLATE TECTONICS AND THE ORIGIN OF CRATONS

THE ATMOSPHERE AND HYDROSPHERE

How Did the Atmosphere Form and Evolve?

Earth's Surface Waters—The Hydrosphere

THE ORIGIN OF LIFE

Experimental Evidence and the Origin of Life

Submarine Hydrothermal Vents and the Origin of Life

The Oldest Known Organisms

ARCHEAN MINERAL RESOURCES

SUMMARY

ThomsonNOW™ Explore interactive tutorials, animations, or practice problems available on the ThomsonNow website at www.thomsonedu.com/login.

[CHAPTER OBJECTIVES]

At the end of this chapter, you will have learned that

- Precambrian time, which accounts for most of geologic time, is divided into two eons: the Archean and the younger Proterozoic.

- The Archean geologic record is difficult to interpret because many of the rocks are metamorphic, deformed, deeply buried, and they contain few fossils.

- Each continent has at least one area, called a shield, in which Precambrian rocks are exposed, and a buried extension of the shield known as a platform. A shield and its platform collectively make up a craton.

- All cratons show evidence of deformation, metamorphism, and emplacement of plutons, but they have been remarkably unaffected by these activities since Precambrian time.

- The two main associations of Archean rocks are granite-gneiss complexes, which are by far the most common, and greenstone belts that consist of igneous and sedimentary rocks.

- Greenstone belts likely formed in several tectonic settings, but many appear to have evolved in back-arc marginal basins and in rifts within continents.

- Plate tectonics was taking place during the Archean, but plates probably moved faster, and igneous activity was more common then because Earth possessed more heat from radioactive decay.

- Gases released during volcanism were responsible for the origin of the hydrosphere and atmosphere, but the atmosphere so formed had little or no free oxygen.

- The oldest known fossil organisms are single-celled bacteria and chemical traces of bacteria-like organisms known as archaea. Bacteria known as blue-green algae produced irregular mats and moundlike structures called stromatolites.

- Resources found in Archean rocks include gold, platinum, copper, zinc, and iron.

Introduction

The duration of geologic time is beyond comprehension. After all, we think of time from the human viewpoint of years or perhaps a few decades, but we have no frame of reference for time measured in millions or billions of years. Accordingly, the Precambrian, which lasted for more than 4 billion years, is longer than we can even imagine, and Earth has existed for 4.6 billion years. To put 4.6 billion in perspective, consider this: Suppose that 1 second equals 1 year. If that were the case, and you were to count to 4.6 billion, the task would take you and your descendants nearly 146 years! Or suppose that a 24-hour clock represented 4.6 billion years. In this case, the Precambrian alone would be more than 21 hours long and constitute about 88% of all geologic time (• Figure 8.1).

Precambrian is a widely used term that refers to both time and rocks. In regard to time, it includes all geologic time from Earth's origin 4.6 billion years ago to the beginning of the Phanerozoic Eon 542 million years ago. Precambrian also encompasses all rocks lying below Cambrian-aged rocks. Unfortunately, no rocks are known for the first 640 million years of geologic time, so our geologic record actually begins 3.96 billion years ago with the oldest known rocks on Earth other than meteorites. The geologic record we do have for the Precambrian is difficult to interpret, particularly for the older part of the Precambrian, because many of the rocks have been metamorphosed and complexly deformed, and in many areas they lie deeply buried beneath younger rocks.

Because of the complexities of Precambrian rocks, and the fact that the few fossils they contain are of little use in stratigraphy, establishing formal subdivisions of the Precambrian is difficult. In 1982, in an effort to standardize terminology, the North American Commission on Stratigraphic Terminology proposed two eons for the Precambrian, the *Archean* and the *Proterozoic*, both of which are now widely used. More recent work in 2004 by the International Commission on Stratigraphy (ICS) has refined the Precambrian subdivisions as shown in • Figure 8.2.

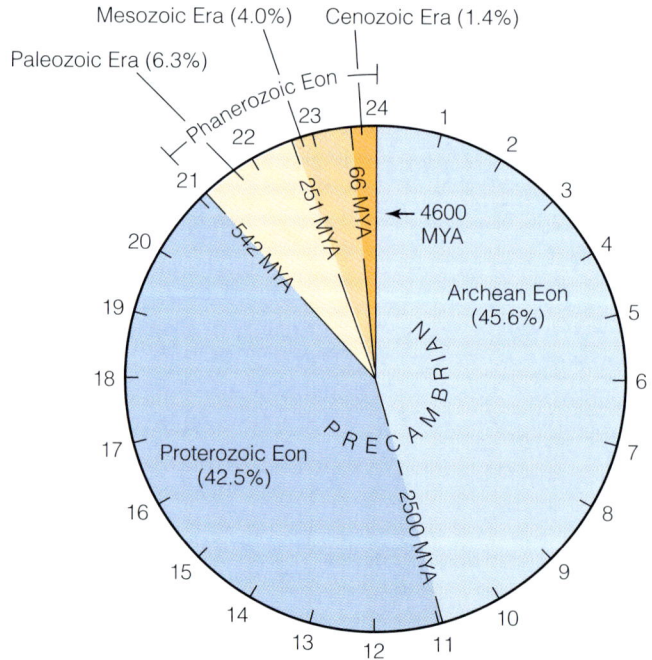

• **Figure 8.1 Geologic Time Represented by a 24-Hour Clock**
If 24 hours represented all geologic time, the Precambrian would be more than 21 hours long, more than 88% of the total.

The Precambrian subdivisions in Figure 8.2 are time units. That is, they are based on absolute ages rather than time-stratigraphic units. This departs from standard practice in which geologic systems based on stratotypes are the basic time-stratigraphic units (see Chapter 5). For example, the Cambrian Period, a term related to time, is based on the Cambrian System, a time-stratigraphic unit with a stratotype in Wales. In contrast, the Precambrian eons, eras, and periods have no stratotypes. They are strictly terms denoting time.

What Happened During the Eoarchean?

According to the data in Figure 8.2, *Eoarchean* refers to all geologic time from Earth's origin until the onset of the Paleoarchean 3.6 billion years ago. Also, remember that the oldest known rocks on Earth are 3.96 billion years old, so we have no geologic record for most of the Eoarchean. Nevertheless, we know of some events that took place during this time. For one thing, it was during the Eoarchean that Earth accreted from planetesimals and differentiated into a core and mantle, and at least some crust was present. Like the other terrestrial planets and the Moon, Earth was bombarded by comets and meteorites, and volcanic activity was ubiquitous. An atmosphere formed, but one that was quite different from the oxygen-rich one we have now, and once the planet cooled sufficiently, the surface waters began to accumulate.

If we could somehow go back and visit Earth shortly after it accreted, we would see a rapidly rotating, hot, barren, waterless planet bombarded by meteorites and comets, no continents, intense cosmic radiation and widespread volcanism (• Figure 8.3). It was not what you would call a hospitable planet, but it evolved and gradually became the way it is now. The oldest known rocks on Earth, the 3.96-billion-year-old Acasta Gneiss in Canada and 3.8-billion-year-old rocks from Montana and Greenland, indicate that some continental crust had evolved by Eoarchean time. In addition, some sedimentary rocks in Australia contain 4.4-billion-year-old detrital zircons ($ZrSiO_4$) so source rocks at least that old must have existed.

We know from studies in geology and astronomy that Earth's rate of rotation is slowing because of tidal effects. That is, friction caused by the Moon on Earth's oceanic waters, as well as its landmasses, causes its rate of rotation to slow very slightly every year. When Earth first formed, it probably rotated in as little as 10 hours. Another effect of the Earth-Moon tidal interaction is the recession of the Moon from Earth at a few centimeters per year. Accordingly, during the Archean the view of the Moon would have been spectacular.

Geologists agree that shortly after Earth formed, it was exceedingly hot and volcanism was widespread (Figure 8.3). However, rather than being a fiery orb for half a billion years as was formerly accepted, some geologists now think that Earth cooled sufficiently by 4.4 billion years ago for surface

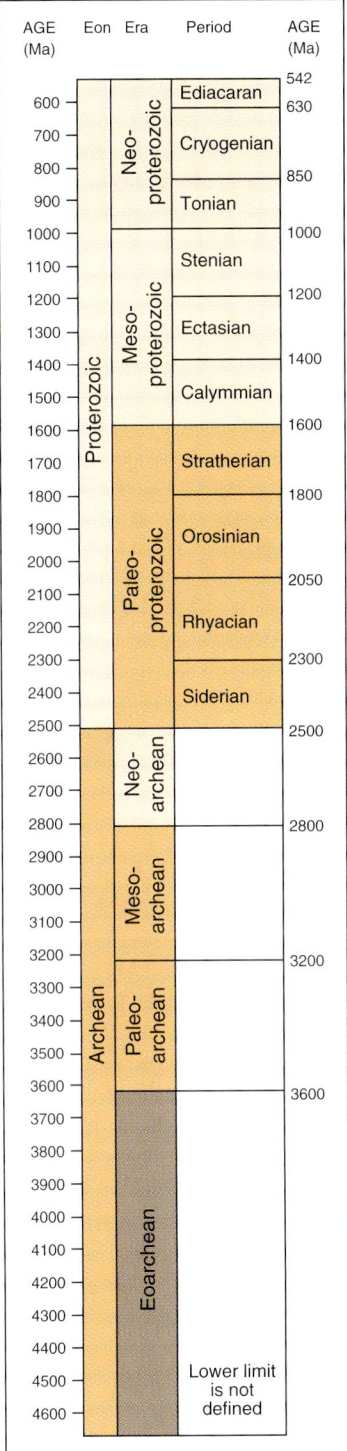

• **Figure 8.2 The Precambrian Geologic Time Scale** This most recent version of the geologic time scale was published by the International Commission on Stratigraphy (ICS) in 2004. Notice the use of the prefixes *eo* (early or dawn), *paleo* (old or ancient), *meso* (middle), and *neo* (new or recent). The age columns on the left and right sides of the time scale are in hundreds and thousands of millions of years (1800 million years = 1.8 billion years, for example).

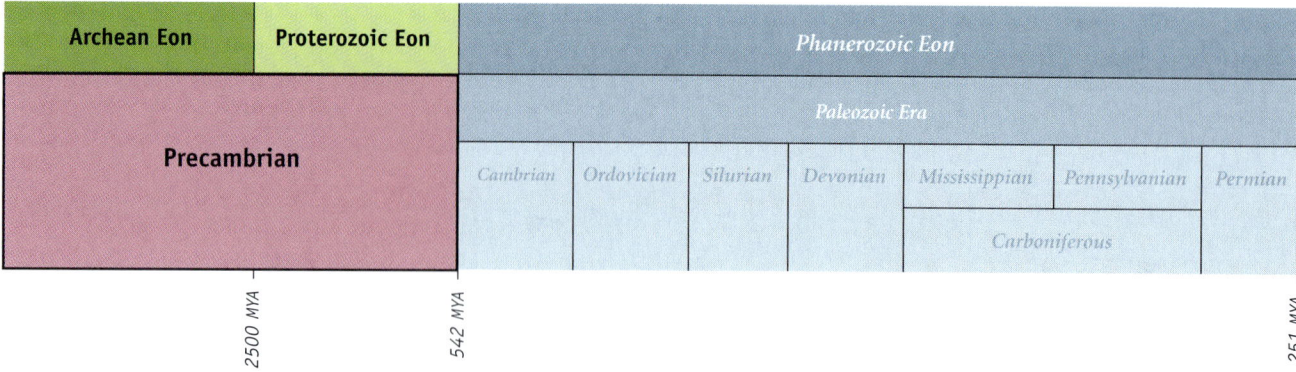

• **Figure 8.3 Earth As It May Have Appeared Soon After It Formed** No rocks are known from this early time in Earth's history, but geologists can make reasonable inferences about the nature of the newly-formed planet.

waters to accumulate. They base this conclusion on oxygen-18 to oxygen-16 ratios in tiny inclusions of oxygen trapped in zircon crystals that indicate reactions with surface water.

The first crust that formed was probably thin and made up of ultramafic rock—that is, igneous rock with less than 45 percent silica. Upwelling mantle currents of mafic magma disrupted this early crust, and numerous subduction zones developed to form the first island arcs (• Figure 8.4a). Weathering of the mafic rocks of island arcs yielded sediments richer in silica, and some of the magma in the arc also became more enriched in silica. Collisions between island arcs eventually formed a few continental cores as silica-rich materials were metamorphosed (Figure 8.4b). Larger groups of merged island arcs, or protocontinents, grew faster by accretion along their margins than smaller ones did, and eventually the first continental nuclei or cratons formed (Figure 8.4c).

In Chapter 1, we introduced the concept of a dynamic Earth and the systems that interact to bring about change. During the Eoarchean these systems became operative, but not all at the same time nor in their present forms. For instance, Earth did not fully differentiate into a core, mantle, and crust until millions of years after it formed. Once it did, though, internal heat was responsible for interactions among plates as they began to diverge, converge, and slide past one another at transform plate boundaries. Indeed, continents began to grow, and continue to do so, by accretion along convergent plate boundaries.

Continental Foundations—Shields, Platforms, and Cratons

Continents are more than simply land areas above sea level. In fact, they consist of rocks with an overall composition similar to that of granite, and continental crust is thicker and less dense than oceanic crust, which is made up of basalt and gabbro. Furthermore, a **Precambrian shield** consisting of a vast area or areas of exposed ancient rocks is found on all continents. Continuing outward from the shields are broad **platforms** of buried Precambrian rocks that underlie much of each continent. Collectively, a shield and platform make up a **craton,** which we can think of as a continent's ancient nucleus.

The cratons are the foundations of continents, and along their margins more continental crust was added as they evolved to their present sizes and shapes. Both Archean and Proterozoic rocks are present in cratons, many of which indicate several episodes of deformation accompanied by metamorphism, igneous activity, and mountain building. However, the cratons have experienced remarkably little deformation since the Precambrian.

In North America, the exposed part of the craton is the **Canadian shield,** which occupies most of northeastern Canada, a large part of Greenland, the Adirondack Mountains

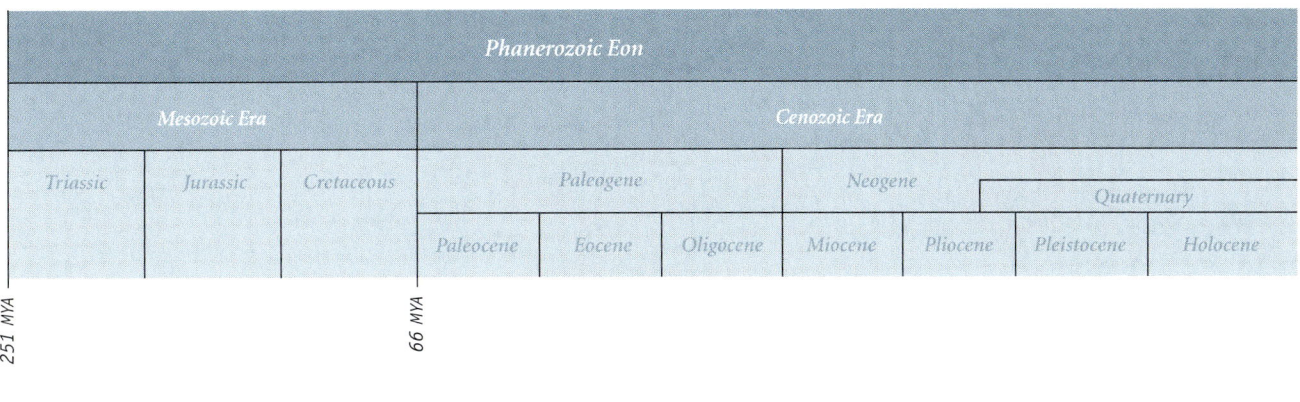

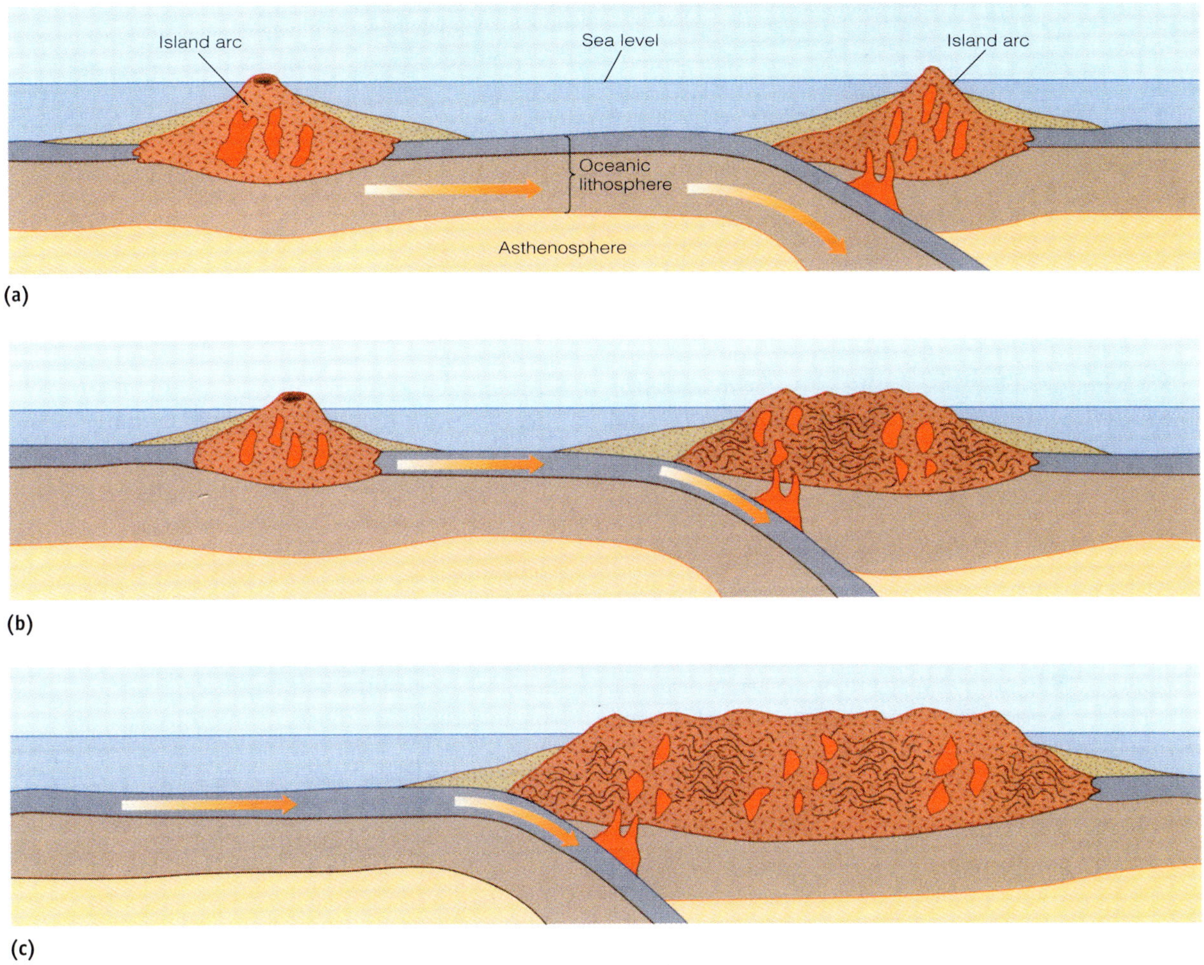

• **Figure 8.4 Origin of Granitic Continental Crust** (a) An andesitic island arc forms by subduction of oceanic lithosphere and partial melting of basaltic oceanic crust. Partial melting of andesite yields granitic magma. (b) The island arc in (a) collides with a previously formed island arc, thereby forming the nucleus of a continent. (c) The processes occur again when the island arc in (b) collides with the evolving continent.

of New York, and parts of the Lake Superior region in Minnesota, Wisconsin, and Michigan (• Figure 8.5). Overall, the Canadian shield is an area of subdued topography, numerous lakes, and exposed Archean and Proterozoic rocks thinly covered in places by Pleistocene glacial deposits. The rocks themselves are plutonic, volcanic, and sedimentary, as well as metamorphic equivalents of all these (• Figure 8.6a).

Actually, the Canadian shield as well as the adjacent platform are made up of numerous units or smaller cratons that amalgamated along deformation belts during the Paleoproterozoic (see Chapter 9). Absolute ages and structural trends differentiate these smaller cratons from one another.

Drilling and geophysical evidence indicate that Precambrian rocks underlie much of North America, but

CONTINENTAL FOUNDATIONS—SHIELDS, PLATFORMS, AND CRATONS **151**

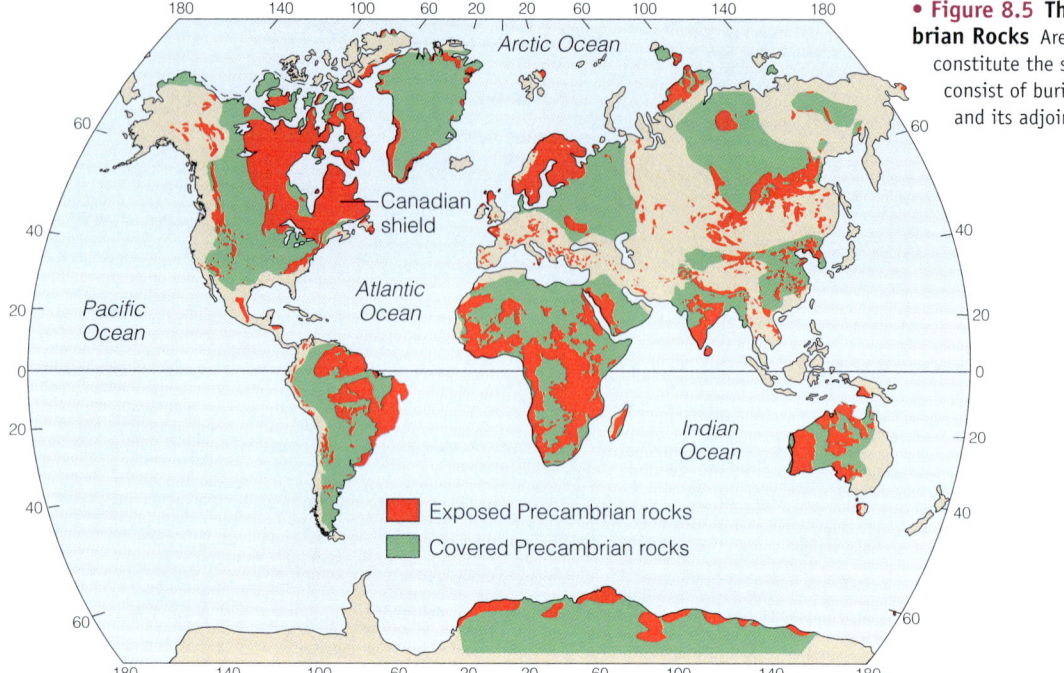

• **Figure 8.5 The Distribution of Precambrian Rocks** Areas of exposed Precambrian rocks constitute the shields, whereas the platforms consist of buried Precambrian rocks. A shield and its adjoining platform make up a craton.

beyond the Canadian shield they are exposed only in areas of deep erosion or uplift. For instance, Archean and Proterozoic rocks are present in the deeper parts of the Grand Canyon of Arizona as well as in the Appalachian Mountains and the Rocky Mountains (Figure 8.6b).

Archean Rocks

Only 22 percent of Earth's exposed Precambrian crust is Archean, with the largest exposures in Africa and North America (see Perspective 8.1). Archean crust is made up of a variety of rocks, but we can characterize most of them as greenstone belts and granite-gneiss complexes, the latter being by far the most common. **Granite-gneiss complexes** are actually composed of a variety of rocks, with granitic gneiss and granitic plutonic rocks being the most common, that were probably derived from plutons emplaced in volcanic island arcs. Nevertheless, there are other rocks ranging from peridotite to sedimentary rocks, all of which have been metamorphosed. Greenstone belts are subordinate, accounting for only 10 percent of Archean rocks, and yet they are important in unraveling some of the complexities of Archean tectonic events.

Greenstone Belts

An ideal **greenstone belt** has three major rock associations; volcanic rocks are most common in the lower and middle

(a)

(b)

• **Figure 8.6 Archean Rocks** (a) Outcrop of Archean gneiss in the Canadian shield. This rock exposure is on the shore of Lake Huron in Ontario, Canada. (b) Beyond the Canadian shield Archean rocks are found in areas of deep erosion and uplift. The rocks making up much of the Teton Range in Wyoming are gneiss, schist, and granite.

Perspective 8.1

Geology of Grand Teton National Park

About 70% of North America is underlain by Precambrian rocks, but outside the Canadian shield they lie deeply buried in most places. That is, they are part of the platform that, along with the shield, constitutes the North American craton. However, Precambrian rocks are well exposed in areas of deep erosion such as the Grand Canyon of Arizona and where Earth's crust has been deformed during mountain-building episodes. One particularly scenic area to view these rocks is in the Teton Range of Grand Teton National Park in northwestern Wyoming (Figure 1a).

The Teton Range is less than 10 million years old, but the rocks that make up the range are predominantly Precambrian. In fact, most of the range consists of Archean layered gneiss and schist, although Proterozoic-aged rocks are also present. The layering in the gneiss results from the different proportions of light minerals (quartz and feldspars) and dark minerals (biotite and hornblende) making up these rocks. In many areas these Archean layered gneisses include layers of schist which is dominated by micas, thereby giving them their flaky to platy appearance. In a few areas, thin layers of marble are found within the gneiss. These Archean-aged metamorphic rocks were probably originally sedimentary and volcanic rocks that formed on the seafloor, perhaps in an island arc system.

In addition to the layered gneiss and schist, the Teton Range also has abundant Proterozoic-aged granite and granite pegmatites that were intruded into the older rocks. In fact, the most prominent peak in the range, the Grand Teton, as well as nearby peaks, is eroded from this huge, irregularly shaped body of granite. The complex of Archean and Proterozoic rocks in the Teton Range are cut by dark green to black basaltic dikes, which formed about 1.3 billion years ago. Paleozoic-, Mesozoic-, and Cenozoic-aged rocks are also present but found only on the west side of the range and in the valley (Jackson Hole) to the east of the range.

Higher and larger mountain ranges are present in the Rocky Mountain region, but few are as scenic or rise as abruptly as the Teton Range, which ascends nearly vertically 2100 m above the valley (Jackson Hole) to the east (Figure 1b). The area of the present-day Teton Range has had a history of deformation and mountain building that goes back at least 90 million years. Nevertheless, the most recent episode of deformation began less than 10 million years ago, when uplift along the Teton fault (a normal fault) began. As up to 6100 m of uplift took place on the Teton fault, the uplifted block was tilted toward the west, and the Paleozoic, Mesozoic, and Cenozoic rocks were mostly stripped away by erosion, exposing the Precambrian rocks. Present-day faults that cut recent sediments along the east side of the range indicate that uplift continues even now.

Uplift and deep erosion, especially by glaciers, has yielded the spectacular, rugged topography we see today. Now the Teton Range supports about a dozen very small glaciers, but during the Pleistocene Epoch it was much more heavily glaciated. These moving bodies of ice deeply scoured the valleys and intricately sculpted the range, thereby forming many classic glacial features such as horn peaks, cirques, arêtes, and end moraines.

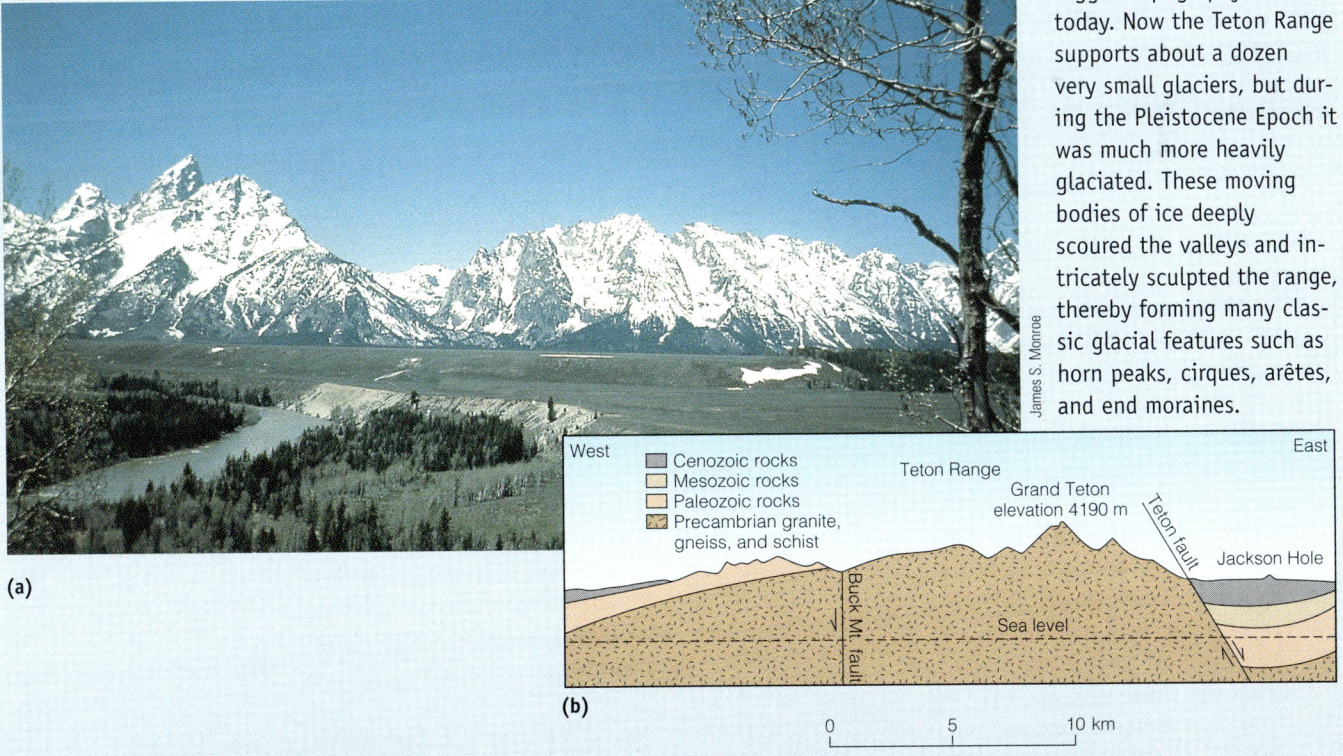

Figure 1 (a) View from the east of the Teton Range. The Grand Teton is the highest peak visible. (b) East–West cross section of the Teton Range showing uplift along the Teton fault. This fault is not visible in (a), but it lies at the base of the mountains.

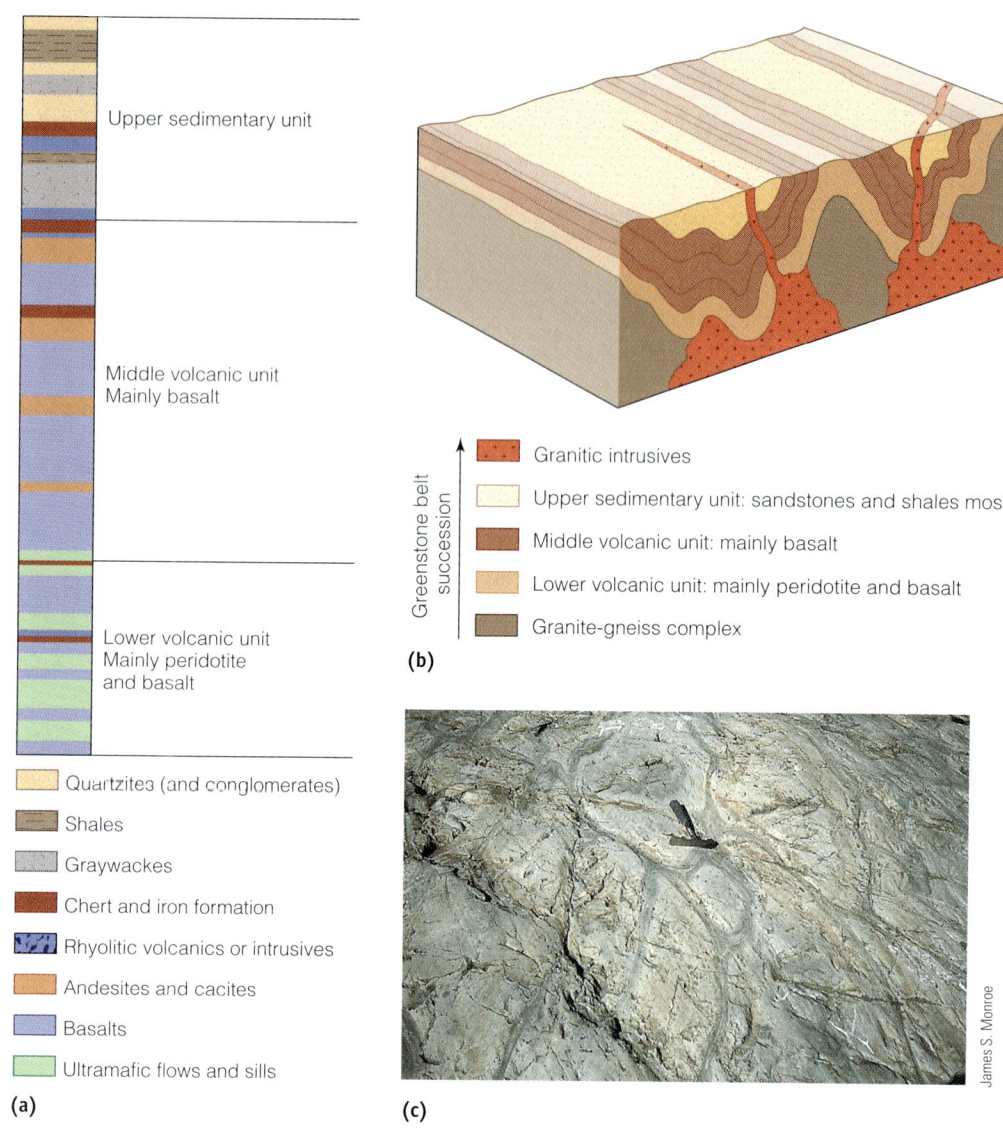

• **Figure 8.7 Greenstone Belts** (a) Idealized column of an Archean greenstone belt. (b) Two adjacent greenstone belts showing their synclinal structure and relationships to granite-gneiss complexes. (c) Pillow lava in the Ispheming greenstone belt at Marquette, Michigan.

parts, whereas the upper rocks are mostly sedimentary (• Figure 8.7a and b). They typically have a synclinal structure, measure anywhere from 40 to 250 km wide and 120 to 800 km long, and have been intruded by granitic magma and cut by thrust faults. Many of the igneous rocks are greenish, because they contain green minerals such as chlorite, actinolite, and epidote that formed during low-grade metamorphism.

Thick accumulations of pillow lava are common in greenstone belts, indicating that much of the volcanism was subaqueous (Figure 8.7c). Pyroclastic materials, in contrast, almost certainly formed by subaerial eruptions where large volcanic centers built above sea level. The most interesting igneous rocks in greenstone belts are *komatiites* that cooled from ultramafic lava flows, which are rare in rocks younger than Archean, and none occur now.

To erupt, ultramafic magma (one with less than 45% silica) requires near-surface magma temperatures of more than 1600°C; the highest recorded surface temperature for recent lava flows is 1350°C. During Earth's early history, however, it possessed more radiogenic heat, and the mantle was as much as 300°C hotter than it is now. Given these conditions, ultramafic magma could reach the surface, but as Earth's production of radiogenic heat decreased, the mantle cooled and ultramafic flows no longer occurred.

Sedimentary rocks are found throughout greenstone belts, but they predominate in the upper unit (Figure 8.7). Many of these rocks are successions of *graywacke* (a sandstone with abundant clay and rock fragments) and *argillite* (slightly metamorphosed mudrocks). Small-scale cross-bedding and graded bedding indicate these rocks represent turbidity current deposition (see Figure 6.4).

Some of the other sedimentary rocks such as quartz sandstone and shale show evidence of deposition in delta, tidal flat, barrier island, and shallow marine environments.

Several other kinds of sedimentary rocks are also present, including conglomerate, chert, and carbonates, although none are very abundant. Iron-rich rocks known as *banded iron formations* are also found, but they are more typical of Proterozoic deposits, and therefore we discuss them in Chapter 9.

Before leaving this topic entirely, we should mention that the oldest large, well-preserved greenstone belts are in South Africa, dating from about 3.6 billion years ago. In North America most greenstone belts are found in the Superior and Slave cratons of the Canadian shield (• Figure 8.8), but they are also found in Michigan, Wyoming, and several other areas. Most formed between 2.7 and 2.5 billion years ago.

Evolution of Greenstone Belts

Greenstone belts probably developed in several tectonic settings, but an appealing model for the origin of some of them relies on Archean plate tectonics and the develop-

• **Figure 8.8 Greenstone Belts in North America** Archean greenstone belts (shown in dark green) of the Canadian shield are mostly in the Superior and Slave cratons.

ment of a greenstone belt in a **back-arc marginal basin** that subsequently closes (• Figure 8.9). Remember that back-arc marginal basins are found between a continent and a volcanic island arc—the Sea of Japan, for instance. Thus there is an early stage of extension when the back-arc marginal basin forms, accompanied by volcanism, emplacement of plutons, and sedimentation, followed by an episode of compression when the basin closes. During this latter stage, the greenstone belt rocks are deformed, metamorphosed, and intruded by granitic magma.

The back-arc basin model for the origin of greenstone belts is well accepted, but there are geologists that think that some greenstone belts form in intracontinental rifts above a rising mantle plume (• Figure 8.10). As the plume rises beneath sialic (silica- and aluminum-rich) crust, it spreads and generates tensional forces that cause rifting. The plume is the source of the lower and middle volcanic units of a developing greenstone belt, and erosion of the volcanic rocks and the flanks of the rift account for the upper sedimentary unit. And finally, there is an episode of subsidence, deformation, low-grade metamorphism, and emplacement of plutons (Figure 8.10).

What Would You Do?

Precambrian rocks have few fossils, and those present are not very useful in stratigraphy. So how would you demonstrate that Archean rocks are the same age as those elsewhere? Also, could you use the principles of superposition, original horizontality, and inclusions to decipher the geologic history at a single location? Explain.

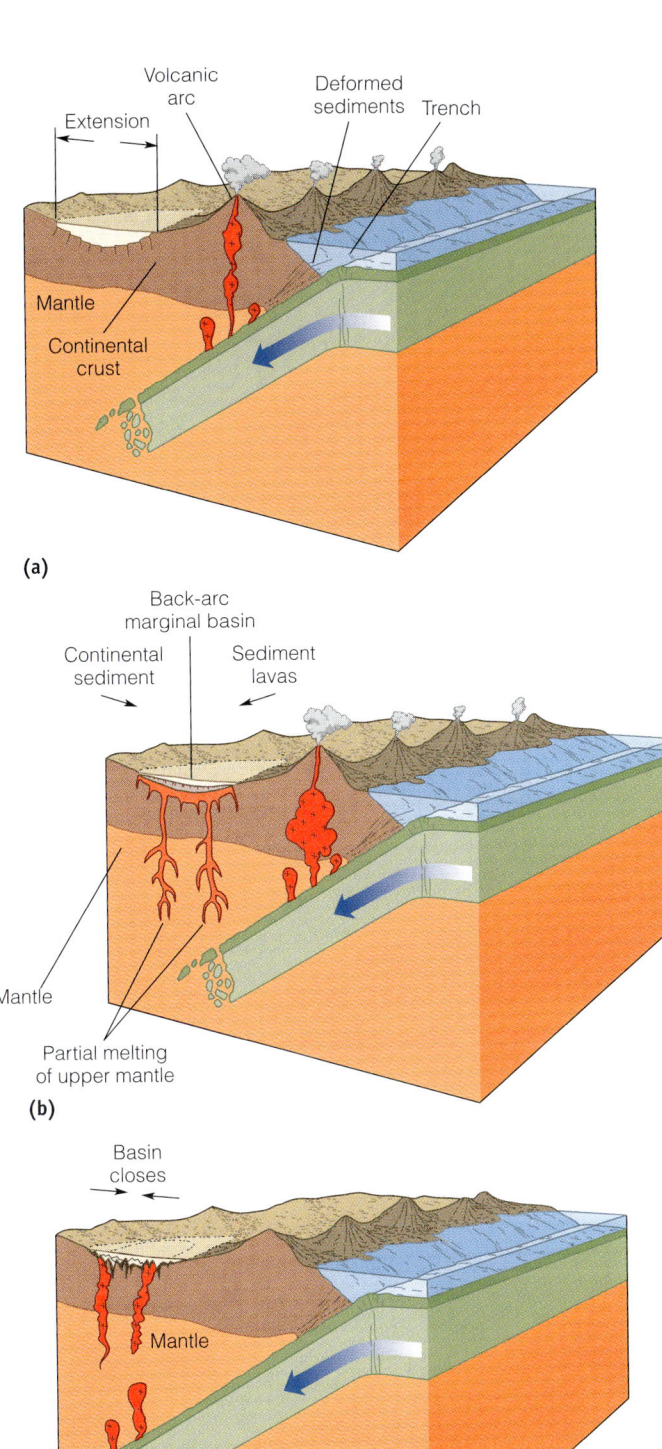

• **Figure 8.9 Origin of a Greenstone Belt in a Back-Arc Marginal Basin** **(a)** Rifting on the continent side of a volcanic arc forms a back-arc marginal basin. Partial melting of subducted oceanic crust supplies andesitic and dioritic magmas to the island arc. **(b)** Basalt lavas and sediment derived from the continent and island arc fill the back-arc marginal basin. **(c)** Closure of the back-arc marginal basin causes compression and deformation. The evolving greenstone belt is deformed into a syncline-like structure into which granitic magmas intrude.

CONTINENTAL FOUNDATIONS—SHIELDS, PLATFORMS, AND CRATONS **155**

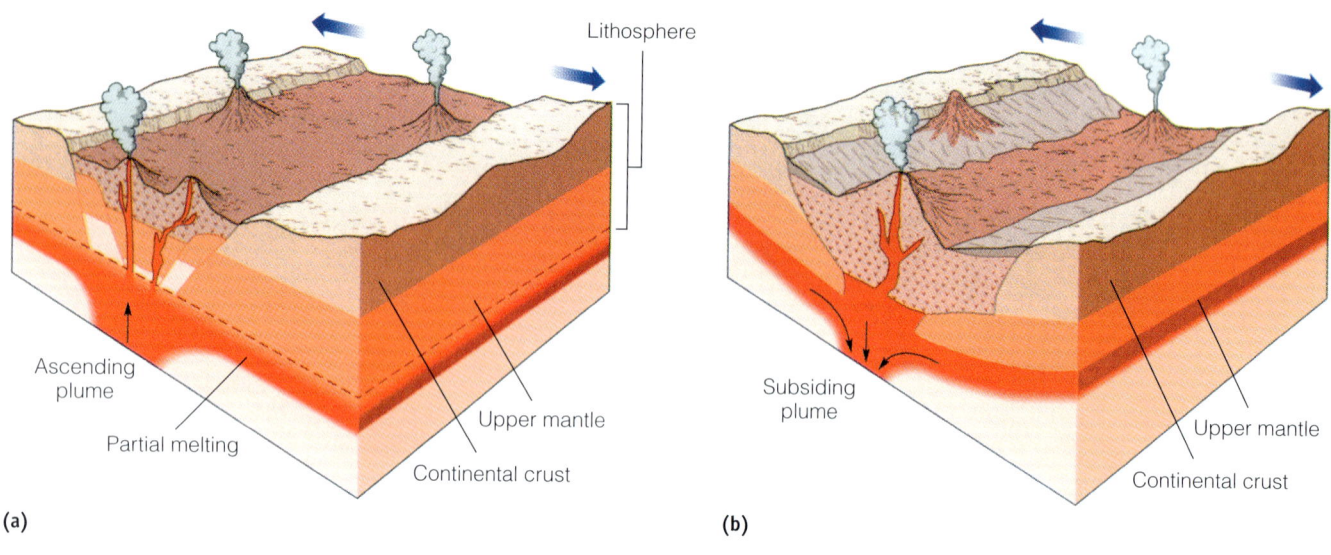

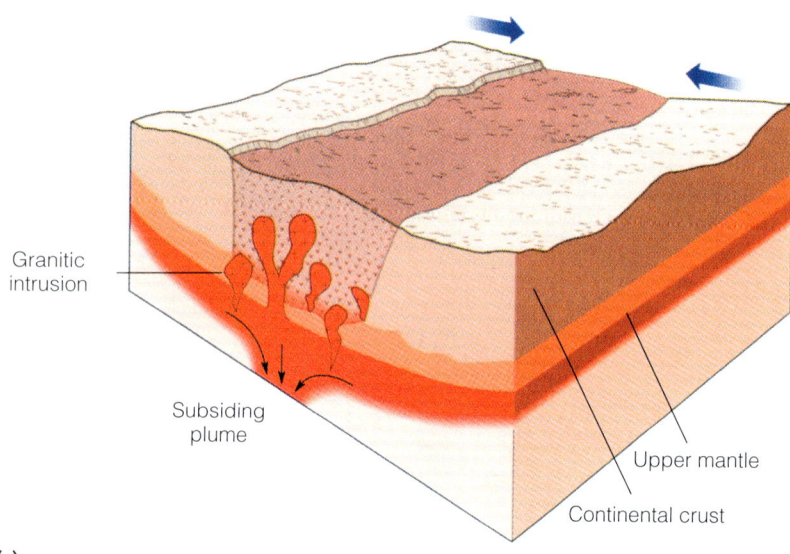

• **Figure 8.10 Model for the Formation of a Greenstone Belt in an Intracontinental Rift** (a) An ascending mantle plume causes rifting and volcanism. (b) As the plume subsides, erosion of the rift flanks accounts for deposition of sediments. (c) Closure of the rift causes compression and deformation. Granitic magma intrudes the greenstone belt.

Archean Plate Tectonics and the Origin of Cratons

The present plate tectonic regime of opening and closing ocean basins has been a primary agent in Earth evolution since the Paleoproterozoic. Now, though, most geologists are convinced that some kind of plate tectonic activity took place during the Archean as well, but it differed in detail from what is going on now. For one thing, with more residual heat from Earth's origin and more radiogenic heat, plates must have moved faster and magma was generated more rapidly. As a result, continents no doubt grew more rapidly along their margins, a process called **continental accretion,** as plates collided with island arcs and other plates. Also, ultramafic lava flows (komatiites) were more common.

There were, however, marked differences between the Archean world and the one that followed. We have little evidence of Archean rocks deposited on broad, passive continental margins, but associations of passive continental margin sediments are widespread in Proterozoic terrains. Deformation belts between colliding cratons indicate that Archean plate tectonics was active, but the *ophiolites* so typical of younger convergent plate boundaries are rare. Nevertheless, Neoarchean ophiolites are now known from several areas, including a 2.5-billion-year-old ophiolite in China and one 2.8 billion years old in Russia.

Certainly several small cratons existed during the Archean and grew by accretion along their margins during the rest of that eon. Remember, though, that they amalgamated into a larger unit during the Proterozoic (see Chapter 9). By the end of the Archean, perhaps 30% to 40% of the present volume of continental crust existed. A plate tectonic model for the Archean crustal evolution of the southern Superior craton of Canada relies on a series of events, including greenstone belt evolution, plutonism, and deformation (• Figure 8.11a). We can take this as a model for Archean crustal evolution in general.

The events leading to the origin of the southern Superior craton (Figure 8.11b) are part of a more extensive orogenic episode that took place near the end of the Archean. Deformation at this time was responsible for the formation of the Superior and Slave cratons, and the origin of some Archean rocks in other parts of the Canadian shield as well

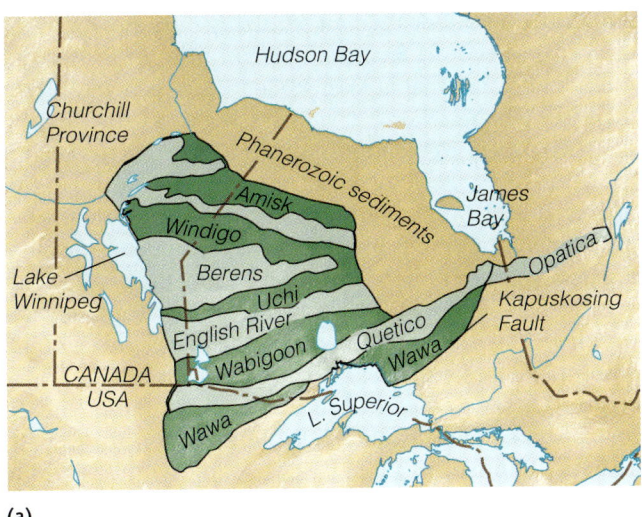
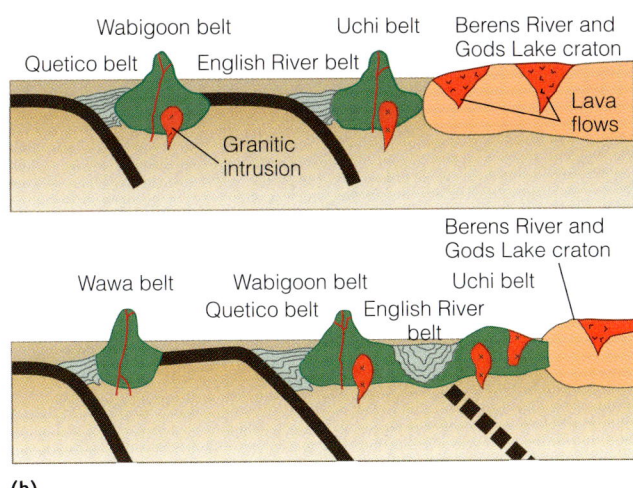

• **Figure 8.11 Origin of the Southern Superior Craton** (a) Geologic map showing greenstone belts (dark green) and areas of granite-gneiss complexes (light green). (b) Plate tectonic model for evolution of the southern Superior craton. The figure is a north–south cross section through the area, and the upper diagram shows an earlier stage of development than the lower one.

as in Wyoming, Montana, and the Mississippi River Valley. This deformation was the last major Archean event in North America. By the time it was over, several sizable cratons existed that are now found in the older parts of the Canadian shield.

What Would You Do?

As a high school science teacher you must explain to your students that an initially hot Earth has continued to cool and at some time in the far distant future its internally driven processes will stop. Put another way, why are internal processes driven by heat, such as volcanism, seismic activity, moving plates, and mountain building, slowing down?

The Atmosphere and Hydrosphere

In Chapter 1 we emphasized the interactions among systems, two of which, the atmosphere and hydrosphere, have had a profound impact on Earth's surface (see Figure 1.1). Shortly after Earth formed, its atmosphere and hydrosphere, although present, were quite different from the way they are now. They did, however, play an important role in the development of the biosphere.

How Did the Atmosphere Form and Evolve?

Today, Earth's atmosphere is quite unlike the noxious one we described earlier. Now it is composed mostly of nitrogen and abundant free oxygen (O_2), meaning oxygen not combined with other elements as in carbon dioxide (CO_2) and water vapor (H_2O). It also has small but important amounts of other gases such as ozone (O_3) (Table 8.1), which, fortunately for us, is common enough in the upper atmosphere to block most of the Sun's ultraviolet radiation.

Earth's very earliest atmosphere was probably composed of hydrogen and helium, the most abundant gases in the universe. If so, it would have quickly been lost into space, for two reasons. First, Earth's gravitational attraction is insufficient to retain gases with such low molecular weights. And second, before Earth differentiated it had no core or magnetic field. Accordingly, it lacked a *magnetosphere*, the area around the planet within which the magnetic field is confined, so a strong solar wind, an outflow of ions from the Sun, would

TABLE 8.1 Composition of Earth's Present-Day Atmosphere

	Symbol	Percentage by Volume*
Nonvariable Gases		
Nitrogen	N_2	78.08
Oxygen	O_2	20.95
Argon	Ar	0.93
Neon	Ne	0.0018
Others		0.001
Variable Gases and Particulates		
Water vapor	H_2O	0.1 to 4.0
Carbon dioxide	CO_2	0.038
Ozone	O_3	0.000004
Other gases		Trace
Particulates		Normally trace

*Percentages, except for water vapor, are for dry air.

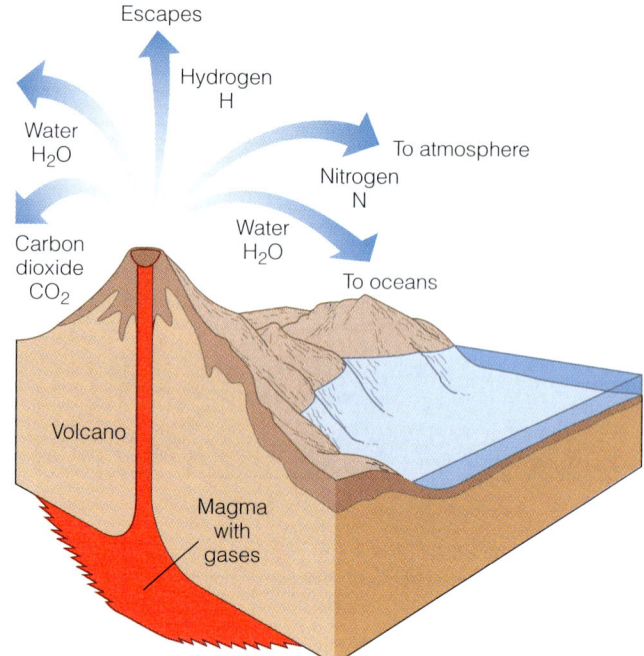

• **Figure 8.12 Outgassing and the Origin of Earth's Early Atmosphere** Erupting volcanoes emit several gases, which consist mostly of water vapor and carbon dioxide. Notice that no free oxygen is discharged. In addition to the gases shown here, chemical reactions in the atmosphere probably yielded methane (CH_4) and ammonia (NH_3).

reducing one rather than an oxidizing one. However, oxidized iron becomes increasingly common in Proterozoic rocks, indicating that at least some free oxygen was present then (see Chapter 9).

Two processes account for introducing free oxygen into the atmosphere, one or both of which began during the Eoarchean. The first, **photochemical dissociation,** involves ultraviolet radiation from the Sun in the upper atmosphere disrupting water molecules, thus releasing their oxygen and hydrogen (• Figure 8.13). This process may eventually have supplied 2% of the present-day oxygen level, but with this amount of free oxygen in the atmosphere ozone forms, creating a barrier against ultraviolet radiation and the formation of more ozone. Even more important were the activities of organisms that practice photosynthesis. **Photosynthesis** is a metabolic process in which organisms use carbon dioxide and water to make organic molecules, and oxygen is released as a waste product (Figure 8.13). Even so, probably no more than 1% of the free oxygen level of today was present by the end of the Archean.

Earth's Surface Waters— The Hydrosphere

Outgassing was responsible for the early atmosphere and also for Earth's surface water, most of which (more than 97%) is in the oceans. However, some but probably not much of our surface water was derived from icy comets.

have swept away any atmospheric gases. Once Earth had differentiated and a magnetosphere was present, though, an atmosphere began accumulating as a result of **outgassing** involving the release of gases from Earth's interior during volcanism (• Figure 8.12).

Water vapor is the most common gas emitted by volcanoes today, but they also emit carbon dioxide, sulfur dioxide, carbon monoxide, sulfur, hydrogen, chlorine, and nitrogen. No doubt Archean volcanoes emitted the same gases, and thus an atmosphere developed, but one lacking free oxygen and an ozone layer. It was, however, rich in carbon dioxide, and gases reacting in this early atmosphere probably formed ammonia (NH_3) and methane (CH_4).

This early oxygen-deficient but carbon dioxide–rich atmosphere persisted throughout the Archean. Some of the evidence for this conclusion comes from detrital deposits containing minerals such as pyrite (FeS_2) and uraninite (UO_2), both of which oxidize rapidly in the presence of free oxygen. So the atmosphere was a chemically

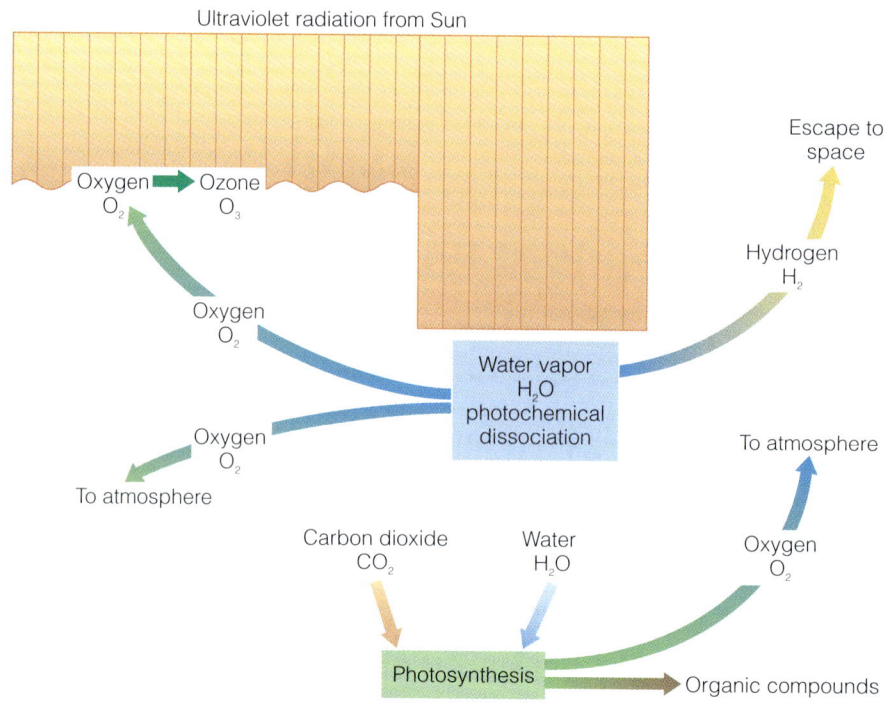

• **Figure 8.13 Photochemical Dissociation and Photosynthesis** Photochemical dissociation and photosynthesis added free oxygen to the atmosphere. Once free oxygen was present, an ozone layer formed in the upper atmosphere and blocked most incoming ultraviolet radiation.

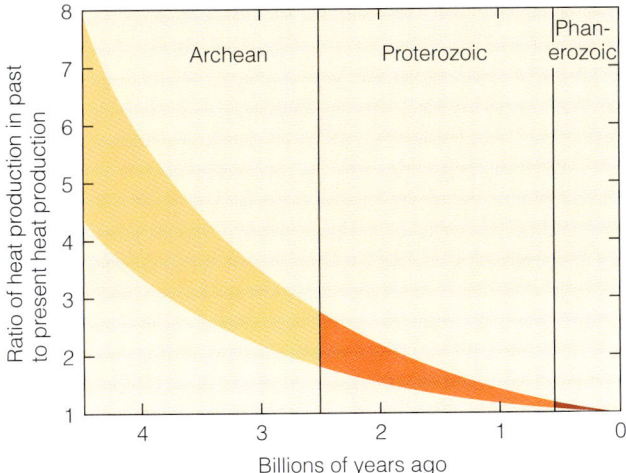

• **Figure 8.14 Ratio of Radiogenic Heat Production in the Past to the Present** The colored band encloses the ratios according to different models, but all show an exponential decay of radioactive elements through time. Heat production 4 billion years ago was three to six times as great as it is now, whereas radiogenic heat production at the beginning of the Phanerozoic Eon was only slightly higher than at present.

Water vapor is the most abundant gas released by volcanoes, so once Earth had cooled sufficiently, water vapor condensed and began to accumulate, perhaps as early as 4.4 billion years ago. We know that oceans were present during Eoarchean time, although their volumes and geographic extent cannot be determined. Nevertheless, we can envision an early Earth with considerable volcanism and a rapid accumulation of surface waters.

Volcanoes continue to erupt and release water vapor, so is the volume of ocean water still increasing? Perhaps, but if so, the rate has decreased considerably because the amount of heat needed to generate magma has diminished (• Figure 8.14). Furthermore, studies of present-day volcanic emissions indicate that much of their water vapor is recycled surface water, so any additions to the oceans are trivial compared to their volumes.

The Origin of Life

Scientists have found fossils in 3.3- to 3.5-billion-year-old Archean rocks, and chemical evidence in 3.85-billion-year-old rocks in Greenland convince some investigators that organisms were present by this early date. There is unequivocal evidence for Archean organisms, but compared to the present, the Archean was biologically impoverished, at least in terms of biotic diversity. Today, the biosphere is made up of millions of species of organisms assigned to five kingdoms—animals, plants, protistans, fungi, and monera (bacteria and archaea),* whereas only

*Archaea includes microscopic organisms that resemble bacteria but differ from them genetically and biochemically.

bacteria and archaea are known from Archean rocks. In Chapter 7 we discussed the theory of evolution, which was proposed to account for the diversification of organisms. Remember that evolutionary theory itself does not address the origin of life—only how organisms have changed through time. Here we are concerned with **abiogenesis**—that is, how life originated from nonliving matter in the first place.

Before discussing abiogenesis, let us be clear on what is living and nonliving. In most cases the distinction is straightforward: dogs and trees are alive, but rocks and water are not. A biologist might use many criteria to make the living-nonliving distinction, such as growth and reaction to stimuli, but minimally a living organism must practice some kind of chemical activity (metabolism) to maintain itself, and it must be capable of reproduction in order to ensure the long-term survival of a group of organisms.

This metabolism/reproduction criterion might seem sufficient to decide whether something is alive or not, and yet the distinction is not always easy to make. Bacteria are living, but under some circumstances, some can go for long periods without showing any signs of life and then go about living again. Are viruses living? They behave like living organisms in the appropriate host cell, but when outside a host cell, they neither metabolize nor reproduce. Some biologists think that viruses represent another way of living, but others disagree. Comparatively simple organic (carbon-based) molecules called *microspheres* form spontaneously and grow and divide in a somewhat organism-like fashion, but these processes are more like random chemical reactions, so they are not living.

So what do viruses and microspheres have to do with the origin of life? First, they show that the living versus nonliving distinction is not always easy to make. And second, if life originated by natural processes from nonliving matter (abiogenesis), it must have passed through prebiotic stages—that is, stages in which the entities would have shown signs of living organisms but were not truly living.

One further note about abiogenesis: It does not hold that a living organism such as a bacterium, or even a complex organic molecule such as a protein, sprang fully formed from nonliving matter. Rather than one huge leap from nonliving to living, abiogenesis holds that several small steps took place, each leading to an increase in organization and complexity.

Experimental Evidence and the Origin of Life

Investigators agree that two requirements were necessary for the origin of life: (1) a source of the appropriate elements for organic molecules and (2) energy sources to promote chemical reactions. All organisms are composed mostly of carbon, hydrogen, nitrogen, and oxygen, all of which were present in Earth's early atmosphere as carbon dioxide (CO_2), water vapor (H_2O), and nitrogen (N_2), and possibly methane (CH_4) and ammonia (NH_3). Energy from lightning and ultraviolet radiation probably promoted chemical reactions during which C, H, N, and O combined to form comparatively simple organic molecules called **monomers,** such as amino acids.

Monomers are the basic building blocks of more complex organic molecules, but is it plausible that they originated in the manner postulated? Experimental evidence indicates it is. During the 1950s, Stanley Miller synthesized several amino acids by circulating gases approximating the early atmosphere in a closed glass vessel (• Figure 8.15). This mixture was subjected to an electric spark to simulate lightning, and in a few days it became cloudy. Analysis showed that several amino acids typical of organisms had formed. In more recent experiments scientists have successfully synthesized all 20 amino acids found in organisms.

Making monomers in a test tube is one thing, but the molecules of organisms are **polymers,** including proteins and nucleic acids such as RNA (ribonucleic acid) and DNA (deoxyribonucleic acid), consisting of monomers linked together in a specific sequence. So how did this phenomenon of polymerization take place? This is more difficult to answer, especially if the event occurred in water, which usually causes depolymerization. However, researchers have synthesized small molecules known as *proteinoids,* some of which consist of more than 200 linked amino acids (• Figure 8.16a), when dehydrated concentrated amino acids were heated. Under these conditions the concentrated amino acids spontaneously polymerized to form

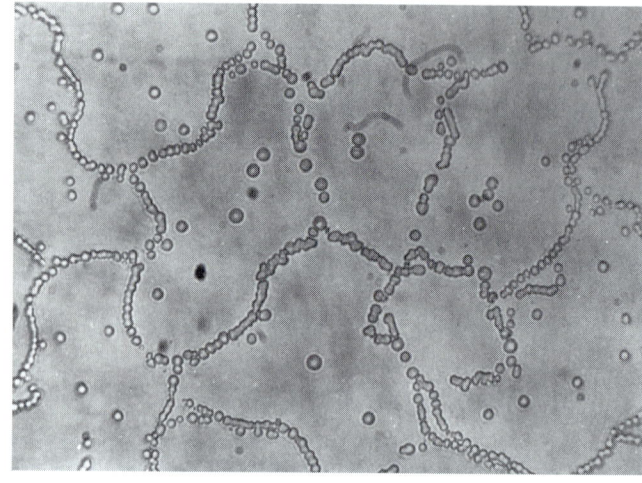

(a)

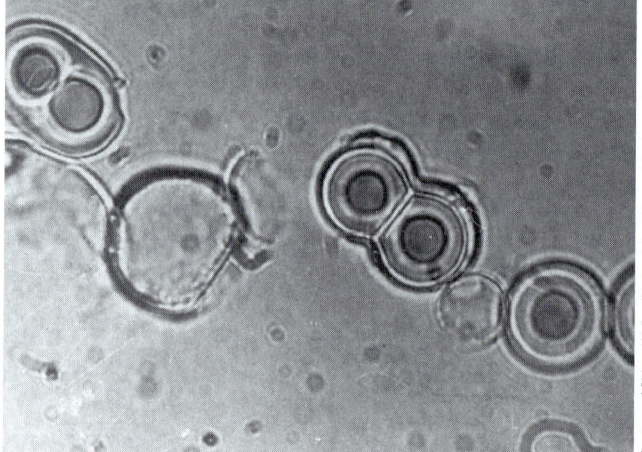

(b)

• **Figure 8.16 Experimental Production of Proteinoids and Microspheres** (a) Bacteria-like proteinoids. (b) Proteinoid microspheres.

proteinoids. Perhaps similar conditions for polymerization existed on early Earth.

At this stage we can refer to these proteinoid molecules as *protobionts,* which are intermediate between inorganic chemical compounds and living organisms. These protobionts would have ceased to exist, though, if some kind of outer membrane had not developed. In other words, they had to be self-contained, as cells are now. Experiments show that proteinoids spontaneously aggregate into microspheres (Figure 8.16b) that are bounded by a cell-like membrane and grow and divide much as bacteria do.

The origin-of-life experiments are interesting, but how can they be related to what may have occurred on Earth more than 3.5 billion years ago? Monomers likely formed continuously and by the billions, and accumulated in the early oceans, which was, as the British biochemist J. B. S. Haldane called it, a hot, dilute soup. The amino acids in this "soup" may have washed up onto a beach or perhaps cinder cones, where they were concentrated by evaporation and polymerized by heat. The poly-

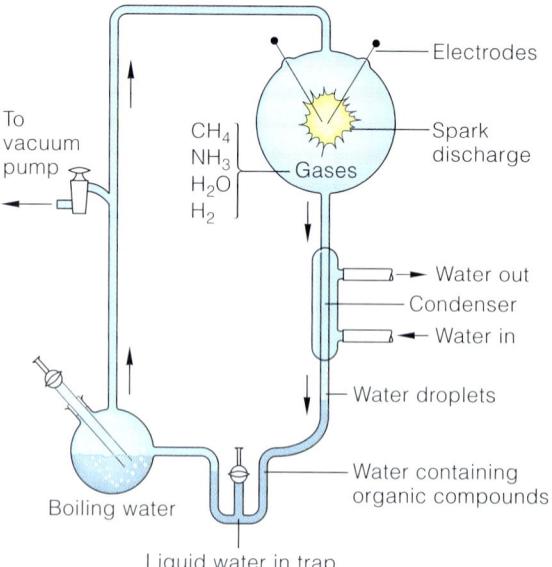

• **Figure 8.15 Stanley Miller's Experimental Apparatus** Miller synthesized several amino acids used by organisms. The spark discharge was to simulate lightning, and the gases approximated Earth's early atmosphere.

mers were then washed back into the ocean, where they reacted further.

Not much is known about the next critical step in the origin of life: the development of a reproductive mechanism. The microspheres shown in Figure 8.16b divide, and some experts think they represent a protoliving system. However, in today's cells nucleic acids, either RNA or DNA, are necessary for reproduction. The problem is that nucleic acids cannot replicate without protein enzymes, and the appropriate enzymes cannot be made without nucleic acids. Or so it seemed until fairly recently.

Now we know that small RNA molecules can replicate without the aid of protein enzymes. In view of this evidence, the first replicating systems may have been RNA molecules. In fact, some researchers propose an early "RNA world" in which these molecules were intermediate between inorganic chemical compounds and the DNA based molecules of organisms. However, RNA cannot be easily synthesized under conditions that probably prevailed on early Earth, so how they were naturally synthesized remains an unsolved problem.

Considering the difficulties facing scientists researching the origin of life, it is remarkable that they have made as much progress as they have. They agree on some of the basic requirements for the origin of life, but the exact steps involved and the significance of some experimental results are still debated. Indeed, our discussion has focused on the role of atmospheric gases in the origin of monomers and polymers, but some scientists think that life originated near hydrothermal vents on the seafloor.

Submarine Hydrothermal Vents and the Origin of Life

Seawater seeps down into the oceanic crust at or near spreading ridges, becomes heated by magma, and rises and discharges onto the seafloor as **submarine hydrothermal vents**. Some of these vents, which are now known from the Atlantic, Pacific, and Indian Oceans, are called **black smokers** because they discharge water saturated with dissolved minerals, giving them the appearance of black smoke (• Figure 8.17a). Because Earth had more radiogenic heat during its early history it is reasonable to infer that submarine hydrothermal vents were more common then.

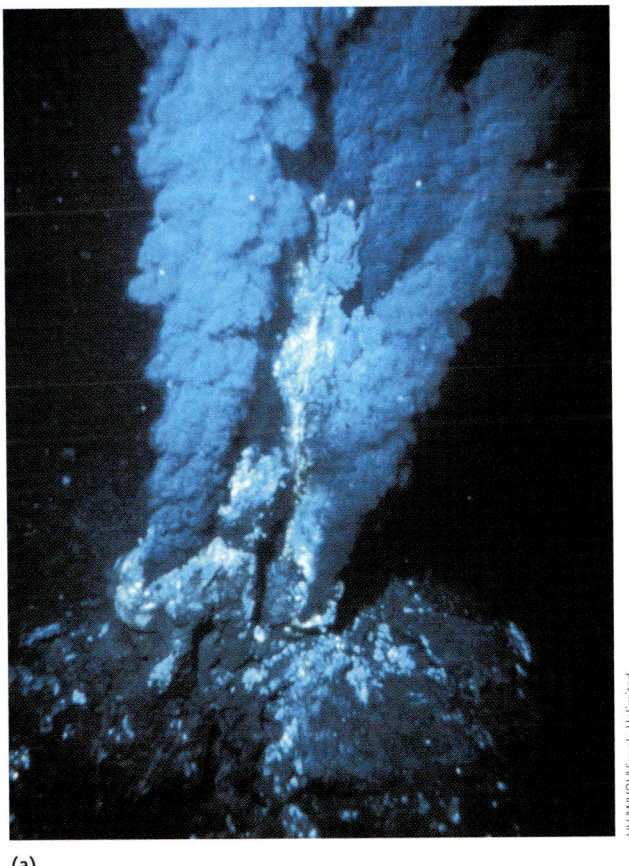

(a)

(b)

• **Figure 8.17 Submarine Hydrothermal Vents** (a) This black smoker on the East Pacific Rise is at a depth of 2800 m. The plume of "black smoke" is heated seawater with dissolved minerals. (b) Several types of organisms including these tube worms, which may be up to 3 m long, live near black smokers. Bacteria that oxidize sulfur compounds lie at the base of the food chain.

Submarine hydrothermal vents are interesting because several minerals containing zinc, copper, and iron are precipitated around them, and they support communities of organisms previously unknown to science (Figure 8.17b). In recent years some investigators have proposed that the first self-replicating molecules came into existence near these vents on the seafloor. The necessary elements (C, H, O, and N) are present in seawater, and heat may have provided the energy necessary for monomers to form. Those endorsing this hypothesis hold that polymerization took place on the surfaces of clay minerals, and, finally, protocells were deposited on the seafloor. In fact, amino acids have been detected in some hydrothermal vent emissions.

Not all scientists are convinced this is a viable environment for the origin of life. They point out that polymerization would be inhibited in this hot, aqueous environment. So this alternate location for the origin of life remains controversial, at least for now.

The Oldest Known Organisms

We now know that the first organisms were members of the kingdom Monera consisting of bacteria and archaea, both of which consist of **prokaryotic cells**—that is, cells that lack an internal, membrane-bounded nucleus and other structures typical of *eukaryotic cells*. (We discuss eukaryotic cells in Chapter 9.) Members of the archaea are single-celled, as are bacteria, and in fact they look much like bacteria but differ from them in the details of their genetics and chemical makeup. In addition, archaea can live in extremely hot, acidic, and saline environments where even bacteria cannot survive. In any case, prior to the 1950s, scientists assumed that the fossils common in Cambrian-aged rocks had a long, earlier history, but the fossil record offered little to support this idea. In fact, it appeared that Precambrian rocks were so devoid of fossils that this interval of time was formerly called the *Azoic*, meaning "without life."

As far back as the early 1900s, Charles Walcott described layered moundlike structures from the Paleoproterozoic-aged Gunflint Iron Formation of Ontario, Canada. He proposed that these structures, now called **stromatolites,** represented reefs constructed by algae (• Figure 8.18), but not until 1954 did paleontologists demonstrate that stromatolites are the product of organic activity. Present-day stromatolites form and grow as sediment grains are trapped on sticky mats of photosynthesizing cyanobacteria (blue-green algae), although now they are restricted to environments where snails cannot live.

Currently, the oldest known undisputed stromatolites are found in 3.0-billion-year-old rocks in South Africa, but probable ones are also known from the 3.3- to 3.5-billion-year-old Warrawoona Group in Australia. And, as noted previously, chemical evidence in rocks 3.85 billion years

(a)

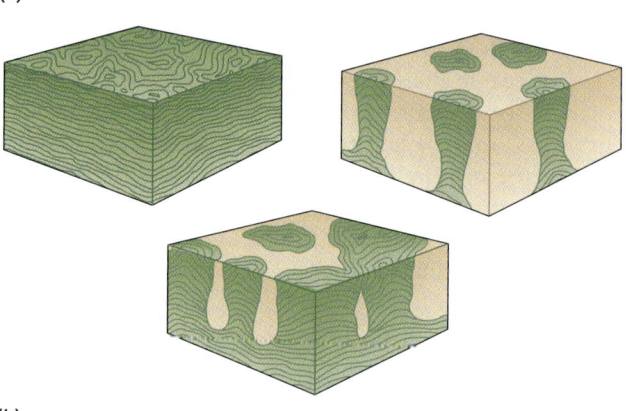

(b)

• **Figure 8.18 Stromatolites** (a) Present-day stromatolites at Shark Bay, Australia. (b) Types of stromatolites include irregular mats, isolated columns, and columns linked by mats.

old in Greenland indicate life was perhaps present then. These oldest known cyanobacteria were photosynthesizing organisms, but photosynthesis is a complex metabolic process that must have been preceded by a simpler type of metabolism.

No fossils are known of these earliest organisms, but they must have resembled tiny bacteria that needed no free oxygen, so they were **anaerobic.** In addition, they must have been totally dependent on an external source of nutrients—that is, they were **heterotrophic,** as opposed to **autotrophic** organisms, which make their own nutrients, as in photosynthesis. And, finally, they were all prokaryotic cells. The nutrient source for these earliest organisms was most likely adenosine triphosphate (ATP), which was used to drive the energy-requiring reactions in cells. ATP can easily be synthesized from simple gases and phosphate, so it was no doubt available in the early Earth environment. These organisms must have acquired their ATP from their surroundings, but this situation could not have persisted for long as more and more cells competed for the same resources. The first organisms to develop a more sophisticated metabolism probably used *fermentation* to meet their energy needs. Fermentation is an anaerobic process in which molecules such as sugars are split, releasing carbon

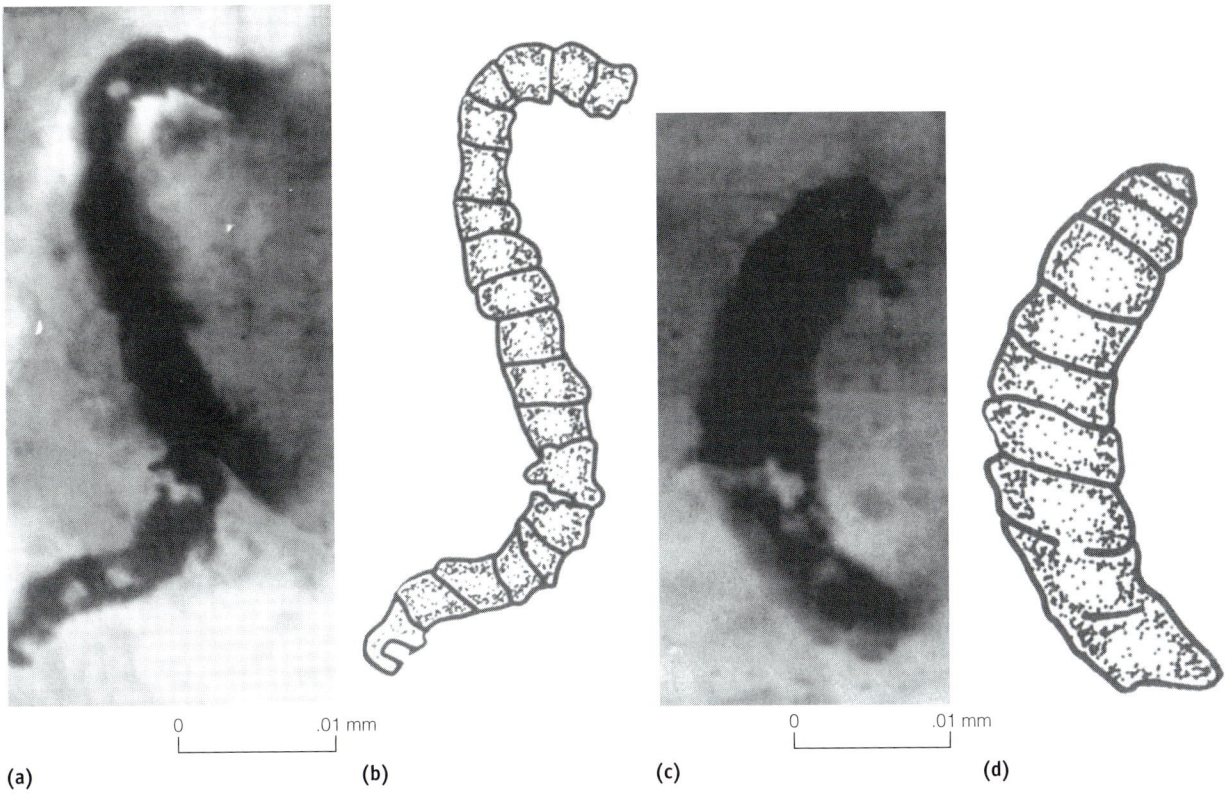

• **Figure 8.19 Archean Fossils** Photomicrographs (a) and (c) and schmatic restorations (b) and (d) of fossil prokaryotes from the 3.3- to 3.5-billion-year-old Warrawoona Group of Western Australia.

dioxide, alcohol, and energy. In fact, most living prokaryotic cells ferment.

Other than the origin of life itself, the most significant biologic event of the Archean was the development of the autotrophic process of photosynthesis as much as 3.5 billion years ago. These prokaryotic cells were still anaerobic, but as autotrophs they were no longer dependent on preformed organic molecules as a source of nutrients. The Archean microfossils in • Figure 8.19 are anaerobic, autotrophic prokaryotes belonging to the kingdom Monera, represented today by bacteria, cyanobacteria, and archaea.

Archean Mineral Resources

Several types of mineral deposits are found in Archean rocks, but gold is the one most commonly associated with them, although it is also found in Proterozoic and Phanerozoic rocks. This soft yellow metal has been the cause of feuds and wars and was one incentive for European exploration of the Americas. It is prized for jewelry, but it is or has also been used as a monetary standard, and it is still used in glass making, electrical circuitry, and the chemical industry. Since 1886, Archean and Proterozoic rocks in South Africa have yielded about half the world's gold. Gold mines are also found in Archean rocks of the Superior craton in Canada.

Archean sulfide deposits of zinc, copper, and nickel are known in several areas, such as Australia, Zimbabwe, and the Abitibi greenstone belt in Ontario, Canada. At least some of these deposits probably formed as mineral accumulations adjacent to hydrothermal vents on the seafloor, much as they do now around black smokers (Figure 8.17a).

About one-fourth of Earth's chrome reserves are in Archean rocks, especially in Zimbabwe. These ore deposits are found in the volcanic units of greenstone belts, where they probably formed when mineral crystals settled and became concentrated in the lower parts of plutons such as mafic and ultramafic sills. Chrome is needed in the steel industry. The United States has very few commercial chrome deposits so must import most of what it uses.

One chrome deposit in the United States is in the Stillwater Complex in Montana. Low-grade ores were mined there during both World Wars as well as during the Korean War, but they were simply stockpiled and never refined for their chrome content. These same rocks are also a source of platinum, one of the precious metals, which in the automotive industry is used in catalytic converters. It is also used in the chemical industry and for cancer chemotherapy. Most of it mined today comes from the Bushveld Complex in South Africa.

Most of Earth's banded iron formations, the primary source of iron ore, were deposited during the Proterozoic, but about 6% of them are of Archean age. Some Archean-aged banded iron formations are mined but they are neither as thick nor as extensive as those of the Proterozoic (see Chapter 9).

Recall that the term *pegmatite* refers to very coarsely crystalline plutonic rock. Most Archean pegmatites approximate the composition of granite and have little commercial value. But some, such as those in the Herb Lake District in Manitoba, Canada, and the Rhodesian Province of Africa, contain valuable minerals. In addition to minerals of gem quality, Archean pegmatites also contain minerals mined for their lithium, beryllium, rubidium, and cesium.

SUMMARY

- All geologic time from Earth's origin to the beginning of the Phanerozoic Eon is included in the Precambrian. Precambrian also refers to rocks lying stratigraphically below Cambrian-aged rocks.
- The Precambrian is divided into two eons, the Archean and the Proterozoic, each of which has further subdivisions.
- Rocks from the latter part of the Eoarchean indicate that crust existed then, but very little of it has been preserved.
- All continents have an ancient craton made up of an exposed shield and a buried platform. In North America, the Canadian shield is made up of smaller units delineated by their ages and structural trends.
- Archean greenstone belts are linear, syncline-like bodies of rock found within much more extensive granite-gneiss complexes.
- An ideal greenstone belts consists of two lower units of mostly igneous rocks and an upper sedimentary unit. They probably formed in back-arc basins and in intracontinental rifts.
- Many geologists are convinced that Archean plate tectonics took place, but plates probably moved faster, and igneous activity was more common then because Earth had more radiogenic heat.
- Outgassing was probably responsible for the early atmosphere and the hydrosphere. However, the atmosphere so formed lacked free oxygen but contained abundant carbon dioxide and water vapor.
- Models for the origin of life by natural processes require an oxygen-deficient atmosphere, the necessary elements for organic molecules, and energy to promote the synthesis of organic molecules.
- The first naturally formed organic molecules were probably monomers, such as amino acids, that linked together to form more complex polymers, including nucleic acids and proteins.
- RNA molecules may have been the first molecules capable of self-replication. However, the method whereby a reproductive system formed is not known.
- The only known Archean fossils are of single-celled, prokaryotic bacteria such as blue-green algae, but chemical compounds in some Archean rocks may indicate the presence of archaea.
- Stromatolites that formed by the activities of photosynthesizing bacteria are found in rocks as much as 3.5 billion years old.
- Archean mineral resources include gold, chrome, zinc, copper, and nickel.

IMPORTANT TERMS

abiogenesis, p. 159
anaerobic, p. 162
autotrophic, p. 162
back-arc marginal basin, p. 155
black smoker, p. 161
Canadian shield, p. 150
continental accretion, p. 156

craton, p. 150
granite-gneiss complex, p. 152
greenstone belt, p. 152
heterotrophic, p. 162
monomer, p. 160
outgassing, p. 158
photochemical dissociation, p. 158

photosynthesis, p. 158
platform, p. 150
polymer, p. 160
Precambrian shield, p. 150
prokaryotic cell, p. 162
stromatolite, p. 162
submarine hydrothermal vent, p. 161

REVIEW QUESTIONS

1. The most widespread Archean-aged rocks are those found in
   a. ____ ultramafic plutons; b. ____ granite-gneiss complexes; c. ____ meteorite impact successions; d. ____ passive continental shelf associations; e. ____ intracontinental rift assemblages.
2. The oldest known organisms had cells we characterize as
   a. ____ felsic; b. ____ cratonic; c. ____ elleptic; d. ____ argillitic; e. ____ prokaryotic.
3. Many scientists think that the first self-replicating system was a(n)
   a. ____ ATP cell; b. ____ stromatolite; c. ____ autotroph; d. ____ RNA molecule; e. ____ proteinoid.
4. Ultramafic lava flows are found mostly in
   a. ____ the lower volcanic units of greenstone belts; b. ____ pyroclastic-gneiss rift deposits; c. ____ sandstone-shale-carbonate deposits of passive continental margins; d. ____ prokaryotic, anaerobic stromatolites; e. ____ banded iron formations.
5. The ancient stable nucleus of a continent is called a
   a. ____ polymer; b. ____ back-arc basin; c. ____ craton; d. ____ rift; e. ____ syncline.
6. One process responsible for at least some of the free oxygen in the Archean atmosphere is
   a. ____ photochemical dissociation; b. ____ heterotrophic metabolism; c. ____ magmatic separation; d. ____ polymerization; e. ____ Eoarchean subduction.
7. Some of the sedimentary rocks in greenstone belts consist of graywacke with graded bedding and argillite, which are interpreted as _____ deposits.
   a. ____ delta; b. ____ braided stream; c. ____ carbonate; d. ____ turbidity current; e. ____ passive shelf.
8. The process whereby Earth's early atmosphere formed by release of gases during volcanic eruptions is known as
   a. ____ dewatering; b. ____ outgassing; c. ____ accretion; d. ____ monomerization; e. ____ photosynthesis.
9. Although the origin of greenstone belts is not fully resolved, many scientists think that some of them formed in
   a. ____ back-arc marginal basins; b. ____ continental shelf environments; c. ____ transform boundary shear zones; d. ____ carbonate-evaporite depositional areas; e. ____ all of the previous answers.
10. The designation for geologic time before the Phanerozoic is
    a. ____ Prearchean; b. ____ Cenozoic; c. ____ Crytozoic; d. ____ Azoic; e. ____ Precambrian.
11. What are stromatolites, and how do they form?
12. What two processes accounted for adding free oxygen to the atmosphere? Which one was most important?
13. Describe the succession of rock types in a typical greenstone belt, and explain how a greenstone belt might form in a back-arc marginal basin.
14. What does *continental accretion* mean?
15. Where is the North American craton exposed other than in the Canadian shield? How do you account for its presence in this (these) area(s)?
16. Summarize the experimental evidence for the origin of monomers and polymers.
17. What are black smokers, and what is their economic and biologic significance?
18. Why is it more difficult to apply the principle of superposition to figure out time-stratigraphic relationships among Archean rocks?

APPLY YOUR KNOWLEDGE

1. What evidence indicates that some Eoarchean crust was present? Why do you think that so few rocks formed during this time have been preserved?
2. How do ultramafic lava flows form, and is their presence in Archean rocks but rarely in younger rocks an exception to the principle of uniformitarianism?

CHAPTER 9
PRECAMBRIAN EARTH AND LIFE HISTORY— THE PROTEROZOIC EON

View of Mesoproterozoic to Neoproterozoic sedimentary rocks of the Belt Supergroup in Glacier National Park, Montana. St. Mary Lake is in the foreground.

David Muench/Corbis

[OUTLINE]

INTRODUCTION

EVOLUTION OF PROTEROZOIC CONTINENTS

Paleoproterozoic History of Laurentia

Paleo- and Mesoproterozoic Igneous Activity

Mesoproterozoic Orogeny and Rifting

Meso- and Neoproterozoic Sedimentation

PROTEROZOIC SUPERCONTINENTS

ANCIENT GLACIERS AND THEIR DEPOSITS

Paleoproterozoic Glaciers

Glaciers of the Neoproterozoic

THE EVOLVING ATMOSPHERE

Banded Iron Formations (BIFs)

Continental Red Beds

IMPORTANT EVENTS IN LIFE HISTORY

Eukaryotic Cells Evolve

Endosymbiosis and the Origin of Eukaryotic Cells

The Dawn of Multicelled Organisms

Neoproterozoic Animals

PROTEROZOIC MINERAL RESOURCES

PERSPECTIVE 9.1 *BIF: From Mine to Steel Mill*

SUMMARY

ThomsonNOW Explore interactive tutorials, animations, or practice problems available on the ThomsonNow website at www.thomsonedu.com/login.

CHAPTER OBJECTIVES

At the end of this chapter, you will have learned that

- A large landmass called Laurentia, made up mostly of North America and Greenland, grew by accretion along its margins during the Proterozoic Eon.

- During the Mesoproterozoic, Laurentia experienced widespread igneous activity, orogenesis, and rifting.

- Proterozoic sandstone-carbonate-shale assemblages of rocks are widespread, and they look much like the rocks that formed on more recent passive continental margins.

- Plate tectonics essentially like that now operating was taking place during the Proterozoic, and at least two Proterozoic supercontinents existed.

- Deposition of banded iron formations and continental red beds indicate that some free oxygen was present in the atmosphere.

- Glacial deposits and associated glacial features such as striations indicate that extensive glaciation took place during the Paleoproterozoic and the Neoproterozoic.

- The first eukaryotic cells, cells with nuclei and that reproduced sexually, evolved and multicelled algae may have evolved as much as 2.1 billion years ago.

- Impressions of multicelled animals are found in Neoproterozoic-aged rocks on all continents except Antarctica.

- Banded iron formations as sources of iron ore are important Proterozoic resources, as are deposits of copper, platinum, and nickel.

Introduction

At 1.958 billion years long, the Proterozoic Eon alone accounts for 42.5% of all geologic time (see Figure 8.1), yet we review this unimaginably long interval of Earth and life history in a single chapter. The Phanerozoic Eon, made up of the more familiar Paleozoic, Mesozoic, and Cenozoic eras, which lasted a comparatively brief 542 million years, is the subject matter for the following 10 chapters of this book. This should not be too surprising, though, because the older parts of the geologic record are more inaccessible and more difficult to interpret. Remember that Precambrian rocks are deeply buried in many areas, and many of those of Archean age have been metamorphosed. Some Proterozoic rocks are also metamorphic, but there are also widespread exposures that show little or no metamorphic effects, but they contain few fossils that are useful in stratigraphy (• Figure 9.1).

Notice in Figure 8.2 that the Proterozoic Eon is subdivided into three eras with the prefixes *Paleo* (old or ancient), *Meso* (middle), and *Neo* (new or recent), and each era is further subdivided into periods. Also remember from Chapter 8 that these subdivisions, just as those for the Archean Eon, are strictly terms denoting time. That is, there are no time-stratigraphic units with stratotypes that are the bases for these time units. Geologists have placed the

(a)

(b)

• **Figure 9.1 Proterozoic Rocks** (a) This view in the Grand Canyon of Arizona shows the Vishnu Schist (black) that was intruded by the Zoraster Granite. The schist, which is about 1.7 billion years old, was originally sedimentary rocks and lava flows. (b) This 1.0-billion-year-old sandstone and mudstone in Glacier National Park in Montana has been only slightly altered by metamorphism.

Archean Eon	Proterozoic Eon	Phanerozoic Eon						
Precambrian		Paleozoic Era						
		Cambrian	Ordovician	Silurian	Devonian	Mississippian	Pennsylvanian	Permian
						Carboniferous		

2500 MYA   542 MYA   251 MYA

Archean–Proterozoic boundary at 2.5 billion years ago (see Figure 8.2), because that marks the approximate time of changes in the style of crustal evolution. However, we must emphasize "approximate," because Archean-type crustal evolution was largely completed in South Africa nearly 3.0 billion years ago, whereas in North America the change took place from 2.95 to 2.45 billion years ago.

What do we mean by a change in the style of crustal evolution? Archean crust-forming processes generated granite-gneiss complexes and greenstone belts that were shaped into cratons (see Figure 8.11). These same rock associations continued to form during the Proterozoic, but they did so at a considerably reduced rate. In addition, many Archean rocks have been metamorphosed, although their degree of metamorphism varies and some are completely unaltered. In contrast, vast exposures of Proterozoic rocks show little or no effects of metamorphism (see the chapter opening photo and Figure 9.1b), and in many areas they are separated from Archean rocks by an unconformity. And finally, the Proterozoic is characterized by widespread sedimentary rock assemblages that are rare or absent in the Archean, and by a plate tectonic style essentially the same as that of the present.

In addition to changes in the style of crustal evolution, the Proterozoic was also an important time in the evolution of the atmosphere and biosphere, as well as the origin of some important mineral resources. It was during the Proterozoic that oxygen-dependent organisms appeared in the fossil record, and the kind of cells that make up most organisms today made their appearance.

Evolution of Proterozoic Continents

In Chapter 8 we noted that Archean cratons assembled during collisions of island arcs and minicontinents (see Figure 8.4). These cratons provided the nuclei around which Proterozoic crust accreted, thereby forming much larger landmasses. Accretion around these nuclei took place much as it does today where plates converge, but it no doubt occurred more rapidly because Earth possessed more radiogenic heat and plates moved faster. Our focus here is on the evolution of **Laurentia,** a landmass made up of what are now North America, Greenland, parts of northwestern Scotland, and perhaps some of the Baltic shield of Scandinavia.

Most greenstone belts formed during the Archean, but they continued to form during the Proterozoic, and at least one is known from Cambrian-aged rocks in Australia. They were not as common, though, and they did differ in one important detail: the near absence of extrusive ultramafic rocks (komatiite). As we discussed in Chapter 8, this detail is no doubt a result of Earth's decreasing rate of radiogenic heat production (see Figure 8.14).

Paleoproterozoic History of Laurentia

The time between 2.0 and 1.8 billion years ago was an important one in the origin and growth of Laurentia. During this time, collisions among plates formed several **orogens,** which are linear or arcuate deformation belts in which many of the rocks have been metamorphosed and intruded by magma, thus forming plutons, especially batholiths. Accordingly, Archean cratons were sutured along these deformation belts, thereby forming a larger landmass (• Figure 9.2). By 1.8 billion years ago, much of what is now Greenland, central Canada, and the north-central United States existed.

Good examples of these craton-forming processes are recorded in rocks of the Thelon orogen in northwestern Canada where the Slave and Rae cratons collided, and the Trans-Hudson orogen, in Canada and the United States, where the Superior, Hearne, and Wyoming cratons were sutured (Figure 9.2a). Note, too, from Figure 9.2a that the southern margin of Laurentia is the site of the Penokian orogen.

Sedimentary rocks of the Wopmay orogen in northwestern Canada are important because they record the opening and closing of an ocean basin, or what is called a **Wilson cycle.** A complete Wilson cycle, named after the Canadian geologist J. Tuzo Wilson, involves rifting of a continent and the opening of an ocean basin with passive

Phanerozoic Eon										
Mesozoic Era			Cenozoic Era							
Triassic	Jurassic	Cretaceous	Paleogene			Neogene		Quaternary		
			Paleocene	Eocene	Oligocene	Miocene	Pliocene	Pleistocene	Holocene	

251 MYA   66 MYA

(a)

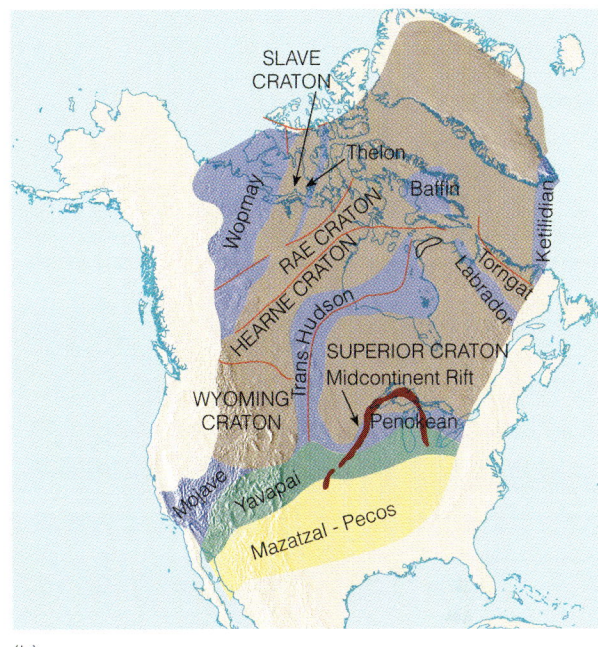

(b)

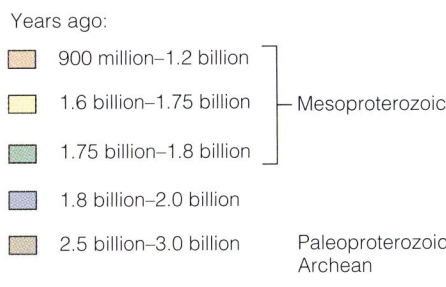

Years ago:
- 900 million–1.2 billion
- 1.6 billion–1.75 billion ⎫ Mesoproterozoic
- 1.75 billion–1.8 billion ⎭
- 1.8 billion–2.0 billion
- 2.5 billion–3.0 billion — Paleoproterozoic Archean

- **Figure 9.2 Proterozoic Evolution of Laurentia** (a) During the Paleoproterozoic, Archean cratons were sutured along deformation belts called *orogens*. (b) Laurentia grew along its southern margin by accretion of the Central Plains, Yavapai, and Mazatzal orogens. Also notice that the Midcontinent rift had formed in the Great Lakes region by this time. (c) A final episode of Proterozoic accretion took place during the Grenville orogeny.

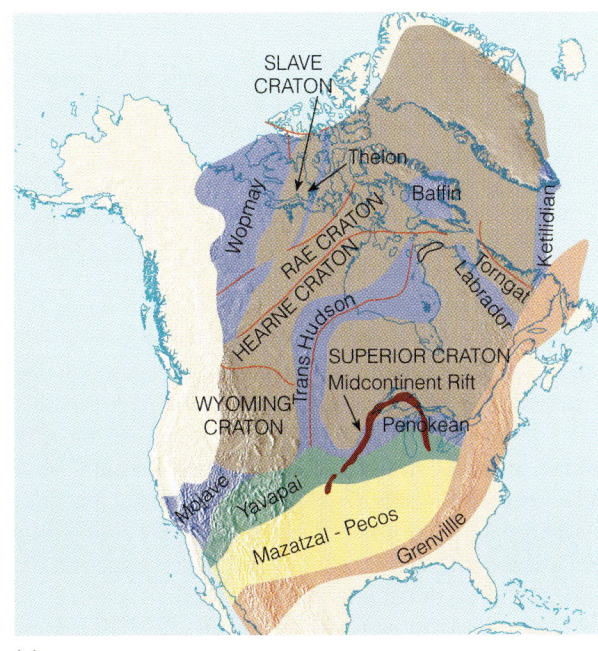

(c)

EVOLUTION OF PROTEROZOIC CONTINENTS

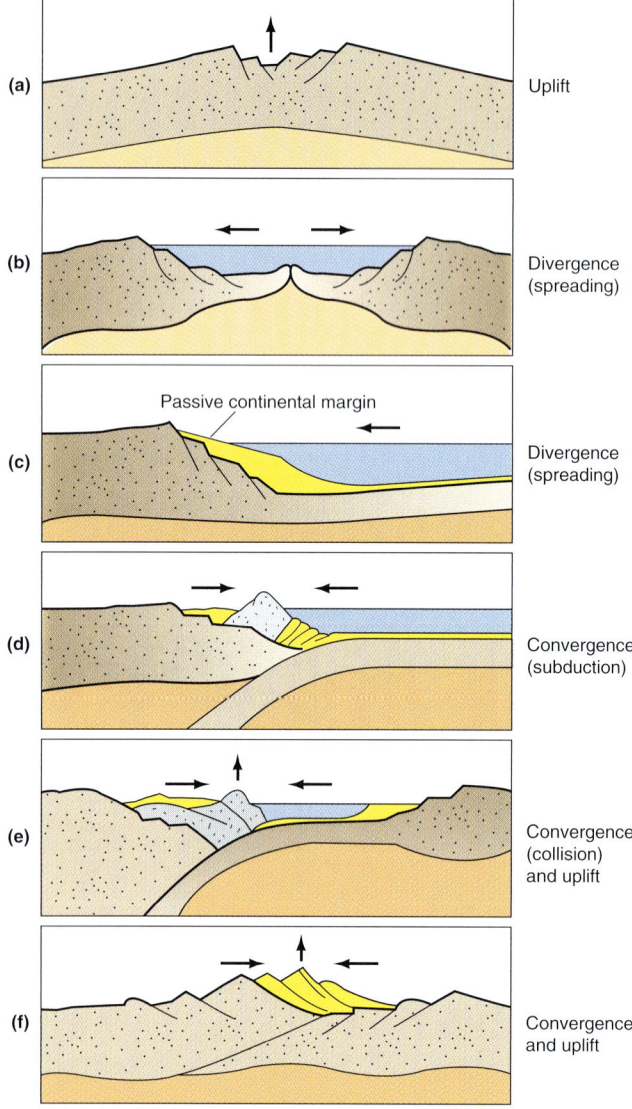

• **Figure 9.3 The Wopmay Orogen and the Wilson Cycle** Some geologists think that the Wopmay orogen in Canada respresents a complete Wilson cycle, which is shown here diagrammatically **(a–f)**. In fact, some of the sedimentary rocks in the Wopmay orogen were probably deposited on a passive continental margin as shown in **(c)**.

continental margins on both sides (• Figure 9.3). As rifting proceeds, an expansive ocean basin forms, but it eventually begins to close, and subduction zones and volcanic island arcs develop on one or both sides of the ocean basin. The final stage involves a continent-continent collision, during which the rocks that formed along the previous passive continental margins and seafloor are deformed during an orogeny (Figure 9.3f). In fact, some of the rocks in the Wopmay orogen belong to a suite of sedimentary rocks called a **sandstone-carbonate-shale assemblage** that is deposited on passive continental margins. These rock assemblages are absent or at least rare in the Archean, but they become common during the Proterozoic.

Paleoproterozoic sandstone-carbonate-shale assemblages are widespread in the Great Lakes region of the United States and Canada (• Figure 9.4). The sandstones (now quartzites) have a variety of sedimentary structures such as ripple marks and cross-beds, whereas some of the carbonate rocks, now mostly dolostone, contain abundant stromatolites (Figure 9.4c). These rocks have been only moderately deformed and are now part of the Penokean orogen.

Following the initial episode of amalgamation of Archean cratons from 2.0 to 1.8 billion years ago, accretion took place along Laurentia's southern margin. From 1.8 to 1.6 billion years ago, continental accretion continued in what is now the southwestern and central United States as successively younger belts were sutured to Laurentia, forming the Yavapai and Mazatzal-Pecos orogens (Figure 9.4b).

The Paleoproterozoic orogenies and rocks just reviewed are certainly important, but this was also the time during which most of Earth's *banded iron formations (BIF)* were deposited. And the first *continental red beds*—sandstone and shale with oxidized iron—were deposited about 1.8 billion years ago. We will have more to say about BIFs and red beds in the section headed "The Evolving Atmosphere." And finally, we should mention that some Paleoproterozoic-aged rocks and associated features provide excellent evidence for widespread glaciation (also covered later in this chapter).

Paleo- and Mesoproterozoic Igneous Activity

During the interval from 1.8 to 1.1 billion years ago, extensive igneous activity took place that was unrelated to orogenic activity (• Figure 9.5). Although quite widespread, this activity did not add to Laurentia's size, because magma was either intruded into or erupted onto already existing continental crust. These igneous rocks are exposed in eastern Canada, extend across Greenland, and are also found in the Baltic shield of Scandinavia. However, they are deeply buried by younger rocks in most areas.

The origins of these granitic and anorthosite* plutons, calderas and their fill, and vast sheets of rhyolite and ash-flows are the subject of debate. According to one hypothesis, large-scale upwelling of magma beneath a Proterozoic supercontinent was responsible for these rocks. According to this hypothesis, the mantle temperature beneath a Proterozoic supercontinent would have been considerably higher than beneath later supercontinents because radiogenic heat production within Earth has decreased. Accordingly, nonorogenic igneous activity would have occurred following the amalgamation of the first supercontinent.

---

*Anorthosite is a plutonic rock composed almost entirely of plagioclase feldspars.

• **Figure 9.4 Paleoproterozoic Sedimentary Rocks**
(a) Geographic distribution of Paleoproterozoic-aged rocks in the Great Lakes region. (b) Outcrop of the Sturgeon Quartzite, and (c) outcrop of the Kona Dolomite showing bulbous structures known as *stromatolites*. Both outcrops are in northern Michigan.

Mesoproterozoic Orogeny and Rifting

The only Mesoproterozoic orogenic event in Laurentia was the 1.3- to 1.0-billion-year-old **Grenville orogeny** in the eastern part of the evolving continent (Figure 9.2c). Grenville rocks are well exposed in the present-day northern Appalachian Mountains, as well as in eastern Canada, Greenland, and Scandinavia. Many geologists think the Grenville orogen resulted from closure of an ocean basin, the final stage in a Wilson cycle. Others disagree, and think that intracontinental deformation or major shearing was responsible for deformation.

Whatever the cause of the Grenville orogeny, it was the final stage in the Proterozoic continental accretion of Laurentia (Figure 9.2c). By then, about 75% of present-day

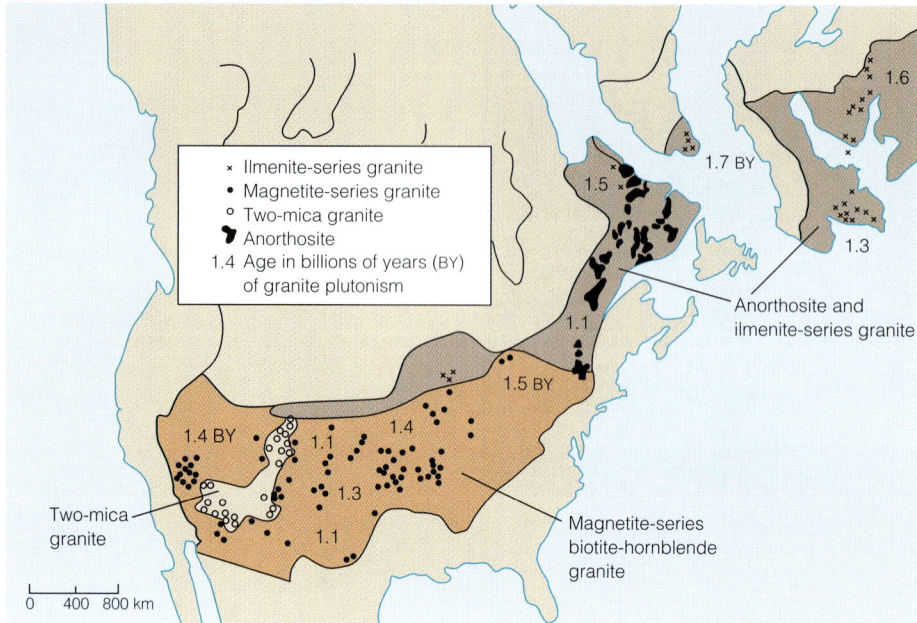

• **Figure 9.5 Proterozoic Igneous Rocks Unrelated to Orogenic Activity** Some of these rocks are 1.7 billion years old, making them Paleoproterozoic, but most are of Mesoproterozoic age. They are deeply buried over most of their extent.

North America existed. The remaining 25% accreted along its margins, particularly its eastern and western margins, during the Phanerozoic Eon.

Beginning about 1.1 billion years ago, tensional forces opened the **Midcontinent rift,** a long, narrow trough bounded by faults that outline two branches; one branch extended southeast as far as Kansas, and one extended southeasterly through Michigan and into Ohio (Figures 9.2b and c and • Figure 9.6). The rift cuts through Archean and Proterozoic rocks and terminates against the Grenville orogen in the east. Although not all geologists agree, many think that the Midcontinent rift is a failed rift where Laurentia began splitting apart. Had this rifting continued, Laurentia would have split into two separate landmasses, but the rifting ceased after about 20 million years.

Most of the Midcontinent rift is buried beneath younger rocks except in the Lake Superior region, where igneous and sedimentary rocks are well exposed (Figure 9.6c and d). The central part of the rift contains numerous overlapping basalt lava flows, forming a volcanic pile several kilometers thick. In fact, the volume of volcanic rocks, between 300,000 and 1,000,000 km^3, is comparable in volume although not area to the great outpourings of lava during the Cenozoic (see Chapter 16). Along the rift's margins, coarse-grained sediments were deposited in large alluvial fans that grade into sandstone and shale with increasing distance from the sediment source.

Meso- and Neoproterozoic Sedimentation

Remember the Grenville orogeny took place between 1.3 and 1.0 billion years ago (Figure 9.2c), the final episode of continental accretion in Laurentia until the Ordovician Period (see Chapter 10). Nevertheless, important geologic events were taking place, such as sediment deposition in what is now the eastern United States and Canada, in the Death Valley region of California and Nevada, and in three huge basins in the west (• Figure 9.7a).

Meso- and Neoproterozoic-aged sedimentary rocks are exceptionally well exposed in the northern Rocky Mountains of Montana, Idaho, and Alberta, Canada. Indeed, their colors, deformation features, and erosion by Pleistocene and recent glaciers have yielded some fantastic scenery. Like the rocks in the Great Lakes region, they are mostly sandstones, shales, and stromatolite-bearing carbonates (Figure 9.7b).

Sedimentary rocks of Proterozoic age are also found in Utah (Figure 9.7c). Proterozoic rocks of the Grand Canyon Supergroup lie unconformably on Archean rocks and in turn are overlain unconformably by Phanerozoic-age rocks (Figure 9.7d). The rocks, consisting mostly of sandstone, shale, and dolostone, were deposited in shallow-water marine and fluvial environments. The presence of stromatolites and carbonaceous impressions of algae in some of these rocks indicate probable marine deposition.

Proterozoic Supercontinents

You have learned that a *continent* is a landmass made up of granitic crust, with much of its surface above sea level. Keep in mind, however, that the geologic margin of a continent—that is, where granitic continental crust changes to basaltic oceanic crust—is beneath sea level. A **supercontinent,** in contrast, consists of at least two continents merged into one, and in the context of Earth history, we usually refer to a supercontinent as one composed of all or most of Earth's landmasses, other than oceanic islands. You are already aware of the supercontinent Pangaea that existed at the end of the Paleozoic Era, but you probably have not heard of previous supercontinents.

Before we specifically address supercontinents, it is important to note that the present style of plate tectonics

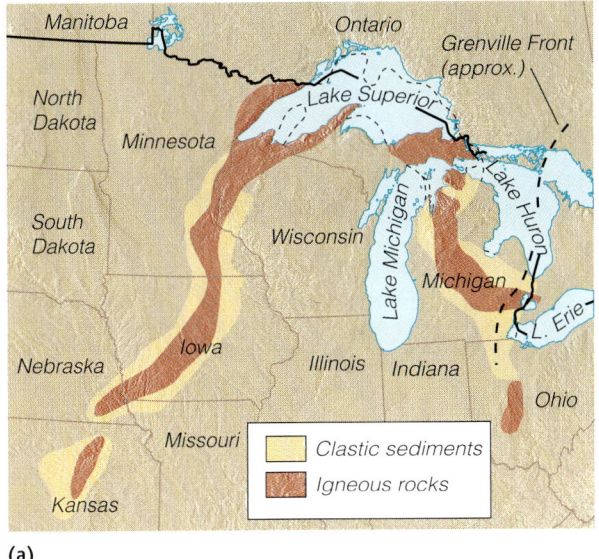
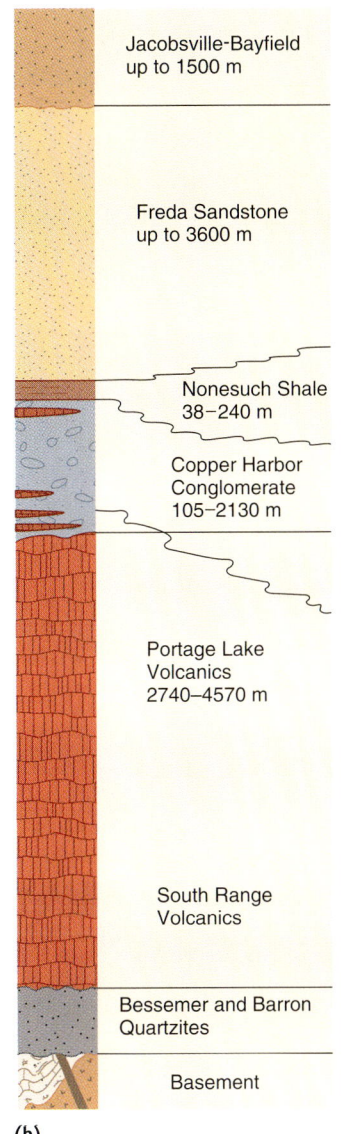

• **Figure 9.6 The Midcontinent Rift** (a) Location of the Midcontinent rift. The rocks filling the rift are well exposed around Lake Superior in the United States and Canada, but they are deeply buried elsewhere. (b) Section showing the vertical relationships among the rocks in the Lake Superior region. (c) The Copper Harbor Conglomerate and (d) the Portage Lake Volcanics, both in Michigan.

involving opening and then closing ocean basins had certainly been established by the Paleoproterozoic. In fact, *ophiolites* that provide evidence for convergent plate boundaries are known from Neoarchean and Paleoproterozoic rocks of Russia and China, respectively. Furthermore, these ancient ophiolites compare closely with younger, well-documented ophiolites, such as the highly deformed but complete Jormua mafic-ultramafic complex in Finland (• Figure 9.8).

Supercontinents may have existed as early as the Neoarchean, but if so we have little evidence of them. The first that geologists recognize with some certainty, known as **Rodinia** (• Figure 9.9), assembled between 1.3 and 1.0 billion years ago and then began fragmenting 750 million years ago. Judging by the large-scale deformation called the *Pan-African orogeny* that took place in what are now the Southern Hemisphere continents, Rodinia's separate

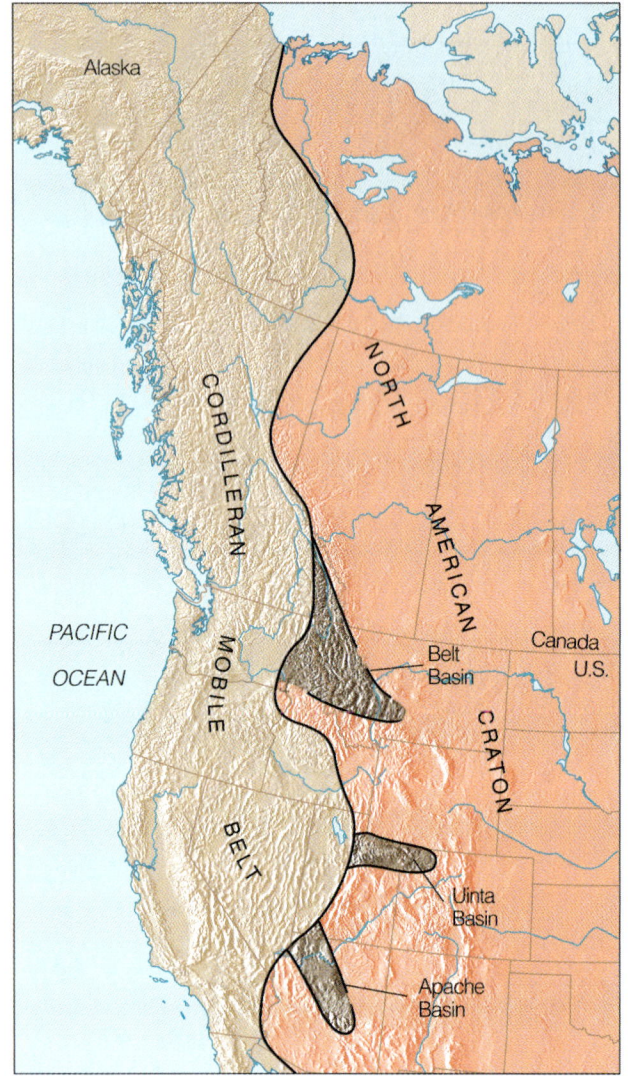

• **Figure 9.7 Proterozoic Rocks in the West** (a) Meso- to Neoproterozoic basins in western North America. (b) Red mudrock in the Belt basin in Montana. (c) Rocks of the Uinta Mountain Group in Utah. (d) Sandstone in the Grand Canyon, Arizona.

pieces reassembled about 650 million years ago and formed another supercontinent, this one known as **Pannotia**. And, finally, by the latest Proterozoic, about 550 million years ago, fragmentation was under way, giving rise to the continental configuration that existed at the onset of the Phanerozoic Eon (see Chapter 10).

What Would You Do?

Suppose you were to visit a planet that, like Earth, has continents and vast oceans. What evidence would indicate that this hypothetical planet's continents formed and evolved like those on Earth?

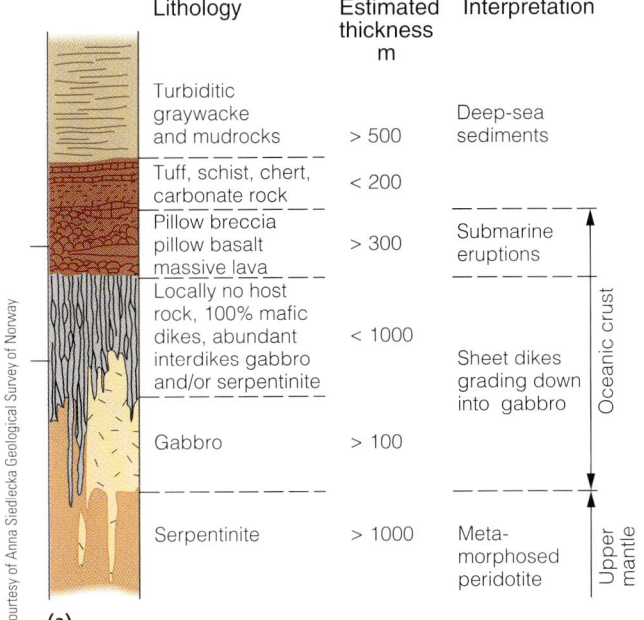

• **Figure 9.8 Proterozoic Ophiolite** (a) Reconstruction of the highly deformed Jorma mafic-ultramafic complex in Finland. This 1.96-billion-year-old sequence of rocks is one of the oldest known complete ophiolites. (b) A metamorphosed basaltic pillow lava. The code plate is 12 cm long. (c) Metamorphosed gabbro between mafic dikes. The hammer shaft is 65 cm long.

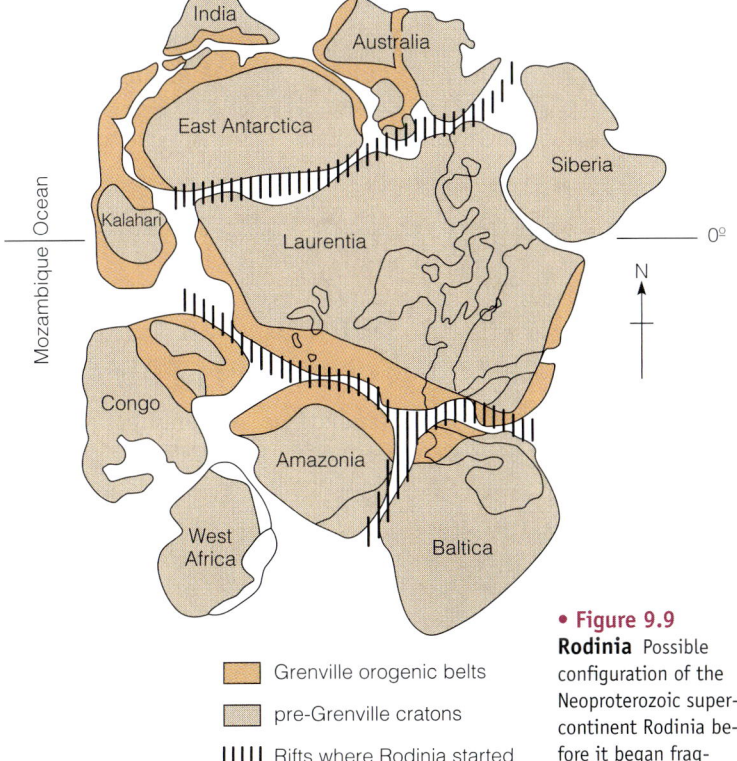

• **Figure 9.9 Rodinia** Possible configuration of the Neoproterozoic supercontinent Rodinia before it began fragmenting about 750 million years ago.

Ancient Glaciers and Their Deposits

Very few instances of widespread glaciation have occurred during Earth history. The most recent one, during the Pleistocene (1.8 million to 10,000 years ago), is certainly the best known, but we also have evidence for Pennsylvanian glaciers (see Chapter 11) and two major episodes of Proterozoic glaciation. But how can we be sure there were Proterozoic glaciers? After all, their most common deposit, called *tillite*, is simply a type of conglomerate that may look much like conglomerate that originated by other processes.

Paleoproterozoic Glaciers

Tillite or tillitelike deposits are known from at least 300 Precambrian localities, and some of these are undoubtedly not glacial deposits. But the extensive geographic distribution of others and their associated glacial features, such as striated and polished bedrock, are distinctive (• Figure 9.10). Based on this kind of evidence, geologists are now

> **What Would You Do?**
>
> As a working geologist you encounter a Proterozoic-aged conglomerate that you are convinced is a glacial deposit. Others, however, think it is simply stream-deposited gravel or perhaps an ancient landslide deposit. What kinds of evidence, if present, would verify your initial interpretation? That is, what attributes of the rock itself and its associated deposits might lend credence to your analysis?

convinced that widespread glaciation took place during the Paleoproterozoic.

Tillites of about the same age in Michigan, Wyoming, and Quebec indicate that North America may have had an ice sheet centered southwest of Hudson Bay (• Figure 9.11a). Tillites of about this age are also found in Australia and South Africa, but dating is not precise enough to determine if there was a single widespread glacial episode or a number of glacial events at different times in different areas. One tillite in the Bruce Formation in Ontario, Canada, is about 2.7 billion years ago, thus making it Neoarchean.

Glaciers of the Neoproterozoic

Tillites and other glacial features dating from between 900 and 600 million years ago are found on all continents except Antarctica. Glaciation was not continuous during this entire time but was episodic, with four major glacial episodes so far recognized. Figure 9.11b shows the approximate distribution of these glaciers, but we must emphasize "approximate," because the actual geographic extent of these glaciers is unknown. In addition, the glaciers covering the area in Figure 9.11b were not all present at the same time. Despite these uncertainties, this Neoproterozoic glaciation was the most extensive in Earth history. In fact, glaciers seem to have been present even in near-equatorial areas.

The Evolving Atmosphere

Geologists agree that the Archean atmosphere contained little or no free oxygen (see Chapter 8), so the atmosphere was not strongly oxidizing as it is now. And even though photochemical dissociation and photosynthesis were adding free oxygen to the atmosphere, the amount present at the beginning of the Proterozoic was probably no more than 1% of that present now. In fact, it might not have exceeded 10% of present levels even at the end of the Proterozoic. Remember from our previous discussions that cyanobacteria (blue-green algae) were present during the Archean, but the structures they formed, called *stromatolites*, did not become common until about 2.3 billion years ago—that is, during the Paleoproterozoic. These photosynthesizing organisms and, to a lesser degree, photochemical dissociation both added free oxygen to the evolving atmosphere (see Figure 8.13).

In Chapter 8 we cited some of the evidence indicating that Earth's early atmosphere had little or no free oxygen but abundant carbon dioxide. Here we contend that more oxygen became available, whereas the amount of carbon dioxide decreased. So what evidence is there indicating that the atmosphere became an oxidizing one, and where is the carbon dioxide now? Of course, a small amount of CO_2 is present in today's atmosphere; it is one of the greenhouse gases partly responsible for global warming. Much of it, however, is now tied up in minerals and rocks, especially the carbonate rocks limestone and dolostone, and in the biosphere. As for evidence that the Proterozoic atmosphere was evolving from a chemically reducing one to an oxidizing one, we must discuss two types of Proterozoic sedimentary rocks that we already alluded to briefly.

Banded Iron Formations (BIFs)

Banded iron formations (BIFs), which consist of alternating layers of iron-rich minerals and chert (• Figure 9.12a and b), are found in Archean rocks, but about 92% of all BIF formed during the interval from 2.5 to 2.0 billion years ago—that is, during the earlier part of the Paleoproterozoic. What do these rocks have to do with the atmosphere? The iron in them is in the form of iron oxides, especially hematite (Fe_2O_3) and magnetite (Fe_3O_4). Iron is a highly reactive element, and in an oxidizing

• **Figure 9.10 Evidence for Proterozoic glaciation in Norway.** This deposit of tillite lies upon a striated bedrock surface of sandstone.

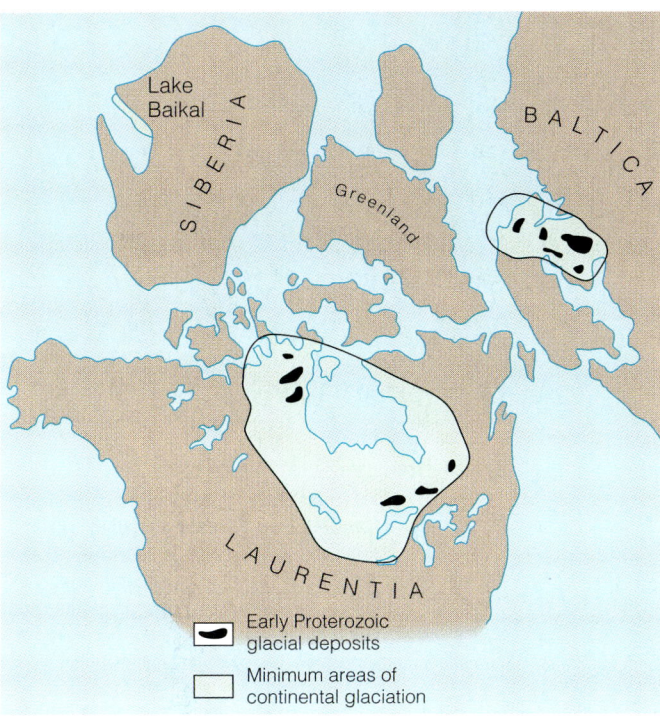

The Archean atmosphere was deficient in free oxygen, so probably little oxygen was dissolved in seawater. However, as photosynthesizing organisms increased in abundance, as indicated by stromatolites, free oxygen released as a metabolic waste product into the oceans caused the precipitation of iron oxides and silica and thus the origin of BIFs.

One model accounting for the details of BIF precipitation involves a Precambrian ocean with an upper oxygenated layer overlying a large volume of oxygen-deficient water that contained reduced iron and silica. Upwelling—that is, transfer of water from depth to the surface—brought iron- and silica-rich waters onto the developing shallow continental shelves and resulted in widespread precipitation of BIFs (Figure 9.12c). A likely source of the iron and silica was submarine volcanism, similar to that now taking place at or near spreading ridges. Huge quantities of dissolved minerals are discharged at submarine hydrothermal vents (see Chapter 8). In any case, the iron and silica combined with oxygen, resulting in the precipitation of large amounts of BIF, and continued until the iron in seawater was largely used up.

Continental Red Beds

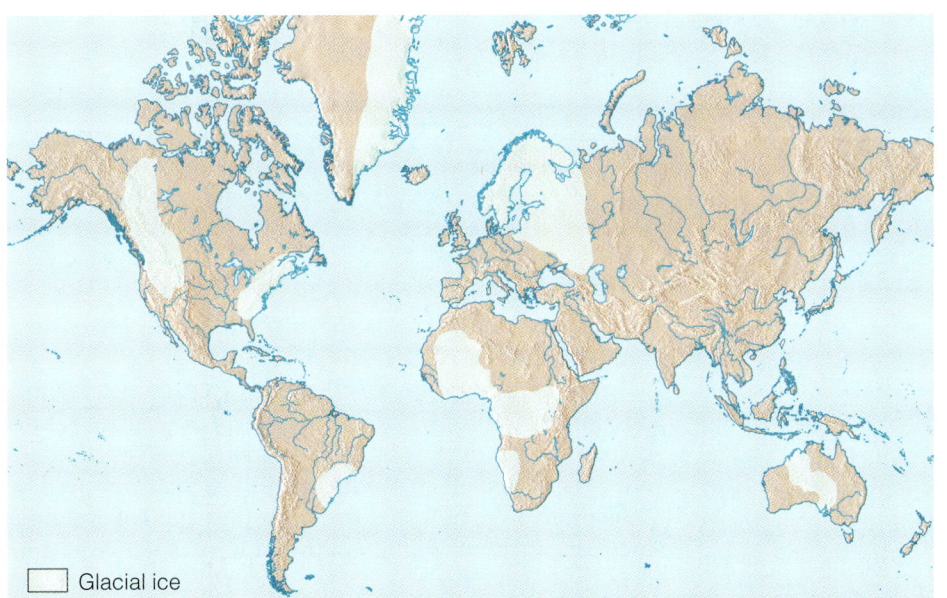

Obviously, the term **continental red beds** refers to red rocks on the continents, but more specifically it means red sandstone or shale colored by iron oxides, especially hematite (Fe_2O_3) (Figure 9.7b and d). These deposits first appear in the geologic record about 1.8 billion years ago, increase in abundance throughout the rest of the Proterozoic, and are quite common in rocks of Phanerozoic age.

The onset of red bed deposition coincides with the introduction of free oxygen into the Proterozoic atmosphere. But the atmosphere at that time may have had only 1% or perhaps 2% of present levels. Is this sufficient to account for oxidized iron in sediment? Probably not, but we must also consider other attributes of this atmosphere.

No ozone (O_3) layer existed in the upper atmosphere before free oxygen (O_2) was present. But as photosynthesizing organisms released free oxygen into the atmosphere, ultraviolet radiation con-

• **Figure 9.11 Proterozoic Glaciation** (a) Paleoproterozoic-aged glacial deposits in the United States and Canada indicate that Laurentia had an extensive ice sheet centered southwest of Hudson Bay. (b) Neoproterozoic glacial centers shown on a map with the continents in their present-day positions. The extent of ice during this time is approximate.

atmosphere it combines with oxygen to form rustlike oxides that are not readily soluble in water. If oxygen is absent, though, iron is easily taken into solution and can accumulate in large quantities in the world's oceans, which it undoubtedly did during the Archean.

verted some of it to elemental oxygen (O) and ozone (O_3), both of which oxidize minerals more effectively than does O_2. Once an ozone layer became established, most ultraviolet radiation failed to penetrate to the surface, and O_2 became the primary agent for oxidizing minerals.

(a)

(b)

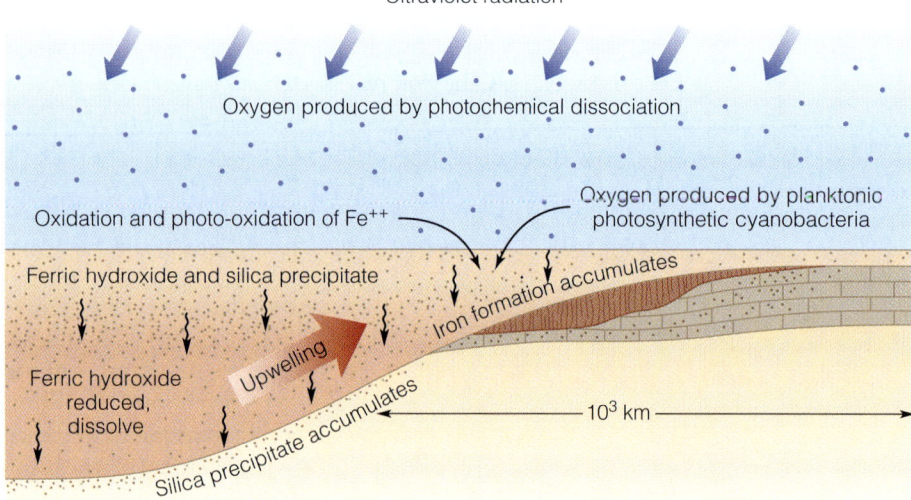
(c)

• **Figure 9.12 Paleoproterozoic-aged Banded Iron Formation (BIF)** (a) At this outcrop in Ishpeming, Michigan, the rocks are brilliantly colored alternating layers of red chert and silver iron minerals. (b) A more typical outcrop of BIF near Negaunee, Michigan. (c) Depositional model for the origin of banded iron formations.

Important Events in Life History

Archean fossils are not very common, and all are varieties of archaea and bacteria, although they undoubtedly existed in profusion. Likewise, the Paleoproterozoic fossil record has mostly bacteria and especially stromatolites (• Figure 9.13a). Even in well-known Early Proterozoic fossil assemblages, such as the Gunflint Iron Formation of Canada, only fossilf bacteria are recognized (Figure 9.13b).

Actually, the lack of organic diversity during this early time in life history is not too surprising, because prokaryotic cells reproduce asexually. Recall from Chapter 7 that most variation in sexually reproducing populations comes from the shuffling of genes and their alleles from generation to generation. Mutations introduce new variation into a population, but in prokaryotes their effects are limited. A beneficial mutation would spread rapidly in sexually reproducing organisms but would have a limited impact in prokaryotic cells, because they do not usually share their genes with other cells.*

Before the appearance of cells capable of sexual reproduction, evolution was a comparatively slow process, thus accounting for the low organic diversity. This situation did not persist. Sexually reproducing cells probably evolved by Paleoproterozoic time, and thereafter the tempo of evolution increased markedly, although from our perspective of time it was still very slow.

Eukaryotic Cells Evolve

The appearance of **eukaryotic cells**—that is, cells with a nucleus and internal organelles—marks a milestone in evolution. But where did these cells come from, and how do they differ from their predecessors the prokaryotic cells? All organisms made up of prokaryotic cells, called prokaryotes,

---

*Prokaryotes usually reproduce by binary fission and give rise to two cells having the same genetic makeup. Under some conditions, they engage in conjugation, during which some genetic material is transferred.

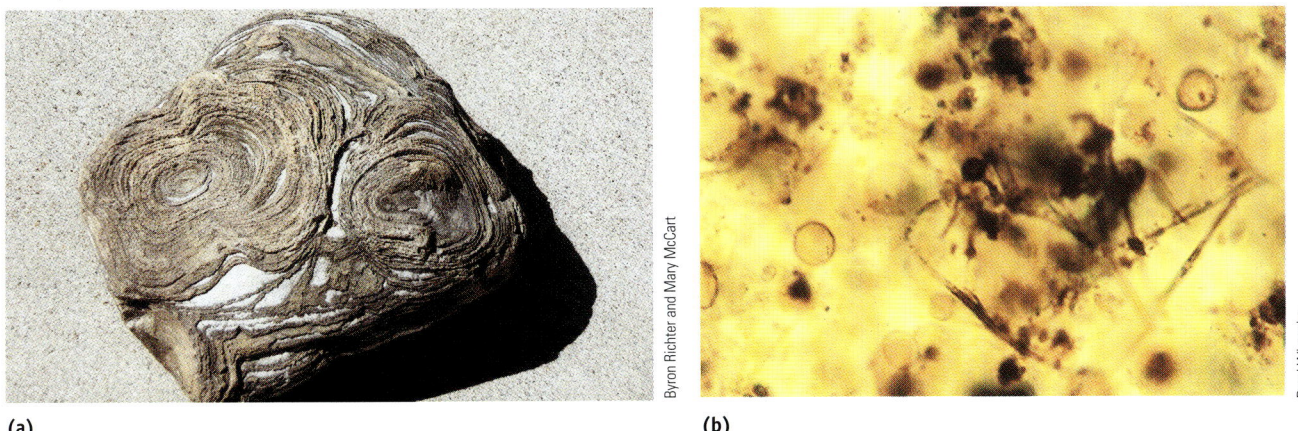

- **Figure 9.13 Proterozoic Stromatolites and Fossil Bacteria** (a) This rock clearly shows two stromatolites that have grown together (see Figure 8.18b). (b) These spherical and filamentous microfossils from the Gunflint Chert of Ontario, Canada, resemble bacteria living today. The filaments measure about 1/1000th of a millimeter across.

are single-celled, but most eukaryotes are multicelled, the notable exception being the protistans (• Figure 9.14, Table 9.1). In marked contrast to prokaryotes, most eukaryotes reproduce sexually, and nearly all are aerobic; that is, they depend on free oxygen to carry out their metabolic processes. Accordingly, they could not have evolved before at least some free oxygen was present in the atmosphere.

The 2.1-billion-year-old Negaunee Iron Formation in Michigan has fossils now generally accepted as the oldest known eukaryotic cells (• Figure 9.15). The Bitter Springs Formation of Australia is much younger (1.0 billion years old), but it has some remarkable fossils of single-celled organisms that show evidence of meiosis and mitosis, processes carried out only by eukaryotic cells.

What kinds of evidence indicate that these microscopic single-celled organisms were eukaryotes rather than prokaryotes? For one thing, eukaryotic cells are larger, commonly much larger, than prokaryotic cells (Figure 9.14). Furthermore, prokaryotic cells are mostly rather simple spherical or platelike structures, whereas eukaryotic cells are much more complex. The latter not only have a well-defined, membrane-bounded cell nucleus, which is lacking in prokaryotes, but they also have internal structures called *organelles,* such as plastids and mitochondria. In short, their organizational complexity is much greater than for prokaryotes (Figure 9.14).

Other organisms that were almost certainly eukaryotes are the *acritarchs* that first appeared about 1.4 billion years ago. They were very common by Neoproterozoic time (• Figure 9.16a and b) and were probably cysts of planktonic (floating) algae. Numerous microfossils of organisms with vase-shaped skeletons have been found in Neoproterozoic-aged rocks in the Grand Canyon (Figure 9.16c). These too have tentatively been identified as cysts of some kind of algae.

Endosymbiosis and the Origin of Eukaryotic Cells

According to a widely accepted theory, eukaryotic cells formed from prokaryotic cells that entered into a symbiotic relationship. Symbiosis involving a prolonged association of two or more dissimilar organisms is quite common today. In many cases both symbionts benefit from the association,

TABLE 9.1	The Five-Kingdom Classification of Organisms (notice that only members of the kingdom Monera have prokaryotic cells)	
Kingdom	**Characteristics**	**Examples**
Monera	Single cells with little internal complexity; prokaryotic cells; reproduce asexually	Archaea; bacteria; cyanobacteria (also called blue-green algae)
Protista	Single-cell or multicell; eukaryotic cells; greater internal complexity than bacteria	Various types of algae, diatoms, protozoans, molds
Fungi	Multicell; eukaryotic cells; major decomposers and nutrient recyclers	Fungi
Plantae	Multicell, eukaryotic cells, obtain nutrients by photosynthesis	Trees, grasses, ferns, palms, roses, broccoli, poison ivy
Animalia	Multicell; eukaryotic cells; obtain nutrients by ingestion of preformed molecules	Worms, sponges, corals, clams, jellyfish, fishes, amphibians, reptiles, birds, mammals

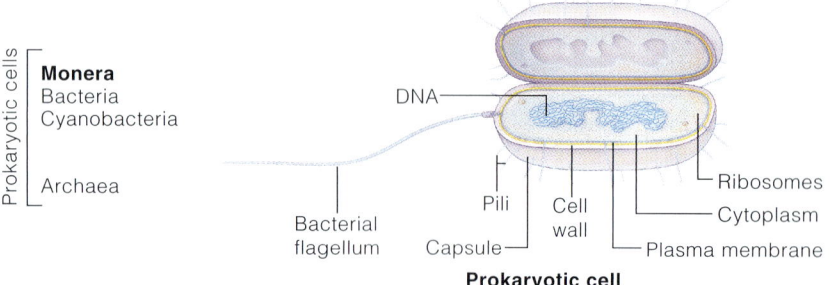

• **Figure 9.14 Prokaryotic and Eukaryotic Cells** Eukaryotic cells have a cell nucleus containing the genetic material and organelles such as mitochondria and plastids, as well as chloroplasts in plant cells. In contrast, prokaryotic cells are smaller and not nearly as complex as eukaryotic cells.

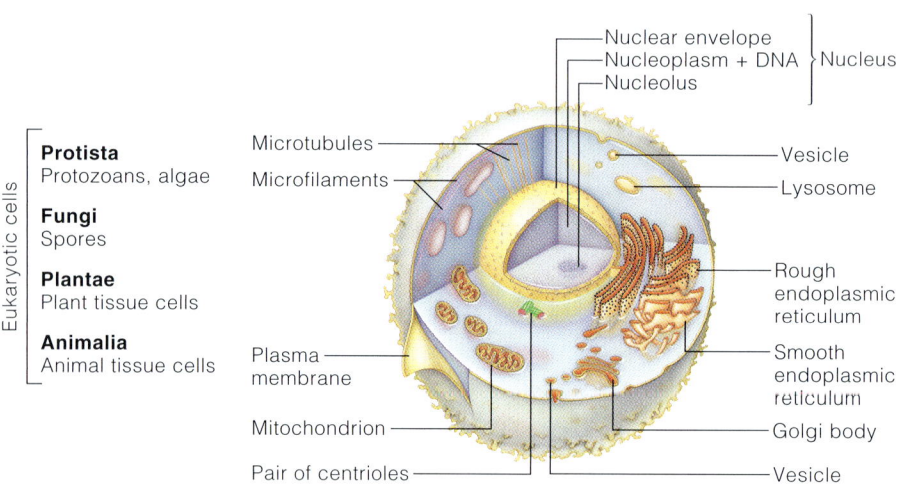

• **Figure 9.15 Oldest Known Fossil Eukaryotic Organism** This fossil from the 2.1-billion-year-old Negaunee Iron Formation at Marquette, Michigan, is probably some kind of multicelled algae.

as in lichens, which were once thought to be plants but actually are symbiotic associations between fungi and algae.

In a symbiotic relationship, each symbiont must be capable of metabolism and reproduction, but the degree of dependence in some relationships is such that one symbiont cannot live independently. This may have been the case with Proterozoic symbiotic prokaryotes that became increasingly interdependent until the unit could exist only as a whole. In this relationship, though, one symbiont lived within the other, which is a special type of symbiosis called **endosymbiosis** (• Figure 9.17).

Supporting evidence for endosymbiosis comes from studies of living eukaryotic cells containing internal structures called organelles, such as mitochondria and plastids, that contain their own genetic material. In addition, prokaryotic cells synthesize proteins as a single system, whereas eukaryotic cells are a combination of protein synthesizing systems. That is, some of the organelles within eukaryotic cells are capable of protein synthesis. These organelles with their own genetic material and protein synthesizing capabilities are thought to have been free-living bacteria that entered into a symbiotic relationship, eventually giving rise to eukaryotic cells.

The Dawn of Multicelled Organisms

Obviously **multicelled organisms** are made up of many cells, perhaps billions, as opposed to a single cell as in bacteria, archaea, and the protestians. In addition, multicelled organisms have cells specialized to perform specific functions such as respiration, food gathering, and reproduction. We know from the fossil record that multicelled organisms

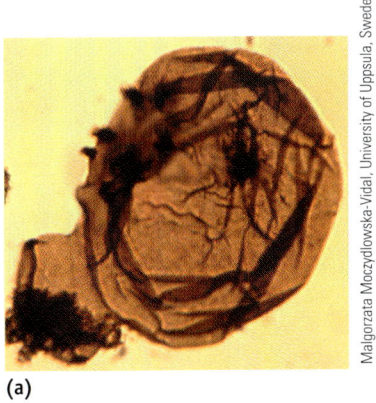 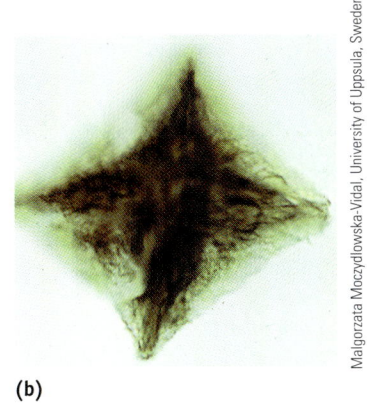 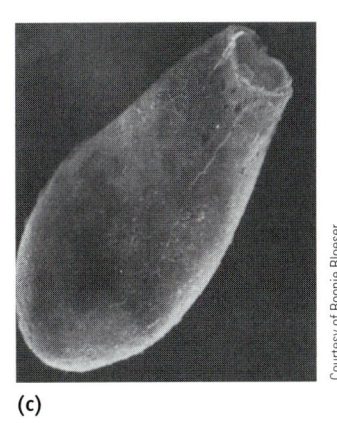

• **Figure 9.16 Proterozoic Fossils** These are probably from eukaryotic organisms and measure only 40 or 50 microns across. **(a)** The acritarch *Tappania plana* is from Mesoproterozoic rocks in China. **(b)** This acritarch, known as *Octoedryxium truncatum*, was found in Neoproterozoic rocks in Sweden. **(c)** This vase-shaped microfossil from Neoproterozoic rocks in the Grand Canyon in Arizona is also a cyst from some kind of algae.

were present during the Proterozoic, but exactly when they appeared has not been resolved. The 2.1-billion-year-old fossils from the Negaunee Iron Formation of Michigan might be some kind of multicelled algae (Figure 9.15). And carbonaceous filaments from 1.8-billion-year-old rocks in China might also be from multicelled algae, and a similar interpretation is likely valid for somewhat younger carbonaceous impressions of filaments and spherical forms (• Figure 9.18).

Studies of present-day organisms give some clues as to how this important transition might have taken place. Perhaps a single-celled organism divided, but the daughter cells formed an association as a colony. Each cell would have been capable of an independent existence, and some cells might have become somewhat specialized, as are the cells of colonial organisms today. Increased specialization of cells might have given rise to comparatively simple multicelled organisms such as algae and sponges.

But is there any particular advantage to being multicelled? After all, for something on the order of 1.5 billion years all organisms were single-celled and life seems to have thrived. In fact, single-celled organisms are quite good at what they do, but what they do is limited. For example, they can't become very large, because as size increases proportionately less of a cell is exposed to the external environment in relation to its volume. So as volume increases, the proportion of surface area decreases and transferring materials from the exterior to the interior becomes less efficient. Also, multicelled organisms live longer, because cells can be replaced and more offspring

• **Figure 9.17 Endosymbiosis and the Origin of Eukaryotic Cells** An aerobic bacterium and a larger host bacterium united to form a mitochondria-containing amoeboid. An amoeboflagellate was formed by a union of the amoeboid and a bacterium of the spirochete group; this amoeboflagellate gave rise to the animal, fungi, and protistan kingdoms. The plant kingdom originated when blue-green algae became plastids within an amoeboflagellate.

• **Figure 9.18 Carbonaceous Impressions in Proterozoic Rocks** These in the Little Belt Mountains in Montana may be impressions of multicelled algae.

IMPORTANT EVENTS IN LIFE HISTORY

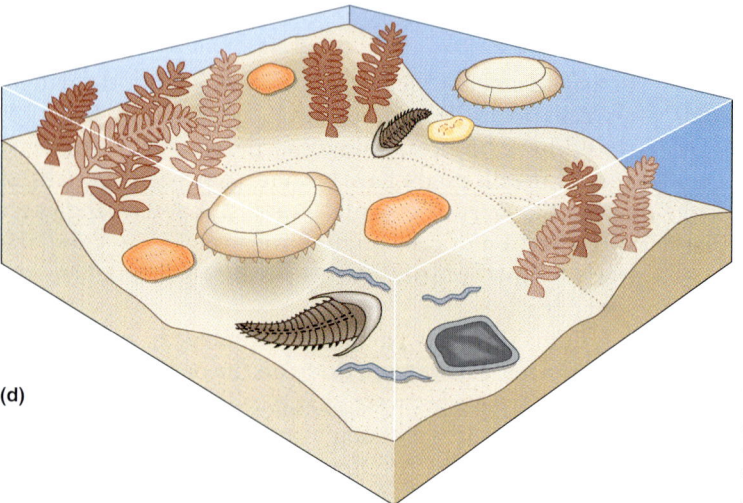

• **Figure 9.19 The Ediacaran Fauna of Australia** (a) The affinities of *Tribrachidium* remain mysterious. It may be either a primitive echinoderm or a cnidarian. (b) *Spriggina* was originally thought to be a segmented worm (annelid), but now it appears to be more closely related to arthropods, possibly even an ancestor of trilobites. (c) Shield-shaped *Parvancorina* is perhaps related to the arthropods. (d) Reconstruction of the Ediacaran environment.

can be produced. And finally, cells have increased functional efficiency when they are specialized into organs with specific functions.

Neoproterozoic Animals

Criteria such as method of reproduction and type of metabolism set forth by biologists allow us to easily distinguish between animals and plants. Or so it would seem—but some present-day organisms blur this distinction, and the same is true for some Proterozoic fossils. Nevertheless, the first fairly controversy-free fossils of animals come from the Ediacaran fauna of Australia and similar faunas of about the same age elsewhere.

The Ediacaran Fauna In 1947, an Australian geologist, R. C. Sprigg, discovered impressions of soft-bodied animals in the Pound Quartzite in the Ediacara Hills of South Australia. Additional discoveries by others turned up what appeared to be impressions of algae and several animals, many bearing no resemblance to any existing now (• Figure 9.19).

Before these discoveries geologists were perplexed by the apparent absence of fossil-bearing rocks predating the Phanerozoic. They had assumed the fossils so common in Cambrian rocks must have had a long previous history but had little evidence to support this conclusion. This discovery and subsequent ones have not answered all questions about pre-Phanerozoic animals, but they have certainly increased our knowledge about this chapter in the history of life.

Some investigators think that three present-day phyla are represented in the Ediacaran fauna: jellyfish and sea pens (phylum Cnidaria), segmented worms (phylum Annelida), and primitive members of the phylum Arthropoda (the phylum with insects, spiders, crabs, and others). One Ediacaran fossil, *Spriggina,* has been cited as a possible ancestor of trilobites (Figure 9.19b), and another may be a primitive member of the phylum Echinodermata. However, some scientists think these Ediacara animals represent an early evolutionary group quite distinct from the ancestry of today's invertebrate animals.

The **Ediacaran fauna**, the collective name for fossil associations similar to those in the Ediacara Hills, is now

• **Figure 9.20 Ediacaran-type Fossils from the Mistaken Point Formation of Newfoundland** (a) These fossils have not been given scientific names. The one at the upper center is informally called *feather duster,* whereas the more obvious ones are called *spindles.* (b) This fossil, known as *Bradgatia,* has also been found in England. These fossils are about 575 million years old.

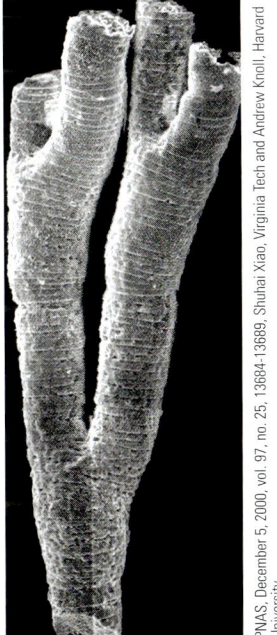

known from all continents except Antarctica. For example, excellent fossils from this 545- to 600-million-year-old fauna have been found in Namibia, Africa, and in Newfoundland, Canada (• Figure 9.20). Thus, Ediacaran animals were widespread, but because they lacked durable skeletons, their fossils are not very common.

Other Proterozoic Animal Fossils Although scarce, a few animal fossils older than those of the Ediacaran fauna are known. A jellyfish-like impression is present in rocks 2000 m below the Pound Quartzite, and, in many areas, burrows, presumably made by worms, are found in rocks at least 700 million years old. Some possible fossil worms from China may be 700 to 900 million years old (• Figure 9.21a).

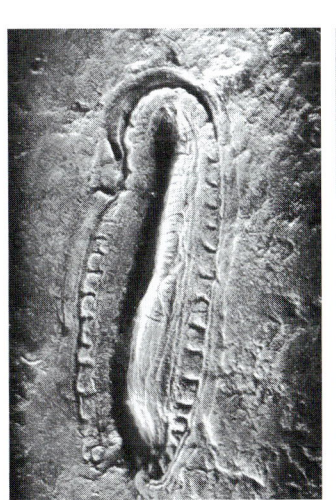

• **Figure 9.21 Neoproterozoic Fossils** (a) Possible fossil worms from 700- to 900-million-year-old rocks in China. (b) These small branching tubes measure only 0.1 to 0.3 mm across. The animals that made these tubes may have been early relatives of corals. (c) All agree that *Kimberella*, from Neoproterozoic-aged rocks in Russia, was an animal, but opinion is divided among those who think it was sluglike or more like a mollusk.

All known Proterozoic animals were soft-bodied, but there is some evidence that the earliest stages in the evolution of skeletons were under way. Even some Ediacaran animals may have had a chitinous carapace, and others appear to have spots or small areas of calcium carbonate. Small branching tubes preserved in 590- to 600-million-year-old rocks in China may be early relatives of corals (Figure 9.21b). The odd creature known as *Kimberella* from the Neoproterozoic of Russia had a tough outer covering similar to that of some present-day marine invertebrates (Figure 9.21c). Exactly what *Kimberella* was is uncertain; some think it was a sluglike creature, but others think it was more like a mollusk.

Minute scraps of shell-like material and small toothlike denticles and spicules, presumably from sponges, indicate that several animals with skeletons, or at least partial skeletons, existed by latest Neoprotoerozoic time. Nevertheless, more durable skeletons of silica, calcium carbonate, and chitin (a complex organic substance) did not appear in abundance until the beginning of the Phanerozoic Eon 542 million years ago (see Chapter 12).

Proterozoic Mineral Resources

In an earlier section we mentioned that most of the world's iron ore comes from Paleoproterozoic banded iron formations (Figure 9.12) and that Canada and the United States have large deposits of these rocks in the Lake Superior region and in eastern Canada (see Perspective 9.1). Both rank among the ten leading nations in iron ore production.

In the Sudbury mining district in Ontario, Canada, nickel and platinum are extracted from Proterozoic rocks. Nickel is essential for the production of nickel alloys such as stainless steel and Monel metal (nickel plus copper), which are valued for their strength and resistance to corrosion and heat. The United States must import more than 50% of all nickel used, mostly from the Sudbury mining district.

Besides its economic importance, the Sudbury Basin, an elliptical area measuring more than 59 by 27 km, is geologically interesting. One hypothesis for the concentration of ores is that they were mobilized from metal-rich rocks beneath the basin following a high-velocity meteorite impact.

Some platinum for jewelry, surgical instruments, and chemical and electrical equipment is also exported to the United States from Canada, but the major exporter is South Africa. The Bushveld Complex of South Africa is a layered complex of igneous rocks from which both platinum and chromite, the only ore of chromium, are mined. Much chromium used in the United States is imported from South Africa; it is used mostly in manufacturing stainless steel.

Economically recoverable oil and gas have been discovered in Proterozoic rocks in China and Siberia, arousing some interest in the Midcontinent rift as a potential source of hydrocarbons (Figure 9.6). So far, land has been leased for exploration, and numerous geophysical studies have been done. However, even though some rocks within the rift are known to contain petroleum, no producing oil or gas wells are operating.

A number of Proterozoic pegmatites are important economically. The Dunton pegmatite in Maine, whose age is generally considered Neoproterozoic, has yielded magnificent gem-quality specimens of tourmaline and other minerals (• Figure 9.22). Other pegmatites are mined for gemstones as well as for tin; industrial minerals, such as feldspars, micas, and quartz; and minerals containing such elements as cesium, rubidium, lithium, and beryllium.

Geologists have identified more than 20,000 pegmatites in the country rocks adjacent to the Harney Peak Granite in the Black Hills of South Dakota. These pegmatites formed about 1.7 billion years ago when the granite was emplaced as a complex of dikes and sills. A few have been mined for gemstones, tin, lithium, and micas, and some of the world's largest known mineral crystals were discovered in these pegmatites.

• Figure 9.22 **Resources from Proterozoic Pegmatites** Gem-quality tourmaline from the Dunton pegmatite mine in Maine.

9.1 Perspective

BIF: From Mine to Steel Mill

The United States and Canada, both highly industrialized nations, owe a good deal of their economic success to abundant natural resources, although both nations, especially the United States, must import some essential commodities. Iron ore, however, is present in great abundance in the Great Lakes region and in the Labrador Trough of eastern Canada. The giant iron-ore freighters that ply the Great Lakes sail from loading facilities in Ontario, Minnesota, and Michigan to iron-producing cities such as Hamilton, Ontario; Cleveland, Ohio; and Gary, Indiana.

Iron ore is produced from rocks of various ages, but by far most of it comes from Paleoproterozoic *banded iron formation* (BIF), consisting of alternating layers of iron-bearing minerals (mostly hematite and magnetite) and chert (Figure 9.12a and b). We already discussed what is known about the origin of BIF, and remember that 92% of it was deposited during the Paleoproterozoic when Earth's atmosphere lacked free oxygen, that is, oxygen not combined with other elements. Most BIF is not particularly attractive but some consists of alternating layers of brilliant, silvery, iron minerals, and red chert as at Jasper Knob in Ishpeming, Michigan (Figure 9.12a).

The richest iron ores in the Great Lakes region had been largely depleted by the time of World War II (1939–1945), but then a technologic innovation was developed that allowed mining of poorer-grade ores. In this process the unusable rock is separated from the iron, which is then shaped into pellets before it is shipped to steel mills (Figure 1). In short, it was now possible to use what otherwise could not be recovered economically. Iron mining now accounts for a large part of the economies of several cities on or near the shores of Lake Superior and in eastern Canada.

The complete process from mine to steel mill consists of hauling the iron-bearing rock in huge dump trucks to a crusher, where it is ground to powder. The powdered ore is then processed to separate the iron from the unusable rock, either by floatation or by magnetic separation. In any case, the iron-rich powdered concentrate is dried and shaped into pellets in large rotating drums, and then finally baked at about 1300°C in huge rotary kilns. From the kilns, the pellets are either stored for later shipment or taken directly to iron ore freighters at Duluth, Minnesota, Marquette, Michigan, or other ports. And then the iron ore freighters make their way through the Great Lakes and deliver the pellets to the steel mills.

The availability of iron ore and a steel-making capacity is essential for large-scale industrialization. Accordingly, industrialized nations and developing nations must either have domestic deposits of iron ore, or they must import it. Japan is industrialized but has no iron ore deposits and thus imports most of what it needs from Australia, which has vast deposits of Proterozoic BIF. Currently, the nations producing the most iron ore for export are Australia and Brazil, each accounting for about a third of the world's total exports, but China is actually the largest iron ore producer, with more than 200 million metric tons per year. During 2003, the United States and Canada produced 54 million and 31 million metric tons of iron ore, respectively.

(a)

(b)

Figure 1 (a) The Empire Mine at Palmer, Michigan, where iron ore from the Paleoproterozoic-aged Negaunee Iron Formation is mined and shaped into (b) pellets containing about 65% iron. The pellets measure about 1 cm in diameter.

SUMMARY

Table 9.2 provides a summary of the Proterozoic geologic and biologic events discussed in the text.

- The crust-forming processes that yielded Archean granite-gneiss complexes and greenstone belts continued into the Proterozoic but at a considerably reduced rate.
- Paleoproterozoic collisions between Archean cratons formed larger cratons that served as nuclei, around which crust accreted. One large landmass so formed was Laurentia, consisting mostly of North America and Greenland.
- Paleoproterozoic amalgamation of cratons, followed by Mesoproterozoic igneous activity, the Grenville orogeny, and the Midcontinent rift, were important events in the evolution of Laurentia.
- Ophiolite sequences marking convergent plate boundaries are first well documented from the Neoarchean and Paleoproterozoic, indicating that a plate tectonic style similar to that operating now had become established.
- Sandstone-carbonate-shale assemblages deposited on passive continental margins were very common by Proterozoic time.
- The supercontinent Rodinia assembled between 1.3 and 1.0 billion years ago, fragmented, and then reassembled to form Pannotia about 650 million years ago, which began fragmenting about 550 million years ago.
- Glaciers were widespread during both the Paleoproterozoic and the Neoproterozoic.
- Photosynthesis continued to release free oxygen into the atmosphere, which became increasingly rich in oxygen through the Proterozoic.
- Fully 92% of Earth's iron ore deposits in the form of banded iron formations were deposited between 2.5 and 2.0 billion years ago.
- Widespread continental red beds dating from 1.8 billion years ago indicate that Earth's atmosphere had enough free oxygen for oxidation of iron compounds.
- Most of the known Proterozoic organisms are single-celled prokaryotes (bacteria). When eukaryotic cells first appeared is uncertain, but they may have been present by 2.1 billion years ago. Endosymbiosis is a widely accepted theory for their origin.
- The oldest known multicelled organisms are probably algae, some of which might date back to the Paleoproterozoic.
- Well-documented multicelled animals are found in several Neoproterozoic localities. Animals were widespread at this time, but because all lacked durable skeletons their fossils are not common.
- Most of the world's iron ore production is from Proterozoic banded iron formations. Other important resources include nickel and platinum.

IMPORTANT TERMS

banded iron formation (BIF), p. 176
continental red beds, p. 177
Ediacaran fauna, p. 182
endosymbiosis, p. 180
eukaryotic cell, p. 178
Grenville orogeny, p. 171
Laurentia, p. 168
Midcontinent rift, p. 172
multicelled organism, p. 180
orogen, p. 168
Pannotia, p. 174
Rodinia, p. 173
sandstone-carbonate-shale assemblage, p. 170
supercontinent, p. 172
Wilson cycle, p. 168

REVIEW QUESTIONS

1. Most of the world's iron ore comes from Proterozoic
   a. _____ continental red beds; b. _____ banded iron formations; c. _____ iron nodules on the seafloor; d. _____ hematitie-magnetite meteorites; e. _____ Ediacaran iron deposits.

2. _____ was a Proterozoic supercontinent made up mostly of North America and Greenland.
   a. _____ Pangaea; b. _____ Grenville; c. _____ Laurentia; d. _____ Sedonia; e. _____ Newfoundland.

3. A plate tectonic cycle involving the opening of an ocean basin and its subsequence closure is known as a(n)
   a. _____ ophiolite sequence; b. _____ intracontinental rift; c. _____ continental amalgamation; d. _____ separation orogeny; e. _____ Wilson cycle.

4. Archean and Proterozoic greenstone belts are similar except that the latter contains little or no _____.
   a. _____ sandstone; b. _____ tillite; c. _____ ultramafic lava flows; d. _____ prokaryotic cells; e. _____ basalt.

5. A sequence of rocks on land made up of mantle rocks overlain by oceanic crust rocks and deep sea sediments is a(n)
   a. _____ ophiolite; b. _____ granite-gneiss complex; c. _____ Archean sedimentary sequence; d. _____ supercontinent; e. _____ midcontinent rift.

6. The most significant biologic event to take place during the Proterozoic was the appearance of
   a. _____ ultramafic organisms; b. _____ autotrophic prokaryotes; c. _____ trilobites and brachiopods; d. _____ eukaryotic cells; e. _____ anaerobic proteinoids.

Table 9.2
Summary of the Proterozoic Geologic and Biologic Events Discussed in the Text

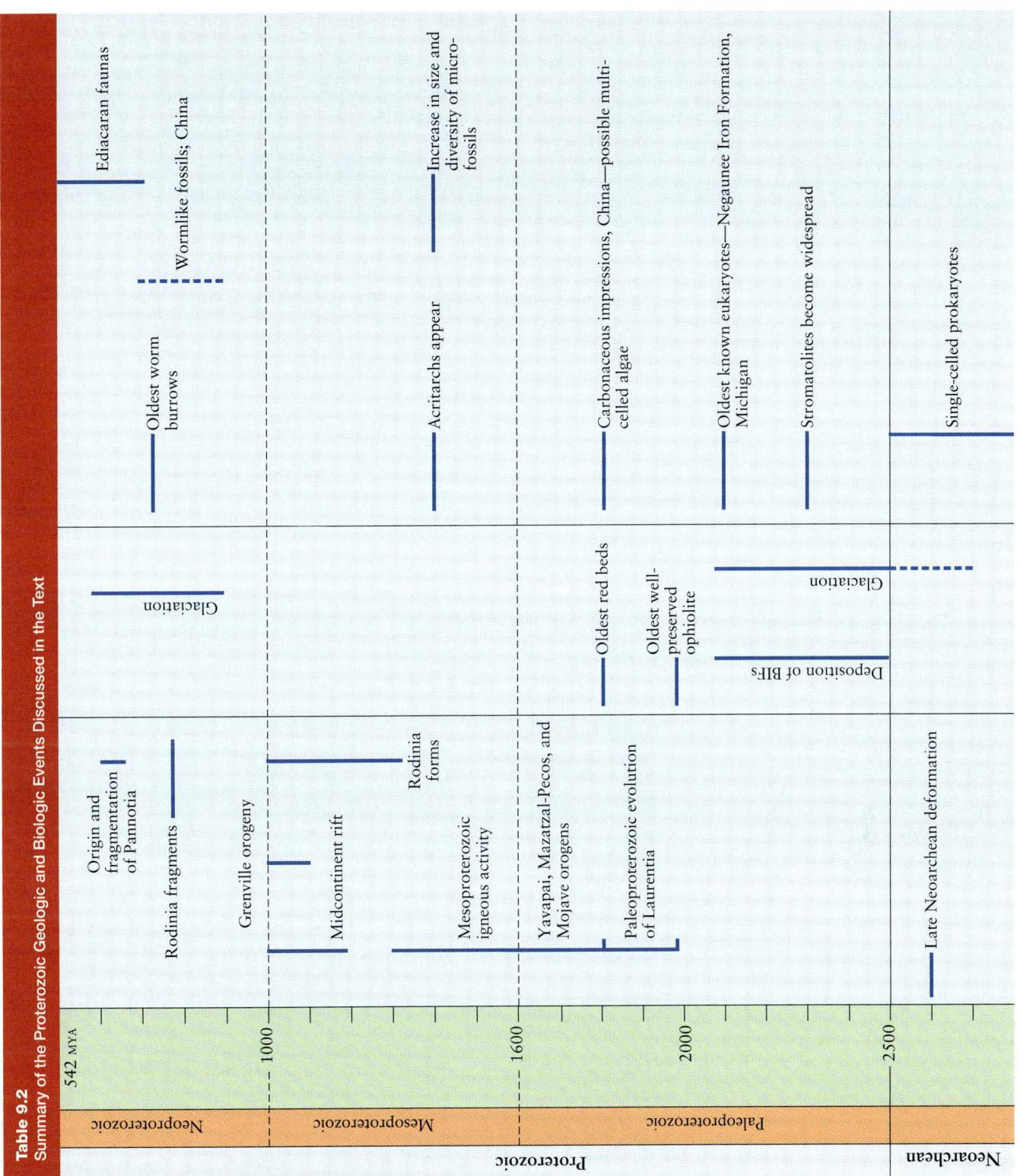

SUMMARY 187

7. Columnar masses of rock produced by the activities of cyanobacteria (blue-green algae) are
a. ____ orogens; b. ____ stromatolites; c. ____ heterotrophs; d. ____ endosymbionts; e. ____ pannotians.

8. The Mesoproterozoic of Laurentia was a time of
a. ____ widespread glaciation; b. ____ origin of animals with skeletons; c. ____ igneous activity unrelated to orogenic activity; d. ____ origin of the oldest-known greenstone belts; e. ____ formation of Pangaea.

9. An elongate, deformed part of the continental crust with metamorphic and plutonic rocks is a(n)
a. ____ ediacaran; b. ____ red bed; c. ____ sandstone-carbonate-shale assemblage; d. ____ orogen; e. ____ eukaryote.

10. Rocks of the Pound Quartzite in the Ediacara Hills of Australia
a. ____ contain important deposits of chrome; b. ____ show evidence of two periods of glaciation; c. ____ have impressions of the first well-documented animals; d. ____ were intensely deformed during the Grenville orogeny; e. ____ formed as alluvial fans in the Midcontinent rift.

11. The transition from prokaryotic cells to eukaryotic cells is not recorded by fossils, so how can we find evidence for how this important event may have occurred?

12. When and where did the Midcontinent rift form, and what kinds of rocks does it contain?

13. Briefly describe the Paleoproterozoic events that led to the formation of Laurentia.

14. When did the first animals appear in the fossil record, and why are their fossils not very common?

15. What evidence indicates that the Proterozoic atmosphere contained some free oxygen?

16. Why are ophiolites and sandstone-carbonate-shale assemblages important in deciphering Proterozoic geologic history?

17. You observe some Proterozoic rocks that might be tillite. How would you confirm or disprove this hypothesis?

18. How did the Mesoproterozoic episode of igneous activity affect the development of Laurentia? What kinds of rocks formed, and where are they found?

19. Describe banded iron formations, explain how geologists think they formed, and indicate why they are important economically.

APPLY YOUR KNOWLEDGE

1. Proterozoic sedimentary rocks in the northern Rocky Mountains are 4000 m thick and were deposited between 1.45 billion and 850 million years ago. What was the average rate of sediment accumulation in millimeters per year? Why is this figure unlikely to represent the actual rate of sedimentation?

FIELD QUESTION

1. Refer to Figure 9.4 and see if you can determine the environment of deposition for these rocks. Give evidence for your conclusion.

CHAPTER 10
EARLY PALEOZOIC EARTH HISTORY

The need to cheaply transport coal from where it was mined to where it was needed resulted in widespread canal building in England during the late 1700s and early 1800s. William Smith, who started his career mapping various coal mines, and later produced the first geologic map of England, was instrumental in helping find the most efficient canal routes to bring coal to market. Canals like the Grand Junction Canal shown here in this 1819 woodcut were critical not only for transporting coal from the mines to market, but also for the movement of people and goods.

[OUTLINE]

INTRODUCTION

CONTINENTAL ARCHITECTURE: CRATONS AND MOBILE BELTS

PALEOZOIC PALEOGEOGRAPHY

Early Paleozoic Global History

EARLY PALEOZOIC EVOLUTION OF NORTH AMERICA

THE SAUK SEQUENCE

PERSPECTIVE 10.1 *Pictured Rocks National Lakeshore*

The Cambrian of the Grand Canyon Region: A Transgressive Facies Model

THE TIPPECANOE SEQUENCE

Tippecanoe Reefs and Evaporites

The End of the Tippecanoe Sequence

THE APPALACHIAN MOBILE BELT AND THE TACONIC OROGENY

EARLY PALEOZOIC MINERAL RESOURCES

SUMMARY

ThomsonNOW Explore interactive tutorials, animations, or practice problems available on the ThomsonNow website at www.thomsonedu.com/login.

[CHAPTER OBJECTIVES]

At the end of this chapter, you will have learned that

- Six major continents were present at the beginning of the Paleozoic Era and that plate movement during the Early Paleozoic resulted in the first of several continental collisions leading to the formation of Pangaea at the end of the Paleozoic.
- The Paleozoic history of North America can be subdivided into six cratonic sequences, which represent major transgressive–regressive cycles.
- During the Sauk Sequence, warm, shallow seas covered most of North America, leaving only a portion of the Canadian shield and a few large islands above sea level.
- Like the Sauk Sequence, the Tippecanoe Sequence began with a major transgression resulting in widespread sandstones, followed by extensive carbonate and evaporite deposition.
- During Tippecanoe time, an oceanic–continental convergent plate boundary formed along the eastern margin of North America (known as the Appalachian mobile belt) resulting in the Taconic orogeny, the first of several orogenies to affect this area.
- Lower Paleozoic rocks contain a variety of important mineral resources.

Introduction

August 1, 1815, is an important date in the history of geology. On that date William Smith, a canal builder, published the world's first true geologic map. Measuring more than eight feet high and six feet wide, Smith's handpainted geologic map of England represented more than 20 years of detailed study of the rocks and fossils of England.

England is a country rich in geologic history. Five of the six Paleozoic geologic systems (Cambrian, Ordovician, Silurian, Devonian, and Carboniferous) were described and named for rocks exposed in England (see Chapter 5). The Carboniferous coal beds of England helped fuel the Industrial Revolution, and the drive to transport coal cheaply from where it was mined to where it was needed helped set off a flurry of canal building during the late 1700s and early 1800s. During this time William Smith, who was mapping various coal mines, first began to notice how rocks and fossils repeated themselves in predictable fashion. During the ensuing years Smith surveyed the English countryside for the most efficient canal routes to bring the coal to market. Much of his success was based on his ability to predict what rocks canal diggers would encounter. Realizing that his observations allowed him to unravel the geologic history of an area and correlate rocks between one region and another, William Smith set out to make the first geologic map of an entire country!

The story of how William Smith came to publish the world's first geologic map is a fascinating tale of determination and perseverance. However, instead of finding fame and success, Smith found himself, slightly less than four years later, in debtors' prison, and—upon his release after spending more than two months in prison—homeless. If such a story can have a happy ending, however, William Smith at least lived long enough to finally be recognized and honored for the seminal contribution he made to the then fledgling science of geology.

Just as William Smith applied basic geologic principles in deciphering the geology of England, we use these same principles in the next two chapters to interpret the geology of the Paleozoic Era. In these chapters we use the geologic principles and concepts discussed in earlier chapters to help explain how Earth's systems and its associated geologic processes have interacted during the Paleozoic to lay the groundwork for the distribution of continental landmasses, ocean basins, and the topography we have today.

The Paleozoic history of most continents involves major mountain-building activity along their margins and numerous shallow-water marine transgressions and regressions over their interiors. These transgressions and regressions were caused by global changes in sea level that most probably were related to plate activity and glaciation.

In the next two chapters, we provide an overview of the geologic history of the world during the Paleozoic Era in order to place in context the geologic events taking place in North America during this time. We then focus our attention on the geologic history of North America—not in a period-by-period chronology but in terms of the major transgressions and regressions taking place on the continent as well as the mountain building activity occurring during this time. Such an approach allows us to place the North American geologic events within a global framework.

Continental Architecture: Cratons and Mobile Belts

During the Precambrian, continental accretion and orogenic activity led to the formation of sizable continents. Movement of these continents during the Late Proterozoic resulted in the formation of a single Pangaea-like supercontinent geologists refer to as Pannotia (see Chapter 9). This supercontinent began breaking apart sometime during the latest Proterozoic, and by the beginning of the Paleozoic Era, six major continents were present. Each continent can be divided into two major components: a craton and one or more mobile belts.

Cratons are the relatively stable and immobile parts of continents and form the foundation on which Phanerozoic sediments were deposited (• Figure 10.1). Cratons typically consist of two parts: a shield and a platform.

Shields are the exposed portion of the crystalline basement rocks of a continent and are composed of Precambrian metamorphic and igneous rocks that reveal a history of extensive orogenic activity during the Precambrian (see Chapters 8 and 9). During the Phanerozoic, however, shields were extremely stable and formed the foundation of the continents.

Extending outward from the shields are buried Precambrian rocks that constitute a *platform,* another part of the craton. Overlying the platform are flat-lying or gently dipping Phanerozoic detrital and chemical sedimentary rocks that were deposited in widespread shallow seas that transgressed and regressed over the craton. These seas, called **epeiric seas,** were a common feature of most Paleozoic cratonic histories. Changes in sea level caused primarily by continental glaciation as well as by plate movement were responsible for the advance and retreat of these epeiric seas.

Whereas most Paleozoic platform rocks are still essentially flat lying, in some places they were gently folded into regional arches, domes, and basins (Figure 10.1). In many cases some of these structures stood out as low islands during the Paleozoic Era and supplied sediments to the surrounding epeiric seas.

Mobile belts are elongated areas of mountain-building activity. They are located along the margins of continents where sediments are deposited in the relatively shallow waters of the continental shelf and the deeper waters at the base of the continental slope. During plate convergence along these margins, the sediments are deformed and intruded by magma, creating mountain ranges.

Four mobile belts formed around the margin of the North American craton during the Paleozoic: the **Franklin, Cordilleran, Ouachita,** and **Appalachian mobile belts** (Figure 10.1). Each was the site of mountain building in response to compressional forces along a convergent plate boundary and formed mountain ranges such as, for example, the Appalachians and Ouachitas.

• **Figure 10.1 Major Cratonic Structures and Mobile Belts** The major cratonic structures and mobile belts of North America that formed during the Paleozoic Era.

Paleozoic Paleogeography

One result of plate tectonics is that Earth's geography is constantly changing. The present-day configuration of the continents and ocean basins is merely a snapshot in time. As the plates move about, the location of continents and ocean basins constantly changes. One of the goals of historical geology is to provide paleogeographic reconstructions of the world during the geologic past. By synthesizing all the pertinent paleoclimatic, paleomagnetic, paleontologic, sedimentologic, stratigraphic, and tectonic data available, geologists can construct paleogeographic maps. Such maps are simply interpretations of the geography of an area for a particular time in the geologic past. The majority of paleogeographic maps show the distribution of land and sea, possible climatic regimes, and such geographic features as mountain ranges, swamps, and glaciers.

The paleogeographic history of the Paleozoic Era, for example, is not as precisely known as for the Mesozoic and Cenozoic eras, in part because the magnetic anomaly patterns preserved in the oceanic crust were destroyed when much of the Paleozoic oceanic crust was subducted during the formation of Pangaea. Paleozoic paleogeographic reconstructions are therefore based primarily on structural relationships, climate-sensitive sediments such as red beds, evaporites, and coals, as well as the distribution of plants and animals.

At the beginning of the Paleozoic, six major continents were present. Besides these large landmasses, geologists have also identified numerous microcontinents such as *Avalonia* (composed of parts of present day Belgium, northern France, England, Wales, Ireland, the Maritime Provinces and Newfoundland of Canada, as well as parts of the New England area of the United States) and various island arcs associated with microplates. We are primarily concerned, however, with the history of the six major continents and their relationship to each other. The six major Paleozoic continents are **Baltica** (Russia west of the Ural Mountains and the major part of northern Europe), **China** (a complex area consisting of at least three Paleozoic continents that were not widely separated and are here considered to include China, Indochina, and the Malay Peninsula), **Gondwana** (Africa, Antarctica, Australia, Florida, India, Madagascar, and parts of the Middle East and southern Europe), **Kazakhstania** (a triangular continent centered on Kazakhstan but considered by some to be an extension of the Paleozoic Siberian continent), **Laurentia** (most of present North America, Greenland, northwestern Ireland, and Scotland), and **Siberia** (Russia east of the Ural Mountains and Asia north of Kazakhstan and south of Mongolia). The paleogeographic reconstructions that follow (• Figure 10.2) are based on the methods

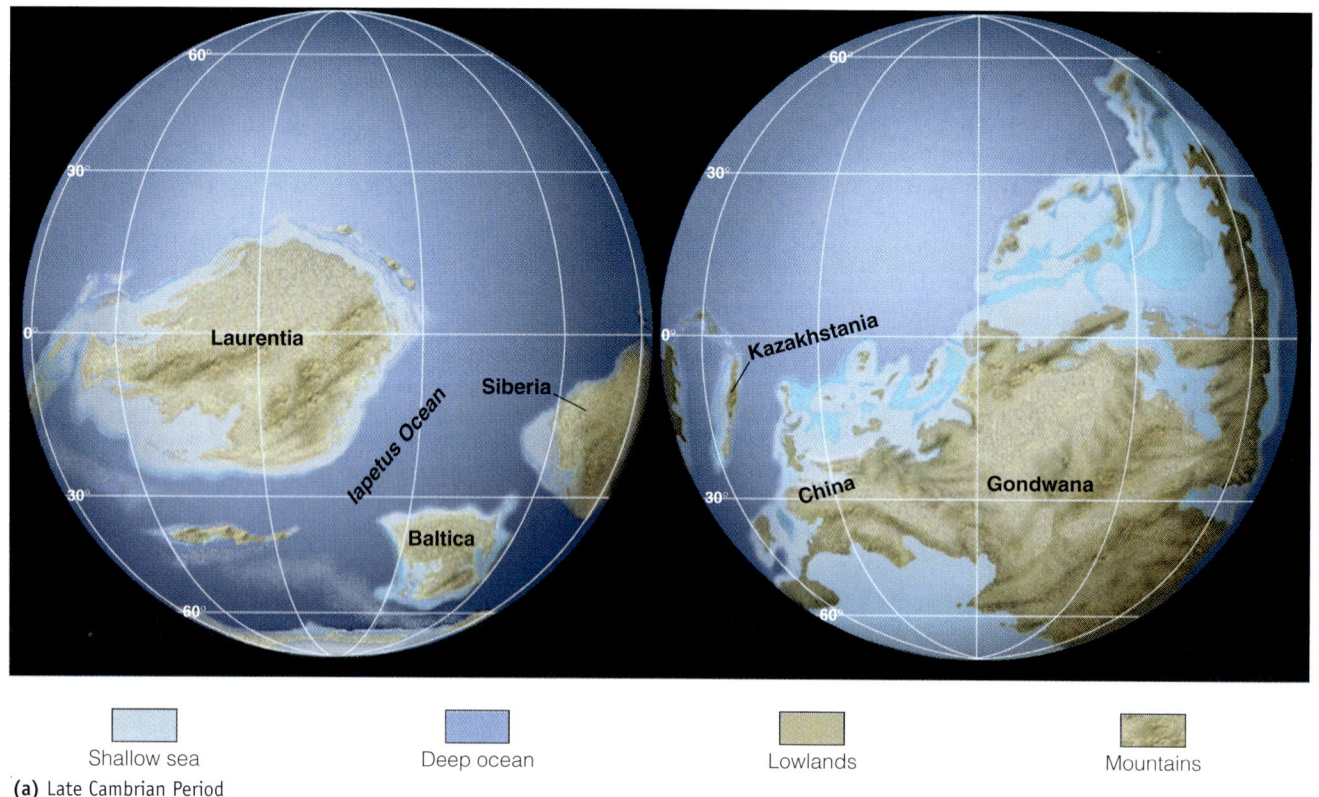

(a) Late Cambrian Period

• **Figure 10.2 Paleozoic Paleogeography** Paleogeography of the world for the **(a)** Late Cambrian Period.

192 CHAPTER 10 EARLY PALEOZOIC EARTH HISTORY

used to determine and interpret the location, geographic features, and environmental conditions on the paleocontinents.

Early Paleozoic Global History

In contrast to today's global geography, the Cambrian world consisted of six major continents dispersed around the globe at low tropical latitudes (Figure 10.2a). Water circulated freely among ocean basins, and the polar regions were mostly ice free. By the Late Cambrian, epeiric seas had covered areas of Laurentia, Baltica, Siberia, Kazakhstania, and China, while highlands were present in northeastern Gondwana, eastern Siberia, and central Kazakhstania.

During the Ordovician and Silurian periods, plate movement played a major role in the changing global geography (Figure 10.2b and c). Gondwana moved southward during the Ordovician and began to cross the South Pole as indicated by Upper Ordovician tillites found today in the Sahara Desert. During the Early Ordovician, the microcontinent Avalonia separated from Gondwana and begin moving northeastward, where it would finally collide with Baltica during the Late Ordovician-Early Silurian. In contrast to the passive continental margin Laurentia exhibited during the Cambrian, an active convergent plate boundary formed along its eastern margin during the Ordovician, as indicated by the Late Ordovician Taconic orogeny that occurred in New England.

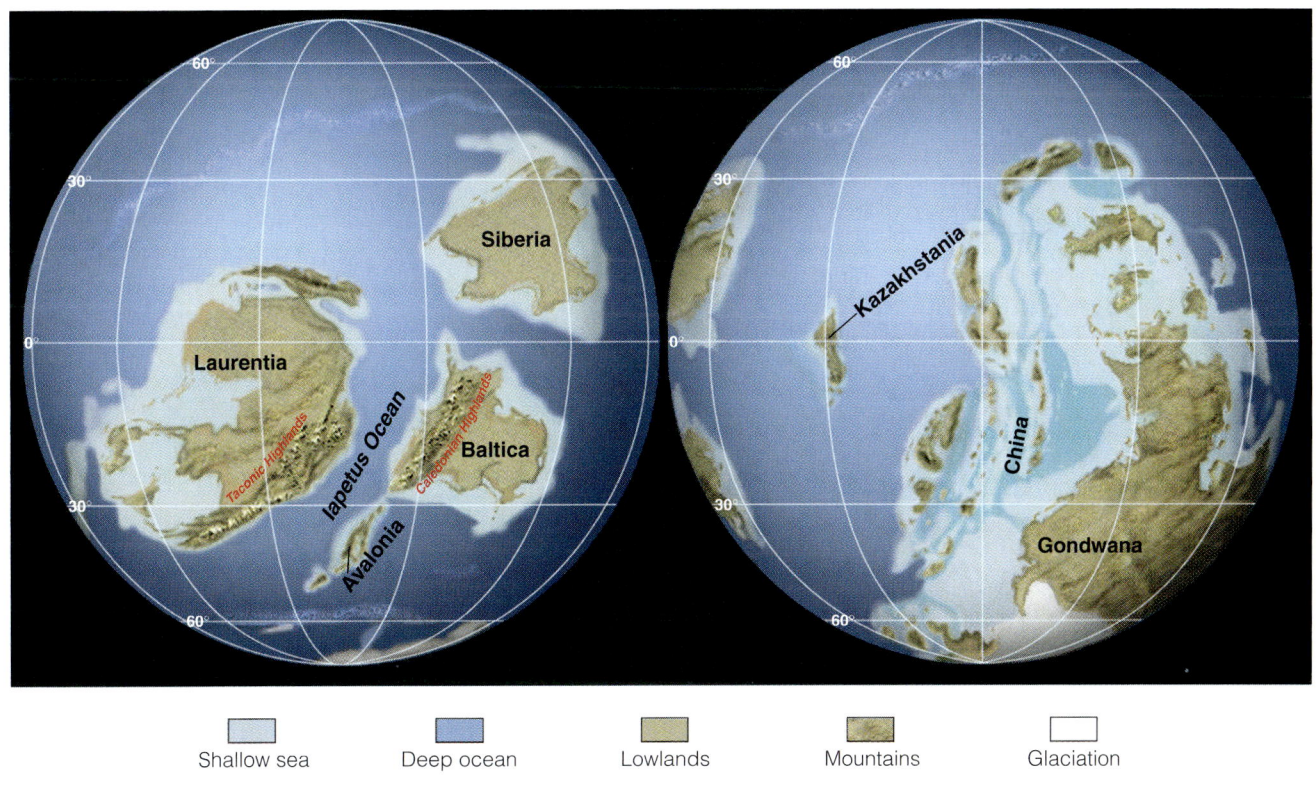

(b) Late Ordovician Period

• Figure 10.2 (cont.) (b) Late Ordovician Period.

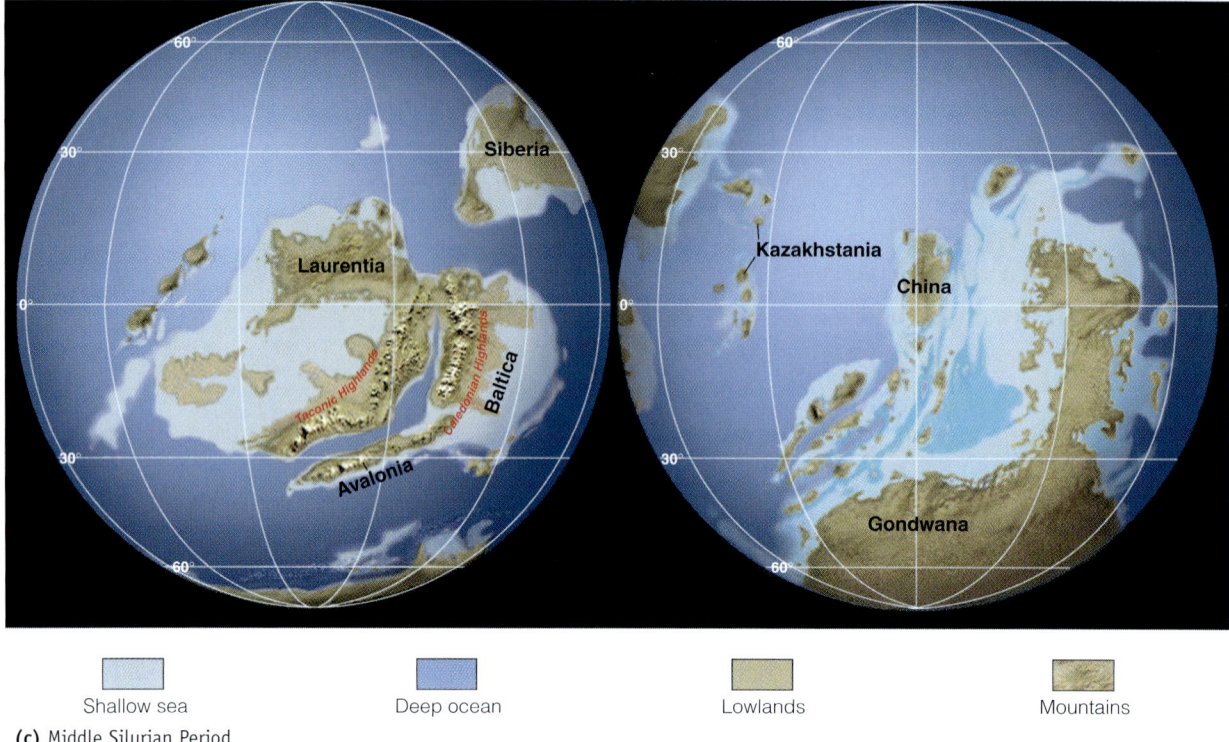

(c) Middle Silurian Period

• Figure 10.2 (cont.) (c) Middle Silurian Period.

During the Silurian, Baltica, along with the newly attached Avalonia, moved northwestward relative to Laurentia and collided with it to form the larger continent of Laurasia. This collision, which closed the northern Iapetus Ocean, is marked by the Caledonian orogeny. After this orogeny, the southern part of the Iapetus Ocean still remained open between Laurentia and Gondwana (Figure 10.2c). Siberia and Kazakhstania moved from a southern equatorial position during the Cambrian to north temperate latitudes by the end of the Silurian Period.

With this plate tectonics overview in mind, we now focus our attention on North America (Laurentia) and its role in the Early Paleozoic geologic history of the world.

What Would You Do?

Attendance at the natural history museum where you work has declined during the past several years. In an effort to increase attendance and make the displays more appealing to the public, the museum director has been asked to create a display showing the geologic history of the region, which just happens to include the Cambrian through Silurian time periods. The director has assigned you to make the display. What kinds of displays, computer animations, or interactive exhibits do you think would appeal to the general public? What events would you focus on, and how would you go about designing exhibits or displays to show the changing geologic panorama of the region during the Early Paleozoic and the impact these events might have on the region today?

Early Paleozoic Evolution of North America

It is convenient to divide the geologic history of the North American craton into two parts: the first dealing with the relatively stable continental interior over which epeiric seas transgressed and regressed and the other with the mobile belts where mountain building occurred.

In 1963, the American geologist Laurence Sloss subdivided the sedimentary rock record of North America into six cratonic sequences. A **cratonic sequence** is a large-scale (greater than supergroup) lithostratigraphic unit representing a major transgressive–regressive cycle bounded by cratonwide unconformities (• Figure 10.3). The transgressive phase, which is usually covered by younger sediments, commonly is well preserved, whereas the regressive phase of each sequence is marked by an unconformity. Where rocks of the appropriate age are preserved, each of the six unconformities can be shown to extend across the various sedimentary basins of the North American craton and into the mobile belts along the cratonic margin.

Geologists have also recognized major unconformity-bounded sequences in cratonic areas outside North America. Such global transgressive and regressive cycles of sea level changes are thought to result from major tectonic and glacial events.

The realization that rock units can be divided into cratonic sequences and that these sequences can be further

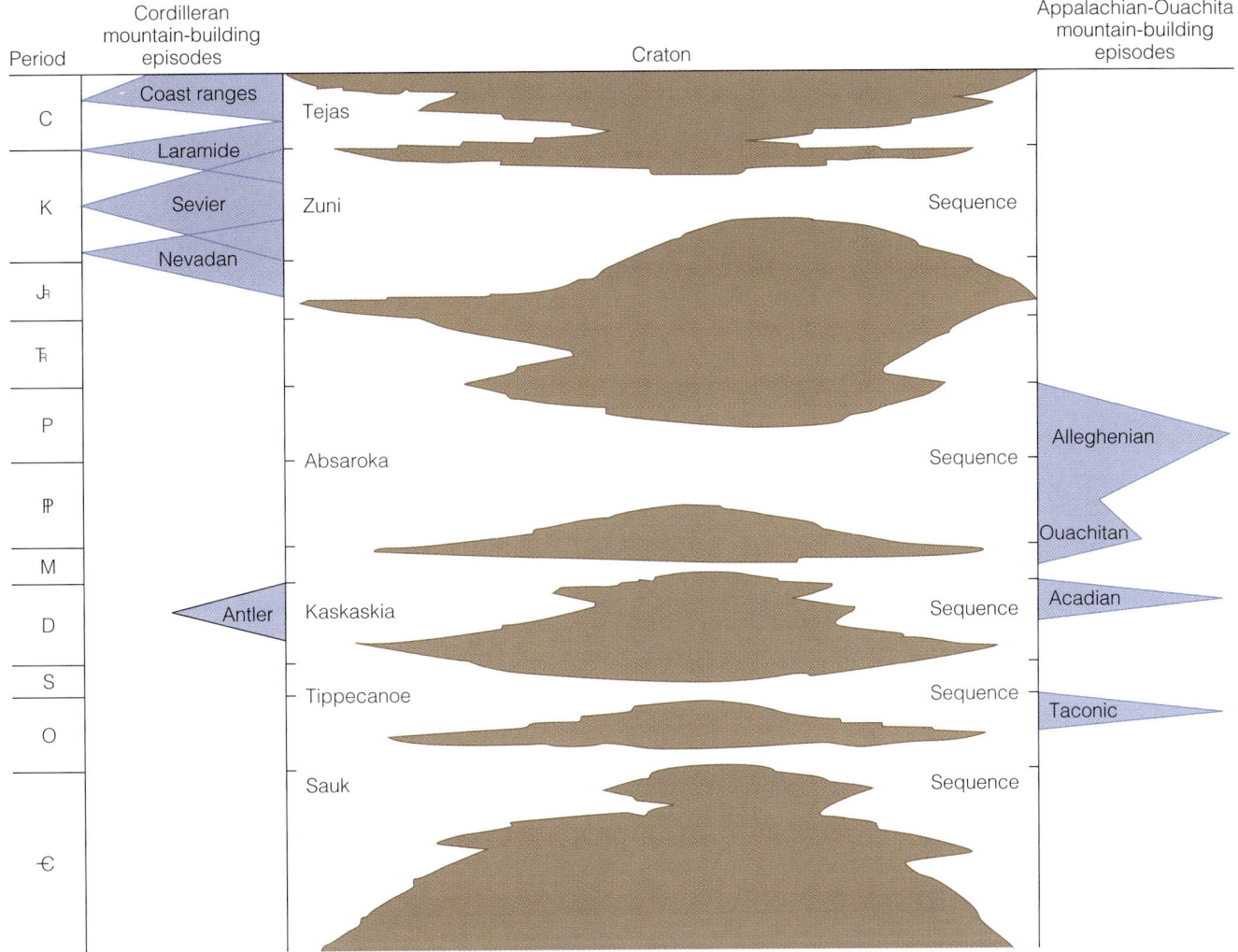

• **Figure 10.3 Cratonic Sequences of North America** A cratonic sequence is a large-scale lithostratigraphic unit representing a major transgressive-regressive cycle and bounded by cratonwide unconformities. The white areas represent sequences of rocks separated by large-scale unconformities (brown areas). The major Cordilleran orogenies are shown on the left side, and the major Appalachian orogenies are on the right.

subdivided and correlated provides the foundation for an important concept in geology that allows high-resolution analysis of time and facies relationships within sedimentary rocks. **Sequence stratigraphy** is the study of rock relationships within a time–stratigraphic framework of related facies bounded by erosional or nondepositional surfaces. The basic unit of sequence stratigraphy is the *sequence*, which is a succession of rocks bounded by unconformities and their equivalent conformable strata. Sequence boundaries result from a relative drop in sea level. Sequence stratigraphy is an important tool in geology because it allows geologists to subdivide sedimentary rocks into related units that are bounded by time–stratigraphically significant boundaries. Geologists use sequence stratigraphy for high-resolution correlation and mapping, as well as for interpreting and predicting depositional environments.

The Sauk Sequence

Rocks of the **Sauk Sequence** (Neoproterozoic–Early Ordovician) record the first major transgression onto the North American craton (Figure 10.3). During the Late Proterozoic and Early Cambrian, deposition of marine sediments was limited to the passive shelf areas of the Appalachian and Cordilleran borders of the craton. The craton itself was above sea level and experiencing extensive weathering and erosion. Because North America was located in a tropical climate at this time and there is no evidence of any terrestrial vegetation, weathering and erosion of the exposed Precambrian basement rocks must have proceeded rapidly.

During the Middle Cambrian, the transgressive phase of the Sauk began with epeiric seas encroaching over the craton (see Perspective 10.1). By the Late Cambrian, these

Perspective 10.1

Pictured Rocks National Lakeshore

Exposed along the south shore of Lake Superior between Au Sable Point and Munising in Michigan's Upper Peninsula is the beautiful and imposing wavecut sandstone called Pictured Rocks cliffs (Figure 1). The rocks exposed in this area, part of which is designated a national lakeshore, comprise the Upper Cambrian Munising Formation, which is divided into two members: the lower Chapel Rock Sandstone and the upper Miner's Castle Sandstone. The Munising Formation unconformably overlies the Neoproterozoic Jacobsville Sandstone and is unconformably overlain by the Middle Ordovician Au Train Formation. The reddish brown, coarse-grained Jacobsville Sandstone was deposited in streams and lakes over an irregular erosion surface. Following deposition, the Jacobsville was slightly uplifted and tilted.

By the Late Cambrian, the transgressing Sauk Sea reached the Michigan area, and the Chapel Rock Sandstone was deposited. The principal source area for this unit was the Northern Michigan highlands, an area that corresponds to the present Upper Peninsula. Following deposition of the Chapel Rock Sandstone, the Sauk Sea retreated from the area.

During a second transgression of the Sauk Sea in this area, the Miner's Castle Sandstone was deposited. This second transgression covered most of the Upper Peninsula of Michigan and drowned the highlands that were the source for the older Chapel Rock Sandstone.

The source area for the Miner's Castle Sandstone was the Precambrian Canadian shield area to the north and northeast. The Miner's Castle Sandstone contains rounder, better sorted, and more abundant quartz grains than the Chapel Rock Sandstone, indicating a different source area. A major unconformity separates the Miner's Castle Sandstone from the overlying Middle Ordovician Au Train Formation.

One of the most prominent features of Pictured Rocks National Lakeshore is Miner's Castle, a wavecut projection along the shoreline (Figure 2). The lower sandstone unit at water level is the Chapel Rock Sandstone, and the rest of the feature is composed of the Miner's Castle Sandstone. The two turrets of the castle formed as sea stacks during a time following the Pleistocene when the water level of Lake Superior was much higher (Figure 2a).

On April 13, 2006, one of the two turrets of Miner's Castle collapsed into Lake Superior (Figure 2b). Witnesses described the collapse as loud and sudden. Although it is disappointing to lose such a prominent landmark, no one should be surprised that it happened. The forces of nature helped shape the structure, and the same forces caused its collapse. It is only a matter of time until the remaining turret meets the same fate as its twin.

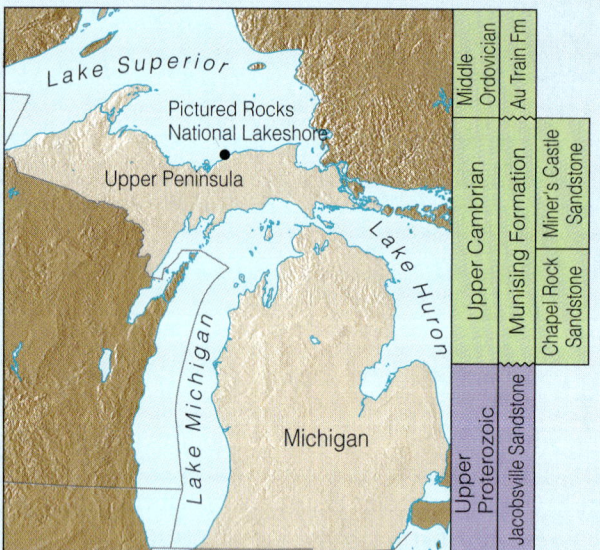

Figure 1 Location of Pictured Rocks National Lakeshore and stratigraphy of rocks exposed in this area.

Figure 2 (a) This photograph of the Munising Formation shows the shoreline projection called Miner's Castle. The two turrets at the top of Miner's Castle formed by wave action when the water level of Lake Superior was higher. The contact between the Miner's Castle and Chapel Rock sandstone members is located just above lake level. (b) This photograph shows the same area, but it was taken shortly after the right-hand turret collapsed into Lake Superior.

epeiric seas had covered most of North America, leaving only a portion of the Canadian shield and a few large islands above sea level (• Figure 10.4). These islands, collectively named the **Transcontinental Arch,** extended from New Mexico to Minnesota and the Lake Superior region.

The sediments deposited on both the craton and along the shelf area of the craton margin show abundant evidence of shallow-water deposition. The only difference between the shelf and craton deposits is that the shelf deposits are thicker. In both areas, the sands are generally clean and well

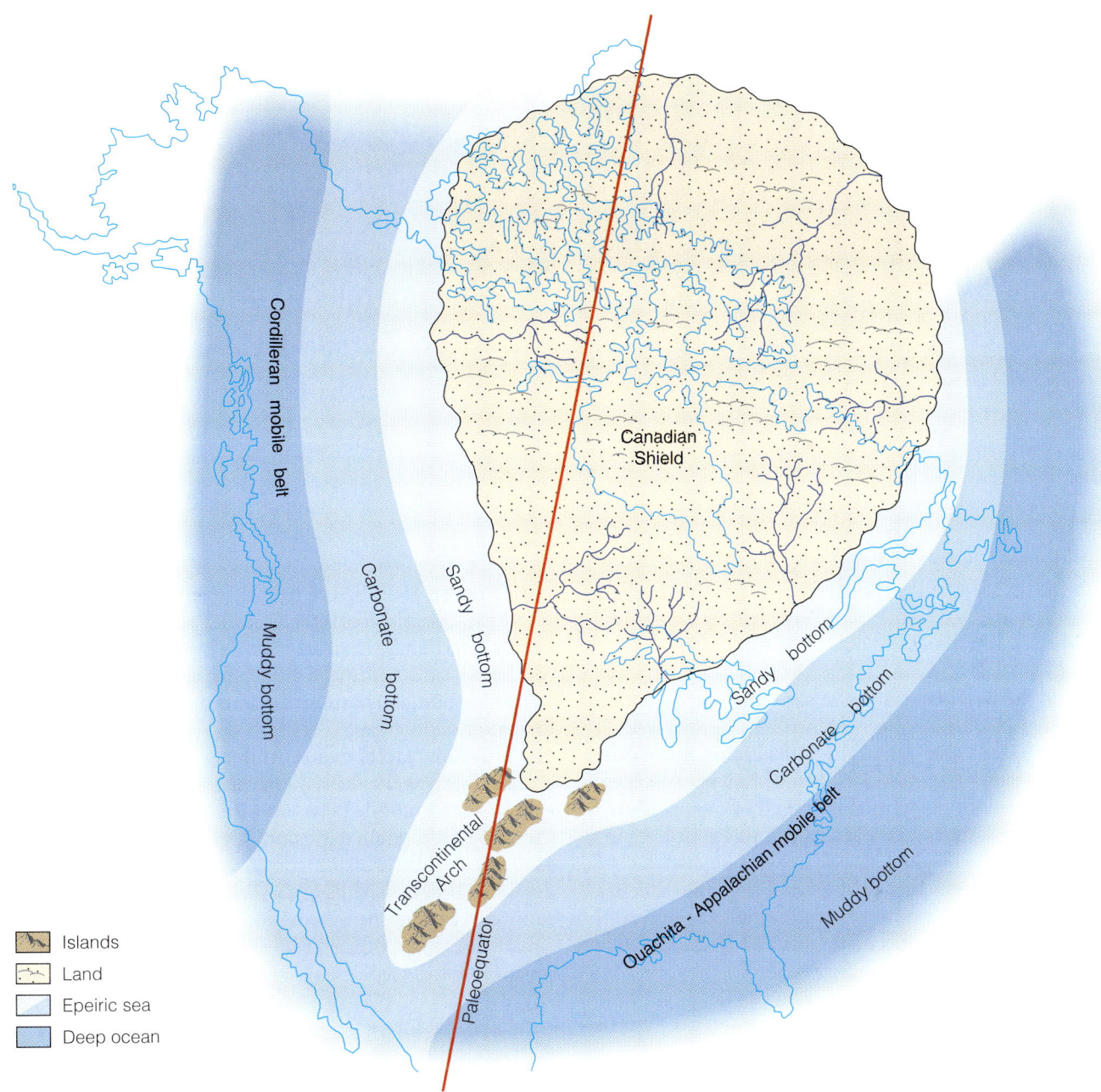

• **Figure 10.4 Cambrian Paleogeography of North America** Note the position of the Cambrian paleoequator. During this time North America straddled the equator as indicated in Figure 10.2a.

sorted and commonly contain ripple marks and small-scale cross-bedding. Many of the carbonates are bioclastic (composed of fragments of organic remains), contain stromatolites, or have oolitic (small, spherical calcium carbonate grains) textures. Such sedimentary structures and textures indicate shallow-water deposition.

The Cambrian of the Grand Canyon Region: A Transgressive Facies Model

Recall from Chapter 5 that sediments become increasingly finer the farther away from land one goes. Therefore, in a stable environment where sea level remains the same, coarse detrital sediments are typically deposited in the nearshore environment, and finer-grained sediments are deposited in the offshore environment. Carbonates form farthest from land in the area beyond the reach of detrital sediments. During a transgression, these facies (sediments that represent a particular environment) migrate in a landward direction (see Figure 5.9).

The Cambrian rocks of the Grand Canyon region (see Chapter 4 opening photo) provide an excellent model of the sedimentation patterns of a transgressing sea. The Grand Canyon region occupied the passive shelf and western margin of the craton during Sauk time. During the

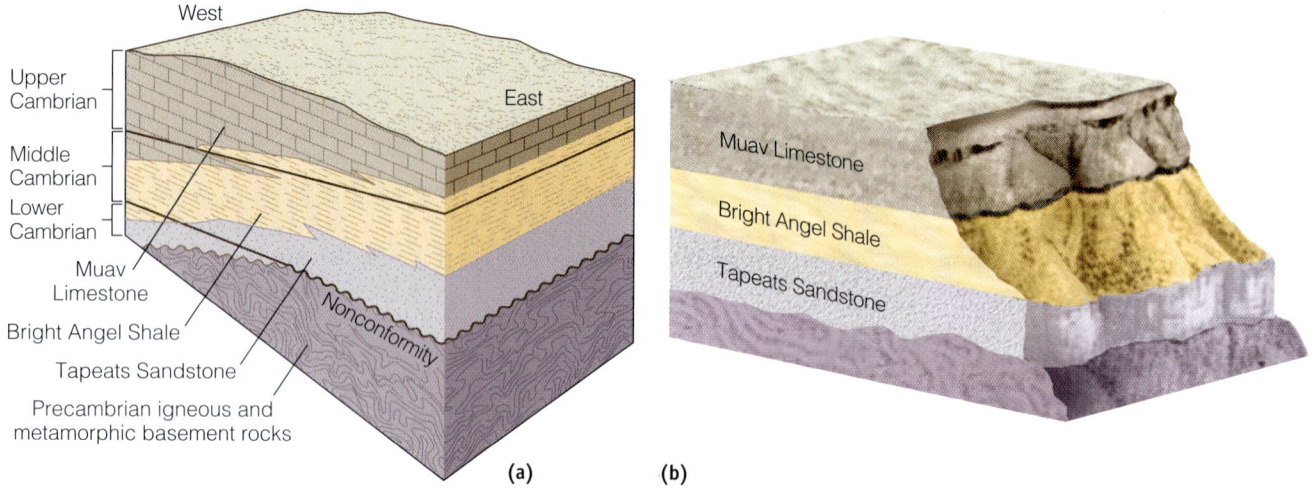

• **Figure 10.5 Cambrian Rocks of the Grand Canyon** (a) Block diagram of Cambrian strata exposed in the Grand Canyon illustrating the transgressive nature of the three formations. (b) Block diagram of Cambrian rocks exposed along the Bright Angel Trail, Grand Canyon, Arizona.

Neoproterozoic and Early Cambrian, most of the craton was above sea level and deposition of marine sediments was mainly restricted to the margins of the craton (continental shelves and slopes).

In the Grand Canyon region, the Tapeats Sandstone represents the basal transgressive shoreline deposits that accumulated as marine waters transgressed across the shelf and just onto the western margin of the craton during the Early Cambrian (• Figure 10.5). These sediments are clean, well-sorted sands of the type one would find on a beach today. As the transgression continued into the Middle Cambrian, muds of the Bright Angel Shale were deposited over the Tapeats Sandstone. By the Late Cambrian, the Sauk Sea had transgressed so far onto the craton that, in the Grand Canyon region, carbonates of the Muav Limestone were being deposited over the Bright Angel Shale.

This vertical succession of sandstone (Tapeats), shale (Bright Angel), and limestone (Muav) forms a typical transgressive sequence and represents a progressive migration of offshore facies toward the craton through time (Figure 10.5; see Figure 5.8).

Cambrian rocks of the Grand Canyon region also illustrate that many formations are time transgressive; that is, their age is not the same in every place they are found. Mapping and correlations based on faunal evidence indicate that deposition of the Muav Limestone had already started on the shelf before deposition of the Tapeats Sandstone was completed on the craton. Faunal analysis of the Bright Angel Shale indicates it is Early Cambrian in age in California and Middle Cambrian in age in the Grand Canyon region, illustrating the time-transgressive nature of formations and facies.

This same facies relationship also occurred elsewhere on the craton as the seas encroached from the Appalachian and Ouachita mobile belts onto the craton interior (• Figure 10.6). Carbonate deposition dominated on the craton as the Sauk transgression continued during the Early Ordovician, and the islands of the Transcontinental Arch were soon covered by the advancing Sauk Sea. By the end of Sauk time, much of the craton was submerged beneath a warm, equatorial epeiric sea (Figure 10.2a).

The Tippecanoe Sequence

As the Sauk Sea regressed from the craton during the Early Ordovician, a landscape of low relief emerged. The rocks exposed were predominantly limestones and dolostones deposited earlier as part of the Sauk transgression. Because North America was still located in a tropical environment when the seas regressed, these carbonates experienced extensive erosion at that time (• Figure 10.7). The resulting cratonwide unconformity thus marks the boundary between the Sauk and Tippecanoe sequences.

Like the Sauk Sequence, deposition of the **Tippecanoe Sequence** (Middle Ordovician–Early Devonian) began with a major transgression onto the craton. This transgressing sea deposited clean, well-sorted quartz sands over most of the craton. The best known of the Tippecanoe basal sandstones is the St. Peter Sandstone, an almost pure

What Would You Do?

You work for a travel agency and are putting together a raft trip down the Colorado River through the Grand Canyon. In addition to the usual information about such a trip, what kind of geologic information would you include in your brochure to make the trip appealing from an educational standpoint as well?

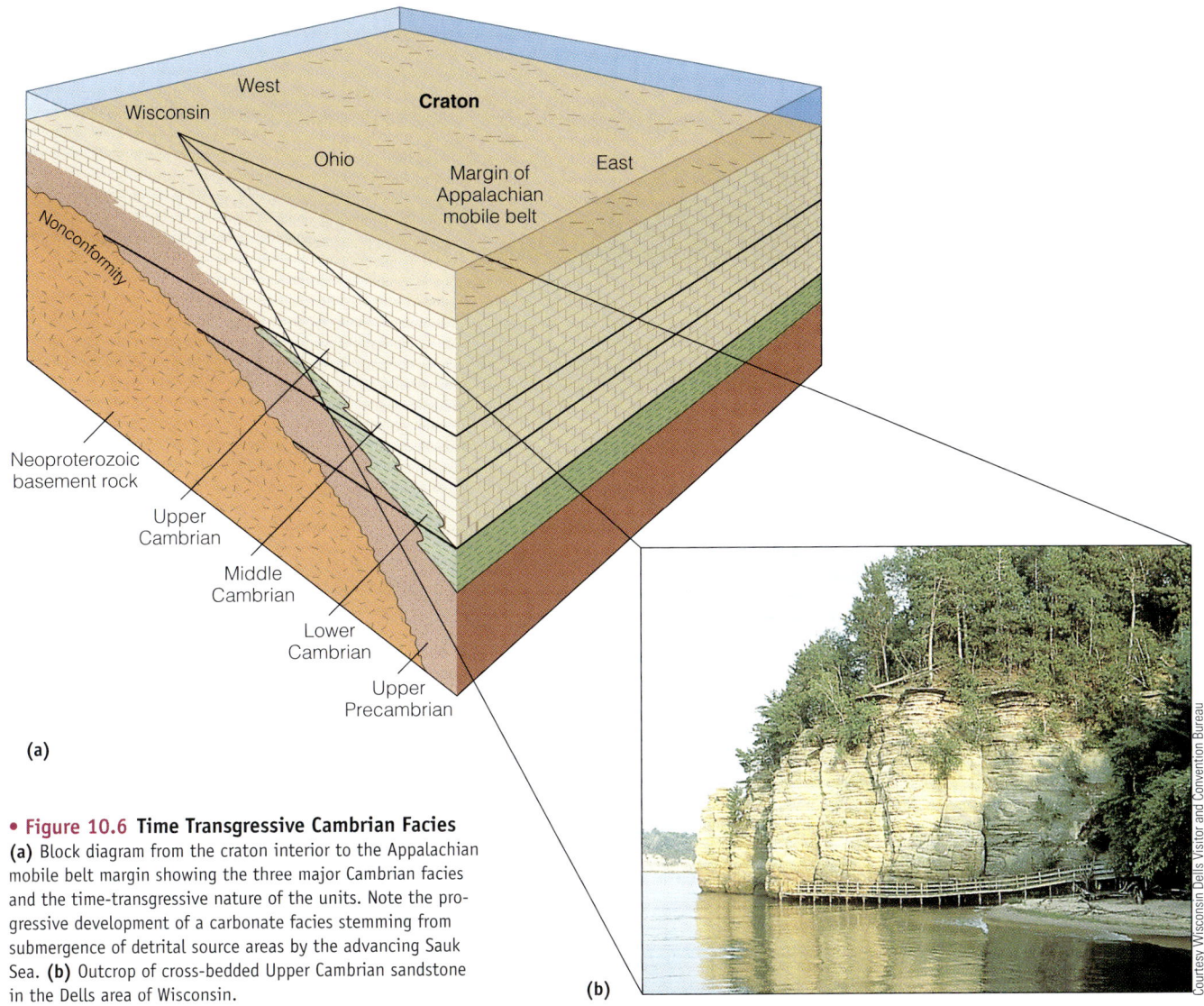

• **Figure 10.6 Time Transgressive Cambrian Facies**
(a) Block diagram from the craton interior to the Appalachian mobile belt margin showing the three major Cambrian facies and the time-transgressive nature of the units. Note the progressive development of a carbonate facies stemming from submergence of detrital source areas by the advancing Sauk Sea. (b) Outcrop of cross-bedded Upper Cambrian sandstone in the Dells area of Wisconsin.

quartz sandstone used in manufacturing glass. It occurs throughout much of the midcontinent and resulted from numerous cycles of weathering and erosion of Proterozoic and Cambrian sandstones deposited during the Sauk transgression (• Figure 10.8).

The Tippecanoe basal sandstones were followed by widespread carbonate deposition (Figure 10.7). The limestones were generally the result of deposition by calcium carbonate–secreting organisms such as corals, brachiopods, stromatoporoids, and bryozoans. Besides the limestones, there were also many dolostones. Most of the dolostones formed as a result of magnesium replacing calcium in calcite, thus converting limestones into dolostones.

In the eastern portion of the craton, the carbonates grade laterally into shales. These shales mark the farthest extent of detrital sediments derived from weathering and erosion of the Taconic Highlands, which resulted from a tectonic event taking place in the Appalachian mobile belt, and which we will discuss later.

Tippecanoe Reefs and Evaporites

Organic reefs are limestone structures constructed by living organisms, some of which contribute skeletal materials to the reef framework (• Figure 10.9). Today, corals and calcareous algae are the most prominent reef builders, but in the geologic past other organisms played a major role.

Regardless of the organisms dominating reef communities, reefs appear to have occupied the same ecologic niche in the geologic past that they do today. Because of the ecologic requirements of reef-building organisms, present-day reefs are confined to a narrow latitudinal belt between approximately 30 degrees north and south of the equator. Corals, the major reef-building organisms today, require warm, clear, shallow water of normal salinity for optimal growth.

The size and shape of a reef are largely the result of interactions among the reef-building organisms, the bottom topography, wind and wave action, and subsidence of

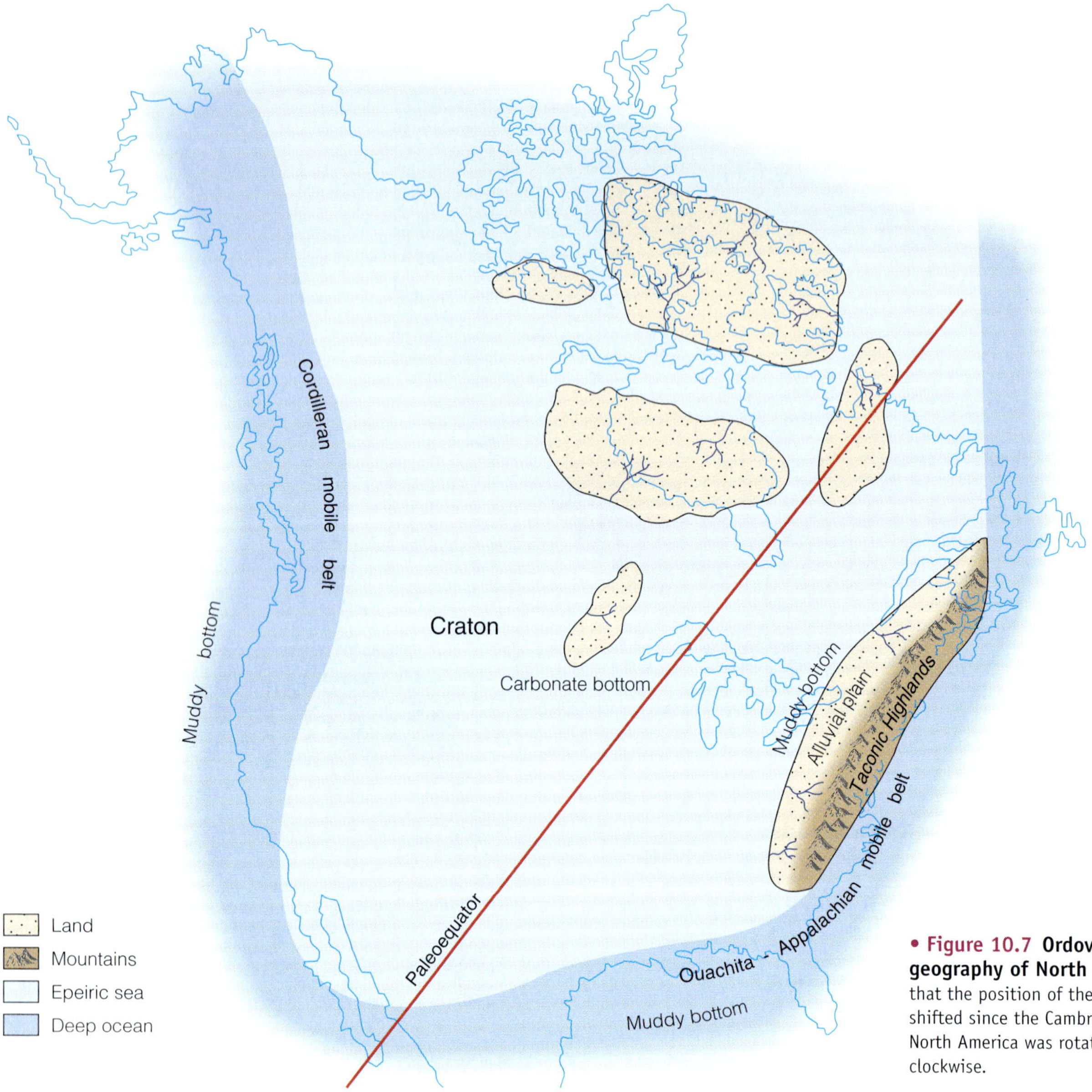

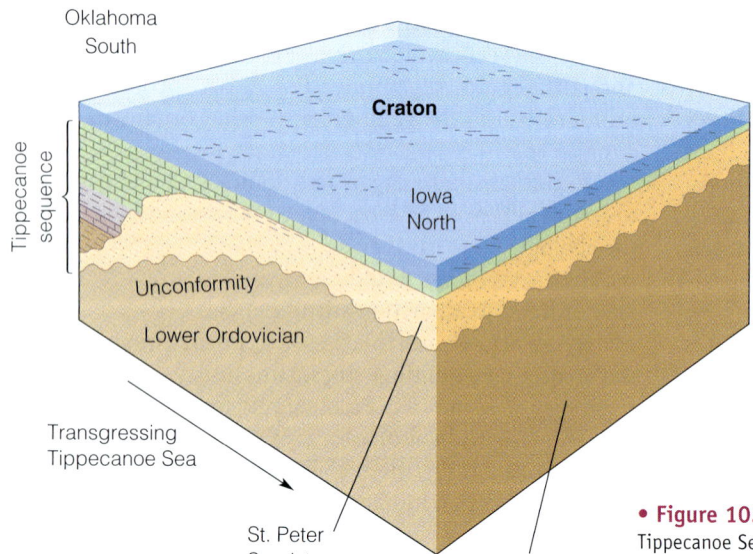

• **Figure 10.7 Ordovician Paleogeography of North America** Note that the position of the equator has shifted since the Cambrian, indicating North America was rotating counterclockwise.

the seafloor. Reefs also alter the area around them by forming barriers to water circulation or wave action.

Reefs typically are long, linear masses forming a barrier between a shallow platform on one side and a comparatively deep marine basin on the other side. Such reefs are known as *barrier reefs* (Figure 10.9). Reefs create and maintain a steep seaward front that absorbs incoming wave energy. As skeletal material breaks off from the reef front, it accumulates as talus along a fore-reef slope. The barrier reef itself is porous and composed of many different reef-building organisms. The lagoon area behind the reef is a low-energy,

• **Figure 10.8 Transgressing Tippecanoe Sea** The transgression of the Tippecanoe Sea resulted in the deposition of the St. Peter Sandstone (Middle Ordovician) over a large area of the craton.

200 CHAPTER 10 EARLY PALEOZOIC EARTH HISTORY

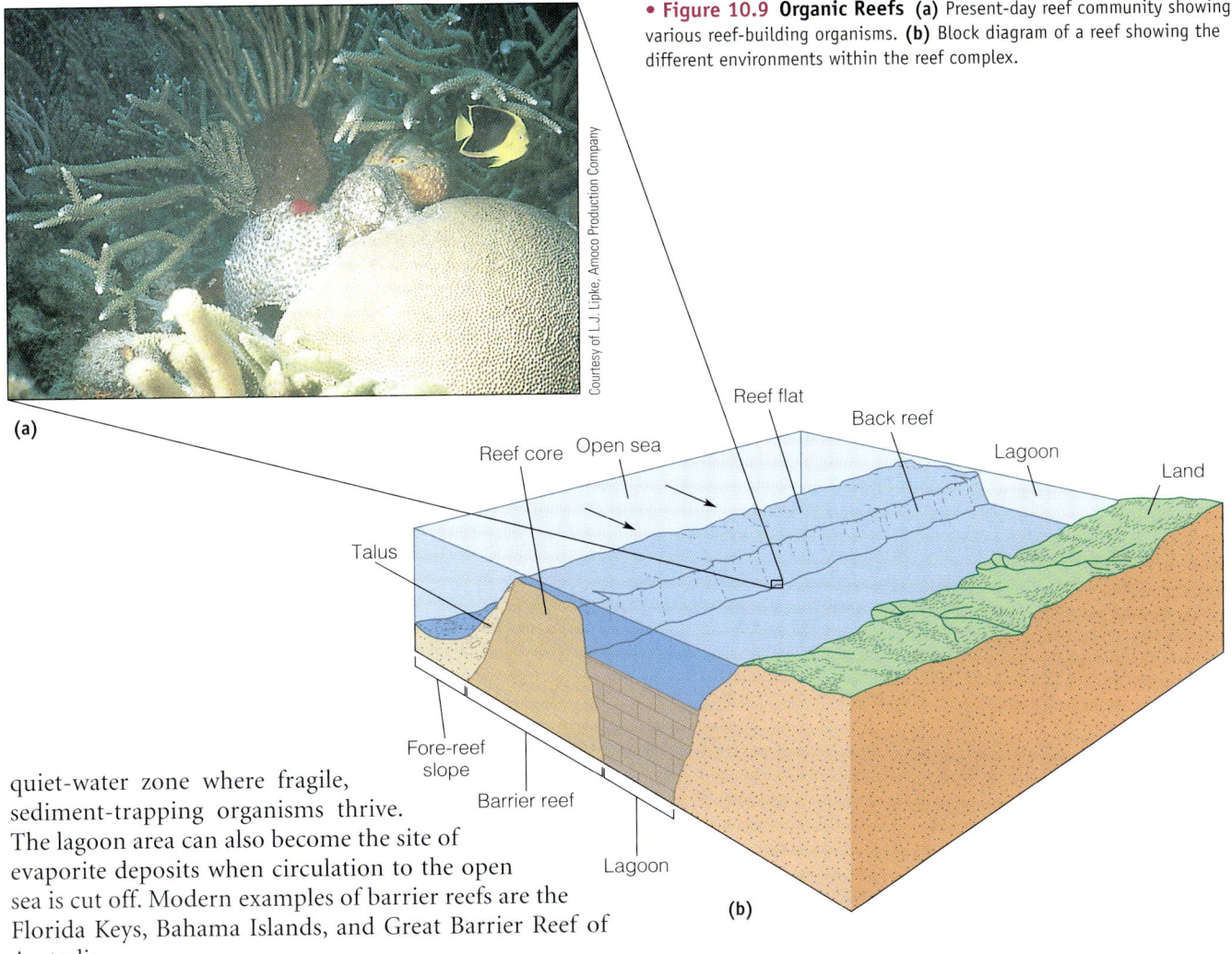

• Figure 10.9 **Organic Reefs** (a) Present-day reef community showing various reef-building organisms. (b) Block diagram of a reef showing the different environments within the reef complex.

quiet-water zone where fragile, sediment-trapping organisms thrive. The lagoon area can also become the site of evaporite deposits when circulation to the open sea is cut off. Modern examples of barrier reefs are the Florida Keys, Bahama Islands, and Great Barrier Reef of Australia.

Reefs have been common features in low latitudes since the Cambrian and have been built by a variety of organisms. The first skeletal builders of reeflike structures were *archaeocyathids*. These conical organisms lived during the Cambrian and had double, perforated, calcareous shell walls. Archaeocyathids built small mounds that have been found on all continents except South America (see Figure 12.6).

Beginning in the Middle Ordovician, stromatoporoid coral reefs became common in the low latitudes, and similar reefs remained so throughout the rest of the Phanerozoic Eon. The burst of reef building seen in the Late Ordovician through Devonian probably occurred in response to evolutionary changes triggered by the appearance of extensive carbonate seafloors and platforms beyond the influence of detrital sediments.

The Middle Silurian rocks (Tippecanoe Sequence) of the present-day Great Lakes region are world famous for their reef and evaporite deposits (• Figure 10.10). The most famous structure in the region, the Michigan Basin, is a broad, circular basin surrounded by large barrier reefs. No doubt these reefs contributed to increasingly restricted circulation and the precipitation of Upper Silurian evaporites within the basin (• Figure 10.11).

Within the rapidly subsiding interior of the basin, other types of reefs are found. *Pinnacle reefs* are tall, spindly structures up to 100 m high. They reflect the rapid upward growth needed to maintain themselves near sea level during subsidence of the basin (Figure 10.11). Besides the pinnacle reefs, bedded carbonates and thick sequences of salt and anhydrite are also found in the Michigan Basin.

As the Tippecanoe Sea gradually regressed from the craton during the Late Silurian, precipitation of evaporite minerals occurred in the Appalachian, Ohio, and Michigan basins (Figure 10.1). In the Michigan Basin alone, approximately 1500 m of sediments were deposited, nearly half of which are halite and anhydrite. How did such thick sequences of evaporites accumulate? One possibility is that when sea level dropped, the tops of the barrier reefs were as high as or above sea level, preventing the influx of new seawater into the basin. Evaporation of the basin seawater would result in the formation of a brine, and as the brine became increasingly concentrated, the precipitation of salts. A second possibility is that the reefs grew upward so close to sea level they formed a sill or barrier that eliminated interior circulation and allowed for the evaporation of the seawater that produced a dense brine that eventually resulted in evaporite deposits (• Figure 10.12).

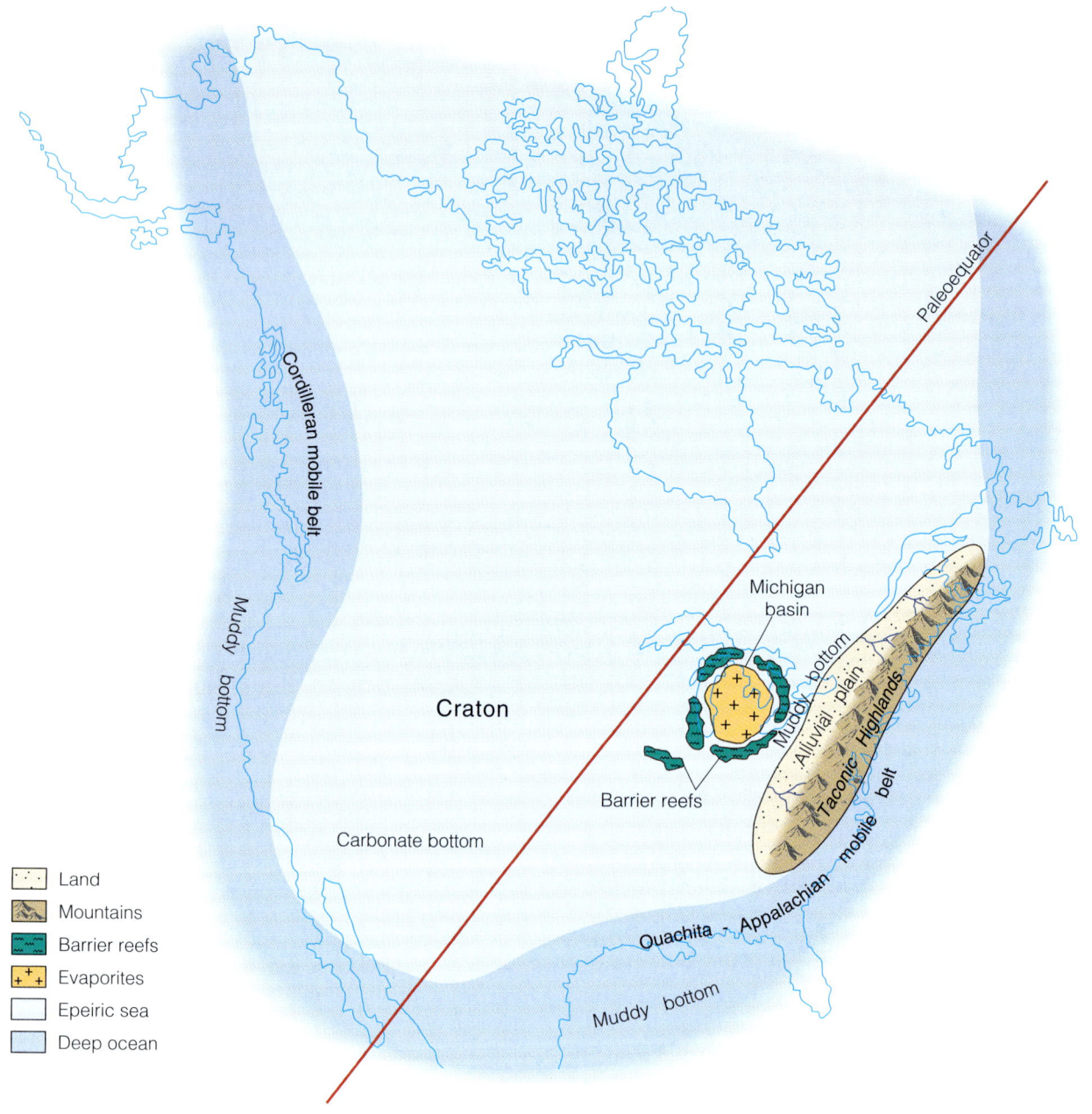

• **Figure 10.10 Silurian Paleogeography of North America** Note the development of reefs in the Michigan, Ohio, and Indiana-Illinois-Kentucky areas.

With North America still near the equator during the Silurian Period (Figure 10.2c), temperatures were probably high. As circulation to the Michigan Basin was restricted or ceased altogether, seawater within the basin evaporated and began forming brine. Because the brine was heavy, it concentrated near the bottom, and minerals precipitated on the basin floor to form evaporite deposits. When seawater flowed back into the Michigan Basin either over the sill and through channels cut in the barrier reefs, this replenishment added new seawater, allowing the process of brine formation and precipitation of evaporites to repeat itself.

The order and type of salts precipitating from seawater depends on their solubility, the original concentration of seawater, and local conditions of the basin. In general, salts precipitate in a sequential order, beginning with the least soluble and ending with the most soluble. Therefore, calcium carbonate usually precipitates out first, followed by gypsum,* and lastly halite. Many lateral shifts and interfingering of the limestone, anhydrite, and halite facies may occur, however, because of variations in the amount of seawater entering the basin and changing geologic conditions.

---

*Recall from Chapter 6 that gypsum ($CaSO_4 \cdot 2H_2O$) is the common sulfate precipitated from seawater, but when deeply buried, gypsum loses its water and is converted to anhydrite ($CaSO_4$).

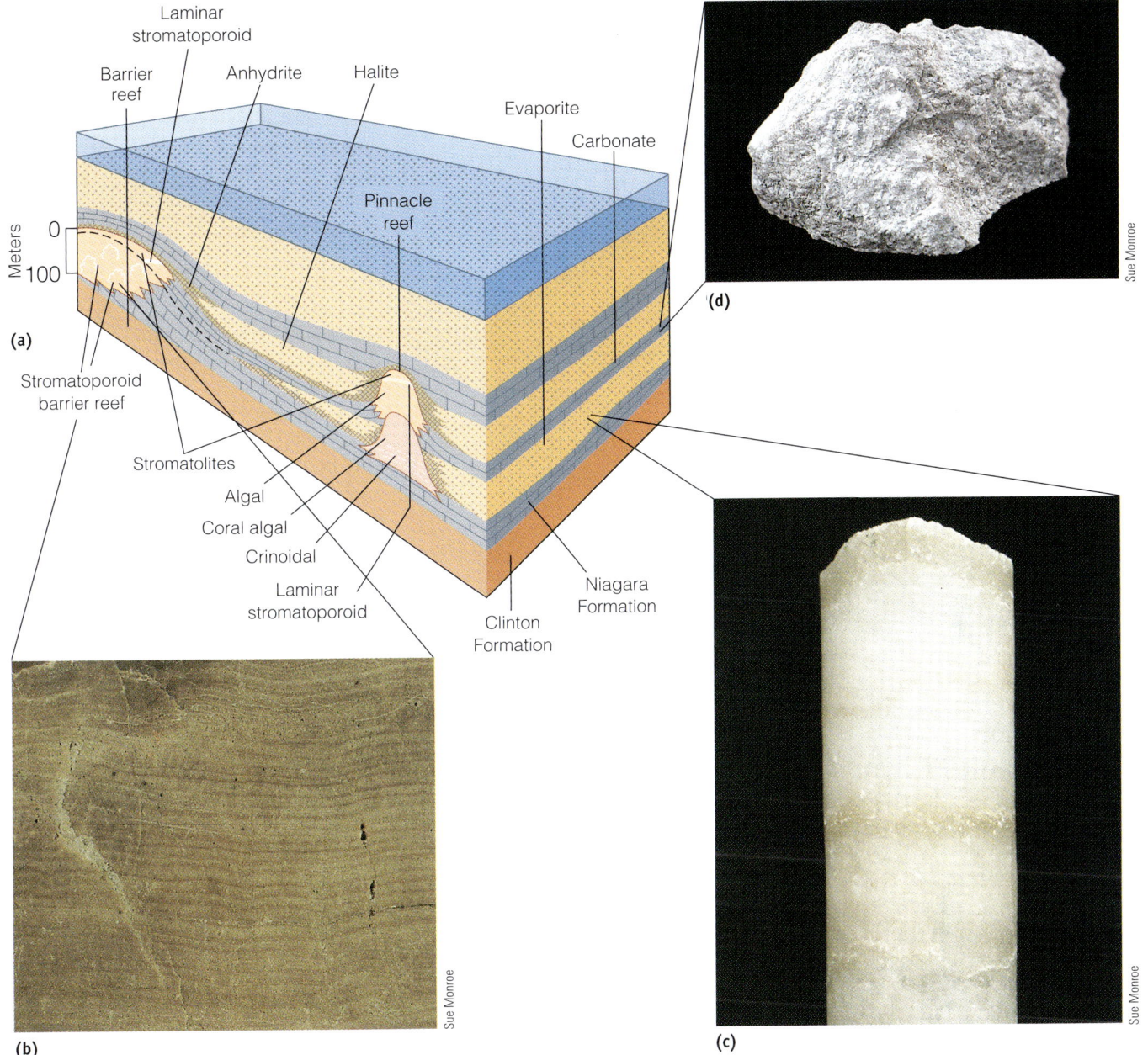

• **Figure 10.11 The Michigan Basin** (a) Generalized block diagram of the northern Michigan Basin during the Silurian Period. (b) Cross section of a stromatoporoid colony from the stromatoporoid barrier reef facies. (c) Core of rock salt from the evaporite facies. (d) Limestone from the carbonate facies.

Thus the periodic evaporation of seawater as just discussed could account for the observed vertical and lateral distribution of evaporites in the Michigan Basin. Associated with those evaporites, however, are pinnacle reefs, and the organisms constructing those reefs could not have lived in such a highly saline environment (Figure 10.11). How, then, can such contradictory features be explained? Numerous models have been proposed, ranging from cessation of reef growth followed by evaporite deposition, to alternation of reef growth and evaporite deposition. Although the Michigan Basin has been studied extensively for years, no model yet proposed completely explains the genesis and relationship of its various reef, carbonate, and evaporite facies.

The End of the Tippecanoe Sequence

By the Early Devonian, the regressing Tippecanoe Sea had retreated to the craton margin, exposing an extensive lowland topography. During this regression, marine deposition was initially restricted to a few interconnected cratonic

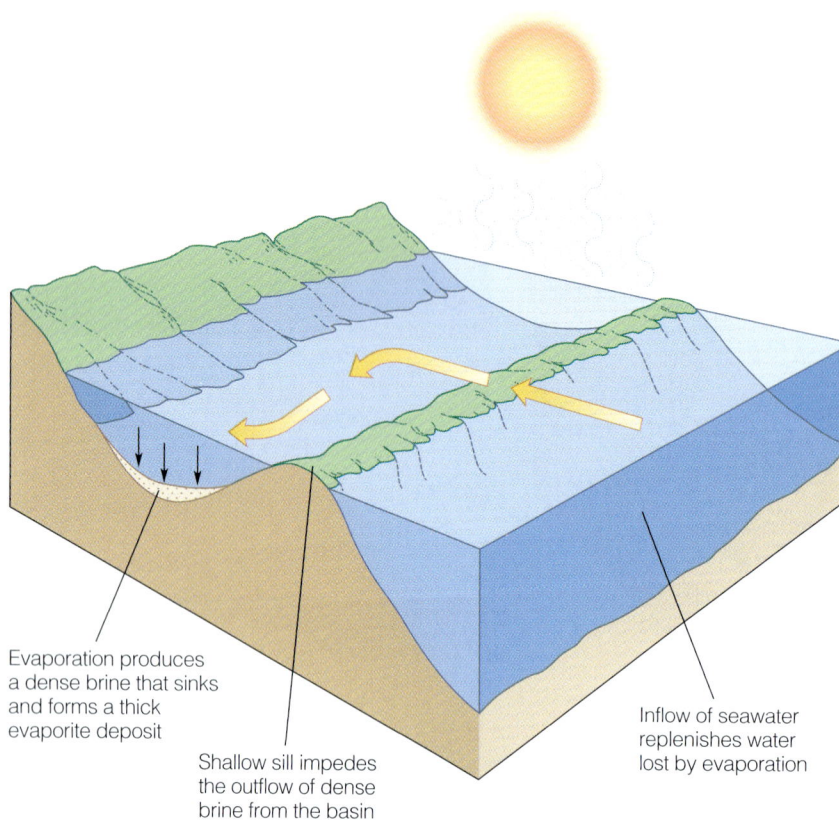

• **Figure 10.12 Evaporite Sedimentation** Silled basin model for evaporite sedimentation by direct precipitation from seawater. Vertical scale is greatly exaggerated.

basins and, finally by the end of the Tippecanoe, to only the mobile belts surrounding the craton.

As the Tippecanoe Sea regressed during the Early Devonian, the craton experienced mild deformation forming many domes, arches, and basins (Figure 10.1). These structures were mostly eroded during the time the craton was exposed, and they were eventually covered by deposits from the ensuing and encroaching Kaskaskia Sea.

The Appalachian Mobile Belt and the Taconic Orogeny

Having examined the Sauk and Tippecanoe geologic history of the craton, we now turn our attention to the Appalachian mobile belt, where the first Phanerozoic orogeny began during the Middle Ordovician. The mountain building occurring during the Paleozoic Era had a profound influence on the climate and sedimentary history of the craton. In addition, it was part of the global tectonic regime that sutured the continents together, forming Pangaea by the end of the Paleozoic.

Throughout Sauk time, the Appalachian region was a broad, passive, continental margin. Sedimentation was closely balanced by subsidence as extensive carbonate deposits succeeded thick, shallow marine sands. During this time, movement along a divergent plate boundary was widening the **Iapetus Ocean** (• Figure 10.13a). Beginning with the subduction of the Iapetus plate beneath Laurentia (an oceanic–continental convergent plate boundary), the Appalachian mobile belt was born (Figure 10.13b). The resulting **Taconic orogeny**—named after the present-day Taconic Mountains of eastern New York, central Massachusetts, and Vermont—was the first of several orogenies to affect the Appalachian region.

The Appalachian mobile belt can be divided into two depositional environments. The first is the extensive, shallow-water carbonate platform that formed the broad eastern continental shelf and stretched from Newfoundland to Alabama (Figure 10.13a). It formed during the transgression of the Sauk Sea onto the craton when carbonates were deposited in a vast shallow sea. The shallow-water depth on the platform is indicated by stromatolites, mud cracks, and other sedimentary structures and fossils.

Carbonate deposition ceased along the East Coast during the Middle Ordovician and was replaced by deepwater deposits characterized by thinly bedded black shales, graded beds, coarse sandstones, graywackes, and associated volcanics. This suite of sediments marks the onset of mountain building, in this case, the Taconic orogeny. The subduction of the Iapetus plate beneath Laurentia resulted in volcanism and downwarping of the carbonate platform (Figure 10.13b). Throughout the Appalachian mobile belt, facies patterns, paleocurrents, and sedimentary structures all indicate that these deposits were derived from the east, where the Taconic Highlands and associated volcanoes were rising.

Additional structural, stratigraphic, petrologic, and sedimentologic evidence has provided much information on the timing and origin of this orogeny. For example, at many locations within the Taconic belt, pronounced angular unconformities occur where steeply dipping Lower Ordovician rocks are overlain by gently dipping or horizontal Silurian and younger rocks.

Other evidence includes volcanic activity in the form of deep-sea lava flows, volcanic ash layers, and intrusive bodies in the area from present-day Georgia to Newfoundland. These igneous rocks show a clustering of radiometric ages corresponding to the Middle to Late Ordovician. In addition, regional metamorphism coincides with the radiometric dates.

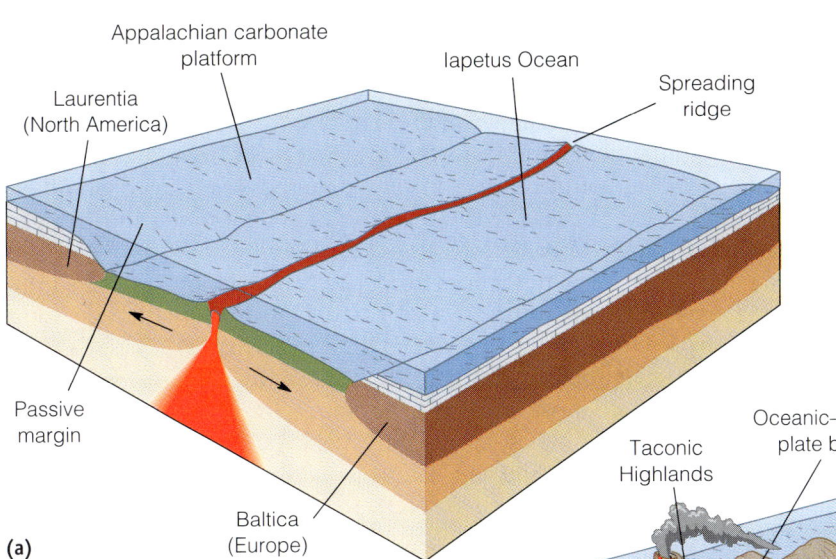

• **Figure 10.13 Neoproterozoic to Late Ordovician Evolution of the Appalachian Mobile Belt** (a) During the Neoproterozoic to the Early Ordovician, the Iapetus Ocean was opening along a divergent plate boundary. Both the east coast of Laurentia and the west coast of Baltica were passive continental margins with large carbonate platforms. (b) Beginning in the Middle Ordovician, the passive margins of Laurentia and Baltica changed to active oceanic–continental plate boundaries, resulting in orogenic activity (see Figure 10.2).

The final piece of evidence for the Taconic orogeny is the development of a large **clastic wedge,** an extensive accumulation of mostly detrital sediments deposited adjacent to an uplifted area. These deposits are thickest and coarsest nearest the highland area and become thinner and finer grained away from the source area, eventually grading into the carbonate cratonic facies (• Figure 10.14). The clastic wedge resulting from the erosion of the Taconic Highlands is referred to as the **Queenston Delta.**

The Taconic orogeny marked the first pulse of mountain building in the Appalachian mobile belt and was a response to the subduction taking place beneath the east coast of Laurentia. As the Iapetus Ocean narrowed and closed, another orogeny occurred in Europe during the Silurian. The *Caledonian orogeny* was essentially a mirror image of the Taconic orogeny and the Acadian orogeny (see Chapter 11) and was part of the global mountain building episode that occurred during the Paleozoic Era (Figure 10.2c). Even though the Caledonian orogeny occurred during Tippecanoe time, we discuss it in the next chapter because it was intimately related to the Acadian orogeny.

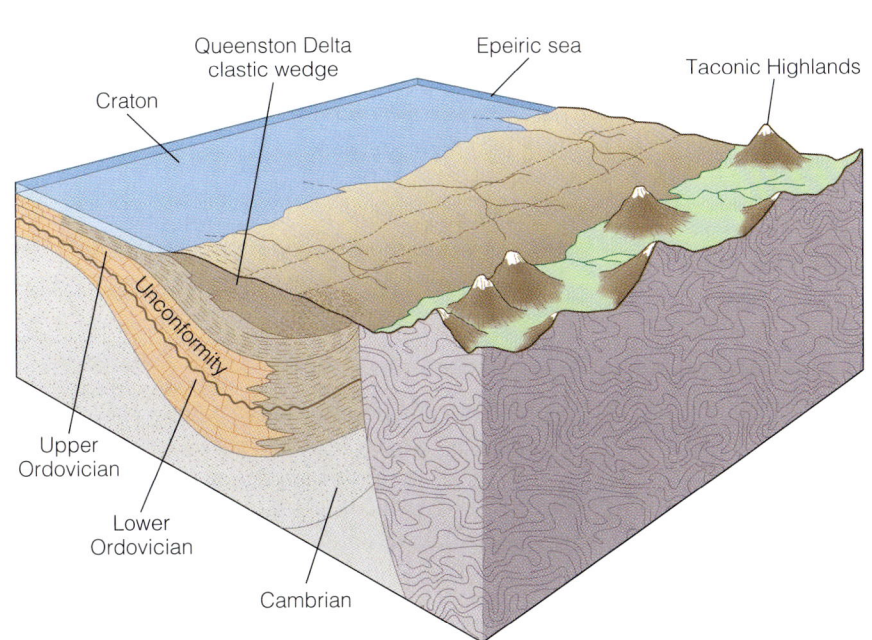

• **Figure 10.14 Reconstruction of the Taconic Highlands and Queenston Delta Clastic Wedge** The Queenston Delta clastic wedge, resulting from the erosion of the Taconic Highlands, consists of thick, coarse-grained detrital sediments nearest the highlands and thins laterally into finer-grained sediments on the craton.

THE APPALACHIAN MOBILE BELT AND THE TACONIC OROGENY   **205**

Early Paleozoic Mineral Resources

Early Paleozoic-aged rocks contain a variety of important mineral resources, including sand and gravel for construction, building stone, and limestone used in the manufacture of cement. Important sources of industrial or silica sand are the Upper Cambrian Jordan Sandstone of Minnesota and Wisconsin, the Lower Silurian Tuscarora Sandstone in Pennsylvania and Virginia, and the Middle Ordovician St. Peter Sandstone. The latter, the basal sandstone of the Tippecanoe Sequence (Figure 10.8), occurs in several states, but the best-known area of production is in La Salle County, Illinois. Silica sand has a variety of uses, including the manufacture of glass, refractory bricks for blast furnaces, and molds for casting iron, aluminum, and copper alloys. Some silica sands, called hydraulic fracturing sands, are pumped into wells to fracture oil- or gas-bearing rocks and provide permeable passageways for the oil or gas to migrate to the well.

Thick deposits of Silurian evaporites, mostly rock salt (NaCl) and rock gypsum ($CaSO_4 \cdot 2H_2O$) altered to rock anhydrite ($CaSO_4$), underlie parts of Michigan, Ohio, New York, and adjacent areas in Ontario, Canada. These rocks are important sources of various salts. In addition, barrier and pinnacle reefs in carbonate rocks associated with these evaporites are the reservoirs for oil and gas in Michigan and Ohio.

The host rocks for deposits of lead and zinc in southeast Missouri are Cambrian dolostones, although some Ordovician rocks contain these metals as well. These deposits have been mined since 1720 but have been largely depleted. Now most lead and zinc mined in Missouri comes from Mississippian-aged sedimentary rocks.

The Silurian Clinton Formation crops out from Alabama north to New York, and equivalent rocks are found in Newfoundland. This formation has been mined for iron in many places. In the United States, the richest ores and most extensive mining occurred near Birmingham, Alabama, but only a small amount of ore is currently produced in that area.

SUMMARY

Table 10.1 summarizes the geologic history of the North American craton and mobile belts as well as global events, sea level changes, and major evolutionary events during the Early Paleozoic.

- Six major continents and numerous microcontinents and island arcs existed at the beginning of the Paleozoic Era; four of the major continents were located near the paleo-equator.

- During the Early Paleozoic (Cambrian-Silurian), Laurentia was moving northward and Gondwana moved to a south polar location, as indicated by tillite deposits.

- Most continents consist of two major components: a relatively stable craton over which epeiric seas transgressed and regressed, surrounded by mobile belts in which mountain building took place.

- The geologic history of North America can be divided into cratonic sequences that reflect cratonwide transgressions and regressions.

- The first major marine transgression onto the craton took place during the Sauk Sequence. At its maximum, the Sauk Sea covered the craton except for parts of the Canadian shield and the Transcontinental Arch, a series of large, northeast–southwest trending islands.

- The Tippecanoe Sequence began with deposition of an extensive sandstone over the exposed and eroded Sauk landscape. During Tippecanoe time, extensive carbonate deposition took place. In addition, large barrier reefs enclosed basins, resulting in evaporite deposition within these basins.

- The eastern edge of North America was a stable carbonate platform during Sauk time. During Tippecanoe time an oceanic–continental convergent plate boundary formed, resulting in the Taconic orogeny, the first of three major orogenies to affect the Appalachian mobile belt.

- The newly formed Taconic Highlands shed sediments into the western epeiric sea, producing a clastic wedge geologists call the Queenston Delta.

- Early Paleozoic-aged rocks contain a variety of mineral resources, including building stone, limestone for cement, silica sand, hydrocarbons, evaporites, and iron ore.

IMPORTANT TERMS

Appalachian mobile belt, p. 191
Baltica, p. 192
China, p. 192
clastic wedge, p. 205
Cordilleran mobile belt, p. 191
craton, p. 191
cratonic sequence, p. 194
epeiric sea, p. 191

Franklin mobile belt, p. 191
Gondwana, p. 192
Iapetus Ocean, p. 204
Kazakhstania, p. 192
Laurentia, p. 192
mobile belt, p. 191
organic reef, p. 199
Ouachita mobile belt, p. 191

Queenston Delta, p. 205
Sauk Sequence, p. 195
sequence stratigraphy, p. 195
Siberia, p. 192
Taconic orogeny, p. 204
Tippecanoe Sequence, p. 198
Transcontinental Arch, p. 196

Table 10.1
Summary of Early Paleozoic Geologic and Evolutionary Events

Age (Millions of Years)	Geologic Period	Sequence	Relative Changes in Sea Level (Rising / Falling)	Cordilleran Mobile Belt	Craton	Ouachita Mobile Belt	Appalachian Mobile Belt	Major Events Outside North America	Major Evolutionary Events
416 —	Silurian	Tippecanoe			Extensive barrier reefs and evaporites common		Acadian orogeny	Caledonian orogeny	First jawed fish evolve
									Early land plants—seedless vascular plants
444 —	Ordovician	Tippecanoe	Present sea level		Queenston Delta clastic wedge		Taconic orogeny	Continental glaciation in Southern Hemisphere	Extinction of many marine invertebrates near end of Ordovician
					Transgression of Tippecanoe Sea				Plants move to land?
					Regression exposing large areas to erosion				Major adaptive radiation of all invertebrate groups
488 —	Cambrian	Sauk			Canadian shield and Transcontinental Arch only areas above sea level				Many trilobites become extinct near end of Cambrian
									Earliest vertebrates—jawless fish evolve
542 —					Transgression of Sauk Sea				

REVIEW QUESTIONS

1. Which was the first major transgressive sequence onto the North American craton?
   a. _____ Absaroka; b. _____ Sauk; c. _____ Zuni; d. _____ Kaskaskia; e. _____ Tippecanoe.
2. What type of plate interaction produced the Taconic orogeny?
   a. _____ divergent; b. _____ transform; c. _____ oceanic–oceanic convergent; d. _____ oceanic–continental convergent; e. _____ continental–continental convergent.
3. During which sequence did the eastern margin of Laurentia change from a passive plate margin to an active plate margin?
   a. _____ Zuni; b. _____ Tippecanoe; c. _____ Sauk; d. _____ Kaskaskia; e. _____ Absaroka.
4. A major transgressive–regressive cycle bounded by cratonwide unconformities is a(n)
   a. _____ biostratigraphic unit; b. _____ cratonic sequence; c. _____ orogeny; d. _____ shallow sea; e. _____ cyclothem.
5. An elongated area marking the site of mountain building is a(n)
   a. _____ cyclothem; b. _____ mobile belt; c. _____ platform; d. _____ shield; e. _____ craton.
6. During which sequence were evaporites and reef carbonates the predominant cratonic rocks?
   a. _____ Kaskaskia; b. _____ Zuni; c. _____ Sauk; d. _____ Absaroka; e. _____ Tippecanoe.
7. What Middle Ordovician formation is an important source of industrial silica sand?
   a. _____ St. Peter; b. _____ Tuscarora; c. _____ Jordan; d. _____ Oriskany; e. _____ Clinton.
8. The ocean separating Laurentia from Baltica is called the
   a. _____ Panthalassa; b. _____ Tethys; c. _____ Iapetus; d: _____ Atlantis; e. _____ Perunica.
9. Which mobile belt is located along the eastern side of North America?
   a. _____ Franklin; b. _____ Cordilleran; c. _____ Ouachita; d. _____ Appalachian; e. _____ answers (b) and (c).
10. During deposition of the Sauk Sequence, the only area above sea level besides the Transcontinental Arch was the
    a. _____ cratonic margin; b. _____ Canadian shield; c. _____ Queenston Delta; d. _____ Appalachian mobile belt; e. _____ Taconic Highlands.
11. The two prominent reef building organisms today are
    a. _____ trilobites and brachiopods; b. _____ corals and bryozoans; c. _____ calcareous algae and oysters; d. _____ corals and calcareous algae; e. _____ oysters and brachiopods.
12. The vertical sequence of the Tapeats Sandstone, Bright Angel Shale, and Muav Limestone represents
    a. _____ a transgression; b. _____ time transgressive formations; c. _____ rocks of the Grand Canyon, Arizona; d. _____ sediments deposited by the Sauk Sea; e. _____ all of the previous answers.
13. At the beginning of the Cambrian, there were _____ major continents.
    a. _____ 3; b. _____ 4; c. _____ 5; d. _____ 6; e. _____ 7.
14. What are some methods geologists can use to determine the locations of continents during the Paleozoic Era?
15. Discuss why cratonic sequences are a convenient way to study the geologic history of the Paleozoic Era.
16. Discuss how the Cambrian rocks of the Grand Canyon illustrate the sedimentation patterns of a transgressive sea.
17. Discuss how sequence stratigraphy can be used to make global correlations and why it is so useful in reconstructing past events.
18. What evidence indicates that the Iapetus Ocean began closing during the Middle Ordovician?
19. Discuss how the evaporites of the Michigan Basin may have formed during the Silurian Period.
20. What evidence in the geologic record indicates that the Taconic orogeny occurred?

APPLY YOUR KNOWLEDGE

1. According to estimates made from mapping and correlation, the Queenston Delta contains more than 600,000 km^3 of rock eroded from the Taconic Highlands. Based on this figure, geologists estimate the Taconic Highlands were at least 4000 m high. It is also estimated that the Catskill Delta (see Chapter 11) contains three times as much sediment as the Queenston Delta. From what you know about the geographic distribution of the Taconic Highlands and the Acadian Highlands (see Chapter 11), can you estimate how high the Acadian Highlands may have been?
2. Paleogeographic maps of what the world looked like can be found in almost every Earth history book and in numerous scientific journals. What criteria are used to determine the location of ancient continents and ocean basins, and why are there minor differences in the location and size of these paleocontinents among the various books and articles?

CHAPTER 11
LATE PALEOZOIC EARTH HISTORY

Tullimonstrum gregarium, also known as the Tully Monster, is Illinois's official state fossil. Left: Specimen from Pennsylvanian rocks, Mazon Creek locality, Illinois. Right: Reconstruction of the Tully Monster (about 30 cm long).

[OUTLINE]

INTRODUCTION

LATE PALEOZOIC PALEOGEOGRAPHY
The Devonian Period
The Carboniferous Period
The Permian Period

LATE PALEOZOIC EVOLUTION OF NORTH AMERICA

THE KASKASKIA SEQUENCE
Reef Development in Western Canada

PERSPECTIVE 11.1 *The Canning Basin, Australia— A Devonian Great Barrier Reef*

Black Shales
The Late Kaskaskia—A Return to Extensive Carbonate Deposition

THE ABSAROKA SEQUENCE

What Are Cyclothems, and Why Are They Important?

Cratonic Uplift—The Ancestral Rockies

The Middle Absaroka—More Evaporite Deposits and Reefs

HISTORY OF THE LATE PALEOZOIC MOBILE BELTS
Cordilleran Mobile Belt
Ouachita Mobile Belt
Appalachian Mobile Belt

WHAT ROLE DID MICROPLATES AND TERRANES PLAY IN THE FORMATION OF PANGAEA?

LATE PALEOZOIC MINERAL RESOURCES

SUMMARY

ThomsonNOW™ Explore interactive tutorials, animations, or practice problems available on the ThomsonNow website at www.thomsonedu.com/login.

CHAPTER OBJECTIVES

At the end of this chapter, you will have learned that

- Movement of the six major continents during the Paleozoic Era resulted in the formation of the supercontinent Pangaea at the end of the Paleozoic.

- In addition to the large-scale plate interactions during the Paleozoic, microplate and terrane activity also played an important role in forming Pangaea.

- Most of the Kaskaskia Sequence is dominated by carbonates and associated evaporites.

- Transgressions and regressions over the low-lying craton during the Absaroka Sequence resulted in cyclothems and the formation of coals.

- During the Late Paleozoic Era, mountain-building activity took place in the Appalachian, Ouachita, and Cordilleran mobile belts.

- The Caledonian, Acadian, Hercynian, and Alleghenian orogenies were all part of the global tectonic activity resulting from the assembly of Pangaea.

- Late Paleozoic-aged rocks contain a variety of mineral resources, including petroleum, coal, evaporites, and various metallic deposits.

Introduction

Approximately 300 million years ago in what is now Illinois, sluggish rivers flowed southwestward through swamps and built large deltas that extended outward into a subtropical shallow sea. These rivers deposited huge quantities of mud, which entombed the plants and animals living in the area. Rapid burial and the formation of ironstone concretions thus preserved many of them as fossils. Known as the Mazon Creek fossils, for the area in northeastern Illinois where most specimens are found, they provide us with significant insights about the soft-part anatomy of the region's biota. Because of the exceptional preservation of this ancient biota, Mazon Creek fossils are known throughout the world and many museums have extensive collections from the area.

During Pennsylvanian time, two major habitats existed in northeastern Illinois. One was a swampy, forested lowland of the subaerial delta, and the other was the shallow marine environment of the actively prograding delta.

In the warm, shallow waters of the delta front lived numerous cnidarians, mollusks, echinoderms, arthropods, worms, and fish. The swampy lowlands surrounding the delta were home to more than 400 plant species, numerous insects, spiders, and other animals such as scorpions and amphibians. In the ponds, lakes, and rivers were many fish, shrimp, and ostracods. Almost all the plants were seedless vascular plants, typical of the kinds that flourished in the coal-forming swamps during the Pennsylvanian Period.

One of the more interesting Mazon Creek fossils is the Tully Monster, which is not only unique to Illinois but also is its official state fossil. Named for Francis Tully, who first discovered it in 1958, *Tullimonstrum gregarium* (its scientific name) was a small (up to 30 cm long), soft-bodied animal that lived in the warm, shallow seas covering Illinois about 300 million years ago.

The Tully Monster had a relatively long proboscis that contained a "claw" with small teeth in it. The round- to oval-shaped body was segmented and contained a crossbar, whose swollen ends some interpreted as the animal's sense organs. The tail had two horizontal fins. It probably swam like an eel, with most of the undulatory movement occurring behind the two sense organs. There presently is no consensus as to what phylum the Tully Monster belongs or to what animals it might be related.

The Late Paleozoic Era was a time not only of interesting evolutionary innovations and novelties such as the Tully Monster but also when the world's continents were colliding along convergent plate boundaries. These collisions profoundly influenced both Earth's geologic and its biologic history and eventually formed the supercontinent Pangaea by the end of the Permian Period.

Late Paleozoic Paleogeography

The Late Paleozoic was a time marked by continental collisions, mountain building, fluctuating sea levels, and varied climates. Coals, evaporites, and tillites testify to the variety of climatic conditions experienced by the different continents during the Late Paleozoic. Major glacial and interglacial episodes took place over much of Gondwana as it continued moving over the South Pole during the Late Mississippian to Early Permian. The growth and retreat of continental glaciers during this time profoundly affected the world's biota as well as contributed to global sea level changes. Collisions between continents not only led to the formation of the supercontinent Pangaea by the end of the Permian but also resulted in mountain building that strongly influenced oceanic and atmospheric circulation patterns. By the end of the Paleozoic, widespread arid and semiarid conditions governed much of Pangaea.

The Devonian Period

During the Silurian, Laurentia and Baltica, which had earlier united with Avalonia, collided along a convergent plate boundary to form the larger continent of **Laurasia.** This collision, which closed the northern Iapetus Ocean, is marked by the **Caledonian orogeny.** During the Devonian, as the southern Iapetus Ocean narrowed between Laurasia and Gondwana, mountain building continued along the eastern margin of Laurasia with the **Acadian orogeny** (• Figure 11.1a). The erosion of the resulting highlands spread vast amounts of reddish fluvial sediments over large areas of northern Europe (Old Red Sandstone) and eastern North America (the Catskill Delta).

Other Devonian tectonic events, probably related to the collision of Laurentia and Baltica, include the Cordilleran **Antler orogeny,** the **Ellesmere orogeny** along the northern margin of Laurentia (which may reflect the collision of Laurentia with Siberia), and the change from a passive continental margin to an active convergent plate boundary in the Uralian mobile belt of eastern Baltica. The distribution of reefs, evaporites, and red beds, as well as the existence of similar floras throughout the world, suggest a rather uniform global climate during the Devonian Period.

The Carboniferous Period

During the Carboniferous Period, southern Gondwana moved over the South Pole, resulting in extensive continental glaciation (Figures 11.1b and • 11.2a). The advance and retreat of these glaciers produced global changes in sea level that affected sedimentation patterns on the cratons. As Gondwana continued moving northward, it began colliding with Laurasia during the Early Carboniferous and continued suturing with it during the rest of the Carboniferous (Figures 11.1b and 11.2a). Because Gondwana rotated clockwise relative to Laurasia, deformation generally progressed in a northeast-to-southwest direction along the Hercynian, Appalachian, and Ouachita mobile belts of the two continents. The final phase of collision between Gondwana and Laurasia is indicated by the Ouachita Mountains of Oklahoma, formed by thrusting during the Late Carboniferous and Early Permian.

Elsewhere, Siberia collided with Kazakhstania and moved toward the Uralian margin of Laurasia (Baltica), colliding with it during the Early Permian. By the end of the Carboniferous, the various continental landmasses were fairly close together as Pangaea began taking shape.

The Carboniferous coal basins of eastern North America, western Europe, and the Donets Basin of the Ukraine all lay in the equatorial zone, where rainfall was high and temperatures were consistently warm. The absence of strong seasonal growth rings in fossil plants from these coal basins indicates such a climate. The fossil plants found in the coals of Siberia, however, show well-developed growth rings, signifying seasonal growth with abundant rainfall and distinct seasons such as in the temperate zones (latitudes 40 degrees to 60 degrees north).

Glacial conditions and the movement of large continental ice sheets in the high southern latitudes are indicated by widespread tillites and glacial striations in southern Gondwana. These ice sheets spread toward the equator and, at their maximum growth, extended well into the middle temperate latitudes.

The Permian Period

The assembly of Pangaea was essentially completed during the Permian as a result of the many continental collisions that began during the Carboniferous (Figure 11.2b). Although geologists generally agree on the configuration and location of the western half of the supercontinent, there is no consensus on the number or configuration of the various terranes and continental blocks that composed the eastern half of Pangaea. Regardless of the exact configuration of the eastern portion, geologists know the supercontinent was surrounded by various subduction zones and moved steadily northward during the Permian. Furthermore, an enormous single ocean, the **Panthalassa,** surrounded Pangaea and spanned Earth from pole to pole (Figure 11.2b).

The formation of a single large landmass had climatic consequences for the terrestrial environment. Terrestrial Permian sediments show that arid and semiarid conditions were widespread over Pangaea. The mountain ranges produced by the **Hercynian, Alleghenian,** and **Ouachita orogenies** were high enough to create rain shadows that blocked the moist, subtropical, easterly winds—much as the southern Andes Mountains do in western South America today. This produced very dry conditions in North America and Europe, as evident from the extensive Permian red beds and evaporites found in western North America, central Europe, and parts of Russia. Permian coals, indicating abundant rainfall, were mostly limited to the northern temperate belts (latitude 40 degrees to 60 degrees north), while the last remnants of the Carboniferous ice sheets continued their recession.

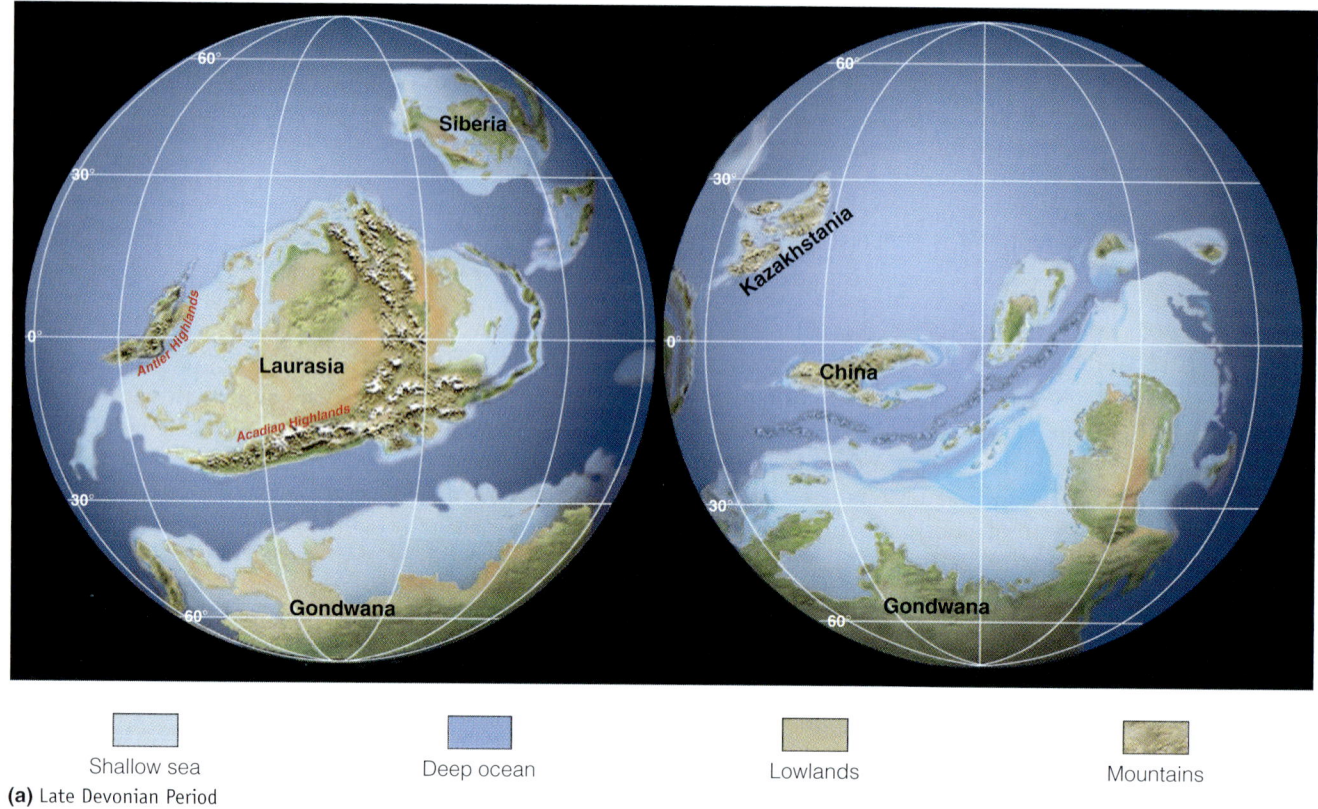

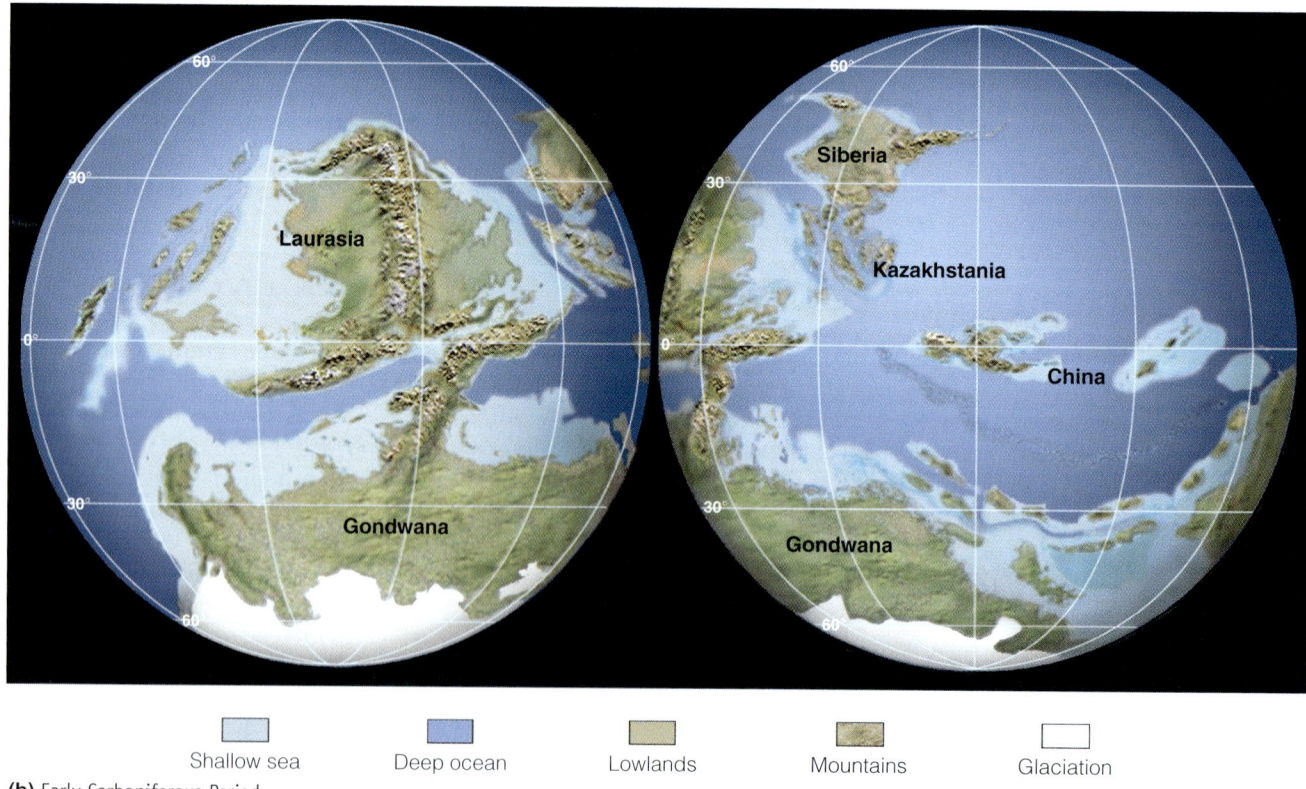

• Figure 11.1 **Paleozoic Paleogeography** Paleogeography of the world for the (a) Late Devonian Period and (b) Early Carboniferous Period.

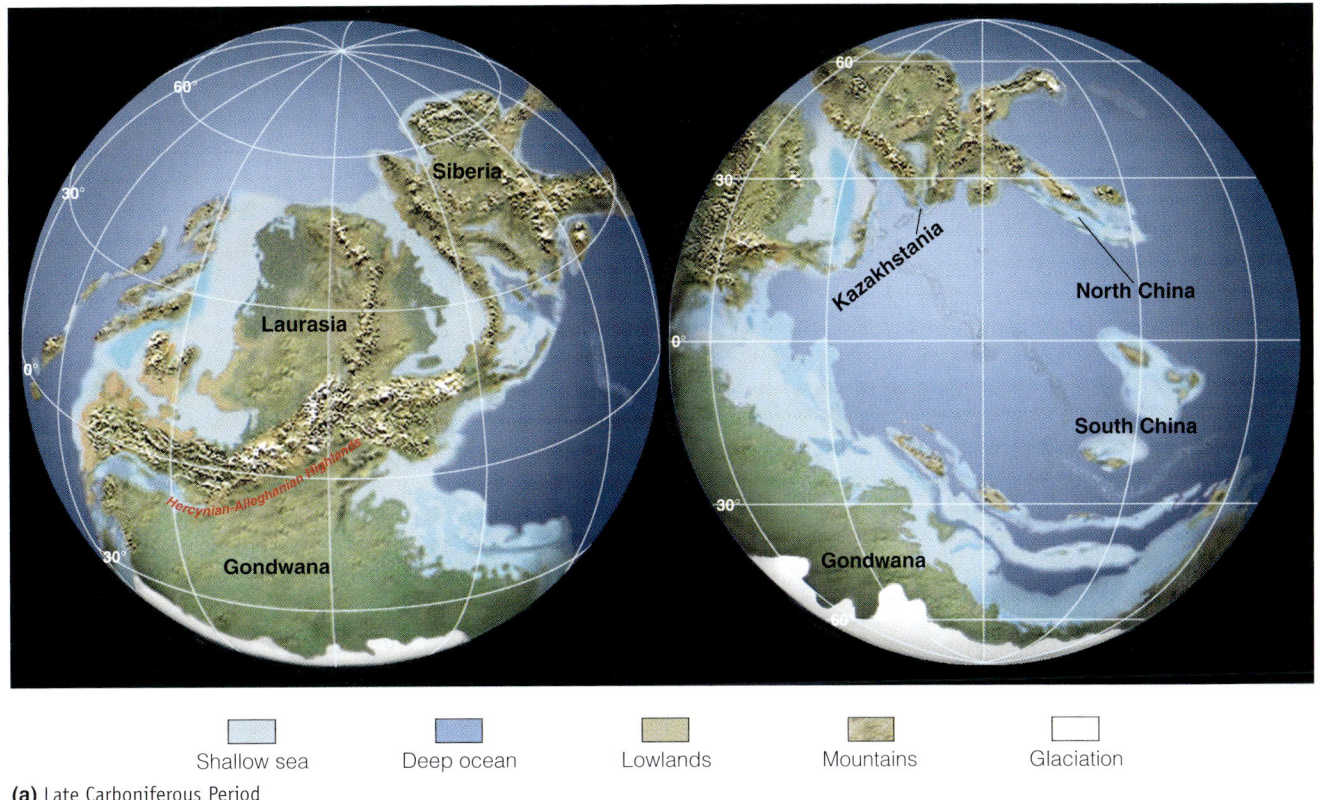

(a) Late Carboniferous Period

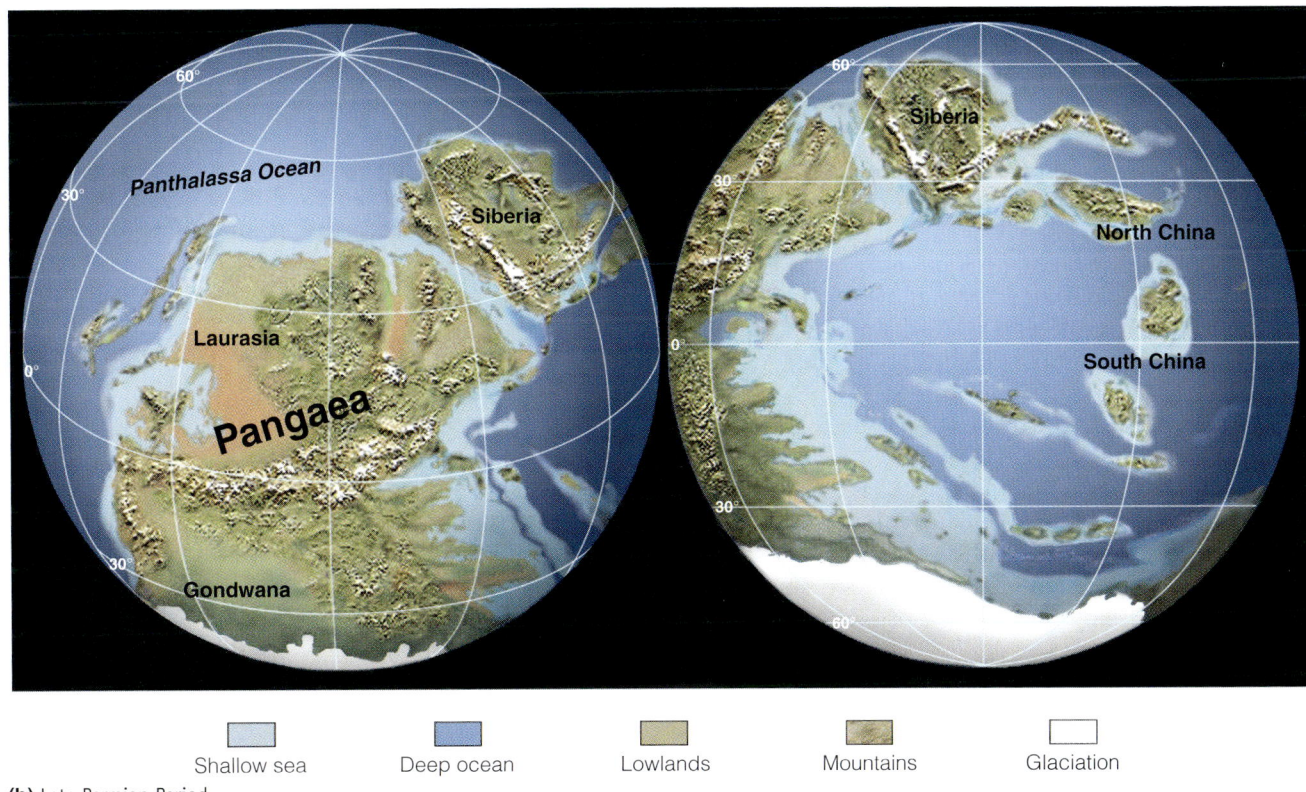

(b) Late Permian Period

• **Figure 11.2 Paleozoic Paleogeography** Paleogeography of the world for the (a) Late Carboniferous Period and (b) Late Permian Period.

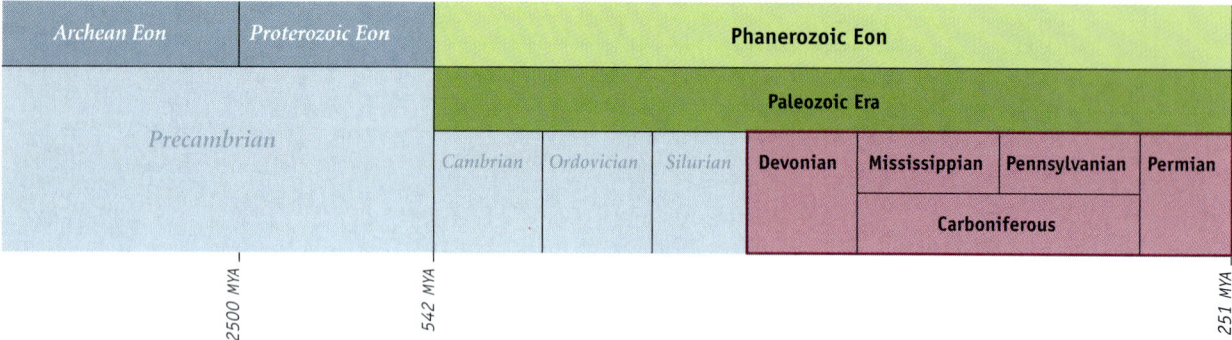

Late Paleozoic Evolution of North America

The Late Paleozoic cratonic history of North America included periods of extensive shallow-marine carbonate deposition and large coal-forming swamps as well as dry, evaporite-forming terrestrial conditions. Cratonic events largely resulted from sea level changes caused by Gondwanan glaciation and tectonic events related to the assemblage of Pangaea.

Mountain building that began with the Ordovician Taconic orogeny continued with the Caledonian, Acadian, Alleghenian, and Ouachita orogenies. These orogenies were part of the global tectonic process that resulted in the formation of Pangaea by the end of the Paleozoic Era.

The Kaskaskia Sequence

The boundary between the Tippecanoe Sequence and the overlying **Kaskaskia Sequence** (Middle Devonian–Late Mississippian) is marked by a major unconformity. As the Kaskaskia Sea transgressed over the low-relief landscape of the craton, most basal beds deposited consisted of clean, well-sorted quartz sandstones. A good example is the Oriskany Sandstone of New York and Pennsylvania and its lateral equivalents (• Figure 11.3). The Oriskany Sandstone, like the basal Tippecanoe St. Peter Sandstone, is an important glass sand as well as a good gas-reservoir rock.

The source areas for the basal Kaskaskia sandstones were primarily the eroding highlands of the Appalachian mobile belt area (• Figure 11.4), exhumed Cambrian and Ordovician sandstones cropping out along the flanks of the Ozark Dome, and exposures of the Canadian Shield in the Wisconsin area. The lack of similar sands in the Silurian carbonate beds below the Tippecanoe–Kaskaskia unconformity indicates that the source areas of the basal Kaskaskia detrital rocks were submerged when the Tippecanoe Sequence was deposited. Stratigraphic studies indicate these source areas were uplifted and the Tippecanoe carbonates removed by erosion before the Kaskaskia transgression.

Kaskaskian basal rocks elsewhere on the craton consist of carbonates that are frequently difficult to differentiate from the underlying Tippecanoe carbonates unless they are fossiliferous.

Except for widespread Upper Devonian and Lower Mississippian black shales, the majority of Kaskaskian rocks are carbonates, including reefs, and associated evaporite deposits. In many other parts of the world, such as southern England, Belgium, central Europe, Australia, and Russia, the Middle and early Late Devonian epochs were times of major reef building (see Perspective 11.1).

Reef Development in Western Canada

The Middle and Late Devonian reefs of western Canada contain large reserves of petroleum and have been widely studied from outcrops and in the subsurface (• Figure 11.5). These reefs began forming as the Kaskaskia Sea transgressed southward into western Canada. By the end of the Middle Devonian, they had coalesced into a large barrier reef system that restricted the flow of oceanic water into the back-reef platform, creating conditions for evaporite

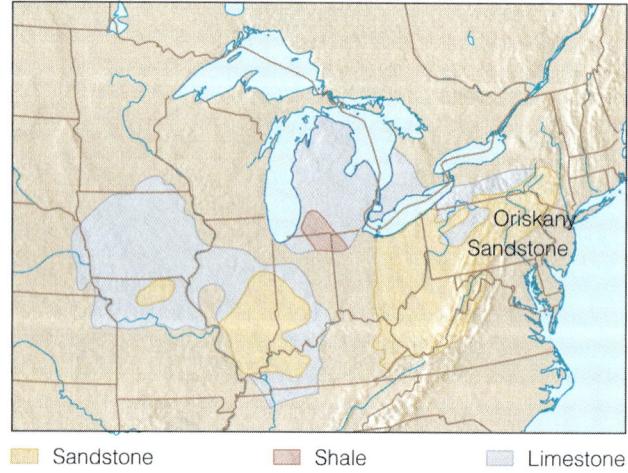

• **Figure 11.3 Basal Rocks of the Kaskaskia Sequence** Extent of the basal rocks of the Kaskaskia Sequence in the eastern and north–central United States.

precipitation (Figure 11.5). In the back-reef area, up to 300 m of evaporites precipitated in much the same way as in the Michigan Basin during the Silurian (see Figure 10.11). More than half the world's potash, which is used in fertilizers, comes from these Devonian evaporites. By the middle of the Late Devonian, reef growth had stopped in the western Canada region, although nonreef carbonate deposition continued.

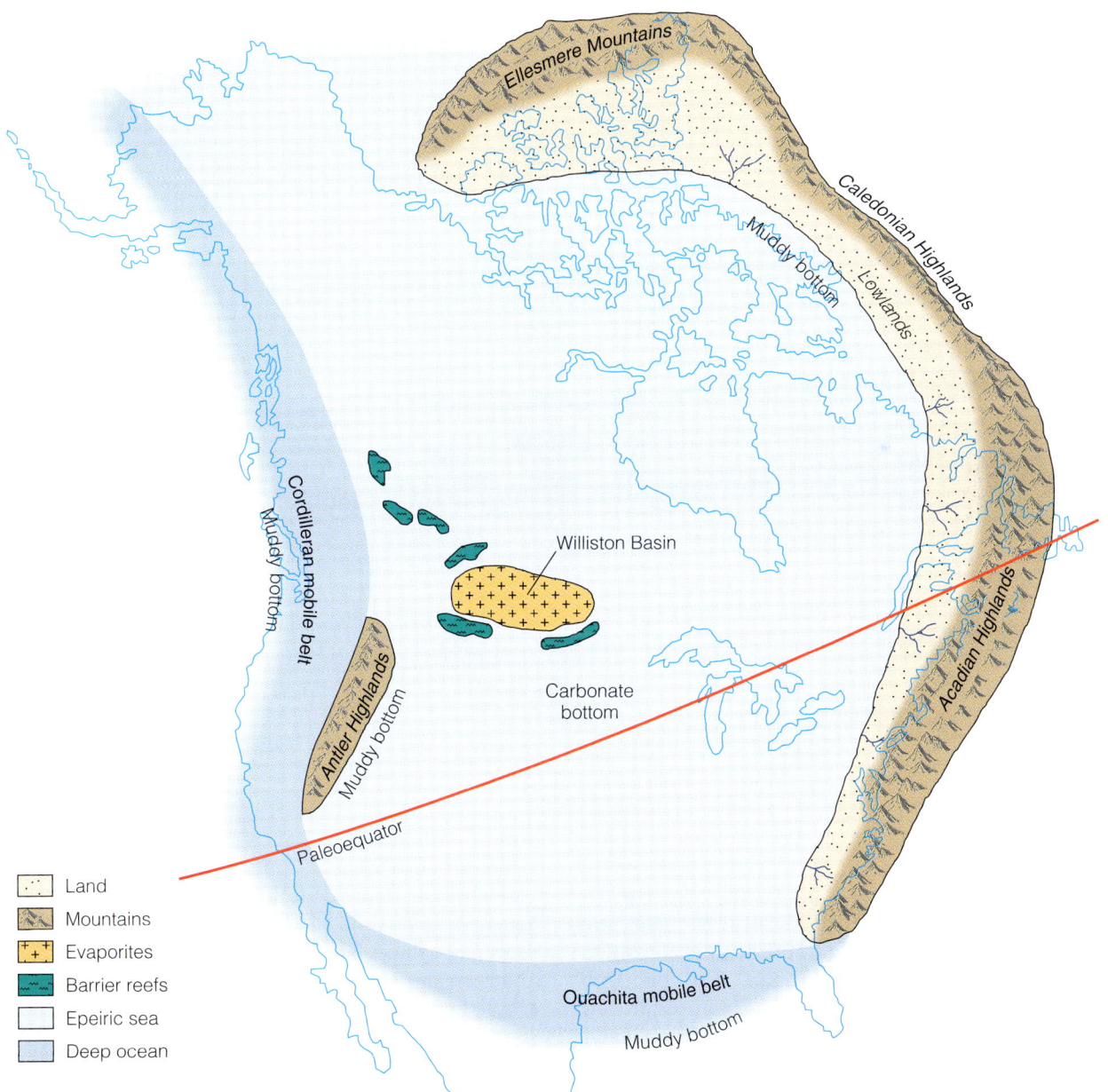

• Figure 11.4 Devonian Paleogeography of North America

Perspective 11.1

The Canning Basin, Australia— A Devonian Great Barrier Reef

Rising majestically 50 to 100 meters above the surrounding plains, the Great Barrier Reef of the Canning Basin, Australia, is one of the largest and most spectacularly exposed fossil reef complexes in the world (Figure 1). This barrier reef complex developed during the Middle and Late Devonian Period, when a tropical epeiric sea covered the Canning Basin (Figure 11.1a).

The reefs themselves were constructed primarily by calcareous algae, stromatoporoids, and various corals, which also were the main components of other large reef complexes in the world at that time. Exposures along Windjana Gorge reveal the various features and facies of the Devonian Great Barrier Reef complex (Figure 2).

On the seaward side of the reef core is a steep fore-reef slope (see Figure 10.9), where such organisms as algae, sponges, and stromatoporoids lived. This facies contains considerable reef talus, an accumulation of debris eroded by waves from the reef front.

The reef core itself consists of unbedded limestones (see Figure 10.9) consisting largely of calcareous algae, stromatoporoids, and corals. The back-reef facies is bedded and is the major part of the total reef complex (see Figure 10.9). In this lagoonal environment lived a diverse and abundant assemblage of calcareous algae, stromatoporoids, corals, bivalves, gastropods, cephalopods, brachiopods, and crinoids.

Near the end of the Late Devonian, almost all the reef-building organisms—as well as much of the associated fauna of the Canning Basin Great Barrier Reef and other large barrier reef complexes—became extinct. As we will discuss in Chapter 12, few massive tabulate-rugose-stromatoporoid reefs are known from latest Devonian or younger rocks anywhere in the world.

Figure 1 Aerial view of Windjana Gorge showing the Devonian Great Barrier Reef.

Figure 2 Outcrop of the Devonian Great Barrier Reef along Windjana Gorge. The talus of the fore-reef can be seen on the left side of the picture sloping away from the reef core, which is unbedded. To the right of the reef core is the back-reef facies that is horizontally bedded.

Black Shales

In North America, many areas of carbonate–evaporite deposition gave way to a greater proportion of shales and coarser detrital rocks beginning in the Middle Devonian and continuing into the Late Devonian. This change to detrital deposition resulted from the formation of new source areas brought on by the mountain-building activity associated with the Acadian orogeny in North America (Figure 11.4).

As the Devonian Period ended, a conspicuous change in sedimentation took place over the North American craton with the appearance of widespread black shales (• Figure 11.6a). These Upper Devonian–Lower Mississippian black shales are typically noncalcareous, thinly bedded, and usually less than 10 m thick (Figure 11.6b). Because most black shales lack body fossils, they are difficult to date and correlate. However, in places where conodonts (microscopic animals), acritarchs (microscopic algae), or plant spores are found, the lower beds are Late Devonian, and the upper beds are Early Mississippian in age.

Although the origin of these extensive black shales is still being debated, the essential features required to produce them include undisturbed anaerobic bottom water, a reduced

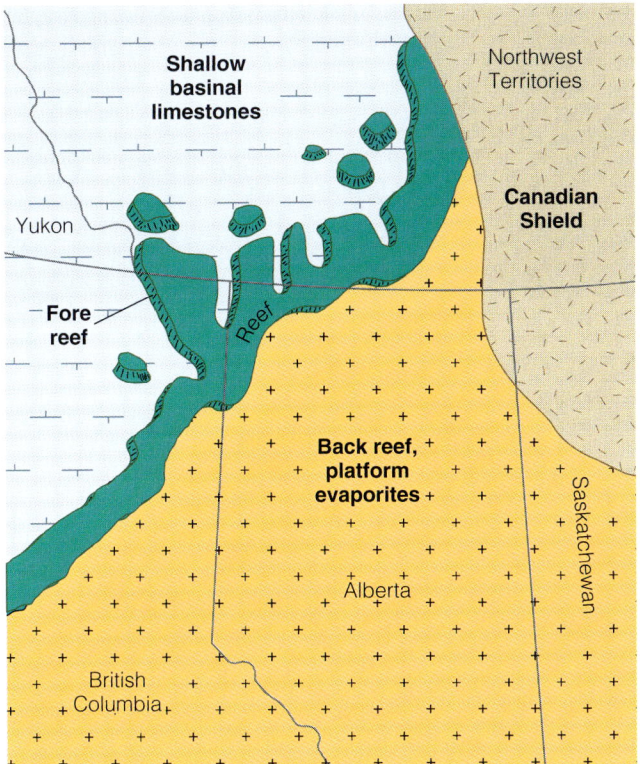

• **Figure 11.5 Devonian Reef Complex of Western Canada** Reconstruction of the extensive Devonian Reef complex of western Canada. These extensive reefs controlled the regional facies of the Devonian epeiric seas.

supply of coarser detrital sediment, and high organic productivity in the overlying oxygenated waters. High productivity in the surface waters leads to a shower of organic material, which decomposes on the undisturbed seafloor and depletes the dissolved oxygen at the sediment–water interface.

The wide extent of such apparently shallow-water black shales in North America remains puzzling. Nonetheless, these shales are rich in uranium and are an important potential source rock for oil and gas in the Appalachian region.

The Late Kaskaskia—A Return to Extensive Carbonate Deposition

Following deposition of the widespread Upper Devonian–Lower Mississippian black shales, carbonate sedimentation on the craton dominated the remainder of the Mississippian Period (• Figure 11.7). During this time, a variety of carbonate sediments was deposited in the epeiric sea, as indicated by the extensive deposits of crinoidal limestones (rich in crinoid fragments), oolitic limestones, and various other limestones and dolostones. These Mississippian carbonates display cross-bedding, ripple marks, and well-sorted fossil fragments, all of which indicate a shallow-water environment. Analogous features can be observed on the present-day Bahama Banks. In addition, numerous small organic reefs occurred throughout the craton during

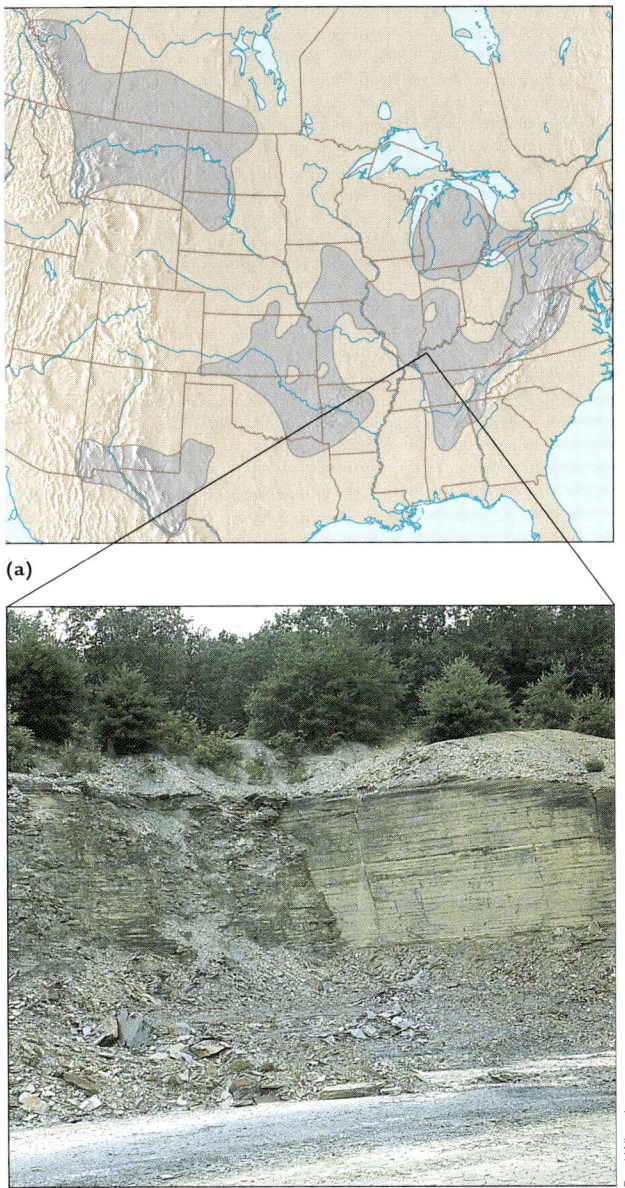

• **Figure 11.6 Upper Devonian–Lower Mississippian Black Shales** (a) The extent of the Upper Devonian to Lower Mississippian Chattanooga Shale and its equivalent units (such as the Antrim Shale and New Albany Shale) in North America. (b) Upper Devonian New Albany Shale, Button Mold Knob Quarry, Kentucky.

the Mississippian. These were all much smaller than the large barrier reef complexes that dominated the earlier Paleozoic seas.

During the Late Mississippian regression of the Kaskaskia Sea from the craton, vast quantities of detrital sediments replaced carbonate deposition. The resulting sandstones, particularly in the Illinois Basin, have been studied in great detail because they are excellent petroleum reservoirs. Before the end of the Mississippian, the epeiric sea had retreated to the craton margin, once again exposing the craton to widespread weathering and erosion resulting in a craton-wide unconformity at the end of the Kaskaskia Sequence.

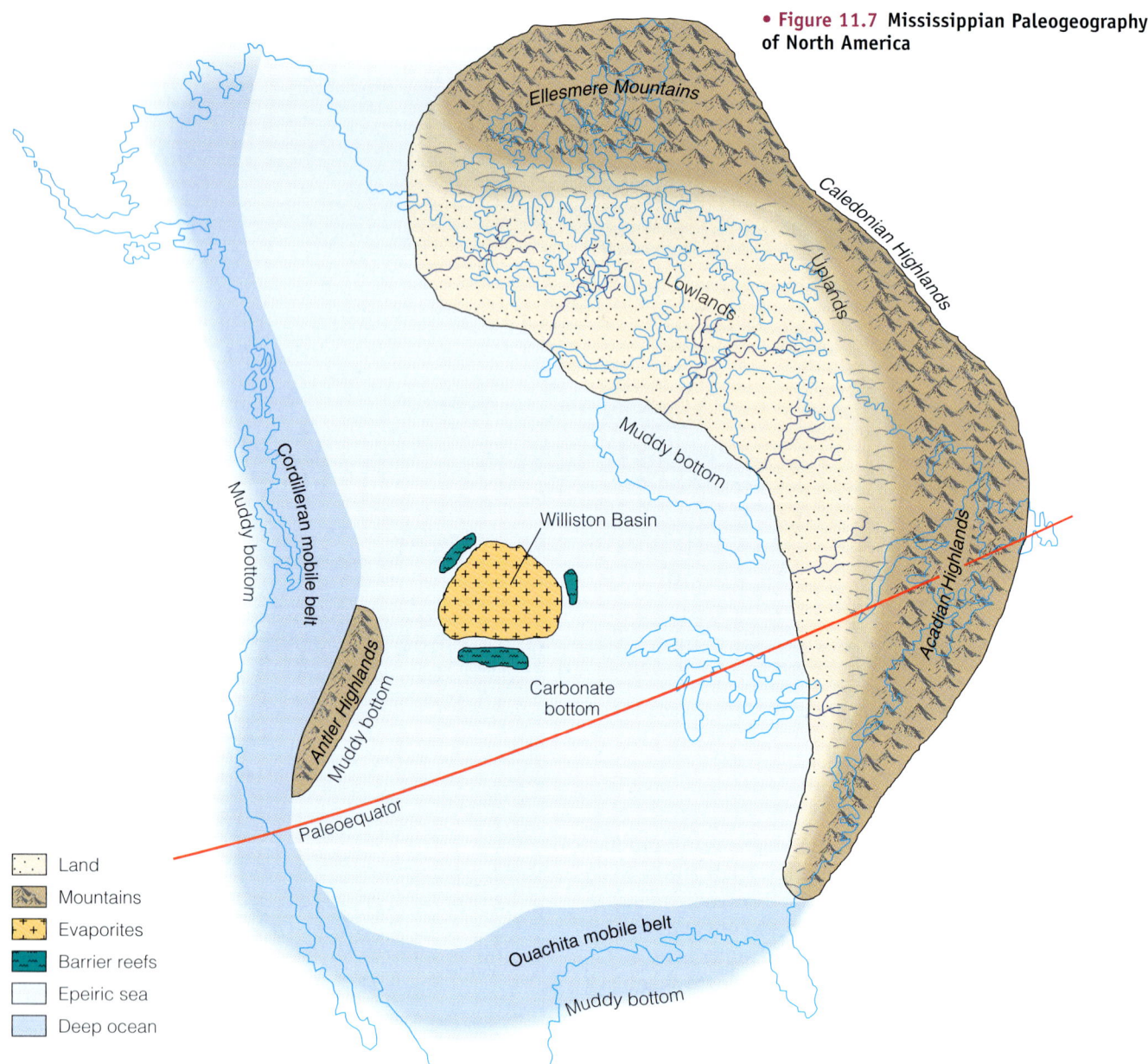

• Figure 11.7 **Mississippian Paleogeography of North America**

The Absaroka Sequence

The **Absaroka Sequence** includes rocks deposited during the Pennsylvanian through Early Jurassic. In this chapter, however, we are concerned only with the Paleozoic rocks of the Absaroka Sequence. The extensive unconformity separating the Kaskaskia and Absaroka sequences essentially divides the strata into the North American Mississippian and Pennsylvanian systems. These two systems are closely equivalent to the European Lower and Upper Carboniferous systems, respectively. The rocks of the Absaroka Sequence not only differ from those of the Kaskaskia Sequence but also result from different tectonic regimes.

The lowermost sediments of the Absaroka Sequence are confined to the margins of the craton. These deposits are generally thickest in the east and southeast, near the emerging highlands of the Appalachian and Ouachita mobile belts, and thin westward onto the craton. The rocks also reveal lateral changes from nonmarine detrital rocks and coals in the east, through transitional marine–nonmarine beds, to largely marine detrital rocks and limestones farther west (• Figure 11.8).

What Are Cyclothems, and Why Are They Important?

One characteristic feature of Pennsylvanian rocks is their repetitive pattern of alternating marine and nonmarine strata. Such repetitive sedimentary sequences are called **cyclothems.** They result from repeated alternations of marine and nonmarine environments, usually in areas of low relief. Although seemingly simple, cyclothems reflect a

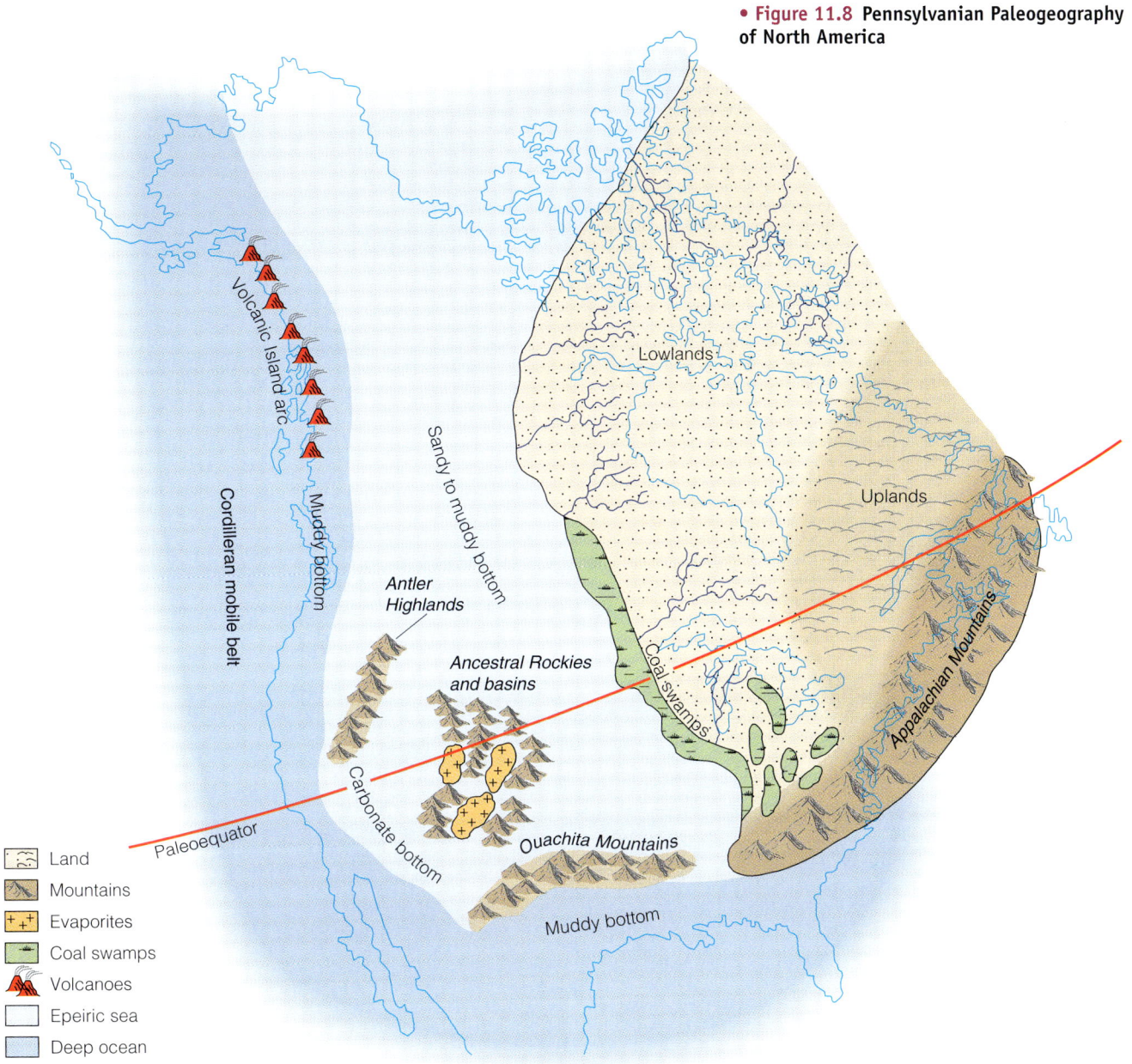

• Figure 11.8 Pennsylvanian Paleogeography of North America

delicate interplay between nonmarine deltaic and shallow-marine interdeltaic and shelf environments.

For illustration, look at a typical coal-bearing cyclothem from the Illinois Basin (• Figure 11.9a). Such a cyclothem contains nonmarine units, capped by a coal and overlain by marine units. Figure 11.9a shows the depositional environments that produced the cyclothem. The initial units represent deltaic and fluvial deposits. Above them is an underclay that frequently contains root casts from the plants and trees that comprise the overlying coal. The coal bed results from accumulations of plant material and is overlain by marine units of alternating limestones and shales, usually with an abundant marine invertebrate fauna. The marine interval ends with an erosion surface. A new cyclothem begins with a nonmarine deltaic sandstone. All the beds illustrated in the idealized cyclothems are not always preserved because of abrupt changes from marine to nonmarine conditions or removal of some units by erosion.

Cyclothems represent transgressive and regressive sequences with an erosional surface separating one cyclothem from another. Thus an idealized cyclothem passes upward from fluvial-deltaic deposits, through coals, to detrital shallow-water marine sediments, and finally to limestones typical of an open marine environment.

Such places as the Mississippi delta, the Florida Everglades, and the Dutch lowlands represent modern coal-forming environments similar to those existing during the Pennsylvanian Period (Figure 11.9d). By studying these modern analogs, geologists can make reasonable deductions about conditions existing in the geologic past.

THE ABSAROKA SEQUENCE   219

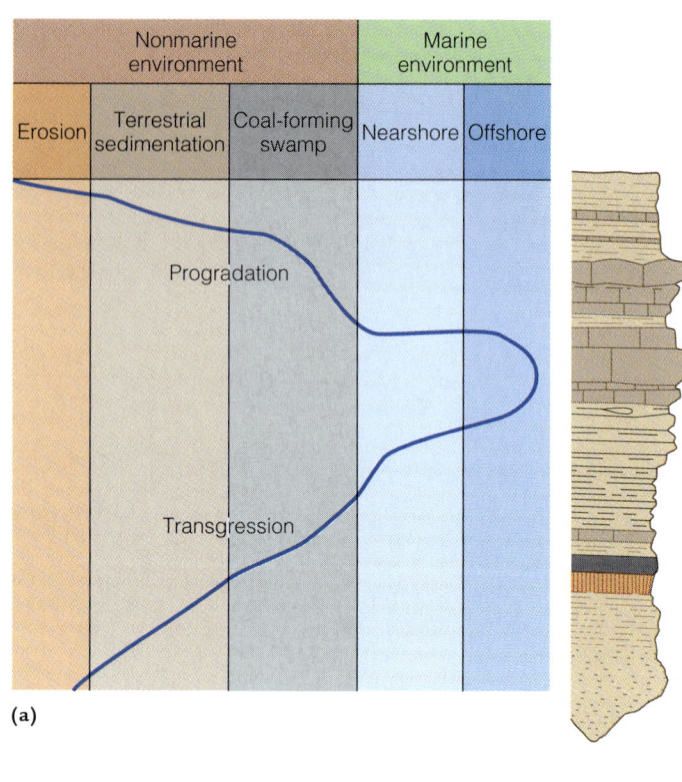

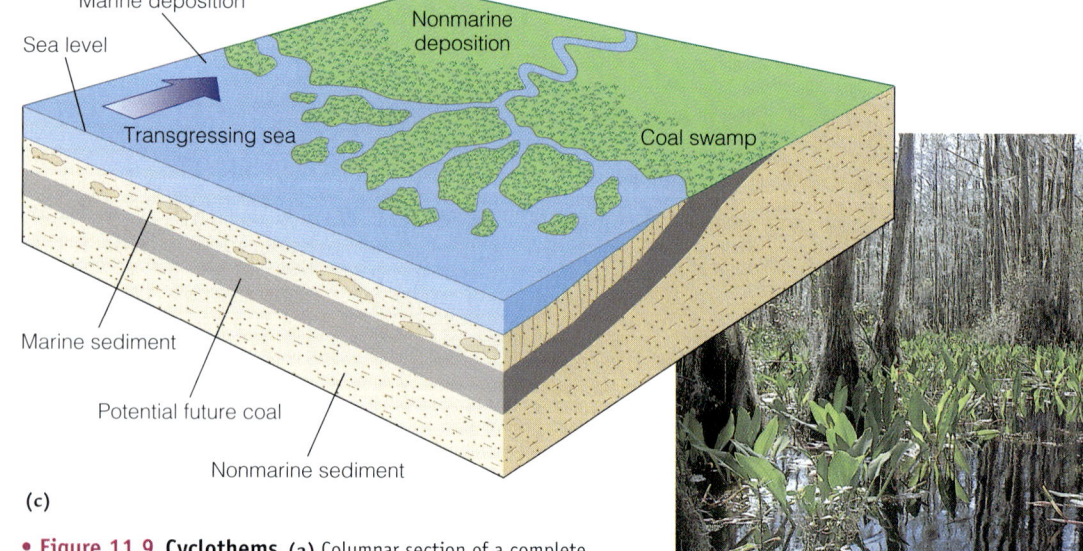

- **Figure 11.9 Cyclothems** (a) Columnar section of a complete cyclothem. (b) Pennsylvanian coal bed, West Virginia. (c) Reconstruction of the environment of a Pennsylvanian coal-forming swamp. (d) The Okefenokee Swamp, Georgia, is a modern example of a coal-forming environment, similar to those occurring during the Pennsylvanian Period.

The Pennsylvanian coal swamps must have been widespread lowland areas with little topographic relief neighboring the sea (Figure 11.9c). In such cases, a very slight rise in sea level would have flooded these large lowland areas, whereas slight drops would have exposed large areas, resulting in alternating marine and nonmarine environments. The same result could also have been caused by a combination of rising sea level and progradation (the seaward extension of a delta by the accumulation of sediment) of a large delta, such as occurs today in Louisiana.

Such repetitious sedimentation over a widespread area requires an explanation. In most cases, local cyclothems of limited extent can be explained by rapid but slight changes in sea level in a swamp-delta complex of low relief near the sea such as by progradation or by localized crustal movement.

Explaining widespread cyclothems is more difficult. The hypothesis currently favored by many geologists is a rise and fall of sea level related to advances and retreats of Gondwanan continental glaciers. When the Gondwanan ice sheets advanced, sea level dropped; when they melted, sea level rose. Late Paleozoic cyclothem activity on all the cratons closely corresponds to Gondwanan glacial–interglacial cycles.

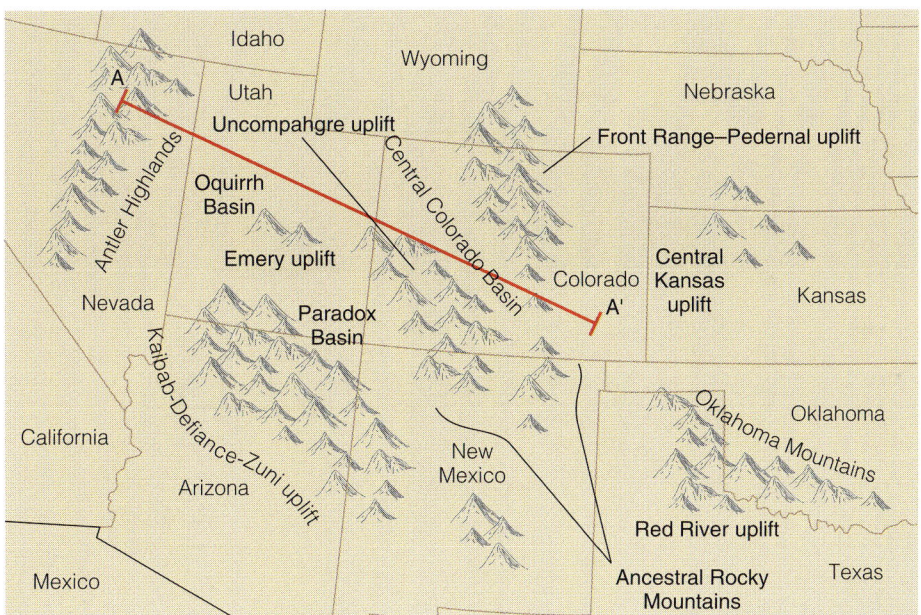

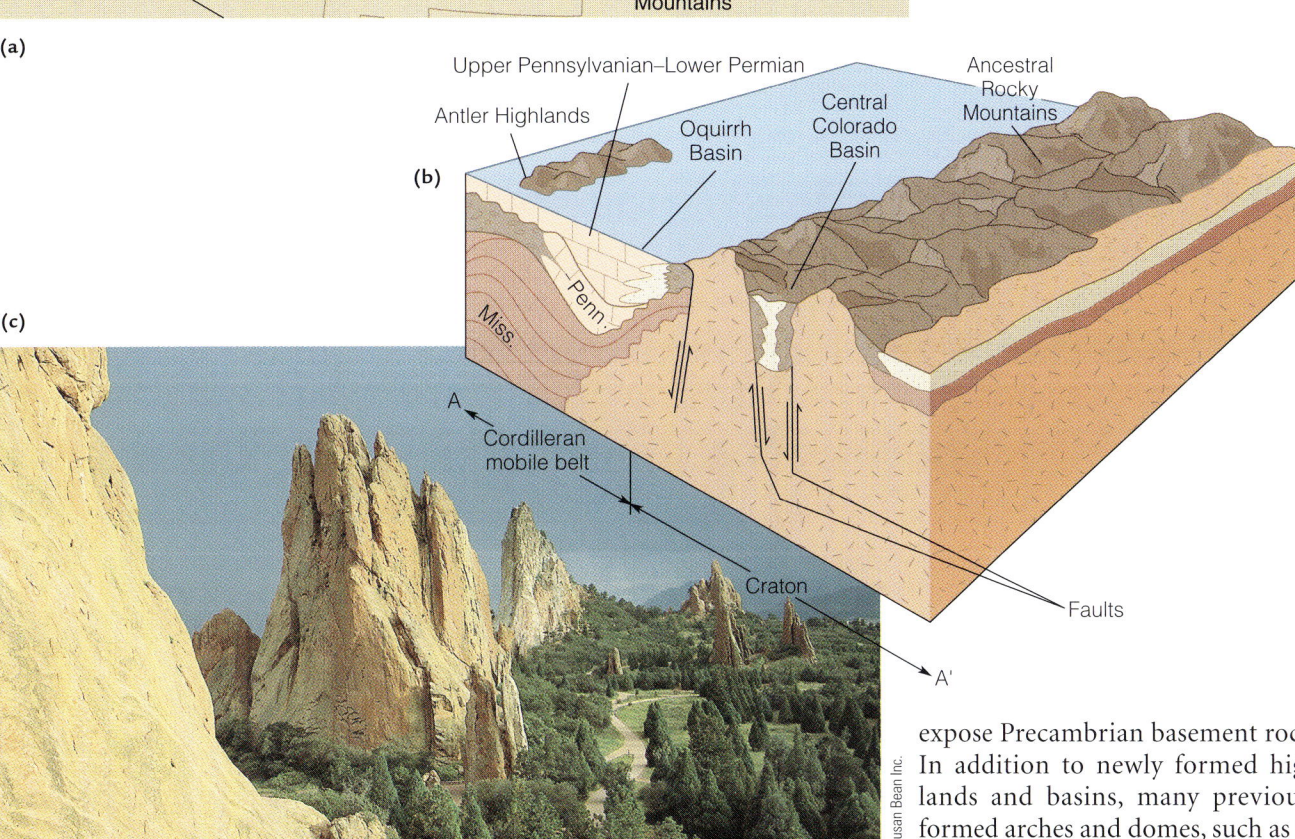

- **Figure 11.10 The Ancestral Rockies** (a) Location of the principal Pennsylvanian highland areas and basins of the southwestern part of the craton. (b) Block diagram of the Ancestral Rockies, elevated by faulting during the Pennsylvanian Period. Erosion of these mountains produced coarse, red sediments deposited in the adjacent basins. (c) Garden of the Gods, storm sky view from Near Hidden Inn, Colorado Springs, Colorado.

Cratonic Uplift— The Ancestral Rockies

Recall that cratons are stable areas, and what deformation they do experience is usually mild. The Pennsylvanian Period, however, was a time of unusually severe cratonic deformation, resulting in uplifts of sufficient magnitude to expose Precambrian basement rocks. In addition to newly formed highlands and basins, many previously formed arches and domes, such as the Cincinnati Arch, Nashville Dome, and Ozark Dome, were also reactivated (see Figure 10.1).

During the Pennsylvanian Period, the area of greatest deformation was in the southwestern part of the North American craton, where a series of fault-bounded uplifted blocks formed the **Ancestral Rockies** (• Figure 11.10a). Uplift of these mountains, some of which were elevated more than 2 km along near-vertical faults, resulted in erosion of overlying Paleozoic sediments and exposure of the Precambrian igneous and metamorphic basement rocks (Figure 11.10b). As the

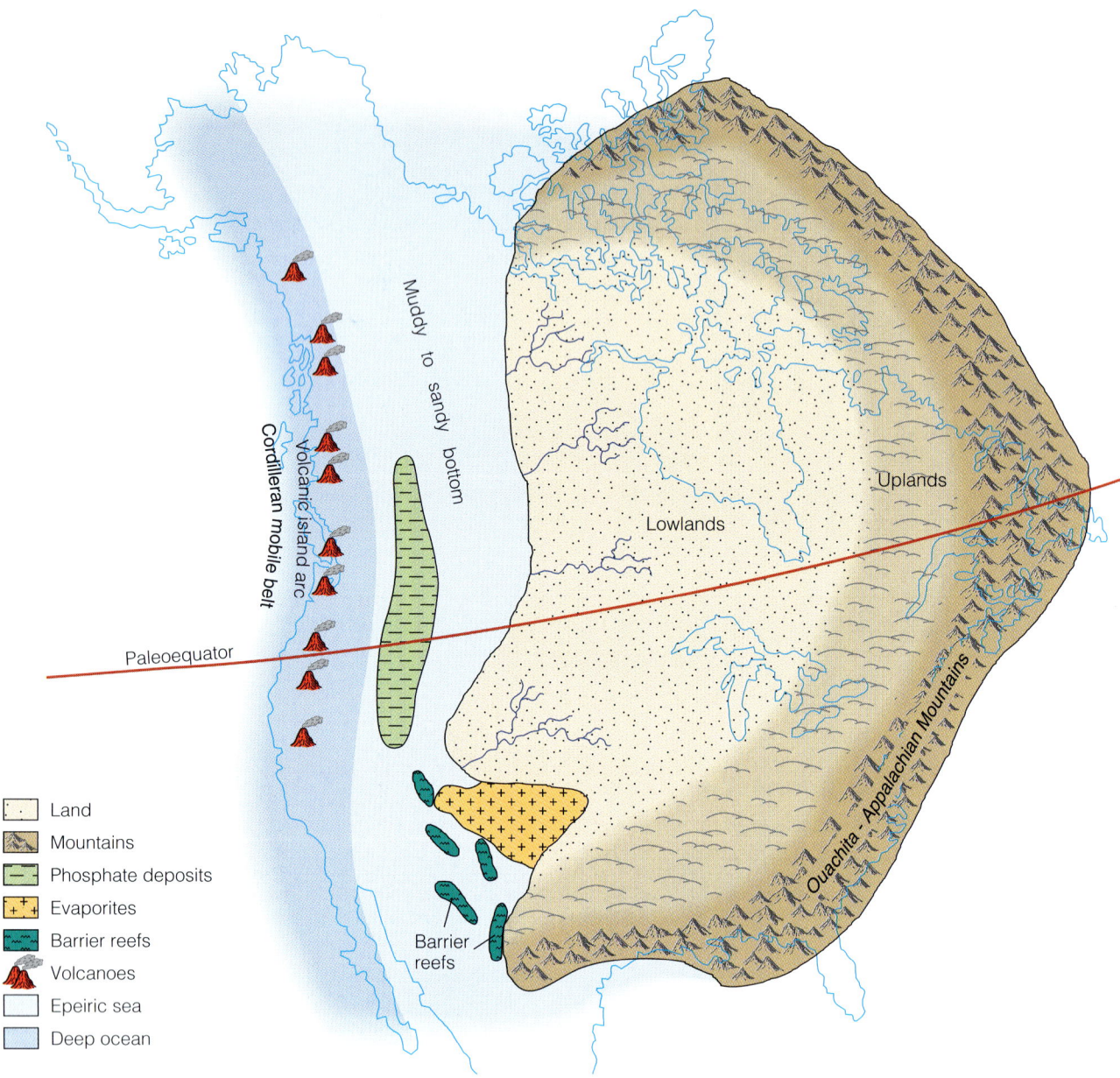

• Figure 11.11 Permian Paleogeography of North America

mountains eroded, tremendous quantities of coarse, red arkosic sand and conglomerate were deposited in the surrounding basins. These sediments are preserved in many areas, including the rocks of the Garden of the Gods near Colorado Springs (Figure 11.10c) and at the Red Rocks Amphitheatre near Morrison, Colorado.

Intracratonic mountain ranges are unusual, and their cause has long been debated. It is currently thought that the collision of Gondwana with Laurasia along the Ouachita mobile belt (Figure 11.2a) generated great stresses in the southwestern region of the North American craton. These crustal stresses were relieved by faulting. Movement along these faults produced uplifted cratonic blocks and downwarped adjacent basins, forming a series of related ranges and basins.

The Middle Absaroka— More Evaporite Deposits and Reefs

While the various intracratonic basins were filling with sediment during the Late Pennsylvanian, the epeiric sea slowly began retreating from the craton. During the Early Permian, the Absaroka Sea occupied a narrow region from Nebraska through west Texas (• Figure 11.11). By the Middle Permian, it had retreated to west Texas and southern

• **Figure 11.12 West Texas Permian Basins and Surrounding Reefs** During the Middle and Late Permian, an interrelated complex of lagoonal, barrier reef, and open-shelf environments formed in the west Texas and southern New Mexico area. Much of the tremendous oil production in this region comes from these reefs.

New Mexico. The thick evaporite deposits in Kansas and Oklahoma show the restricted nature of the Absaroka Sea during the Early and Middle Permian and its southwestward retreat from the central craton.

During the Middle and Late Permian, the Absaroka Sea was restricted to west Texas and southern New Mexico, forming an interrelated complex of lagoonal, reef, and open-shelf environments (• Figure 11.12). Three basins separated by two submerged platforms developed in this area during the Permian. Massive reefs grew around the basin margins (• Figure 11.13), and limestones, evaporites, and red beds were deposited in the lagoonal areas behind the reefs. As the barrier reefs grew and the passageways between the basins became more restricted, Late Permian evaporites gradually filled the individual basins.

Spectacular deposits representing the geologic history of this region can be seen today in the Guadalupe Mountains of Texas and New Mexico where the Capitan Limestone forms the caprock of these mountains (• Figure 11.14). These reefs have been extensively studied because tremendous oil production comes from this region.

By the end of the Permian Period, the epeiric sea had retreated from the craton, exposing continental red beds over most of the southwestern and eastern region (Figure 11.2b).

• **Figure 11.13 Middle Permian Capitan Limestone Reef Environment** A reconstruction of the Middle Permian Capitan Limestone reef environment. Shown are brachiopods, corals, bryozoans, and large glass sponges.

• **Figure 11.14 Guadalupe Mountains, Texas** The prominent Capitan Limestone forms the caprock of the Guadalupe Mountains. The Capitan Limestone is rich in fossil corals and associated reef organisms.

History of the Late Paleozoic Mobile Belts

Having examined the Kaskaskian and Absarokian history of the craton, let's now look at the orogenic activity in the mobile belts. The mountain building during this time profoundly influenced the climatic and sedimentary history of the craton. In addition, it was part of the global tectonic regime that formed Pangaea.

Cordilleran Mobile Belt

During the Neoproterozoic and Early Paleozoic, the Cordilleran area was a passive continental margin along which extensive continental shelf sediments were deposited (see Figures 9.7 and 10.5). Thick sections of marine sediments graded laterally into thin cratonic units as the Sauk Sea transgressed onto the craton. Beginning in the Middle Paleozoic, an island arc formed off the western margin of the craton. A collision between this eastward-moving island arc and the western border of the craton took place during the Late Devonian and Early Mississippian, producing a highland area (• Figure 11.15).

This orogenic event, the *Antler orogeny,* was caused by subduction resulting in deep water deposits and oceanic crustal rocks being thrust eastward over shallow-water continental shelf sediments, thus closing the narrow ocean basin separating the island arc from the craton. Erosion of the resulting Antler Highlands produced large quantities of sediment that were deposited to the east in the epeiric sea covering the craton and to the west in the deep sea. The Antler orogeny was the first in a series of orogenic events to affect the Cordilleran mobile belt. During the Mesozoic and Cenozoic, this area was the site of major tectonic activity caused by oceanic–continental convergence and accretion of various terranes.

Ouachita Mobile Belt

The Ouachita mobile belt extends for approximately 2100 km from the subsurface of Mississippi to the Marathon region of Texas. Approximately 80% of the former mobile belt is buried beneath a Mesozoic and Cenozoic sedimentary cover. The two major exposed areas in this region are the Ouachita Mountains of Oklahoma and Arkansas and the Marathon Mountains of Texas.

During the Late Proterozoic to Early Mississippian, shallow-water detrital and carbonate sediments were deposited on a broad continental shelf, and in the deeper-water portion of the adjoining mobile belt, bedded cherts and shales were accumulating (• Figure 11.16a). Beginning in the Mississippian Period, the sedimentation rate increased dramatically as the region changed from a passive continental margin to an active convergent plate boundary, marking the beginning of the Ouachita orogeny (Figure 11.16b).

Thrusting of sediments continued throughout the Pennsylvanian and Early Permian, driven by the compressive forces generated along the zone of subduction as Gondwana collided with Laurasia (Figure 11.16c). The collision of Gondwana and Laurasia formed a large mountain range, most of which eroded during the Mesozoic Era. Only the rejuvenated Ouachita and Marathon Mountains remain of this once lofty mountain range.

The Ouachita deformation was part of the general worldwide tectonic activity that occurred when Gondwana united with Laurasia. The Hercynian, Appalachian, and Ouachita mobile belts were continuous and marked the southern boundary of Laurasia (Figure 11.2). The tectonic activity that uplifted the Ouachita mobile belt was very complex and involved not only the collision of Laurasia and Gondwana but also several microplates and terranes between the continents that eventually became part of Central America. The compressive forces impinging on the Ouachita mobile belt also affected the craton by broadly uplifting the southwestern part of North America.

Appalachian Mobile Belt

Caledonian Orogeny The Caledonian mobile belt stretches along the western border of Baltica and includes the present-day countries of Scotland, Ireland, and Norway (see Figure 10.2c). During the Middle Ordovician, subduction along the boundary between the Iapetus plate and Baltica (Europe) began, forming a mirror image of the convergent plate boundary off the east coast of Laurentia (North America).

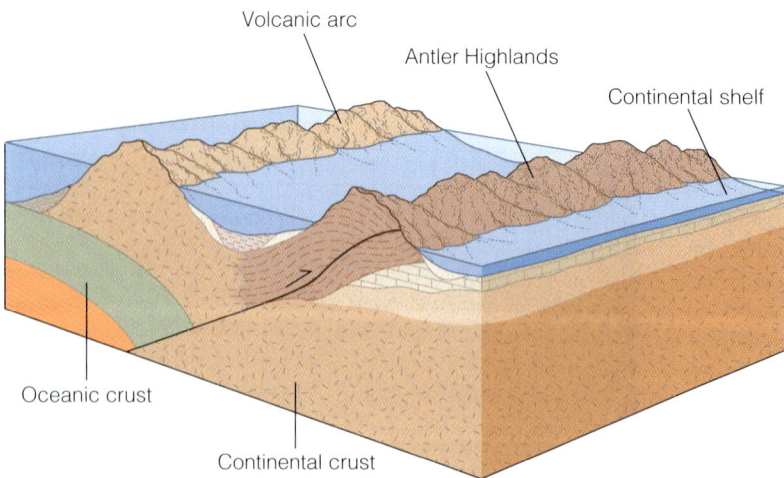

• **Figure 11.15 Antler Orogeny** Reconstruction of the Cordilleran mobile belt during the Early Mississippian in which deep-water continental slope deposits were thrust eastward over shallow-water continental shelf carbonates, forming the Antler Highlands.

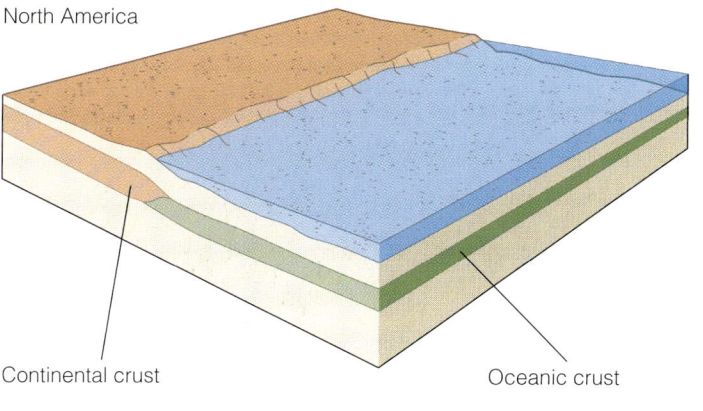
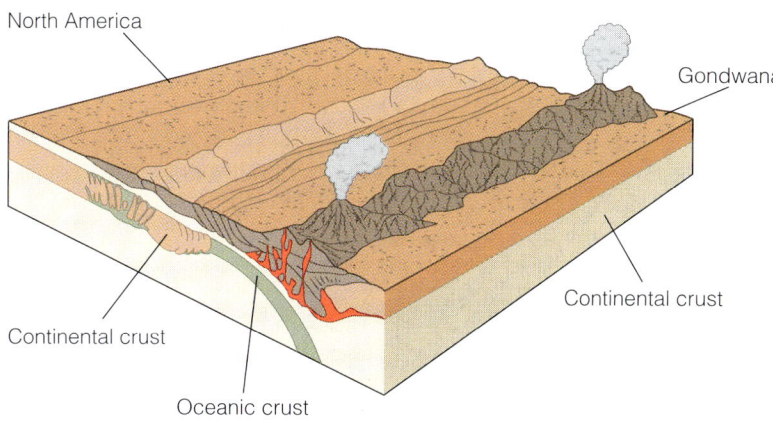
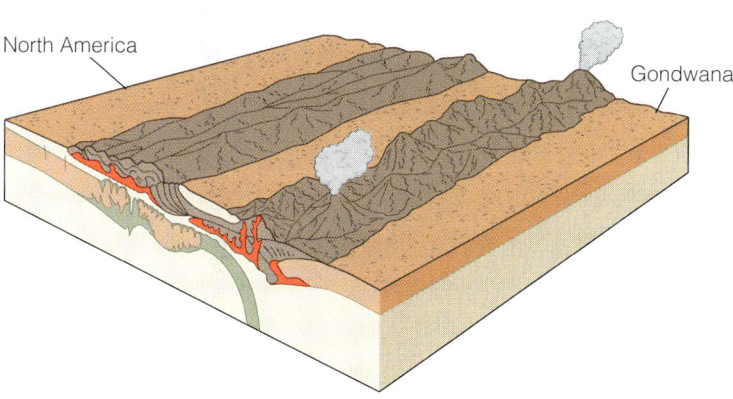

• **Figure 11.16 Ouachita Mobile Belt** Plate tectonic model for deformation of the Ouachita mobile belt. **(a)** Depositional environment prior to the beginning of orogenic activity. **(b)** Incipient continental collision between North America and Gondwana began during the Mississippian Period. **(c)** Continental collision continued during the Pennsylvanian and Permian periods.

The *Caledonian orogeny* (mentioned earlier) culminated during the Late Silurian and Early Devonian with the formation of a mountain range along the western margin of Baltica (see Figure 10.2c).

Acadian Orogeny The third Paleozoic orogeny to affect Laurentia and Baltica began during the Late Silurian and concluded at the end of the Devonian Period. The *Acadian orogeny* (mentioned earlier) affected the Appalachian mobile belt from Newfoundland to Pennsylvania as sedimentary rocks were folded and thrust against the craton.

As with the preceding Taconic and Caledonian orogenies, the Acadian orogeny occurred along an oceanic–continental convergent plate boundary. As the northern Iapetus Ocean continued to close during the Devonian, the plate carrying Baltica finally collided with Laurentia, forming a continental–continental convergent plate boundary along the zone of collision (Figure 11.1a).

As the increased metamorphic and igneous activity indicates, the Acadian orogeny was more intense and lasted longer than the Taconic orogeny. Radiometric dates from the metamorphic and igneous rocks associated with the Acadian orogeny cluster between 360 and 410 million years ago. Just as with the Taconic orogeny, deepwater sediments were folded and thrust northwestward during the Acadian orogeny, producing angular unconformities separating Upper Silurian from Mississippian rocks.

Weathering and erosion of the Acadian Highlands produced the **Catskill Delta,** a thick clastic wedge named for the Catskill Mountains in upstate New York, where it is well exposed. The Catskill Delta, composed of red, coarse conglomerates, sandstones, and shales, contains nearly three times as much sediment as the Queenston Delta (see Figure 10.14).

The Devonian rocks of New York are among the best studied on the continent. A cross section of the Devonian strata clearly reflects an eastern source (Acadian Highlands) for the Catskill facies (• Figure 11.17). These detrital rocks can be traced from eastern Pennsylvania, where the coarse-grained deposits are approximately 3 km thick, to Ohio, where the deltaic facies are only about 100 m thick and consist of cratonic shales and carbonates.

What Would You Do?

Your daughter has asked you to come to her high school science class and explain how the Appalachian Mountains formed as part of the class unit on local geology. The class is just beginning their geology studies and has not yet been introduced to plate tectonics as the unifying theory of geology. How will you explain the Paleozoic history of the Appalachians in a manner that is not only entertaining to high school students but also informative?

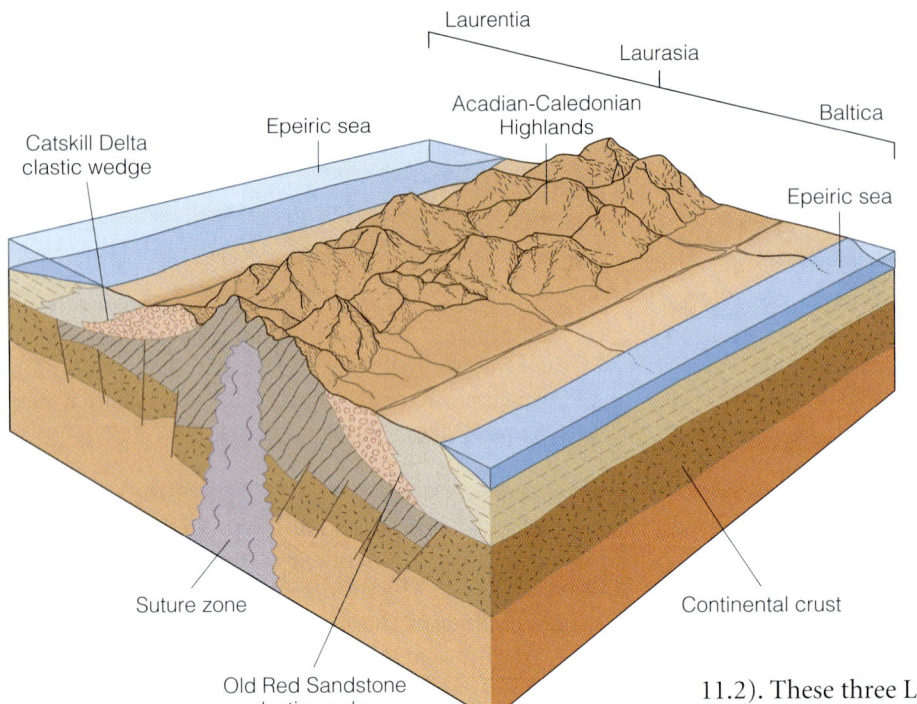

• **Figure 11.17 Formation of Laurasia** Block diagram showing the area of collision between Laurentia and Baltica. Note the bilateral symmetry of the Catskill Delta clastic wedge and the Old Red Sandstone and their relationship to the Acadian and Caledonian Highlands.

Hercynian–Alleghenian Orogeny The Hercynian mobile belt of southern Europe and the Appalachian and Ouachita mobile belts of North America mark the zone along which Europe (part of Laurasia) collided with Gondwana (Figure 11.1). While Gondwana and southern Laurasia collided during the Pennsylvanian and Permian in the area of the Ouachita mobile belt, eastern Laurasia (Europe and southeastern North America) joined together with Gondwana (Africa) as part of the *Hercynian–Alleghenian orogeny* (Figure 11.2). These three Late Paleozoic orogenies (Hercynian, Alleghenian, and Ouachita) represent the final joining of Laurasia and Gondwana into the supercontinent Pangaea during the Permian.

The red beds of the Catskill Delta derive their color from the hematite in the sediments. Plant fossils and oxidation of the hematite indicate the beds were deposited in a continental environment.

The Old Red Sandstone The red beds of the Catskill Delta have a European counterpart in the Devonian Old Red Sandstone of the British Isles (Figure 11.17). The Old Red Sandstone was a Devonian clastic wedge that grew eastward from the Caledonian Highlands onto the Baltica craton. The Old Red Sandstone, just like its North American Catskill counterpart, contains numerous fossils of freshwater fish, early amphibians, and land plants.

By the end of the Devonian Period, Baltica and Laurentia were sutured together, forming Laurasia (Figure 11.17). The red beds of the Catskill Delta can be traced north, through Canada and Greenland, to the Old Red Sandstone of the British Isles and into northern Europe. These beds were deposited in similar environments along the flanks of developing mountain chains formed at convergent plate boundaries.

The Taconic, Caledonian, and Acadian orogenies were all part of the same major orogenic event related to the closing of the Iapetus Ocean (see Figures 10.13b and 11.17). This event began with paired oceanic–continental convergent plate boundaries during the Taconic and Caledonian orogenies and culminated along a continental–continental convergent plate boundary during the Acadian orogeny as Laurentia and Baltica became sutured. After this, the Hercynian–Alleghenian orogeny began, followed by orogenic activity in the Ouachita mobile belt.

What Role Did Microplates and Terranes Play in the Formation of Pangaea?

We have presented the geologic history of the mobile belts bordering the Paleozoic continents in terms of subduction along convergent plate boundaries. It is becoming increasingly clear, however, that accretion along the continental margins is more complicated than the somewhat simple, large-scale plate interactions we have described. Geologists now recognize that numerous terranes or microplates existed during the Paleozoic and were involved in the orogenic events that occurred during that time.

In this chapter and the previous one, we have been concerned only with the six major Paleozoic continents. However, terranes and microplates of varying size were present during the Paleozoic and participated in the formation of Pangaea. For example, as we mentioned in the previous chapter, the microcontinent of *Avalonia* consisted of some coastal parts of New England, southern New Brunswick, much of Nova Scotia, the Avalon Peninsula of eastern Newfoundland, southeastern Ireland, Wales, England, and parts of Belgium and northern France. This microcontinent separated from Gondwana in the Early Ordovician and existed as a separate continent until it collided with Baltica during the Late Ordovician–Early Silurian and then with Laurentia (as part of Baltica) during the Silurian (see Figures 10.2b, 10.2c, and 11.1a).

Other terranes and microplates include *Iberia-Armorica* (a portion of southern France, Sardinia, and most of the Iberian peninsula), *Perunica* (Bohemia), numerous Alpine fragments (especially in Austria), as well as many other bits and pieces of island arcs and suture zones. Not only did these terranes and microplates separate and move away from the larger continental landmasses during the Paleozoic, but they usually developed their own unique faunal and floral assemblages.

Thus, although the basic history of the formation of Pangaea during the Paleozoic remains essentially the same, geologists now realize that microplates and terranes also played an important role in the formation of Pangaea. Furthermore, they help explain some previously anomalous geologic and paleontologic situations.

What Would You Do?

You are the geology team leader of an international mining company. Your company holds the mineral rights on large blocks of acreage in various countries along the west coast of Africa. The leases on these mineral rights will shortly expire, and you've been given the task of evaluating which leases are the most promising. How do you think your knowledge of Paleozoic plate tectonics can help you in these evaluations?

Late Paleozoic Mineral Resources

Late Paleozoic-aged rocks contain a variety of important mineral resources including energy resources and metallic and nonmetallic mineral deposits. Petroleum and natural gas are recovered in commercial quantities from rocks ranging from the Devonian through Permian. For example, Devonian rocks in the Michigan Basin, Illinois Basin, and the Williston Basin of Montana, South Dakota, and adjacent parts of Alberta, Canada, have yielded considerable amounts of hydrocarbons. Permian reefs and other strata in the western United States, particularly Texas, have also been prolific producers.

Although Permian coal beds are known from several areas, including Asia, Africa, and Australia, much of the coal in North America and Europe comes from Pennsylvanian (Upper Carboniferous) deposits. Large areas in the Appalachian region and the midwestern United States are underlain by coal deposits (• Figure 11.18). These coal deposits formed from the lush vegetation that flourished in Pennsylvanian coal-forming swamps (Figure 11.9).

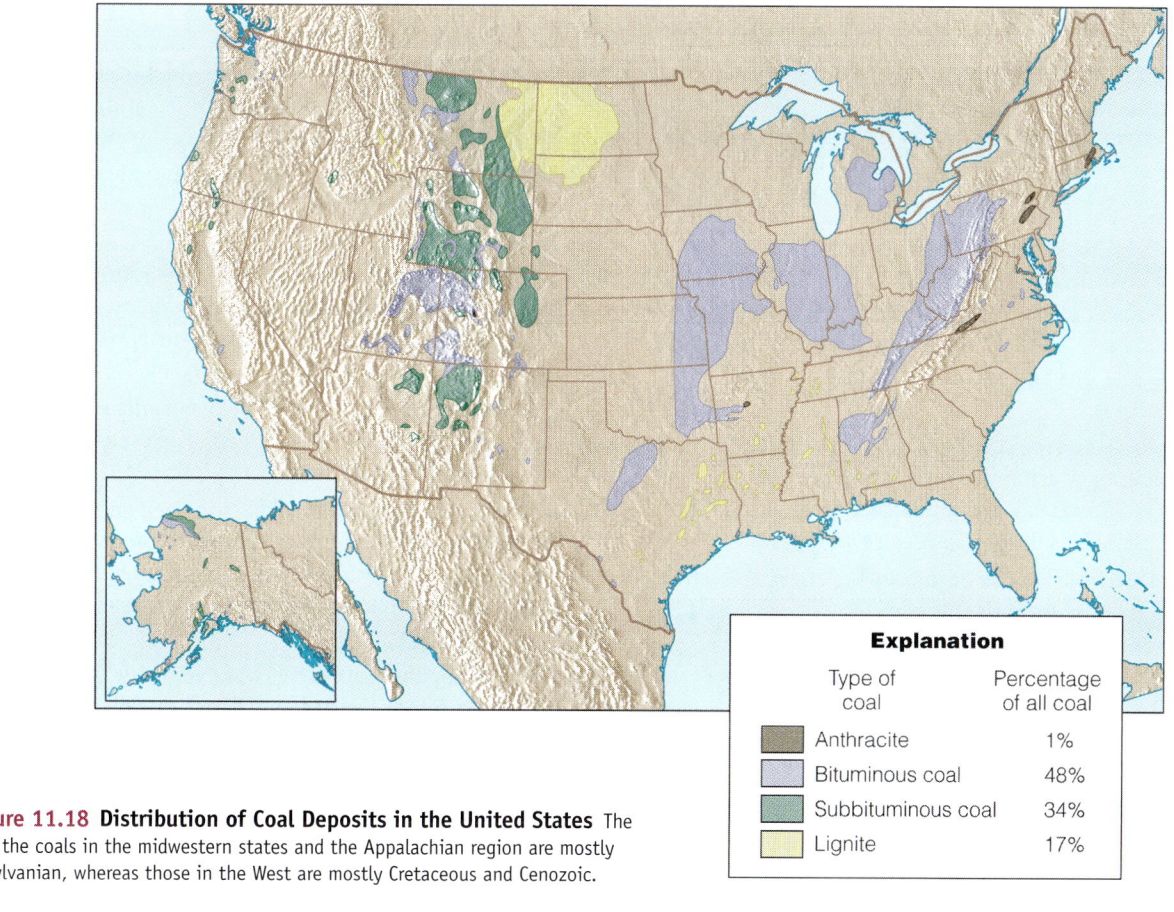

• **Figure 11.18 Distribution of Coal Deposits in the United States** The age of the coals in the midwestern states and the Appalachian region are mostly Pennsylvanian, whereas those in the West are mostly Cretaceous and Cenozoic.

Much of this coal is bituminous coal, which contains about 80% carbon. It is a dense, black coal that has been so thoroughly altered that plant remains can be seen only rarely. Bituminous coal is used to make *coke,* a hard, gray substance made up of the fused ash of bituminous coal. Coke is used to fire blast furnaces for steel production.

Some Pennsylvanian coal from North America is *anthracite,* a metamorphic type of coal containing up to 98% carbon. Most anthracite is in the Appalachian region (Figure 11.18). It is especially desirable because it burns with a smokeless flame and yields more heat per unit volume than other types of coal. Unfortunately, it is the least common type—much of the coal used in the United States is bituminous.

A variety of Late Paleozoic-aged evaporite deposits are important nonmetallic mineral resources. The Zechstein evaporites of Europe extend from Great Britain across the North Sea and into Denmark, the Netherlands, Germany, and eastern Poland and Lithuania. Besides the evaporites themselves, Zechstein deposits form the caprock for the large reservoirs of the gas fields of the Netherlands and part of the North Sea region.

Other important evaporite mineral resources include those of the Permian Delaware Basin of West Texas and New Mexico and Devonian evaporites in the Elk Point Basin of Canada. In Michigan, gypsum is mined and used in the construction of sheetrock.

Most silica sand mined in the United States comes from east of the Mississippi River, and much of this comes from Late Paleozoic-aged rocks. For example, the Devonian Ridgeley Formation is mined in West Virginia, Maryland, and Pennsylvania, and the Devonian Sylvania Sandstone is mined near Toledo, Ohio. Recall from Chapter 10 that silica sand is used in manufacturing glass; for refractory bricks in blast furnaces; for molds for casting aluminum, iron, and copper alloys; and for a variety of other uses.

Late Paleozoic-aged limestones from many areas in North America are used in manufacturing cement. Limestone is also mined and used in blast furnaces for steel production.

Metallic mineral resources including tin, copper, gold, and silver are also known from Late Paleozoic-aged rocks, especially those deformed during mountain building. Although the precise origin of the Missouri lead and zinc deposits remains unresolved, much of the ores of these metals come from Mississippian rocks. In fact, mines in Missouri account for a substantial amount of all domestic production of lead ores.

SUMMARY

Table 11.1 summarizes the geologic history of the North American craton and mobile belts as well as global events, sea level changes, and evolutionary events during the Late Paleozoic.

- During the Late Paleozoic, Baltica and Laurentia collided, forming Laurasia. Siberia and Kazakhstania collided and finally were sutured to Laurasia. Gondwana moved over the South Pole and experienced several glacial–interglacial periods, resulting in global sea level changes and transgressions and regressions along the low-lying craton margins.
- Laurasia and Gondwana underwent a series of collisions beginning in the Carboniferous. During the Permian, the formation of Pangaea was completed. Surrounding the supercontinent was the global ocean Panthalassa.
- The Late Paleozoic history of the North American craton can be deciphered from the rocks of the Kaskaskia and Absaroka sequences.
- The basal beds of the Kaskaskia Sequence that were deposited on the exposed Tippecanoe surface consisted either of sandstones derived from the eroding Taconic Highlands, or of carbonate rocks.
- Most of the Kaskaskia Sequence is dominated by carbonates and associated evaporites. The Devonian Period was a time of major reef building in western Canada, southern England, Belgium, Australia, and Russia.
- Widespread black shales were deposited over large areas of the craton during the Late Devonian and Early Mississippian.
- The Mississippian Period was dominated for the most part by carbonate deposition.
- Transgressions and regressions, probably caused by advancing and retreating Gondwanan ice sheets, over the low-lying North American craton, resulted in cyclothems and the formation of coals during the Pennsylvanian Period.
- Cratonic mountain building, specifically the Ancestral Rockies, occurred during the Pennsylvanian Period and resulted in thick nonmarine detrital sediments and evaporites being deposited in the intervening basins.
- By the Early Permian, the Absaroka Sea occupied a narrow zone of the south–central craton. Here, several large reefs and associated evaporites developed. By the end of the Permian Period, this epeiric sea had retreated from the craton.
- The Cordilleran mobile belt was the site of the Antler orogeny, a minor Devonian orogeny during which deep-water sediments were thrust eastward over shallow-water sediments.
- During the Pennsylvanian and Early Permian, mountain building occurred in the Ouachita mobile belt. This tectonic activity was partly responsible for the cratonic uplift in the southwest, resulting in the Ancestral Rockies.

Table 11.1
Summary of Late Paleozoic Geologic and Evolutionary Events

Age (millions of years)	Geologic Period	Sequence	Relative Changes in Sea Level (Rising / Falling)	Cordilleran Mobile Belt	Craton	Ouachita Mobile Belt	Appalachian Mobile Belt	Major Events Outside North America	Major Evolutionary Events
251	Permian	Absaroka			Deserts, evaporites, and continental red beds in southwestern United States. Extensive reefs in Texas area		Allegheny orogeny	Formation of Pangaea / Hercynian orogeny	Largest mass extinction event to affect the invertebrates. Many vertebrates go extinct. Gymnosperms diverse and abundant
299	Carboniferous – Pennsylvanian	Absaroka			Coal swamps common. Formation of Ancestral Rockies. Transgression of Absaroka Sea	Ouachita orogeny		Continental glaciation in Southern Hemisphere	Amphibians diverse and abundant. Abundant coal swamps with seedless vascular plants
318	Carboniferous – Mississippian	Kaskaskia			Widespread black shales				Reptiles evolve. Gymnosperms evolve
359	Devonian	Kaskaskia / Tippecanoe	←Present sea level	Antler orogeny	Widespread black shales. Catskill Delta clastic wedge. Extensive barrier reef formation in Western Canada. Transgression of Kaskaskia Sea		Acadian orogeny	Old Red Sandstone clastic wedge in British Isles / Caledonian orogeny	Extinction of many reef-building invertebrates. Amphibians evolve. All major groups of fish present—Age of Fish. Early land plants–seedless vascular plants
416									

- The Caledonian, Acadian, Hercynian, and Alleghenian orogenies were all part of the global tectonic activity that assembled Pangaea.
- During the Paleozoic Era, numerous terranes, such as Avalonia, existed and played an important role in forming Pangaea.
- Late Paleozoic-aged rocks contain a variety of mineral resources, including petroleum, coal, evaporites, silica sand, lead, zinc, and other metallic deposits.

IMPORTANT TERMS

Absaroka Sequence, p. 218
Acadian orogeny, p. 211
Alleghenian orogeny, p. 211
Ancestral Rockies, p. 221
Antler orogeny, p. 211
Caledonian orogeny, p. 211
Catskill Delta, p. 225
cyclothem, p. 218
Ellesmere orogeny, p. 211
Hercynian orogeny, p. 211
Kaskaskia Sequence, p. 214
Laurasia, p. 211
Ouachita orogeny, p. 211
Panthalassa Ocean, p. 211

REVIEW QUESTIONS

1. Which of the following resulted from intracratonic deformation?
   a. _____ Antler Highlands; b. _____ Ancestral Rockies; c. _____ Acadian Highlands; d. _____ Caledonian Highlands; e. _____ Taconic Highlands.

2. The Catskill Delta clastic wedge resulted from weathering and erosion of the _____ highlands.
   a. _____ Taconic; b. _____ Nevadan; c. _____ Transcontinental Arch; d. _____ Acadian; e. _____ Sevier.

3. The European Old Red Sandstone is the equivalent of the North American
   a. _____ Queenston Delta; b. _____ Capitan Limestone; c. _____ Phosphoria Formation; d. _____ Oriskany Sandstone; e. _____ Catskill Delta.

4. During which Paleozoic cratonic sequence were cyclothems common?
   a. _____ Sauk; b. _____ Absaroka; c. _____ Kaskaskia; d. _____ Zuni; e. _____ Tippecanoe.

5. During which period did extensive continental glaciation of the Gondwana continent occur?
   a. _____ Cambrian; b. _____ Silurian; c. _____ Devonian; d. _____ Carboniferous; e. _____ Permian.
6. Repetitive sedimentary sequences of alternating marine and nonmarine sedimentary rocks are
   a. _____ cyclothems; b. _____ reefs; c. _____ orogenies; d. _____ evaporites; e. _____ tillites.
7. The Antler orogeny was the first Paleozoic orogeny to occur in the _____ mobile belt.
   a. _____ Franklin; b. _____ Cordilleran; c. _____ Ouachita; d. _____ Appalachian; e. _____ Caledonian.
8. In what two areas can Late Paleozoic barrier reefs be found?
   a. _____ Michigan and Ohio; b. _____ Western Canada and Michigan; c. _____ Western Canada and Texas–New Mexico; d. _____ Colorado and Texas–New Mexico; e. _____ Montana and Utah.
9. Following deposition of the black shales during the Late Devonian–Early Mississippian, what type of deposition predominated on the craton during the remainder of the Mississippian Period?
   a. _____ carbonates; b. _____ clastics; c. _____ evaporites; d. _____ volcanics; e. _____ cherts and graywackes.
10. The Ancestral Rockies formed during which geologic period?
    a. _____ Permian; b. _____ Pennsylvanian; c. _____ Mississippian; d. _____ Devonian; e. _____ Silurian.
11. The economically valuable deposit in a cyclothem is
    a. _____ gravel; b. _____ metallic ore; c. _____ coal; d. _____ carbonates; e. _____ evaporites.
12. Which orogeny was *not* involved in the closing of the Iapetus Ocean?
    a. _____ Alleghenian; b. _____ Acadian; c. _____ Taconic; d. _____ Caledonian; e. _____ Antler.
13. Discuss how plate movement during the Paleozoic Era affected worldwide weather patterns.
14. What was the relationship between the Ouachita orogeny and the cratonic uplifts on the craton during the Pennsylvanian Period?
15. Based on the discussion of Milankovitch cycles and their role in causing glacial–interglacial cycles (Chapter 17), could these cycles be partly responsible for the transgressive–regressive cycles that resulted in cyclothems during the Pennsylvanian Period?
16. How did the formation of Pangaea and Panthalassa affect the world's climate at the end of the Paleozoic Era?
17. What were the major differences between the Appalachian, Ouachita, and Cordilleran mobile belts during the Paleozoic Era?
18. Describe the geologic history for the Iapetus Ocean during the Paleozoic Era.
19. How are the Caledonian, Acadian, Ouachita, Hercynian, and Alleghenian orogenies related to modern concepts of plate tectonics?
20. How does the origin of evaporite deposits of the Kaskaskia Sequence compare with the origin of evaporites of the Tippecanoe Sequence?

APPLY YOUR KNOWLEDGE

1. In your travels you notice that many buildings in the eastern United States as well as numerous castles in the United Kingdom seem to be constructed of the same coarse-grained red sandstones and conglomerates. How would you account for such a coincidence. Or is it a coincidence? Explain.
2. What is the economic benefit to the automobile industry in having Paleozoic silica sand deposits nearby in and around Toledo, Ohio?

FIELD QUESTION

1. This closeup of a Devonian red rock from a building in Glasgow, Scotland, shows a distinctive sedimentary structure. Identify the sedimentary structure, indicate what type of environment you think it was deposited in, and give the name of the formation this rock comes from.

CHAPTER 12

PALEOZOIC LIFE HISTORY: INVERTEBRATES

Diorama of the environment and biota of the Phyllopod bed of the Middle Cambrian Burgess Shale, British Columbia, Canada. In the background is a vertical wall of a submarine escarpment with algae growing on it. The large, cylindrical ribbed organisms on the muddy bottom in the foreground are sponges.

Courtesy of Smithsonian Institution Libraries, Washington, DC

[OUTLINE]

INTRODUCTION
WHAT WAS THE CAMBRIAN EXPLOSION?
THE EMERGENCE OF A SHELLY FAUNA
PALEOZOIC INVERTEBRATE MARINE LIFE
The Present Marine Ecosystem
Cambrian Marine Community
The Burgess Shale Biota

Ordovician Marine Community
Silurian and Devonian Marine Communities
PERSPECTIVE 12.1 *Mass Extinctions and Their Possible Causes*
Carboniferous and Permian Marine Communities
The Permian Mass Extinction
SUMMARY

ThomsonNOW Explore interactive tutorials, animations, or practice problems available on the ThomsonNow website at **www.thomsonedu.com/login**.

CHAPTER OBJECTIVES

At the end of this chapter, you will have learned that

- Animals with skeletons appeared abruptly at the beginning of the Paleozoic Era and experienced a short period of rapid evolutionary diversification.

- The present marine ecosystem is a complex organization of organisms that interrelate and interact not only with each other, but also with the physical environment.

- The Cambrian Period was a time of many evolutionary innovations during which almost all the major invertebrate phyla evolved.

- The Ordovician Period witnessed striking changes in the marine community, resulting in a dramatic increase in diversity of the shelly fauna, followed by a mass extinction at the end of the Ordovician.

- The Silurian and Devonian periods were a time of rediversification and recovery for many of the invertebrate phyla as well as a time of major reef building.

- Following the Late Devonian extinctions, the marine community again experienced renewed adaptive radiation and diversification.

- The greatest recorded mass extinction in Earth's history occurred at the end of the Permian Period.

Introduction

On August 30 and 31, 1909, near the end of the summer field season, Charles D. Walcott, geologist and head of the Smithsonian Institution, was searching for fossils along a trail on Burgess Ridge between Mount Field and Mount Wapta, near Field, British Columbia, Canada. On the west slope of this ridge, he discovered the first soft-bodied fossils from the Burgess Shale, a discovery of immense importance in deciphering the early history of life. During the following week, Walcott and his collecting party split open numerous blocks of shale, many of which yielded carbonized impressions of a number of soft-bodied organisms beautifully preserved on bedding planes. Walcott returned to the site the following summer and located the shale stratum that was the source of his fossil-bearing rocks in the steep slope above the trail. He quarried the site and shipped back thousands of fossil specimens to the United States National Museum of Natural History, where he later catalogued and studied them.

The importance of Walcott's discovery is not that it was another collection of well-preserved Cambrian fossils, but rather that it allowed geologists a rare glimpse into a world previously almost unknown: that of the soft-bodied animals that lived some 530 million years ago. The beautifully preserved fossils from the Burgess Shale present a much more complete picture of a Middle Cambrian community than do deposits containing only fossils of the hard parts of organisms. In fact, 60% of the total fossil assemblage of more than 100 genera is composed of soft-bodied animals, a percentage comparable to present-day marine communities.

What conditions led to the remarkable preservation of the Burgess Shale fauna? The depositional site of the Burgess Shale lay at the base of a steep submarine escarpment. The animals whose exquisitely preserved fossil remains are found in the Burgess Shale lived in and on mud banks that formed along the top of this escarpment. Periodically, this unstable area would slump and slide down the escarpment as a turbidity current. At the base, the mud and animals carried with it were deposited in a deep-water anaerobic environment devoid of life. In such an environment, bacterial degradation did not destroy the buried animals, and thus they were compressed by the weight of the overlying sediments and eventually preserved as carbonaceous films.

In the following two chapters, we examine the history of Paleozoic life as a system in which its parts consist of a series of interconnected biologic and geologic events. The underlying processes of evolution and plate tectonics are the forces that drove this system. The opening and closing of ocean basins, transgressions and regressions of epeiric seas, the formation of mountain ranges, and the changing positions of the continents profoundly affected the evolution of the marine and terrestrial communities.

A time of tremendous biologic change began with the appearance of skeletonized animals near the Precambrian–Cambrian boundary. Following this event, marine invertebrates began a period of adaptive radiation and evolution, during which the Paleozoic marine invertebrate community greatly diversified. Indeed, the history of the Paleozoic marine invertebrate community was one of diversification and extinction, culminating at the end of the Paleozoic Era in the greatest mass extinction in Earth history.

What Was the Cambrian Explosion?

At the beginning of the Paleozoic Era, animals with skeletons appeared rather abruptly in the fossil record. In fact, their appearance is described as an explosive development of new types of animals and is referred to as the "Cambrian explosion" by most scientists. This sudden appearance of new animals in the fossil record is rapid, however, only in the context of geologic time, having taken place over millions of years during the Early Cambrian Period.

This seemingly sudden appearance of animals in the fossil record is not a recent discovery. Early geologists observed that the remains of skeletonized animals appeared rather abruptly in the fossil record. Charles Darwin addressed this problem in *On the Origin of Species by Means of Natural Selection* (1859) and observed that, without a convincing explanation, such an event was difficult to reconcile with his newly expounded evolutionary theory.

The sudden appearance of shelled animals during the Early Cambrian contrasts sharply with the biota living during the preceding Proterozoic Eon. Up until the evolution of the Ediacaran fauna, Earth was populated primarily by single-celled organisms. Recall from Chapter 9 that the Ediacaran fauna, which is found on all continents except Antarctica, consists primarily of multicelled soft-bodied organisms. Microscopic calcareous tubes, presumably housing wormlike suspension-feeding organisms, have also been found at some localities. In addition, trails and burrows, which represent the activities of worms and other sluglike animals, are also found associated with Ediacaran faunas throughout the world. The trails and burrows are similar to those made by present-day soft-bodied organisms.

Until recently, it appeared there was a fairly long period of time between the extinction of the Ediacaran fauna and the first Cambrian fossils. That gap has been considerably narrowed in recent years with the discovery of new Proterozoic fossiliferous localities. Now, Proterozoic fossil assemblages continue right to the base of the Cambrian.

Nonetheless, the cause of the sudden appearance of so many different animal phyla during the Early Cambrian is still a hotly debated topic. Newly developed molecular techniques that allow evolutionary biologists to compare the similarity of molecular sequences of the same gene from different species is being applied to the phylogeny of many organisms. In addition, new fossil sites and detailed stratigraphic studies are shedding light on the early history and ancestry of the various invertebrate phyla.

The Cambrian explosion probably had its roots firmly planted in the Proterozoic. However, the mechanism or mechanisms that triggered this event are still being investigated. Although some would argue for a single causal event, it is more likely that the Cambrian explosion was a combination of factors, both biological and geological. For example, geologic evidence indicates Earth was glaciated one or more times during the Proterozoic, followed by global warming during the Cambrian. These global environmental changes may have stimulated evolution and contributed to the Cambrian explosion.

Others would argue that a change in the chemistry of the oceans favored the evolution of a mineralized skeleton. In this scenario, an increase in the concentration of calcium from the Neoproterozoic through the Early Cambrian allowed for the precipitation of calcium carbonate and calcium phosphate, compounds that comprise the shells of most invertebrates.

Another hypothesis gaining favor is that the rapid evolution of a skeletonized fauna was a response to the evolution of predators. A shell or mineralized covering would provide protection against predation by the various predators evolving during this time.

An interesting line of research related to the Cambrian explosion involves *Hox* genes, which are sequences of genes that control the development of individual regions of the body. Studies indicate that the basic body plans for all animals were apparently established by the end of the Cambrian explosion, and only minor modifications have occurred since then. Whatever the ultimate cause of the Cambrian explosion, the appearance of a skeletonized fauna and the rapid diversification of that fauna during the Early Cambrian was a major event in life history.

The Emergence of a Shelly Fauna

The earliest organisms with hard parts are Proterozoic calcareous tubes found associated with Ediacaran faunas from several locations throughout the world. These are followed by other microscopic skeletonized fossils from the Early Cambrian (• Figure 12.1) and the appearance of large skeletonized animals during the Cambrian explosion. Along with the question of why did animals appear so suddenly in the fossil record is the equally intriguing one of why they initially acquired skeletons and what selective

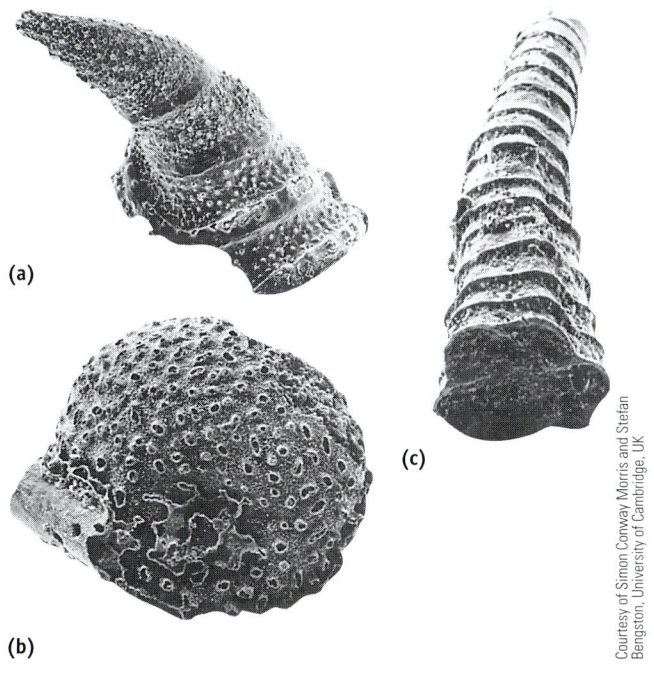

• **Figure 12.1 Lower Cambrian Shelly Fossils** Three small (several millimeters in size) Lower Cambrian shelly fossils. **(a)** A conical sclerite (a piece of the armor covering) of *Lapworthella* from Australia. **(b)** *Archaeooides,* an enigmatic spherical fossil from the Mackenzie Mountains, Northwest Territories, Canada. **(c)** The tube of an anabaritid from the Mackenzie Mountains, Northwest Territories, Canada.

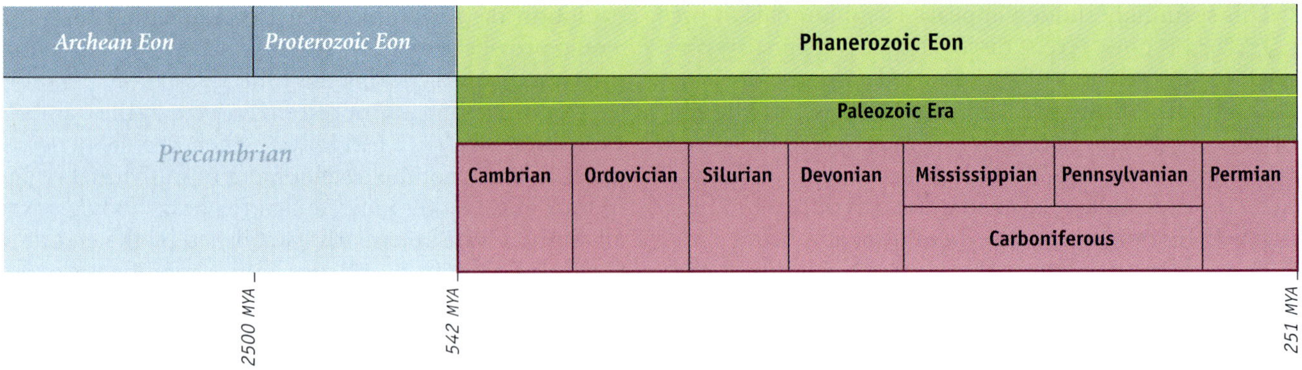

advantage this provided. A variety of explanations about why marine organisms evolved skeletons have been proposed, but none is completely satisfactory or universally accepted.

The formation of an exoskeleton, or shell, confers many advantages on an organism: (1) It provides protection against ultraviolet radiation, allowing animals to move into shallower waters; (2) it helps prevent drying out in an intertidal environment; (3) a supporting skeleton, whether an exo- or endoskeleton, allows animals to increase their size and provides attachment sites for muscles; and (4) it provides protection against predators. Evidence of actual fossils of predators and specimens of damaged prey, as well as antipredatory adaptations in some animals, indicates that the impact of predation during the Cambrian was great, leading some scientists to hypothesize that the rapid evolution of a shelly invertebrate fauna was a response to the rise of predators (• Figure 12.2). With predators playing an important role in the Cambrian marine ecosystem, any mechanism or feature that protected an animal would certainly be advantageous and confer an adaptive advantage to the organism.

Scientists currently have no clear answer about why marine organisms evolved mineralized skeletons during the Cambrian explosion and shortly thereafter. They undoubtedly evolved because of a variety of biologic and environmental factors. Whatever the reason, the acquisition of a mineralized skeleton was a major evolutionary innovation allowing invertebrates to successfully occupy a wide variety of marine habitats.

Paleozoic Invertebrate Marine Life

Having considered the origin, differentiation, and evolution of the Precambrian–Cambrian marine biota, we now examine the changes that occurred in the marine invertebrate community during the Paleozoic Era. Rather than focusing on the history of each invertebrate phylum (Table 12.1), we will survey the evolution of the marine invertebrate communities through time, concentrating on the major features and changes that took place. To do that, we need to briefly examine the nature and structure of living marine communities so that we can make a reasonable interpretation of the fossil record.

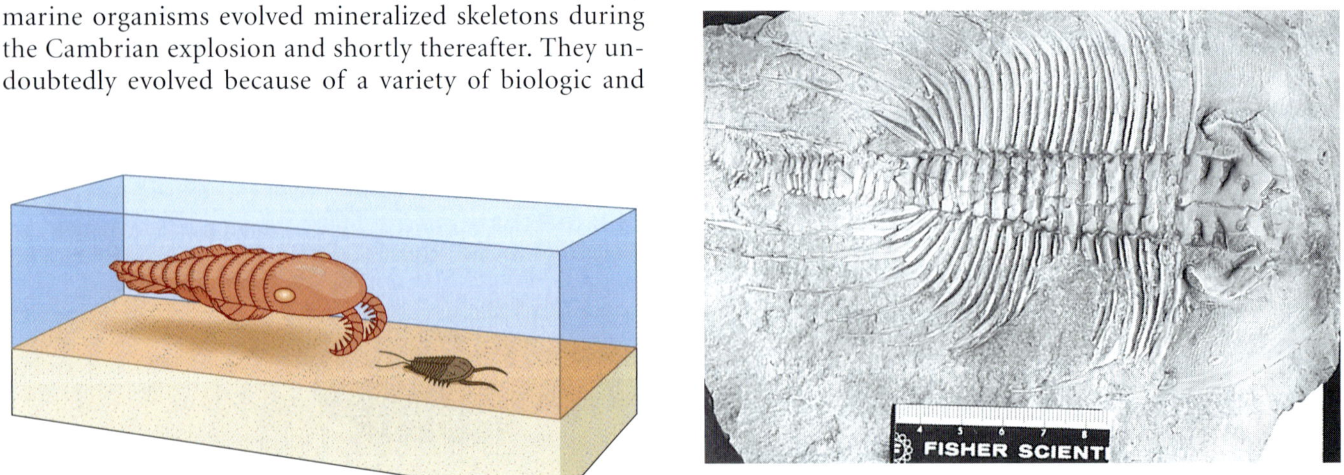

• **Figure 12.2 Cambrian Predation** (a) Reconstruction of *Anomalocaris*, a predator from the Early and Middle Cambrian Period. It was about 45 cm long and probably fed on trilobites. Its gripping appendages presumably carried food to its mouth. (b) Wounds (area just above the ruler) to the body of the trilobite *Olenellus robsonensis*. The wounds have healed, demonstrating that they occurred when the animal was alive and were not inflicted on an empty shell.

	Phanerozoic Eon										
	Mesozoic Era			Cenozoic Era							
Triassic	Jurassic	Cretaceous	Paleogene			Neogene		Quaternary			
			Paleocene	Eocene	Oligocene	Miocene	Pliocene	Pleistocene	Holocene		

251 MYA · 66 MYA

TABLE 12.1 The Major Invertebrate Groups and Their Stratigraphic Ranges

Phylum Protozoa	Cambrian–Recent		**Phylum Mollusca**	Cambrian–Recent
Class Sarcodina	Cambrian–Recent		Class Monoplacophora	Cambrian–Recent
Order Foraminifera	Cambrian–Recent		Class Gastropoda	Cambrian–Recent
Order Radiolaria	Cambrian–Recent		Class Bivalvia	Cambrian–Recent
Phylum Porifera	Cambrian–Recent		Class Cephalopoda	Cambrian–Recent
Class Demospongea	Cambrian–Recent		**Phylum Annelida**	Precambrian–Recent
Order Stromatoporoida	Cambrian–Oligocene		**Phylum Arthropoda**	Cambrian–Recent
Phylum Archaeocyatha	Cambrian		Class Trilobita	Cambrian–Permian
Phylum Cnidaria	Cambrian–Recent		Class Crustacea	Cambrian–Recent
Class Anthozoa	Ordovician–Recent		Class Insecta	Silurian–Recent
Order Tabulata	Ordovician–Permian		**Phylum Echinodermata**	Cambrian–Recent
Order Rugosa	Ordovician–Permian		Class Blastoidea	Ordovician–Permian
Order Scleractinia	Triassic–Recent		Class Crinoidea	Cambrian–Recent
Phylum Bryozoa	Ordovician–Recent		Class Echinoidea	Ordovician–Recent
Phylum Brachiopoda	Cambrian–Recent		Class Asteroidea	Ordovician–Recent
Class Inarticulata	Cambrian–Recent		**Phylum Hemichordata**	Cambrian–Recent
Class Articulata	Cambrian–Recent		Class Graptolithina	Cambrian–Mississippian

The Present Marine Ecosystem

In analyzing the present-day marine ecosystem, we must look at where organisms live, how they get around, and how they feed (• Figure 12.3). Organisms that live in the water column above the seafloor are called *pelagic*. They can be divided into two main groups: the floaters, or **plankton,** and the swimmers, or **nekton.**

Plankton are mostly passive and go where currents carry them. Plant plankton such as diatoms, dinoflagellates, and various algae are called *phytoplankton* and are mostly microscopic. Animal plankton are called *zooplankton* and are also mostly microscopic. Examples of zooplankton include foraminifera, radiolarians, and jellyfish. The nekton are swimmers and are mainly vertebrates such as fish; the invertebrate nekton include cephalopods.

Organisms that live on or in the seafloor make up the **benthos.** They are characterized as *epifauna* (animals) or *epiflora* (plants)—those that live on the seafloor—or as *infauna*—animals that live in and move through the sediments. The benthos are further divided into those organisms that stay in one place, called *sessile*, and those that move around on or in the seafloor, called *mobile*.

The feeding strategies of organisms are also important in terms of their relationships with other organisms in the marine ecosystem. There are basically four feeding groups: **Suspension-feeding** animals remove or consume microscopic plants and animals as well as dissolved nutrients from the water, **herbivores** are plant eaters, **carnivore-scavengers** are meat eaters, and **sediment-deposit feeders** ingest sediment and extract the nutrients from it.

We can define an organism's place in the marine ecosystem by where it lives and how it eats. For example, an articulate brachiopod is a benthic, epifaunal suspension feeder, whereas a cephalopod is a nektonic carnivore.

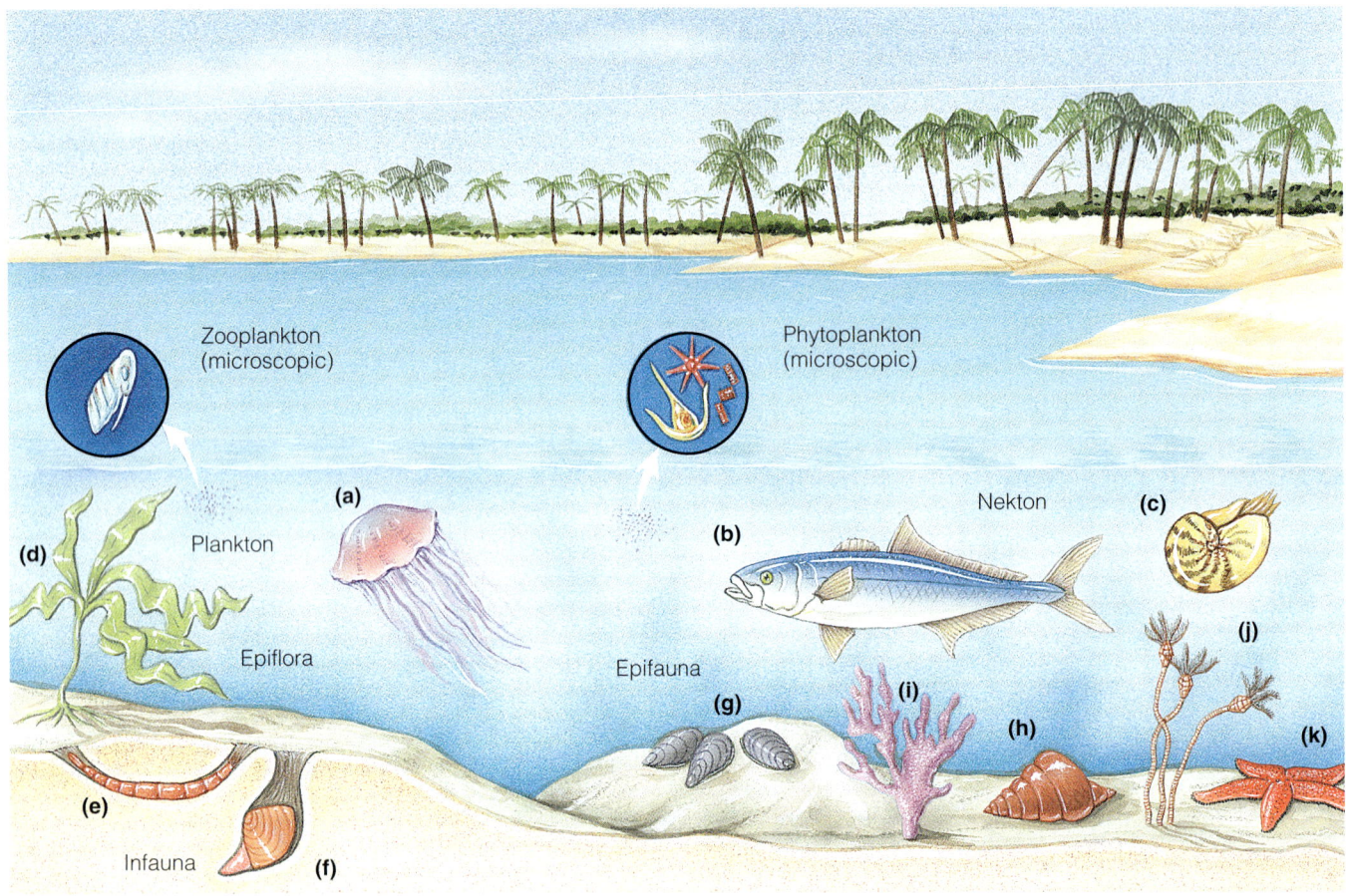

• **Figure 12.3 Marine Ecosystem** Where and how animals and plants live in the marine ecosystem. Plankton: **(a)** jellyfish. Nekton: **(b)** fish and **(c)** cephalopod. Benthos: **(d)** through **(k)**. Sessile epiflora: **(d)** seaweed. Sessile epifauna: **(g)** bivalve, **(i)** coral, and **(j)** crinoid. Mobile epifauna: **(k)** starfish and **(h)** gastropod. Infauna: **(e)** worm and **(f)** bivalve. Suspension feeders: **(g)** bivalve, **(i)** coral, and **(j)** crinoid. Herbivores: **(h)** gastropod. Carnivore-scavengers: **(k)** starfish. Sediment-deposit feeders: **(e)** worm.

An ecosystem includes several **trophic levels,** which are tiers of food production and consumption within a feeding hierarchy. The feeding hierarchy, and hence energy flow, in an ecosystem comprise a food web of complex interrelationships among the producers, consumers, and decomposers (• Figure 12.4). The **primary producers,** or *autotrophs,* are those organisms that manufacture their own food. Virtually all marine primary producers are phytoplankton. Feeding on the primary producers are the primary consumers, which are mostly suspension feeders. Secondary consumers feed on the primary consumers and thus are predators, whereas tertiary consumers, which are also predators, feed on the secondary consumers. Besides the producers and consumers, there are also transformers and decomposers. These are bacteria that break down the dead organisms that have not been consumed into organic compounds, which are then recycled.

When we look at the marine realm today, we see a complex organization of organisms interrelated by trophic interactions and affected by changes in the physical environment. When one part of the system changes, the whole structure changes, sometimes almost insignificantly, other times catastrophically.

As we examine the evolution of the Paleozoic marine ecosystem, keep in mind how geologic and evolutionary changes can have a significant impact on its composition and structure. For example, the major transgressions onto the craton opened up vast areas of shallow seas that

What Would You Do?

Congress has just voted a large cut in funding for the National Science Foundation (NSF). Because of reduced funding, some earth science areas will not be funded this year. One area slated for major funding reduction or no funding this year is paleontology. As a paleontology professor whose specialty is Paleozoic marine ecosystems and whose research depends on NSF funds, you will be testifying against cutting funding for paleontology.

What arguments would you use to persuade the members of Congress to keep funding paleontologic research? What possible outcomes from your research could you point to that might benefit society today?

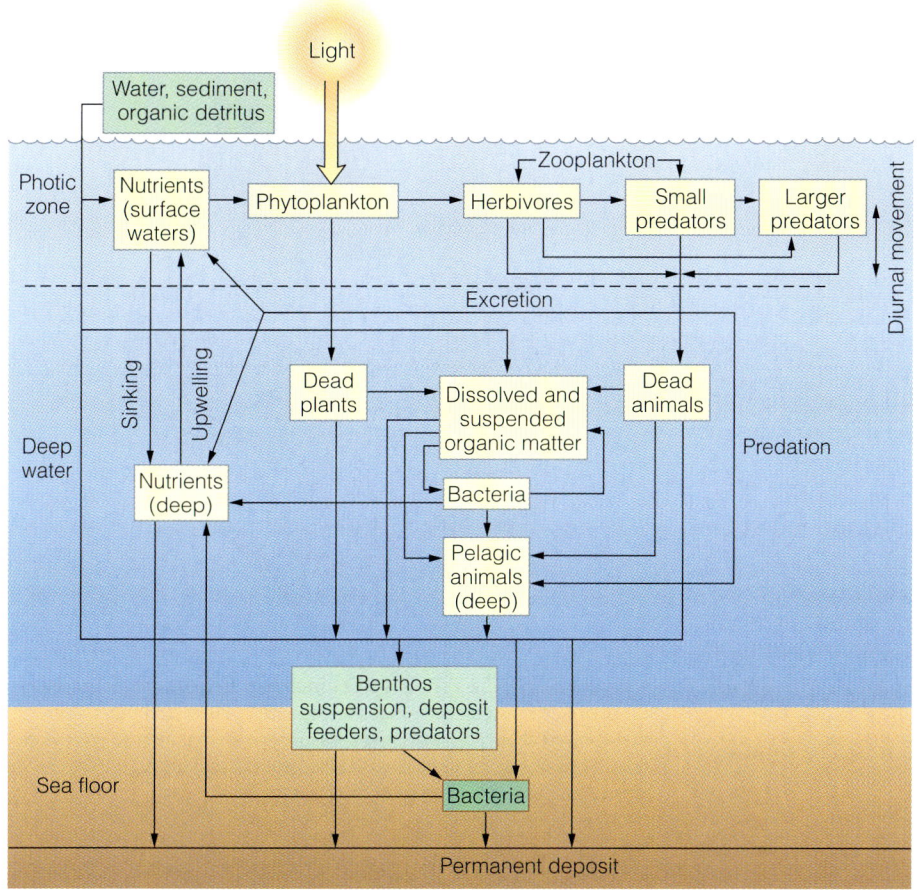

• Figure 12.4 **Marine Food Web** Marine food web showing the relationships among the producers, consumers, and decomposers.

• Figure 12.5 **Cambrian Marine Community** Reconstruction of a Cambrian marine community showing floating jellyfish, swimming arthropods, benthonic sponges, and scavenging trilobites.

Cambrian Marine Community

The Cambrian Period was a time during which many new body plans evolved and animals moved into new niches. As might be expected, the Cambrian witnessed a higher percentage of such experiments than any other period of geologic history.

Although almost all the major invertebrate phyla evolved during the Cambrian Period (Table 12.1), many were represented by only a few species. Whereas trace fossils are common and echinoderms diverse, trilobites, inarticulate brachiopods, and archaeocyathids comprised the majority of Cambrian skeletonized life (• Figure 12.5).

Trilobites were by far the most conspicuous element of the Cambrian marine invertebrate community and made up about half of the total fauna. Trilobites were benthic, mobile, sediment-deposit feeders that crawled or swam along the seafloor. They first appeared in the Early Cambrian, rapidly diversified, reached their maximum diversity in the Late Cambrian, and then suffered mass extinctions near the end of the Cambrian from which they never fully recovered.

As yet no consensus exists on what caused the trilobite extinctions, but a combination of factors were likely involved, possibly including a reduction of shelf space, increased competition, and a rise in predators. Some scientists have also suggested that a cooling of the seas may have played a role, particularly for the extinctions at the end of the Ordovician Period.

Cambrian **brachiopods** were mostly types called *inarticulates*. They secreted a chitinophosphate shell, composed of the organic compound chitin combined with calcium phosphate. Inarticulate brachiopods also lacked a tooth-and-socket arrangement along the hinge line where the two shells pivot. The *articulate* brachiopods, which have a tooth-and-socket arrangement, were also present but did not become abundant until the Ordovician Period.

could be inhabited. The movement of continents affected oceanic circulation patterns as well as causing environmental changes.

• **Figure 12.6 Archaeocyathids** Restoration of a Cambrian reeflike structure built by archaeocyathids.

The third major group of Cambrian organisms were the **archaeocyathids** (• Figure 12.6). These organisms were benthic, sessile, presumably suspension feeders that constructed reeflike structures beginning in the Early Cambrian. Archaeocyathids went extinct at the end of the Cambrian.

The rest of the Cambrian fauna consisted of representatives of the other major phyla, including many organisms that were short-lived evolutionary experiments (• Figure 12.7). As might be expected during times of adaptive radiation and evolutionary experimentation, many of the invertebrates that evolved during the Cambrian soon became extinct.

The Burgess Shale Biota

No discussion of Cambrian life would be complete without mentioning one of the best examples of a preserved soft-bodied fauna and flora: the Burgess Shale biota. As the Sauk Sea transgressed from the Cordilleran shelf onto the western edge of the craton, Early Cambrian-aged sands were covered by Middle Cambrian-aged black muds that allowed a diverse soft-bodied benthic community to be preserved. As we discussed in the Introduction, these fossils were discovered in 1909 by Charles Walcott near Field, British Columbia. They represent one of the most significant fossil finds of the 20th century because they consist of carbonized impressions of soft-bodied animals and plants that are rarely preserved in the fossil record (• Figure 12.8). This discovery therefore provides us with a valuable glimpse of rarely preserved organisms as well as the soft-part anatomy of many extinct groups.

In recent years, the reconstruction, classification, and interpretation of many of the Burgess Shale fossils have undergone a major change that has led to new theories and explanations of the Cambrian explosion of life. Recall that during the Neoproterozoic, multicelled organisms evolved, and shortly thereafter animals with hard parts made their first appearance. These were followed by an explosion of invertebrate groups during the Cambrian, many of which are now extinct. These Cambrian organisms represent the root stock and basic body plans from which all present-day invertebrates evolved.

• **Figure 12.7 The Primitive Echinoderm** *Helicoplacus* *Helicoplacus,* a primitive echinoderm that became extinct 20 million years after it first evolved about 510 million years ago. Such an organism (a representative of one of several short-lived echinoderm classes) illustrates the "experimental" nature of the Cambrian invertebrate fauna.

The question that paleontologists are still debating is how many phyla arose during the Cambrian, and at the center of that debate are the Burgess Shale fossils. For years, most paleontologists placed the bulk of the Burgess Shale organisms into existing phyla, with only a few assigned to phyla that are now extinct. Thus the phyla of the Cambrian world were viewed as being essentially the same in number as the phyla of the present-day world but with fewer species in each phylum. According to this view, the history of life has been simply a gradual increase in the diversity of species within each phylum through time. The number of basic body plans has therefore remained more or less constant since the initial radiation of multicelled organisms.

This view, however, has been challenged by other paleontologists who think the initial explosion of varied life-forms in the Cambrian was promptly followed by a short period of experimentation and then extinction of many phyla. The richness and diversity of modern life-forms are the result of repeated variations of the basic body plans that survived the Cambrian extinctions. In other words, life was much more diverse in terms of phyla during the Cambrian than it is today. The reason why members of the Burgess Shale biota look so strange to us is that no living organisms possess their basic body plan, and therefore many of them have been reassigned into new phyla.

Discoveries of Cambrian fossils at localities such as Sirius Passet, Greenland, and Yunnan, China, have resulted in reassignment of some Burgess Shale specimens back

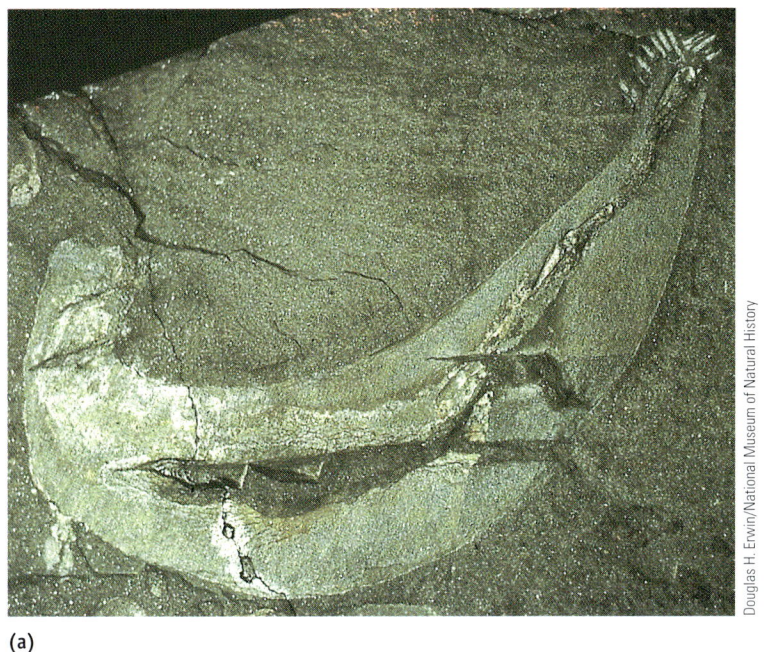

• Figure 12.8 **Fossils from the Burgess Shale** Some of the fossil animals preserved in the Burgess Shale. (a) *Ottoia*, a carnivorous worm. (b) *Wiwaxia*, a scaly armored sluglike creature whose affinities remain controversial. (c) *Hallucigenia*, a velvet worm. (d) *Waptia*, an anthropod.

into extant phyla. If these reassignments to known phyla prove correct, then no massive extinction event followed the Cambrian explosion, and life has gradually increased in diversity through time. Currently, there is no clear answer to this debate, and the outcome will probably be decided as more fossil discoveries are made.

Ordovician Marine Community

A major transgression that began during the Middle Ordovician (Tippecanoe Sequence) resulted in widespread inundation of the craton. This vast epeiric sea, which had a uniformly warm climate during this time, opened many new marine habitats that were soon filled by a variety of organisms.

Not only did sedimentation patterns change dramatically from the Cambrian to the Ordovician, but the fauna underwent equally striking changes. Whereas the Cambrian invertebrate community was dominated by trilobites, inarticulate brachiopods, and archaeocyathids, the Ordovician was characterized by the adaptive radiation of many other animal phyla (such as articulate brachiopods, bryozoans, and corals), with a consequent dramatic increase

PALEOZOIC INVERTEBRATE MARINE LIFE **239**

• **Figure 12.9 Middle Ordovician Marine Community** Re-creation of a Middle Ordovician seafloor fauna. Cephalopods, crinoids, colonial corals, bryozoans, trilobites, and brachiopods are shown.

in the diversity of the total shelly fauna (• Figure 12.9). The Ordovician was also a time of increased diversity and abundance of the **acritarchs** (organic-walled phytoplankton of unknown affinity), which were the major phytoplankton group of the Paleozoic Era and the primary food source of the suspension feeders (• Figure 12.10).

During the Cambrian, archaeocyathids were the main builders of reeflike structures, but bryozoans, stromatoporoids, and tabulate and rugose corals assumed that role beginning in the Middle Ordovician. Many of these reefs were small patch reefs similar in size to those of the Cambrian but of a different composition, whereas others were quite large. As with present-day reefs, Ordovician reefs showed a high diversity of organisms and were dominated by suspension feeders.

Three Ordovician fossil groups have proved particularly useful for biostratigraphic correlation—the *articulate brachiopods, graptolites,* and *conodonts.* The articulate brachiopods, present since the Cambrian, began a period of major diversification in the shallow-water marine environment during the Ordovician (• Figure 12.11a). They became a conspicuous element of the invertebrate fauna during the Ordovician and in succeeding Paleozoic periods.

Most **graptolites** were planktonic animals carried about by ocean currents. Because most graptolites were planktonic and most individual species existed for less than a million years, graptolites are excellent guide fossils. They were especially abundant during the Ordovician and Silurian periods. Graptolites are most commonly found in black shales where they are preserved as carbonaceous impressions (Figure 12.11b).

Conodonts are a group of well-known, small, toothlike fossils composed of the mineral apatite (calcium phosphate), the same mineral that composes bone (• Figure 12.12a). Although conodonts have been known for more than 150 years, their affinity was debated until the discovery of the conodont animal in 1983 (Figure 12.12b). Several specimens of the carbonized impression of the conodont animal from Lower Carboniferous rocks of Scotland show it is a member of a group of primitive jawless animals assigned to the phylum Chordata. Study of the specimens indicates the conodont animal was probably an elongate swimming organism. The wide distribution and short stratigraphic range of individual conodont species make them excellent fossils for biostratigraphic zonation and correlation.

The end of the Ordovician was a time of mass extinctions in the marine realm. More than 100 families of marine invertebrates became extinct, and in North America alone, about half of the brachiopods and bryozoans died out. What caused such an event? Many geologists think these extinctions were the result of extensive glaciation in Gondwana at the end of the Ordovician Period (see Chapter 10).

Mass extinctions, those geologically rapid events in which an unusually high percentage of the fauna and/or flora become extinct, have occurred throughout geologic time (at or near the end of the Ordovician, Devonian, Permian, and Cretaceous periods) and are the focus of much research and debate (see Perspective 12.1).

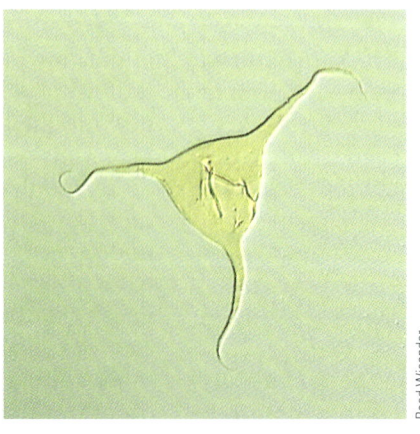
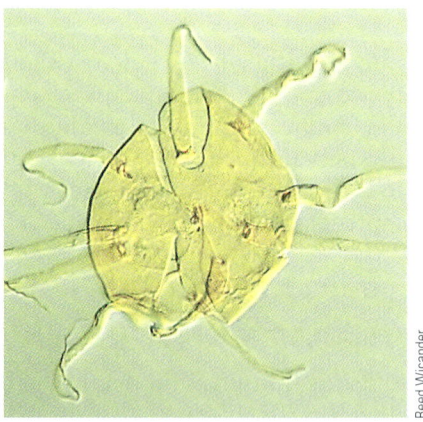

• **Figure 12.10 Late Ordovician Acritarchs** Acritarchs from the Upper Ordovician Sylvan Shale, Oklahoma. Acritarchs are organic-walled phytoplankton and were the primary food source for suspension feeders during the Paleozoic Era.

(a)

(b)

• **Figure 12.11 Representative Brachiopods and Graptolites** (a) Brachiopods are benthic, sessile, suspension feeders. (b) Graptolites are planktonic suspension feeders. Shown is *Phyllograptus angustifolius* from Norway.

(a)

(b)

• **Figure 12.12 Conodonts and the Conodont Animal** (a) Conodonts are microscopic toothlike fossils. *Cahabagnathus sweeti*, Copenhagen Formation (Middle Ordovician), Monitor Range, Nevada (*left*); *Scolopodus*, sp., Shingle Limestone, Single Pass, Nevada (*right*). (b) The conodont animal preserved as a carbonized impression in the Lower Carboniferous Granton Shrimp Bed in Edinburgh, Scotland. The animal measures about 40 mm long and 2 mm wide.

Silurian and Devonian Marine Communities

The mass extinction at the end of the Ordovician was followed by rediversification and recovery of many of the decimated groups. Brachiopods, bryozoans, gastropods, bivalves, corals, crinoids, and graptolites were just some of the groups that rediversified beginning during the Silurian.

As we discussed in Chapters 10 and 11, the Silurian and Devonian were times of major reef building. Whereas most of the Silurian radiations of invertebrates represented repopulating of niches, organic reef builders diversified in new ways, building massive reefs larger than any produced during the Cambrian or Ordovician. This repopulation was probably caused in part by renewed transgressions over the craton, and although a major drop in sea level occurred at the end of the Silurian, the Middle Paleozoic sea level was generally high (see Table 10.1).

The Silurian and Devonian reefs were dominated by tabulate and colonial rugose corals and stromatoporoids (• Figure 12.13). Although the fauna of these Silurian and Devonian reefs was somewhat different from that of earlier reefs and reeflike structures, the general composition and structure are the same as in present-day reefs.

The Silurian and Devonian periods were also the time when *eurypterids* (arthropods with scorpion-like bodies and impressive pincers) were abundant, and unlike many other marine invertebrates, eurypterids expanded into brackish and freshwater habitats (• Figure 12.14). *Ammonoids*, a subclass of the cephalopods, evolved from nautiloids during the Early Devonian and rapidly diversified. With their distinctive suture patterns, short stratigraphic ranges, and widespread distribution, ammonoids are excellent guide fossils for the Devonian through Cretaceous periods (• Figure 12.15).

Near the end of the Devonian, another mass extinction occurred that resulted in a worldwide near-total collapse of the massive marine reef communities. On land, however, the seedless vascular plants were seemingly unaffected. Thus extinctions at this time were most extensive among marine life, particularly in the reef and pelagic communities.

Perspective 12.1

Mass Extinctions and Their Possible Causes

Throughout geologic history, various plant and animal species have become extinct. In fact, extinction is a common feature of the fossil record, and the rate of extinction through time has fluctuated only slightly. Just as new species evolve, others become extinct. There have, however, been brief intervals in the geologic past during which mass extinctions have eliminated large numbers of species. Extinctions of this magnitude could only occur due to radical changes in the environment on a regional or global scale.

When we look at the different mass extinctions that have occurred during the geologic past, several common themes stand out. The first is that mass extinctions typically have affected life both in the sea and on land. Second, tropical organisms, particularly in the marine realm, apparently are more affected than organisms from the temperate and high-latitude regions. Third, some animal groups repeatedly experience mass extinctions.

During the first mass extinction, such groups are severely affected but not wiped out. Following the initial crisis, the survivors diversify, only to have their numbers reduced further by another mass extinction. Three marine invertebrate groups in particular display this characteristic: the trilobites, graptolites, and ammonoids. Each of these groups experienced high rates of extinction, followed by high rates of speciation.

When we examine the mass extinctions for the last 650 million years, we see that the first major extinction event involved only the acritarchs. Several extinction events occurred during the Cambrian, and these affected only marine invertebrates, particularly trilobites. Three other marine mass extinctions took place during the Paleozoic Era: one at the end of the Ordovician, involving many invertebrates; one near the end of the Devonian, affecting the major barrier reef–building organisms as well as the primitive armored fish; and the most severe at the end of the Permian, when about 90% of all marine invertebrate species and more than 65% of all land animals became extinct.

The Mesozoic Era experienced several mass extinctions, the most devastating occurring at the end of the Cretaceous, when almost all large animals, including dinosaurs, flying reptiles, and seagoing animals such as plesiosaurs and ichthyosaurs, became extinct. Many scientists think the terminal Cretaceous mass extinction was caused by a meteorite impact.

Several mass extinctions occurred during the Cenozoic Era. The most severe was near the end of the Eocene Epoch and is correlated with global cooling and climatic change.

Although many scientists think of the marine mass extinctions as sudden events from a geologic perspective, they were rather gradual from a human perspective, occurring over hundreds of thousands and even millions of years. Furthermore, many geologists think that climatic changes, rather than a single catastrophic event, were primarily responsible for the extinctions, particularly in the marine realm. Evidence of glacial episodes or other signs of climatic change, such as global warming, have been correlated with the extinction events recorded in the fossil record.

The demise of the Middle Paleozoic reef communities highlights the geographic aspects of the Late Devonian mass extinction. The tropical groups were most severely affected; in contrast, the higher latitude communities were seemingly little affected. Apparently, an episode of global cooling was largely responsible for the extinctions near the end of the Devonian. During such a cooling, the disappearance of tropical conditions would have had a severe effect on reef and other warm-water organisms. Cool-water species, in contrast, could have simply migrated toward the equator. Although cooling temperatures certainly played an important role in the Late Devonian extinctions, the closing of the Iapetus Ocean and the orogenic events of this time undoubtedly also played a role by

• Figure 12.13 **Middle Devonian Marine Reef Community** Reconstruction of a Middle Devonian reef from the Great Lakes area. Shown are corals, bryozoans, cephalopods, trilobites, crinoids, and brachiopods.

• Figure 12.14 **Silurian Brackish Water Community** Restoration of a Silurian brackish water scene near Buffalo, New York. Shown are algae, eurypterids, gastropods, worms, and shrimp.

reducing the area of shallow shelf environments where many marine invertebrates lived.

Carboniferous and Permian Marine Communities

The Carboniferous invertebrate marine community responded to the Late Devonian extinctions in much the same way the Silurian invertebrate marine community responded to the Late Ordovician extinctions—that is, by renewed adaptive radiation and rediversification. The brachiopods and ammonoids quickly recovered and again assumed important ecologic roles. Other groups, such as the lacy bryozoans and crinoids, reached their greatest diversity during the Carboniferous. With the decline of the stromatoporoids and the tabulate and rugose corals, large organic reefs such as those existing earlier in the Paleozoic virtually disappeared and were replaced by small patch reefs. These reefs were dominated by crinoids, blastoids, lacy bryozoans, brachiopods, and calcareous algae and flourished during the Late Paleozoic (• Figure 12.16). In addition, bryozoans and crinoids contributed large amounts of skeletal debris to the formation of the vast bedded limestones that constitute the majority of Mississippian sedimentary rocks.

The Permian invertebrate marine faunas resembled those of the Carboniferous but were not as widely distributed because of the restricted size of the shallow seas on the cratons and the reduced shelf space along the continental margins (see Figure 11.11). The spiny and odd-shaped productids dominated the brachiopod assemblage and constituted an important part of the reef complexes that formed in the Texas region during the Permian (• Figure 12.17). The fusulinids (spindle-shaped foraminifera), which first evolved during the Late Mississippian and greatly diversified during the Pennsylvanian (• Figure 12.18),

• Figure 12.15 **Ammonoid Cephalopod** A Late Devonian-aged ammonoid cephalopod from Erfoud, Morocco. The distinctive suture pattern, short stratigraphic range, and wide geographic distribution make ammonoids excellent guide fossils.

What Would You Do?

One concern of environmentalists is that environmental degradation is leading to vast reductions in global biodiversity. As a paleontologist, you are aware mass extinctions have taken place throughout Earth history. What facts and information can you provide from your geological perspective that will help focus the debate as to whether or not Earth's biota is being adversely affected by such human activities as industrialization, and what the possible outcome(s) might be if global biodiversity is severely reduced?

• Figure 12.16 **Late Mississippian Marine Community** Reconstruction of marine life during the Mississippian, based on an Upper Mississippian fossil site at Crawfordville, Indiana. Invertebrate animals shown include crinoids, blastoids, lacy bryozoans, brachiopods, and small corals.

experienced a further diversification during the Permian. Because of their abundance, diversity, and worldwide occurrence, fusulinids are important guide fossils for Pennsylvanian and Permian strata. Bryozoans, sponges, and some types of calcareous algae also were common elements of the Permian invertebrate fauna.

• Figure 12.17 **Permian Patch-Reef Marine Community** Reconstruction of a Permian patch-reef community from the Glass Mountains of West Texas. Shown are algae, productid brachiopods, cephalopods, sponges, and corals.

The Permian Mass Extinction

The greatest recorded mass extinction to affect Earth occurred at the end of the Permian Period (• Figure 12.19). Before the Permian ended, roughly 50% of all marine invertebrate families and about 90% of all marine invertebrate species became extinct. Fusulinids, rugose and tabulate corals, several bryozoan and brachiopod orders, as well as trilobites and blastoids did not survive the end of the Permian. All of these groups had been very successful during the Paleozoic Era. In addition, more than 65% of all amphibians and reptiles, as well as nearly 33% of insects on land also became extinct.

What caused such a crisis for both marine and land-dwelling organisms? Various hypotheses have been proposed, but no completely satisfactory answer has yet been found. Because the extinction event extended over many millions of years at the end of the Permian, a meteorite impact such as occurred at the end of the Cretaceous Period (see Chapter 15) can be reasonably discounted.

A reduction in the habitable marine shelf area caused by the formation of Pangaea and a widespread marine regression resulting from glaciation can also be rejected as the primary cause of the extinctions. By the end of the Permian, most collisions of the continents had already taken place, such that the reduction in the shelf area had already occurred before the mass extinctions began in earnest. Furthermore, the widespread glaciation that took place during the Carboniferous was now waning in the Permian.

Currently, many scientists think an episode of deep-sea anoxia resulted in a highly stratified ocean during the Late Permian. In other words, there was very little, if any, circulation of oxygen-rich surface waters into the deep ocean. During this time, stagnant waters also covered the shelf regions, thus affecting the shallow marine fauna.

• **Figure 12.18 Fusulinids** Fusulinids are spindle-shaped, microscopic benthonic foraminifera that are excellent guide fossils for the Pennsylvanian and Permian periods. Shown here are three natural cross-sections of fusulinids from the Lower Permian Owens Valley Group, Inyo County, California. The two elongated specimens are *Parafusulina* sp. and the circular specimen in the lower right (view is a cross-section perpendicular to the long axis of the specimen) is an unidentified fusulinid.

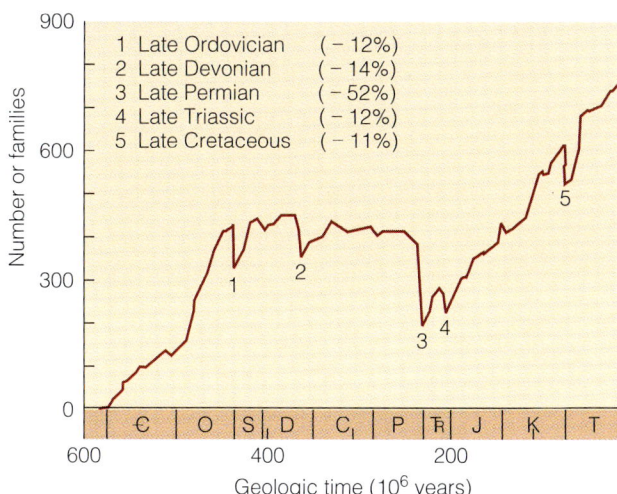

• **Figure 12.19 Phanerozoic Marine Diversity** Phanerozoic diversity for marine invertebrate and vertebrate families. Note the three episodes of Paleozoic mass extinctions, with the greatest occurring at the end of the Permian Period.

In addition, there is also evidence of increased global warming during the Late Permian. This would also contribute to a stratified global ocean because warming of the high latitudes would significantly reduce or eliminate the downwelling of cold, dense, oxygenated waters from the polar areas into the deep oceans at lower latitudes, as occurs today. This would result in stagnant, stratified oceans, rather than a well-mixed, oxygenated oceanic system.

During this time (Late Permian), widespread volcanic and continental fissure eruptions were also taking place, releasing additional carbon dioxide into the atmosphere and contributing to increased climatic instability and ecologic collapse. By the end of the Permian, a near collapse of both the marine and terrestrial ecosystem had occurred. Although the ultimate cause of such devastation is still being debated and investigated, it is safe to say that it was probably a combination of interconnected and related geologic and biologic events.

SUMMARY

Table 12.2 summarizes the major evolutionary and geologic events of the Paleozoic Era and shows their relationships to each other.

- Multicelled organisms presumably had a long Precambrian history, during which they lacked hard parts. Invertebrates with hard parts suddenly appeared during the Early Cambrian in what is called the Cambrian explosion. Skeletons provided such advantages as protection against predators and support for muscles, enabling organisms to grow large and increase locomotor efficiency. Hard parts probably evolved as a result of various geologic and biologic factors rather than a single cause.

- Marine organisms are classified as plankton if they are floaters, nekton if they swim, and benthos if they live on or in the seafloor.

- Marine organisms are divided into four basic feeding groups: suspension feeders, which consume microscopic plants and animals as well as dissolved nutrients from water; herbivores, which are plant eaters; carnivore-scavengers, which are meat eaters; and sediment-deposit feeders, which ingest sediment and extract nutrients from it.

- The marine ecosystem consists of various trophic levels of food production and consumption. At the base are primary producers, on which all other organisms are dependent. Feeding on the primary producers are the primary consumers, which in turn are fed on by higher levels of consumers. The decomposers are bacteria that break down the complex organic compounds of dead organisms and recycle them within the ecosystem.

- The Cambrian invertebrate community was dominated by three major groups: the trilobites, inarticulate brachiopods, and archaeocyathids. Little specialization existed among the invertebrates, and most phyla were represented by only a few species. The Middle Cambrian Burgess Shale contains one of the finest examples of a well-preserved soft-bodied biota in the world.

Table 12.2
Major Evolutionary and Geologic Events of the Paleozoic Era

Age (millions of years)	Geologic Period		Invertebrates	Vertebrates	Plants	Major Geologic Events
251	Permian		Largest mass extinction event to affect the invertebrates	Acanthodians, placoderms, and pelycosaurs become extinct Therapsids and pelycosaurs the most abundant reptiles	Gymnosperms diverse and abundant	Formation of Pangaea Alleghenian orogeny Hercynian orogeny
299	Carboniferous	Pennsylvanian	Fusulinids diversify	Amphibians abundant and diverse	Coal swamps with flora of seedless vascular plants and gymnosperms	Coal-forming swamps common Formation of Ancestral Rockies Continental glaciation in Gondwana
318		Mississippian	Crinoids, lacy bryozoans, blastoids become abundant Renewed adaptive radiation following extinctions of many reef-builders	Reptiles evolve	Gymnosperms appear (may have evolved during Late Devonian)	Ouachita orogeny Widespread deposition of black shale
359	Devonian		Extinctions of many reef-building invertebrates near end of Devonian Reef building continues Eurypterids abundant	Amphibians evolve All major groups of fish present—Age of Fish	First seeds evolve Seedless vascular plants diversify	Widespread deposition of black shale Acadian orogeny Antler orogeny
416	Silurian		Major reef building Diversity of invertebrates remains high	Ostracoderms common Acanthodians, the first jawed fish, fish evolve	Early land plants— seedless vascular plants	Caledonian orogeny Extensive barrier reefs and evaporites
444	Ordovician		Extinctions of a variety of marine invertebrates near end of Ordovician Major adaptive radiation of all invertebrate groups Suspension feeders dominant	Ostracoderms diversify	Plants move to land?	Continental glaciation in Gondwana Taconic orogeny
488	Cambrian		Many trilobites become extinct near end of Cambrian Trilobites, brachiopods, and archaeocyathids are most abundant	Earliest vertebrates— jawless fish called ostracoderms		First Phanerozoic transgression (Sauk) onto North American craton
542						

- The Ordovician marine invertebrate community marked the beginning of the dominance by the shelly fauna and the start of large-scale reef building. The end of the Ordovician Period was a time of major extinction for many invertebrate phyla.
- The Silurian and Devonian periods were times of diverse faunas dominated by reef-building animals, whereas the Carboniferous and Permian periods saw a great decline in invertebrate diversity.
- A major extinction occurred at the end of the Paleozoic Era, affecting the invertebrates as well as the vertebrates. Its cause is still the subject of debate.

IMPORTANT TERMS

acritarch, p. 242
archaeocyathid, p. 240
benthos, p. 237
brachiopod, p. 239
carnivore-scavenger, p. 237
conodont, p. 242
graptolite, p. 242
herbivore, p. 237
nekton, p. 237
plankton, p. 237
primary producer, p. 238
sediment-deposit feeder, p. 237
suspension feeder, p. 237
trilobite, p. 239
trophic level, p. 238

REVIEW QUESTIONS

1. Organisms that manufacture their own food are
   a. _____ autotrophs; b. _____ herbivores; c. _____ benthos; d. _____ epifaunal; e. _____ none of the previous answers.
2. The major organic-walled phytoplankton group of the Paleozoic Era was
   a. _____ acritarchs; b. _____ coccolithophorids; c. _____ diatoms; d. _____ dinoflagellates; e. _____ graptolites.
3. Which group of planktonic invertebrates that were especially abundant during the Ordovician and Silurian periods are excellent guide fossils?
   a. _____ brachiopods; b. _____ cephalopods; c. _____ fusulinids; d. _____ graptolites; e. _____ trilobites.
4. Organisms living in the water column above the seafloor are
   a. _____ benthic; b. _____ epifaunal; c. _____ infaunal; d. _____ epifloral; e. _____ pelagic.
5. The Burgess Shale fauna is significant because it contains the
   a. _____ first shelled animals; b. _____ carbonized impressions of many extinct soft-bodied animals; c. _____ fossils of rare marine plants; d. _____ earliest known benthic community; e. _____ conodont animal.
6. Brachiopods are
   a. _____ benthic mobile carnivores; b. _____ benthic mobile scavengers; c. _____ benthic suspension feeders; d. _____ nektonic carnivore-scavengers; e. _____ planktonic primary producers.
7. What type of invertebrates dominated the Ordovician invertebrate community?
   a. _____ epifloral planktonic primary producers; b. _____ infaunal nektonic carnivores; c. _____ infaunal benthic sessile suspension feeders; d. _____ epifaunal benthic mobile suspension feeders; e. _____ epifaunal benthic sessile suspension feeders.
8. An exoskeleton is advantageous because it
   a. _____ prevents drying out in an intertidal environment; b. _____ provides protection against ultraviolet radiation; c. _____ provides protection against predators; d. _____ provides attachment sites for development of strong muscles; e. _____ all of the previous answers.
9. Mass extinctions occurred at or near the end of which three periods?
   a. _____ Cambrian, Ordovician, Permian; b. _____ Cambrian, Silurian, Devonian; c. _____ Ordovician, Devonian, Permian; d. _____ Silurian, Devonian, Permian; e. _____ Cambrian, Devonian, Permian.
10. The earliest reeflike structures were constructed by
    a. _____ bryozoans; b. _____ mollusks; c. _____ archaeocyathids; d. _____ sponges; e. _____ corals.
11. The age of the Burgess Shale is
    a. _____ Cambrian; b. _____ Ordovician; c. _____ Silurian; d. _____ Devonian; e. _____ Mississippian.
12. The greatest recorded mass extinction in Earth history took place at the end of which period?
    a. _____ Cambrian; b. _____ Ordovician; c. _____ Devonian; d. _____ Permian; e. _____ Cretaceous.
13. The three invertebrate groups that comprised the majority of Cambrian skeletonized life were
    a. _____ trilobites, archaeocyathids, brachiopods; b. _____ echinoderms, corals, bryozoans; c. _____ brachiopods, archaeocyathids, corals; d. _____ trilobites, echinoderms, corals; e. _____ trilobites, brachiopods, corals.
14. The _____ and _____ were times of major reef building.
    a. _____ Cambrian, Ordovician; b. _____ Ordovician, Silurian; c. _____ Silurian, Devonian; d. _____ Devonian, Mississippian; e. _____ Mississippian, Pennsylvanian.
15. Discuss how changing geologic conditions affected the evolution of invertebrate life during the Paleozoic Era.

16. If the Cambrian explosion of life was partly the result of filling unoccupied niches, why don't we see such rapid evolution following mass extinctions such as those that occurred at the end of the Permian and Cretaceous periods?
17. Discuss the significance of the appearance of the first shelled animals and possible causes for the acquisition of a mineralized exoskeleton.
18. Discuss how the incompleteness of the fossil record may play a role in what is known as the Cambrian explosion.
19. What are the major differences between the Cambrian marine community and the Ordovician marine community?
20. Discuss some of the possible causes for the Permian mass extinction.

APPLY YOUR KNOWLEDGE

1. Draw a marine food web that shows the relationships among the producers, consumers, and decomposers.

FIELD QUESTION

1. What are the majority of fossils in this slab of Paleozoic limestone? Assuming you answered the first part of this question correctly, what would you estimate the age of this rock to be? In other words, what geologic period would be a reasonable answer based on the fossils in this rock?

CHAPTER 13
PALEOZOIC LIFE HISTORY: VERTEBRATES AND PLANTS

Courtesy of Ken Higgs, Department of Geology, University College, Cork, Ireland

Tetrapod trackway at Valentia Island, Ireland. These fossilized footprints, which are more than 365 million years old, are evidence of one of the earliest four-legged animals on land.

[OUTLINE]

INTRODUCTION

VERTEBRATE EVOLUTION

FISH

AMPHIBIANS—VERTEBRATES INVADE THE LAND

EVOLUTION OF THE REPTILES— THE LAND IS CONQUERED

PLANT EVOLUTION

PERSPECTIVE 13.1 *Palynology: A Link between Geology and Biology*

Silurian and Devonian Floras

Late Carboniferous and Permian Floras

SUMMARY

ThomsonNOW™ Explore interactive tutorials, animations, or practice problems available on the ThomsonNow website at **www.thomsonedu.com/login**.

[CHAPTER OBJECTIVES]

At the end of this chapter, you will have learned that

- Vertebrates first evolved during the Cambrian Period, and fish diversified rapidly during the Paleozoic Era.

- Amphibians first appear in the fossil record during the Late Devonian, having made the transition from water to land, and became extremely abundant during the Pennsylvanian Period.

- The evolution of the amniote egg allowed reptiles to colonize all parts of the land beginning in the Late Mississippian.

- The pelycosaurs or finback reptiles were the dominant reptiles during the Permian and were the ancestors to the therapsids or mammal-like reptiles.

- The earliest land plants are known from the Ordovician Period, whereas the oldest known vascular land plants first appear in the Middle Silurian.

- Seedless vascular plants were very abundant during the Pennsylvanian Period.

- With the onset of arid conditions during the Permian Period, the gymnosperms became the dominant element of the world's flora.

Introduction

The discovery in 1992 of fossilized tetrapod footprints more than 365 million years old has forced paleontologists to rethink how and when animals emerged onto land. The Late Devonian trackway that Swiss geologist Iwan Stössel discovered that year on Valentia Island, off the southwest coast of Ireland, has helped shed light on the early evolution of tetrapods (from the Greek *tetra,* meaning "four," and *podos,* meaning "foot"). Given these footprints, geologists estimate that the creature was longer than three feet and had fairly large back legs. Furthermore, instead of walking on dry land, this animal was probably walking or wading around in a shallow, tropical stream, filled with aquatic vegetation and predatory fish. This hypothesis is based on the fact the trackway showed no evidence of a tail being dragged behind it. Unfortunately, no bones are associated with the tracks to help reconstruct what this primitive tetrapod looked like.

One of the intriguing questions paleontologists ask is, Why did limbs evolve in the first place? It was probably not for walking on land. In fact, many scientists think aquatic limbs made it easier to move around in streams, lakes, or swamps that were choked with water plants or other debris. The scant fossil evidence also seems to support this hypothesis. Fossils of *Acanthostega*, a tetrapod found in 360-million-year-old rocks from Greenland, reveals an animal that had limbs but was clearly unable to walk on land. Paleontologist Jennifer Clack, who has recovered and analyzed hundreds of specimens of *Acanthostega*, points out that *Acanthostega*'s limbs were not strong enough to support its weight on land, and its rib cage was too small for the necessary muscles needed to hold its body off the ground. In addition, *Acanthostega* had gills and lungs, meaning it could survive on land, but it was more suited for the water. Clack thinks *Acanthostega* used its limbs to maneuver around in swampy, plant-filled waters, where swimming would be difficult and limbs would be an advantage.

Presently, there are many more questions about the evolution of the earliest tetrapods than there are answers. However, the publication in 2006 of a Late Devonian tetrapod-like fish from Canada's Ellesmere Island provides important insights into the transition between lobe-finned fish and tetrapods. It is discoveries like this that make studying prehistoric life so interesting and exciting.

In the previous chapter, we examined the Paleozoic history of invertebrates, beginning with the acquisition of hard parts and concluding with the massive Permian extinctions that claimed about 90% of all invertebrates and more than 65% of all amphibians and reptiles. In this chapter we examine the Paleozoic evolutionary history of vertebrates and plants.

What Would You Do?

Because of the recent controversy concerning the teaching of evolution in the public schools, your local school board has asked you to make a 30-minute presentation on the history of life and how such a history is evidence that evolution is a valid scientific theory. With so much material to cover and so little time, you decide to focus on the Paleozoic evolutionary history of vertebrates. You have chosen the Paleozoic Era because during this time the stage was set, so to speak, for the later evolution of dinosaurs, birds, and mammals, groups with which most citizens are familiar. What features of the Paleozoic vertebrate fossil record would you emphasize, and how would you go about convincing the school board that evolution has taken place? How would the recent discovery of *Tiktaalik roseae* help your argument?

One of the striking parallels between plants and animals is that in making the transition from water to land, both plants and animals had to solve the same basic problems. For both groups, the method of reproduction proved the major barrier to expansion into the various terrestrial environments. With the evolution of the seed in plants and the amniote egg in animals, this limitation was removed, and both groups expanded into all terrestrial habitats.

Vertebrate Evolution

A **chordate** (phylum Chordata) is an animal that has, at least during part of its life cycle, a notochord, a dorsal hollow nerve cord, and gill slits (• Figure 13.1). **Vertebrates**, which are animals with backbones, are simply a subphylum of chordates.

The ancestors and early members of the phylum Chordata were soft-bodied organisms that left few fossils (• Figure 13.2). Consequently, we know little about the early evolutionary history of the chordates or vertebrates. Surprisingly, a close relationship exists between echinoderms (see Table 12.1) and chordates, and they may even have shared a common ancestor. This is because in the developing embryo of echinoderms and chordates, cells divide by radial cleavage so that the cells are aligned directly above each other (• Figure 13.3a). In all other invertebrates, cells undergo spiral cleavage, which results in having cells nested between each other in successive rows (Figure 13.3b). Furthermore, the biochemistry of muscle activity and blood proteins, and the larval stages are similar in both echinoderms and chordates.

The evolutionary pathway to vertebrates thus appears to have taken place much earlier and more rapidly than many scientists have long thought. Based on fossil evidence and recent advances in molecular biology, one scenario suggests that vertebrates evolved shortly after an ancestral chordate, probably resembling *Yunnanozoon*, and acquired a second set of genes. According to this hypothesis, a random mutation produced a duplicate set of genes, letting the ancestral vertebrate animal evolve entirely new body structures that proved to be evolutionarily advantageous. Not all scientists accept this hypothesis, and the origin of vertebrates is still hotly debated.

• **Figure 13.2** *Yunnanozoon lividum* Found in 525-million-year-old rocks in Yunnan province, China, *Yunnanozoon lividum*, a 5-cm-long animal, is one of the oldest known chordates.

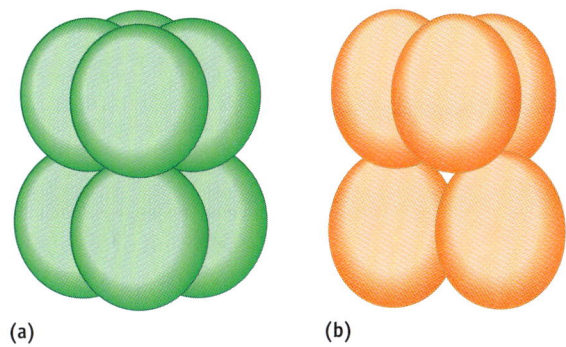

• **Figure 13.3 Cell Cleavage** (a) Arrangement of cells resulting from radial cleavage is characteristic of chordates and echinoderms. In this configuration, cells are directly above each other. (b) Arrangement of cells resulting from spiral cleavage. In this arrangement, cells in successive rows are nested between each other. Spiral cleavage is characteristic of all invertebrates except echinoderms.

Fish

The most primitive vertebrates are fish, and some of the oldest fish remains are found in the Upper Cambrian Deadwood Formation in northeastern Wyoming (• Figure 13.4). Here phosphatic scales and plates of *Anatolepis*, a primitive member of the class Agnatha (jawless fish), have been recovered from marine sediments. All known Cambrian and Ordovician fossil fish have been found in shallow,

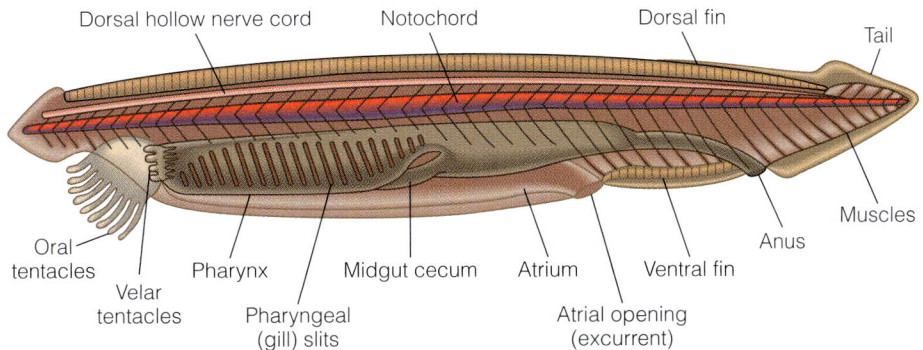

• **Figure 13.1 Three Characteristics of a Chordate** The structure of the lancelet *Amphioxus* illustrates the three characteristics of a chordate: a notochord, a dorsal hollow nerve cord, and gill slits.

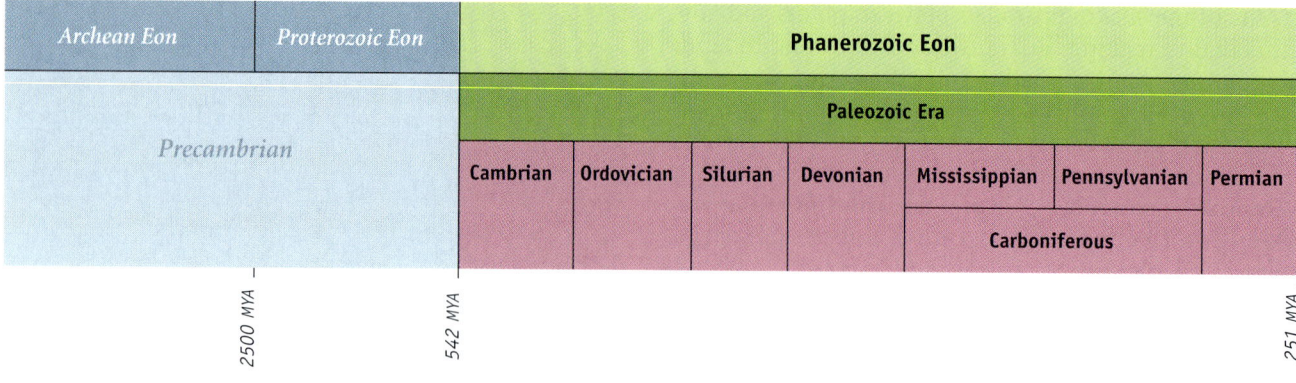

• Figure 13.4 **Ostracoderm Fish Plate** A fragment of a plate from *Anatolepis* cf. *A. heintzi* from the Upper Cambrian Deadwood Formation of Wyoming. *Anatolepis* is one of the oldest known fish.

nearshore marine deposits, whereas the earliest nonmarine (freshwater) fish remains have been found in Silurian strata. This does not prove that fish originated in the oceans, but it does lend strong support to the idea.

As a group, fish range from the Late Cambrian to the present (• Figure 13.5). The oldest and most primitive of the class Agnatha are the **ostracoderms,** whose name means "bony skin" (Table 13.1). These are armored, jawless fish that first evolved during the Late Cambrian, reached their zenith during the Silurian and Devonian, and then became extinct.

The majority of ostracoderms lived on the seafloor. *Hemicyclaspis* is a good example of a bottom-dwelling ostracoderm (• Figure 13.6a). Vertical scales allowed *Hemicyclaspis* to wiggle sideways, propelling itself along the seafloor, and the eyes on the top of its head allowed it to see such predators as cephalopods and jawed fish approaching from above. While moving along the sea bottom, it probably sucked up small bits of food and sediments through its jawless mouth. Another type of ostracoderm, represented by *Pteraspis,* was more elongated and probably an active

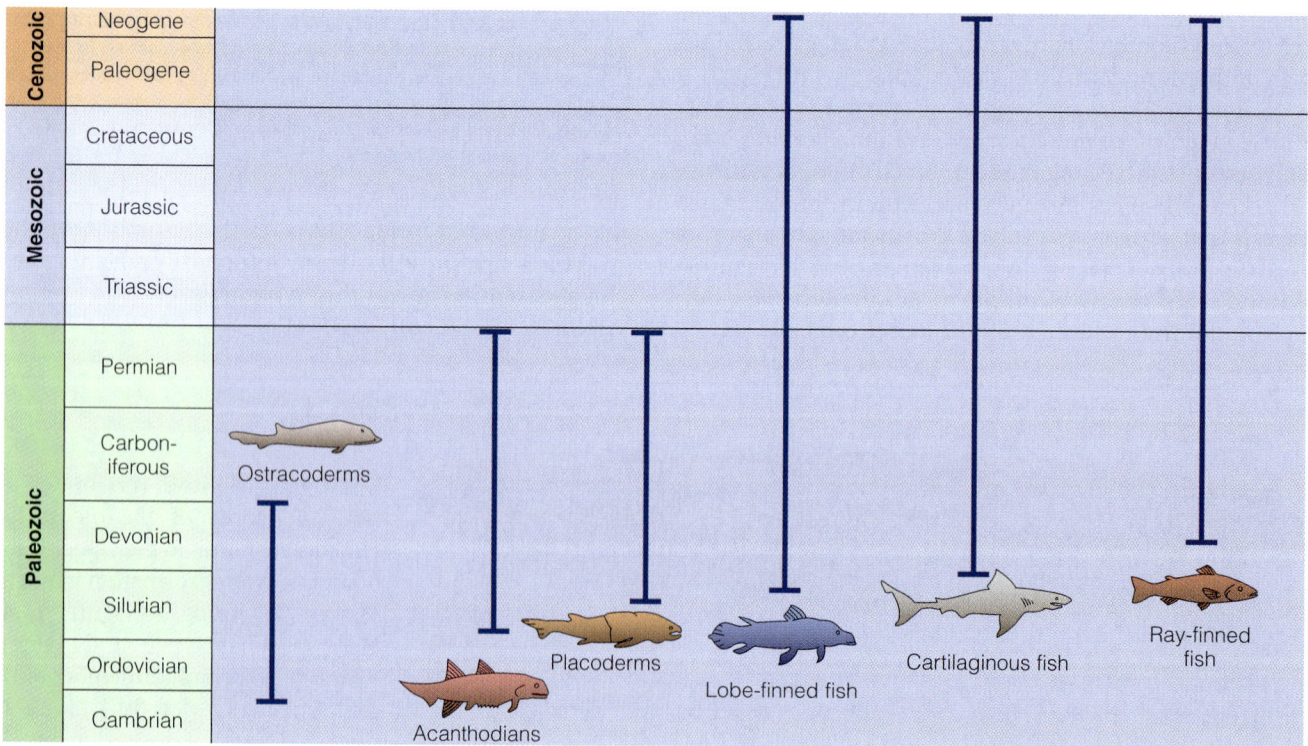

• Figure 13.5 **Geologic Ranges of the Major Fish Groups** Ostracoderms are early members of the class Agnatha (jawless fish). Acanthodians (class Acanthodii) are the first fish with jaws. Placoderms (class Placodermii) are armored, jawed fish. Lobe-finned (subclass Sarcopterygii) and ray-finned (subclass Actinopterygii) fish are members of the class Osteichthyes (bony fish), whereas cartilaginous fish belong to the class Chondrichthyes.

Phanerozoic Eon											
Mesozoic Era			Cenozoic Era								
Triassic	Jurassic	Cretaceous	Paleogene			Neogene		Quaternary	
			Paleocene	Eocene	Oligocene	Miocene	Pliocene	Pleistocene	Holocene

251 MYA 66 MYA

swimmer, although it also seemingly fed on small pieces of food that it was able to suck up.

The evolution of jaws was a major evolutionary advance among primitive vertebrates. Although their jawless ancestors could only feed on detritus, jawed fish could chew food and become active predators, thus opening many new ecologic niches.

The vertebrate jaw is an excellent example of evolutionary opportunism. Various studies suggest that the jaw originally evolved from the first three gill arches of jawless fish. Because the gills are soft, they are supported by gill arches of bone or cartilage. The evolution of the jaw may thus have been related to respiration rather than to feeding (• Figure 13.7). By evolving joints in the forward gill arches, jawless fish could open their mouths wider. Every time a fish opened and closed its mouth, it would pump more water past the gills, thereby increasing the oxygen intake. The modification from rigid to hinged forward gill arches let fish increase both their food consumption and oxygen intake, and the evolution of the jaw as a feeding structure rapidly followed.

The fossil remains of the first jawed fish are found in Lower Silurian rocks and belong to the **acanthodians** (class Acanthodii), a group of small, enigmatic fish characterized by large spines, paired fins, scales covering much of the body, jaws, teeth, and greatly reduced body armor (Figure

• **Figure 13.6 Devonian Seafloor** Recreation of a Devonian seafloor showing **(a)** an ostracoderm (*Hemicyclaspis*), **(b)** a placoderm (*Bothriolepis*), **(c)** an acanthodian (*Parexus*), and **(d)** a ray-finned fish (*Cheirolepis*).

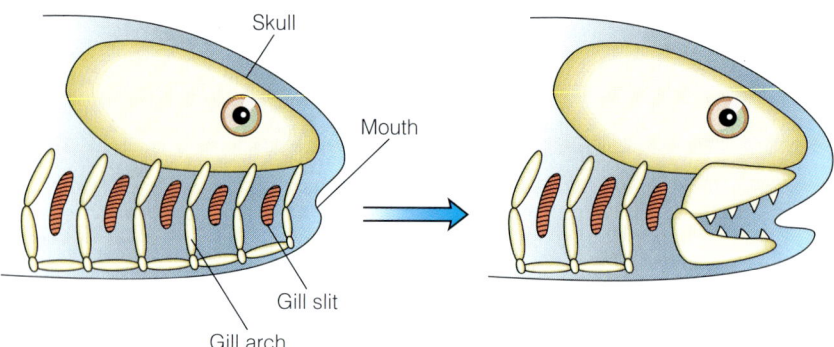

• **Figure 13.7 Evolution of the Vertebrate Jaw** The evolution of the vertebrate jaw is thought to have begun from the modification of the first two or three anterior gill arches. This theory is based on the comparative anatomy of living vertebrates.

13.6c, Table 13.1). Although their relationship to other fish is not well established, many scientists think the acanthodians included the probable ancestors of the present-day bony and cartilaginous fish groups. The acanthodians were most abundant during the Devonian, declined in importance through the Carboniferous, and became extinct during the Permian.

The other jawed fish, the **placoderms** (class Placodermii), whose name means "plate-skinned," evolved during the Late Silurian. Placoderms were heavily armored, jawed fish that lived in both freshwater and the ocean, and, like the acanthodians, reached their peak of abundance and diversity during the Devonian.

The placoderms showed considerable variety, including small bottom dwellers (Figure 13.6b), as well as large major predators such as *Dunkleosteus,* a Late Devonian fish that lived in the midcontinental North American epeiric seas (• Figure 13.8a). It was by far the largest fish of the time, reaching a length of more than 12 m. It had a heavily armored head and shoulder region, a huge jaw lined with razor-sharp bony teeth, and a flexible tail, all features consistent with its status as a ferocious predator.

Besides the abundant acanthodians, placoderms, and ostracoderms, other fish groups, such as the cartilaginous and bony fish, also evolved during the Devonian Period. Small wonder, then, that the Devonian is informally called the "Age of Fish," because all major fish groups were present during this time period.

The **cartilaginous fish** (class Chrondrichthyes) (Table 13.1), represented today by sharks, rays, and skates, first evolved during the Early Devonian, and by the Late Devonian, primitive marine sharks such as *Cladoselache* were quite abundant (Figure 13.8b). Cartilaginous fish have never been as numerous or as diverse as their cousins, the bony fish, but they were, and still are, important members of the marine vertebrate fauna.

Along with cartilaginous fish, the **bony fish** (class Osteichthyes) (Table 13.1) also first evolved during the Devonian. Because bony fish are the most varied and numerous of all the fishes, and because the amphibians evolved from them, their evolutionary history is particularly important. There are two groups of bony fish: the common **ray-finned fish** (subclass Actinopterygii) (Figure 13.8d) and the less familiar **lobe-finned fish** (subclass Sarcopterygii) (Table 13.1).

The term *ray-finned* refers to the way the fins are supported by thin bones that spread away from the body (• Figure 13.9a). From a modest freshwater beginning during the Devonian, ray-finned fish, which include most of the familiar fish such as trout, bass, perch, salmon, and tuna, rapidly diversified to dominate the Mesozoic and Cenozoic seas.

Present-day lobe-finned fish are characterized by muscular fins. The fins do not have radiating bones but rather have articulating bones with the fin attached to the body by a fleshy shaft (Figure 13.9b). Such an arrangement allows for a powerful stroke of the fin, making the fish an effective

TABLE 13.1 Brief Classification of Fish Groups Referred to in the Text

Classification	Geologic Range	Living Example
Class Agnatha (jawless fish)	Late Cambrian–Recent	Lamprey, hagfish
Early members of the class are called ostracoderms		No living ostracoderms
Class Acanthodii (the first fish with jaws)	Early Silurian–Permian	None
Class Placodermii (armored jawed fish)	Late Silurian–Permian	None
Class Chondrichthyes (cartilaginous fish)	Devonian–Recent	Sharks, rays, skates
Class Osteichthyes (bony fish)	Devonian–Recent	Tuna, perch, bass, pike, catfish, trout, salmon, lungfish, *Latimeria*
Subclass Actinopterygii (ray-finned fish)	Devonian–Recent	Tuna, perch, bass, pike, catfish, trout, salmon
Subclass Sarcopterygii (lobe-finned fish)	Devonian–Recent	Lungfish, *Latimeria*
Order Coelacanthimorpha	Devonian–Recent	*Latimeria*
Order Dipnoi	Devonian–Recent	Lungfish
Order Crossopterygii	Devonian–Permian	None
Suborder Rhipidistia	Devonian–Permian	None

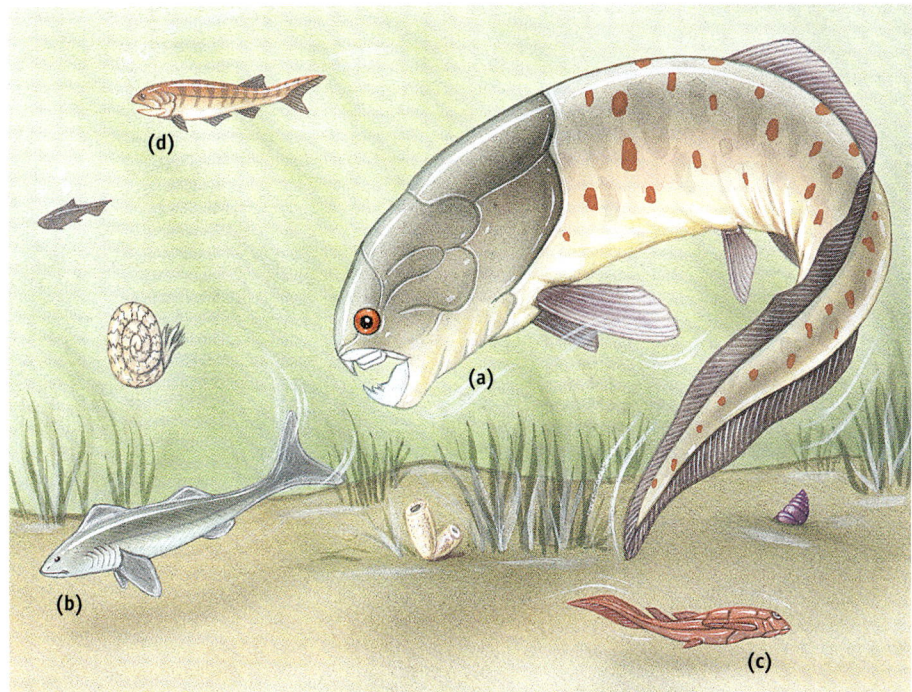

• **Figure 13.8 Late Devonian Seascape** A Late Devonian marine scene from the midcontinent of North America. **(a)** The giant placoderm *Dunkleosteus* (length more than 12 m) is pursuing **(b)** the shark *Cladoselache* (length up to 1.2 m). Also shown are **(c)** the bottom-dwelling placoderm *Bothriolepis* and **(d)** the swimming ray-finned fish *Cheirolepsis,* both of which reached a length of 40–50 cm.

swimmer. Three orders of lobe-finned fish are recognized: *coelacanths*, *lungfish*, and *crossopterygians* (Table 13.1).

Coelacanths (order Coelacanthimorpha) are marine lobe-finned fish that evolved during the Middle Devonian and were thought to have gone extinct at the end of the Cretaceous. In 1938, however, a fisherman caught a coelacanth in the deep waters off Madagascar (see Figure 7.17a), and since then, several dozen more have been caught, both there and in Indonesia.

Lungfish (order Dipnoi) were fairly abundant during the Devonian, but today only three freshwater genera exist, one each in South America, Africa, and Australia. Their present-day distribution presumably reflects the Mesozoic breakup of Gondwana.

The "lung" of a modern-day lungfish is actually a modified swim bladder that most fish use for buoyancy in swimming. In lungfish, this structure absorbs oxygen, allowing them to breath air when the lakes or streams in which they live become stagnant and dry up. During such times, they burrow into the sediment to prevent dehydration and breath through their swim bladder until the stream begins flowing or the lake they were living in fills with water. When they are back in the water, lungfish then rely on gill respiration.

The **crossopterygians** (order Crossopterygii) are an important group of lobe-finned fish, because it is probably from them that amphibians evolved. However, the transition between crossopterygians and true amphibians is not as simple as it was once portrayed. The group of crossopterygians that appears to be ancestral to amphibians are *rhipidistians* (Table 13.1). These fish, reaching lengths of over 2 m, were the dominant freshwater predators during the Late Paleozoic. *Eusthenopteron,* a good example of a rhipidistian crossopterygian and the classic example of the transitional form between fish and amphibians, had an elongated body that helped it move swiftly through the water and paired, muscular fins that many scientists thought could be used for moving on land (• Figure 13.10). The structural similarity between crossopterygian fish and the earliest amphibians is striking and one of the most widely cited examples of a transition from one major group to another (• Figure 13.11). However, recent discoveries of older lobe-finned fish and tetrapods like *Acanthostega* (see the Introduction), and newly published findings of tetrapod-like fish, are filling in the gaps in the time of the evolution between fish and tetrapods.

Before discussing this transition and the evolution of amphibians, it is useful to place the evolutionary history of Paleozoic fish in the larger context of Paleozoic evolutionary events. Certainly, the evolution and diversification of jawed fish as well as eurypterids and ammonoids had a profound

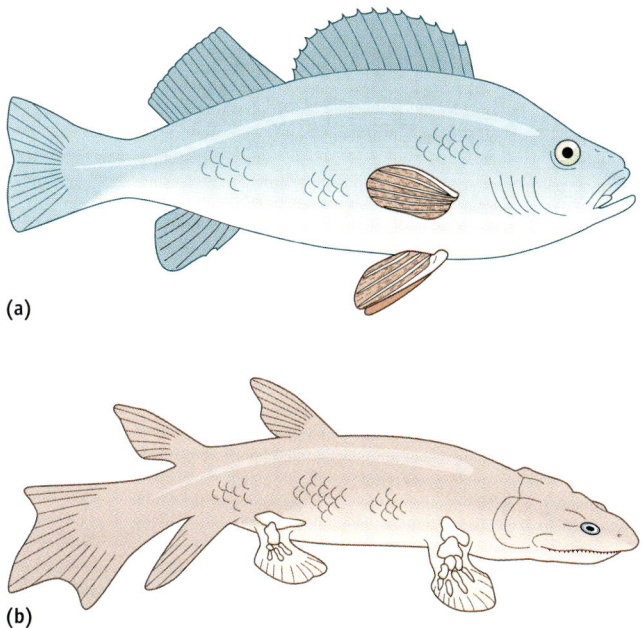

• **Figure 13.9 Ray-finned and Lobe-finned Fish** Arrangement of fin bones for **(a)** a typical ray-finned fish and **(b)** a lobe-finned fish. The muscles extend into the fin of the lobe-finned fish, allowing greater flexibility of movement than that of the ray-finned fish.

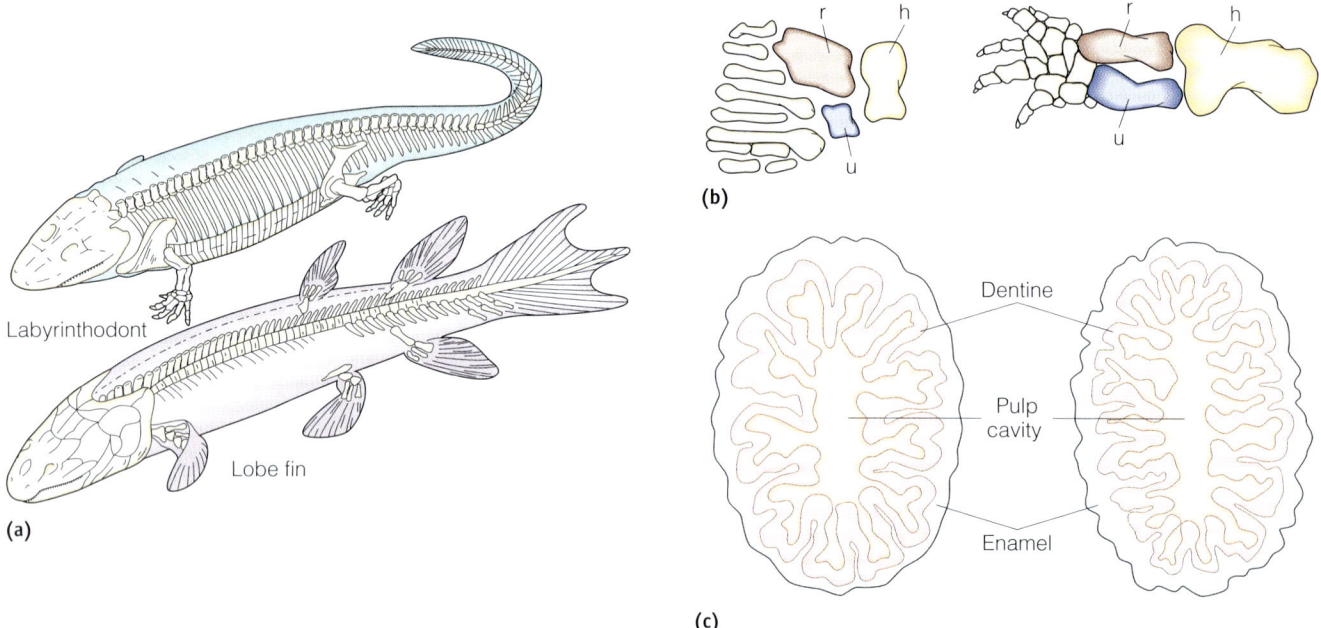

• **Figure 13.10 Rhipidistian Crossopterygian** The crossopterygians are the group from which the amphibians are thought to have evolved. *Eusthenopteron*, a member of the rhipidistian crossopterygians, had an elongated body and paired fins that could be used for moving about on land.

• **Figure 13.11 Similarities Between Crossopterygians and Labyrinthodonts** Similarities between the crossopterygian lobe-finned fish and the labyrinthodont amphibians. **(a)** Skeletal similarity. **(b)** Comparison of the limb bones of a crossopterygian *(left)* and amphibian *(right)*; color identifies the bones (u = ulna, shown in blue, r = radius, mauve, h = humerus, gold) that the two groups have in common. **(c)** Comparison of tooth cross sections shows the complex and distinctive structure found in both the crossopterygians *(left)* and labyrinthodont amphibians *(right)*.

effect on the marine ecosystem. Previously defenseless organisms either evolved defensive mechanisms or suffered great losses, possibly even extinction.

Recall from Chapter 12 that trilobites experienced major extinctions at the end of the Cambrian, recovered slightly during the Ordovician, and then declined greatly from the end of the Ordovician to final extinction at the end of the Permian. Perhaps their lightly calcified external covering made them easy prey for the rapidly evolving jawed fish and cephalopods.

Ostracoderms, although armored, would also have been easy prey for the swifter jawed fishes. Ostracoderms became extinct by the end of the Devonian, a time that coincides with the rapid evolution of jawed fish. Placoderms, like acanthodians, greatly decreased in abundance after the Devonian and became extinct by the end of the Paleozoic Era. In contrast, cartilaginous and ray-finned bony fish expanded during the Late Paleozoic, as did the ammonoid cephalopods, the other major predators of the Late Paleozoic seas.

Amphibians—Vertebrates Invade the Land

Although amphibians were the first vertebrates to live on land, they were not the first land-living organisms. Land plants, which probably evolved from green algae, first evolved during the Ordovician. Furthermore, insects, millipedes, spiders, and even snails invaded the land before amphibians. Fossil evidence indicates that such land-dwelling arthropods as scorpions and flightless insects had evolved by at least the Devonian.

The transition from water to land required animals to surmount several barriers. The most critical were desiccation, reproduction, the effects of gravity, and the extraction of oxygen from the atmosphere by lungs rather than from water by gills. Up until the 1990s, the traditional evolutionary sequence had a Rhipidistian crossopterygian, like *Eusthenopteron*, evolving into a primitive amphibian like *Ichthyostega*. At that time, fossils of those two genera were about all paleontologists had to work with, and although there were gaps in morphology, the link between crossopterygians and these earliest amphibians was easy to see (Figure 13.11).

Crossopterygians already had a backbone and limbs that could be used for walking and lungs that could extract oxygen (Figure 13.11). The oldest amphibian fossils, on the other hand, found in the Upper Devonian Old Red Sandstone of eastern Greenland and belonging to such genera as *Ichthyostega,* had streamlined bodies, long tails, and fins along their backs, in addition to four legs, a strong backbone, a rib cage, and pelvic and pectoral girdles, all of which were structural adaptations for walking on land (• Figure 13.12). These earliest amphibians thus appear to have inherited many characteristics from the crossopterygians with little modification (Figure 13.11). However, with the discovery of such fossils as *Acanthostega* and others like it, the transition between fish and amphibians involves a number of new genera that are intermediary between the two groups.

Panderichthys, a large (up to 1.3 m long), Late Devonian (~380 million years ago) lobe-finned fish from Latvia, was essentially a contemporary of *Eusthenopteron*. It had a large tetrapod-like head with a pointed snout, dorsally located eyes, and modifications to that part of the skull related to the ear region. From paleoenvironmental evidence, *Panderichthys* lived in shallow tidal flats or estuaries, using its lobe fins to maneuver around in the shallow waters.

We've already discussed in the Introduction *Acanthostega*, a Late Devonian (365 million years ago) tetrapod that seemed to be the perfect intermediary between fish and true land-dwelling tetrapods. However, its limbs could not support its weight on land, and thus it was an aquatic animal, using its limbs to navigate in water, rather than walking on land.

In 2006, an exciting discovery of a 1.2–2.8 m long, 375-million-year-old (Late Devonian) "fishapod" was announced. Discovered on Ellesmere Island, Canada, *Tiktaalik roseae*, from the Inuktitut meaning "large fish in

• **Figure 13.12 Late Devonian Landscape in Eastern Greenland** Shown is *Ichthyostega,* an amphibian that grew to a length of about 1 m. The flora of the time was diverse, consisting of a variety of small and large seedless vascular plants.

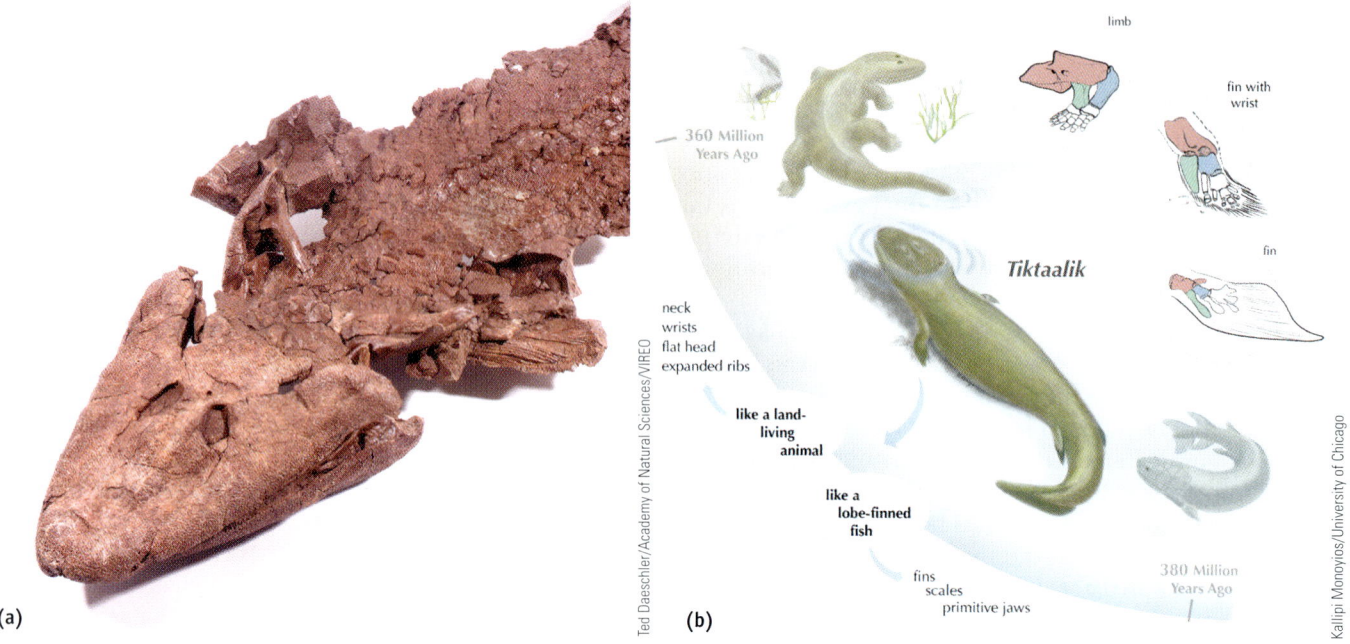

• **Figure 13.13 Tiktaalik roseae** (a) Skeleton of *Tiktaalik roseae*, a "fishapod" hailed as an intermediary between lobe-finned fish and tetrapods, has characteristics of both fish and tetrapods. (b) Diagram illustrating how *Tiktaalik roseae* is a transitional species between lobe-finned fish and tetrapods.

a stream," was hailed as an intermediary between the lobe-finned fish like *Panderichthys* and the earliest tetrapod, *Acanthostega* (• Figure 13.13a).

Tiktaalik roseae is truly a "fishapod" in that it has a mixture of both fish and tetrapod characteristics (• Figure 13.13b). For example, it has gills and fish scales but also a broad skull, eyes on top of its head, a flexible neck and large ribcage that could support its body on land or in shallow water, and lungs, all of which are tetrapod features. What really excited scientists, however, was that *Tiktaalik roseae* has the beginnings of a true tetrapod forelimb, complete with functional wrist bones and five digits, as well as a modified ear region. Sedimentological evidence suggests *Tiktaalik roseae* lived in a shallow water habitat associated with Late Devonian floodplains of Laurasia.

As previously mentioned, the oldest known amphibian, *Ichthyostega*, had skeletal features that allowed it to spend its life on land. Because amphibians did not evolve until the Late Devonian, they were a minor element of the Devonian terrestrial ecosystem. Like other groups that moved into new and previously unoccupied niches, amphibians underwent rapid adaptive radiation and became abundant during the Carboniferous and Early Permian.

The Late Paleozoic amphibians did not at all resemble the familiar frogs, toads, newts, and salamanders that make up the modern amphibian fauna. Rather, they displayed a broad spectrum of sizes, shapes, and modes of life (• Figure 13.14). One group of amphibians were the **labyrinthodonts**, so named for the labyrinthine wrinkling and folding of the chewing surface of their teeth (Figure 13.11c). Most labyrinthodonts were large animals, as much as 2 m in length. These typically sluggish creatures lived in swamps and streams, eating fish, vegetation, insects, and other small amphibians (Figure 13.14).

Labyrinthodonts were abundant during the Carboniferous when swampy conditions were widespread (see Chapter 11) but soon declined in abundance during the Permian, perhaps in response to changing climatic conditions. Only a few species survived into the Triassic.

Evolution of the Reptiles— The Land Is Conquered

Amphibians were limited in colonizing the land because they had to return to water to lay their gelatinous eggs. The evolution of the **amniote egg** (• Figure 13.15) freed reptiles from this constraint. In such an egg, the developing embryo is surrounded by a liquid-filled sac called the *amnion* and provided with both a yolk, or food sac, and an allantois, or waste sac. In this way the emerging reptile is in essence a miniature adult, bypassing the need for a larval stage in the water. The evolution of the amniote egg allowed vertebrates to colonize all parts of the land, because they no longer had to return to the water as part of their reproductive cycle.

Many of the differences between amphibians and reptiles are physiologic and are not preserved in the fossil record. Nevertheless, amphibians and reptiles differ sufficiently in skull structure, jawbones, ear location, and limb and vertebral construction to suggest that reptiles evolved from labyrinthodont ancestors by the Late Mississippian. This assessment is based on the discovery of a well-preserved fossil skeleton of the oldest known reptile, *Westlothiana*, and

• **Figure 13.14 Carboniferous Coal Swamp** The varied amphibian fauna of the time is shown, including the large labyrinthodont amphibian *Eryops* (foreground), the larval *Branchiosaurus* (center), and the serpentlike *Dolichosoma* (background).

other fossil reptile skeletons from Late Mississippian-aged rocks in Scotland.

Other early reptile fossils occur in the Lower Pennsylvanian Joggins Formation in Nova Scotia, Canada. Here remains of *Hylonomus* are found in the sediments filling in tree trunks (• Figure 13.16). These earliest reptiles from Scotland and Canada were small and agile and fed largely on grubs and insects. They are loosely grouped together as **protorothyrids,** whose members include the earliest known reptiles (• Figure 13.17). During the Permian Period, reptiles diversified and began displacing many amphibians. The reptiles succeeded partly because of their advanced method of reproduction and their more advanced jaws and teeth, as well as their ability to move rapidly on land.

The **pelycosaurs,** or finback reptiles, evolved from the protorothyrids during the Pennsylvanian and were the dominant reptile group by the Early Permian. They evolved into a diverse assemblage of herbivores, exemplified by the

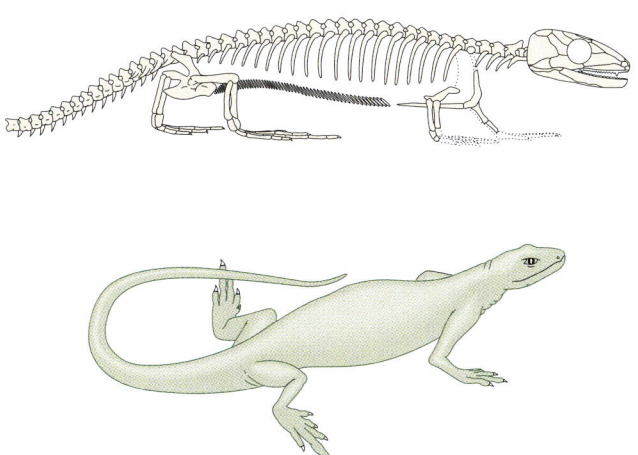

• **Figure 13.15 Amniote Egg** In an amniote egg, the embryo is surrounded by a liquid sac (amnion cavity) and provided with a food source (yolk sac) and waste sac (allantois). The evolution of the amniote egg freed reptiles from having to return to the water for reproduction and let them inhabit all parts of the land.

• **Figure 13.16 *Hylonomus lyelli*** Reconstruction and skeleton of one of the oldest known reptiles, *Hylonomus lyelli,* from the Pennsylvanian Period. Fossils of this animal have been collected from sediments that filled tree stumps. *Hylonomus lyelli* was about 30 cm long.

EVOLUTION OF THE REPTILES— THE LAND IS CONQUERED **259**

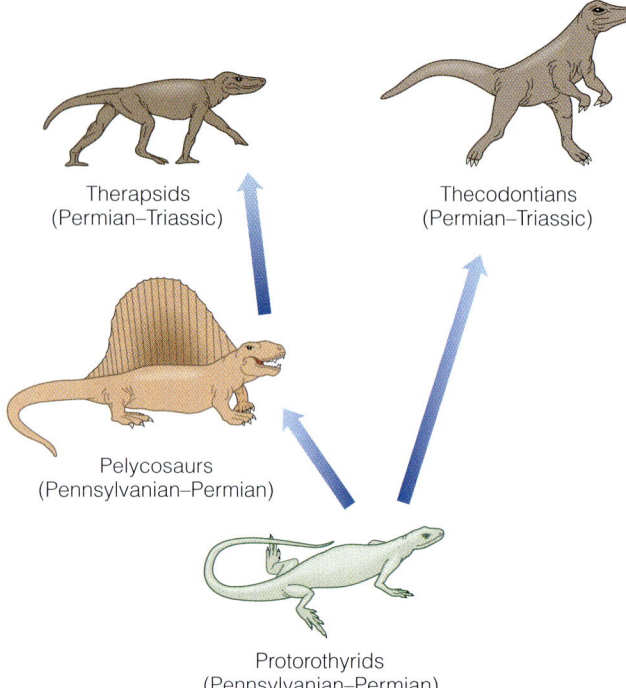

• Figure 13.17 **Evolutionary Relationship Among the Paleozoic Reptiles**

herbivore *Edaphosaurus* and carnivores such as *Dimetrodon* (• Figure 13.18). An interesting feature of the pelycosaurs is their sail. It was formed by vertebral spines that, in life, were covered with skin. The sail has been variously explained as a type of sexual display, a means of protection, and a display to look more ferocious. The current consensus seems to be that the sail served as some type of thermoregulatory device, raising the reptile's temperature by catching the sun's rays or cooling it by facing the wind. Because pelycosaurs are considered the group from which therapsids evolved, it is interesting that they may have had some sort of body-temperature control.

The pelycosaurs became extinct during the Permian and were succeeded by the **therapsids,** mammal-like reptiles that evolved from the carnivorous pelycosaur lineage and rapidly diversified into herbivorous and carnivorous lineages (• Figure 13.19). Therapsids were small- to medium-sized animals that displayed the beginnings of many mammalian features: fewer bones in the skull, because many of the small skull bones were fused; enlarged lower jawbone; differentiation of teeth for various functions such as nipping, tearing, and chewing food; and more vertically placed legs for greater flexibility, as opposed to the way the legs sprawled out to the side in primitive reptiles.

Furthermore, many paleontologists think therapsids were *endothermic,* or warm-blooded, enabling them to maintain a constant internal body temperature. This characteristic would have let them expand into a variety of habitats, and indeed, the Permian rocks in which their fossil remains are found are distributed not only in low latitudes but in middle and high latitudes as well.

As the Paleozoic Era came to an end, the therapsids constituted about 90% of the known reptile genera and occupied a wide range of ecologic niches. The mass extinctions that decimated the marine fauna at the close of the Paleozoic had an equally great effect on the terrestrial population. By the end of the Permian, about 90% of all marine invertebrate species were extinct, compared with more than two-thirds of all amphibians and reptiles. Plants, in contrast, apparently did not experience as great a turnover as animals.

Plant Evolution

When plants made the transition from water to land, they had to solve most of the same problems that animals did: desiccation, support, and the effects of gravity. Plants did so by evolving a variety of structural adaptations that were

• Figure 13.18 **Pelycosaurs** Most pelycosaurs or finback reptiles have a characteristic sail on their back. One hypothesis explains the sail as a type of thermoregulatory device. Other hypotheses are that it was a type of sexual display or a device to make the reptile look more intimidating. Shown here are **(a)** the carnivore *Dimetrodon* and **(b)** the herbivore *Edaphosaurus*.

• **Figure 13.19 Therapsids** A Late Permian scene in southern Africa showing various therapsids including *Dicynodon* (left foreground) and *Moschops* (right). Many paleontologists think therapsids were endothermic and may have had a covering of fur, as shown here.

fundamental to the subsequent radiations and diversification that occurred during the Silurian, Devonian, and later periods (see Perspective 13.1) (Table 13.2). Most experts agree that the ancestors of land plants first evolved in a marine environment and then moved into a freshwater environment before finally moving onto land. In this way, the differences in osmotic pressures between salt and freshwater were overcome while the plant was still in the water.

The higher land plants are divided into two major groups: nonvascular and vascular. Most land plants are **vascular,** meaning they have a tissue system of specialized cells for the movement of water

TABLE 13.2 Major Events in the Evolution of Land Plants. The Devonian Period was a time of rapid evolution for the land plants. Major events were the appearance of leaves, heterospory, secondary growth, and the emergence of seeds.

PLANT EVOLUTION 261

Perspective 13.1

Palynology: A Link between Geology and Biology

Palynology is the study of organic microfossils called *palynomorphs*. These include such familiar things as spores and pollen (both of which cause allergies for many people) (see Figure 17.7), but also such unfamiliar organisms as acritarchs (see Figure 12.10), dinoflagellates (marine and freshwater single-celled phytoplankton, some species of which in high concentrations make shellfish toxic to humans) (see Figure 15.5c), chitinozoans (vase-shaped microfossils of unknown origin), scolecodonts (jaws of marine annelid worms), and microscopic colonial algae. Fossil palynomorphs are extremely resistant to decay and are extracted from sedimentary rocks by dissolving the rocks in various acids.

A specialty of palynology that attracts many biologists and geologists is the study of spores and pollen. By examining the fossil spores and pollen preserved in sedimentary rocks, palynologists can tell when plants colonized Earth's surface (Figure 13.20), which in turn influenced weathering and erosion rates, soil formation, and changes in the composition of atmospheric gases. Furthermore, because plants are not particularly common as fossils, the study of spores and pollen can frequently reveal the time and region for the origin and extinction of various plant groups.

Analysis of fossil spores and pollen is used to solve many geologic and biologic problems. One of the more important uses of fossil spores and pollen is determining the geologic age of sedimentary rocks. Because spores and pollen are microscopic, resistant to decay, deposited in both marine and terrestrial environments, extremely abundant, and are part of the life cycle of plants (Figure 13.23), they are very useful for determining age. Many spore and pollen species have narrow time ranges that make them excellent guide fossils.

Rocks considered lacking in fossils by paleontologists who were looking only for megafossils often actually contain thousands, even millions, of fossil spores or pollen grains that allow palynologists to date these so-called unfossiliferous rocks.

Fossil spores and pollen are also useful in determining the environment and climate in the past. Their presence in sedimentary rocks helps palynologists determine what plants and trees were living at the time, even if the fossils of those plants and trees are not preserved. Plants are very sensitive to climatic changes, and by plotting the abundance and types of vegetation present, based on their preserved spores and pollen, palynologists can determine past climates and changes in climates through time (see Figure 17.7).

An interesting study by C. B. Foster and S. A. Afonin published in 2005 related morphologic abnormalities in gymnosperm pollen grains to deteriorating atmospheric conditions around the Permian-Triassic boundary—that is, at the time of the global Permian extinction event. One cause of morphologic abnormalities in living gymnosperms and angiosperms is environmental stress on the parent plant. Plants are sensitive indicators of environmental change, and studies have shown that pollen wall abnormalities are caused by atmospheric pollution, UV radiation, or a combination of both.

Processing of samples from nonmarine sedimentary rock sequences that span the Permian-Triassic boundary from the Junggar Basin, Xinjiang Province, China, and Nedubrovo, Russia, yielded a diverse, abundant, and well-preserved pollen assemblage. Examination of the assemblage revealed that greater than 3% of the pollen showed morphologic abnormalities (Figure 1). Based on this finding, the authors concluded these abnormalities were the result of atmospheric pollution, including increased UV radiation, caused by extensive volcanism. Recall from Chapter 12 that one of the possible causes of the Permian extinction was increased global warming caused by higher atmospheric carbon dioxide levels and volcanic

---

and nutrients. The **nonvascular** plants, such as bryophytes (liverworts, hornworts, and mosses) and fungi, do not have these specialized cells and are typically small and usually live in low, moist areas.

The earliest land plants from the Middle to Late Ordovician were probably small and bryophyte-like in their overall organization (but not necessarily related to bryophytes). The evolution of vascular tissue in plants was an important step because it allowed transport of food and water.

Discoveries of probable vascular plant megafossils and characteristic spores indicate to many paleontologists that vascular plants evolved well before the Middle Silurian. Sheets of cuticle-like cells—that is, the cells that cover the surface of present-day land plants—and tetrahedral clusters that closely resemble the spore tetrahedrals of primitive land plants have been reported from Middle to Upper Ordovician rocks from western Libya and elsewhere (• Figure 13.20).

The ancestor of terrestrial vascular plants was probably some type of green algae. Although no fossil record of the transition from green algae to terrestrial vascular plants exists, comparison of their physiology reveals a strong link. Primitive *seedless vascular plants* (discussed later in this chapter) such as ferns resemble green algae in their pigmentation, important metabolic enzymes, and type of reproductive cycle. Furthermore, green algae are one of the few plant groups to have made the transition from saltwater to freshwater.

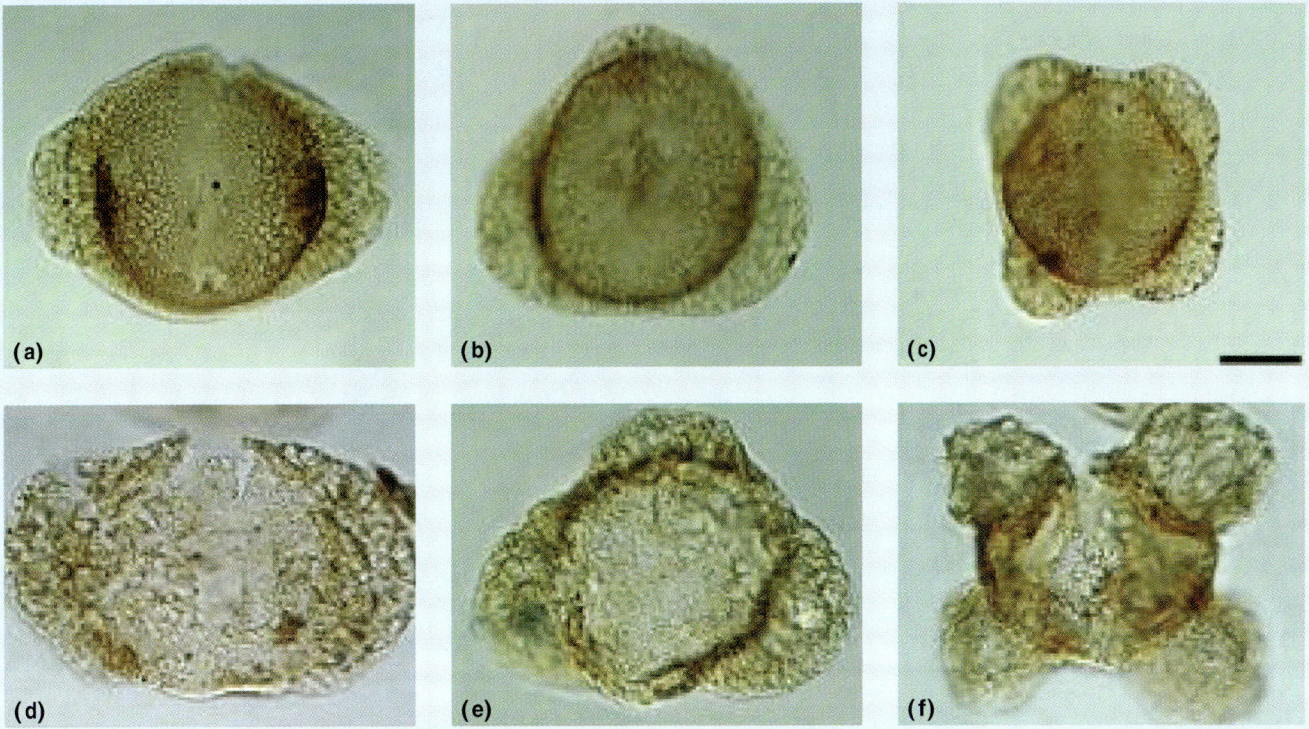

Figure 1 Normal and abnormal pollen grains of *Klausipollenites schaubergeri* from Nedubrovo, Russia (a-c) and the Junggar Basin, China (d-f). Normal pollen grains of *Klausipollenites schaubergeri* are bisaccate, that is they have a central body with two air sacs (a and d), whereas abnormal forms have three or four air sacs (b and c; e and f). The high percentage of abnormal pollen grains of *Klausipollenites schaubergeri* recovered from sedimentary rocks spanning the Permian-Triassic boundary in Russia and China have been used as evidence of increased atmospheric pollution relating to the Permian mass extinction event. The scale bar represents 25 µm (25/1000 mm). Photos courtesy of Clinton B. Foster, Petroleum and Marine Division, Geoscience Australia, and Sergey A. Afonin, formerly of the Palaeontological Institute, Moscow, Russia.

activity. This study is another example of the usefulness of palynology in solving geologic and evolutionary problems. It will be interesting to see if other rocks from other parts of the world yield morphologically abnormal pollen during this time interval.

From this short survey of palynology, it can be seen that the study of spores and pollen provides a tremendous amount of information about the vegetation in the past, its evolution, the type of climate, and changes in climate. In addition, spores and pollen are very useful for dating rocks and correlating marine and terrestrial rocks, both regionally and globally.

The evolution of terrestrial vascular plants from an aquatic, probably green alga ancestry was accompanied by various modifications that let them occupy this new, harsh environment. Besides the primary function of transporting water and nutrients throughout a plant, vascular tissue also provides some support for the plant body. Additional strength is derived from the organic compounds *lignin* and *cellulose,* found throughout a plant's walls.

The problem of desiccation was circumvented by the evolution of *cutin,* an organic compound found in the outer-wall layers of plants. Cutin also provides additional resistance to oxidation, the effects of ultraviolet light, and the entry of parasites.

Roots evolved in response to the need to collect water and nutrients from the soil and to help anchor the plant in the ground. The evolution of *leaves* from tiny outgrowths on the stem or from branch systems provided plants with an efficient light-gathering system for photosynthesis.

Silurian and Devonian Floras

The earliest known vascular land plants are small Y-shaped stems assigned to the genus *Cooksonia* from the Middle Silurian of Wales and Ireland. Together with Upper Silurian and Lower Devonian species from Scotland, New York State, and the Czech Republic, these earliest

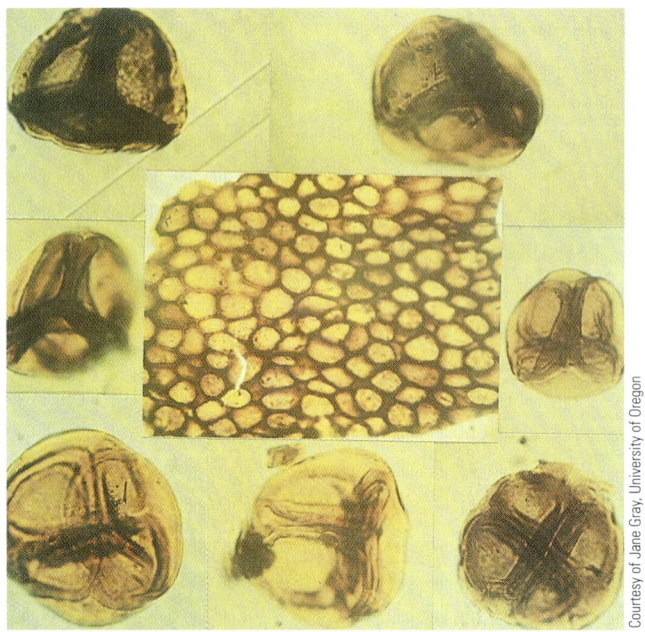

• **Figure 13.20 Upper Ordovician Plant Spores and Cells** Fossils that closely resemble the spore tetrahedrals of primitive land plants. The sheet of cuticle-like cells *(center)* is from the Upper Ordovician Melez Chograne Formation of Libya. The others are from the Upper Ordovician Djeffara Formation of Libya.

• **Figure 13.21 *Cooksonia*** The earliest known fertile land plant was *Cooksonia*, seen in this fossil from the Upper Silurian of South Wales. *Cooksonia* consisted of upright, branched stems terminating in sporangia (spore-producing structures). It also had a resistant cuticle and produced spores typical of a vascular plant. These plants probably lived in moist environments such as mud flats. This specimen is 1.49 cm long.

plants were small, simple, leafless stalks with a spore-producing structure at the tip (• Figure 13.21). They are known as **seedless vascular plants** because they did not produce seeds. They also did not have a true root system. A *rhizome*, the underground part of the stem, transferred water from the soil to the plant and anchored the plant to the ground. The sedimentary rocks in which these plant fossils are found indicate that they lived in low, wet, marshy, freshwater environments.

An interesting parallel can be seen between seedless vascular plants and amphibians. When they made the transition from water to land, both plants and animals had to overcome the same problems such a transition involved. Both groups, while successful, nevertheless required a source of water in order to reproduce. In the case of amphibians, their gelatinous egg had to remain moist, and the seedless vascular plants required water for the sperm to travel through to reach the egg.

From this simple beginning, the seedless vascular plants evolved many of the major structural features characteristic of modern plants such as leaves, roots, and secondary growth. These features did not all evolve simultaneously but rather at different times, a pattern known as *mosaic evolution*. This diversification and adaptive radiation took place during the Late Silurian and Early Devonian and resulted in a tremendous increase in diversity (• Figure 13.22). During the Devonian, the number of plant genera remained about the same, yet the composition of the flora changed. Whereas the Early Devonian landscape was dominated by relatively small, low-growing, bog-dwelling types of plants, the Late Devonian witnessed forests of large, tree-sized plants up to 10 m tall.

In addition to the diverse seedless vascular plant flora of the Late Devonian, another significant floral event took place. The evolution of the seed at this time liberated land plants from their dependence on moist conditions and allowed them to spread over all parts of the land.

Seedless vascular plants require moisture for successful fertilization because the sperm must travel to the egg on the surface of the gamete-bearing plant (gametophyte) to produce a successful spore-generating plant (sporophyte). Without moisture, the sperm would dry out before reaching the egg (• Figure 13.23a). In the seed method of reproduction, the spores are not released to the environment, as they are in the seedless vascular plants, but are retained on the spore-bearing plant, where they grow into the male and female forms of the gamete-bearing generation.

In the case of the **gymnosperms**, or flowerless seed plants, these are male and female cones (Figure 13.23b). The male cone produces pollen, which contains the sperm and has a waxy coating to prevent desiccation, and the egg, or embryonic seed, is contained in the female cone. After fertilization, the seed then develops into a mature, cone-bearing plant. In this way the need for a moist environment

• **Figure 13.22 Early Devonian Landscape** Reconstruction of an Early Devonian landscape showing some of the earliest land plants. **(a)** *Dawsonites*. **(b)** *Protolepidodendron*. **(c)** *Bucheria*.

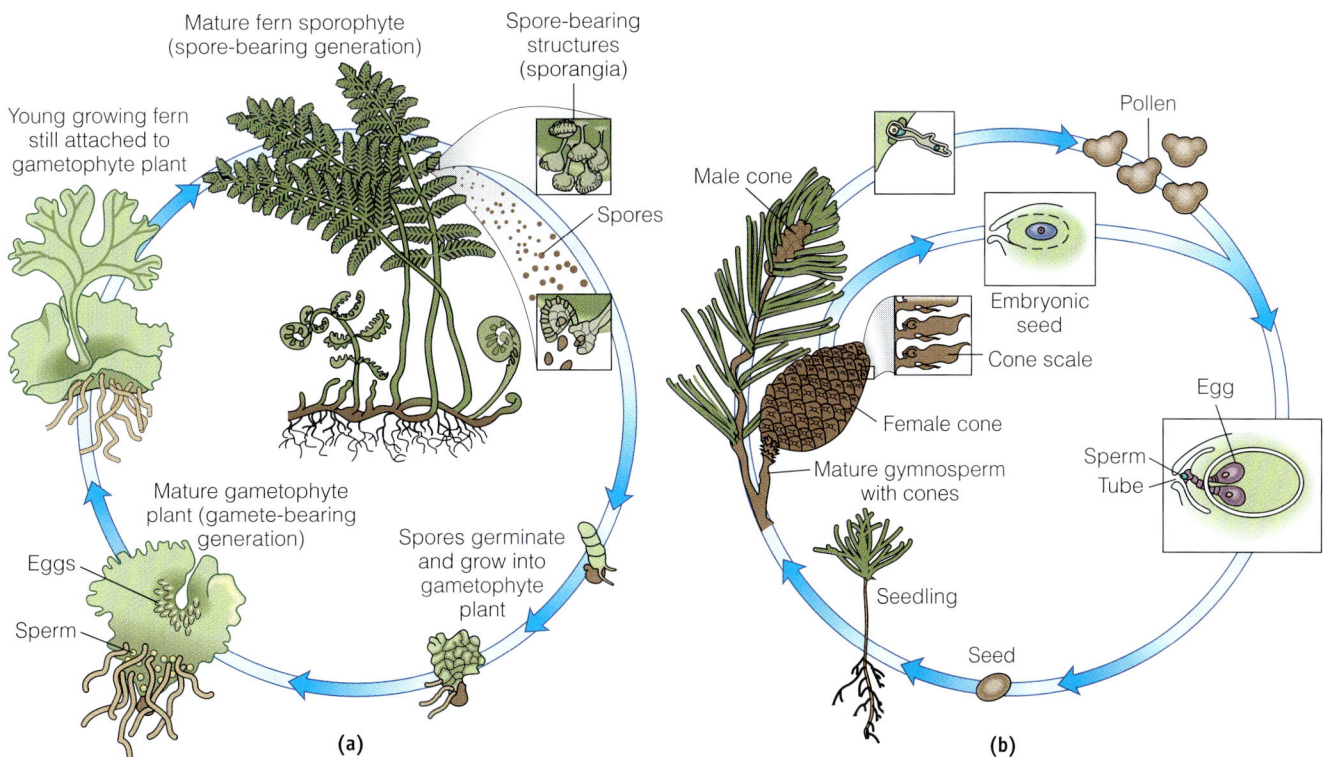

• **Figure 13.23 Generalized Life History of a Seedless Vascular Plant and Gymnosperm Plant** **(a)** Generalized life history of a seedless vascular plant. The mature sporophyte plant produces spores, which on germination grow into small gametophyte plants that produce sperm and eggs. The fertilized eggs grow into the spore-producing mature plant, and the sporophyte–gametophyte life cycle begins again. **(b)** Generalized life history of a gymnosperm plant. The mature plant bears both male cones that produce sperm-bearing pollen grains and female cones that contain embryonic seeds. Pollen grains are transported to the female cones by the wind. Fertilization occurs when the sperm moves through a moist tube growing from the pollen grain and unites with the embryonic seed, which then grows into a cone-bearing mature plant.

(a)

• **Figure 13.24** *Chaleuria cirrosa* (a) Specimen and (b) reconstruction of the Early Devonian plant *Chaleuria cirrosa* from New Brunswick, Canada. This plant was heterosporous, meaning that it produced two sizes of spores.

(b)

for the gametophyte generation is solved. The significance of this development is that seed plants, like reptiles, were no longer restricted to wet areas but were free to migrate into previously unoccupied dry environments.

Before seed plants evolved, an intermediate evolutionary step was necessary. This was the development of *heterospory*, whereby a species produces two types of spores: a large one (megaspore) that gives rise to the female gamete-bearing plant and a small one (microspore) that produces the male gamete-bearing plant. The earliest evidence of heterospory is found in the Early Devonian plant *Chaleuria cirrosa*, which produced spores of two distinct sizes (• Figure 13.24). The appearance of heterospory was followed several million years later by the emergence of progymnosperms—Middle and Late Devonian plants with fernlike reproductive habits and a gymnosperm anatomy—which gave rise in the Late Devonian to such other gymnosperm groups as the seed ferns and conifer-type seed plants.

Although the seedless vascular plants dominated the flora of the Carboniferous coal-forming swamps, the gymnosperms made up an important element of the Late Paleozoic flora, particularly in the nonswampy areas.

Late Carboniferous and Permian Floras

As discussed earlier, the rocks of the Pennsylvanian Period (Late Carboniferous) are the major source of the world's coal. Coal results from the alteration of plant remains accumulating in low, swampy areas. The geologic and geographic conditions of the Pennsylvanian were ideal for the growth of seedless vascular plants, and consequently these coal swamps had a very diverse flora (• Figure 13.25).

It is evident from the fossil record that whereas the Early Carboniferous flora was similar to its Late Devonian counterpart, a great deal of evolutionary experimentation was taking place that would lead to the highly successful Late Paleozoic flora of the coal swamps and adjacent habitats. Among the seedless vascular plants, the lycopsids and sphenopsids were the most important coal-forming groups of the Pennsylvanian Period.

The lycopsids were present during the Devonian, chiefly as small plants, but by the Pennsylvanian, they were the dominant element of the coal swamps, achieving

What Would You Do?

You are an elementary school teacher, and your sixth-grade class is studying prehistoric life. To make this project as interesting as possible, you decide to have the class construct a "living diorama" of life during the Pennsylvanian Period. Although you can't clone extinct plants and animals, you can collect representatives of the different plant, animal, and insect groups that lived in the coal-forming swamps during that time. What types of present-day plants, animals, and insects would you have your class gather to recreate a Pennsylvanian Period ecosystem?

• **Figure 13.25 Pennsylvanian Coal Swamp** Reconstruction of a Pennsylvanian coal swamp with its characteristic vegetation. The amphibian is *Eogyrinus*.

heights up to 30 m in such genera as *Lepidodendron* and *Sigillaria*. The Pennsylvanian lycopsid trees are interesting because they lacked branches except at their top, which had elongate leaves similar to the individual palm leaf of today. As the trees grew, the leaves were replaced from the top, leaving prominent and characteristic rows or spirals of scars on the trunk. Today, the lycopsids are represented by small temperate-forest ground pines.

The sphenopsids, the other important coal-forming plant group, are characterized by being jointed and having horizontal underground stem-bearing roots. Many of these plants, such as *Calamites,* average 5 to 6 m tall. Living sphenopsids include the horsetail *(Equisetum)* or scouring rushes (• Figure 13.26). Small, seedless vascular plants and seed ferns formed a thick undergrowth or ground cover beneath these treelike plants.

Not all plants were restricted to the coal-forming swamps. Among those plants that occupied higher and

• **Figure 13.26 Horsetail *Equisetum*** Living sphenopsids include the horsetail *Equisetum*.

PLANT EVOLUTION 267

• **Figure 13.27 Late Carboniferous Cordaite Forest** A cordaite forest from the Late Carboniferous. Cordaites were a group of gymnosperm trees that grew up to 50 m tall.

drier ground were some of the cordaites, a group of tall gymnosperm trees that grew up to 50 m high and probably formed vast forests (• Figure 13.27). Another important nonswamp dweller was *Glossopteris*, the famous plant so abundant in Gondwana (see Figure 3.1), whose distribution is cited as critical evidence that the continents have moved through time.

The floras that were abundant during the Pennsylvanian persisted into the Permian, but because of climatic and geologic changes resulting from tectonic events (see Chapter 11), they declined in abundance and importance. By the end of the Permian, the cordaites became extinct, and the lycopsids and sphenopsids were reduced to mostly small, creeping forms. Gymnosperms with lifestyles more suited to the warmer and drier Permian climates diversified and came to dominate the Permian, Triassic, and Jurassic landscapes.

SUMMARY

Table 13.3 summarizes the major evolutionary and geologic events of the Paleozoic Era and shows their relationships to each other.

- Chordates are characterized by a notochord, dorsal hollow nerve cord, and gill slits. The earliest chordates were soft-bodied organisms that were rarely fossilized. Vertebrates are a subphylum of the chordates.
- Fish are the earliest known vertebrates with their first fossil occurrence in Upper Cambrian rocks. They have had a long and varied history, including jawless and jawed armored forms (ostracoderms and placoderms), cartilaginous forms, and bony forms. It is from the lobe-finned fish that amphibians evolved.
- The link between crossopterygian lobe-finned fish and the earliest amphibians is convincing and includes a close similarity of bone and tooth structures. The transition from fish to amphibians occurred during the Late Devonian. During the Carboniferous, the labyrinthodont amphibians were the dominant terrestrial vertebrate animals.
- The earliest fossil record of reptiles is from the Late Mississippian. The evolution of an amniote egg was the critical factor that allowed reptiles to completely colonize the land.
- Pelycosaurs were the dominant reptile group during the Early Permian, whereas therapsids dominated the landscape for the rest of the Permian Period.
- In making the transition from water to land, plants had to overcome the same basic problems as animals—namely, desiccation, reproduction, and gravity.
- The earliest fossil record of land plants is from Middle to Upper Ordovician rocks. These plants were probably small and bryophyte-like in their overall organization.
- The evolution of vascular tissue was an important event in plant evolution as it allowed food and water to be transported throughout the plant and provided the plant with additional support.

Table 13.3
Major Evolutionary and Geologic Events of the Paleozoic Era

Age (millions of years)	Geologic Period		Invertebrates	Vertebrates	Plants	Major Geologic Events
251	Permian		Largest mass extinction event to affect the invertebrates	Acanthodians, placoderms, and pelycosaurs become extinct. Therapsids and pelycosaurs the most abundant reptiles	Gymnosperms diverse and abundant	Formation of Pangaea. Alleghenian orogeny. Hercynian orogeny
299	Carboniferous	Pennsylvanian	Fusulinids diversify	Amphibians abundant and diverse	Coal swamps with flora of seedless vascular plants and gymnosperms	Coal-forming swamps common. Formation of Ancestral Rockies. Continental glaciation in Gondwana
318	Carboniferous	Mississippian	Crinoids, lacy bryozoans, blastoids become abundant. Renewed adaptive radiation following extinctions of many reef-builders	Reptiles evolve	Gymnosperms appear (may have evolved during Late Devonian)	Ouachita orogeny. Widespread deposition of black shale
359	Devonian		Extinctions of many reef-building invertebrates near end of Devonian. Reef building continues. Eurypterids abundant	Amphibians evolve. All major groups of fish present—Age of Fish	First seeds evolve. Seedless vascular plants diversify	Widespread deposition of black shale. Acadian orogeny. Antler orogeny
416	Silurian		Major reef building. Diversity of invertebrates remains high	Ostracoderms common. Acanthodians, the first jawed fish, evolve	Early land plants—seedless vascular plants	Caledonian orogeny. Extensive barrier reefs and evaporites
444	Ordovician		Extinctions of a variety of marine invertebrates near end of Ordovician. Major adaptive radiation of all invertebrate groups. Suspension feeders dominant	Ostracoderms diversify	Plants move to land?	Continental glaciation in Gondwana. Taconic orogeny
488	Cambrian		Many trilobites become extinct near end of Cambrian. Trilobites, brachiopods, and archaeocyathids are most abundant	Earliest vertebrates—jawless fish called ostracoderms		First Phanerozoic transgression (Sauk) onto North American craton
542						

- The ancestor of terrestrial vascular plants was probably some type of green alga based on such similarities as pigmentation, metabolic enzymes, and the same type of reproductive cycle.
- The earliest seedless vascular plants were small, leafless stalks with spore-producing structures on their tips. From this simple beginning, plants evolved many of the major structural features characteristic of today's plants.
- By the end of the Devonian Period, forests with tree-sized plants up to 10 m had evolved. The Late Devonian also witnessed the evolution of the flowerless seed plants (gymnosperms) whose reproductive style freed them from having to stay near water.
- The Carboniferous Period was a time of vast coal swamps, where conditions were ideal for the seedless vascular plants. With the onset of more arid conditions during the Permian, the gymnosperms became the dominant element of the world's flora.

IMPORTANT TERMS

acanthodian, p. 253
amniote egg, p. 258
bony fish, p. 254
cartilaginous fish, p. 254
chordate, p. 251
crossopterygian, p. 255
gymnosperm, p. 264
labyrinthodont, p. 258
lobe-finned fish, p. 254
nonvascular plant, p. 262
ostracoderm, p. 252
pelycosaur, p. 259
placoderm, p. 254
protorothyrid, p. 259
ray-finned fish, p. 254
seedless vascular plant, p. 264
therapsid, p. 260
vascular plant, p. 261
vertebrate, p. 251

REVIEW QUESTIONS

1. Labyrinthodonts are
   a. _____ plants; b. _____ fish; c. _____ amphibians; d. _____ reptiles; e. _____ none of the previous answers.
2. The first fish group to evolve jaws were the
   a. _____ bony; b. _____ acanthodians; c. _____ placoderms; d. _____ ostracoderms; e. _____ lobe-finned.
3. Based on similarity of embryo cell division, which invertebrate phylum is most closely allied with the chordates?
   a. _____ Mollusca; b. _____ Echinodermata; c. _____ Porifera; d. _____ Annelida; e. _____ Arthropoda.
4. Which of the following groups did amphibians evolve from?
   a. _____ coelacanths; b. _____ ray-finned fish; c. _____ lobe-finned fish; d. _____ pelycosaurs; e. _____ therapsids.
5. Which was the first plant group that did not require a wet area for the reproductive part of its life cycle?
   a. _____ seedless vascular; b. _____ naked seedless; c. _____ gymnosperms; d. _____ angiosperms; e. _____ flowering.
6. Which plant group first successfully invaded land?
   a. _____ seedless vascular; b. _____ gymnosperms; c. _____ naked seed bearing; d. _____ angiosperms; e. _____ flowering.
7. An organism must possess which of the following during at least part of its life cycle to be classified a chordate?
   a. _____ notochord, dorsal solid nerve cord, lungs; b. _____ vertebrae, dorsal hollow nerve cord, gill slits; c. _____ vertebrae, dorsal hollow nerve cord, lungs; d. _____ notochord, ventral solid nerve cord, lungs; e. _____ notochord, dorsal hollow nerve cord, gill slits.
8. Which of the following is thought by many scientists to be endothermic?
   a. _____ crossopterygians; b. _____ therapsids; c. _____ amphibians; d. _____ rhipidistians; e. _____ labyrinthodonts.
9. Which reptile group gave rise to the mammals?
   a. _____ labyrinthodonts; b. _____ acanthodians; c. _____ pelycosaurs; d. _____ protothyrids; e. _____ therapsids.
10. The Age of Fish is which period?
    a. _____ Cambrian; b. _____ Silurian; c. _____ Devonian; d. _____ Pennsylvanian; e. _____ Permian.
11. What evolutionary innovation allowed reptiles to colonize all of the land?
    a. _____ tear ducts; b. _____ additional bones in the jaw; c. _____ the middle-ear bones; d. _____ an egg that contained a food-and-waste sac and surrounded the embryo in a fluid-filled sac; e. _____ limbs and a backbone capable of supporting the animals on land.
12. Pelycosaurs are
    a. _____ jawless fish; b. _____ jawed armored fish; c. _____ reptiles; d. _____ amphibians; e. _____ plants.
13. Which algal group was the probable ancestor to vascular plants?
    a. _____ yellow; b. _____ blue-green; c. _____ red; d. _____ brown; e. _____ green.
14. In which period were amphibians and seedless vascular plants most abundant?
    a. _____ Permian; b. _____ Pennsylvanian; c. _____ Mississippian; d. _____ Silurian; e. _____ Cambrian.

15. The discovery of *Tiktaalik roseae* is significant because it is a. _____ the ancestor of modern reptiles; b. _____ an intermediary between lobe-finned fish and amphibians; c. _____ the first vascular land plant; d. _____ the "missing-link" between amphibians and reptiles; e. _____ the oldest known fish.
16. What are the major differences between the seedless vascular plants and the gymnosperms, and why are these differences significant in terms of exploiting the terrestrial environment?
17. Outline the evolutionary history of fish.
18. Describe the problems that had to be overcome before organisms could inhabit and completely colonize the land.
19. Discuss the significance and possible advantages of the pelycosaur sail.
20. Discuss how changing geologic conditions affected the evolution of plants and vertebrates.

APPLY YOUR KNOWLEDGE

1. Based on what you know about Carboniferous geology (Chapter 12), why was this time period so advantageous to the evolution of both plants and amphibians?

CHAPTER 14

MESOZOIC EARTH HISTORY

By 1852, during the California gold rush, mining operations were well under way on the American River near Sacramento.

Bettmann/Corbis

[OUTLINE]

INTRODUCTION

THE BREAKUP OF PANGAEA

The Effects of the Breakup of Pangaea on Global Climates and Ocean Circulation Patterns

MESOZOIC HISTORY OF NORTH AMERICA

CONTINENTAL INTERIOR

EASTERN COASTAL REGION

GULF COASTAL REGION

WESTERN REGION

Mesozoic Tectonics

Mesozoic Sedimentation

PERSPECTIVE 14.1 *Petrified Forest National Park*

WHAT ROLE DID ACCRETION OF TERRANES PLAY IN THE GROWTH OF WESTERN NORTH AMERICA?

MESOZOIC MINERAL RESOURCES

SUMMARY

ThomsonNOW™ Explore interactive tutorials, animations, or practice problems available on the ThomsonNow website at www.thomsonedu.com/login.

[CHAPTER OBJECTIVES]

At the end of this chapter, you will have learned that

- The Mesozoic breakup of Pangaea profoundly affected geologic and biologic events.
- During the Triassic Period most of North America was above sea level.
- During the Jurassic Period a seaway flooded the interior of western North America.
- A global rise in sea level during the Cretaceous Period resulted in an enormous interior seaway that extended from the Gulf of Mexico to the Arctic Ocean and divided North America into two large landmasses.
- Western North America was affected by four orogenies that took place along an oceanic–continental plate boundary.
- Microplate tectonics played an important role in the geologic history of Western North America.
- Coal, petroleum, uranium, and copper deposits are major Mesozoic mineral resources.

Introduction

Approximately 150 to 210 million years after the emplacement of massive plutons created the Sierra Nevada (Nevadan orogeny), gold was discovered at Sutter's Mill on the South Fork of the American River at Coloma, California. On January 24, 1848, James Marshall, a carpenter building a sawmill for John Sutter, found bits of the glittering metal in the mill's tailrace. Soon settlements throughout the state were completely abandoned as word of the chance for instant riches spread throughout California.

Within a year after the news of the gold discovery reached the East Coast, the Sutter's Mill area was swarming with more than 80,000 prospectors, all hoping to make their fortune. At least 250,000 gold seekers prospected the Sutter's Mill area, and although most were Americans, they came from all over the world, even as far away as China. Most thought the gold was simply waiting to be taken and didn't realize prospecting was very hard work.

No one thought much about the consequences of so many people converging on the Sutter's Mill area, all intent on making easy money. In fact, life in the mining camps was extremely hard and expensive. The shop owners and traders frequently made more money than the prospectors. In reality, only a few prospectors ever hit it big or were even moderately successful. The rest barely eked out a living until they eventually abandoned their dream and went home.

The gold these prospectors sought was mostly in the form of placer deposits. Weathering of gold-bearing igneous rocks and mechanical separation of minerals by density during stream transport forms placer deposits. Although many prospectors searched for the mother lode, all the gold recovered during the gold rush came from placers (deposits of sand and gravel containing gold particles large enough to be recovered by panning). Panning is a common method of mining placer deposits. In this method, a shallow pan is dipped into a streambed, the material is swirled around and the lighter material is poured off. Gold, being about six times heavier than most sand grains and rock chips, concentrates on the bottom of the pan and can then be picked out.

Although some prospectors dug $30,000 worth of gold dust a week out of a single claim and some gold was found practically sitting on the surface of the ground, most of this easy gold was recovered very early during the gold rush. Most prospectors barely made a living wage working their claims. Nevertheless, during the five years from 1848 to 1853 that constituted the gold rush proper, more than $200 million in gold was extracted.

The Mesozoic Era (251 to 66 million years ago) was an important time in Earth history. The major geologic event was the breakup of Pangaea, which affected oceanic and climatic circulation patterns and influenced the evolution of the terrestrial and marine biotas. Other important Mesozoic geologic events resulting from plate movement include the origin of the Atlantic Ocean basin and the Rocky Mountains, the accumulation of vast salt deposits that eventually formed salt domes adjacent to which oil and natural gas were trapped, and the emplacement of huge batholiths accounting for the origin of various mineral resources.

The Breakup of Pangaea

Just as the formation of Pangaea influenced geologic and biologic events during the Paleozoic, the breakup of this supercontinent had profound geologic and biologic effects during the Mesozoic. The movement of continents affected the global climatic and oceanic regimes as well as the climates of the individual continents. Populations became isolated or were brought into contact with other populations, leading to evolutionary changes in the biota. So great was the effect of this breakup on the world that it forms the central theme of this chapter.

Pangaea's breakup began with rifting between Laurasia and Gondwana during the Triassic (• Figure 14.1a). By the end of the Triassic, the expanding Atlantic Ocean separated North America from Africa. This change was followed by the rifting of North America from South America sometime during the Late Triassic and Early Jurassic.

Separation of the continents allowed water from the Tethys Sea to flow into the expanding central Atlantic Ocean, whereas Pacific Ocean waters flowed into the newly formed Gulf of Mexico, which at that time was little more than a restricted bay (• Figure 14.2). During that time, these areas were located in the low tropical latitudes, where high temperatures and high rates of evaporation were ideal for the formation of thick evaporite deposits.

During the Late Triassic and Jurassic periods, Antarctica and Australia, which remained sutured together, began separating from South America and Africa. Also during this time, India began rifting from the Gondwana continent.

South America and Africa began rifting apart during the Jurassic (Figure 14.1b) and the subsequent separation of these two continents formed a narrow basin where thick evaporite deposits accumulated from the evaporation of southern ocean waters (Figure 14.2). During this time, the eastern end of the Tethys Sea began closing as a result of the clockwise rotation of Laurasia and the northward movement of Africa. This narrow Late Jurassic and Cretaceous seaway between Africa and Europe was the forerunner of the present Mediterranean Sea.

By the end of the Cretaceous, Australia and Antarctica had detached from each other, and India had moved into the low southern latitudes and was nearly to the equator. South America and Africa were widely separated, and Greenland was essentially an independent landmass with only a shallow sea between it and North America and Europe (Figure 14.1c).

A global rise in sea level during the Cretaceous resulted in worldwide transgressions onto the continents. These transgressions resulted from higher heat flow along oceanic ridges that were initially caused by increased rifting and the consequent expansion of oceanic crust. By the Middle Cretaceous, sea level was probably as high as at any time since the Ordovician, and about one-third of the present land area was inundated by epeiric seas (Figure 14.1c).

The final stage in Pangaea's breakup occurred during the Cenozoic. During this time, Australia continued moving northward, and Greenland was completely separated from Europe and North America and formed a separate landmass.

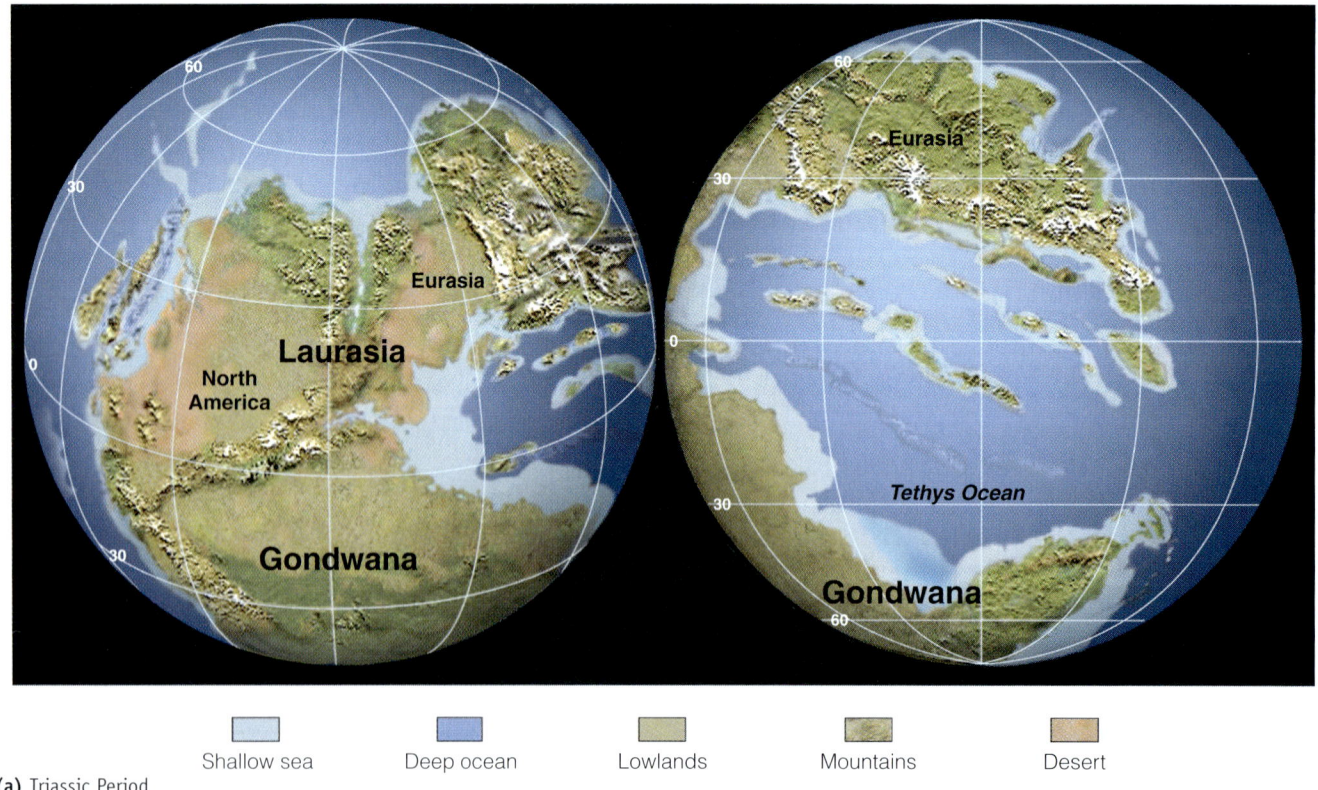

(a) Triassic Period

• Figure 14.1 **Mesozoic Paleogeography** Paleogeography of the world for the (a) Triassic Period, (b) Jurassic Period, and (c) Late Cretaceous Period.

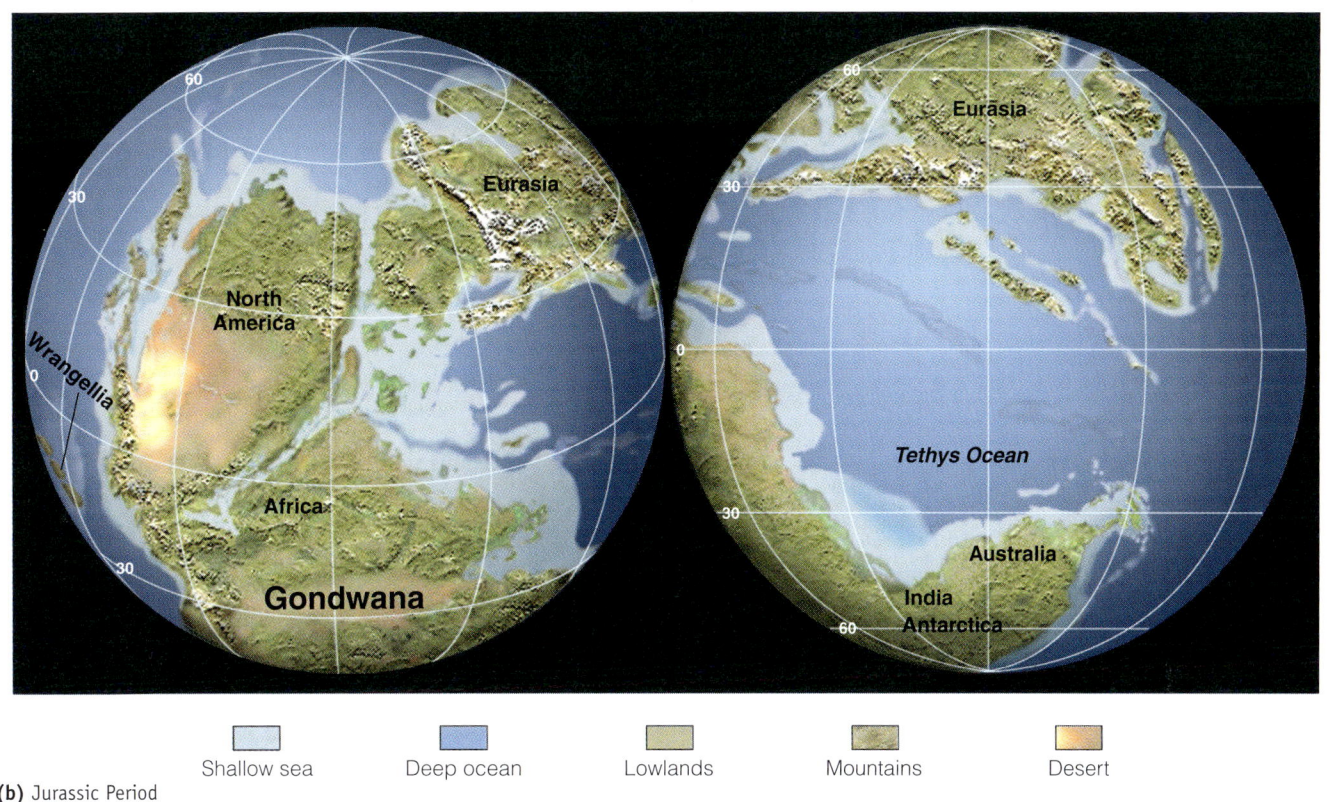

(b) Jurassic Period

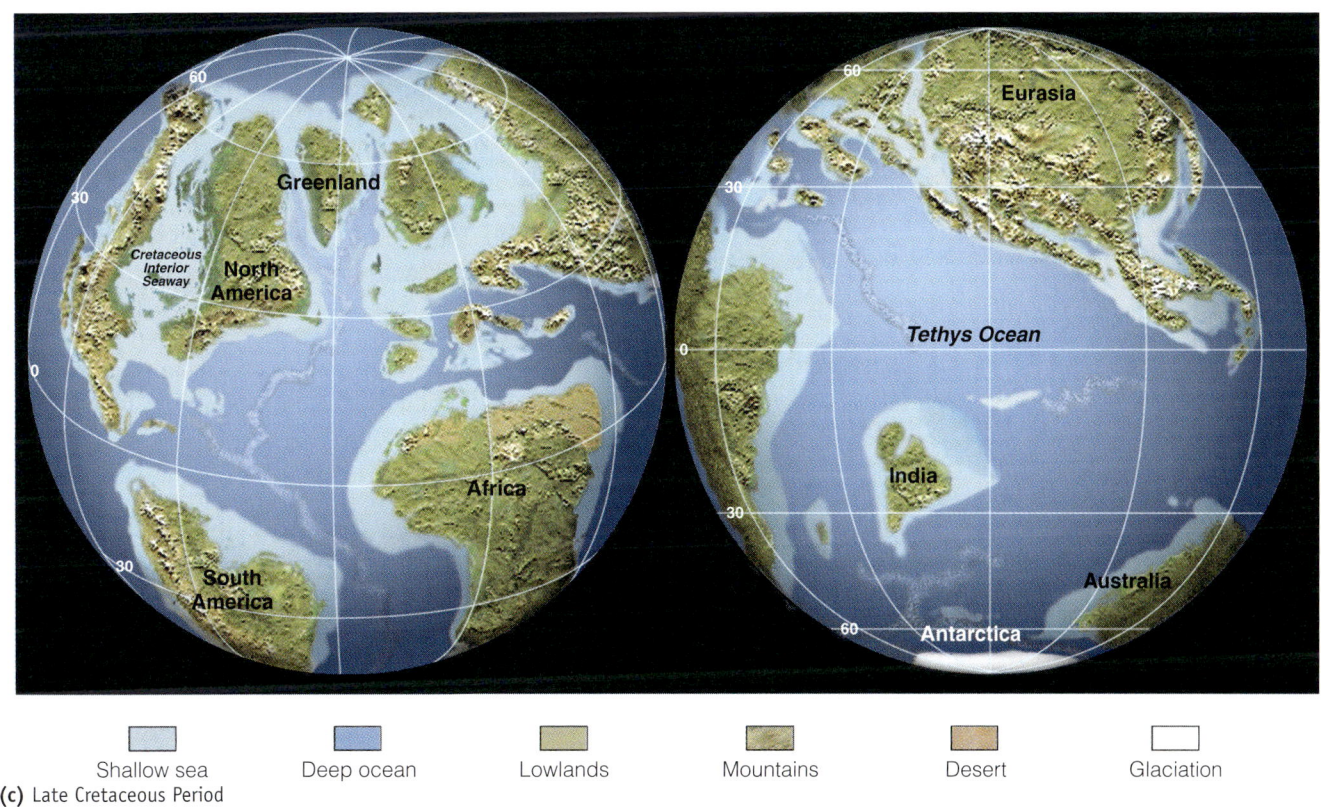

(c) Late Cretaceous Period

THE BREAKUP OF PANGAEA 275

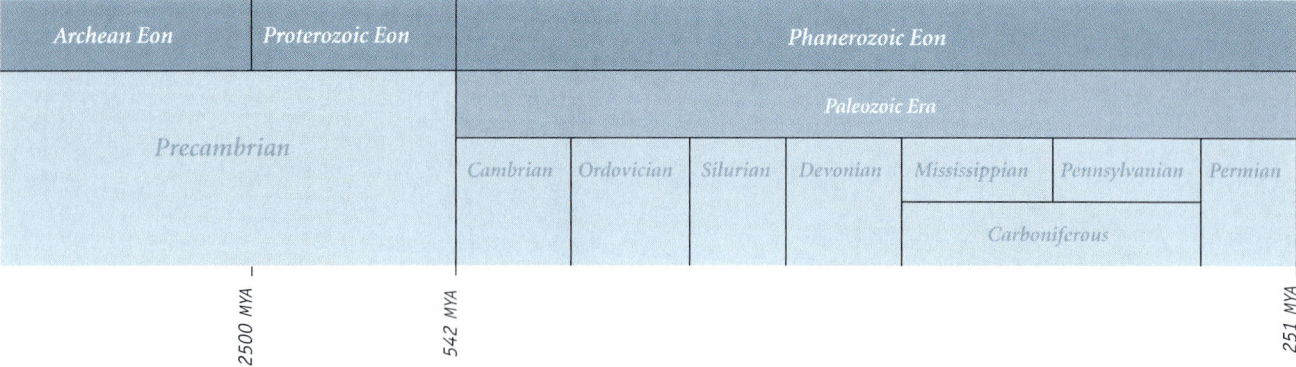

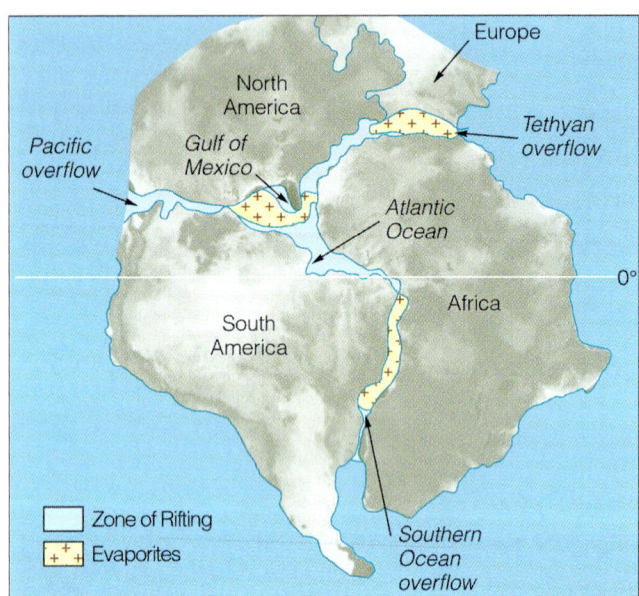

• **Figure 14.2 Evaporite Formation** Evaporites accumulated in shallow basins as Pangaea broke apart during the Early Mesozoic. Water from the Tethys Sea flowed into the central Atlantic Ocean, and water from the Pacific Ocean flowed into the newly formed Gulf of Mexico. Marine water from the south flowed into the southern Atlantic Ocean.

The Effects of the Breakup of Pangaea on Global Climates and Ocean Circulation Patterns

By the end of the Permian Period, Pangaea extended from pole to pole, covered about one-fourth of Earth's surface, and was surrounded by Panthalassa, a global ocean that encompassed about 300 degrees of longitude (see Figure 11.2b). Such a configuration exerted tremendous influence on the world's climate and resulted in generally arid conditions over large parts of Pangaea's interior.

The world's climates result from the complex interaction between wind and ocean currents and the location and topography of the continents. In general, dry climates occur on large landmasses in areas remote from sources of moisture and where barriers to moist air, such as mountain ranges, exist. Wet climates occur near large bodies of water or where winds can carry moist air over land.

Past climatic conditions can be inferred from the distribution of climate-sensitive deposits. Evaporite deposits result when evaporation exceeds precipitation. Although sand dunes and red beds may form locally in humid regions, they are characteristic of arid regions. Coal forms in both warm and cool humid climates. Vegetation that is eventually converted into coal requires at least a good seasonal water supply; thus coal deposits are indicative of humid conditions.

Widespread Triassic evaporites, red beds, and desert dunes in the low and middle latitudes of North and South America, Europe, and Africa indicate dry climates in those regions, whereas coal deposits are found mainly in the high latitudes, indicating humid conditions (Figure 14.1a). These high latitude coals are analogous to today's Scottish peat bog or Canadian muskeg. The lands bordering the Tethys Sea were probably dominated by seasonal monsoon rains resulting from the warm, moist winds and warm oceanic currents impinging against the east-facing coast of Pangaea.

The temperature gradient between the tropics and the poles also affects oceanic and atmospheric circulation. The greater the difference in temperature between the tropics and the poles, the steeper the temperature gradient and the faster the circulation of the oceans and atmosphere. Oceans absorb about 90% of the solar radiation they receive, whereas continents absorb only about 50%, even less if they are snow covered. The rest of the solar radiation is reflected back into space. Areas dominated by seas are thus warmer than those dominated by continents. By knowing the distribution of continents and ocean basins, geologists can generally estimate the average annual temperature for any region on Earth, as well as determine a temperature gradient.

The breakup of Pangaea during the Late Triassic caused the global temperature gradient to increase because the Northern Hemisphere continents moved farther northward, displacing higher-latitude ocean waters. Because of the steeper global temperature gradient produced by a

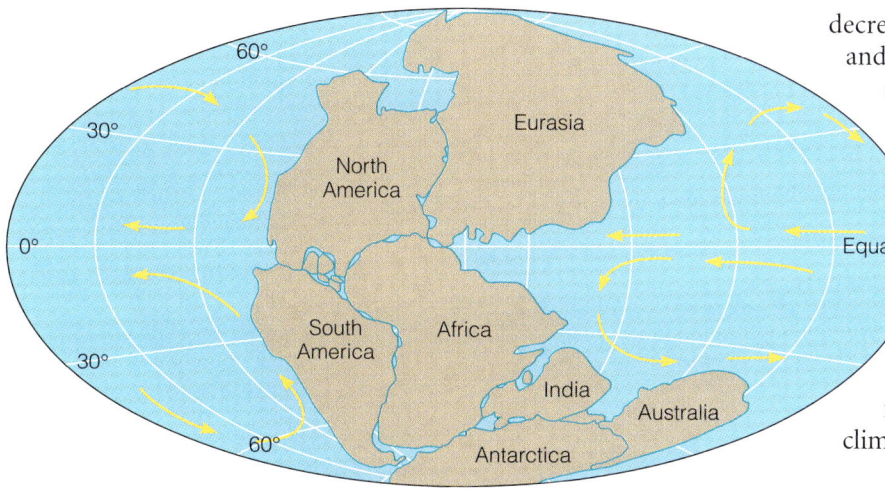

decrease in temperature in the high latitudes and the changing positions of the continents, oceanic and atmospheric circulation patterns greatly accelerated during the Mesozoic (• Figure 14.3). Although the temperature gradient and seasonality on land were increasing during the Jurassic and Cretaceous, the middle- and higher-latitude oceans were still quite warm because warm waters from the Tethys Sea were circulating to the higher latitudes. The result was a relatively equable worldwide climate through the end of the Cretaceous.

Mesozoic History of North America

The beginning of the Mesozoic Era was essentially the same in terms of tectonism and sedimentation as the preceding Permian Period in North America (see Figure 11.11). Terrestrial sedimentation continued over much of the craton, and block faulting and igneous activity began in the Appalachian region as North America and Africa began to separate (• Figure 14.4). The newly forming Gulf of Mexico experienced extensive evaporite deposition during the Late Triassic and Jurassic as North America separated from South America (Figures 14.2 and • 14.5).

A global rise in sea level during the Cretaceous resulted in worldwide transgressions onto the continents such that marine deposition was continuous over much of the North American Cordilleran (• Figure 14.6).

• **Figure 14.3 Mesozoic Oceanic Circulation Patterns** Oceanic circulation evolved from **(a)** a simple pattern in a single ocean (Panthalassa) with a single continent (Pangaea) to **(b)** a more complex pattern in the newly formed oceans of the Cretaceous Period.

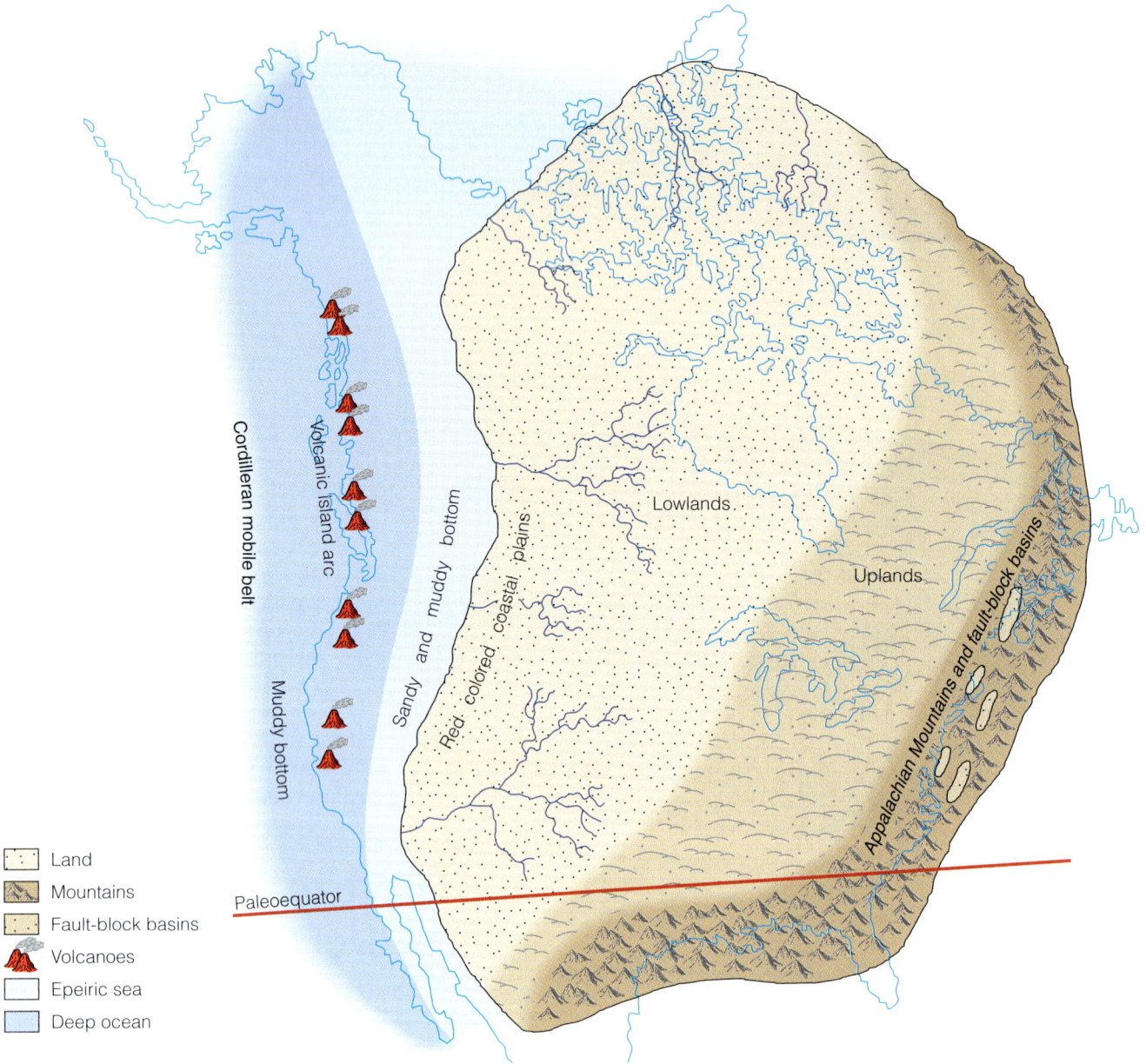

• Figure 14.4 **Triassic Paleogeography of North America**

A volcanic island arc system that formed off the western edge of the craton during the Permian was sutured to North America sometime later during the Permian or Triassic. This event is referred to as the *Sonoma orogeny* and will be discussed later in the chapter. During the Jurassic, the entire Cordilleran area was involved in a series of major mountain-building episodes resulting in the formation of the Sierra Nevada, the Rocky Mountains, and other lesser mountain ranges. Although each orogenic episode has its own name, the entire mountain-building event is simply called the *Cordilleran orogeny* (also discussed later in this chapter). With this simplified overview of the Mesozoic history of North America in mind, we will now examine the specific regions of the continent.

Continental Interior

Recall that the history of the North American craton is divided into unconformity-bound sequences reflecting advances and retreats of epeiric seas over the craton (see Figure 10.3). Although these transgressions and regressions played a major role in the Paleozoic geologic history of the continent, they were not as important during the Mesozoic. Most of the continental interior during the Mesozoic was well above sea level and was not inundated by epeiric seas. Consequently, the two Mesozoic cratonic sequences, the Absaroka Sequence (Late Mississippian to Early Jurassic) and **Zuni Sequence** (Early Jurassic to Early Paleocene) (see Figure 10.3), are not

278 CHAPTER 14 MESOZOIC EARTH HISTORY

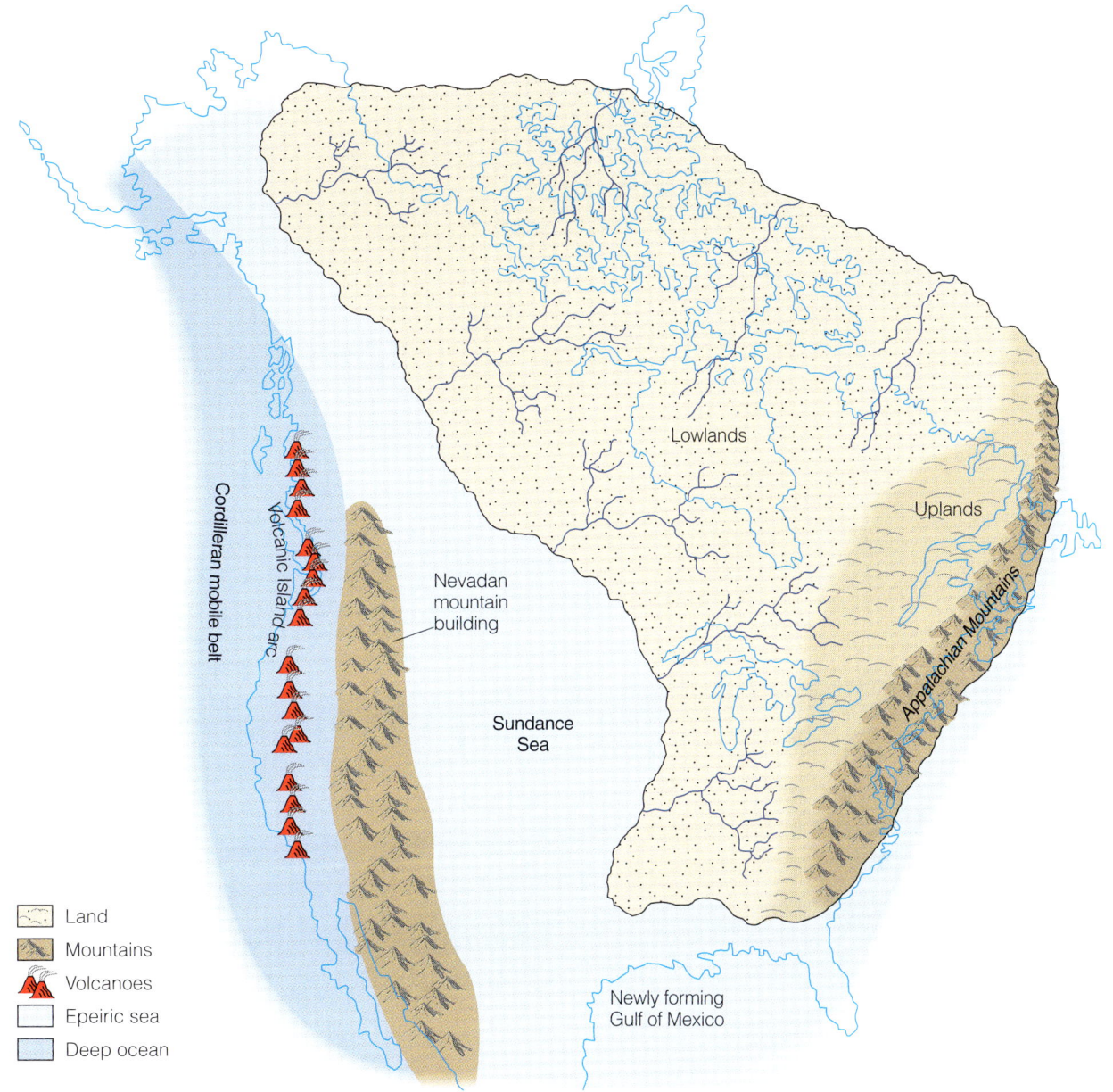

• Figure 14.5 Jurassic Paleogeography of North America

treated separately here; instead, we will examine the Mesozoic history of the three continental margin regions of North America.

Eastern Coastal Region

During the Early and Middle Triassic, coarse detrital sediments derived from erosion of the recently uplifted Appalachians (Alleghenian orogeny) filled the various intermontane basins and spread over the surrounding areas. As weathering and erosion continued during the Mesozoic, this once lofty mountain system was reduced to a low-lying plain.

During the Late Triassic, the first stage in the breakup of Pangaea began with North America separating from Africa. Fault-block basins developed in response to upwelling magma beneath Pangaea in a zone stretching from present-day Nova Scotia to North Carolina (• Figure 14.7). Erosion of the adjacent fault-block mountains filled these basins with great quantities (up to 6000 m) of poorly sorted red nonmarine detrital sediments known as the *Newark Group*.

Reptiles roamed along the margins of the various lakes and streams that formed in these basins, leaving their footprints and trackways in the soft sediments (• Figure 14.8). Although the Newark Group rocks contain numerous dinosaur footprints, they are almost completely devoid of dinosaur bones! The Newark Group is mostly Late Triassic

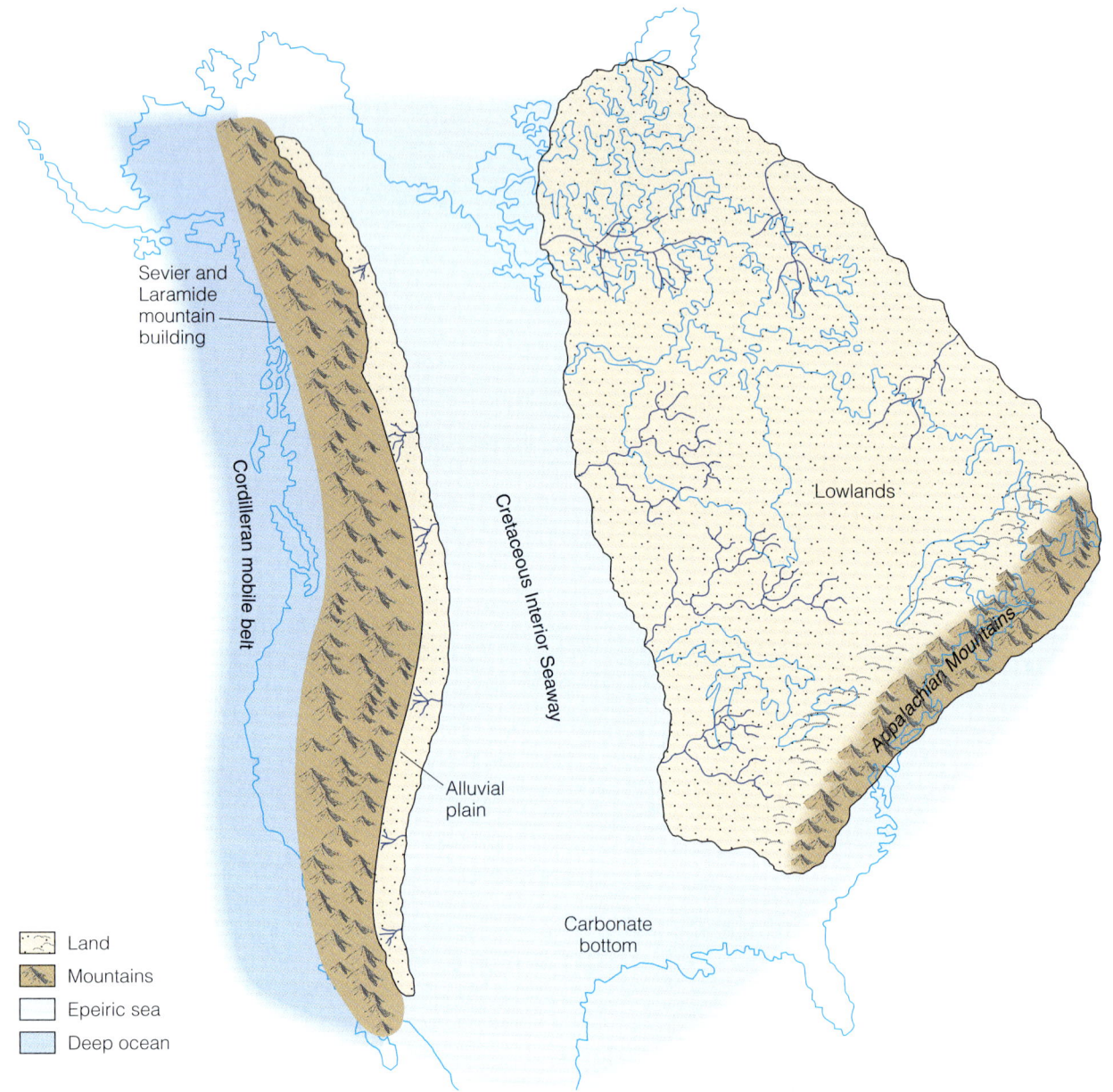

• **Figure 14.6 Cretaceous Paleogeography of North America**

in age, but in some areas deposition did not begin until the Early Jurassic.

Concurrent with sedimentation in the fault-block basins were extensive lava flows that blanketed the basin floors, as well as intrusions of numerous dikes and sills. The most famous intrusion is the prominent Palisades sill along the Hudson River in the New York–New Jersey area (Figure 14.7d).

As the Atlantic Ocean grew, rifting ceased along the eastern margin of North America, and this once active convergent plate margin became a passive, trailing continental margin. The fault-block mountains produced by this rifting continued eroding during the Jurassic and Early Cretaceous until all that was left was an area of low-relief. The sediments produced by this erosion contributed to the growing eastern continental shelf. During the Cretaceous Period, the Appalachian region was reelevated and once again shed sediments onto the continental shelf, forming a gently dipping, seaward-thickening wedge of rocks up to 3000 m thick. These rocks are currently exposed in a belt extending from Long Island, New York, to Georgia.

Gulf Coastal Region

The Gulf Coastal region was above sea level until the Late Triassic (Figure 14.4). As North America separated from South America during the Late Triassic and Early Jurassic,

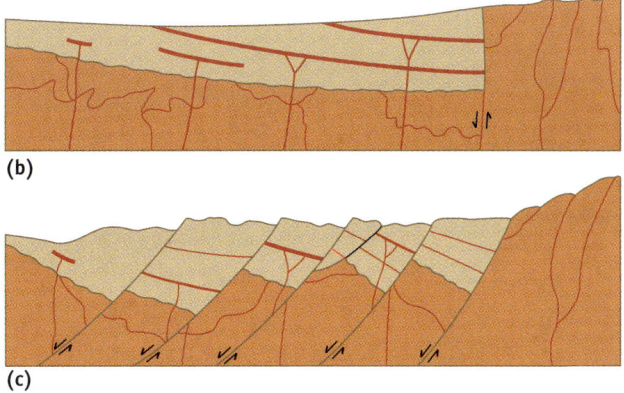

• **Figure 14.7 North American Triassic Fault-Block Basins**
(a) Areas where Triassic fault-block basin deposits crop out in eastern North America. (b) After the Appalachians were eroded to a low-lying plain by the Middle Triassic, fault-block basins formed as a result of Late Triassic rifting between North America and Africa. (c) These valleys accumulated tremendous thickness of sediments and were themselves broken by a complex of normal faults during rifting. (d) Palisades of the Hudson River. This sill was one of many intruded into the Newark sediments during the Late Triassic rifting that marked the separation of North America from Africa.

the Gulf of Mexico began to form (Figure 14.5). With oceanic waters flowing into this newly formed, shallow, restricted basin, conditions were ideal for evaporite

• **Figure 14.8 Triassic Newark Group Reptile Footprints** Reptile tracks in the Triassic Newark Group were uncovered during the excavation for a new state building in Hartford, Connecticut. Because the tracks were so spectacular, the building site was moved, and the excavation was designated as a state park.

formation. These Jurassic evaporites are thought to be the source for the Paleogene salt domes found today in the Gulf of Mexico and southern Louisiana. The history of these salt domes and their associated petroleum accumulations is discussed in Chapter 16.

By the Late Jurassic, circulation in the Gulf of Mexico was less restricted, and evaporite deposition ended. Normal marine conditions returned to the area with alternating transgressing and regressing seas. The earlier formed evaporites were covered and buried by thousands of meters of Cretaceous and Cenozoic sediments.

During the Cretaceous, the Gulf Coastal region, like the rest of the continental margin, was flooded by northward-transgressing seas (Figure 14.6). As a result, nearshore sandstones are overlain by finer sediments characteristic of deeper waters. Following an extensive regression at the end of the Early Cretaceous, a major transgression began, during which a wide seaway extended from the Arctic Ocean to the Gulf of Mexico (Figure 14.6). Sediments deposited in the Gulf Coastal region formed a seaward-thickening wedge.

Reefs were also widespread in the Gulf Coastal region during the Cretaceous and were composed primarily of bivalves called *rudists* (see Chapter 15). Because of their high porosity and permeability, rudistoid reefs make excellent petroleum reservoirs. A good example of a Cretaceous reef

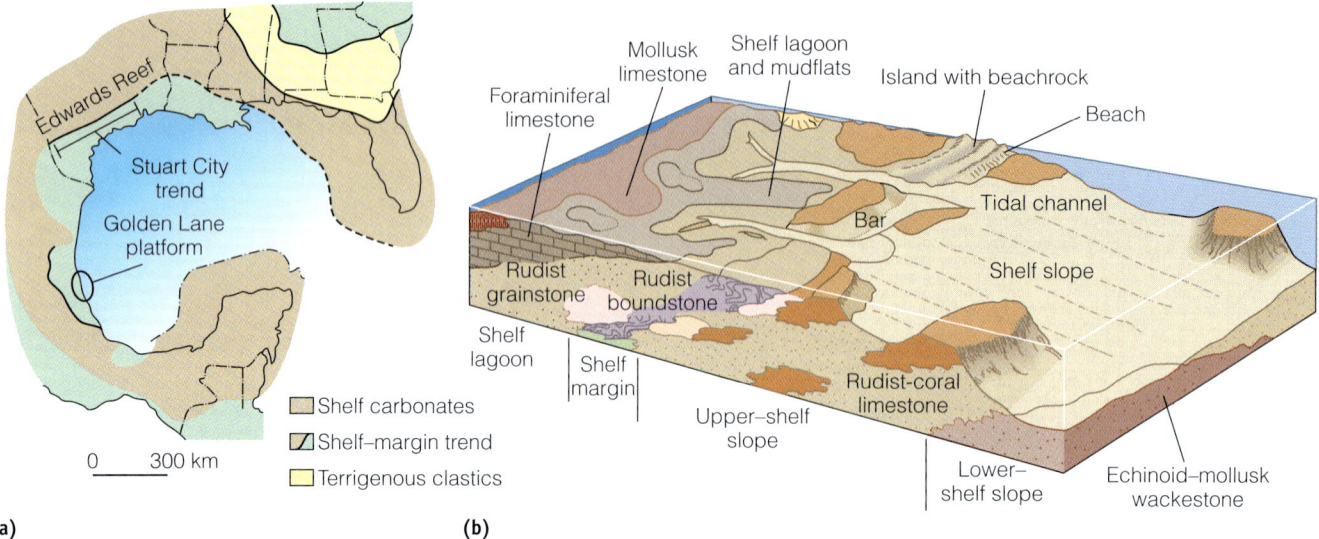

- **Figure 14.9 Early Cretaceous Shelf-Margin Sedimentary Facies** (a) Early Cretaceous shelf-margin facies around the Gulf of Mexico Basin. The reef trend is shown as a black line. (b) Reconstruction of the depositional environment and facies changes across the Stuart City reef trend, South Texas.

complex occurs in Texas where the reef trend strongly influenced the carbonate platform deposition of the region (• Figure 14.9). The facies patterns of these Cretaceous carbonate rocks are as complex as those in the major barrier-reef systems of the Paleozoic Era.

Western Region

Mesozoic Tectonics

The Mesozoic geologic history of the North American Cordilleran mobile belt is very complex, involving the eastward subduction of the oceanic Farallon plate under the continental North American plate. Activity along this oceanic–continental convergent plate boundary resulted in an eastward movement of deformation. This orogenic activity progressively affected the trench and continental slope, the continental shelf, and the cratonic margin, thickening the continental crust. In addition, the accretion of terranes and microplates along the western margin of North America also played a significant role in the Mesozoic tectonic history of this area.

Except for the Late Devonian–Early Mississippian Antler orogeny (see Figure 11.15), the Cordilleran region of North America experienced little tectonism during the Paleozoic. However, an island arc and ocean basin formed off the western North American craton during the Permian (Figure 14.4), followed by subduction of an oceanic plate beneath the island arc and the thrusting of oceanic and island arc rocks eastward against the craton margin (• Figure 14.10). This event, known as the **Sonoma orogeny,** occurred at or near the Permian–Triassic boundary and resulted in the suturing of island-arc terranes along the western edge of North America.

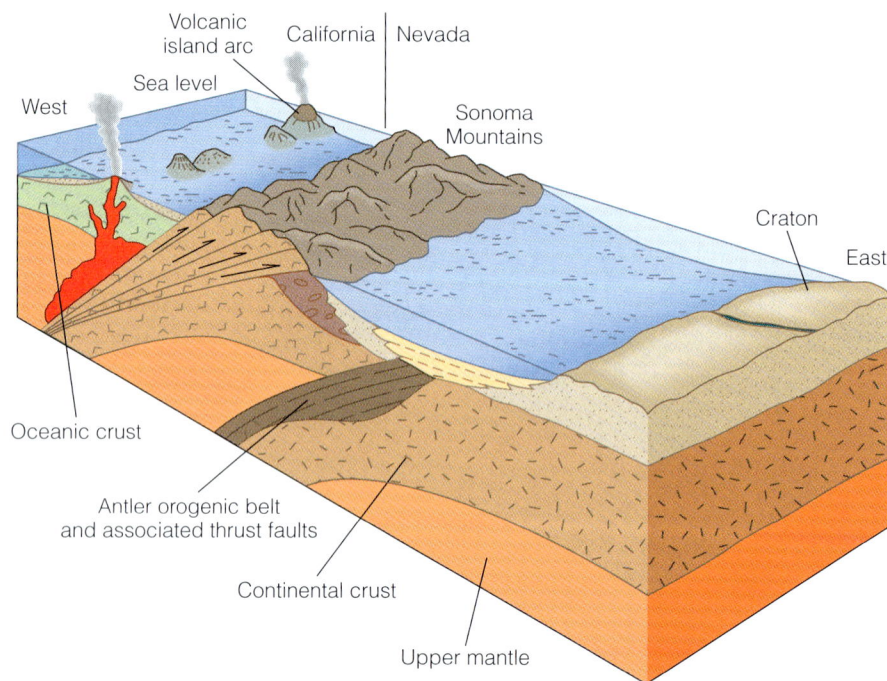

- **Figure 14.10 Sonoma Orogeny** Tectonic activity that culminated in the Permian–Triassic Sonoma orogeny in western North America. The Sonoma orogeny was the result of a collision between the southwestern margin of North America and an island arc system.

282 CHAPTER 14 MESOZOIC EARTH HISTORY

Following the Late Paleozoic–Early Mesozoic destruction of the volcanic island arc during the Sonoma orogeny, the western margin of North America became an oceanic–continental convergent plate boundary. During the Late Triassic, a steeply dipping subduction zone developed along the western margin of North America in response to the westward movement of North America over the Farallon plate. This newly created oceanic–continental plate boundary controlled Cordilleran tectonics for the rest of the Mesozoic Era and for most of the Cenozoic Era; this subduction zone marks the beginning of the modern circum-Pacific orogenic system.

Two subduction zones, dipping in opposite directions from each other, formed off the west coast of North America during the Middle and early Late Jurassic (• Figure 14.11). The more westerly subduction zone was eliminated by the westward-moving North American plate, which overrode the oceanic Farallon plate.

The *Franciscan Complex,* which is up to 7000 m thick, is an unusual rock unit consisting of a chaotic mixture of rocks that accumulated during the Late Jurassic and Cretaceous. The various rock types—graywacke, volcanic breccia, siltstone, black shale, chert, pillow basalt, and blueschist metamorphic rocks—suggest that continental-shelf, slope, and deep-sea environments were brought together in a submarine trench when North America overrode the subducting Farallon plate (• Figure 14.12). This complex was squeezed against the edge of the North American craton along with the *Great Valley Group* to the east which are currently separated from each other by a major thrust fault. The Great Valley Group consists of more than 16,000 m of conglomerates, sandstones, siltstones, and shales and were deposited on the continental shelf and slope at the same time the Franciscan deposits were accumulating in the submarine trench (Figure 14.12).

The general term **Cordilleran orogeny** is applied to the mountain-building activity that began during the Jurassic and continued into the Cenozoic (• Figure 14.13). The Cordilleran orogeny consisted of a series of individual mountain-building events that occurred in different regions at different times but overlapped to some extent.

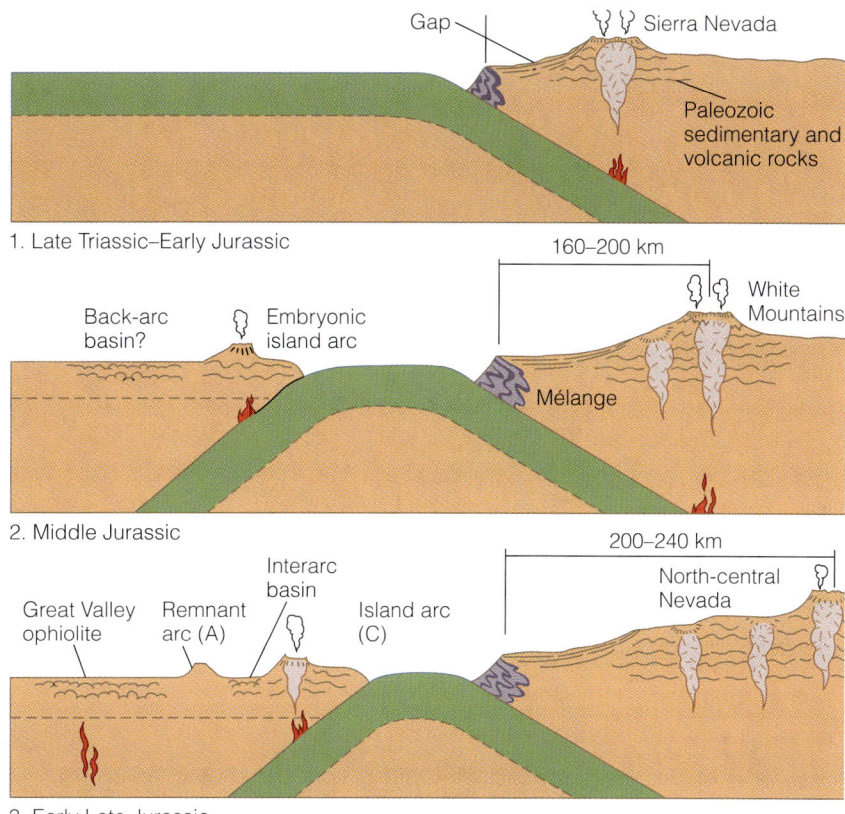
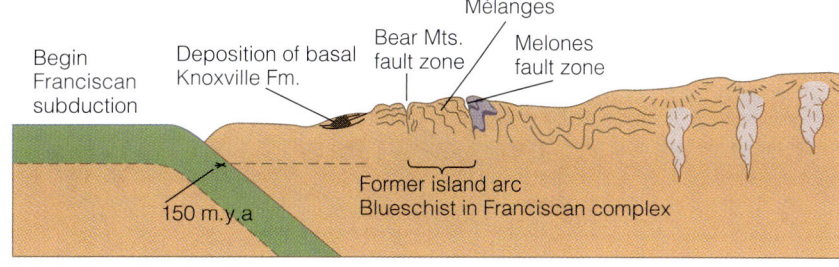
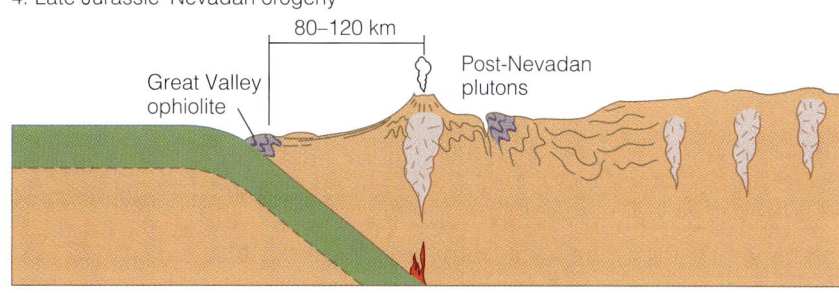

• **Figure 14.11 Mesozoic Tectonic Evolution of the Sierra Nevada** Interpretation of the tectonic evolution of the Sierra Nevada during the Mesozoic Era.

Most of this Cordilleran orogenic activity is related to the continued westward movement of the North American plate as it overrode the Farallon plate and its history is highly complex.

The first phase of the Cordilleran orogeny, the **Nevadan orogeny** (Figure 14.13), began during the Late Jurassic and continued into the Cretaceous as large

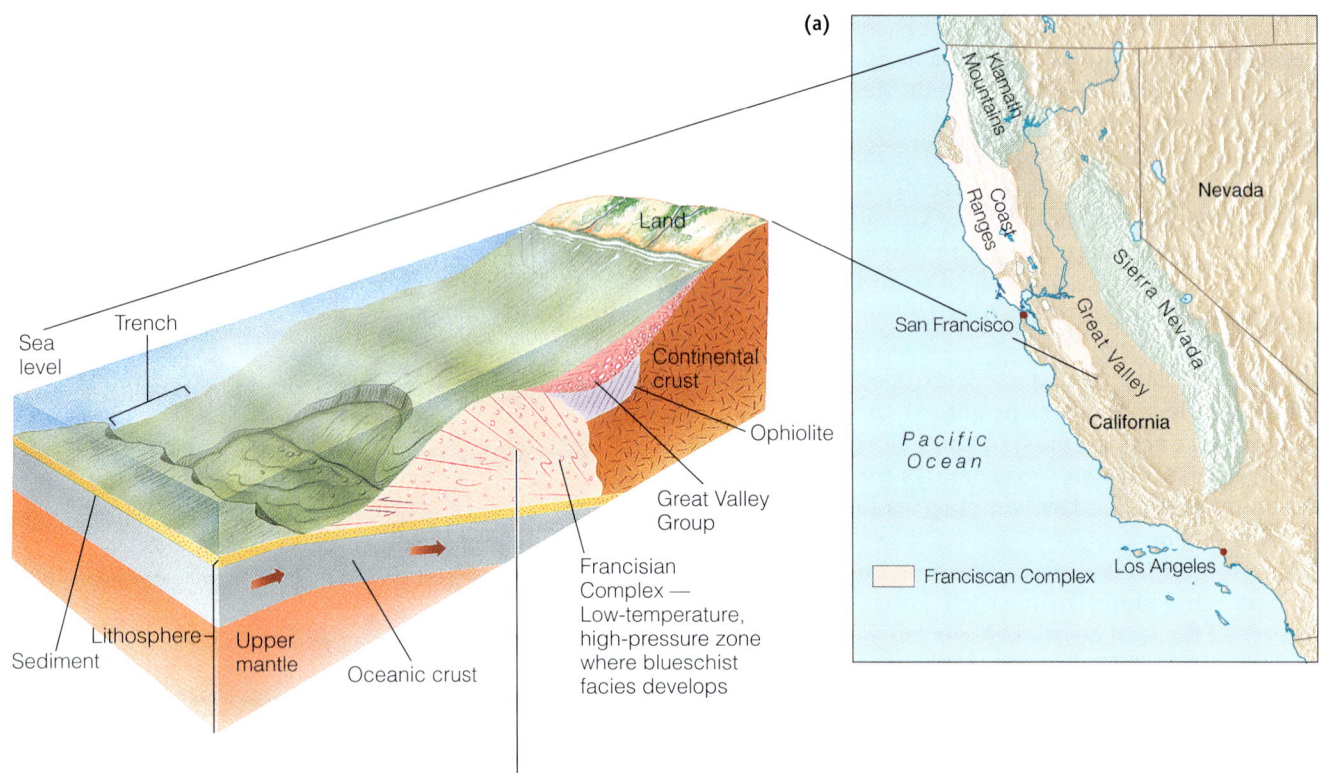

• Figure 14.12 **The Franciscan Complex** (a) Reconstruction of the depositional environment of the Franciscan Complex during the Late Jurassic and Cretaceous. (b) Exposures of the Franciscan Complex along the central California coast.

volumes of granitic magma were generated at depth beneath the western edge of North America. These granitic masses ascended as huge batholiths that are now recognized as the Sierra Nevada, Southern California, Idaho, and Coast Range batholiths (• Figure 14.14). It was also during this time that the Franciscan Complex and Great Valley Group were being deposited and deformed as part of the Nevadan orogeny within the Cordilleran mobile belt.

By the Late Cretaceous, most of the volcanic and plutonic activity had migrated eastward into Nevada and Idaho. This migration was probably caused by a change from high-angle to low-angle subduction, resulting in the subducting oceanic plate reaching its melting depth farther east (• Figure 14.15). Thrusting occurred progressively farther east so that by the Late Cretaceous it extended all the way to the Idaho–Washington border.

The second phase of the Cordilleran orogeny, the **Sevier orogeny**, was mostly a Cretaceous event even though it began in the Late Jurassic and is associated with the tectonic activity of the earlier Nevadan orogeny (Figure 14.13). Subduction of the Farallon plate beneath the North American plate continued during this time, resulting in numerous overlapping, low-angle thrust faults in which blocks of older strata were thrust eastward on top of younger strata (• Figure 14.16). This deformation produced generally north–south-trending mountain ranges that stretch from Montana to western Canada.

During the Late Cretaceous to Early Cenozoic, the final pulse of the Cordilleran orogeny took place (Figure 14.13). The **Laramide orogeny** developed east of the Sevier orogenic belt in the present-day Rocky Mountain areas of New Mexico, Colorado, and Wyoming. Most features of the present-day Rocky Mountains resulted from the Cenozoic phase of the Laramide orogeny, and for that reason it is discussed in Chapter 16.

Mesozoic Sedimentation

Concurrent with the tectonism in the Cordilleran mobile belt, Early Triassic sedimentation on the western continental shelf consisted of shallow-water marine sandstones, shales, and limestones. During the Middle and Late Triassic, the western shallow seas regressed farther west, exposing large areas of former seafloor to erosion. Marginal marine and nonmarine Triassic rocks, particularly red beds, contribute to the spectacular and colorful scenery of the region.

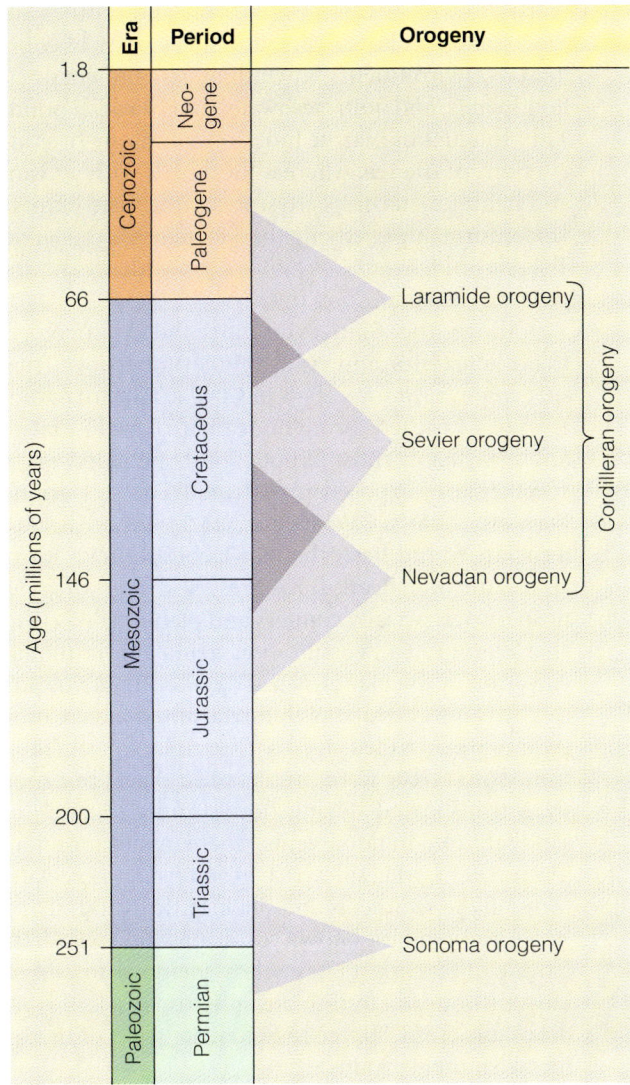

• **Figure 14.13 Mesozoic Cordilleran Orogenies** Mesozoic orogenies occurring in the Cordilleran mobile belt.

• **Figure 14.14 Cordilleran Batholiths** Location of Jurassic and Cretaceous batholiths in western North America.

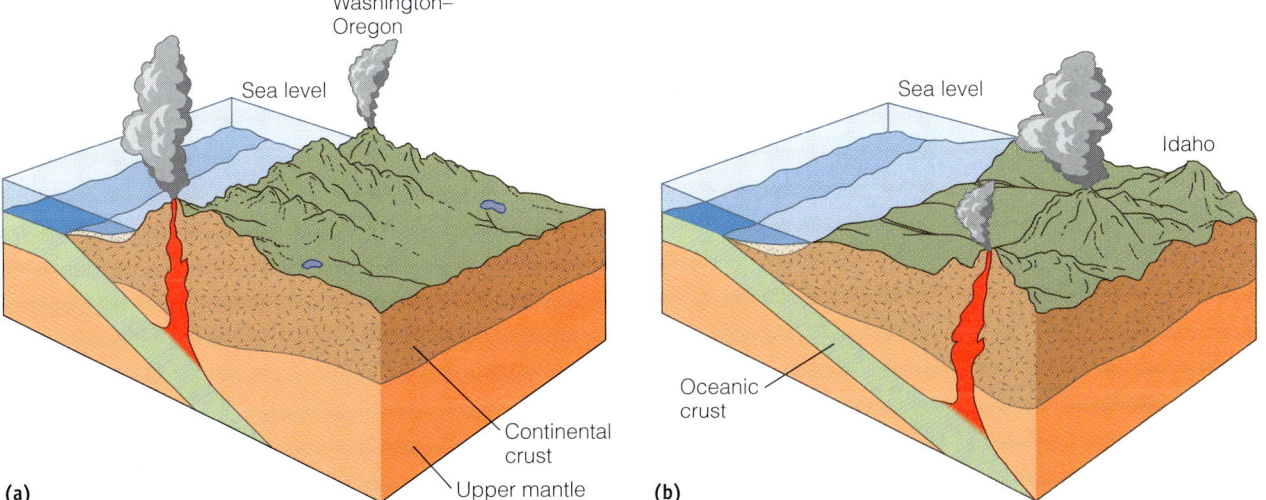

• **Figure 14.15 Cretaceous Subduction in the Cordilleran Mobile Belt** A possible cause for the eastward migration of igneous activity in the Cordilleran region during the Cretaceous Period was a change from **(a)** high-angle to **(b)** low-angle subduction. As the subducting plate moved downward at a lower angle, the depth of melting moved farther east.

WESTERN REGION 285

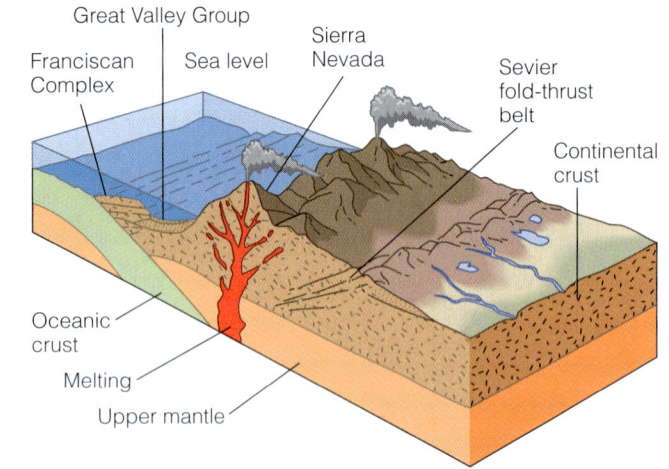

• **Figure 14.16 Sevier Orogeny (a)** Restoration showing the associated tectonic features of the Late Cretaceous Sevier orogeny caused by subduction of the Farallon plate under the North American plate. **(b)** The Keystone thrust fault, a major fault in the Sevier overthrust belt, is exposed west of Las Vegas, Nevada. The sharp boundary between the light-colored Mesozoic rocks and the overlying dark-colored Paleozoic rocks marks the trace of the Keystone thrust fault.

The Lower Triassic *Moenkopi Formation* of the southwestern United States consists of a succession of brick-red and chocolate-colored mudstones (• Figure 14.17). Such sedimentary structures as desiccation cracks and ripple marks, as well as fossil amphibians and reptiles and their tracks, indicate deposition in a variety of continental environments, including stream channels, floodplains, and both freshwater and brackish ponds. Thin tongues of marine limestones indicate brief incursions of the sea, whereas local beds with gypsum and halite crystal casts attest to a rather arid climate. Unconformably overlying the Moenkopi is the Upper Triassic *Shinarump Conglomerate*, a widespread unit generally less than 50 m thick.

Above the Shinarump are the multicolored shales, siltstones, and sandstones of the Upper Triassic *Chinle Formation* (Figure 14.17). This formation is widely exposed throughout the Colorado Plateau and is probably most famous for its petrified wood, spectacularly exposed in Petrified Forest National Park, Arizona (see Perspective 14.1). Whereas fossil ferns are found here, the park is best known for its abundant and beautifully preserved logs of gymnosperms, especially conifers and plants called *cycads* (see Figure 15.6a). Fossilization resulted from the silicification of the plant tissues. Weathering of volcanic ash beds interbedded with fluvial and deltaic Chinle sediments provided most of the silica for silicification. Some trees were preserved in place, but most were transported during floods and deposited on sandbars and on floodplains, where fossilization took place. After burial, silica-rich groundwater percolated through the sediments and silicified the wood.

Although best known for its petrified wood, the Chinle Formation has also yielded fossils of labyrinthodont amphibians, phytosaurs, and small dinosaurs (see Chapter 15 for a discussion of the latter two animal groups).

Early Jurassic-aged deposits in a large part of the western region consist mostly of clean, cross-bedded sandstones indicative of windblown deposits. The lowermost unit is the *Wingate Sandstone*, a desert dune deposit, and it is overlain by the *Kayenta Formation*, a stream and lake deposit (Figure 14.17a). These two formations are well exposed in southwestern Utah. The thickest and most prominent of the Jurassic cross-bedded sandstones is the *Navajo Sandstone* (Figure 14.17b), a widespread formation that accumulated in a coastal dune environment along the southwestern margin of the craton. The sandstone's most distinguishing feature is its large-scale cross-beds, some of which are more than 25 m high (• Figure 14.18). The upper part of the Navajo contains smaller cross-beds as well as dinosaur and crocodilian fossils.

Marine conditions returned to the region during the Middle Jurassic when a seaway called the **Sundance Sea** twice flooded the interior of western North America (Figure 14.5). The resulting deposits, the *Sundance Formation*, were produced from erosion of tectonic highlands to the west that paralleled the shoreline. These highlands resulted from intrusive igneous activity and associated volcanism that began during the Triassic.

What Would You Do?

The U.S. economy has been in a recession for several years, and to reduce spending, the president has proposed that funding for the National Park system be significantly reduced. As director of the National Park system, you have been called to testify at the upcoming budget hearings. What arguments would you use against the proposed reduction in funding? Would a knowledge of the geology and ecology of the different parks be helpful? Explain.

14.1 Perspective

Petrified Forest National Park

Petrified Forest National Park is located in eastern Arizona about 42 km east of Holbrook. The park consists of two sections: the Painted Desert, which is north of Interstate 40, and the Petrified Forest, which is south of the Interstate.

The Painted Desert is a brilliantly colored landscape whose colors and hues change constantly throughout the day. The multicolored rocks of the Triassic Chinle Formation have been weathered and eroded to form a badlands topography of numerous gullies, valleys, ridges, mounds, and mesas. The Chinle Formation is composed predominantly of various-colored shale beds. These shales and associated volcanic ash layers are easily weathered and eroded. Interbedded locally with the shales are lenses of conglomerates, sandstones, and limestones, which are more resistant to weathering and erosion than the shales and form resistant ledges.

The Petrified Forest was originally set aside as a national monument to protect the large number of petrified logs that lay exposed in what is now the southern part of the park (Figure 1). When the transcontinental railroad constructed a coaling and watering stop in Adamana, Arizona, passengers were encouraged to take excursions to "Chalcedony Park," as the area was then called, to see the petrified forests. In a short time, collectors and souvenir hunters hauled off tons of petrified wood, quartz crystals, and Native American relics. Not until a huge rock crusher was built to crush the logs for the manufacture of abrasives was the area declared a national monument and the petrified forests preserved and protected.

During the Triassic Period, the climate of the area was much wetter than today, with many rivers, streams, and lakes. About 40 different fossil plant species have been identified from the Chinle Formation. These include numerous seedless vascular plants such as rushes and ferns, as well as gymnosperms such as cycads and conifers. Such plants thrive in floodplains and marshes. Most logs are conifers and belong to the genus *Araucarioxylon*. Some of these trees were more than 60 m tall and up to 4 m in diameter. Apparently, most of the conifers grew on higher ground or riverbanks. Although many trees were buried in place, most seem to have been uprooted and transported by raging streams during times of flooding. Burial of the logs was rapid, and groundwater saturated with silica from the ash of nearby volcanic eruptions quickly permineralized the trees.

Deposition continued in the Colorado Plateau region during the Jurassic and Cretaceous, further burying the Chinle Formation. During the Laramide orogeny, the Colorado Plateau area was uplifted and eroded, exposing the Chinle Formation. Because the Chinle is mostly shales, it was easily eroded, leaving the more resistant petrified logs and log fragments exposed on the surface—much as we see them today.

Figure 1 Petrified Forest National Park, Arizona. The petrified log shown here is *Araucarioxylon*, which is the most abundant tree in the park. The petrified logs have been weathered from the Chinle Formation, where many were buried in place some 200 million years ago.

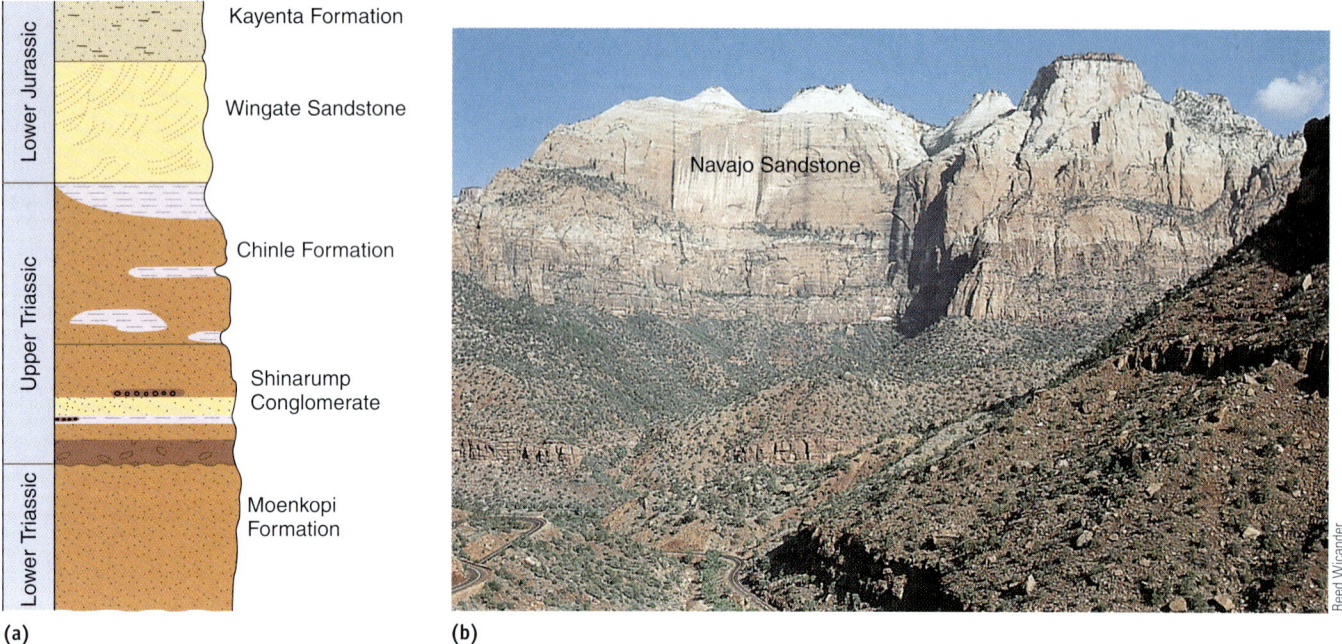

• **Figure 14.17 Triassic and Jurassic Formations in the Western United States** (a) Stratigraphic section of Triassic and Lower Jurassic formations in the western United States. (b) View of East Entrance of Zion Canyon, Zion National Park, Utah. The massive light-colored rocks are the Jurassic Navajo Sandstone, and the slope-forming rocks below the Navajo are the Lower Jurassic Kayenta Formation.

During the Late Jurassic, a mountain chain formed in Nevada, Utah, and Idaho as a result of the deformation produced by the Nevadan orogeny. As the mountain chain grew and shed sediments eastward, the Sundance Sea began retreating northward. A large part of the area formerly occupied by the Sundance Sea was then covered by multicolored sandstones, mudstones, shales, and occasional lenses of conglomerates that comprise the world-famous *Morrison Formation* (• Figure 14.19a).

The Morrison Formation contains one of the world's richest assemblages of Jurassic dinosaur remains. Although most of the skeletons are broken up, as many as 50 individuals have been found together in a small area. Such a concentration indicates that the skeletons were brought together during times of flooding and deposited on sandbars in stream channels. Soils in the Morrison indicate that the climate was seasonally dry.

Although most major museums have either complete dinosaur skeletons or at least bones from the Morrison Formation, the best place to see the bones still embedded in the rocks is the visitors' center at Dinosaur National Monument near Vernal, Utah (Figure 14.19b).

Shortly before the end of the Early Cretaceous, Arctic waters spread southward over the craton, forming a large inland sea in the Cordilleran region. Mid-Cretaceous transgressions also occurred on other continents, and all were part of the global mid-Cretaceous rise in

• **Figure 14.18 Navajo Sandstone** Large cross-beds of the Jurassic Navajo Sandstone in Zion National Park, Utah.

• **Figure 14.19 Morrison Formation** (a) Panoramic view of the Jurassic Morrison Formation as seen from the visitors' center at Dinosaur National Monument, Utah. (b) North wall of visitors' center showing dinosaur bones in bas relief, just as they were deposited 140 million years ago.

sea level that resulted from accelerated seafloor spreading as Pangaea continued to fragment. These Middle Cretaceous transgressions are marked by widespread black shale deposition within the oceanic areas, the shallow sea shelf areas, and the continental regions that were inundated by the transgressions.

By the beginning of the Late Cretaceous, this incursion joined the northward-transgressing waters from the Gulf area to create an enormous **Cretaceous Interior Seaway** that occupied the area east of the Sevier orogenic belt. Extending from the Gulf of Mexico to the Arctic Ocean and more than 1500 km wide at its maximum extent, this seaway effectively divided North America into two large landmasses until just before the end of the Late Cretaceous (Figure 14.6).

Cretaceous deposits less than 100 m thick indicate that the eastern margin of the Cretaceous Interior Seaway subsided slowly and received little sediment from the emergent, low-relief craton to the east. The western shoreline, however, shifted back and forth, primarily in response to fluctuations in the supply of sediment from the Cordilleran Sevier orogenic belt to the west. The facies relationships show lateral changes from conglomerate and coarse sandstone adjacent to the mountain belt through finer sandstones, siltstones, shales, and even limestones and chalks in the east (• Figure 14.20). During times of particularly active mountain building, these coarse clastic wedges of gravel and sand prograded even further east.

As the Mesozoic Era ended, the Cretaceous Interior Seaway withdrew from the craton. During this regression, marine waters retreated to the north and south, and marginal marine and continental deposition formed widespread coal-bearing deposits on the coastal plain.

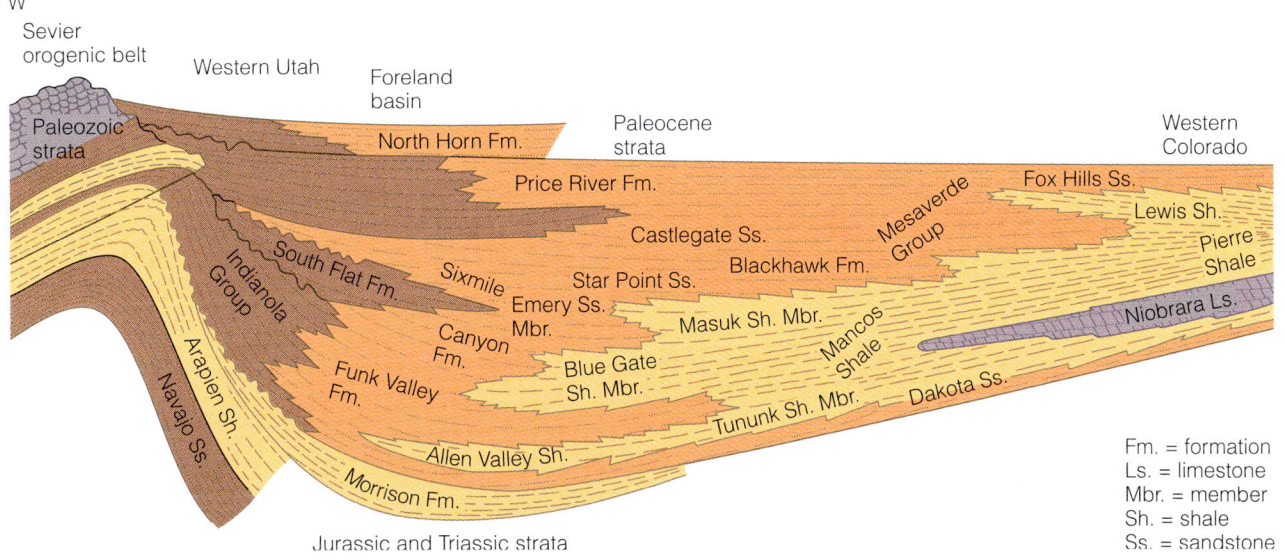

• **Figure 14.20 Cretaceous Facies of the Western Cretaceous Interior Seaway** This restored west–east cross section of Cretaceous facies of the western Cretaceous Interior Seaway shows their relationship to the Sevier orogenic belt.

What Role Did Accretion of Terranes Play in the Growth of Western North America?

In the preceding sections, we discussed orogenies along convergent plate boundaries resulting in continental accretion. Much of the material accreted to continents during such events is simply eroded older continental crust, but a significant amount of new material is added to continents as well—for example, igneous rocks that formed as a consequence of subduction and partial melting. Although subduction is the predominant influence on the tectonic history in many regions of orogenesis, other processes are also involved in mountain building and continental accretion, especially the accretion of terranes.

Geologists now know that portions of many mountain systems are composed of small accreted lithospheric blocks that are clearly of foreign origin. These **terranes** differ completely in their fossil content, stratigraphy, structural trends, and paleomagnetic properties from the rocks of the surrounding mountain system and adjacent craton. In fact, these terranes are so different from adjacent rocks that most geologists think they formed elsewhere and were carried great distances as parts of other plates until they collided with other terranes or continents.

Geologic evidence indicates that more than 25% of the entire Pacific Coast from Alaska to Baja California consists of accreted terranes. These accreting terranes are composed of volcanic island arcs, oceanic ridges, seamounts, volcanic plateaus, hot spot tracks, and small fragments of continents that were scraped off and accreted to the continent's margin as the oceanic plate with which they were carried was subducted under the continent. It is estimated that more than 100 different-sized terranes have been added to the western margin of North America during the last 200 million years (• Figure 14.21). The Wrangellian terranes (Figure 14.1b) are a good example of terranes that have been accreted to North America's western margin (Figure 14.21).

The basic plate tectonic reconstruction of orogenies and continental accretion remains unchanged, but the details of such reconstructions are decidedly different in view of terrane tectonics. For example, growth along active continental margins is faster than along passive continental margins because of the accretion of terranes. Furthermore, these accreted terranes are often new additions to a continent, rather than reworked older continental material.

So far, most terranes have been identified in mountains of the North American Pacific Coast region, but a number of such plates are suspected to be present in other

• **Figure 14.21 Mesozoic Terranes** Some of the accreted lithospheric blocks, called terranes, that form the western margin of the North American Craton. The dark brown blocks probably originated as terranes and were accreted to North America. The light green blocks are possibly displaced parts of North America. The North American craton is shown in dark green. See Figure 14.1b for position of Wrangellian terranes during the Jurassic.

mountain systems as well. They are more difficult to recognize in older mountain systems, such as the Appalachians, however, because of greater deformation and erosion. Thus terranes provide another way of viewing Earth and gaining a better understanding of the geologic history of the continents.

> **What Would You Do?**
>
> Because of political events in the Middle East, the oil-producing nations of this region have reduced the amount of petroleum they export, resulting in shortages in the United States. To alleviate U.S. dependence on overseas oil, the major oil companies want Congress to let them explore for oil in many of our national parks. As director of the National Park system, you have been called to testify at the congressional hearing addressing this possibility. What arguments would you use to discourage such exploration? Would a knowledge of the geology of the area be helpful in your testimony? Explain.

Mesozoic Mineral Resources

Although much of the coal in North America is Pennsylvanian or Paleogene in age, important Mesozioc coals occur in the Rocky Mountains states. These are mostly lignite and bituminous coals, but some local anthracites are present as well. Particularly widespread in western North America are coals of Cretaceous age. Mesozoic coals are also known from Alberta and British Columbia, Canada, as well as from Australia, Russia, and China.

Large concentrations of petroleum occur in many areas of the world, but more than 50% of all proven reserves are in the Persian Gulf region (see Perspective 3.1). During the Mesozoic Era, what is now the Gulf region was a broad passive continental margin conducive for the formation of petroleum. Similar conditions existed in what is now the Gulf Coast region of the United States and Central America. Here, petroleum and natural gas also formed on a broad shelf over which transgressions and regressions occurred. In this region, the hydrocarbons are largely in reservoir rocks that were deposited as distributary channels on deltas and as barrier-island and beach sands. Some of these hydrocarbons are associated with structures formed adjacent to rising salt domes. The salt, called the *Louann Salt,* initially formed in a long, narrow sea when North America separated from Europe and North Africa during the fragmentation of Pangaea (Figure 14.2).

The richest uranium ores in the United States are widespread in Mesozoic rocks of the Colorado Plateau area of Colorado and adjoining parts of Wyoming, Utah, Arizona, and New Mexico. These ores, consisting of fairly pure masses of a complex potassium-, uranium-, vanadium-bearing mineral called *carnotite,* are associated with plant remains in sandstones that were deposited in ancient stream channels.

As noted in Chapter 9, Proterozoic banded iron formations are the main sources of iron ores. There are, however, important exceptions. For example, the Jurassic-aged "Minette" iron ores of western Europe, composed of oolitic limonite and hematite, are important ores in France, Germany, Belgium, and Luxembourg. In Great Britain, low-grade iron ores of Jurassic age consist of oolitic siderite, which is an iron carbonate. And in Spain, Cretaceous rocks are the host rocks for iron minerals.

South Africa, the world's leading producer of gem-quality diamonds and among the leaders in industrial diamond production, mines these minerals from kimberlite pipes, conical igneous intrusions of dark gray or blue igneous rock. Diamonds, which form at great depth where pressure and temperature are high, are brought to the surface during the explosive volcanism that forms kimberlite pipes. Although kimberlite pipes have formed throughout geologic time, the most intense episode of such activity in South Africa and adjacent countries was during the Cretaceous Period. Emplacement of Triassic and Jurassic diamond-bearing kimberlites also occurred in Siberia.

In the Introduction we noted that the mother lode or source for the placer deposits mined during the California gold rush is in Jurassic-aged intrusive rocks of the Sierra Nevada. Gold placers are also known in Cretaceous-aged conglomerates of the Klamath Mountains of California and Oregon.

Porphyry copper was originally named for copper deposits in the western United States mined from porphyritic granodiorite, but the term now applies to large, low-grade copper deposits disseminated in a variety of rocks. These porphyry copper deposits are an excellent example of the relationship between convergent plate boundaries and the distribution, concentration, and exploitation of valuable metallic ores. Magma generated by partial melting of a subducting plate rises toward the surface, and as it cools, it precipitates and concentrates various metallic ores. The world's largest copper deposits were formed during the Mesozoic and Cenozoic in a belt along the western margins of North and South America (see Figure 3.29).

SUMMARY

Table 14.1 provides a summary of the geologic history of North America, as well as global events and sea-level changes during the Mesozoic.

- The breakup of Pangaea can be summarized as follows.
  a. During the Late Triassic, North America began separating from Africa. This was followed by the rifting of North America from South America.
  b. During the Late Triassic and Jurassic periods, Antarctica and Australia—which remained sutured together—began separating from South America and Africa, and India began rifting from Gondwana.
  c. South America and Africa began separating during the Jurassic, and Europe and Africa began converging during this time.
  d. The final stage in Pangaea's breakup occurred during the Cenozoic when Greenland completely separated from Europe and North America.
- The breakup of Pangaea influenced global climatic and atmospheric circulation patterns. Although the temperature gradient from the tropics to the poles gradually increased during the Mesozoic, overall global temperatures remained equable.
- An increased rate of seafloor spreading during the Cretaceous Period caused sea level to rise and transgressions to occur.
- Except for incursions along the continental margin and two major transgressions (the Sundance Sea and the Cretaceous Interior Seaway), the North American craton was above sea level during the Mesozoic Era.
- The Eastern Coastal Plain was the initial site of the separation of North America from Africa that began during the Late Triassic. During the Cretaceous Period, it was inundated by marine transgressions.
- The Gulf Coastal region was the site of major evaporite accumulation during the Jurassic as North America rifted from South America. During the Cretaceous, it was inundated by a transgressing sea, which, at its maximum, connected with a sea transgressing from the north to create the Cretaceous Interior Seaway.
- Mesozoic rocks of the western region of North America were deposited in a variety of continental and marine environments. One of the major controls of sediment distribution patterns was tectonism.
- Western North America was affected by four orogenies: the Sonoma, Nevadan, Sevier, and Laramide. Each involved igneous intrusions, as well as eastward thrust faulting and folding.
- The cause of the Nevadan, Sevier, and Laramide orogenies was the changing angle of subduction of the oceanic Farallon plate under the continental North American plate. The timing, rate, and, to some degree, the direction of plate movement were related to seafloor spreading and the opening of the Atlantic Ocean.
- Orogenic activity associated with the oceanic–continental convergent plate boundary in the Cordilleran mobile belt explains the structural features of the western margin of North America. It is thought, however, that more than 25% of the North American western margin originated from the accretion of terranes.
- Mesozoic rocks contain a variety of mineral resources, including coal, petroleum, uranium, gold, and copper.

IMPORTANT TERMS

Cordilleran orogeny, p. 283
Cretaceous Interior Seaway, p. 289
Laramide orogeny, p. 284
Nevadan orogeny, p. 283
Sevier orogeny, p. 284
Sonoma orogeny, p. 282
Sundance Sea, p. 286
terrane, p. 290
Zuni Sequence, p. 278

REVIEW QUESTIONS

1. The formation or complex responsible for the spectacular scenery of the Painted Desert and Petrified Forest is the
   a. _____ Franciscan; b. _____ Morrison; c. _____ Chinle; d. _____ Wingate; e. _____ Navajo.
2. A possible cause for the eastward migration of igneous activity in the Cordilleran region during the Cretaceous was a change from
   a. _____ high-angle to low-angle subduction; b. _____ divergent plate margin activity to subduction; c. _____ subduction to divergent plate margin activity; d. _____ oceanic–oceanic convergence to oceanic–continental convergence; e. _____ divergent to convergent plate margin activity.
3. The first Mesozoic orogeny in the Cordilleran region was the
   a. _____ Sevier; b. _____ Laramide; c. _____ Sonoma; d. _____ Antler; e. _____ Nevadan.
4. During the Jurassic, the newly forming Gulf of Mexico was the site of primarily what type of deposition?
   a. _____ evaporites; b. _____ siliciclastics; c. _____ volcaniclastics; d. _____ detrital; e. _____ answers b and c.
5. What is the evidence for the breakup of Pangaea?
   a. _____ rift valleys; b. _____ dikes; c. _____ great quantities of poorly sorted nonmarine detrital sediments; d. _____ sills; e. _____ all of the previous answers.

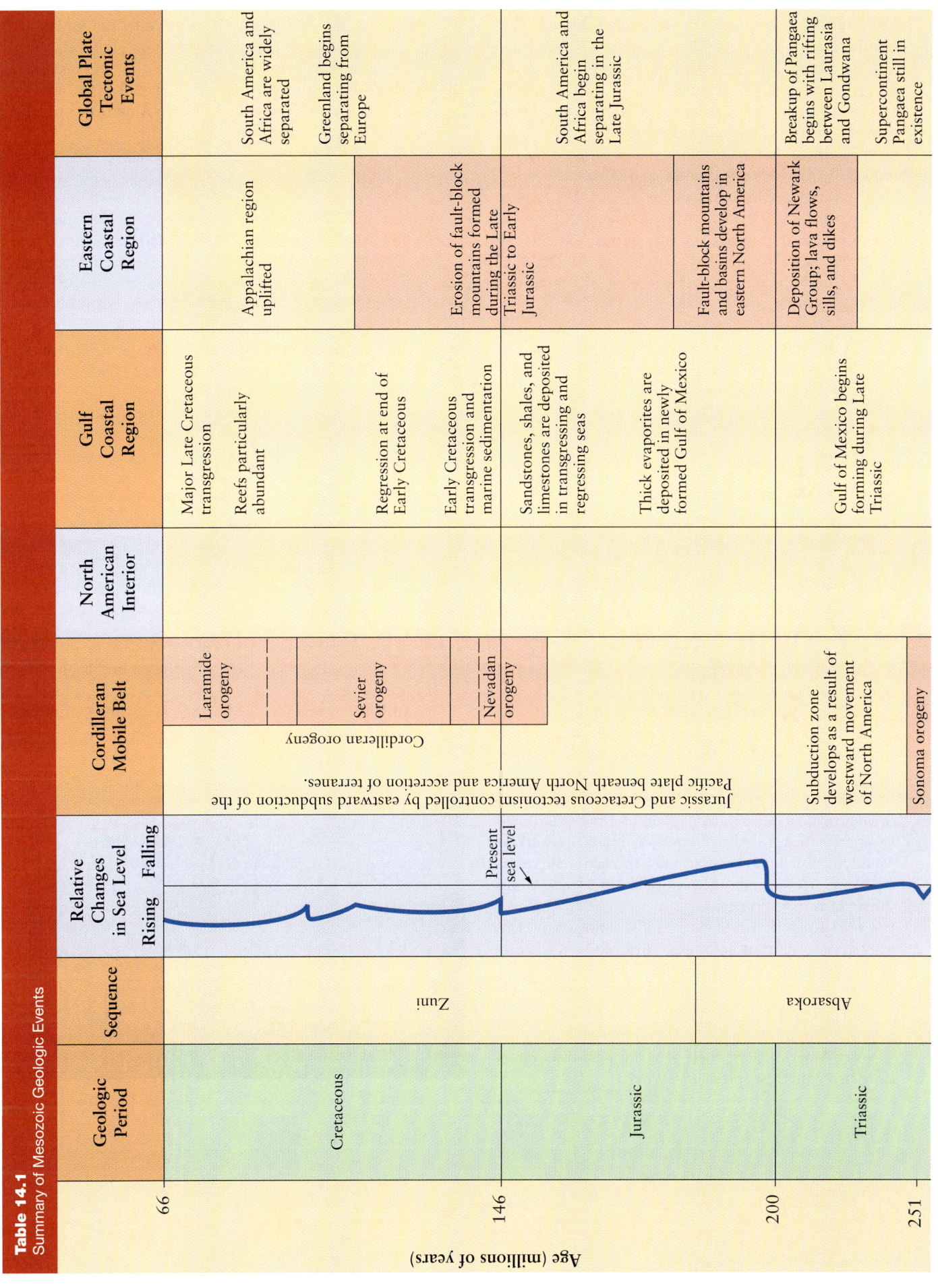

Table 14.1 Summary of Mesozoic Geologic Events

6. The Jurassic formation or complex famous for dinosaur fossils is the
   a. _____ Chinle; b. _____ Morrison; c. _____ Franciscan; d. _____ Navajo; e. _____ Sundance.
7. Triassic rifting between which two continental landmasses initiated the breakup of Pangaea?
   a. _____ India and Australia; b. _____ Antarctica and India; c. _____ South America and Africa; d. _____ North America and Eurasia; e. _____ Laurasia and Gondwana.
8. The orogeny responsible for the present-day Rocky Mountains is the
   a. _____ Sonoma; b. _____ Sevier; c. _____ Laramide; d. _____ Nevadan; e. _____ Antler.
9. The Mesozoic tectonic history of the North American Cordilleran region is very complex and involves
   a. _____ oceanic–continental convergence; b. _____ continental–continental convergence; c. _____ terrane accretion; d. _____ answers a and b; e. _____ answers a and c.
10. The time of greatest post-Paleozoic inundation of the craton occurred during which geologic period?
    a. _____ Triassic; b. _____ Jurassic; c. _____ Cretaceous; d. _____ Paleogene; e. _____ Neogene.
11. Which orogeny produced the Sierra Nevada, Southern California, Idaho, and Coast Range batholiths?
    a. _____ Laramide; b. _____ Sonoma; c. _____ Nevadan; d. _____ Sevier; e. _____ none of the previous answers.
12. The first major seaway to flood North America was the
    a. _____ Cretaceous Interior Seaway; b. _____ Sundance; c. _____ Cordilleran; d. _____ Zuni; e. _____ Newark.
13. Which formation or group filled the Late Triassic fault-block basins of the east coast of North America with red nonmarine sediments?
    a. _____ Morrison; b. _____ Chinle; c. _____ Navajo; d. _____ Franciscan; e. _____ Newark.
14. Discuss the similarities and differences between the orogenic activity that occurred in the Appalachian mobile belt during the Paleozoic Era and that which occurred in the Cordilleran mobile belt during the Mesozoic Era.
15. How did the Mesozoic rifting that took place on the East Coast of North America affect the tectonics in the Cordilleran mobile belt?
16. Compare the tectonic setting and depositional environment of the Gulf of Mexico evaporites with the evaporite sequences of the Paleozoic Era.
17. How did the breakup of Pangaea affect oceanic and climatic circulation patterns?
18. How does terrane accretion change our interpretations of the geologic history of the western margin of North America, and how does it relate to the Mesozoic orogenies that took place in that area?
19. Compare the tectonics of the Sonoma and Antler orogenies.
20. Explain and diagram how increased seafloor spreading can cause a rise in sea level along the continental margins.

APPLY YOUR KNOWLEDGE

1. The breakup of Pangaea influenced the distribution of continental landmasses, ocean basins, and oceanic and atmospheric circulation patterns, which in turn affected the distribution of natural resources, landforms, and the evolution of the world's biota. Reconstruct a hypothetical history of the world for a different breakup of Pangaea—one in which the continents separate in a different order or rift apart in a different configuration. How would such a scenario affect the distribution of natural resources? Would the distribution of coal and petroleum reserves be the same? How might evolution be affected? Would human history be different?
2. The gold discovered at Sutter's Mill, California, in 1848 sparked the California gold rush. This gold is widely disseminated throughout the granitic rocks of the Sierra Nevada batholith and was concentrated in placer deposits. During the Nevadan orogeny, several other large granitic batholiths were intruded in the Cordillera region of North America. Have there been any gold discoveries associated with these intrusions? Why?

FIELD QUESTION

1. What is the sedimentary structure shown in this photo? In what type of environment did this formation form? From your knowledge of Mesozoic formations, what is the name of this formation, and where is it located?

© John Sibbick

CHAPTER 15 | LIFE OF THE MESOZOIC ERA

This Mesozoic scene shows the 10-m-long carnivorous dinosaur *Baryonyx* and two large herbivorous dinosaurs in the background. Dinosaurs are the most popular of all extinct organisms, and restorations such as this one bring them to life at least in our mind's eye.

[OUTLINE]

INTRODUCTION

MARINE INVERTEBRATES AND PHYTOPLANKTON

AQUATIC AND SEMIAQUATIC VERTEBRATES—FISH AND AMPHIBIANS

PLANTS—PRIMARY PRODUCERS ON LAND

THE DIVERSIFICATION OF REPTILES

Archosaurs and the Origin of Dinosaurs

Dinosaurs

Warm-Blooded Dinosaurs?

Flying Reptiles

Mesozoic Marine Reptiles

Crocodiles, Turtles, Lizards, and Snakes

PERSPECTIVE 15.1 *Mary Anning and Her Contributions to Paleontology*

FROM REPTILES TO BIRDS

ORIGIN AND EVOLUTION OF MAMMALS

Cynodonts and the Origin of Mammals

Mesozoic Mammals

MESOZOIC PALEOBIOGEOGRAPHY

MASS EXTINCTIONS—A CRISIS IN THE HISTORY OF LIFE

SUMMARY

ThomsonNOW™ Explore interactive tutorials, animations, or practice problems available on the ThomsonNow website at www.thomsonedu.com/login.

[CHAPTER OBJECTIVES]

At the end of this chapter, you will have learned that

- Marine invertebrates that survived the Paleozoic extinctions diversified and repopulated the seas.
- Land plant communities changed considerably when flowering plants evolved during the Cretaceous.
- Reptile diversification began during the Mississippian and continued throughout the Mesozoic Era.
- Among the Mesozoic reptiles, dinosaurs had evolved by the Late Triassic and soon became the dominant land-dwelling vertebrate animals.
- In addition to dinosaurs, the Mesozoic was also the time of flying reptiles and marine reptiles, as well as turtles, lizards, snakes, and crocodiles.
- Birds evolved from reptiles, probably from some kind of small carnivorous dinosaur.
- Mammals evolved from reptiles only distantly related to dinosaurs, and they existed as contemporaries with dinosaurs.
- The transition from reptiles to mammals is very well supported by fossil evidence.
- Several varieties of Mesozoic mammals are known, all of which were small, and their diversity remained low.
- The proximity of continents and generally mild Mesozoic climates allowed many plants and animals to spread over extensive geographic areas.
- Extinctions at the end of the Mesozoic Era were second in magnitude only to the Paleozoic extinctions. These extinctions have received more attention than any other because dinosaurs were among the victims.

Introduction

In 1842, Sir Richard Owen coined the term *dinosaur*, and since then, these animals have been the object of intense curiosity and the subject matter for countless articles and books, TV specials, and movies. Indeed, with the release of the movie *Jurassic Park* in 1993 and its sequels *The Lost World* (1997) and *Jurassic Park III* (2001), the interest in dinosaurs has probably been its greatest ever (see the chapter opening photo). No other group of animals has so thoroughly captured the public imagination, but dinosaurs, although certainly captivating, were only one type of Mesozoic reptile. Equally remarkable were the flying reptiles (pterosaurs), several types of marine reptiles, and crocodiles, turtles, lizards, and snakes. In fact, geologists refer informally to the Mesozoic Era as "The Age of Reptiles," alluding to the importance of these creatures among land-dwelling animals.

The origin and evolution of dinosaurs and other reptiles were certainly important in the history of life, but so were several other Mesozoic biologic events. For instance, mammals evolved from mammal-like reptiles during the Triassic and thus were contemporaries of dinosaurs, although they were not nearly as diverse and none were very large. Birds also made their appearance, probably evolving from small carnivorous dinosaurs during the Jurassic. Remarkable discoveries of feathered dinosaurs in China have important implications not only about dinosaur biology (at least some were probably warm-blooded), but they are also important in evaluating dinosaur relationships with birds.

Important changes also took place in Mesozoic land plant communities as the flowering plants evolved during the Cretaceous and soon became widespread and numerous.

Certainly the major groups of Paleozoic land plants persisted, but now they constitute less than 10% of all species. Recall the Paleozoic extinctions that decimated the marine invertebrate faunas, causing a decrease in biotic diversity. The survivors of this crisis in life history diversified during the Triassic and repopulated the seas, accounting for the success of several types of cephalopods, bivalves, and many other invertebrates.

Throughout this book, we have emphasized the systems approach to Earth and life history, and in this chapter we do so again. Remember that the distribution of land and sea profoundly influences oceanic circulation, which in turn partly controls climate, and that the proximity or separation of landmasses partly determines the geographic distribution of organisms. Pangaea began fragmenting during the Triassic and continues to do so. As it did, intercontinental migrations became increasingly difficult. In fact, South America and Australia became isolated island continents, and their faunas became quite different from those elsewhere.

Mass extinctions at the end of the Mesozoic, second in magnitude only to the Paleozoic extinctions, had a tremendous impact on the biosphere. Given that dinosaurs were among the victims, these extinctions have received much more attention than any other extinction. So just as at the end of the Paleozoic Era, biotic diversity was sharply reduced, but once again many of the survivors evolved rapidly, giving rise to the Cenozoic fauna (see Chapter 18).

Marine Invertebrates and Phytoplankton

Following the Paleozoic mass extinctions, the Mesozoic was a time when marine invertebrates repopulated the seas. The Early Triassic invertebrate fauna was not very diverse, but by the Late Triassic the seas were once again swarming with invertebrates from planktonic foraminifera to a wide variety of cephalopods. The brachiopods that had been so abundant during the Paleozoic never completely recovered from their near extinction, and although still present in modern seas, the bivalves have largely taken over their ecologic niche.

Mollusks such as cephalopods, bivalves, and gastropods were the most important elements of the Mesozoic marine invertebrate fauna. Their rapid evolution and the fact that many cephalopods were nektonic make them excellent guide fossils (• Figure 15.1). The ammonoids (Order Ammonoidea), cephalopods with wrinkled sutures, constitute three groups: the goniatites, ceratites, and ammonites. The latter, while present during the entire Mesozoic, were most prolific during the Jurassic and Cretaceous. Most ammonites were coiled, with some attaining diameters of 2 m, whereas others were uncoiled and led a near benthonic existence. Ammonites became extinct at the end of the Cretaceous, but two related groups of cephalopods survived into the Cenozoic: the *nautiloids* (Order Nautiloidea), which include the living pearly nautilus, and the *coleoids* (Order Coleoidea), represented by the extinct belemnoids as well as the extant squid and octopus. Belemnoids, which were squidlike in appearance, were particularly abundant during the Cretaceous Period and are excellent guide fossils for the Jurassic and Cretaceous (• Figure 15.2).

Mesozoic bivalves diversified to inhabit many epifaunal and infaunal niches (• Figure 15.3). Oysters and clams (epifaunal suspension feeders) became particularly diverse and abundant, and despite a reduction in diversity at the end of the Cretaceous, they remain important animals in the marine fauna today.

As is true now, where shallow marine waters were warm and clear, coral reefs proliferated. However, during the Mesozoic, reefs did not rebound from the Permian extinctions until the Middle Triassic. An important reef-builder throughout the Mesozoic was a group of bivalves known as *rudists*. Rudists are important because they displaced corals as the main reef-builders during the later Mesozoic

• **Figure 15.1 Cretaceous Seascape** Cephalopods such as the Late Cretaceous ammonoids *Baculites* (foreground) and *Helioceros* (background) were present throughout the Mesozoic, but they were most abundant during the Jurassic and Cretaceous. They were important predators, and they are excellent guide fossils.

• **Figure 15.2 Belemnoids** Belemnoids are extinct squidlike cephalopods that were particularly abundant during the Cretaceous. They are excellent guide fossils for the Jurassic and Cretaceous. Shown here are several belemnoids swimming in a Cretaceous sea.

Archean Eon	Proterozoic Eon	Phanerozoic Eon							
			Paleozoic Era						
Precambrian		Cambrian	Ordovician	Silurian	Devonian	Mississippian	Pennsylvanian	Permian	
						Carboniferous			

2500 MYA — 542 MYA — 251 MYA

• **Figure 15.3 Cretaceous Hard Chalk Seafloor** A recreation of a Cretaceous hard chalk seafloor. Shown on the seafloor are crinoids, echinoids, brachiopods, molluscs, and a lobster. Swimming above the seafloor is an ammonite. Molluscs were major elements of the Mesozoic marine invertebrate fauna, particularly during the Cretaceous Period.

and are excellent guide fossils for the Late Jurassic and Cretaceous.

A new and familiar type of coral also first appeared during the Triassic: the *scleractinians*. Whether scleractinians evolved from rugose corals or from an as yet unknown soft-bodied ancestor with no known fossil record is still unresolved. In addition to the familiar present-day reef-building colonial scleractinian corals, solitary or individual scleractinian corals inhabited relatively deep waters during the Mesozoic.

Another invertebrate group that prospered during the Mesozoic was the echinoids. Echinoids were exclusively epifaunal during the Paleozoic but branched out into the infaunal habitat during the Mesozoic, and both groups began a major adaptive radiation during the Late Triassic that continued throughout the remainder of the Mesozoic and into the Cenozoic.

The foraminifera (single-celled consumers) diversified rapidly during the Jurassic and Cretaceous and continue to be diverse and abundant to the present (• Figure 15.4). The planktonic forms in particular diversified rapidly, but most genera became extinct at the end of the Cretaceous. The planktonic foraminifera are excellent guide fossils for the Cretaceous.

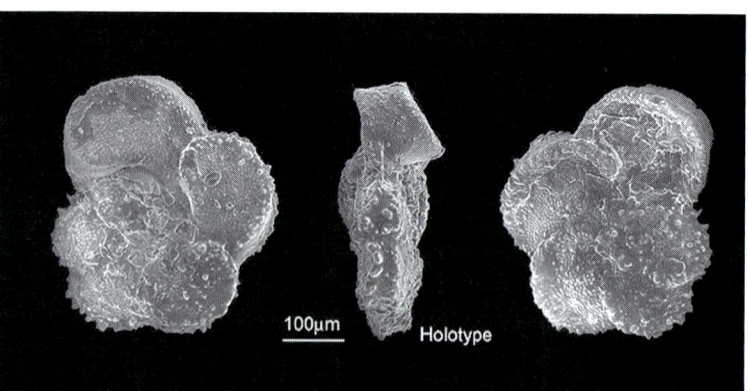

• **Figure 15.4 Planktonic Foraminifera** Planktonic foraminifera diversified and became abundant during the Jurassic and Cretaceous, and continued to be diverse and abundant throughout the Cenozoic. Many planktonic foraminifera are excellent guide fossils for the Cretaceous, such as species of the genus *Globotruncana*, which is restricted to the Upper Cretaceous. Shown here are three views of the holotype (the specimen that defines the species) of *Globotruncana loeblichi*, a Late Cretaceous planktonic foraminifera.

Phanerozoic Eon									
Mesozoic Era			Cenozoic Era						
Triassic	Jurassic	Cretaceous	Paleogene			Neogene		Quaternary	
			Paleocene	Eocene	Oligocene	Miocene	Pliocene	Pleistocene	Holocene

251 MYA — 66 MYA

A major difference between the Paleozoic and Mesozoic marine invertebrate faunas was the increased abundance and diversity of burrowing organisms. With few exceptions, Paleozoic burrowers were soft-bodied animals such as worms. The bivalves and echinoids, which were epifaunal animals during the Paleozoic, evolved various means of entering infaunal habitats. This trend toward an infaunal existence may have been an adaptive response to increasing predation from the rapidly evolving fish and cephalopods. Bivalves, for instance, expanded into the infaunal niche during the Mesozoic, and they could escape predators by burrowing.

The primary producers in the Mesozoic seas included the microscopic *coccolithophores*, *diatoms*, and *dinoflagellates* (• Figure 15.5). Coccolithophores are an important group of calcareous phytoplankton (Figure 15.5a) that first evolved during the Jurassic and became extremely common during the Cretaceous. Diatoms (Figure 15.5b), which build skeletons of silica, made their appearance during the Cretaceous, but they are more important as primary producers during the Cenozoic. Diatoms are presently most abundant in cooler oceanic waters, and some species inhabit freshwater lakes. Dinoflagellates, which are organic-walled phytoplankton, were common during the Mesozoic and today are the major primary producers in warm water (Figure 15.5c).

In general terms, we can think of the Mesozoic as a time of increasing complexity among the marine invertebrate fauna. At the beginning of the Triassic, diversity was low and food chains were short. Near the end of the Cretaceous, though, the marine invertebrate fauna was highly complex, with interrelated food chains. This evolutionary history reflects changing geologic conditions influenced by plate tectonic activity (see Chapter 3).

Aquatic and Semiaquatic Vertebrates—Fish and Amphibians

Sharks and the other cartilaginous fishes became more abundant during the Mesozoic, but even so, they never came close to matching the diversity of the bony fishes. Nevertheless, sharks, although an evolutionarily conservative group,

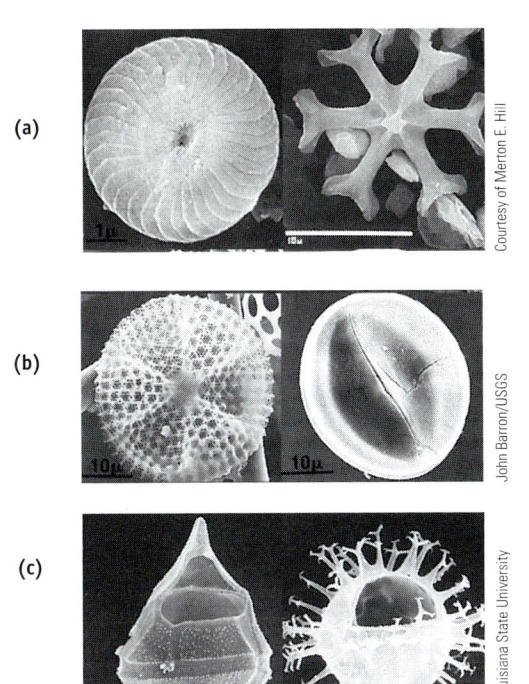

• **Figure 15.5 Primary Producers** (a) A Miocene (left) and a Miocene-Pliocene coccolithophore from the Gulf of Mexico. (b) Diatoms from Upper Miocene rocks in Java. (c) An Eocene dinoflagellate from Alabama (left) and a Miocene-Pliocene dinoflagellate from the Gulf of Mexico.

were and still are important in the marine fauna, especially among predators. Few species of lungfishes and crossopterygians existed during the Mesozoic, and the latter declined and were nearly extinct by the end of the era. Only one crossopterygian species exists now (see Figure 7.17a), and the group has no known Cenozoic fossil record.

All bony fish, except lungfishes and crossopterygians, belong to three groups, which for convenience we call *primitive*, *intermediate*, and *advanced*. The primitive bony fishes existed mostly during the Paleozoic, but by Middle Mesozoic time the intermediate group predominated. The advanced group, more formally known as *teleosts*, was dominant by Cretaceous time in both marine and freshwater environments. With about 20,000 living species, they are by far the most diverse and numerous of all living vertebrate animals.

During 2003, a rather complete fossil of the largest known fish was found in England. This Jurassic-aged fish measured 22 m long, nearly twice the length of today's largest fish: the whale shark. Like the whale shark, though, this giant fish probably ate plankton.

The labyrinthodont amphibians were common during the latter part of the Paleozoic, but the few surviving Mesozoic species died out by the end of the Triassic. Since their greatest abundance during the Pennsylvanian Period, amphibians have made up only a small part of the total vertebrate fauna. Frogs and salamanders evolved during the Mesozoic, but both have poor fossil records.

Plants—Primary Producers on Land

Just as during the Late Paleozoic, seedless vascular plants and gymnosperms dominated Triassic and Jurassic land plant communities, and, in fact, representatives of both groups are still common (• Figure 15.6). Among the gymnosperms, the large seed ferns became extinct by the end of the Triassic, but *ginkgos* (see Figure 7.17b) remained abundant and still exist in isolated regions, and *conifers* continued to diversify and are now widespread, particularly at high elevations and high latitudes. A new group of gymnosperms known as *cycads* made their appearance during the Triassic. They became widespread and now exist in tropical and semitropical areas (Figure 15.6b).

The long dominance of seedless plants and gymnosperms ended during the Early Cretaceous, perhaps the Late Jurassic, when many were replaced by **angiosperms**, or flowering plants (• Figure 15.7). Studies of fossil and living gymnosperms show that some have close relationships to angiosperms, but unfortunately, the early fossil record of angiosperms is sparse, so their precise ancestors remain obscure.

Since they first evolved, angiosperms have adapted to nearly every terrestrial habitat from mountains to deserts, and some have even adapted to shallow coastal waters. Several factors account for their phenomenal success, but chief among them were the evolution of flowers, which attract animal pollinators, especially insects, and the evolution of enclosed seeds (Figure 15.7a).

Seedless vascular plants and gymnosperms are important and still flourish in many environments; in fact, many botanists regard ferns and conifers as emerging groups. Nevertheless, a measure of the angiosperms' success is that today with 250,000 to 300,000 species they account for more than 90% of all land plant species, and they occupy some habitats in which other land plants do poorly or cannot exist.

The Diversification of Reptiles

Reptile diversification began during the Mississippian Period when the protothyrids, the first animals to lay amniotic eggs, evolved (see Chapter 13). From this basic stock

(a)

(b)

• Figure 15.6 **Mesozoic Plants** (a) This Jurassic landscape was dominated by seedless vascular plants, particularly ferns, as well as gymnosperms such as conifers, tree ferns, and cycads. (b) These living cycads look much like some of those depicted in (a).

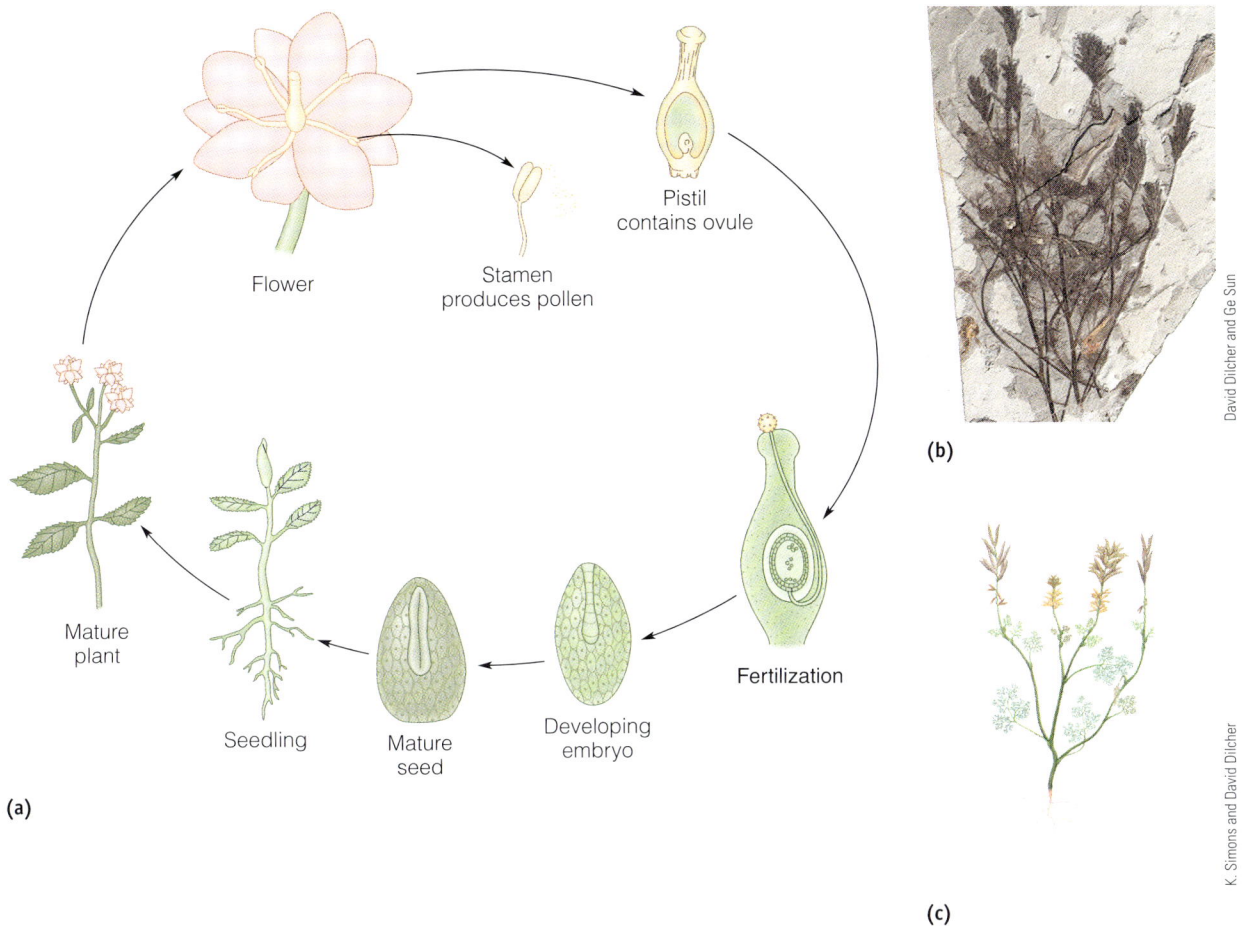

• **Figure 15.7 The Angiosperms or Flowering Plants (a)** The reproductive cycle in angiosperms. **(b)** *Archaefructus sinensis* from Lower Cretaceous rocks in China is among the oldest known angiosperms. **(c)** Restoration of *Archaefructus sinensis*.

of so-called *stem reptiles*, all other reptiles as well as birds and mammals evolved (• Figure 15.8). Recall from Chapter 13 that pelycosaurs were the dominant land vertebrates during the Pennsylvanian and Permian. Here we continue our story of reptile diversification with a group called *archosaurs*.

Archosaurs and the Origin of Dinosaurs

Crocodiles, flying reptiles (pterosaurs), dinosaurs, and birds are all **archosaurs** (*archo,* meaning "ruling," and *sauros,* meaning "lizard"). Including such diverse animals in a single group implies that they share a common ancestor (Figure 15.8), and in fact, they possess several characteristics that unite them and distinguish them from other reptiles. Of course, they lay amniote eggs, as do all reptiles and birds, but they also have teeth set in individual sockets, except today's birds, but even the early birds had this feature and several other skeletal characteristics that set them apart. One of the most important is features of the limbs that indicate semi-upright or upright posture. We will have more to say about crocodiles, pterosaurs, and birds later in this chapter, but now we turn to a discussion of dinosaurs.

Dinosaurs

Sir Richard Owen proposed the term **dinosaur** in 1842 to mean "fearfully great lizard," although now "fearfully" has come to mean "terrible," thus the characterization of dinosaurs as "terrible lizards." Of course, they were not terrible, or at least no more terrible than animals living today, and they were not lizards. Dinosaurs more than any other kind of animal have inspired awe, but unfortunately, their popularization in many cartoons, books, and movies has been inaccurate and has contributed to many misunderstandings. For instance, many people think that all dinosaurs were very large and that because they are extinct, they must have been poorly adapted. It is true that many were large, but dinosaurs varied from giants weighing several tens of metric tons to those that weighed no more than 2 or 3 kg. And to consider them poorly adapted is to ignore

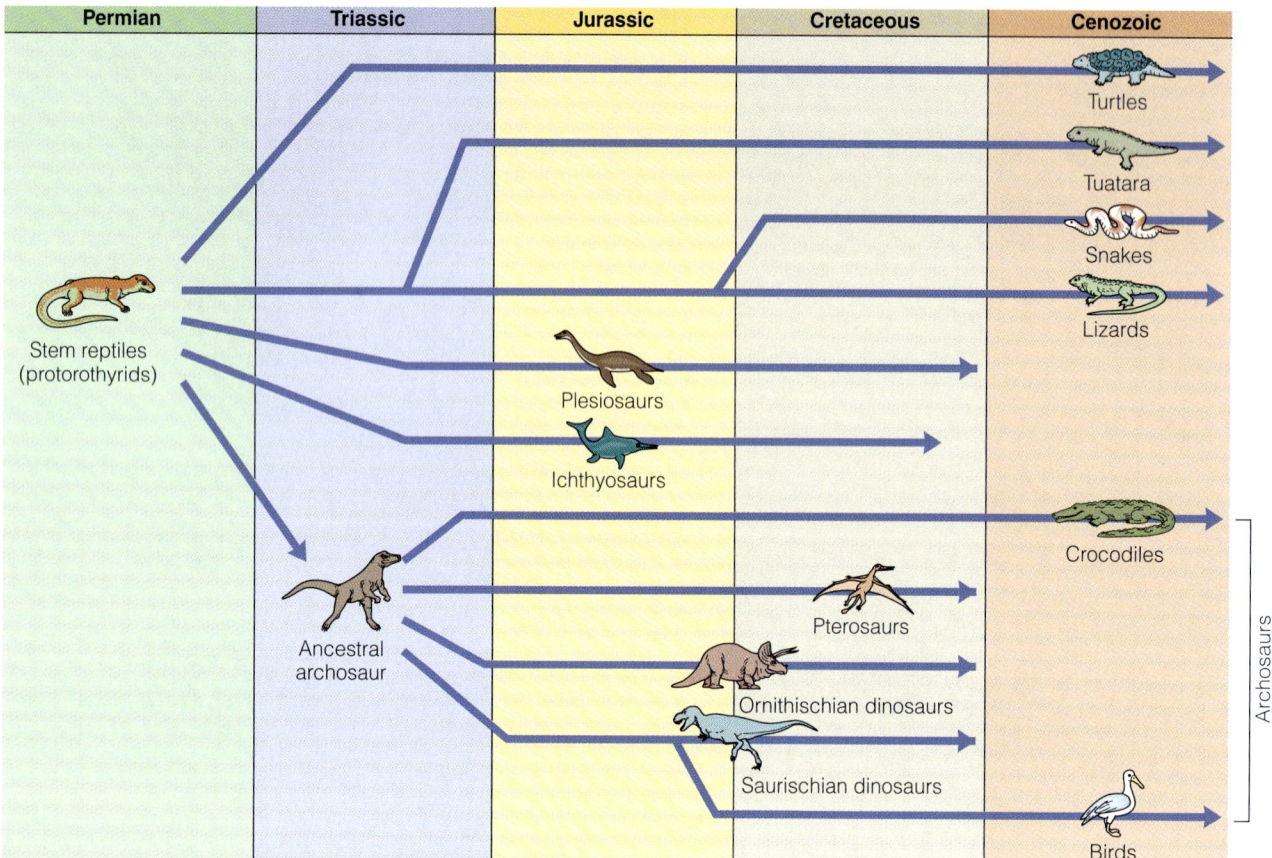

• **Figure 15.8 Relationships Among Fossil and Living Reptiles and Birds**

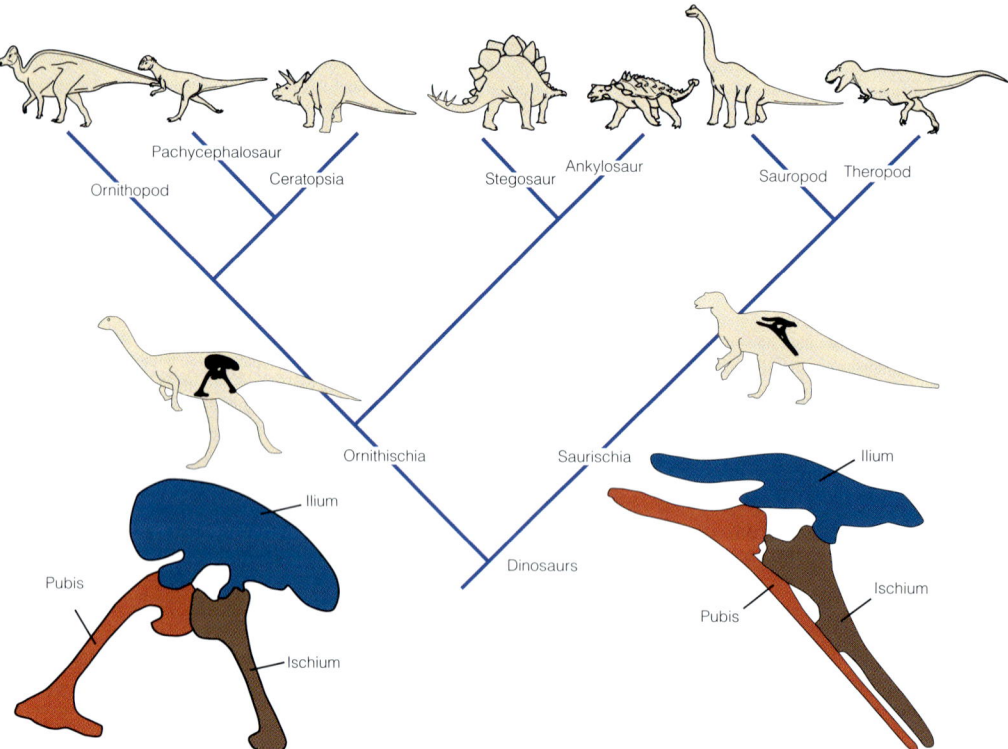

• **Figure 15.9 Cladogram Showing Relationships Among Dinosaurs** Pelvises of ornithischian and saurischian dinosaurs are shown for comparison. All the dinosaurs shown here were herbivores except for theropods. Notice that bipedal and quadrupedal dinosaurs are fond in both ornithischians and saurischians.

302 CHAPTER 15 LIFE OF THE MESOZOIC ERA

the fact that dinosaurs were extremely diverse and widespread for more than 140 million years!

Although various media now portray dinosaurs as more active animals, the misconception persists that they were lethargic, dimwitted beasts. Evidence now available indicates that some were very active and perhaps even warm-blooded. Furthermore, some species probably cared for their young long after hatching, a behavioral characteristic most often found in birds and mammals. Many questions remain unanswered about dinosaurs, but their fossils and the rocks containing them are revealing more and more about their evolutionary relationships and behavior. All dinosaurs share several characteristics such as fully upright posture with the limbs directly beneath their bodies as in mammals and birds, yet they differ enough for us to recognize two distinct orders: the **Saurischia** and the **Ornithischia**. Saurischian dinosaurs have a lizardlike pelvis and accordingly are called *lizard-hipped dinosaurs*, whereas a birdlike pelvis is present in the ornithiscians, so they are called *bird-hipped dinosaurs* (• Figure 15.9). Paleontologists are convinced that both orders of dinosaurs share a common ancestor, much like the archosaurs from Middle Triassic rocks in Argentina. These dinosaur ancestors were small (less than 1 m long), long-legged carnivores that walked and ran on their hind limbs, so they were **bipedal**, as opposed to **quadrupedal** animals that move on all four limbs.

Saurischian Dinosaurs The saurischians include two distinct groups known as *theropods* and *sauropods* (Figure 15.9, Table 15.1). All theropods were carnivorous bipeds that varied from tiny *Compsognathus* to giants such as *Tyrannosaurus* and similar but even larger species (• Figure 15.10a and b). Since 1996, Chinese scientists have discovered several

TABLE 15.1 Summary Chart for the Orders and Suborders of Dinosaurs (lengths and weights approximate, from several sources)

Order	Suborder	Familiar Genera	Comments
Saurischia	Theropoda	*Allosaurus, Coelophysis, Compsognathus, Deinonychus, Tyrannosaurus,* * *Velociraptor*	Bipedal carnivores. Late Triassic to end of Cretaceous. Size from 0.6 to 15 m long, 2 or 3 kg to 7.3 metric tons. Some smaller genera may have hunted in packs.
	Sauropoda	*Apatosaurus, Brachiosaurus, Camarasaurus, Diplodocus, Titanosaurus*	Giant quadrupedal herbivores. Late Triassic to Cretaceous, but most common during Jurassic. Size up to 27 m long, 75 metric tons.** Trackways indicate sauropods lived in herds. Preceded in fossil record by the smaller prosauropods.
Ornithischia	Ornithopoda	*Anatosaurus, Camptosaurus, Hypsilophodon, Iguanodon, Parasaurolophus*	Some ornithopods, such as *Anatosaurus*, had a flattened bill-like mouth and are called duck-billed dinosaurs. Size from a few meters long up to 13 m and 3.6 metric tons. Especially diverse and common during the Cretaceous. Primarily bipedal herbivores, but could also walk on all fours.
	Pachycephalosauria	*Stegoceras*	*Stegoceras* only 2 m long and 55 kg, but larger species known. Thick bones of skull cap might have aided in butting contests for dominance and mates. Bipedal herbivores of Cretaceous.
	Ankylosauria	*Ankylosaurus*	*Ankylosaurus* more than 7 m long and about 2.5 metric tons. Heavily armored with bony plates on top of head, back, and sides. Quadrupedal herbivore.
	Stegosauria	*Stegosaurus*	A variety of stegosaurs are known, but *Stegosaurus*, with bony plates on its back and a spiked tail, is best known. Plates probably were for absorbing and dissipating heat. Quadrupedal herbivores that were most common during the Jurassic. *Stegosaurus* 9 m long, 1.8 metric tons.
	Ceratopsia	*Triceratops*	Numerous genera known. Some early ones bipedal, but later large animals were quadrupedal herbivores. Much variation in size; *Triceratops* to 7.6 m long and 5.4 metric tons, with large bony frill over top of neck, three horns on skull, and beaklike mouth. Especially common during the Cretaceous.

*Until recently *Tyrannosaurus* at 4.5 metric tons was the largest known theropod, but now similar and larger animals are known from Argentina and Africa.
**Partial remains indicate even larger brachiosaurs existed, perhaps measuring 30 m long and weighing 100 metric tons.

• **Figure 15.10 Theropod Dinosaurs Were Bipedal Carnivores (a)** *Compsognathus* was about 60 cm long and probably weighed no more than 2 or 3 kg. Bones found within its ribcage indicate that it ate lizards. **(b)** *Tyrannosaurus* weighed several metric tons. Its skull measured more than 1 m long. **(c)** Lifelike restoration of *Deinonychus*, a 3 m long theropod that may have weighed 80 kg.

theropods with feathers.* No one doubts that these dinosaurs had feathers, and molecular evidence indicates they were composed of the same material as bird feathers.

The movie *Jurassic Park* and its sequels popularized some of the smaller theropods, especially *Velociraptor*, a 1.8 m long predator with large sickle-like claws on the hind feet. This dinosaur and its somewhat larger relative, *Deinonychus*, probably used its claws in a slashing type of attack (Figure 15.10c). A particularly interesting fossil from Mongolia shows a *Velociraptor* grasping a herbivorous dinosaur called *Protoceratops*, which both perished during a struggle. Despite what you might see in movies about dinosaurs, theropods, like predators today, probably avoided large, dangerous prey and went for the easy kill, preying on the young, old, or disabled, or they dined on carrion if available. No doubt the larger theropods simply chased smaller predators away from their kills. From the evidence available, theropods such as diminutive *Coelophysis* and medium-sized *Deinonychus* probably hunted in packs.

The second group of saurischians, the sauropods, includes the truly giant, quadrupedal herbivorous dinosaurs such as *Apatosaurus*, *Diplodocus*, and *Brachiosaurus*, the largest land animals ever (Table 15.1). *Brachiosaurus*, a giant even by sauropod standards, may have weighed 75 metric tons, and partial remains from several areas indicate that even larger sauropods may have existed. Trackways show that sauropods moved in herds.

Sauropods were preceded in the fossil record by the smaller Late Triassic to Early Jurassic *prosauropods*, which were undoubtedly related to sauropods but probably not their ancestors. Sauropods were most common during the Jurassic; only a few genera existed during the Cretaceous.

Ornithischian Dinosaurs Scientists recognize five groups of ornithischians: ornithopods, pachycephalosaurs, ankylosaurs, stegosaurs, and ceratopsians (Figure 15.9, Table 15.1). Ornithopods consist of several subgroups, including the familiar duck-billed dinosaurs or hadrosaurs. Hadrosaurs were especially numerous during the Cretaceous, and several species had head crests that may have been used to amplify bellowing, for sexual display, or for species recognition (• Figure 15.11). All ornithopods were herbivorous and primarily bipedal, but they had well-developed front limbs that allowed them to walk in a quadrupedal fashion, too.

The Late Cretaceous ornithopod *Miasaura* ("good mother dinosaur") nested in colonies and used the same

---
*One feathered specimen purchased in Utah that was reported to have come from China turned out to be a forgery. It even appeared in *National Geographic* before scientists exposed its fraudulent nature.

(a)

(b)

(c)

• **Figure 15.11 Duck-billed Dinosaurs** Among the ornithopods the duck-billed dinosaurs, or hadrosaurs, all had flattened bill-like mouths, and some had crests on their heads. Crests are obvious on **(a)** *Corythosaurus* and **(b)** *Parasaurolophus*, but **(c)** *Edmontosaurus* had no crest. The crests were hollow, bony extensions of the skull.

nesting area repeatedly where 2 m diameter nests were placed 7 m apart or about the length of an adult. Some nests contain juveniles up to 1 m long, which is much larger than when they hatched, so they probably stayed in the nest area where adults protected them and perhaps fed them. The fact that these animals lived in vast herds is demonstrated by the fossils of an estimated 10,000 individuals in a single bone bed in Montana. Apparently they were overcome by toxic gases from a volcano and later buried in flood deposits.

The most distinctive feature of the pachycephalosaurs is their thick-boned, dome-shaped skull (Table 15.1). However, the traditional view of these as animals that butted heads for dominance or mates has been challenged. Now some paleontologists note that the thick skull bones are found only in juveniles but not in adults. In any case, pachycephalosaurs were bipedal herbivores that varied from 1 to 4.5 m long. Their fossils are known only from Late Cretaceous-aged rocks.

The fossil record of ceratopsians (horned dinosaurs) shows that small Early Cretaceous animals were the ancestors of large Late Cretaceous genera such as *Triceratops* (• Figure 15.12a). *Triceratops* and related genera with huge heads, a large bony frill over the neck, and a horn or horns on the skull were very common in North America. Fossil trackways show that these large, quadrupedal herbivores moved in herds. Furthermore, bone beds with fossils from a single species indicate that large numbers of ceratopsians perished quickly, probably during river crossings.

The most distinctive features of *Stegosaurus*, a medium-sized, herbivorous quadruped from the Jurassic Period, are a spiked tail, almost certainly used for defense, and plates on its back (Figure 15.12b, Table 15.1). The exact arrangement of these plates is uncertain, although they are usually depicted in two rows with the plates on one side offset from those on the other. In any case, most paleontologists think the plates functioned to absorb and dissipate heat.

All ankylosaurs were quadrupedal herbivores and more heavily armored than any other dinosaur (Figure 15.12c, Table 15.1). Bony armor protected the animal's back, flanks, and top of the head. The tail of some species such as *Ankylosaurus* ended in a bony club that undoubtedly could deliver a crippling blow to an attacking predator.

Various media depict dinosaurs as aggressive, dangerous beasts, but we have every reason to think that they behaved much as land animals do today. Certainly some lived in herds and no doubt interacted by bellowing, snorting, grunting, and foot stomping in defense, territorial disputes, or attempts to attract mates. Since the renaissance in dinosaur study that began in the 1970s, our view of dinosaurs has changed so that we now think that they were active animals that behaved much like animals today.

Warm-Blooded Dinosaurs?

Were dinosaurs **endotherms** (warm-blooded) like today's mammals and birds, or were they **ectotherms** (cold-blooded), as are all of today's reptiles? Almost everyone now agrees some compelling evidence exists for dinosaur endothermy, but opinion is divided among (1) those holding that all dinosaurs were endotherms, (2) those who think only some were endotherms, and (3) those proposing that dinosaur metabolism, and thus their ability to regulate body temperature, changed as they matured.

Bones of endotherms typically have numerous passageways that contain blood vessels, but far fewer passageways

(a)

(c)

(b)

• **Figure 15.12 Representatives of Three of the Five Types of Ornithischians (a)** Skeleton of the rhinoceros-sized, Late Cretaceous ceratopsian *Triceratops*. **(b)** *Stegosaurus* from the Late Jurassic was about 9 m long. Notice the large plates along the back and the bony spikes on the tail. **(c)** This 6 m long ankylosaur known as *Sauropelta* lived in Western North America during the Early Cretaceous. Some ankylosaurs had a bony club at the end of the tail.

are present in bones of ectotherms. Dinosaur bones are more similar to those of living endotherms. Crocodiles and turtles, however, have a so-called endothermic bone, even though they are ectotherms, and some small mammals have bones more typical of ectotherms. Perhaps bone structure is related more to body size and growth patterns than to endothermy, so this evidence is not conclusive.

Endotherms must eat more than comparably sized ectotherms because their metabolic rates are so much higher. Consequently, endothermic predators require large prey populations and thus constitute a much smaller proportion of the total animal population than their prey, usually only a few percent. In contrast, the proportion of the ectothermic predators to prey may be as high as 50%.

Where data are sufficient to allow an estimate, dinosaur predators made up 3% to 5% of the total population. Nevertheless, uncertainties in the data make this argument less than convincing for many paleontologists.

A large brain in comparison to body size requires a rather constant body temperature and thus implies endothermy. And some dinosaurs were rather brainy, especially the small and medium-sized theropods, so brain size might be a convincing argument for these dinosaurs.

But even more compelling evidence for theropod endothermy comes from their probable relationship to birds and from the discoveries in China of theropods with feathers or a featherlike covering (• Figure 15.13). Today, only endotherms have hair, fur, or feathers for insulation.

Scientists point out that some duck-billed dinosaurs grew and reached maturity much more quickly than would be expected for ectotherms and conclude that they must have been warm-blooded. Furthermore, a fossil ornithopod discovered in 1993 has a preserved four-chambered heart much like that of living mammals and birds. Three-dimensional imaging of this structure, now on display at the North Carolina Museum of Natural Sciences, has convinced many scientists that this animal was an endotherm.

There are good arguments for endothermy for several types of dinosaurs, although the large sauropods were probably not endothermic but nevertheless were capable of maintaining a rather constant body temperature. Large animals heat up and cool down more slowly than smaller

• **Figure 15.13 Feathered Dinosaurs** Restoration of the feathered dinosaur *Caudipteryx* from China. The fact that *Caudipteryx* had short forelimbs, symmetric feathers, and was larger than the oldest known bird indicate that it was flightless.

ones, because they have a small surface area compared to their volume. With their comparatively smaller surface area for heat loss, sauropods probably retained heat more effectively than their smaller relatives.

Flying Reptiles

Paleozoic insects were the first animals to achieve flight, but the first among vertebrates were **pterosaurs**, or flying reptiles, which were common in the skies from the Late Triassic until their extinction at the end of the Cretaceous (• Figure 15.14). Adaptations for flight include a wing membrane supported by an elongated fourth finger (Figure 15.14c), light, hollow bones; and development of those parts of the brain that controlled muscular coordination and sight. Because at least one pterosaur species had a coat of hair or hairlike feathers, possibly it, and perhaps all pterosaurs, were endotherms.

Pterosaurs are generally depicted in movies as large, aggressive creatures, but some were no bigger than today's sparrows, robins, and crows. However, a few species had wingspans of several meters, and the wingspan of one Cretaceous pterosaur was at least 12 m! Nevertheless, even the very largest species probably weighed no more than a few tens of kilograms.

Experiments and studies of fossils indicate that the wing bones of large pterosaurs such as *Pteranodon* (Figure 15.14b) were too weak for sustained flapping. These comparatively large animals probably took advantage of rising air currents to stay airborne, mostly by soaring but occasionally flapping their wings for maneuvering. In contrast, smaller pterosaurs probably stayed aloft by vigorously flapping their wings just as present-day small birds do.

Mesozoic Marine Reptiles

Several types of Mesozoic reptiles adapted to a marine environment, including turtles and some crocodiles, as well as the Triassic mollusk-crushing placodonts. Here, though, we concentrate on the ichthyosaurs and plesiosaurs and the less familiar mosasaurs. All were thoroughly aquatic marine predators, but other than all being reptiles, they were not closely related to one another. Furthermore, none were dinosaurs, although some popular media depict them as such.

The streamlined, rather porpoise-like **ichthyosaurs** varied from species measuring only 0.7 m long to giants more than 15 m long (• Figure 15.15a). Details of their ancestry are still not clear, but fossil ichthyosaurs from Japan prompted researcher Ryosuke Motani to say, "I knew *Utatsusaurus* was exactly what paleontologists had been expecting to find for years: an ichthyosaur that looked like a lizard with legs."*

Ichthyosaurs used their powerful tail for propulsion and maneuvered with their flipperlike forelimbs. They had numerous sharp teeth, and preserved stomach contents reveal a diet of fish, cephalopods, and other marine organisms. It is doubtful that ichthyosaurs could come onto land, so females must have retained eggs within their bodies and given birth to live young. A few fossils with small ichthyosaurs in the appropriate part of the body cavity support this interpretation.

An interesting side note in the history of paleontology is the story of Mary Anning (see Perspective 15.1), who, when she was only about 11 years old, discovered and directed the excavation of a nearly complete ichthyosaur in southern England.

The **plesiosaurs** belonged to one of two subgroups: short-necked and long-necked (Figure 15.15b). Most were modest-sized animals 3.6 to 6 m long, but one species found in Antarctica measures 15 m. Short-necked plesiosaurs might have been bottom feeders, but their long-necked cousins may have used their necks in a snakelike fashion, and their numerous sharp teeth, to capture fish. These animals probably came ashore to lay their eggs.

Mosasaurs were Late Cretaceous marine lizards related to the present-day Komodo dragon or monitor lizard. Some species measured no more than 2.5 m long, but a few such as *Tylosaurus* were large, measuring up to 9 m (• Figure 15.16). Mosasaur limbs resemble paddles and were used mostly for maneuvering, whereas the long tail provided propulsion. All

---

*Ryosuke Motani, 2004. Rulers of the Jurassic Seas. *Scientific American*, v. 14, no. 2, p. 7.

• **Figure 15.14 Flying Reptiles (a)** *Pterodactylus*, a long-tailed pterosaur, is well known from the Late Jurassic. Among the several known species, wingspans ranged from 50 cm to 2.5 m. **(b)** The short-tailed pterosaur known as *Pteranodon* was a large, Cretaceous animal with a wingspan of more than 6 m. **(c)** In all flying vertebrates, the forelimb has been modified into a wing. A long fourth finger supports the pterosaur wing, whereas in birds the second and the third fingers are fused together, and in bats fingers 2 through 5 support the wing. Are these wings analogous, homologous, or both?

What Would You Do?

In your high school science class a student notices that ichthyosaurs and porpoises look somewhat similar, and she speculates that the former is the ancestor of the latter. After all, they look similar, and ichthyosaurs lived before porpoises, so, she reasons, there must be a relationship. How would you explain that there is no evidence to indicate that porpoises evolved from ichthyosaurs? (Hint: See the discussion on whales in Chapter 18.)

were predators, and preserved stomach contents indicate they ate fish, birds, smaller mosasaurs, and a variety of invertebrates, including ammonoids.

Crocodiles, Turtles, Lizards, and Snakes

By Jurassic time, crocodiles had become the most common freshwater predators. All crocodiles are amphibious, spending much of their time in water, but they are well equipped for walking on land. Overall, crocodile evolution has been conservative, involving changes mostly in size from a meter or so in Jurassic forms to 15 m in some Cretaceous species.

Turtles, too, have been evolutionarily conservative since their appearance during the Triassic. The most remarkable feature of turtles is their heavy, bony armor; turtles are more thoroughly armored than any other vertebrate animal, living or fossil. Turtle ancestry is uncertain. One Permian animal had eight broadly expanded ribs, which may represent the first stages in the development of turtle armor.

Lizards and snakes are closely related, and lizards were in fact ancestral to snakes. The limbless condition in snakes (some lizards are limbless, too) and skull modifications that allow snakes to open their mouths very wide are the main difference between these two groups. Lizards are known from Upper Permian strata, but they were not abundant until the Late Cretaceous. Snakes first appear during the Cretaceous, but the families to which most living snakes belong differentiated since the Early Miocene. One Early Cretaceous genus from Israel shows characteristics intermediate between snakes and their lizard ancestors.

15.1 Perspective

MARY ANNING AND HER CONTRIBUTIONS TO PALEONTOLOGY

Men and women from many countries contribute to our understanding of prehistoric life, but this has not always been the case. The early history of paleontology was dominated by Western European males, but there was a notable exception: Mary Anning (1799–1847), who began a remarkable career as a fossil collector when she was only 11 years old (Figure 1).

Mary Anning was born in Lyme Regis on England's south coast. When only 15 months old, she survived a lightning strike that, according to one report, killed three girls, and according to another, killed a nurse tending her. Of the ten or so children in the Anning family, only Mary and her brother Joseph reached maturity.

In 1810 Mary's father, a cabinetmaker who also sold fossils part time, died leaving the family nearly destitute. Mary Anning expanded the fossil business and became a professional fossil collector known to the paleontologists of her time, some of whom visited her shop to buy fossils or gather information. She collected fossils from the Dorset coast near Lyme Regis and is reported to have been the inspiration for the tongue twister "She sells seashells by the sea shore."

Soon after her father's death, she made her first important discovery: a nearly complete skeleton of a Jurassic ichthyosaur, which was described in 1814 by Sir Everard Home. The sale of this fossil specimen provided considerable financial relief for her family. In 1821 she made a second major discovery and excavated the remains of another Mesozoic marine reptile, a plesiosaur. And in 1818 she excavated the remains of the first Mesozoic flying reptile (pterosaur) found in England, which was sent to the eminent geologist William Buckland of Oxford University.

By 1830 Mary Anning's fortunes began declining as collectors and museums had less money to spend on fossils. Indeed, she may once again have become destitute if it had not been for her friend Henry Thomas de la Beche, also a resident of Lyme Regis. De la Beche drew a fanciful scene called *Duria antiquior*, meaning "An earlier Dorset," in which he brought to life the fossils Mary Anning had collected. The scene was printed and sold widely, the proceeds of which went directly to Mary Anning.

Mary Anning died of cancer in 1847, and although only 48 years old, she had a fossil-collecting career that spanned 36 years. Her contributions to paleontology are now widely recognized, but, unfortunately, soon after her death she was mostly forgotten. Apparently, people who purchased her fossils were credited with finding them. So even though Mary Anning became a respected fossil collector, many scientists of that time could not accept that an untutored girl could possess such knowledge and

Figure 1 Mary Anning

skill. "It didn't occur to them to credit a woman from the lower classes with such astonishing work. So an uneducated little girl, with a quick mind and an accurate eye, played a key role in setting the course of the 19th-century geologic revolution. Then we simply forgot about her."*

* John Lienhard, University of Houston.

From Reptiles to Birds

Long ago, scientists were aware of a number of characteristics shared by reptiles and birds. For example, birds and reptiles lay shelled, yolked eggs, and both share a number of skeletal features, such as the way the jaw articulates with the skull. But, of course, birds have feathers, whereas reptiles have scales or a tough, beaded skin. Furthermore, birds do not closely resemble any living reptile, so why do scientists think that birds evolved from reptiles? Several fossils with feather impressions have been discovered in the Solnhofen Limestone of Germany. They definitely have feathers and a wishbone, consisting of fused clavicle bones so typical of birds, and yet in most other skeletal features they most closely resemble small theropod dinosaurs. These animals, known as ***Archaeopteryx*** (• Figure 15.17), are birds by definition, but their numerous reptilian features convince scientists that they evolved from small theropods. Even fused clavicles are found in several theropods, and recent discoveries in China of theropods with feathers provide more evidence of this relationship.

Opponents of the theropod–bird scenario point out that theropods are typically found in Cretaceous-aged

• Figure 15.15 **Ichthyosaurs and Plesiosaurs (a)** The fully aquatic ichthyosaurs evolved from land-dwelling ancestors. Most ichthyosaurs were a few meters long, but some exceeded 15 m in length. **(b)** Plesiosaurs were also aquatic, but their flipperlike forelimbs probably allowed them to come out onto land. Both long-necked and short-necked plesiosaurs are known.

rocks, whereas *Archaeopteryx* is Jurassic. However, some of the fossils coming from China are about the same age as *Archaeopteryx*, narrowing the gap between presumed ancestor and descendant.

Some other Mesozoic fossils shed more light on bird evolution. One specimen, from China, is slightly younger than *Archaeopteryx* and possesses both primitive and advanced features. It retains abdominal ribs similar to those of *Archaeopteryx* and theropods, but it has a reduced tail more typical of present-day birds. Another Mesozoic bird from Spain is also a mix of primitive and advanced characteristics, but it appears to lack abdominal ribs. And finally, discoveries in 2004 and 2005 in China of five specimens of an Early Cretaceous bird indicate that today's birds may have had an aquatic ancestor. With few exceptions, the bones of these birds known as *Gansus yumenensis* are much like those of present-day birds.

The early fossil records of birds are not good enough to resolve whether *Archaeopteryx* is the ancestor of today's birds or an animal that died out without leaving any descendants. Of course, that in no way diminishes the fact that it had both reptile and bird characteristics. However, some claim that fossils of two crow-sized individuals known as *Protoavis* represent an even earlier bird than *Archaeopteryx*. These Late Triassic-aged fossils have hollow bones and the breastbone structure of birds, but because no feather impressions were found, many paleontologists think they are small theropods.

One hypothesis for the origin of bird flight—*from the ground up*—holds that bird ancestors were bipedal, fleet-footed ground dwellers. Wings enabled these animals to leap into the air and glide, at least for short distances, to catch flying insects or to escape predators. According to another hypothesis, flight evolved *from the trees down*, meaning that bird ancestors were bipeds that climbed trees and used their rudimentary wings for gliding or parachuting. According to this idea, this ancestral animal increased its time in the air by wing flapping. The from-the-ground-up hypothesis is probably better supported by evidence in that a bipedal theropod ancestor is reasonable because some small theropods had forelimbs much like those of *Archaeopteryx*. On the other hand, from-the-trees-down has the advantage because takeoff from an elevated position is easier, whereas landing is a challenge.

Origin and Evolution of Mammals

Recall from Chapter 13 that mammal-like reptiles called **therapsids** diversified into many species of herbivores and carnivores during the Permian Period. In fact, they were the most numerous and diverse land-dwelling vertebrates at that time. Among the therapsids one group known as **cynodonts** was the most mammal-like of all, and by Late Triassic time true mammals evolved from them.

Cynodonts and the Origin of Mammals

We can easily recognize living mammals as warm-blooded animals that have hair or fur and mammary glands and, except for the platypus and spiny anteater, give birth to live

• Figure 15.16 **Mosasaurs Were Marine Lizards (a)** *Tylosaurus* was a large, Late Cretaceous mosasaur. It measured up to 9 m long. **(b)** Mosasaur skull on display in the Museum of Geology and Paleontology, University of Florence, Italy.

young. However, these criteria are not sufficient for recognizing fossil mammals; for them, we must rely on skeletal structure only. Several skeletal modifications took place during the transition from mammal-like reptiles to mammals, but distinctions between the two groups are based mostly on details of the middle ear, the lower jaw, and the teeth (Table 15.2). Fortunately, the evolution of mammals from cynodonts is so well documented by fossils that classification of some fossils as reptile or mammal is difficult.

Reptiles have one small bone in the middle ear (the stapes), whereas mammals have three: the incus, the malleus, and the stapes. Also, the lower jaw of a mammal is composed of a single bone called the dentary, but a reptile's jaw is composed of several bones (• Figure 15.18). In addition, a reptile's jaw is hinged to the skull at a contact between the articular and quadrate bones, whereas in mammals the dentary contacts the squamosal bone of the skull (Figure 15.18).

During the transition from cynodonts to mammals, the quadrate and articular bones that had formed the joint between the jaw and skull in reptiles were modified into the incus and malleus of the mammalian middle ear (Figure 15.18, Table 15.2). Fossils document the progressive enlargement of the dentary until it became the only element in the mammalian jaw. Likewise, a progressive change from the reptile to mammal jaw joint is documented by fossil evidence. In fact, some of the most advanced cynodonts were truly transitional, because they had a compound jaw joint consisting of (1) the articular and quadrate bones typical of reptiles and (2) the dentary and squamosal bones as in mammals (Table 15.2).

In Chapter 7 we noted that the study of embryos provides some of the evidence for evolution. Opossum embryos show that the middle-ear bones of mammals were originally part of the jaw. In fact, when opossums are born, the middle ear elements are still attached to the dentary, but as they develop further, these elements migrate to the middle ear, and a typical mammal jaw joint develops.

Several other aspects of cynodonts also indicate they were ancestors of mammals. Their teeth were becoming double-rooted as they are in mammals, and they were somewhat differentiated into distinct types that performed specific functions. In mammals the teeth are fully differentiated into incisors, canines, and chewing teeth, but typical reptiles do not have differentiated teeth (• Figure 15.19). In addition, mammals have only two sets of teeth during their lifetimes—a set of baby teeth and the permanent adult

TABLE 15.2 Summary Chart Showing Some Characteristics and How They Changed during the Transition from Reptiles to Mammals

Features	Typical Reptile	Cynodont	Mammal
Lower Jaw	Dentary and several other bones	Dentary enlarged, other bones reduced	Dentary bone only, except in earliest mammals
Jaw-Skull Joint	Articular-quadrate	Articular-quadrate; some advanced cynodonts had both the reptile jaw-skull joint and the mammal jaw-skull joint	Dentary-squamosal
Middle-Ear Bones	Stapes	Stapes	Stapes, incus, malleus
Secondary Palate	Absent	Partly developed	Well developed
Teeth	No differentiation; chewing teeth single rooted	Some differentiation; chewing teeth partly double rooted	Fully differentiated into incisors, canines, and chewing teeth; chewing teeth double rooted
Tooth Replacement	Teeth replaced continuously	Only two sets of teeth in some advanced cynodonts	Two sets of teeth
Occipital condyle	Single	Partly divided	Double
Occlusion (chewing teeth meet surface to surface to allow grinding)	No occlusion	Occlusion in some advanced cynodonts	Occlusion
Endothermic vs. Ectothermic	Ectothermic	Probably endothermic	Endothermic
Body Covering	Scales	One fossil shows it had skin similar to that of mammals	Skin with hair or fur

teeth. Typical reptiles have teeth replaced continuously throughout their lives, the notable exception being some cynodonts who in mammal fashion had only two sets of teeth. Another important feature of mammal teeth is occlusion; that is, the chewing teeth meet surface to surface to allow grinding. Thus mammals chew their food, but reptiles, amphibians, and fish do not. However, tooth occlusion is known in some advanced cynodonts (Table 15.2).

Reptiles and mammals have a bony protuberance from the skull that fits into a socket in the first vertebra: the atlas. This structure, called the *occipital condyle*, is a single feature in typical reptiles, but in cynodonts it is partly divided into a double structure typical of mammals (Table 15.2). Another mammalian feature, the secondary palate, was partially developed in advanced cynodonts (• Figure 15.20). This bony shelf separating the nasal passages from the mouth cavity is an adaptation for eating and breathing at the same time, a necessary requirement for endotherms with their high demands for oxygen.

Mesozoic Mammals

Mammals evolved during the Late Triassic not long after the first dinosaurs appeared, but for the rest of the Mesozoic Era most of them were small. There were, however, a few exceptions. One is a Middle Jurassic-aged aquatic mammal found in China that measures about 50 cm long, which also has the distinction of being the oldest known fossil with fur. The other is an Early Cretaceous-aged mammal called *Repenomamus giganticus*, also from China, that was about 1 m long, weighed 12 to 14 kg, and had the remains of a juvenile dinosaur in its stomach. Nevertheless, most other Mesozoic mammals were about the size of mice and rats, and they were not very diverse—certainly not as diverse as they were during the Cenozoic Era. Furthermore, they retained reptile characteristics but had mammalian features, too. The Triassic triconodonts, for instance, had the fully differentiated teeth typical of mammals, but they also had both the reptile and the mammal types of jaw

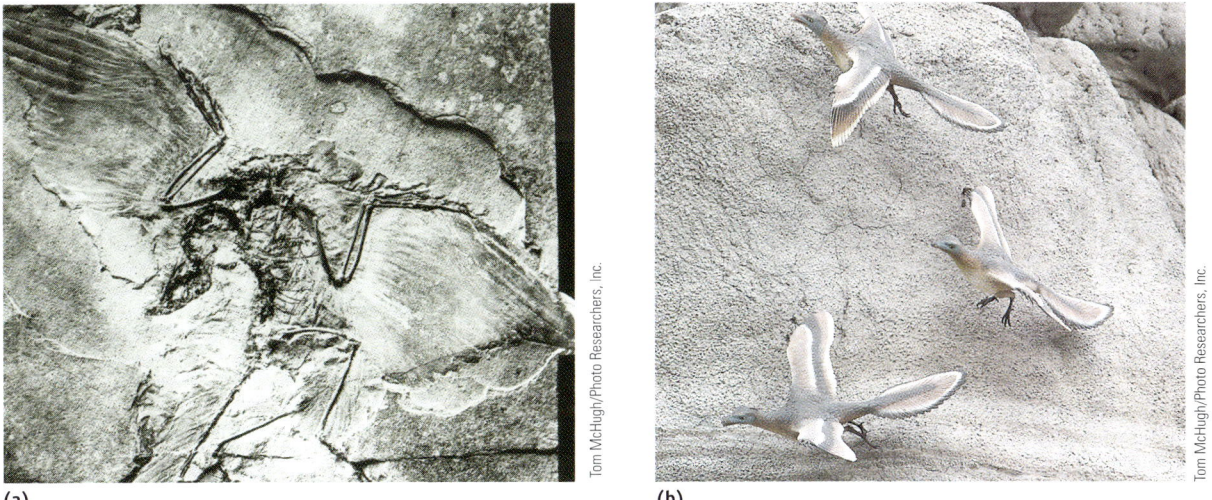

• **Figure 15.17** *Archaeopteryx* **from the Jurassic-aged Solnhofen Limestone of Germany.** (a) Fossil showing the feather impressions on the wings and a long tail. This animal had feathers and a wishbone, so it is a bird, but in most details of its anatomy it more closely resembles small theropod dinosaurs: It had reptile-like teeth, claws on its wings, and a long tail. (b) Restoration of *Archaeopteryx*. It was about the size of a present-day crow.

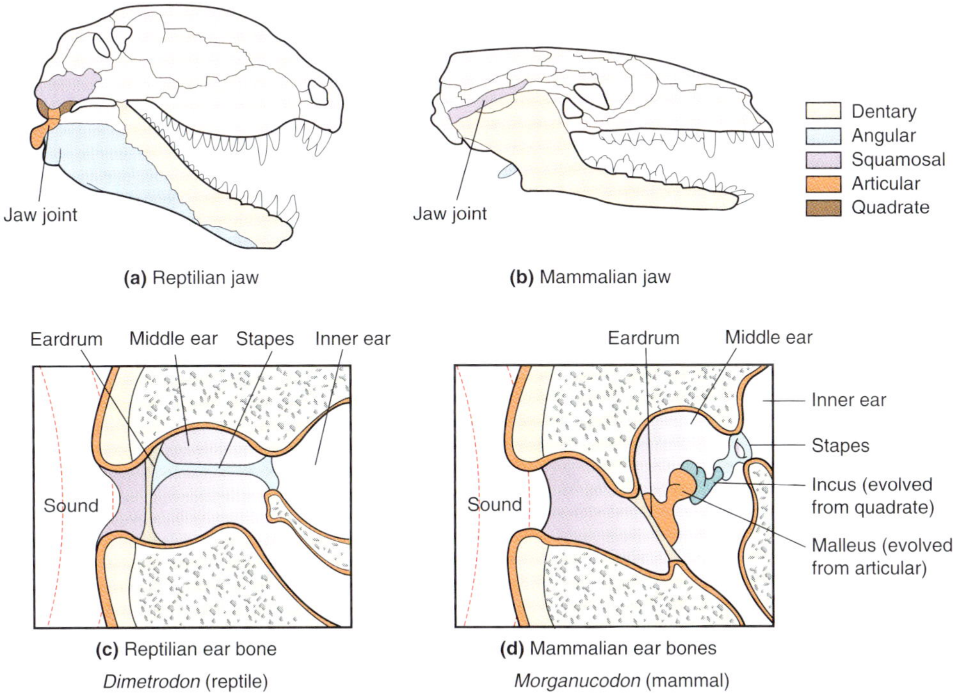

• **Figure 15.18 Evolution of the Mammal Jaw and Middle Ear** (a) Cynodonts had four bones in the jaw, whereas (b) mammals have only one. Note that the cynodont jaw-skull joint is between the articular and the quadrate bones, but in mammals it is between the dentary and squamosal bones. (c) Cynodonts had one middle ear bone—the stapes—but (d) mammals have three: the malleus, the incus, and the stapes. The malleus and incus were derived from the cynodont articular and quadrate bones.

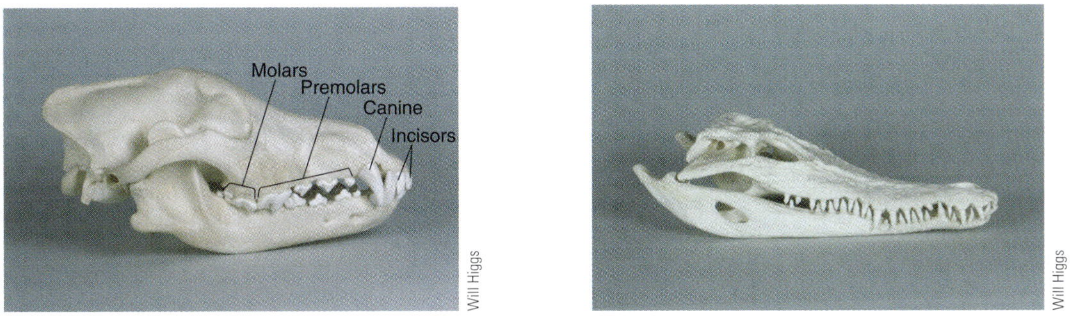

• **Figure 15.19 Comparison of the Teeth of a Mammal and a Reptile** (a) This wolf skull shows that mammal teeth are differentiated into incisors, canines, premolars, and molars. (b) Reptiles, represented here by a crocodile, may have teeth that vary somewhat in size, but otherwise they all look the same. The only exception is among some mammal-like reptiles.

ORIGIN AND EVOLUTION OF MAMMALS

joints. In short, some mammal features appeared sooner than others (remember the concept of *mosaic evolution* from Chapter 7).

The early mammals diverged into two distinct branches. One branch includes the triconodonts (• Figure 15.21) and their probable descendants, the **monotremes**,* or egg-laying mammals such as the spiny anteater and platypus of the Australian region. The other branch includes the **marsupial** (pouched) **mammals** and the **placental mammals** and their ancestors, the eupantotheres (Figure 15.21).

Although the history of the monotremes is uncertain, fossils of several Mesozoic animals are relevant to the evolution of marsupials and placentals. In fact, the divergence of marsupials and placental mammals from a common ancestor took place during the Early Cretaceous (• Figure 15.22).

Mesozoic Paleobiogeography

Fragmentation of the supercontinent Pangaea began by the Late Triassic, but during much of the Mesozoic, close connections existed between the various landmasses. The proximity of these landmasses alone, however, is not enough to explain Mesozoic biogeographic distributions, because climates are also effective barriers to wide dispersal. During much of the Mesozoic, though, climates were more equable and lacked the strong north and south zonation characteristic of the present. In short, Mesozoic plants and animals had greater opportunities to occupy much more extensive geographic ranges.

Pangaea persisted as a supercontinent through most of the Triassic (see Figure 14.1a), and the Triassic climate was warm temperate to tropical, although some areas, such as the present southwestern United States, were arid. Mild

---

*The fossil record for monotremes is poor, so their relationship to Mesozoic mammals is uncertain.

temperatures extended 50 degrees north and south of the equator, and even the polar regions may have been temperate. The fauna had a truly worldwide distribution. Some dinosaurs had continuous ranges across Laurasia and Gondwana, the peculiar gliding lizards lived in New Jersey and England, and reptiles known as phytosaurs lived in North America, Europe, and Madagascar.

By the Late Jurassic, Laurasia had become partly fragmented by the opening North Atlantic, but a connection still existed. The South Atlantic had begun to open so that a long, narrow sea separated the southern parts of Africa and South America. Otherwise the southern continents were still close together.

The mild Triassic climate persisted into the Jurassic. Ferns whose living relatives are now restricted to the tropics of southeast Asia lived as far as 63 degrees south and 75 degrees north. Dinosaurs roamed widely across Laurasia and Gondwana. For example, the giant sauropod *Brachiosaurus* is found in western North America and eastern Africa. Stegosaurs and some families of carnivorous dinosaurs lived throughout Laurasia and in Africa.

By the Late Cretaceous, the North Atlantic had opened further, and Africa and South America were completely separated (see Figure 14.1c). South America remained an island continent until late in the Cenozoic, and its fauna became increasingly different from faunas of the other continents (see Chapter 18). Marsupial mammals reached Australia from South America via Antarctica, but the South American connection was eventually severed. Placentals, other than bats and a few rodents, never reached Australia, explaining why marsupials continue to dominate the continent's fauna even today.

Cretaceous climates were more strongly zoned by latitude, but they remained warm and equable until the close of that period. Climates then became more seasonal and cooler, a trend that persisted into the Cenozoic. Dinosaur and mammal fossils demonstrate that interchange was still possible, especially between the various components of Laurasia.

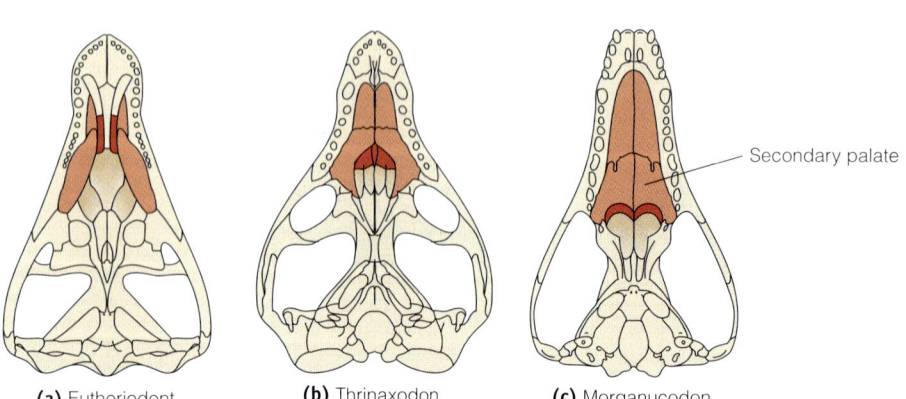

(a) Eutheriodont   (b) Thrinaxodon   (c) Morganucodon

• **Figure 15.20 Evolution of the Secondary Palate** Views of the bottoms of the skulls of **(a)** an early therapsid, **(b)** a cynodont, and **(c)** an early mammal showing the development of the bony secondary palate. In mammals, the secondary palate is a bony shelf that separates the nasal passages from the mouth cavity.

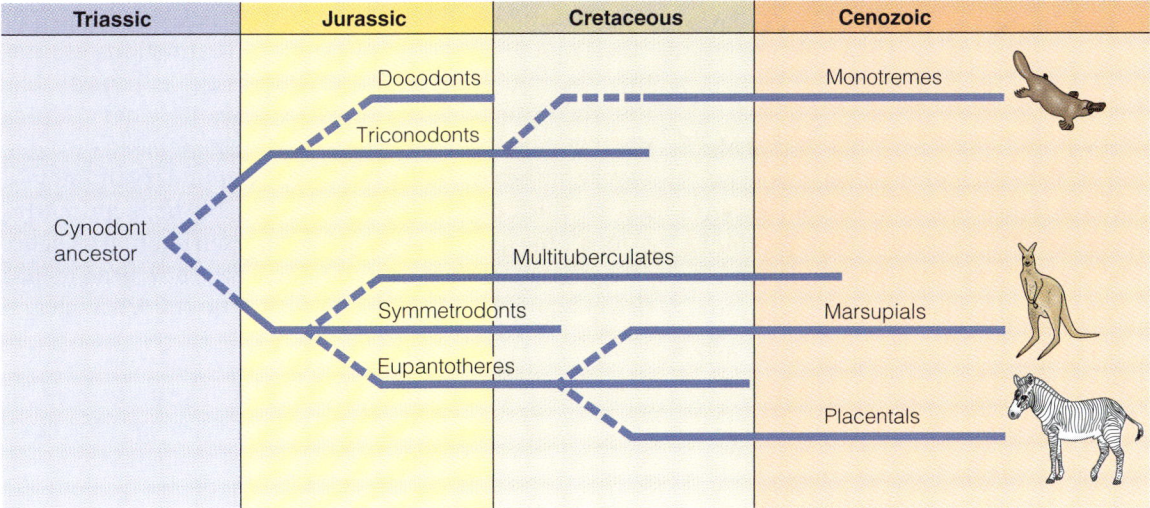

• **Figure 15.21 Relationships Among the Various Recognized Groups of Early Mammals and Their Descendants** Mammal evolution proceeded along two branches, one leading to today's monotremes, or egg-laying mammals, and the other to marsupial and placental animals.

(a) (b)

• **Figure 15.22 Mesozoic Mammals (a)** Restoration of the oldest known marsupial mammal, *Sinodelphys*, which measures 15 cm long. **(b)** This restoration shows *Eomaia*, the oldest known placental mammal. It was only 12 or 13 cm long. Both fossils come from Early Cretaceous-aged rocks in China.

MESOZOIC PALEOBIOGEOGRAPHY

Mass Extinctions—A Crisis in the History of Life

The greatest mass extinction took place at the end of the Paleozoic Era, but the one at the close of the Mesozoic has attracted more attention because among its casualties were dinosaurs, flying reptiles, and marine reptiles. Several kinds of marine invertebrates also went extinct, including ammonites, which had been so abundant during the Mesozoic, rudistid bivalves, and some planktonic organisms.

Hypotheses to explain Mesozoic extinctions are numerous, but most have been dismissed as improbable, untestable, or inconsistent with the available data. A proposal that has become popular since 1980 is based on a discovery at the Cretaceous–Paleogene boundary in Italy—a 2.5 cm thick clay layer with a remarkably high concentration of the platinum-group element iridium (• Figure 15.23). High iridium concentrations have now been identified at many other Cretaceous–Paleogene boundary sites.

What Would You Do?

While you are giving a speech on dinosaurs at a local civic club, an audience member asks how it was possible for some dinosaurs to live so far north and south of the equator and have such large geographic ranges. What is your answer?

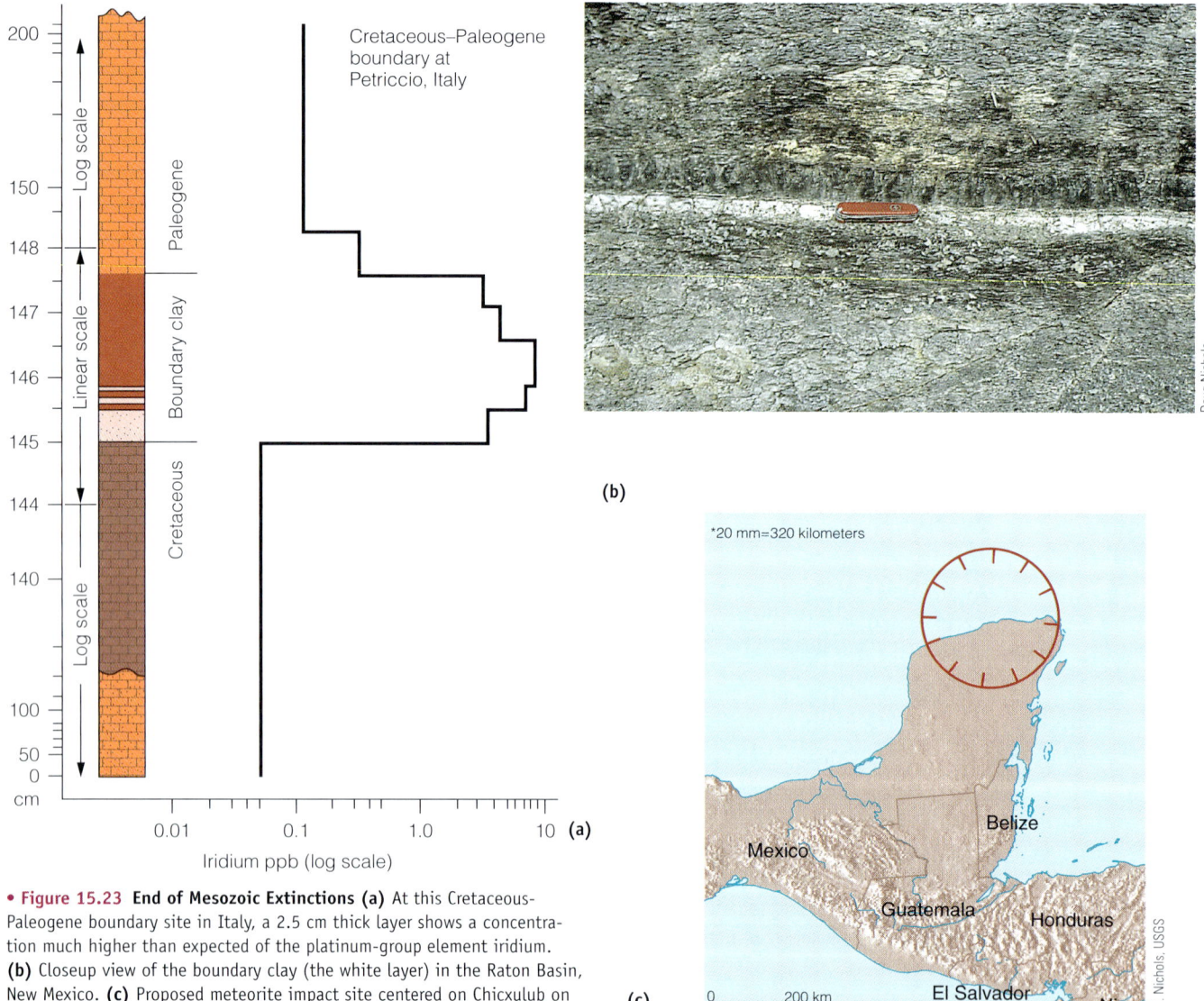

• **Figure 15.23 End of Mesozoic Extinctions** (a) At this Cretaceous-Paleogene boundary site in Italy, a 2.5 cm thick layer shows a concentration much higher than expected of the platinum-group element iridium. (b) Closeup view of the boundary clay (the white layer) in the Raton Basin, New Mexico. (c) Proposed meteorite impact site centered on Chicxulub on the Yucatan Peninsula of Mexico.

The significance of this **iridium anomaly** is that iridium is rare in crustal rocks but is found in much higher concentrations in some meteorites. Accordingly, some investigators propose a meteorite impact to explain the anomaly and further postulate that the meteorite, perhaps 10 km in diameter, set in motion a chain of events leading to extinctions. Some Cretaceous–Paleogene boundary sites also contain soot and shock-metamorphosed quartz grains, both of which are cited as additional evidence of an impact.

According to the impact hypothesis, about 60 times the mass of the meteorite was blasted from the crust high into the atmosphere, and the heat generated at impact started raging forest fires that added more particulate matter to the atmosphere. Sunlight was blocked for several months, temporarily halting photosynthesis; food chains collapsed; and extinctions followed. Furthermore, with sunlight greatly diminished, Earth's surface temperatures were drastically reduced, adding to the biologic stress. Another consequence of the impact was that vaporized rock and atmospheric gases produced sulfuric acid (H_2SO_4) and nitric acid (HNO_3). Both would have contributed to strongly acid rain that may have had devastating effects on vegetation and marine organisms.

Now some geologists point to a probable impact site centered on the town of Chicxulub on the Yucatán Peninsula of Mexico (Figure 15.23c). The 180 km diameter structure lies beneath layers of sedimentary rock and appears the right age. Evidence that supports the conclusion that the Chicxulub structure is an impact crater includes shocked quartz, the deposits of huge waves, and tektites—small pieces of rock melted during the impact and hurled into the atmosphere.

Even if a meteorite did hit Earth, did it lead to these extinctions? If so, both terrestrial and marine extinctions must have occurred at the same time. To date, strict time equivalence between terrestrial and marine extinctions has not been demonstrated. The selective nature of the extinctions is also a problem. In the terrestrial realm, large animals were the most affected, but not all dinosaurs were large, and crocodiles, close relatives of dinosaurs, survived, although some species died out. Some paleontologists think dinosaurs, some marine invertebrates, and many plants were already on the decline and headed for extinction before the end of the Cretaceous. A meteorite impact may have simply hastened the process.

In the final analysis, Mesozoic extinctions have not been explained to everyone's satisfaction. Most geologists now concede a large meteorite impact occurred, but we also know vast outpourings of lava were taking place in what is now India. Perhaps these brought about detrimental atmospheric changes. Furthermore, the vast, shallow seas that covered large parts of the continents had mostly withdrawn by the end of the Cretaceous, and the mild equable Mesozoic climates became harsher and more seasonal by the end of that era. Nevertheless, these extinctions were selective, and no single explanation accounts for all aspects of this crisis in life history.

SUMMARY

Table 15.3 summarizes many Mesozoic biologic and plate tectonic events.

- Invertebrate survivors of the Paleozoic extinctions diversified and gave rise to increasingly diverse marine communities.
- Some of the most abundant invertebrates were cephalopods, especially ammonoids, foraminifera, and the reef-building rudists.
- Land plant communities of the Triassic and Jurassic consisted of seedless vascular plants and gymnosperms. The angiosperms, or flowering plants, evolved during the Early Cretaceous, diversified rapidly, and were soon the most abundant land plants.
- Dinosaurs evolved from small, bipedal archosaurs during the Late Triassic, but they were most common during the Jurassic and Cretaceous periods.
- All dinosaurs evolved from a common ancestor but differ enough that two distinct orders are recognized: the Saurischia and the Ornithischia.
- Bone structure, predator–prey relationships, and other features have been cited as evidence of dinosaur endothermy. Although there is still no solid consensus, many paleontologists think some dinosaurs were indeed endotherms.
- That some theropods had feathers indicates they were warm-blooded and provides further evidence of their relationship to birds.
- Pterosaurs, the first flying vertebrates, varied from sparrow size to comparative giants. The larger pterosaurs probably depended on soaring to stay aloft, whereas smaller ones flapped their wings. At least one species had hair or hairlike feathers.
- The fish-eating, porpoise-like ichthyosaurs were thoroughly adapted to an aquatic environment, whereas the plesiosaurs with their paddle-like limbs could most likely come out of the water to lay their eggs. The marine reptiles known as mosasaurs were most closely related to lizards.
- During the Jurassic, crocodiles became the dominant freshwater predators. Turtles and lizards were present during most of the Mesozoic. By the Cretaceous, snakes had evolved from lizards.
- Jurassic-aged *Archaeopteryx*, the oldest known bird, possesses so many theropod characteristics that it has convinced most paleontologists the two are closely related.

Table 15.3
Summary of Mesozoic Geologic and Evolutionary Events

Age (millions of years)	Geologic Period	Invertebrates	Vertebrates	Plants	Climate	Plate Tectonics
66 —	Cretaceous	Extinction of ammonites, rudists, and most planktonic foraminifera at end of Cretaceous. Continued diversification of ammonites and belemnoids. Rudists become major reef builders.	Extinctions of dinosaurs, flying reptiles, marine reptiles, and some marine invertebrates. Placental and marsupial mammals diverge.	Angiosperms evolve and diversify rapidly. Seedless plants and gymnosperms still common but less varied and abundant.	North–south zonation of climates more marked but remains equable. Climate becomes more seasonal and cooler at end of Cretaceous.	Further fragmentation of Pangaea. South America and Africa have separated. Australia separated from South America but remains connected to Antarctica. North Atlantic continues to open.
146 —	Jurassic	Ammonites and belemnoid cephalopods increase in diversity. Scleractinian coral reefs common. Appearance of rudist bivalves.	First birds (may have evolved in Late Triassic). Time of giant sauropod dinosaurs. *Greatest Diversity of Dinosaurs*	Seedless vascular plants and gymnosperms only.	Much like Triassic. Ferns with living relatives restricted to tropics live at high latitudes, indicating mild climates.	Fragmentation of Pangaea continues, but close connections exist among all continents.
200 —	Triassic	The seas are repopulated by invertebrates that survived the Permian extinction event. Bivalves and echinoids expand into the infaunal niche.	Mammals evolve from cynodonts. Cynodonts become extinct. Ancestral archosaur gives rise to dinosaurs. Flying reptiles and marine reptiles evolve.	Land flora of seedless vascular plants and gymnosperms as in Late Paleozoic.	Warm temperate to tropical. Mild temperatures extend to high latitudes; polar regions may have been temperate. Local areas of aridity.	Fragmentation of Pangaea begins in Late Triassic.
251 —						

- Mammals evolved by the Late Triassic, but they differed little from their ancestors, the cynodonts. Minor differences in the lower jaw, teeth, and middle ear differentiate one group of fossils from the other.
- Several types of Mesozoic mammals existed, but most were small, and their diversity was low. Both marsupial and placental mammals evolved during the Cretaceous from a group known as eupantotheres.
- Because during much of the Mesozoic the continents were close together and climates were mild, plants and animals occupied much larger geographic ranges than they do now.
- Among the victims of the Mesozoic mass extinction were dinosaurs, flying reptiles, marine reptiles, and several groups of marine invertebrates. A huge meteorite impact may have caused these extinctions, but some paleontologists think other factors were important, too.

IMPORTANT TERMS

angiosperm, p. 300
Archaeopteryx, p. 309
archosaur, p. 301
bipedal, p. 303
cynodont, p. 310
dinosaur, p. 301
ectotherm, p. 305
endotherm, p. 305
ichthyosaur, p. 307
iridium anomaly, p. 317
marsupial mammal, p. 314
monotreme, p. 314
mosasaur, p. 307
Ornithischia, p. 303
placental mammal, p. 314
plesiosaur, p. 307
pterosaur, p. 307
quadrupedal, p. 303
Saurischia, p. 303
therapsid, p. 310

REVIEW QUESTIONS

1. Which one of the following was a Mesozoic marine reptile?
   a.____mosasaur; b.____pelycosaur; c.____pterodactyl; d.____saurischian; e.____monotreme.
2. Occlusion means that mammal teeth
   a.____are simple cones; b.____meet surface to surface for chewing; c.____consist of parietals, occipitals, and dentals; d.____evolved from the incus and malleus; e.____are replaced continuously.
3. Which one of the following statements is correct?
   a.____the first flying vertebrate animals were placodonts; b.____ichthyosaurs look much like living opossums; c.____the first mammals evolved during the Cretaceous; d.____rudists were important Mesozoic reef-building animals; e.____most Triassic and Jurassic plants were angiosperms.
4. The bipedal, carnivorous dinosaurs were
   a.____marsupials; b.____gymnosperms; c.____theropods; d.____pachycephalosaurs; e.____plesiosaurs.
5. The typical mammal jaw joint is between the _____ and _____ bones.
   a.____occipital/incus; b.____articular/jugular; c.____stapes/quadrate; d.____ethmoid/zygomatic; e.____dentary/squamosal.
6. Because of their rapid evolution and nektonic lifestyle, the _____ are good guide fossils.
   a.____amphibians; b.____cephalopods; c.____burrowing worms; d.____therapsids; e.____bivalves.
7. The eupantotheres were the
   a.____probable ancestors of placental and marsupial mammals; b.____most dolphin-like of all ichthyosaurs; c.____largest of all sauropods; d.____first vertebrate animals capable of flight; e.____most diverse bipedal angiosperms.
8. Of all the mammal-like reptiles, the _____ are the most probable ancestors of mammals.
   a.____cephalopods; b.____endotherms; c.____cynodonts; d.____ornithopods; e.____quadrupeds.
9. Modification of the hand yielding an elongated fourth finger for wing support is found in
   a.____birds; b.____insects; c.____pterosaurs; d.____theropods; e.____bats.
10. Which one of the following is evidence of a meteorite impact at the end of the Cretaceous Period?
    a.____evolution of ammonites; b.____iridium anomaly; c.____breakup of Pangaea; d.____mild climates near the poles; e.____extinction of angiosperms.
11. During the transition from mammal-like reptiles to mammals what changes took place in the jaw and middle ear?
12. What changes first seen in angiosperms account for their remarkable success?
13. Why do paleontologists think that birds evolved from theropod dinosaurs?
14. Explain how plate positions and climate influenced the geographic distribution of Mesozoic plants and animals.
15. What modifications for flight are found in pterosaurs, and how were ichthyosaurs modified for an aquatic environment?
16. How do paleontologists explain the use of stegosaur plates and the crests seen in some ornithopods?
17. What features of *Archaeopteryx* are birdlike and what features are reptilelike?

18. Summarize the evidence that convinces many geologists that an asteroid impact occurred at the end of the Mesozoic Era.
19. What evidence indicates that dinosaurs were endotherms? Were they all endotherms?
20. What are the two main groups of dinosaurs, and how do they differ from one another?

APPLY YOUR KNOWLEDGE

1. Someone tells you that any animal with three middle ear bones and a dentary-squamosal jaw joint is without a doubt a mammal, whereas any animal with one middle ear bone and an articular-quadrate jaw joint is a reptile. This person also claims that there are no fossils and no evidence from living animals to show that mammals evolved from reptiles. What can you cite from the fossil record as well as from living animals to refute these assertions?

FIELD QUESTION

1. You observe limestone beds with fossil trilobites and brachiopods dipping at 50 degrees, but an overlying layer of volcanic ash followed upward by sandstone beds with dinosaur fossils dips at only 15 degrees. A basalt dike cuts through all of the strata. How can you explain the sequence of events that took place? What basic geologic principles did you use to make your interpretations? Is it possible to determine the absolute ages of any of the events? If so, explain.

CHAPTER 16

CENOZOIC GEOLOGIC HISTORY: THE PALEOGENE AND NEOGENE

Outcrop of the Miocene Dove Springs Formation in Red Rock Canyon State Park at the western end of the El Paso Mountains in California. The rocks in the park have served as a backdrop for many movies, especially westerns. The red rocks are sandstones that were deposited in stream channels, whereas the gray rocks are tuffaceous sandstones, and the lightest layers are floodplain deposits.

[OUTLINE]

INTRODUCTION
CENOZOIC PLATE TECTONICS—AN OVERVIEW
CENOZOIC OROGENIC BELTS
The Alpine–Himalayan Orogenic Belt
The Circum-Pacific Orogenic Belt
THE NORTH AMERICAN CORDILLERA
The Laramide Orogeny
Cordilleran Igneous Activity
Basin and Range Province
Colorado Plateau
The Rio Grande Rift

Pacific Coast
THE CONTINENTAL INTERIOR
PERSPECTIVE 16.1: *Shiprock, New Mexico*
CENOZOIC HISTORY OF THE APPALACHIAN MOUNTAINS
NORTH AMERICA'S SOUTHERN AND EASTERN CONTINENTAL MARGINS
The Gulf Coastal Plain
The Atlantic Continental Margin
PALEOGENE AND NEOGENE MINERAL RESOURCES
SUMMARY

ThomsonNOW™ Explore interactive tutorials, animations, or practice problems available on the ThomsonNow website at **www.thomsonedu.com/login**.

CHAPTER OBJECTIVES

At the end of this chapter, you will have learned that

- The 66-million-year-long Cenozoic Era is divided into two periods, the Paleogene and the Neogene, each of which consists of several epochs.

- The breakup of Pangaea that began during the Triassic Period continued to the present, giving rise to the present distribution of land and sea.

- Cenozoic orogenies were concentrated in two belts, one that nearly encircles the Pacific Ocean basin, and another that trends east-west through the Mediterranean basin and on into southeast Asia.

- The Late Cretaceous to Eocene Laramide orogeny resulted in deformation of a large area in the west, called the North American Cordillera, which extends from Alaska to Mexico.

- Following the Laramide orogeny, the North American Cordillera continued to evolve as it experienced volcanism, uplift of broad plateaus, large-scale block faulting, and deep erosion.

- A subduction zone was present along the western margin of the North American plate until the plate collided with a spreading ridge, producing the San Andreas and Queen Charlotte transform faults.

- An epeiric sea briefly occupied North America's interior lowlands during the Paleogene.

- Thick deposits of sediment accumulated along the Gulf and Atlantic Coastal plains.

- Renewed uplift and erosion account for the present-day Appalachian Mountains.

- Paleogene and Neogene rocks contain mineral resources such as oil, oil shale, coal, phosphorus, and gold.

Introduction

At 66 million years long, the Cenozoic Era is only 1.5% of all geologic time, or just 20 minutes on our hypothetical 24-hour clock for geologic time (see Figure 8.1). So the Cenozoic Era was comparatively brief when considered in the context of geologic time, and yet it was extremely long by any other measure. It was certainly long enough for significant changes to occur as plates changed position, mountains and landscapes continued to developed, an ice age took place, and the biota evolved. When the Cenozoic began, semitropical forests covered much of North America, and many of the mammals that dwelled in these forests are completely unfamiliar to us (see Chapter 18). Many events that began during the Cenozoic continue to the present, including the ongoing erosion of the Grand Canyon, continued uplift and erosion of the Himalayas in Asia and the Andes in South America, the origin and evolution of the San Andreas fault, and the origin of the volcanoes that make the Cascade Range.

Geologists divide the Cenozoic Era into two periods of unequal duration. The Paleogene Period (66 to 23 million years ago) includes the Paleocene, Eocene, and Oligocene epochs, and the Neogene Period (23 million years ago to the present) includes the Miocene, Pliocene, Pleistocene, and Holocene or Recent epochs. Although the terms *Tertiary Period* (66 to 1.8 million years ago) and *Quaternary Period* (for the last 2.6 million years) are still used by some geologists, they are no longer recommended as subdivisions of the Cenozoic Era (• Figure 16.1). However, many geologists continue to use the term

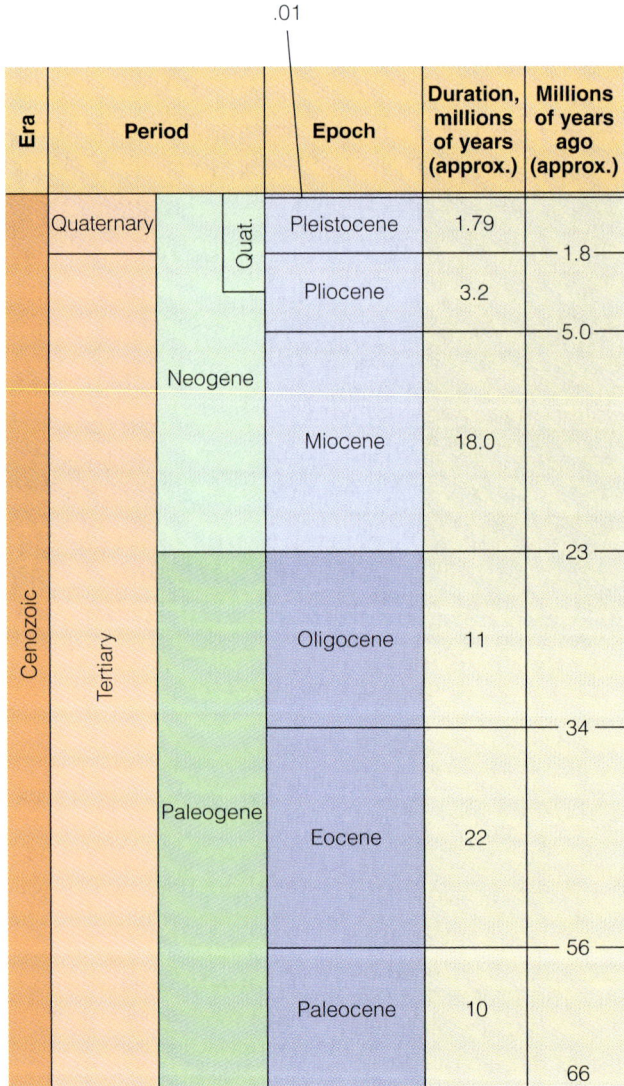

• **Figure 16.1 The Geological Time Scale for the Cenozoic Era**
In this book we use the terms *Paleogene* and *Neogene* rather than *Tertiary* and *Quaternary*.

Quaternary, and we will discuss the use of this term in the next chapter.

Geologists know more about Cenozoic Earth and life history than for any of the other intervals of geologic time because Cenozoic rocks, being the youngest, are the most accessible at the surface or in the shallow subsurface. Vast exposures of Cenozoic sedimentary and igneous rocks in western North America record the presence of a shallow sea in the continental interior, terrestrial depositional environments, lava flows (• Figure 16.2), and volcanism on a huge scale in the Pacific Northwest. Exposures of Cenozoic rocks in eastern North America are limited, except for Ice Age deposits, but one notable exception is Florida where fossil-bearing rocks of Middle to Late Cenozoic age are present.

One reason to study Cenozoic Earth history is that the present distribution of land and sea, climatic and oceanic circulation patterns, and Earth's present-day distinctive topography resulted from systems interactions during this time. In this chapter our concern is Earth history of the Paleogene and the Neogene periods (except for the Pleistocene and Holocene epochs). The latter part of the Neogene was unusual because it was one of the few times in Earth history when widespread glaciers were present, so we consider the Pleistocene and Holocene epochs in Chapter 17.

Cenozoic Plate Tectonics— An Overview

The progressive fragmentation of Pangaea, the supercontinent that existed at the end of the Paleozoic (see Figure 14.1), accounts for the present distribution of Earth's landmasses. Moving plates also directly affect the biosphere because the geographic locations of continents profoundly influence the atmosphere and hydrosphere. As we examine Cenozoic life history, you will see that some important biological events are related to isolation and/or connections between landmasses (see Chapter 18).

Notice from • Figure 16.3 that as the Americas separated from Europe and Africa, the Atlantic Ocean basin opened, first in the south and later in the north. Spreading ridges such as the Mid-Atlantic Ridge and East Pacific Rise were established, along which new oceanic crust formed and continues to form. However, the age of the oceanic crust in the Pacific is very asymmetric, because much of the crust in the eastern Pacific Ocean basin has been subducted beneath the westerly moving North and South American plates (see Figure 3.12).

Another important plate tectonic event was the northward movement of the Indian plate and its eventual collision with Asia (Figure 16.3). Simultaneous northward

(a)

(b)

(c)

• **Figure 16.2 Cenozoic Sedimentary and Volcanic Rocks in the Western United States** (a) The Paleocene Fort Union Formation in Montana. (b) The Wasatch Formation in Bryce Canyon National Park in Utah was deposited during the Eocene. (c) The Miocene Lovejoy Basalt near Orland, California, caps the small hill in the distance.

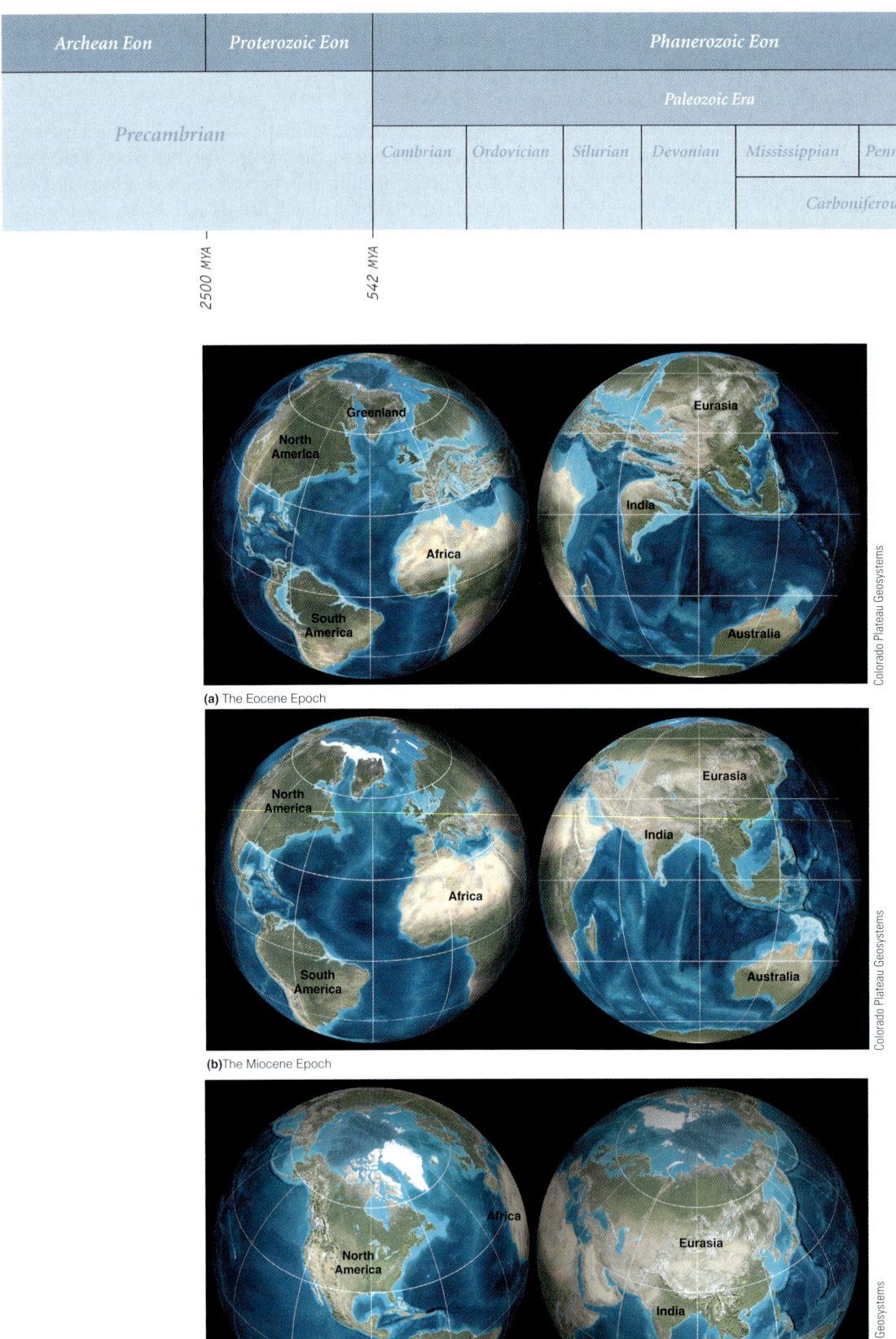

- **Figure 16.3 Paleogeography of the World During the Cenozoic** (a) The Eocene Epoch. (Compare with Figure 14.1c.) (b) The Miocene Epoch. (c) The world today.

Phanerozoic Eon										
Mesozoic Era			Cenozoic Era							
Triassic	Jurassic	Cretaceous	Paleogene			Neogene		Quaternary		
			Paleocene	Eocene	Oligocene	Miocene	Pliocene	Pleistocene	Holocene	

251 MYA • 66 MYA

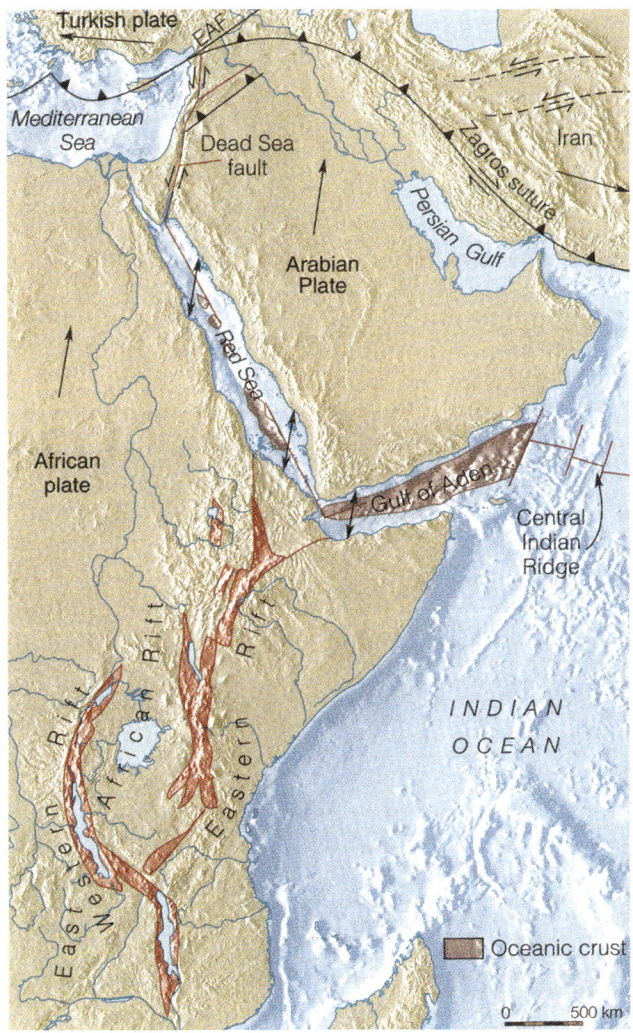

• **Figure 16.4 Triple Junction at the Junction of the East African Rift System and the Rifts in the Gulf of Aden and the Red Sea** Oceanic crust began forming in the Gulf of Aden about 10 million years ago. Rifting began later in the Red Sea, and oceanic crust is now forming. In East Africa, the continental crust has not yet stretched and thinned enough for oceanic crust to form from below. Much of the tectonic activity in the Middle East is caused by the Arabian plate moving northward against Eurasia.

movement of the African plate caused the closure of the Tethys Sea and initiated the tectonic activity that currently takes place throughout an east–west zone from the Mediterranean through northern India. Erupting volcanoes in Italy and Greece as well as seismic activity in Italy, Turkey, Greece, and Pakistan remind us of the continuing plate interactions in this part of the world.

Rifting and the separation of landmasses is not restricted to the Triassic. In fact, Neogene rifting began in East Africa, the Red Sea, and the Gulf of Aden (• Figure 16.4). Rifting in East Africa is in its early stages, because the continental crust has not yet stretched and thinned enough for new oceanic crust to form from below. Nevertheless, this area shows seismic activity and considerable volcanism. In the Red Sea, rifting and the Late Pliocene origin of oceanic crust followed vast eruptions of basalt, and in the Gulf of Aden, Earth's crust had stretched and thinned enough by Late Miocene time for upwelling basaltic magma to form new oceanic crust. Notice in Figure 16.4 that the Arabian plate is moving north, so it too causes some of the deformation taking place from the Mediterranean through India.

In the meantime, the North and South American plates continued their westerly movement as the Atlantic Ocean basin widened. Subduction zones bounded both continents on their western margins, but the situation changed in North America as it moved over the northerly extension of the East Pacific Rise and it now has a transform plate boundary, a topic we discuss more fully in a later section.

Cenozoic Orogenic Belts

Remember that an *orogeny* is an episode of mountain building, during which deformation takes place over an elongate area. In addition, most orogenies involve volcanism, the emplacement of plutons, and regional metamorphism as Earth's crust is locally thickened and stands higher than adjacent areas. Cenozoic orogenic activity took place largely in two major zones or belts: the *Alpine–Himalayan orogenic belt* and the *circum-Pacific orogenic belt* (• Figure 16.5). Both belts are made up of smaller segments known as **orogens,** each of which shows the characteristics of an orogeny.

The Alpine–Himalayan Orogenic Belt

Volcanism, seismicity, and deformation remind us that the **Alpine–Himalayan orogenic belt** remains active. It extends eastward from Spain through the Mediterranean region as well as the Middle East and India and on into Southeast Asia (Figure 16.5). Remember that during Mesozoic time the Tethys Sea separated much of Gondwana from Eurasia. Closure of this sea took place during the Cenozoic as the African and Indian plates collided with the huge landmass to the north (Figure 16.3).

The Alps During the **Alpine orogeny,** deformation occurred in a linear zone in southern Europe extending from Spain eastward through Greece and Turkey. Concurrent deformation also occurred along Africa's northwest coast (Figure 16.5). Many details of this long, complex event are still poorly understood, but the overall picture is now becoming clear.

Events leading to Alpine deformation began during the Mesozoic, yet Eocene to Late Miocene deformation was also important. Northward movements of the African and Arabian plates against Eurasia caused compression and deformation, but the overall picture is complicated by the collision of several smaller plates with Europe. These small plates were also deformed and are now incorporated in the mountains in the Alpine orogen.

Mountain building in this region produced the Pyrenees between Spain and France, the Apennines of Italy, as well as the Alps of mainland Europe (Figure 16.5). Indeed, the compressional forces generated by colliding plates resulted in complex thrust faults and huge overturned folds known as *nappes* (• Figure 16.6). As a result, the geology of such areas in France, Switzerland, and Austria is extremely complex. Plate convergence also produced an almost totally isolated sea in the Mediterranean basin, which had previously been part of the Tethys Sea. Late Miocene deposition in this sea, which was then in an arid environment, accounts for evaporite deposits up to 2 km thick.

The collision of the African plate with Eurasia also accounts for the Atlas Mountains of northwest Africa, and further to the east, in the Mediterranean basin, Africa continues to force oceanic lithosphere northward beneath Greece and Turkey. Active volcanoes in Italy and Greece as well as seismic activity throughout this region indicate that southern Europe and the Middle East remain geologically active. In 2005, for instance, an earthquake with a magnitude of 7.6 on the Richter scale killed more than 86,000 people in Pakistan, and Mount Vesuvius in Italy has erupted 80 times since it destroyed Pompeii in A.D. 79.

The Himalayas—Roof of the World During the Early Cretaceous, India broke away from Gondwana and began moving north, and oceanic lithosphere was consumed at a subduction zone along the southern margin of Asia (• Figure 16.7a). The descending plate partially melted, yielding magma that rose to form a volcanic chain and large granitic plutons in what is now Tibet. The Indian plate eventually approached these volcanoes and destroyed them as it collided with Asia. As a result, two continental plates were sutured along a zone now recognized as the Himalayan orogen.

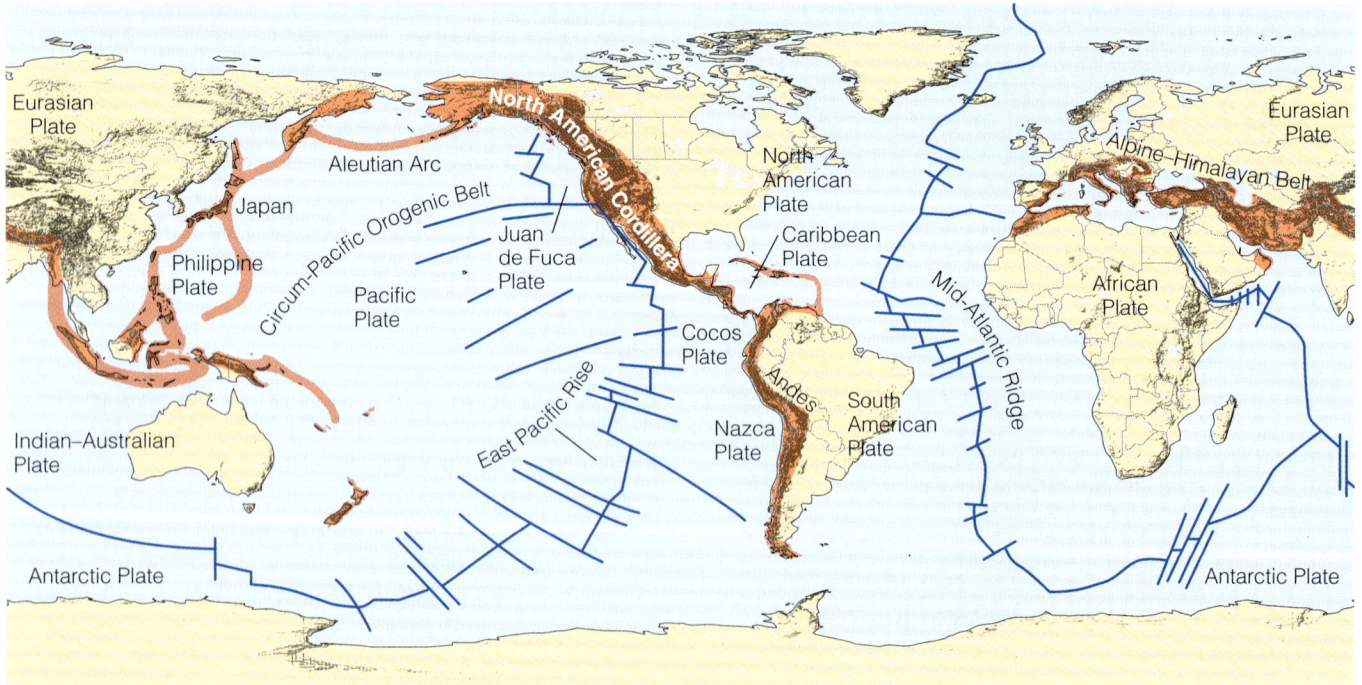

• Figure 16.5 **Earth's Present-Day Orogenic Belts** Most of Earth's geologically recent and present-day orogenic activity takes place in the circum-Pacific orogenic belt and the Alpine–Himalayan orogenic belt. Each belt is made up of smaller units known as *orogens*.

• **Figure 16.6 The Alps in Europe Are Part of the Alpine–Himalayan Orogenic Belt** (a) View of the Alps from the Lauterbrunnen Valley near Interlaken, Switzerland. The waterfall is about 300 m high. (b) Folded rocks in the Alps of Switzerland.

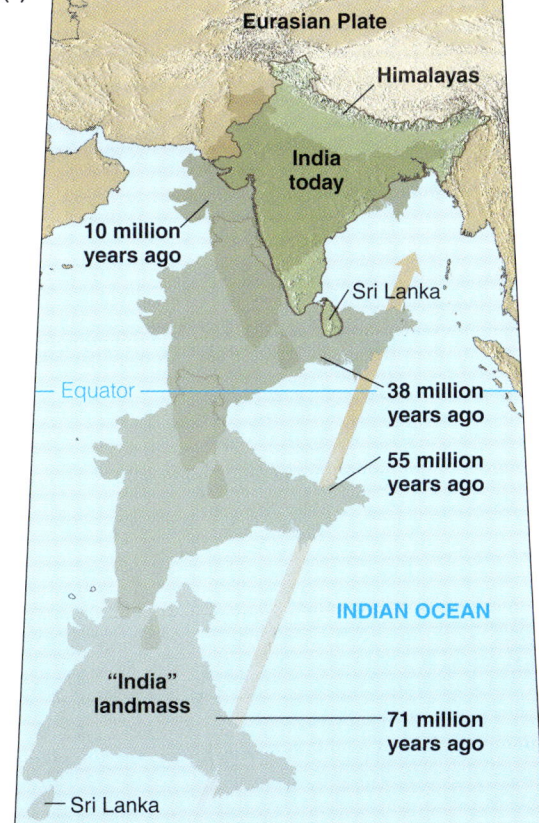

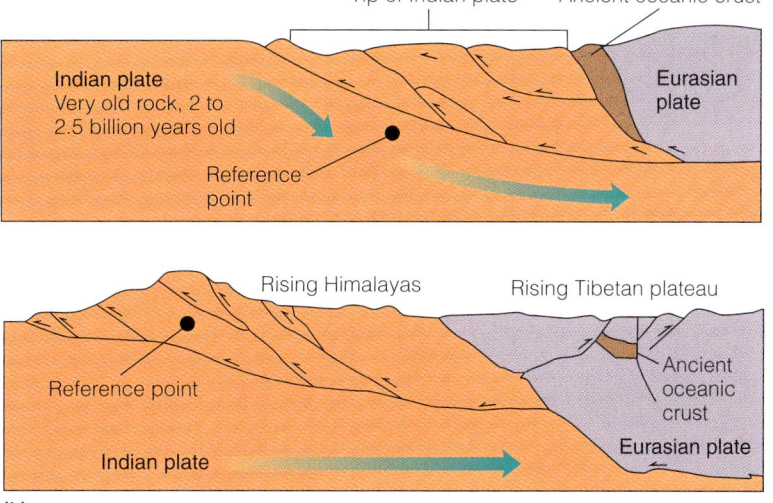

• **Figure 16.7 Plate Tectonics and the Origin of the Himalayas** (a) The Indian plate moved northward for millions of years until it collided with the Eurasian plate, causing crustal thickening and uplift of the Himalayas. (b) These diagrams show the collision of the two plates. The top illustration is earlier than the lower one. The reference point shows the uplift of an imaginary point during this orogeny.

CENOZOIC OROGENIC BELTS **327**

Just when this collision took place is not certain, but sometime between 40 and 50 million years ago India's drift rate decreased abruptly from 15 to 20 cm/year to about 5 cm/year. Because continental lithosphere is not dense enough to be subducted, this decrease most likely marks the time of collision and India's resistance to subduction (Figure 16.7b). As a result of India's low density and resistance to subduction, it was underthrust about 2000 km beneath Asia, causing crustal thickening and uplift, a process that continues at about 5 cm/year. Furthermore, sedimentary rocks formed in the sea south of Asia were thrust northward into Tibet, and two huge thrust faults carried Paleozoic and Mesozoic rocks of Asian origin onto the Indian plate (Figure 16.7c). Seismic activity continues in the Himalayan orogen, but there is no volcanism because the Indian plate does not penetrate deeply enough to generate magma.

The Circum-Pacific Orogenic Belt

The **circum-Pacific orogenic belt** is an area of active tectonism consisting of several orogens along the western margins of South, Central, and North America as well as the eastern margin of Asia and the islands north of Australia and New Zealand (Figure 16.5). Subduction of oceanic lithosphere accompanied by deformation and igneous activity characterize the orogens in the western and northern Pacific. Japan, for instance, is bounded on the east by the Japan Trench, where the Pacific plate is subducted, and the Sea of Japan, a **back-arc marginal basin,** lies between Japan and mainland Asia. According to some geologists, Japan was once part of mainland Asia and was separated when back-arc spreading took place (• Figure 16.8). Separation began during the Cretaceous as Japan moved westward over the Pacific plate and oceanic crust formed in the Sea of Japan.

Japan's geology is complex, and much of its deformation predates the Cenozoic, but considerable deformation, metamorphism, and volcanism occurred during the Cenozoic and continues to the present.

In the eastern part of the Pacific, the Cocos and Nazca plates moved west from the East Pacific Rise only to be consumed at subduction zones in Central and South America, respectively (Figure 16.3). Volcanism and seismic activity indicate these orogens in both Central and South America remain active. One manifestation of ongoing tectonic activity in South America is the Andes Mountains, with more than 49 peaks higher than 6000 m. The Andes formed, and continue to do so, as Mesozoic-Cenozoic plate convergence resulted in crustal thickening as sedimentary rocks were deformed, uplifted, and intruded by huge granitic plutons (• Figure 16.9).

The North American Cordillera

The **North American Cordillera,** a complex mountainous region in western North America, is one large segment of the circum-Pacific orogenic belt extending from Alaska to central Mexico. In the United States it widens to 1200 km, stretching east–west from the eastern flank of the Rocky Mountains to the Pacific Ocean (• Figure 16.10).

The geologic evolution of the North American Cordillera actually began during the Neoproterozoic when huge quantities of sediment accumulated along a westward-facing continental margin (see Figure 9.7). Deposition continued into the Paleozoic, and during the Devonian part of the region was deformed at the time of the Antler orogeny (see Chapter 11). A protracted episode

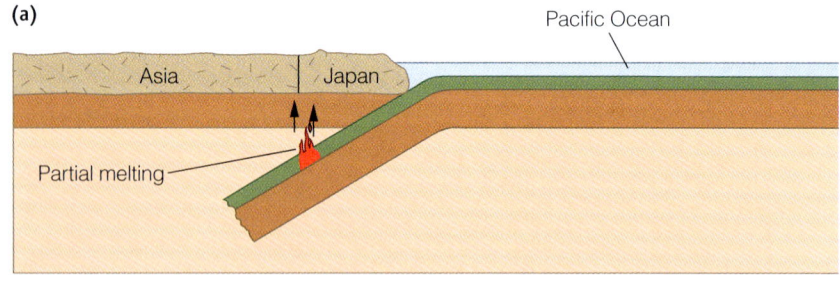

• Figure 16.8 Back-Arc Spreading and the Sea of Japan (a) Model showing the initial stage in the origin of the Sea of Japan. (b) A more advanced stage in which the Sea of Japan, a back-arc marginal basin, has opened.

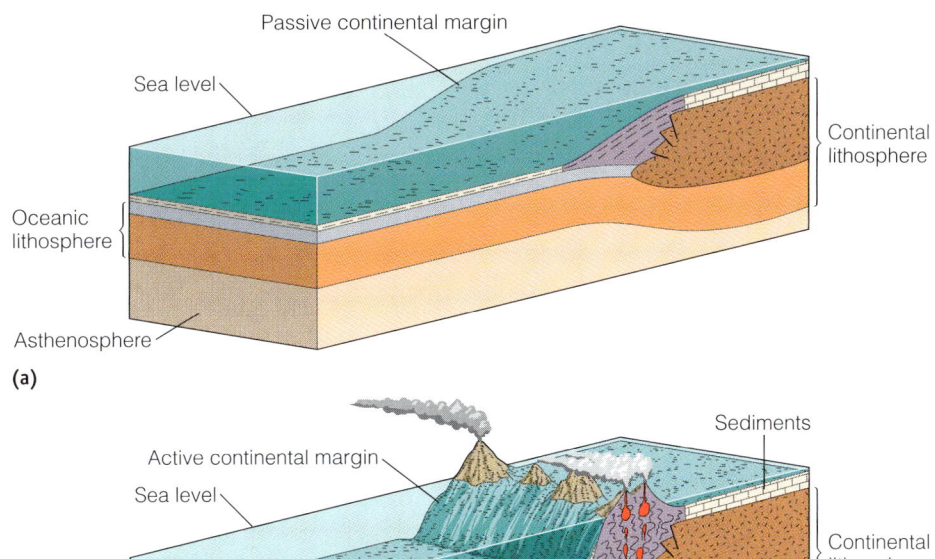
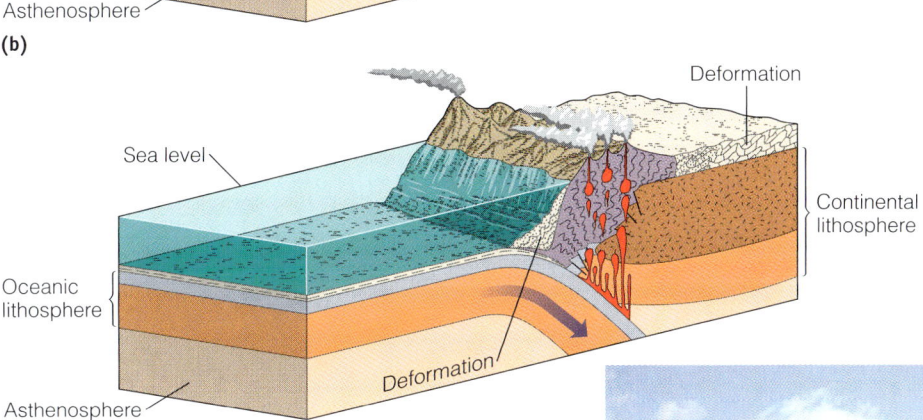

• **Figure 16.9 The Andes Mountains in South America** (a) Prior to 200 million years ago, the western margin of South America was a passive continental margin. (b) Orogeny began when this area became an active continental margin as the South American plate moved to the west and collided with oceanic lithosphere. (c) Continued deformation, plutonism, and volcanism. (d) View of the Andes in Chile.

of deformational events known as the *Cordilleran orogeny* began during the Late Jurassic as the Nevadan, Sevier, and Laramide orogenies progressively affected areas from west to east (see Figure 14.13). The first two of these orogenies were discussed in Chapter 14; the Laramide orogeny, a Late Cretaceous to Eocene episode of deformation, is discussed in the following section.

After Laramide deformation, the North American Cordillera continued to evolve as parts of it experienced large-scale block-faulting, extensive volcanism, and vertical uplift and deep erosion. Furthermore, during about the first half of the Cenozoic Era a subduction zone was present along the entire western margin of the Cordillera, but now most of it is a transform plate boundary. Present-day seismic activity and volcanism indicate plate interactions continue in the Cordillera, especially near its western margin.

The Laramide Orogeny

We have already mentioned that the **Laramide orogeny** was the third in a series of deformational events in the Cordillera beginning during the Late Jurassic. However, this orogeny was Late Cretaceous to Eocene and it differed from the previous orogenies in important ways. First, it

• **Figure 16.10** The North American Cordillera and the Major Provinces in the United States and Canada

occurred much further inland from a convergent plate boundary, and neither volcanism nor emplacement of plutons was very common. Second, deformation mostly took the form of vertical, fault-bounded uplifts rather than the compression-induced folding and thrust faulting typical of most orogenies. To account for these differences, geologists have had to modify their model for orogenies at convergent plate boundaries. During the preceding Nevadan and Sevier orogenies, the Farallon plate was subducted at about a 50-degree angle along the western margin of North America. Volcanism and plutonism took place 150 to 200 km inland from the oceanic trench, and sediments of the continental margin were compressed and deformed. Most geologists agree that by Early Paleogene time there was a change in the subduction angle from steep to gentle and the Farallon plate moved nearly horizontally beneath North America, but they disagree on what caused the change in the angle of subduction.

According to one hypothesis, a buoyant oceanic plateau that was part of the Farallon plate that descended beneath North America resulted in shallow subduction. Another hypothesis holds that North America overrode the Farallon plate, beneath which was the deflected head of a mantle plume (• Figure 16.11). The lithosphere above the mantle plume was buoyed up, accounting for the change from steep to shallow subduction. As a result, igneous activity shifted farther inland and finally ceased because the descending plate no longer penetrated to the mantle.

This changing angle of subduction also caused a change in the type of deformation—the fold-thrust deformation of the Sevier orogeny gave way to large-scale buckling and fracturing, which yielded fault-bounded vertical uplifts. Erosion of the uplifted blocks yielded rugged mountainous topography and supplied sediments to the intervening basins.

The Laramide orogen is centered in the middle and southern Rocky Mountains of Wyoming and Colorado, but deformation also took place far to the north and south. In the northern Rocky Mountains of Montana and Alberta, Canada, huge slabs of pre-Laramide strata moved eastward

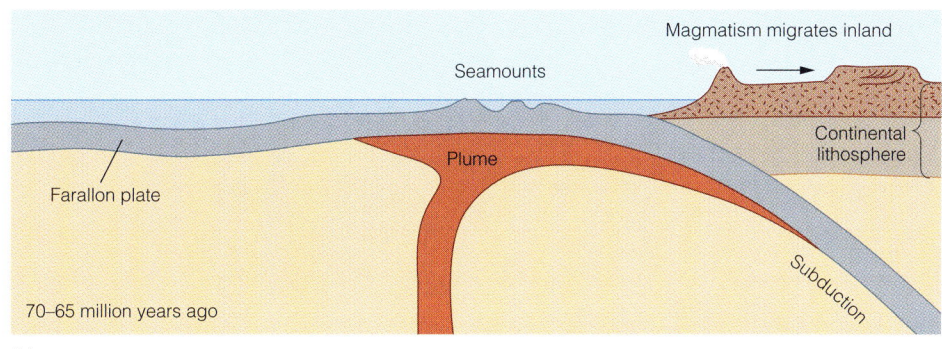

(a)

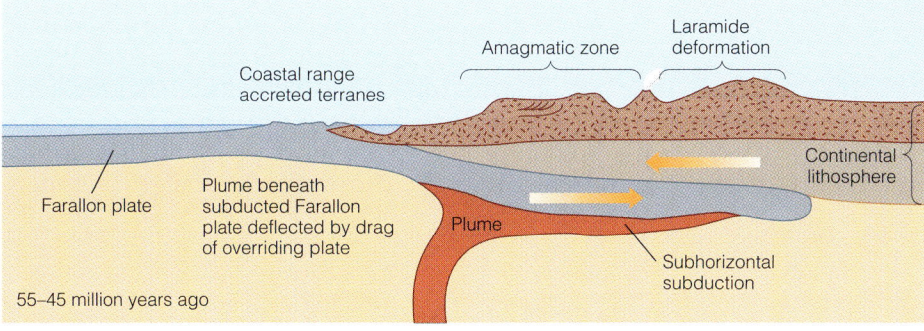

(b)

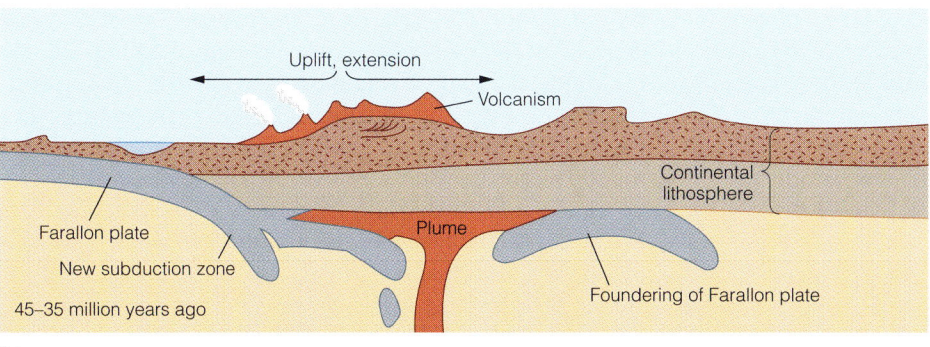

(c)

• **Figure 16.11 Laramide Orogeny** The Late Cretaceous to Eocene Laramide orogeny took place as the Farallon plate was subducted beneath North America. **(a)** As North America moved westward over the Farallon plate, beneath which was the deflected head of a mantle plume, the angle of the subduction decreased and the igneous activity shifted inland. **(b)** With nearly horizontal subduction, igneous activity ceased and the continental crust was deformed, mostly by vertical uplift. **(c)** Disruption of the oceanic plate by the mantle plume marked the onset of renewed igneous activity.

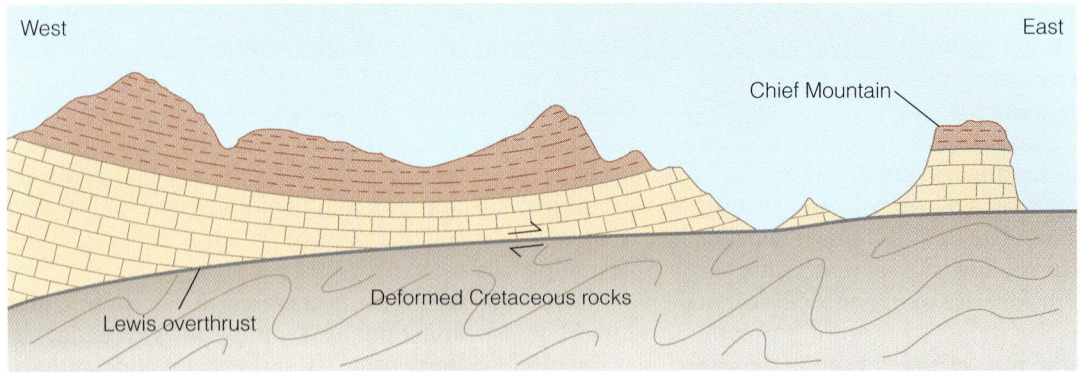

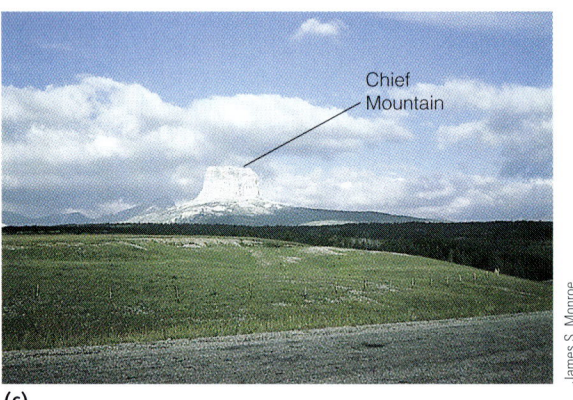

• **Figure 16.12 Laramide Deformation in the Northern Rocky Mountains (a)** Cross section of the Lewis overthrust in the Glacier National Park, Montana. Meso- and Neoproterozoic-aged rocks of the Belt Supergroup rest on deformed Cretaceous rocks. **(b)** The trace of the fault is visible in this image as a light-colored line on the mountainside. **(c)** Erosion has isolated Chief Mountain from the rest of the slab of overthrust rock.

along overthrust faults.* On the Lewis overthrust in Montana, for instance, a slab of Precambrian rocks was displaced eastward about 75 km (• Figure 16.12), and similar deformation can be seen in the Canadian Rocky Mountains. Far to the south of the main Laramide orogen, sedimentary rocks in the Sierra Madre Oriental of east-central Mexico are now part of a major fold-thrust belt.

By Middle Eocene time, Laramide deformation ceased and igneous activity resumed in the Cordillera when the mantle plume beneath the lithosphere disrupted the overlying oceanic plate (Figure 16.11c). The uplifted blocks of the Laramide orogen continued to erode, and by the Neogene the rugged, eroded mountains had been nearly buried in their own debris, forming a vast plain across which streams flowed. During a renewed cycle of erosion, these streams removed much of the basin fill sediments and incised their valleys into the uplifted blocks. Late Neogene uplift accounts for the present ranges, uplift that continues in some areas.

Cordilleran Igneous Activity

The enormous batholiths in Idaho, British Columbia, Canada, and the Sierra Nevada of California were emplaced during the Mesozoic (see Chapter 14), but intrusive activity continued into the Paleogene Period. Numerous small plutons formed, including copper- and molybdenum-bearing stocks in Utah, Nevada, Arizona, and New Mexico.

Volcanism was more or less continuous in the Cordillera, but it varied in location, intensity, and eruptive style, and it ceased temporarily in the area of the Laramide orogen (Figure 16.11b). In the Pacific Northwest, the Columbia Plateau (Figure 16.10) is underlain by 200,000 km^3 of Miocene lava flows of the Columbia River basalts that have an aggregate thickness of about 2500 m. These vast lava flows—the Rosa flow alone covers 40,000 km^2 and flowed about 300 km from its source—are now well exposed in the walls of the canyons eroded by the Columbia and Snake rivers (• Figure 16.13a and b). The relationship of this huge outpouring of lava to plate tectonics remains unclear, but some geologists think it resulted from a mantle plume beneath western North America.

The Snake River Plain (Figure 16.10), which is mostly in Idaho, is actually a depression in the crust that was filled by Miocene and younger rhyolite, volcanic ash, and basalt (Figure 16.13a and c). These rocks are youngest in the southwest part of the area and become older toward the northeast, leading some geologists to propose that North America has migrated over a mantle plume that now lies beneath Yellowstone National Park in Wyoming. Other geologists disagree, thinking that these volcanic rocks erupted along an intracontinental rift zone.

---

*An overthrust fault is a large-scale, low-angle thrust fault with movement measured in kilometers.

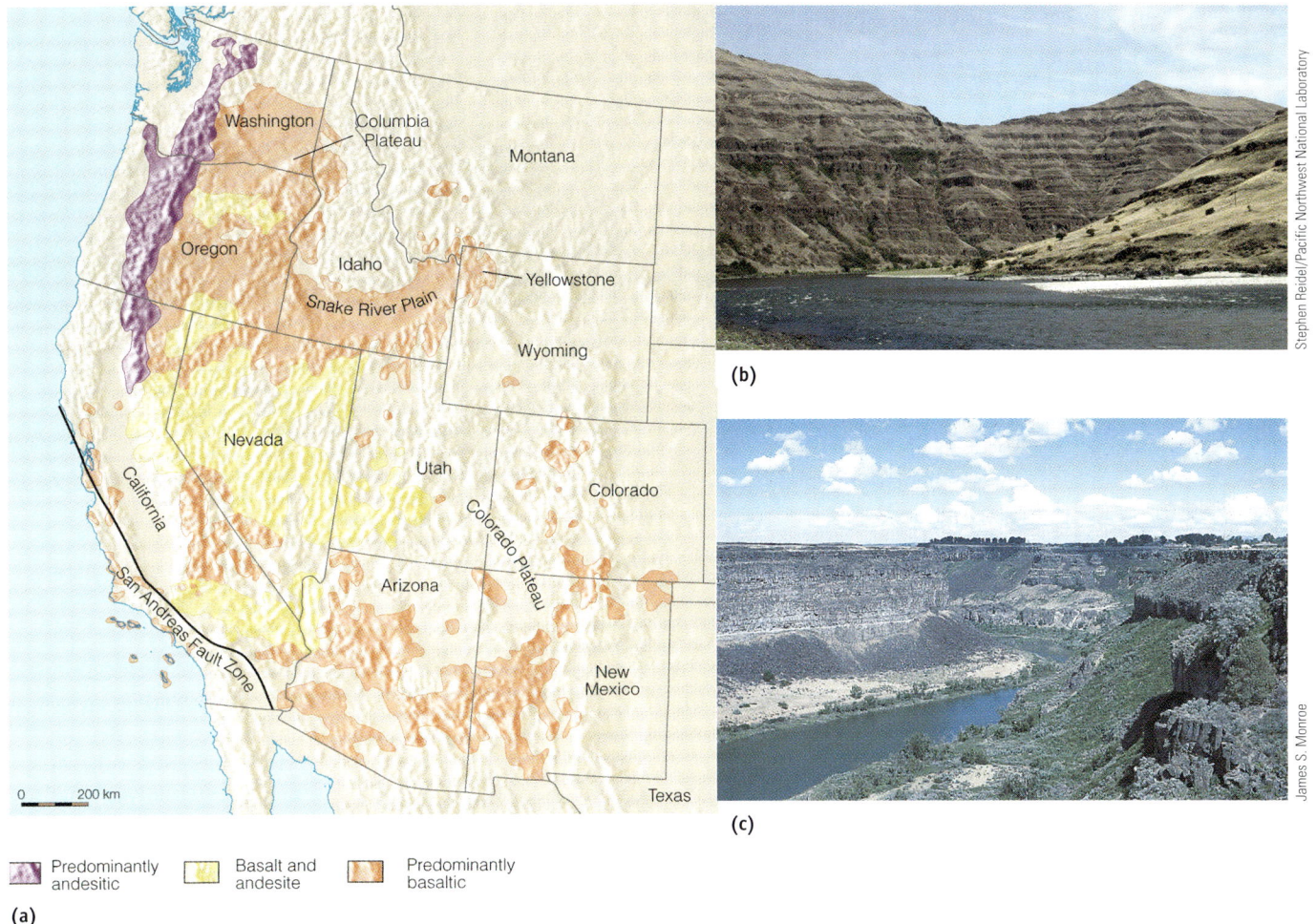

• **Figure 16.13 Cenozoic Volcanism (a)** Distribution of Cenozoic volcanic rocks in the western United States. **(b)** About 20 lava flows of the Columbia River basalts are exposed in the canyon of the Grand Ronde River in Washington. **(c)** Basalt lava flows of the Snake River Plain near Twin Falls, Idaho.

Bordering the Snake River Plain on the northeast is the Yellowstone Plateau (Figure 16.10), an area of Pliocene and Pleistocene volcanism. Perhaps a mantle plume lies beneath the area, as just noted, that accounts for the ongoing hydrothermal activity there, but the heat may come from an intruded body of magma that has not yet completely cooled.

Elsewhere in the Cordillera, andesite, volcanic breccia, and welded tuffs (ignimbrites), mostly of Oligocene age, cover more than 25,000 km^2 in the San Juan volcanic field in Colorado. In Arizona, Pliocene and Pleistocene eruptions built up the San Francisco Mountains where volcanism may have ceased as little as 1200 years ago. Eruptions in the Coso volcanic field in California began during the Pliocene and continued until only a few thousands of years ago (• Figure 16.14a), and in Arizona, the San Francisco volcanic field formed during the Pliocene and Pleistocene (Figure 16.14b).

Some of the most majestic and highest mountains in the Cordillera are in the **Cascade Range** of northern California, Oregon, Washington, and southern British Columbia, Canada (• Figure 16.15). Thousands of volcanic vents are present, the most impressive of which are the dozen or so large composite volcanoes and Lassen Peak in California, the world's largest lava dome. Volcanism in this region is related to subduction of the Juan de Fuca plate beneath North America. Volcanism in the Cascade Range goes back at least to the Oligocene, but the most recent episode began during the Late Miocene or Early Pliocene. The eruption of Lassen Peak in California from 1914 to 1917 and the eruptions of Mount St. Helens in Washington in 1980 and again in 2004—and continuing—indicate that Cascade volcanoes remain active.

Basin and Range Province

Earth's crust in the **Basin and Range Province** (Figure 16.10)—an area of nearly 780,000 km^2 centered on Nevada but extending into adjacent states and northern Mexico—has been stretched and thinned yielding north–south oriented mountain ranges with intervening valleys or

• **Figure 16.14 Cenozoic Volcanism in California and Oregon (a)** Pliocene to Pleistocene volcanism took place in the Coso volcanic field in California. The cinder cone, called Red Hill, formed no more than a few tens of thousands of years ago. **(b)** This view shows Sunset Crater in Sunset Crater National Monument in Arizona. It is only a little more than 1000 years old, but it is part of the extensive San Francisco volcanic field that formed mostly during the Pliocene and Pleistocene.

basins. The ranges are bounded on one or both sides by steeply dipping normal faults that probably curve and dip less steeply with depth. In any case the faults outline blocks that show displacement and rotation (• Figure 16.16a).

Before faulting began, the region was deformed during the Nevadan, Sevier, and Laramide orogenies. Then during the Paleogene, the entire area was highlands undergoing extensive erosion, but Early Miocene eruptions of rhyolitic lava flows and pyroclastic materials covered large areas. By the Late Miocene, large-scale faulting had begun, forming the basins and ranges. Sediment derived from the ranges was transported into the adjacent basins and accumulated as alluvial fan and playa lake deposits.

At its western margin the Basin and Range Province is bounded by normal faults along the east face of the Sierra Nevada (Figure 16.16b). Pliocene and Pleistocene uplift tilted the Sierra Nevada toward the west, and its crest now stands 3000 m above the basins to the east. Before this uplift took place, the Basin and Range had a subtropical climate, but the rising mountains created a rain shadow, making the climate increasingly arid.

Geologists have proposed several models to account for basin-and-range structure but have not reached a consensus. Among these are back-arc spreading, spreading at the East Pacific Rise, the northern part of which is thought to now lie beneath this region; spreading above a mantle plume; and deformation related to movements along the San Andreas fault.

Colorado Plateau

The vast, elevated region in Colorado, Utah, Arizona, and New Mexico known as the **Colorado Plateau** has volcanic mountains rising above it, brilliantly colored rocks, and deep canyons. In Chapters 11 and 14, we noted that during the Permian and Triassic the Colorado Plateau region was the site of extensive red bed deposition; many of these rocks are now exposed in the uplifts and canyons.

Cretaceous-aged marine sedimentary rocks indicate the Colorado Plateau was below sea level, but during the Paleogene Period Laramide deformation yielded broad anticlines and arches and basins, and a number of large normal faults. However, deformation was far less intense than elsewhere in the Cordillera. Neogene uplift elevated the region from near sea level to the 1200 to 1800 m elevations seen today, and as uplift proceeded, streams and rivers began eroding deep canyons.

Geologists disagree on the details of just how the deep canyons so typical of the region developed—such as the

• **Figure 16.15 Cascade Range Volcanism** Volcanic activity in the Cascade Range dates back at least to the Oligocene, but the large volcanoes of the range formed more recently. This view shows Mount Shasta in northern California, which last erupted during the late 1700s. The main peak is on the left; the one on the right is Shastina, a smaller cone on the flank of Mount Shasta.

• **Figure 16.16 The Basin and Range Province (a)** The Basin and Range is mostly in Nevada, but it extends into adjacent states and northern Mexico. The ranges and the intervening basins trend north–south. Each range is bounded on one or both sides by normal faults. **(b)** The Sierra Nevada at the western margin of the Basin and Range has risen along normal faults so that it is more than 3000 m above the valley to the east.

Grand Canyon (see Chapter 4 opening photo). Some think the streams were *antecedent*, meaning they existed before the present topography developed, in which case they simply eroded downward as uplift proceeded. Others think the streams were *superposed*, implying that younger strata covered the area, on which streams were established. During uplift, the streams stripped away these younger rocks and eroded down into the underlying strata. In either case, the landscape continues to evolve as erosion of the canyons and their tributaries deepens and widens them.

The Rio Grande Rift

A major rift in Earth's crust known as the **Rio Grande rift** extends north to south about 1000 km from central Colorado through New Mexico and into northern Mexico. Recall our discussions of the Mesoproterozoic Midcontinent rift in the Great Lakes region (see Chapter 9) and the present-day rifting in the Gulf of Aden, the Red Sea, and East Africa (Figure 16.4). The Rio Grande rift is similar in that Earth's crust has been stretched and thinned, the rift is bounded on both sides by normal faults, seismic activity continues, and volcanoes and calderas are present (• Figure 16.17). Actually the Rio Grande rift consists of several basins through which the present-day Rio Grande flows, although the river simply exploited an easy route to the sea but was not responsible for the rift itself.

Rifting in this area began about 29 million years ago and persisted for 10 to 12 million years, from the Late Oligocene into the Early Miocene. A second period of rifting began during the Middle Miocene, about 17 million years ago, and it continues to the present. The displacement on some of the faults is as much as 8000 m, but concurrent with faulting, the basins along the rift filled with huge quantities of sediments and volcanic rocks. Some of the volcanic features such as Valles caldera, which measures 19 by 24 km, and the Bandelier tuff are prominent features in New Mexico (Figure 16.17). Rifting continues, but it is progressing very slowly—only 2 mm or less per year. So even though ongoing rifting may eventually completely split the area so that it resembles the Red Sea, it will be in the far distant future.

Pacific Coast

Before the Eocene, the entire Pacific Coast was a convergent plate boundary where the **Farallon plate** was consumed at a subduction zone that stretched from Mexico to Alaska. Now there are only two small remnants of the Farallon plate—the Juan de Fuca and Cocos plates (• Figure 16.18). Continuing subduction of these small plates accounts for the

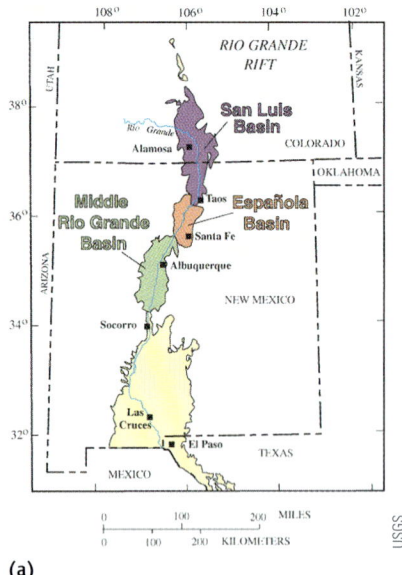

• **Figure 16.17 The Rio Grande Rift (a)** Location of the basins making up the Rio Grande rift. A complex of normal faults is present on both sides of the rift. **(b)** The Bandelier Tuff in Bandelier National Monument, New Mexico, erupted in the Jemez volcanic field 1.14 million years ago.

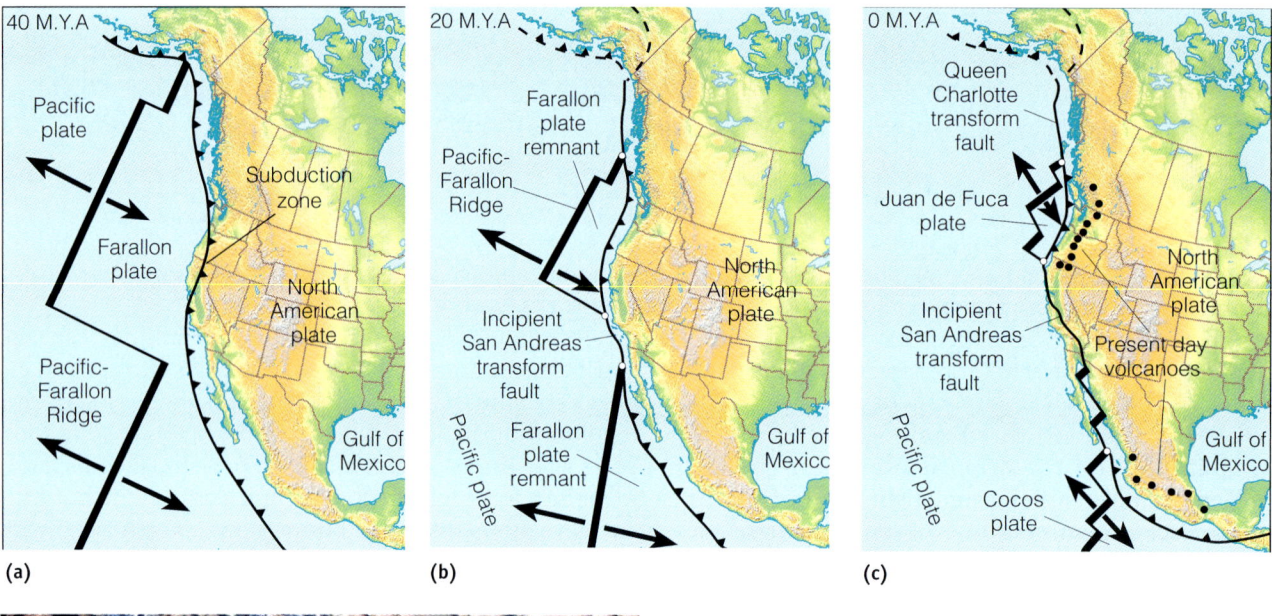

• **Figure 16.18 Origin of the San Andreas and Queen Charlotte Faults (a–c)** Three stages in the westward movement of North America and its collision with the Pacific-Farallon Ridge. As North America overrode the ridge, its margin became bounded by transform faults except in the Pacific Northwest. **(d)** Aerial view of the San Andreas fault today. On land we call it a right-lateral strike-slip fault.

present seismic activity and volcanism in the Pacific Northwest and Central America, respectively.

Another consequence of plate interactions in this region was the westward movement of the North American plate and its collision with the **Pacific–Farallon ridge.** Because the Pacific–Farallon ridge was oriented at an angle to the margin of North America, the continent–ridge collision took place first during the Eocene in northern Canada and only later during the Oligocene in southern California (Figure 16.18). In southern California, two triple junctions formed, one at the intersection of the North American, Juan de Fuca, and Pacific plates, the other at the intersection of the North American, Cocos, and Pacific plates. Continued westward movement of the North American plate over the Pacific plate caused the triple junctions to migrate, one to the north and the other to the south, giving rise to the **San Andreas transform fault** (Figure 16.18). A similar occurrence along Canada's west coast produced the *Queen Charlotte transform fault.*

Seismic activity on the San Andreas fault results from continuing movements of the Pacific and North American plates along this complex zone of shattered rocks. Indeed, where the fault cuts through coastal California it is actually a zone as much as 2 km wide, and it has numerous branches. Movements on such complex fault systems subject blocks of rocks adjacent to and within the fault zone to extensional and compressive stresses forming basins and elevated areas, the latter supplying sediments to the former. Many of the fault-bounded basins in the southern California area have subsided below sea level and soon filled with turbidites and other deposits. A number of these basins are areas of prolific oil and gas production.

What Would You Do?

Suppose you teach Earth science in high school. A curious student wants to know why western North America has volcanoes, earthquakes, numerous mountain ranges, and many small glaciers, whereas these same features or phenomena are absent or nearly so in the eastern part of the continent. How would you explain this disparity, and further, can you think of how the situation might be reversed? That is, what kinds of events would lead to these kinds of geologic events in the east?

The Continental Interior

Notice in Figure 16.10 that much of central North America is a vast area called the **Interior Lowlands,** which in turn are made up of the *Great Plains* and the *Central Lowlands.* During the Cretaceous, the Great Plains were covered by the **Zuni epeiric sea,** but by Early Paleogene time, this sea had largely withdrawn except for a sizable remnant that remained in North Dakota. Sediments eroded from the Laramide highlands were transported to this sea and deposited in transitional and marine environments.

The Paleocene Cannonball Formation, and its equivalents, mark the transition from marine to terrestrial deposition in this region. Following this brief episode of marine deposition, all other sedimentation in the Great Plains took place in terrestrial environments, especially fluvial systems. These formed eastward-thinning wedges of sediment that now underlie the entire region (• Figure 16.19).

(a)

(b)

• **Figure 16.19 Vast Amounts of Sediment Shed from the Laramide Highlands Were Deposited on the Great Plains (a)** Paleocene sedimentary rocks exposed in Theodore Roosevelt National Park in North Dakota. **(b)** Oligocene-Miocene sedimentary rocks, mostly siltstone and sandstone, and volcanic ash layers at Scott's Bluff National Monument in Nebraska.

Perspective 16.1

Shiprock, New Mexico

According to Navajo legend, a young man named Nayenezgani asked his grandmother where the mythical birdlike creatures known as Tse'na'hale lived. She replied, "They dwell at Tsae-bidahi," which means Winged Rock or Rock with Wings. We know Winged Rock as Shiprock, a volcanic neck rising nearly 550 m above the surrounding plain. Radiating outward from this conical volcanic neck are three dikes (Figure 1). Navajo legend holds that Winged Rock represents a giant bird that brought the Navajo people from the north, and the dikes are snakes that have turned to stone.

Shiprock is the most impressive of many volcanic necks exposed in the Four Corners region of the southwestern United States. (Four Corners is a designation for the point where the boundaries of Colorado, Utah, Arizona, and New Mexico converge.) Shiprock is visible from 160 km and was a favorite with rock climbers for many years until the Navajos put a stop to all climbing on the reservation. The country rock penetrated by this volcanic neck includes ancient metamorphic and igneous rocks and about 1000 m of overlying sedimentary rocks. The rock unit exposed at the surface, the Mancos Shale, is sedimentary rock composed mostly of mud that was deposited in an arm of the sea that existed in North America during the Cretaceous Period. Absolute dating of one of the dikes indicates that the magma that solidified to form Shiprock was emplaced about 27 million years ago, during the Late Oligocene.

Shiprock is one of several volcanic necks in the Navajo volcanic field that formed as a result of explosive volcanic eruptions. During these eruptions, volcanic materials along with large pieces of country rock torn from the vent walls were hurled high into the air and fell randomly around the area. The material composing Shiprock itself is characterized as a tuff-breccia consisting of fragmental volcanic debris along with inclusions of various sedimentary rocks and some granite and metamorphic rocks. Because Shiprock now stands about 550 m above the surrounding plain, at least that much erosion must have occurred to expose it in its present form. We can only speculate as to how much higher and larger it was when it was part of an active volcano.

The dikes radiating from Shiprock formed when magma ascended rather quietly and was emplaced in the country rocks. However, the fractures along which this magma rose may have formed as a result of the explosive emplacement of the tuff-breccia that filled the volcanic vent. The dike on the northeast side of Shiprock extends more than 2900 m outward from the vent and averages 2.3 m thick. Because the dike rock, like the material composing the volcanic neck, is more resistant to erosion than the adjacent Mancos Shale, the dikes stand as near-vertical walls above the surrounding plain.

The only local sediment source within the Great Plains was the Black Hills in South Dakota. This area has a history of marine deposition during the Cretaceous followed by the origin of terrestrial deposits derived from the Black Hills that are now well exposed in Badlands National Park, South Dakota. Judging from the sedimentary rocks and their numerous fossil mammals and other animals, the area was initially covered by semitropical forest, but grasslands replaced the forests as the climate became more arid (see Chapter 18).

Igneous activity was not widespread in the Interior Lowlands but was significant in some parts of the Great Plains. For instance, igneous activity in northeastern New Mexico was responsible for volcanoes and numerous lava flows, and several small plutons were emplaced in Colorado, Wyoming, Montana, South Dakota, and New Mexico. Indeed, one of the most widely recognized igneous bodies in the entire continent, Devil's Tower in northeastern Wyoming, is probably an Eocene volcanic neck, although some geologists think it is an eroded laccolith (• Figure 16.20). Another prominent volcanic feature is Shiprock in New Mexico, a volcanic neck that dates from the Late Oligocene (see Perspective 16.1).

Our discussion thus far has focused on the Great Plains, but what about the Central Lowlands to the east? Pleistocene glacial deposits are present in the northern part

• **Figure 16.20 Cenozoic Intrusive Activity in the Great Plains** At 650 m high, Devil's Tower in Wyoming can be seen from a distance of 48 km. It was emplaced as a small pluton during the Eocene Epoch between 45 and 50 million years ago.

Figure 1 Shiprock in northwest New Mexico is a volcanic neck that rises 550 m above the surrounding plain. It formed during the Oligocene Epoch, about 27 million years ago.

of this region, as well as in the northern Great Plains (see Chapter 17), but during most of the Cenozoic Era nearly all the Central Lowlands was an area of active erosion rather than deposition. Of course, the eroded materials had to be deposited somewhere, and that was on the Gulf Coastal Plain (Figure 16.10).

Cenozoic History of the Appalachian Mountains

Deformation and mountain building in the area of the present Appalachian Mountains began during the Neoproterozoic with the Grenville orogeny (see Chapter 9). The area was deformed again during the Taconic and Acadian orogenies, and during the Late Paleozoic closure of the Iapetus Ocean, which gave rise to the Hercynian-Alleghenian orogeny (see Chapters 10 and 11). Then during Late Triassic time, the entire region experienced block-faulting as Pangaea fragmented (see Figure 14.7). By the end of the Mesozoic, though, erosion had reduced the mountains to a plain across which streams flowed eastward to the ocean.

The present distinctive aspect of the Appalachian Mountains developed as a result of Cenozoic uplift and erosion (• Figure 16.21). As uplift proceeded, upturned resistant rocks formed northeast–southwest trending ridges with intervening valleys eroding into less resistant rocks. The preexisting streams eroded downward while uplift took place, were superposed on resistant rocks, and cut large canyons across the ridges, forming *water gaps* (• Figure 16.22), deep passes through which streams flow, and *wind gaps,* which are water gaps no longer containing streams.

Erosion surfaces at different elevations in the Appalachians are a source of continuing debate among geologists. Some are convinced these more or less planar surfaces show evidence of uplift followed by extensive erosion and then renewed uplift and another cycle of erosion. Others think that each surface represents differential response to weathering and erosion. According to this view, a low–elevation erosion surface developed on softer strata that eroded more or less uniformly, whereas higher surfaces represent weathering and erosion of more resistant rocks.

North America's Southern and Eastern Continental Margins

In a previous section we mentioned that much of the Interior Lowlands eroded during the Cenozoic. Even in the Great

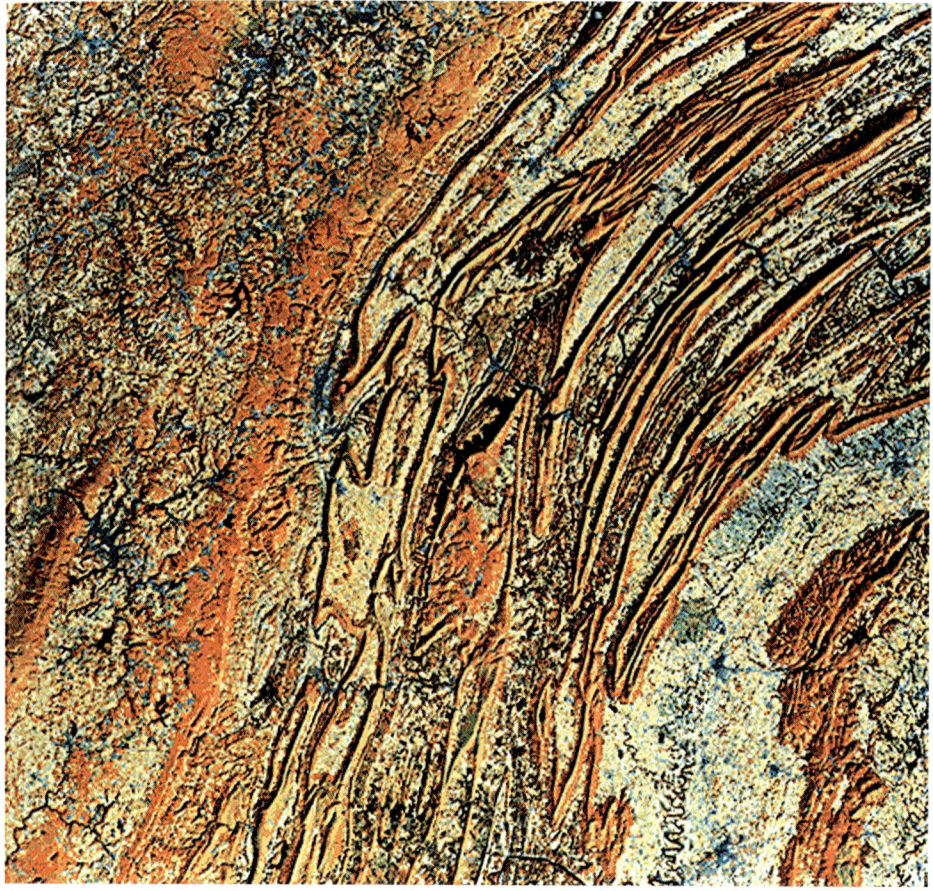

• **Figure 16.21 Landsat Image of the Appalachian Mountains** This view of the Appalachians shows the central part of Pennsylvania. Notice the long ridges with intervening valleys.

Plains where vast deposits of Cenozoic rocks are present, sediment was carried across the region and into the drainage systems that emptied into the Gulf of Mexico. Likewise sediment eroded from the western margin of the Appalachian Mountains ended up in the Gulf, but these mountains also shed huge quantities of sediment eastward that was deposited along the Atlantic Coastal Plain. Notice in Figure 16.10 that the **Atlantic Coastal Plain** and the **Gulf Coastal Plain** form a continuous belt extending from the northeastern United States to Texas. Both areas have horizontal or gently seaward-dipping strata deposited mostly by streams. Seaward of the coastal plains lie the continental shelf, slope, and rise, also areas of notable Mesozoic and Cenozoic deposition.

The Gulf Coastal Plain

After the withdrawal of the Cretaceous to Early Paleogene Zuni Sea, the Cenozoic **Tejas epeiric sea** made a brief appearance on the continent. But even at its maximum extent it was largely restricted to the Atlantic and Gulf Coastal plains and parts of coastal California. It did, however, extend up the Mississippi River Valley, where it reached as far north as southern Illinois.

The overall Gulf Coast sedimentation pattern was established during the Jurassic and persisted throughout the Cenozoic. Sediments derived from the Cordillera, western Appalachians, and the Interior Lowlands were transported toward the Gulf of Mexico, where they were deposited in terrestrial, transitional, and marine environments. In general, the sediments form seaward-thickening wedges grading from terrestrial facies in the north to marine facies in the south (• Figure 16.23).

Sedimentary facies development was controlled mostly by regression of the Tejas epeiric sea. After its maximum extent onto the continent during the Paleogene, this sea began its long withdrawal toward the Gulf of Mexico. Its regression, however, was periodically reversed by minor transgressions—eight transgressive–regressive episodes are recorded in Gulf Coastal Plain sedimentary rocks, accounting for the intertonguing among the various facies.

Many sedimentary rocks in the Gulf Coastal Plain are either source rocks or reservoirs for hydrocarbons, a topic we discuss more fully in the section on Paleogene and Neogene mineral resources.

Most of the Gulf Coastal Plain was dominated by detrital deposition, but in the Florida section of the region

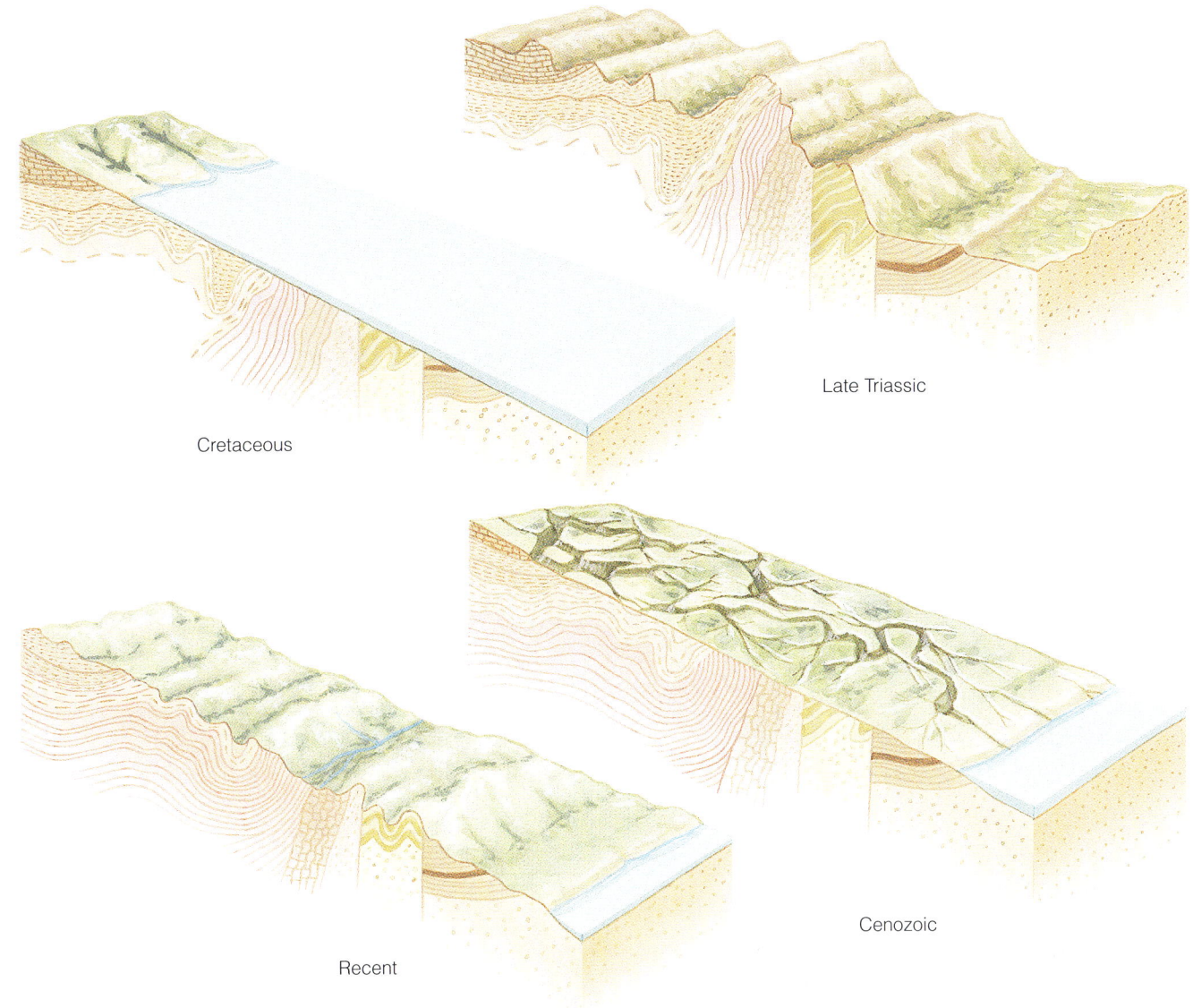

• **Figure 16.22 Evolution of the Present Topography of the Appalachian Mountains** Although these mountains have a long history, their present topographic expression resulted mainly from Cenozoic uplift and erosion.

and the Gulf Coast of Mexico significant carbonate deposition took place. Florida was a carbonate platform during the Cretaceous and continued as an area of carbonate deposition into the Early Paleogene; carbonate deposition continues even now in Florida Bay and the Florida Keys. Southeast of Florida, across the 85-km-wide Florida Strait, lies the Great Bahama Bank, an area of carbonate deposition from the Cretaceous to the present (see Figure 6.19b).

The Atlantic Continental Margin

The east coast of North America includes the Atlantic Coastal Plain and extends seaward across the continental shelf, slope, and rise (• Figure 16.24). It is a classic example of a passive continental margin. When Pangaea began fragmenting during the Triassic, continental crust rifted, and a new ocean basin began to form. Remember that the North American plate moved westerly, so its eastern margin was within the plate, where a passive continental margin developed.

The Atlantic continental margin has a number of Mesozoic and Cenozoic basins, formed as a result of rifting, in which sedimentation began by Jurassic time. Even though Jurassic-aged rocks have been detected in only a few deep wells, geologists assume they underlie the entire continental margin. The distribution of Cretaceous and Cenozoic rocks is better known, because both are exposed on the Atlantic Coastal Plain, and both have been penetrated by wells on the continental shelf.

Sediments—now sedimentary rocks—on the broad Atlantic Coastal Plain as well as those underlying the continental

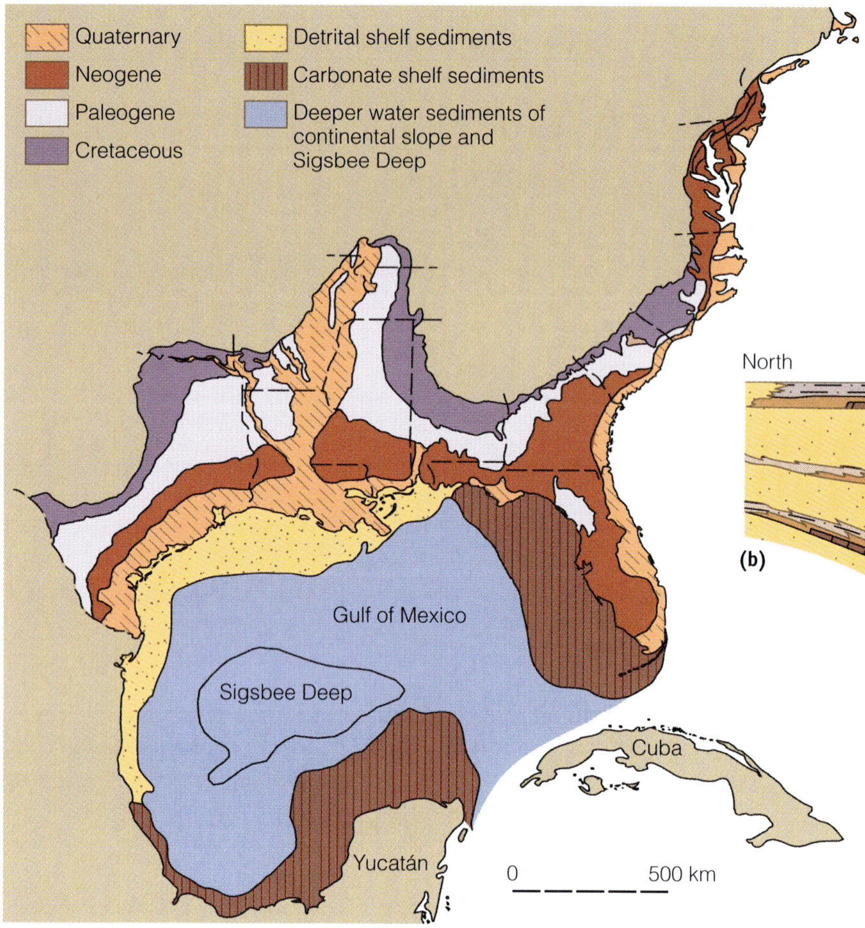

• **Figure 16.23 Cenozoic Deposition on the Gulf Coastal Plain** (a) Areas of deposition in the Gulf of Mexico and surface geology on the Gulf and Atlantic Coastal Plains. (b) Cross section of the Eocene Claiborn Group of the Gulf Coastal Plain, showing facies and the seaward thickening of the deposits.

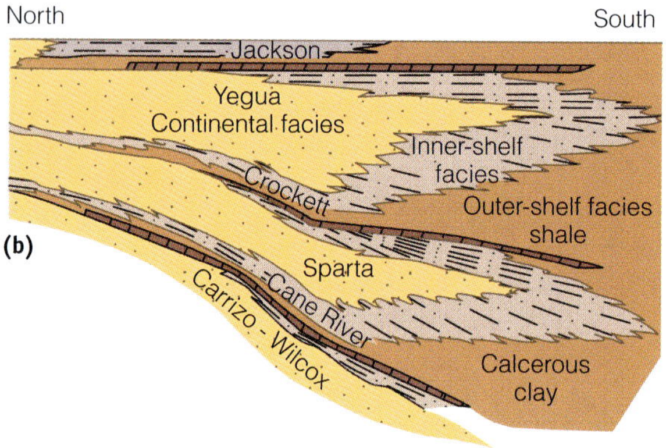

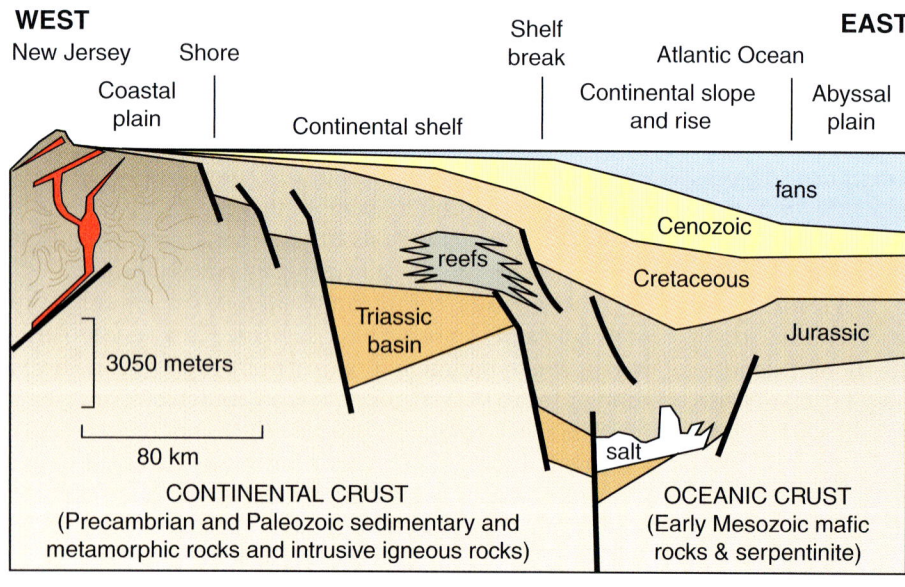

• **Figure 16.24 The Continental Margin in Eastern North America** The coastal plain and continental margin of New Jersey are covered mostly by Cenozoic sandstones and shales. Beneath these rocks lie Cretaceous and probably Jurassic sedimentary rocks.

342 CHAPTER 16 CENOZOIC GEOLOGIC HISTORY: THE PALEOGENE AND NEOGENE

shelf, slope, and rise, were derived from the Appalachian Mountains. Numerous rivers and streams transported sediments toward the east, where they were deposited in seaward thickening wedges (up to 14 km thick) that grade from terrestrial deposits on the west to marine deposits further east. For instance, the Calvert Cliffs in Maryland consist of rocks deposited in marginal marine environments (• Figure 16.25).

An interesting note regarding the geologic evolution of the Atlantic Coastal Plain is the evidence convincing some geologists that a 3- to 5-km-diameter comet or asteroid impact occurred in the present-day area of Chesapeake Bay. This postulated event took place about 35 million years ago, during the Late Eocene, and left an impact crater measuring 85 km in diameter and 1.3 km deep. Now buried beneath 300 to 500 m of younger sedimentary rocks, it has been detected by drilling and geophysical surveys.

Paleogene and Neogene Mineral Resources

The Eocene Green River Formation of Wyoming, Utah, and Colorado, well known for fossils, also contains huge quantities of oil shale and evaporites of economic interest. Oil shale consists of clay particles, carbonate minerals, and an organic compound called *kerogen* from which liquid oil and combustible gases can be extracted. No oil is currently derived from these rocks, but according to one estimate 80 billion barrels of oil could be recovered with present technology. The evaporite mineral *trona* is mined from Green River rocks for sodium compounds.

Mining of phosphorus-rich sedimentary rocks in Central Florida accounts for more than half that state's mineral production (• Figure 16.26). The phosphorus from these rocks has a variety of uses in metallurgy, preserved foods, ceramics, matches, fertilizers, and animal feed supplements. Some of these phosphate rocks also contain interesting assemblages of fossil mammals (see Chapter 18).

Diatomite is a soft, low-density sedimentary rock made up of microscopic shells of diatoms, single-celled marine and freshwater plants with skeletons of silicon dioxide (SiO_2). In fact, diatomite is so porous and light that when dry it will float. Diatomite is used mostly to purify gas and to filter liquids such as molasses, fruit juices, and sewage. The United States leads the world in diatomite production, mostly from Cenozoic deposits in California, Oregon, and Washington.

Historically, most coal mined in the United States (Canada has very little coal) has been Pennsylvanian-aged bituminous coal from mines in Pennsylvania, West Virginia, Kentucky, and Ohio. Now, though, huge deposits of lignite and subbituminous coal in the Northern Great Plains are becoming important resources. These Late

• **Figure 16.25 Sedimentary Rocks of the Atlantic Coastal Plain** These Miocene- and Pliocene-aged sedimentary rocks at Rocky Point in the Calvert Cliffs of Maryland were deposited in marine environments. In addition to fossils of marine microorganisms, the rocks also contain fossil invertebrates, sharks, and marine mammals.

deposits are found. A placer is an accumulation resulting from the separation and concentration of minerals of greater density from those of lesser density in streams or on beaches. The gold in these placers was weathered and eroded from Mesozoic-aged quartz veins in the Sierra Nevada batholith and adjacent rocks (see Chapter 14 Introduction).

Hydrocarbons are recovered from the fault-bounded basins in Southern California and from many rocks of the Gulf Coastal Plain. Many rocks in the latter region form reservoirs for petroleum and natural gas because of different physical properties of the strata, and are thus called *stratigraphic traps*. Hydrocarbons are also found in geologic structures, such as folds, particularly those adjacent to salt domes, and such reservoirs are accordingly called *structural traps*. Because rock salt is a low-density sedimentary rock, when deeply buried and under pressure it rises toward the surface, and in doing so it penetrates and deforms the overlying rocks.

Another potential resource is methane hydrate, which consists of single methane molecules bound up in networks formed by frozen water. Huge deposits of methane hydrate are present along the eastern continental margin of North America, but so far it is not known whether they can be effectively recovered and used as an energy source. According to one estimate, the amount of carbon in methane hydrates worldwide is double that in all coal, oil, and conventional natural gas reserves.

• **Figure 16.26 Neogene Sedimentary Rocks in Florida** Miocene-aged phosphorus-rich rocks in the Bone Valley Member of the Peace River Formation in the IMC Four Corners Mine, Polk County, Florida.

Cretaceous to Early Paleogene-aged coal deposits are most abundant in the Williston and Powder River basins of North Dakota, Montana, and Wyoming. Besides having a low sulfur content, which makes them desirable, some of these coal beds are more than 30 m thick!

Gold from the Pacific Coast states, particularly California, comes largely from stream gravel in which placer

What Would You Do?

According to one estimate, present-day technology is capable of extracting 80 billion barrels of oil from the Eocene-aged Green River Formation in Wyoming, Colorado, and Utah, yet none is currently being produced. Given that U.S. domestic oil production is about 7.6 million barrels per day and we must import about 64% of what we consume, what contribution would the Green River Formation make, assuming its entire potential could be realized? Do you see any problems with your projections? Also, what kinds of economic, technologic, and environmental problems might be encountered?

SUMMARY

- The Late Triassic rifting of Pangaea continued through the Cenozoic and accounts for the present distribution of continents and oceans.
- Cenozoic orogenic activity was concentrated mostly in two major belts: the Alpine–Himalayan orogenic belt and the circum-Pacific orogenic belt. Each belt is composed of smaller units called orogens.
- The Alpine orogeny resulted from convergence of the African and Eurasian plates. Mountain building took place in southern Europe, the Middle East, and North Africa. Plate motions also caused the closure of the Mediterranean basin, which became a site of evaporite deposition.
- India separated from Gondwana, moved north, and eventually collided with Asia, causing deformation and uplift of the Himalayas.
- Orogens characterized by subduction of oceanic lithosphere and volcanism took place in the western and northern Pacific Ocean basin. Back-arc spreading produced back-arc marginal basins such as the Sea of Japan.

- Subduction of oceanic lithosphere occurred along the western margins of the Americas during much of the Cenozoic.
- Subduction continues beneath Central and South America, but the North American plate is now bounded mostly by transform faults, except in the Pacific Northwest.
- The North American Cordillera is a complex mountainous region extending from Alaska into Mexico. Its Cenozoic evolution included deformation during the Laramide orogeny, extensional tectonics that formed the Basin-and-Range structures, intrusive and extrusive igneous activity, and uplift and erosion.
- Shallow angle subduction of the Farallon plate beneath North America resulted in the vertical uplifts of the Laramide orogeny. The Laramide orogen is centered in the middle and southern Rockies, but deformation occurred from Alaska to Mexico.
- Cordilleran volcanism was more or less continuous through the Cenozoic. The Columbia River basalts represent one of the world's greatest eruptive events. Volcanism continues in the Cascade Range of the Pacific Northwest.
- Crustal extension in the Basin and Range Province yielded north–south oriented, normal faults. Differential movement on these faults produced uplifted ranges separated by broad, sediment-filled basins.
- The Colorado Plateau was deformed less than other areas in the Cordillera. Late Neogene uplift and erosion were responsible for the present topography of the region.
- The westward drift of North America resulted in its collision with the Pacific–Farallon ridge. Subduction ceased, and the continental margin became bounded by major transform faults, except where the Juan de Fuca plate continues to collide with North America.
- The Rio Grande rift formed as north–south oriented rifting took place in an area extending from Colorado into Mexico. The basins within this rift filled with sediments and volcanic rocks.
- Sediments eroded from Laramide uplifts were deposited in intermontane basins, on the Great Plains, and in a remnant of the Cretaceous epeiric sea in North Dakota.
- Deposition on the Gulf Coastal Plain and Atlantic Coastal Plain took place throughout the Cenozoic, resulting in seaward-thickening wedges of rocks grading from terrestrial facies to marine facies.
- Cenozoic uplift and erosion were responsible for the present topography of the Appalachian Mountains. Much of the sediment eroded from the Appalachians was deposited on the Atlantic Coastal Plain.
- Paleogene and Neogene mineral resources include oil and natural gas, gold, and phosphorus-rich sedimentary rocks.

IMPORTANT TERMS

Alpine–Himalayan orogenic belt, p. 326
Alpine orogeny, p. 326
Atlantic Coastal Plain, p. 340
back-arc marginal basin, p. 328
Basin and Range Province, p. 333
Cascade Range, p. 333
circum-Pacific orogenic belt, p. 328

Colorado Plateau, p. 334
Farallon plate, p. 335
Gulf Coastal Plain, p. 340
Interior Lowlands, p. 337
Laramide orogeny, p. 329
North American Cordillera, p. 328
orogen, p. 325

Pacific–Farallon ridge, p. 337
Rio Grande rift, p. 335
San Andreas transform fault, p. 337
Tejas epeiric sea, p. 340
Zuni epeiric sea, p. 337

REVIEW QUESTIONS

1. One area of Cenozoic mountain building was in the _____.
   a.____Atlantic coastal plain mountain chain; b.____Alpine–Himalayan orogenic belt; c.____Colorado Plateau–Cascades volcanic arc; d.____Great Plains deformational zone; e.____Pacific-Farallon ridge.
2. Shallow angle subduction of the Farallon plate beneath North America may have been responsible for the
   a.____Hercynian orogeny; b.____Antler orogeny; c.____Caledonian orogeny; d.____Atlas orogeny; e.____Laramide orogeny.
3. A broad area mostly in the United States that was deformed into numerous north–south oriented mountain ranges with intervening valleys is the
   a.____Basin and Range Province; b.____Tejas epeiric sea; c.____circum-Pacific orogenic belt; d.____Alpine convergence zone; e.____Atlantic Coastal Plain.
4. A vast area of overlapping lava flows mostly in Washington state is known as the
   a.____Coast Ranges; b.____San Juan volcanic field; c.____Columbia River basalts; d.____Gulf Coastal Plain; e.____Zuni epeiric sea.
5. The topographic expression of the present-day Appalachian Mountains resulted from
   a.____a collision between the Cocos plate and Central America; b.____volcanism in the Basin and Range Province; c.____intense folding and thrust faulting in the Colorado Plateau; d.____Cenozoic uplift

and erosion in eastern North America; e.____erosion by glaciers.

6. Ongoing subduction of the Juan de Fuca plate beneath North America is the cause of
a.____volcanism in the Cascade Range; b.____uplift of the Central Lowlands; c.____regression of the Tejas epeiric sea; d.____the Alpine orogeny; e.____seismic activity on the Atlantic Coastal Plain.

7. The Himalayas formed when the _____ plate collided with the _____ plate.
a.____North American / Pacific; b.____Nazca / Cocos; c.____Indian / Asian; d.____Farallon / African; e.____Atlantic / Middle Eastern

8. The Cenozoic Era consists of two periods: the _____ and _____.
a.____Miocene and Triassic; b.____Paleogene and Neogene; c.____Jurassic and Cretaceous; d.____Proterozoic and Archean; e.____Cordilleran and Tejas.

9. The _____ formed when the North American plate collided with the Pacific–Farallon ridge.
a.____California back-arc basin; b.____Alpine orogen; c.____San Andreas fault; d.____Interior Lowlands thrust zone; e.____Appalachian Mountains.

10. Which one of the following statements is correct?
a.____The Cenozoic Era began 1.8 million years ago; b.____The Colorado Plateau rocks are mostly lava flows; c.____The Laramide orogeny began during the Neogene Period; d.____The Sea of Japan is a back-arc marginal basin; e.____Epeic seas covered most of North America during the Paleogene.

11. Where is the Cascade Range, what kinds of volcanoes are found there, and what accounts for ongoing volcanism in this area?

12. Describe the sequence of events leading to the origin of the Himalayas of Asia.

13. How did the Laramide orogeny differ from more typical orogenies at convergent plate boundaries?

14. Explain how the San Andreas fault evolved.

15. Where are the Interior Lowlands, and what geologic events took place there during the Cenozoic Era?

16. What features are shared by both the Gulf Coastal Plain and the Atlantic Coastal Plain? Are there any differences between these two areas?

17. Describe the Basin and Range Province, and explain what may have been responsible for its origin.

18. Briefly summarize the events leading to the Alpine orogeny. What kinds of geologic evidence indicate that this region remains tectonically active?

19. Give a brief summary of the events that took place in the North American Cordillera during the Cenozoic Era.

20. How does the Cenozoic geologic history of the Colorado Plateau differ from the history of other parts of the North American Cordillera?

APPLY YOUR KNOWLEDGE

1. The United States uses about 860 million metric tons of coal from its reserve of 243 billion metric tons. Assuming that all of this coal could be mined, how long will it last at the current rate of consumption? Is there any reason to think that the current rate of consumption will be the same in the future, and is it even likely that all of this coal reserve could be mined?

FIELD QUESTION

1. The outcrops in these images show Paleogene rocks in Northern Ireland that are made up of lava flows lying below and above a deep red soil (laterite). How can you explain the events that took place here, including why the lava flows have vertical columns? Can you make any inferences about the ancient climate from the soil?

CHAPTER 17

CENOZOIC GEOLOGIC HISTORY: THE PLEISTOCENE AND HOLOCENE EPOCHS

Peter Essick/Aurora/Getty Images

The Little Ice Age began in about 1500 and lasted into the 1800s. During this time of cooler temperatures, glaciers in Europe and elsewhere extended much farther down their valleys than they do now. This image shows the Unteraar Glacier in Switzerland which is a shrunken remnant of the much larger glacier that occupied this valley during the Little Ice Age.

[OUTLINE]

INTRODUCTION

PLEISTOCENE AND HOLOCENE TECTONISM AND VOLCANISM

PERSPECTIVE 17.1: *Supervolcanoes and the Origin of the Yellowstone Caldera*

PLEISTOCENE STRATIGRAPHY

Terrestrial Stratigraphy

Deep-Sea Stratigraphy

ONSET OF THE ICE AGE

Climate of the Pleistocene

Glaciers—What Are They and How Do They Form?

GLACIATION AND ITS EFFECTS

Glacial Landforms

Changes in Sea Level

Glaciers and Isostasy

Pluvial and Proglacial Lakes

WHAT CAUSED PLEISTOCENE GLACIATION?

The Milankovitch Theory

Short-Term Climatic Changes

GLACIERS TODAY

PLEISTOCENE MINERAL RESOURCES

SUMMARY

ThomsonNOW Explore interactive tutorials, animations, or practice problems available on the ThomsonNow website at **www.thomsonedu.com/login**.

CHAPTER OBJECTIVES

At the end of this chapter, you will have learned that

- The Pleistocene and the Holocene or Recent epochs encompass only the most recent 1.8 million years of Earth history.

- The Pleistocene Epoch, lasting from 1.8 million years to 10,000 years ago, is best known for widespread glaciers but was also a time of continuing orogeny and volcanism.

- Much of our information about Pleistocene climates comes from oxygen isotope ratios, pollen analyses, and the distribution and coiling directions of planktonic foraminifera.

- Pleistocene continental glaciers were present on the Northern Hemisphere continents as well as Antarctica, and thousands of small valley glaciers occupied valleys in mountain ranges on all continents.

- Sea level fell and rose during the several Pleistocene advances and retreats of glaciers, depending on how much water from the ocean was frozen on land.

- The tremendous weight of continental glaciers caused Earth's crust to subside into the mantle and to rise again when the glaciers wasted away.

- Many now-arid regions far from glaciated areas supported large lakes as a result of greater precipitation during the Pleistocene, and numerous other lakes formed along the margins of glaciers.

- A current widely accepted theory explaining the onset of ice ages relies on irregularities in Earth's rotation and orbit.

- Important Pleistocene mineral resources include sand and gravel, diatomite, peat, and placer deposits of gold.

Introduction

The most recent 1.8 million years of geologic time consists of the *Pleistocene Epoch*, better known as the Ice Age, and the *Holocene* or *Recent Epoch* (see Figure 16.1). The Pleistocene Epoch, from 1.8 million to 10,000 years ago, constitutes most of what has traditionally been considered the Quaternary Period and is thus the main focus of this chapter. Recall our analogy of all geologic time represented by a 24-hour clock (see Figure 8.1). In this context the Pleistocene is only 38 seconds long, but they are certainly an important 38 seconds, at least from our perspective, because during this time our species (*Homo sapiens*) evolved (see Chapter 19). Furthermore, the Pleistocene deserves special attention because it is one of the few times in Earth history when vast glaciers were present.

A **glacier** is a body of ice on land that moves as a result of *plastic flow* (internal deformation in response to pressure) and by *basal slip* (sliding over its underlying surface). The most important for our consideration in this chapter are **continental glaciers** (also known as ice sheets) that by definition cover at least 50,000 km² and are unconfined by topography (• Figure 17.1a); that is, they flow outward in all directions from a central point or points of accumulation. An **ice cap** is similar to a continental glacier but covers less than 50,000 km² (Figure 17.1b). **Valley glaciers** (also called alpine glaciers and mountain glaciers) are long, narrow tongues of ice confined to mountain valleys in which they flow from higher to lower elevations (Figure 17.1c).

Geologists have traditionally divided the Cenozoic Era into the *Tertiary Period* and *Quaternary Period*. However, as mentioned in the previous chapter, these terms are falling into disuse. Because a formal decision on the chronostratigraphic status of the Quaternary has not yet been resolved, we place it in the Geologic Time Scale (see Figures 1.9, 4.1, 16.1) as occurring from the late Pliocene to the present.

In hindsight, it is difficult to believe that many naturalists of 165 years ago refused to accept the evidence indicating that widespread glaciation had occurred during the recent past. Many invoked the biblical deluge to explain the large boulders throughout Europe far from their source, whereas others thought the boulders were rafted by ice during vast floods. In 1837 Swiss naturalist Louis Agassiz argued convincingly that the large, displaced boulders, as well as polished and striated bedrock and U-shaped valleys found throughout Europe and elsewhere resulted from huge masses of ice moving over the land (Figure 17.1).

We now know that the Ice Age was a time of several intervals of glacial expansion separated by warmer interglacial periods. Furthermore, during glacial expansions more precipitation fell in regions now arid, such as the Sahara Desert of North Africa and Death Valley in California, both of which supported lush vegetation, streams, and lakes. An unresolved question is whether the Ice Age is truly over or whether we are in an interglacial period that will be followed by renewed glaciation.

Climatic fluctuations have occurred since the Pleistocene, the most recent significant one being the Little Ice Age from about A.D. 1500 until sometime in the 1800s (see the chapter opening photo). During the Little Ice Age glaciers in mountain valleys expanded, and the summers were cooler and wetter with shorter growing seasons. In Europe and Iceland, glaciers reached their greatest historic extent by the early 1800s, and glaciers in the western United States, Alaska, and Canada also expanded.

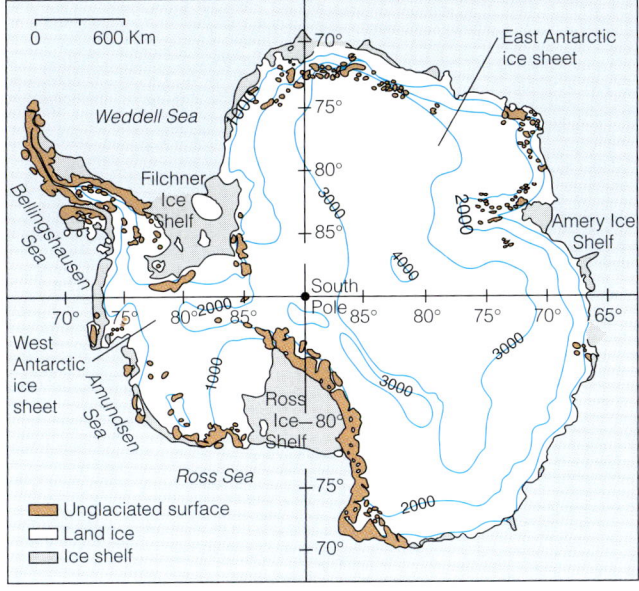

(a)

(b)

(c)

• **Figure 17.1 Types of Glaciers (a)** Two continental glaciers called the East and West Antarctic ice sheets merge to form a nearly continuous ice cover that averages 2160 m thick. The blue lines are lines of equal thickness. **(b)** The Penny Ice Cap on Baffin Island, Canada, covers about 6000 km². **(c)** A valley glacier such as this one in Alaska is a long, narrow tongue of moving ice confined to a mountain valley.

Pleistocene and Holocene Tectonism and Volcanism

The Pleistocene Epoch is best known for glaciation, but it was also a time of volcanism and tectonic activity. For instance, continuing orogeny took place in the Himalayas of Asia and the Andes Mountains in South America, and deformation at convergent plate boundaries proceeded unabated in the Aleutian Islands, Japan, the Philippines, and elsewhere. Interactions between the North American and Pacific plates along the San Andreas transform plate boundary produced folding, faulting, and a number of basins and uplifts. Marine terraces covered with Pleistocene sediments attest to periodic uplift in southern California (• Figure 17.2a).

Archean Eon	Proterozoic Eon	Phanerozoic Eon							
Precambrian		Paleozoic Era							
		Cambrian	Ordovician	Silurian	Devonian	Mississippian	Pennsylvanian	Permian	
						Carboniferous			

2500 MYA — 542 MYA — 251 MYA

17.1 Perspective

Supervolcanoes and the Origin of the Yellowstone Caldera

Geologists have no formal definition for the term *supervolcano,* but we can take it to mean an eruption resulting in the explosive ejection of hundreds of cubic kilometers of pyroclastic materials and the origin of a huge caldera. Fortunately, such voluminous eruptions are rare. Two supervolcanoes of particular interest in North America are the ones that yielded the Yellowstone Caldera, Wyoming, and the Long Valley Caldera, California (Table 1). In both cases, continuing earthquakes and hydrothermal activity (hot springs and geysers) remind us that renewed volcanism in these areas is possible.

The Yellowstone Caldera lies within the confines of Yellowstone National Park, the first area in the United States set aside as a national park. This area is noted for its scenery, wildlife, boiling mud pots, hot springs, and geysers, especially Old Faithful, but few tourists are aware of the region's volcanic history. The fact that a large body of magma is still present beneath the surface is well accepted by geologists, many of whom are convinced the Yellowstone Caldera remains active.

On three separate occasions, supervolcano eruptions followed accumulation of rhyolitic magma beneath the surface of the region. Each eruption yielded a widespread blanket of volcanic ash and pumice as well as collapse of the surface and origin of a large caldera. We can summarize Yellowstone's volcanic history by noting that supervolcano eruptions took place 2 million years ago, 1.3 million years ago, and 600,000 years ago, each responsible for the explosive ejection of hundreds to thousands of cubic kilometers of pyroclastic materials (Table 1). It was during this last huge eruption that the present-day Yellowstone Caldera originated. Actually, the present caldera is part of a larger composite caldera that resulted from the three cataclysmic events just noted (Figure 1a).

The magnitude of the supervolcano eruptions yielding the three calderas is difficult to imagine. In fact, the pyroclastic flows and airborne ash cover not only the areas within and adjacent to the Yellowstone region but are also found over a large part of the western United States and northern Mexico. Nothing in the immediate areas of the eruptions could have survived the intense heat and choking clouds of volcanic ash.

Since the last supervolcano eruption, rhyolitic magma continued to accumulate beneath the caldera, elevating part of its floor

TABLE 1 Supervolcano Eruptions

Name	Location	Volume of Material Erupted	Size of Caldera	Last Erupted
Long Valley Caldera	California	600 km³	15 x 30 km	760,000 YA
Yellowstone Caldera	Wyoming	1000-2000	45 x 76	600,000 YA
Toba Caldera	Sumatra	1500	30 x 100	74,000 YA
La Garita Caldera	Colorado	5000	35 x 75	27.8 MYA
Cerro Galán Caldera	Argentina	2000	25 x 35	2.2 MYA

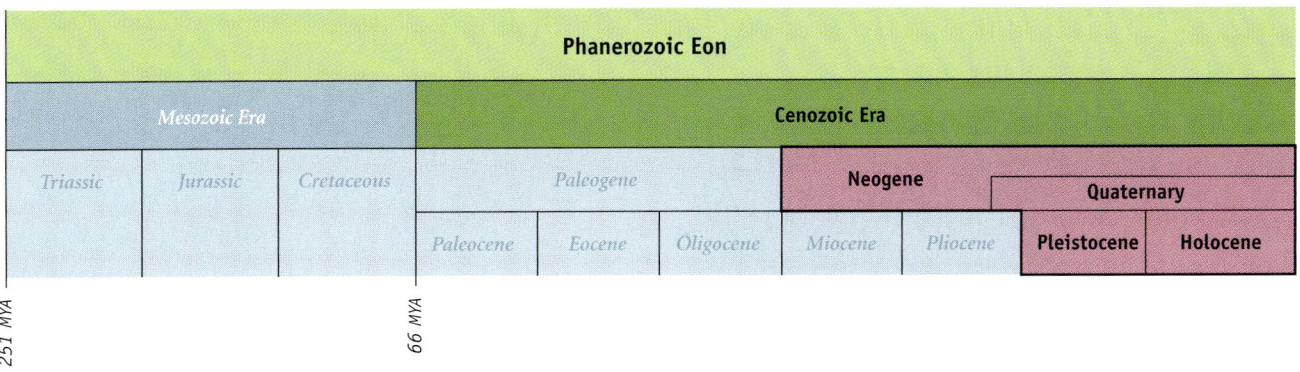

in what is known as a *resurgent dome*. Between 150,000 and 75,000 years ago an additional 1000 km³ of pyroclastic materials were erupted, but these pyroclastic layers are confined to the caldera. An excellent example is the Yellowstone Tuff, into which the picturesque Grand Canyon of the Yellowstone River is incised (Figure 1b). Precise leveling indicates the caldera floor continues to rise as magma moves beneath the surface; more than 80 cm of uplift has taken place since 1923.

Many geologists are convinced that a *mantle plume*, a cylindrical mass of magma rising from the mantle, underlies the Yellowstone region. As this rising magma nears the surface, it triggers volcanic eruptions, and because the magma is rhyolitic and thus viscous the eruptions are particularly explosive. The Yellowstone hot spot, as it is called, is one of only a few dozen hot spots recognized on Earth.

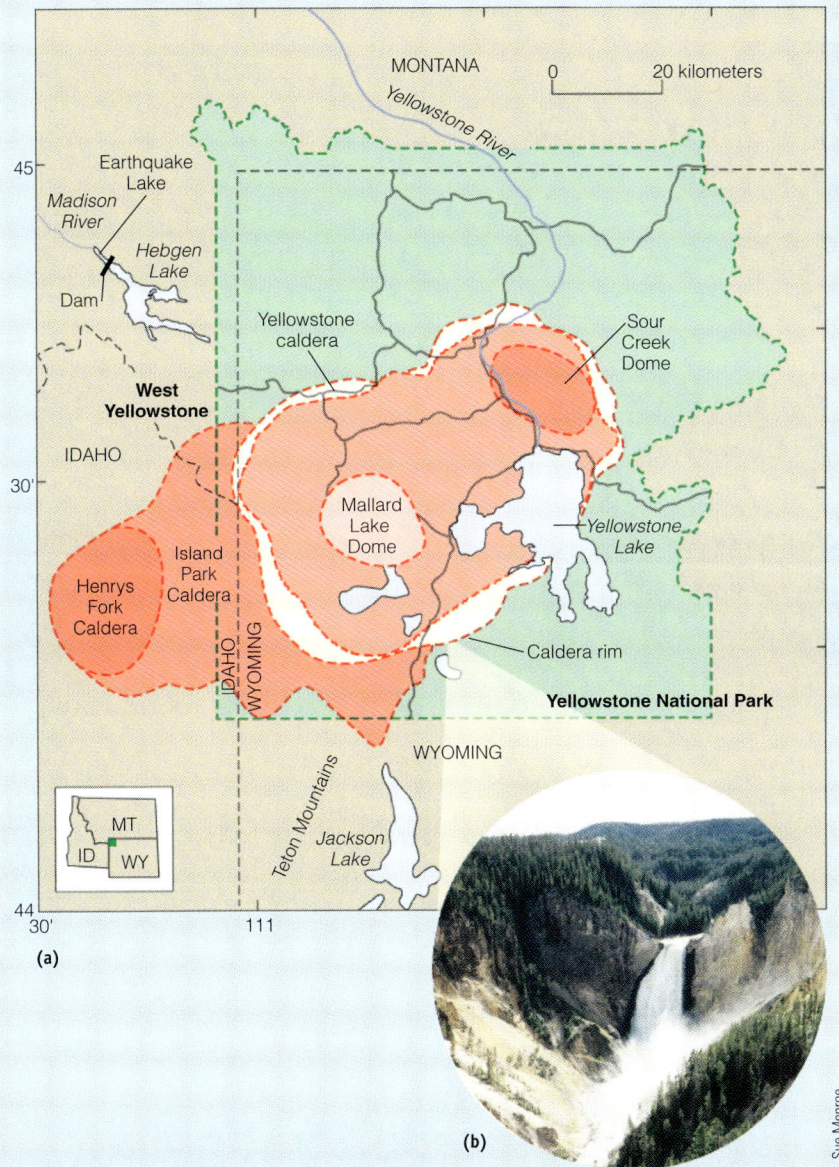

Figure 1 Pleistocene Supervolcano Eruptions (a) This huge caldera is a composite made up of the Henry's Fork, Island Park, and Yellowstone calderas. The Yellowstone caldera (shown in yellow) has been partly filled with younger volcanic rocks. (b) The walls of the Grand Canyon of the Yellowstone River are made up of the hydrothermally altered Yellowstone Tuff that partly fills the Yellowstone caldera.

John Shelton

(a)

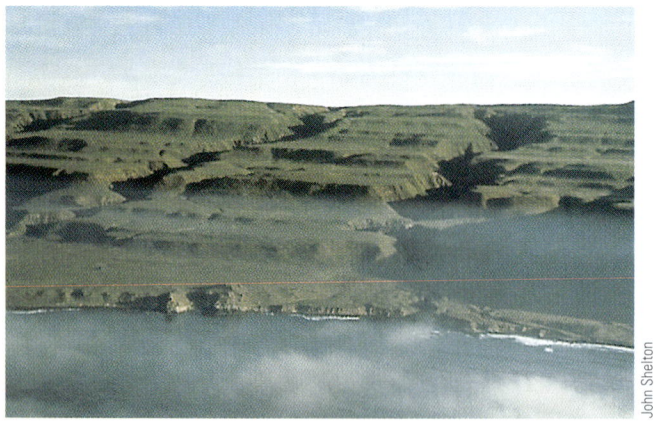

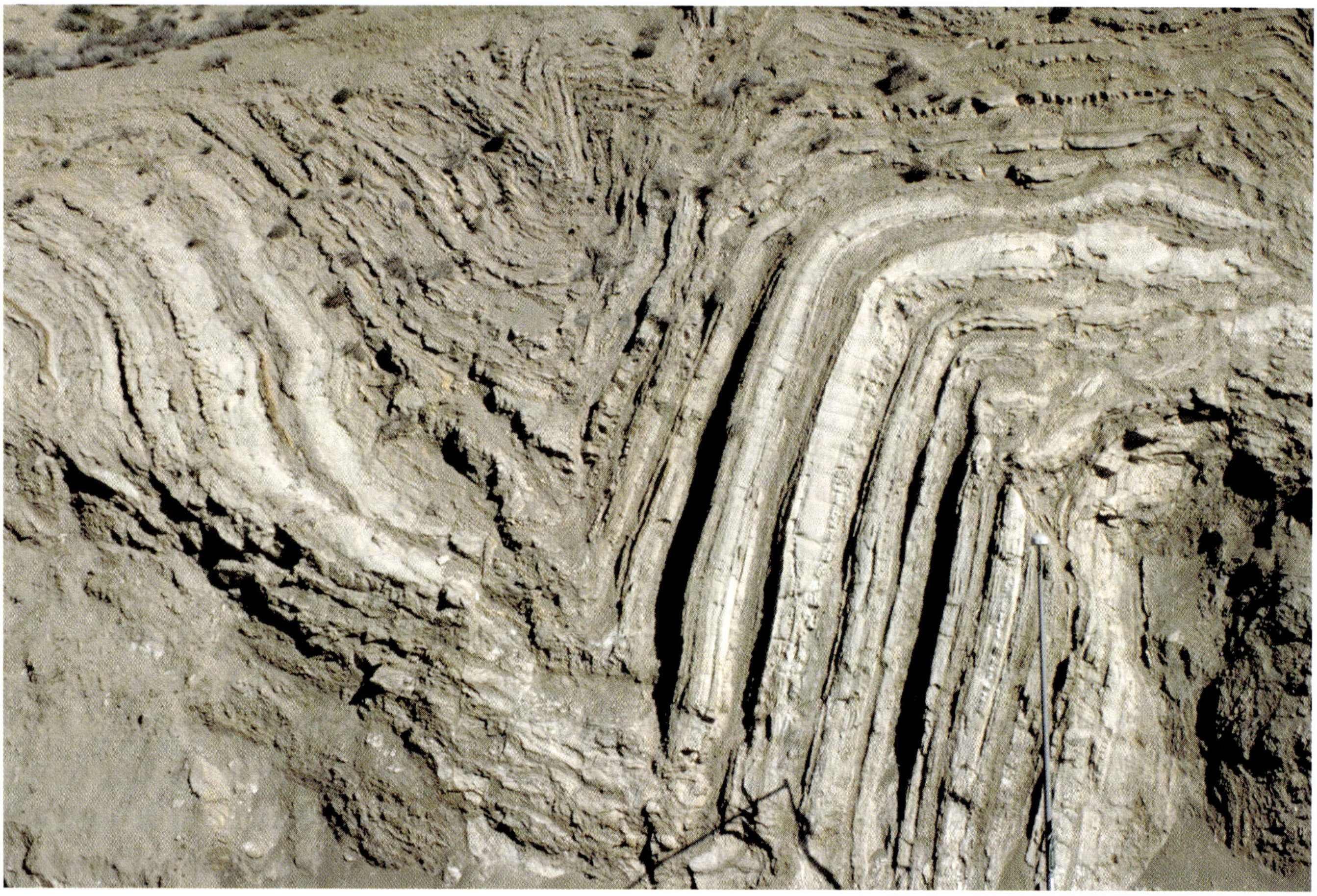

Marli Bryant Miller

(b)

• **Figure 17.2 Pleistocene Uplift and Tectonism (a)** Marine terraces on San Clemente Island, California. Each terrace represents a time when the area was at sea level. The highest terrace is now about 400 m above sea level. **(b)** These deformed Pliocene-aged rocks are only a few hundred meters from the San Andreas fault in southern California.

Ongoing subduction of remnants of the Farallon plate beneath Central America and the Pacific Northwest account for volcanism in these two areas. The Cascade Range of California, Oregon, Washington, and British Columbia has a history dating back to the Oligocene, but the large volcanoes now present formed during the last 1.8 million years (• Figure 17.3a). Volcanism also occurred in many other areas in the western United States, including Idaho, Arizona, and California (Figure 17.3b). Following colossal eruptions, huge calderas formed in the area of Yellowstone National Park, Wyoming (see Perspective 17.1). Elsewhere, volcanoes erupted in South America, Japan, the Philippines, and East Indies, as well as in Iceland, Spitzbergen, and the Azores.

Pleistocene Stratigraphy

Although geologists still debate which rocks should serve as the Pleistocene stratotype,* they agree the Pleistocene

*Recall from Chapter 5 that a stratotype is a section of rocks where a named stratigraphic unit such as a system or series was defined—for example, the stratotype for the Cambrian system.

• **Figure 17.3 Pleistocene and Recent Volcanism** (a) Some of the volcanoes in the Cascade Range in central Oregon. Mount Bachelor at 11,000 to 15,000 years old is the youngest volcano in the range. (b) View of the Cima volcanic field in Mojave National Preserve in California which was active between 7.6 million (Late Miocene) and 10,000 years ago (Late Pleistocene). Basalt lava flows and about 40 cinder cones are present here.

PLEISTOCENE STRATIGRAPHY

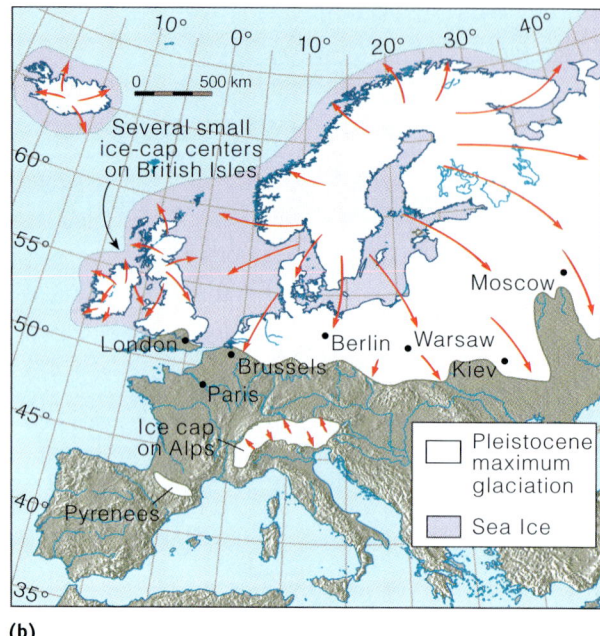

• **Figure 17.4 Areas Covered by Pleistocene Glaciers (a)** Centers of ice accumulation and extent of glaciers in North America. **(b)** Centers of ice accumulation and directions of ice movement (red arrows) during the maximum extent of glaciers in Europe.

Epoch began 1.8 million years ago. The Pleistocene–Holocene boundary at 10,000 years ago is based on climatic change from cold to warmer conditions concurrent with the melting of the most recent ice sheets. Changes in vegetation, as well as oxygen isotope ratios determined from shells of marine organisms, provide ample evidence for this climatic change.

Terrestrial Stratigraphy

Soon after Louis Agassiz proposed his theory for glaciation, research focused on deciphering the history of the Ice Age. This work involved recognizing and mapping terrestrial glacial features and placing them in a stratigraphic sequence. From such glacial features as the distribution of moraines, erratic boulders, and glacial striations, geologists have determined that at their greatest extent Pleistocene glaciers as much as 3 km thick covered about three times as much of Earth's surface as they do now, or about 45,000,000 km^2 (• Figure 17.4). Furthermore, detailed mapping of glacial features reveals that several glacial advances and retreats occurred.

By mapping the distribution of glacial deposits, geologists have determined that North America has had at least four major episodes of Pleistocene glaciation. Each glacial advance was followed by a glacial retreat and warmer climates. The four **glacial stages**, the *Wisconsinan, Illinoian, Kansan,* and *Nebraskan*, are named for the states representing the southernmost advance where deposits are well exposed. The three **interglacial stages**, the *Sangamon, Yarmouth,* and *Aftonian*, are named for localities of well-exposed interglacial soil and other deposits (• Figure 17.5). Recent detailed studies of glacial deposits indicate, however, that there were an as yet undetermined number of pre-Illinoian glacial events and that the history of glacial advances and retreats in North America is more complex than previously thought.

Six or seven major glacial advances and retreats are recognized in Europe, and at least 20 major warm–cold cycles can be detected in deep-sea cores. Why isn't there better correlation among the different areas if glaciation was so widespread? Part of the problem is that glacial deposits are typically chaotic mixtures of coarse materials that are difficult to correlate. Furthermore, glacial advances and retreats usually destroy the sediment left by the previous advances, obscuring older evidence. Even within a single major glacial advance, several minor advances and retreats may have occurred. For example, careful study of deposits from the Wisconsinan glacial stage reveals at least four distinct fluctuations of the ice margin during the last 70,000 years in Wisconsin and Illinois.

Deep-Sea Stratigraphy

Until recently, the traditional view of Pleistocene chronology was based on sequences of glacial sediments on land. During the early 1960s, however, new evidence from ocean sediment samples indicated numerous climatic fluctuations during the Pleistocene. Evidence for these climatic fluctuations comes from changes in surface ocean temperature recorded in the shells of planktonic foraminifera, which after they die sink to the seafloor and accumulate as sediment.

One way to determine past changes in ocean surface temperatures is to resolve whether planktonic foraminifera were warm- or cold-water species. Many are sensitive to

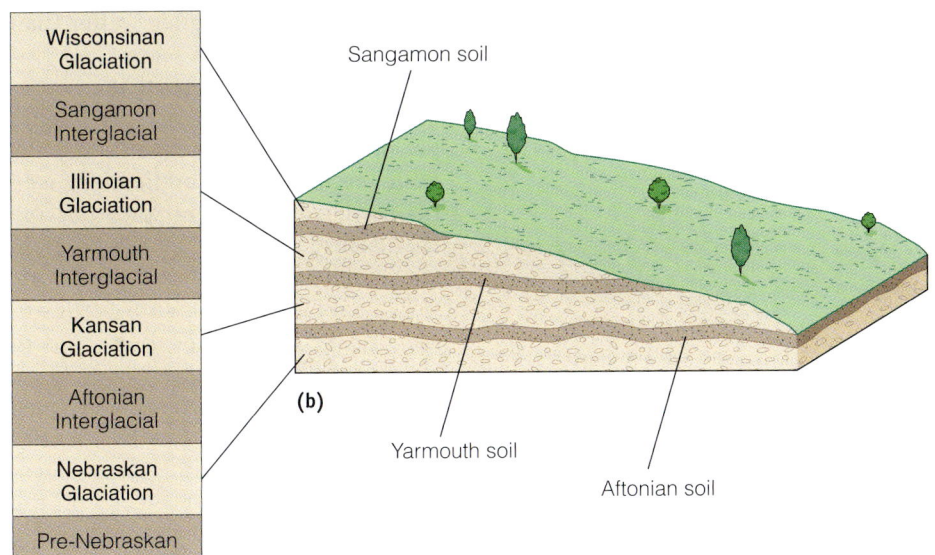

• **Figure 17.5 Pleistocene Glaciers in North America (a)** Traditional terminology for Pleistocene glacial and interglacial stages in North America. **(b)** Idealized succession of deposits and soils developed during the glacial and interglacial stages.

variations in temperature and migrate to different latitudes when the surface water temperature changes. For example, the tropical species *Globorotalia menardii* is present or absent within Pleistocene sediment samples, depending on what the surface water temperature was at the time. During periods of cooler climate, it is found only near the equator, whereas during times of warming its range extends into the higher latitudes.

Some planktonic foraminifera species change the direction they coil during growth in response to temperature fluctuations. The Pleistocene species *Globorotalia truncatulinoides* coils predominantly to the right in water temperatures above 10°C but coils mostly to the left in water below 8°–10°C. On the basis of changing coiling ratios, geologists have constructed detailed climatic curves for the Pleistocene and earlier epochs.

Changes in the O^{18}-to-O^{16} ratio in the shells of planktonic foraminifera also provide data about climatic events. The abundance of these two oxygen isotopes in their calcareous ($CaCO_3$) shells is a function of the oxygen isotope ratio in water molecules and water temperature when the shell forms. The ratio of these two isotopes reflects the amount of ocean water stored in glacial ice. Seawater has a higher O^{18}-to-O^{16} ratio than glacial ice, because water containing the lighter O^{16} isotope is more easily evaporated than water containing the O^{18} isotope. Therefore, Pleistocene glacial ice was enriched in O^{16} relative to O^{18}, whereas the heavier O^{18} isotope is concentrated in seawater. The declining percentage of O^{16} and consequent rise of O^{18} in seawater during times of glaciation is preserved in the shells of planktonic foraminifera. Consequently, oxygen isotope fluctuations indicate surface water temperature changes and thus climatic changes.

Unfortunately, geologists have not yet been able to correlate these detailed climatic changes with corresponding changes recorded in the sedimentary record on land. The time lag between the onset of cooling and any resulting glacial advance produces discrepancies between the marine and terrestrial records. Thus, it is unlikely that all the minor climatic fluctuations recorded in deep-sea sediments will ever be correlated with continental deposits.

Onset of the Ice Age

Glacial conditions actually set in about 40 million years ago when surface ocean waters at high southern latitudes rapidly cooled, and the water in the deep ocean became much colder than it was previously. The gradual closure of the Tethys Sea during the Oligocene limited the flow of warm water to higher latitudes, and by Middle Miocene time an Antarctic ice sheet had formed, accelerating the formation of very cold oceanic waters. After a brief Pliocene warming trend, continental glaciers began forming in the Northern Hemisphere about 1.8 million years ago—the Pleistocene Ice Age was underway.

Climate of the Pleistocene

The climatic conditions leading to Pleistocene glaciation were, as you would expect, worldwide. Contrary to popular belief and depictions in cartoons and movies, Earth was not as cold as commonly portrayed. In fact, evidence of various kinds indicates the world's climate cooled gradually from Eocene through Pleistocene time (• Figure 17.6). Oxygen isotope ratios (O^{18} to O^{16}) from deep-sea cores reveal that during the last 2 million years Earth has had 20 major warm–cold cycles during which the temperature fluctuated by as much as 10 degrees C (see the section on deep-sea stratigraphy). And studies of glacial deposits

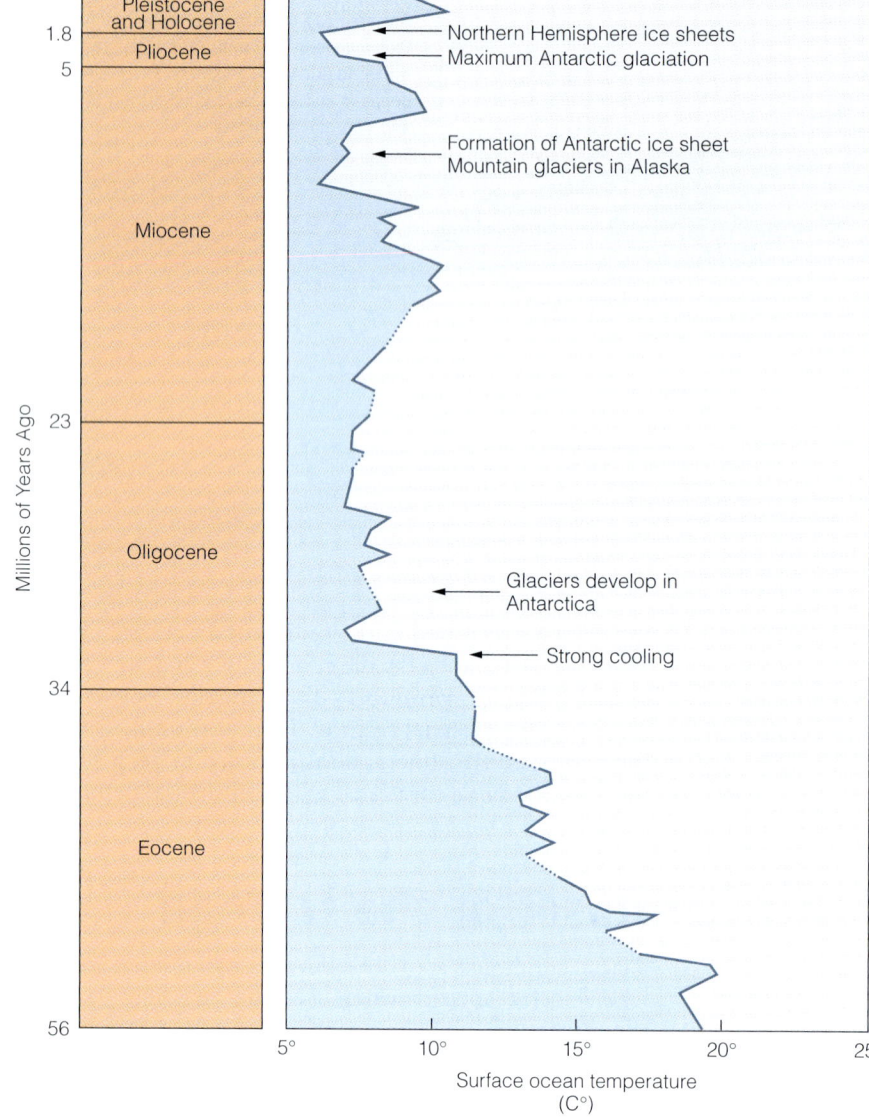

• **Figure 17.6 Oxygen Isotope Ratios and Climate** Fluctuations in O^{18}-to-O^{16} isotope ratios from a sediment core in the western Pacific Ocean reveal changes in ocean surface temperatures during the last 56 million years. A change from warm surface waters to colder conditions took place about 32 million years ago.

attest to at least four major episodes of glaciation in North America and six or seven similar events in Europe.

During glacial growth, those areas covered by or near glaciers experienced short, cool summers and long, wet winters. Areas distant from glaciers had varied climates. When glaciers grew and advanced, lower ocean temperatures reduced evaporation rates, so most of the world was drier than now. Some areas now arid were much wetter during the Ice Age. For instance, the expansion of the cold belts at high latitudes compressed the temperate, subtropical, and tropical zones toward the equator. Consequently, the rain that now falls on the Mediterranean then fell farther south on the Sahara of North Africa, enabling lush forests to grow in what is now desert. In North America a high-pressure zone over the northern ice sheets deflected storms south, so the arid Southwest was much wetter than today.

Pollen analysis is particularly useful in paleoclimatology (• Figure 17.7). Pollen grains, produced by the male reproductive bodies of seed plants, have a resistant waxy coating that ensure many will be preserved in the fossil record. Most seed plants disperse pollen by wind, so it settles in streams, lakes, swamps, bogs, and in nearshore marine environments. Once paleontologists recover pollen from sediments, they can usually identify the type of plant it came from, determine the floral composition of the area, and make climatic inferences (Figure 17.7).

Pollen diagrams (Figure 17.7b), tree-ring analysis (see Chapter 4), and studies of the advances and retreats of valley glaciers have yielded a wealth of information about the Northern Hemisphere climate for the last 10,000 years— that is, since the time the last major continental glaciers retreated and disappeared. Data from pollen analysis indicate a continuous trend toward a warmer climate until about 6000 years ago. In fact, between 8000 to 6000 years ago temperatures were very warm. Then the climate became cooler and moister, favoring the growth of valley glaciers on the Northern Hemisphere continents. Three episodes of glacial expansion took place during this **neoglaciation,** as it is called. The most recent one, the **Little Ice Age** between 1500 and the mid- to late 1800s, was a time of generally cooler temperatures, glacial expansion, and cooler, wetter summers. It had a profound effect on the social and economic fabric of human society, accounting for several famines as well as migrations of many Europeans to the New World.

Glaciers—What Are They, and How Do They Form?

We have already defined the terms *continental glacier, ice cap,* and *valley glacier* as moving bodies of ice on land (see the Introduction) (Figure 17.1). During the Pleistocene, all types of glaciers were much more widespread than now. For example, the only continental glaciers today are the ones in

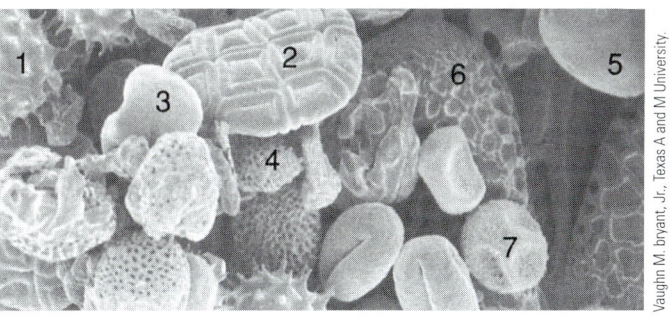

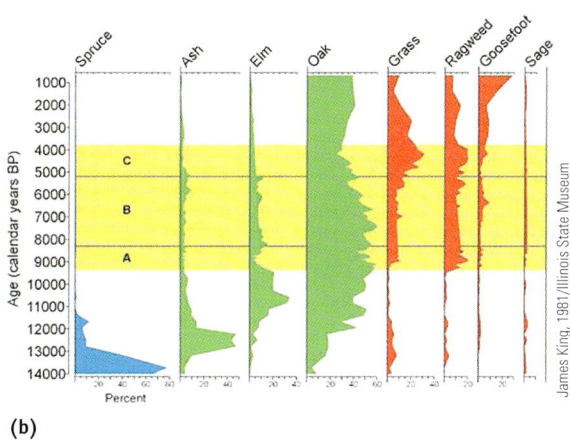

• **Figure 17.7 Pollen and Climate (a)** Scanning electron microscope view of present-day pollen: (1) sunflower, (2) acacia, (3) oak, (4) white mustard, (5) little walnut, (6) agave, and (7) juniper. **(b)** Pollen diagrams and the climate for the last 14,000 years for Chatsworth Bog in Livingston County, Illinois.

Antarctica and Greenland (• Figure 17.8), but during the Pleistocene they covered about 30% of Earth's land surface, especially on the Northern Hemisphere continents. These continental glaciers formed, advanced, and then retreated several times, forming much of the present topography of the glaciated regions and nearby areas. The Pleistocene was also a time when small valley glaciers were more common in mountain ranges. Indeed, much of the spectacular scenery in such areas as Grand Teton National Park, Wyoming, resulted from erosion by valley glaciers (see Perspective 8.1).

The question "How do glaciers form?" is rather easily answered, unlike "What causes the onset of an ice age?" Any area that receives more snow in the cold season than melts in the warm season has a net accumulation over the years. As accumulation takes place, the snow at depth is converted to glacial ice and when a critical thickness of about 40 m is reached, flow in response to pressure begins. Once a glacier forms, it moves from a zone of accumulation, where additions exceed losses, toward its zone of wastage, where losses exceed additions. As long as a balance exists between the two, the glacier has a *balanced budget*, but the budget may be *negative* or *positive*, depending on any imbalances that exist in these two zones. Consequently, a glacier's terminus may advance, retreat, or remain stationary, depending on its budget.

Glaciation and Its Effects

Huge glaciers moving over Earth's surface reshaped the previously existing topography and yielded many distinctive glacial landforms. And as glaciers formed and wasted away, sea level fell and rose, depending on how much water was frozen on land, and the continental margins were alternately exposed and water covered. In addition, the climatic changes that initiated glacial growth had effects far beyond the glaciers themselves. Another legacy of the Pleistocene is that areas once covered by thick glaciers are still rising as a result of isostatic rebound.

Glacial Landforms

Remember that glaciers by definition are moving masses of ice on land, and as such continental and valley glaciers yield a number of easily recognized erosional and depositional landforms. A large part of Canada and parts of some northern states have subdued topography, little or no soil, striated and polished bedrock exposures, and poor surface drainage, characteristics of an **ice-scoured plain** (• Figure 17.9a). Pleistocene valley glaciers also yielded several distinctive

What Would You Do?

As a local expert on geology, you are asked what evidence other than glacial deposits indicated that Earth had a number of cold–warm cycles during the Pleistocene. How would you respond?

systems approach to Earth history, and here we have an excellent opportunity to see interactions among systems at work. Cape Cod, Massachusetts, is a distinctive landform that resembles a human arm extending into the Atlantic Ocean. It and nearby Martha's Vineyard and Nantucket Island owe their existence to deposition by Pleistocene glaciers and modification of these deposits by wind-generated waves and nearshore currents (• Figure 17.12).

Changes in Sea Level

Today, between 28 and 35 million km^3 of water is frozen in glaciers, all of which came from the oceans. During the maximum extent of Pleistocene glaciers, though, more than 70 million km^3 of ice was present on the continents. These huge masses of ice themselves had a tremendous impact on the glaciated areas (see the next section), and they contained enough frozen water to lower sea level by 130 m. Accordingly, large areas of today's continental shelves were exposed and quickly blanketed by vegetation. In fact, at the Bering Strait, Alaska connected with Siberia via a broad land bridge across which Native Americans and various mammals such as the bison migrated (• Figure 17.13). The shallow floor of the North Sea was also above sea level so Great Britain and mainland Europe formed a single landmass. When the glaciers melted, these areas were flooded, drowning the plants and forcing the animals to migrate.

Lower sea level during the several Pleistocene glacial intervals also affected the *base level*, the lowest level to which running water can erode, of rivers and streams flowing into the oceans. As sea level dropped, rivers eroded deeper valleys and extended them across the emergent continental shelves. During times of lower sea level, rivers transported huge quantities of sediment across the exposed continental shelves and onto the continental slopes, where the sediment contributed to the growth of submarine fans. As the glaciers melted, however, sea level rose, and the lower ends of these river valleys along North America's East Coast were flooded, whereas those along the West Coast formed impressive submarine canyons.

What would happen if the world's glaciers all melted? Obviously, the water stored in them would return to the oceans, and sea level would rise about 70 m. If this were to happen, many of the world's large population centers would be flooded.

• **Figure 17.8 The Greenland Ice Sheet** Greenland is mostly covered by a continental glacier that is more than 3000 m thick. Notice that only a few high mountains are not ice covered. Continental glaciers were much more widespread during the Pleistocene, but today only Greenland and Antarctica have continental glaciers.

landforms such as bowl-shaped depressions on mountainsides known as **cirques** and broad valleys called **U-shaped glacial troughs** (Figure 17.9b).

The most important deposits of continental and valley glaciers are various **moraines**, which are chaotic mixtures of poorly sorted sediment deposited directly by glacial ice, and **outwash** consisting of stream-deposited sand and gravel (• Figure 17.10). Any *moraine* deposited at a glacier's terminus is an *end moraine*, but notice from Figure 17.10 that *terminal* and *recessional moraines* are types of end moraines. Terminal moraines and outwash in southern Ohio, Indiana, and Illinois mark the greatest southerly extent of Pleistocene continental glaciers in the midcontinent region. Recessional moraines indicate various positions where the ice front stabilized temporarily during a general retreat to the north (• Figure 17.11).

Glaciers are of course made up of frozen water and thus constitute an important part of the hydrosphere, one of Earth's major systems. In Chapter 1 we emphasized the

Glaciers and Isostasy

In a manner of speaking, Earth's crust floats on the denser mantle below, a phenomenon geologists call **isostasy**. An analogy can help you understand this concept, which is certainly counterintuitive; after all, how can rock float in rock? Consider an iceberg. Ice is slightly less dense than water, so an iceberg sinks to its equilibrium position in water with

No one doubts that Earth's crust subsided from the great weight of glaciers during the Pleistocene or that it has rebounded and continues to do so in some areas. Indeed, the surface in some places was depressed as much as 300 m below preglacial elevations. But as the glaciers melted and eventually wasted away, the downwarped areas gradually rebounded to their former positions. Evidence of isostatic rebound can be found in formerly glaciated areas such as Scandinavia and the North American Great Lakes Region (• Figure 17.14). Some coastal cities in Scandinavia have rebounded enough so that docks built only a few centuries ago are now far inland from the shore. And in Canada as much as 100 m of isostatic rebound has taken place during the last 6000 years.

(a)

(b)

• **Figure 17.9 Erosion by Continental and Valley Glaciers Yields Distinctive Landscapes** (a) A continental glacier eroded this ice-scoured plain, a subdued surface with extensive bedrock exposures, in the Northwest Territories of Canada. (b) Valley glaciers erode mountains and leave sharp, angular peaks and ridges and broad, smooth valleys as seen here in the Chigmit Mountains in Alaska.

only about 10% of its volume above the surface. Earth's crust is a bit more complicated, but it sinks into the mantle, which behaves like a fluid, until it reaches its equilibrium position depending on its thickness and density. Remember, oceanic crust is thinner but denser than continental crust, which varies considerably in thickness.

If the crust has more mass added to it, as when thick layers of sediment accumulate or vast glaciers form, it sinks lower into the mantle until it once again achieves equilibrium. However, if erosion or melting ice reduces the load, the crust slowly rises by **isostatic rebound.** Think of the iceberg again. If some of it above sea level were to melt, it would rise in the water until it regained equilibrium.

Pluvial and Proglacial Lakes

During the Wisconsinan glacial stage, many now arid parts of the western United States supported large lakes when glaciers were present far to the north. These **pluvial lakes,** as they are called, existed because of the greater precipitation and overall cooler temperatures, especially during the summer, which lowered the evaporation rate. Wave-cut cliffs, beaches, deltas, and various lake deposits along with fossils of freshwater organisms attest to the presence of these lake (• Figure 17.15). Lake Bonneville, with a maximum size of about 50,000 km^2 and at least 335 m deep, was a large pluvial lake; the vast salt deposits of the Bonneville Salt Flats west of Salt Lake City, Utah, formed when parts of this ancient lake dried up. The present Great Salt Lake is a shrunken remnant of this once much larger lake.

Death Valley on the California–Nevada border is the hottest, driest place in North America, yet during the Wisconsinan it supported Lake Manly, another large pluvial lake (• Figure 17.15). It was 145 km long, nearly 180 m deep, and when it dried up dissolved salts precipitated on the valley floor. Borax, one of the minerals in these lake deposits, is mined for use in ceramics, fertilizers, glass, solder, and pharmaceuticals.

In contrast to pluvial lakes, which are far from areas of glaciation, **proglacial lakes** form where meltwater accumulates along a glacier's margin. Lake Agassiz, named in honor of the French naturalist Louis Agassiz, was a proglacial lake formed in this manner. It covered about 250,000 km^2 in North Dakota, Manitoba, Saskatchewan, and Ontario and persisted until the ice along its northern margin melted, at which time it drained northward into Hudson Bay.

Deposits in lakes adjacent to or near glaciers vary considerably from gravel to mud, but of special interest are the finely laminated mud deposits consisting of alternating dark and light layers. Each dark-light couplet is a

GLACIATION AND ITS EFFECTS **359**

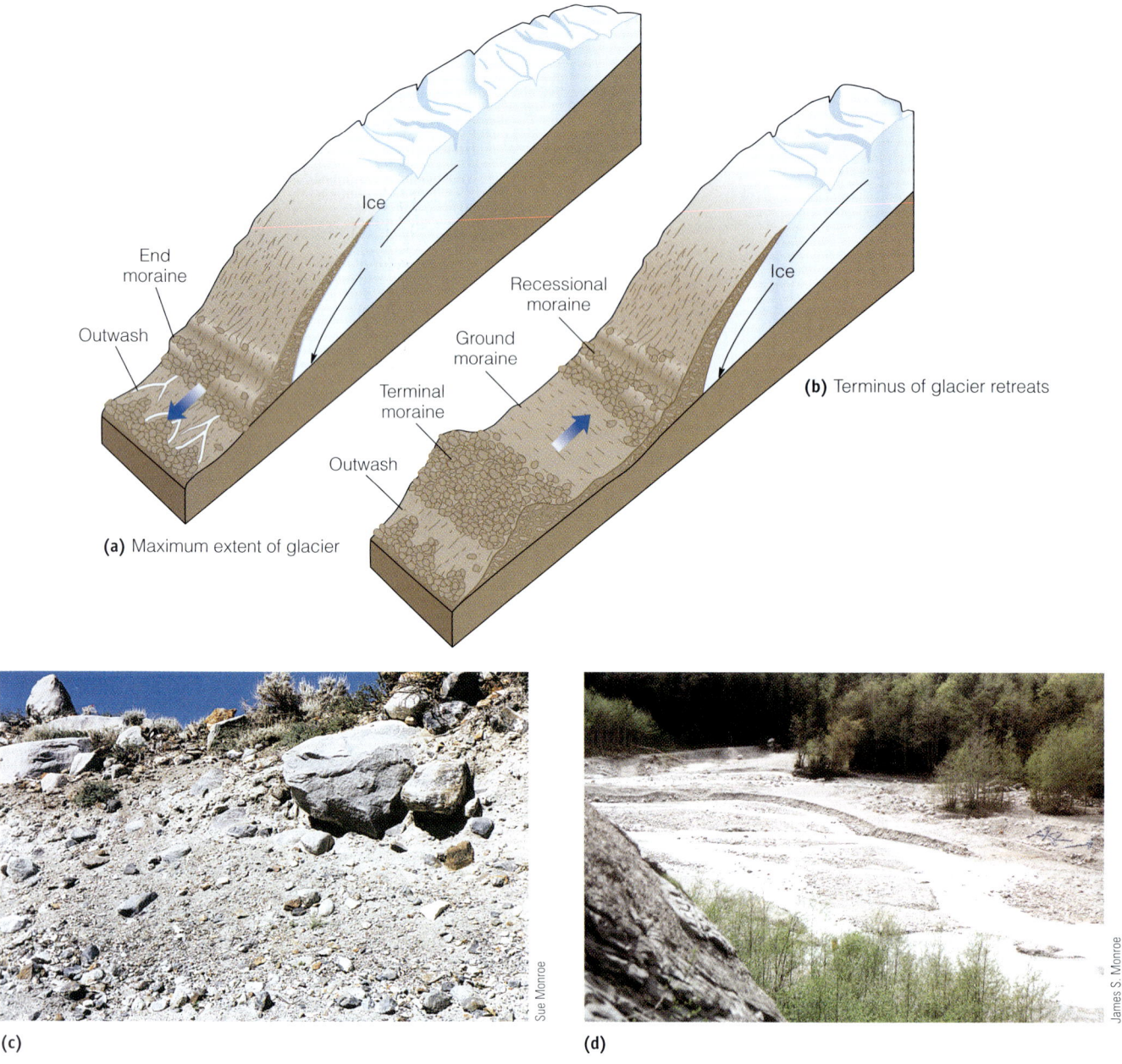

• **Figure 17.10 End Moraines and Outwash (a)** The end moraine deposited at the maximum extent of a glacier is a terminal moraine, whereas a recessional moraine **(b)** forms when the glacier's terminus retreats and becomes stabilized again. **(a)** and **(b)** also show the origin of outwash, which is sand and gravel deposited by streams that discharge from glaciers. **(c)** This terminal moraine in California is typical; it is unsorted and not stratified. **(d)** This outwash in Switzerland is made up of sand and gravel.

varve (see Figure 6.13d), representing an annual deposit. The light-colored layer of silt and clay formed during the spring and summer, and the dark layer is made up of smaller particles and organic matter formed during winter when the lake froze over. These varved deposits may also contain gravel-sized particles known as *dropstones*, released from melting ice.

Glacial Lake Missoula In 1923 geologist J. Harlan Bretz proposed that a Pleistocene lake in what is now western Montana periodically burst through its ice dam and flooded a large area in the Pacific Northwest. He further claimed that these huge floods had made the giant ripple marks and other fluvial features seen in Montana and Idaho as well as created the *scablands* of eastern Washington, an area in which the surface deposits were scoured, exposing underlying bedrock (• Figure 17.16).

Bretz's hypothesis initially met with considerable opposition, but he marshaled his evidence and eventually convinced geologists these huge floods had taken place, the most recent one probably no more than 18,000 to 20,000 years ago. It now is well accepted that Lake Missoula, a large proglacial lake covering about 7800 km^2, was impounded by an ice dam in Idaho that periodically failed. In fact, the

• **Figure 17.11 Terminal and Recessional Moraines in the Mid-Continent Region** These moraines were deposited during the latter part of the Wisconsinan. The oldest ones, those farthest south, are about 16,000 years old.

shorelines of this ancient lake are still clearly visible on the mountainsides around Missoula, Montana. When the ice dam failed, the water rushed out at tremendous velocity, accounting for the various fluvial features seen in Montana and Idaho and the scablands in eastern Washington (• Figure 17.16).

A Brief History of the Great Lakes Before the Pleistocene, the Great Lakes region was a rather flat lowland with broad stream valleys. As the continental glaciers advanced southward from Canada, the entire area was ice covered and deeply eroded. Indeed, four of the five Great Lakes basins were eroded below sea level; glacial erosion is not restricted by base level, as erosion by running water is. In any case, the glaciers advanced far to the south, but eventually began retreating north, depositing numerous recessional moraines as they did so (Figure 17.11).

By about 14,000 years ago, parts of the Lake Michigan and Lake Erie basins were ice free, and glacial meltwater began forming proglacial lakes (• Figure 17.17). As the ice front resumed its retreat northward—although interrupted by minor readvances—the Great Lakes basins eventually became ice free, and the lakes expanded until they reached their present size and shape.

This brief history of the Great Lakes is generally correct, but oversimplified. The minor readvances of the ice front mentioned earlier caused the lakes to fluctuate widely, and as they filled they overflowed their margins and partly drained. In addition, once the glaciers were gone, isostatic rebound took place, and this too has affected the Great Lakes.

GLACIATION AND ITS EFFECTS **361**

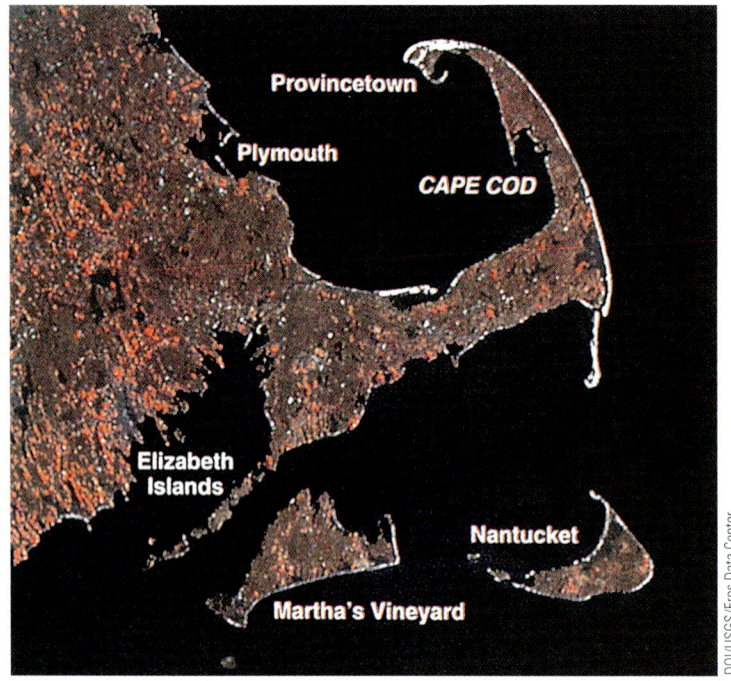

(a)

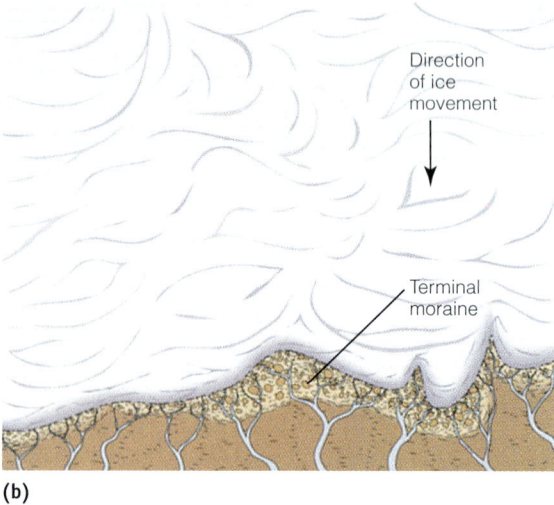

(b)

• **Figure 17.12 The Geologic Evolution of Cape Cod, Massachusetts, and Nearby Areas During the Ice Age.** (a) Cape Cod and the nearby islands are made up of mostly end moraines, although the deposits have been modified by waves since they were deposited 14,000 to 23,000 years ago. (b) Position of the glacier when it deposited a terminal moraine that would become Martha's Vineyard and Nantucket Island. (c) Position of the glacier when it deposited a recessional moraine that now forms much of Cape Cod.

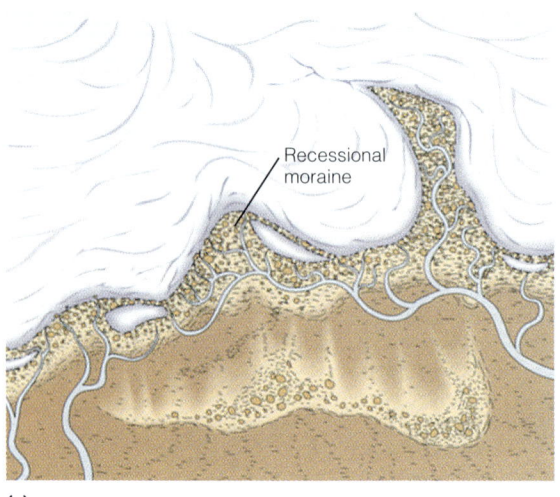

(c)

What Caused Pleistocene Glaciation?

We know how glaciers move, erode, transport, and deposit sediment, and we even know the conditions necessary for them to originate—more winter snowfall than melts during the following warmer seasons. But this really does not address the broader question of what caused large-scale glaciation during the Ice Age, and why so few episodes of glaciation have occurred. Geologists, oceanographers, climatologists, and others have tried for more than a century to develop a comprehensive theory explaining all aspects of ice ages, but so far have not been completely successful. One reason for their lack of success is that the climatic changes responsible for glaciation, the cyclic occurrence of glacial–interglacial stages, and short-term events such as the Little Ice Age operate on vastly different time scales.

The few periods of glaciation recognized in the geologic record are separated from one another by long intervals of mild climate. Slow geographic changes related to plate tectonic activity are probably responsible for such long-term climatic changes. Plate movements may carry continents into latitudes where glaciers are possible, provided they receive enough snowfall. Long-term climatic changes also take place as plates collide, causing uplift of vast areas far above sea level, and of course the distribution of land and sea has an important influence on oceanic and atmospheric circulation patterns.

One proposed mechanism for the onset of the cooling trend that began following the Mesozoic and culminated with Pleistocene glaciation is decreased levels of carbon dioxide (CO_2) in the atmosphere. Carbon dioxide is a greenhouse gas, so if less were present to trap sunlight Earth's overall temperature would perhaps be low enough for glaciers to form. The problem is, no hard data exist to demonstrate that such a decrease in CO_2 levels actually

• **Figure 17.13 The Bering Land Bridge** During the Pleistocene, sea level was as much as 130 m lower than it is now, and a broad area called the Bering land bridge (Beringia) connected Asia and North America. It was exposed above sea level during times of glacial advances and served as a corridor for the migration of people, animals, and plants.

(a)

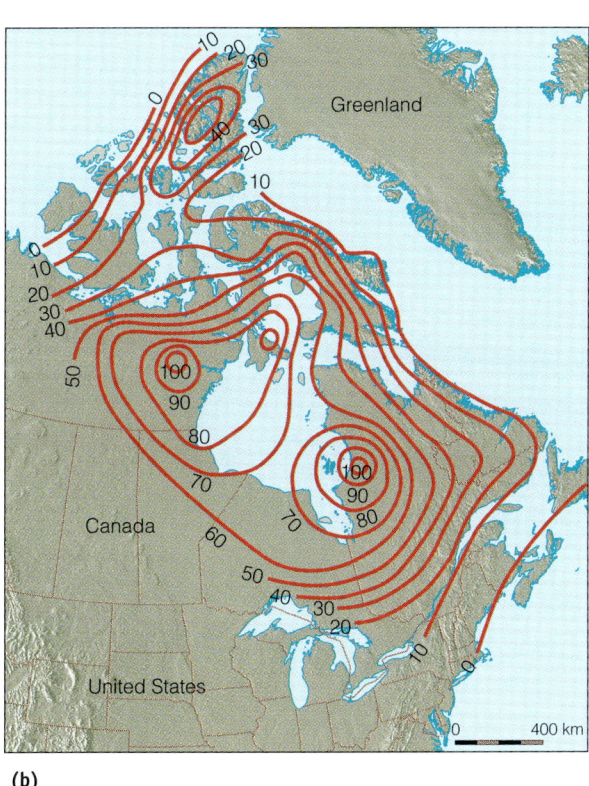
(b)

• **Figure 17.14 Glaciers and Isostasy** (a) Isostatic rebound in Scandinavia. The lines show rates of uplift in centimeters per century. (b) Isostatic rebound in eastern Canada in meters during the last 6000 years.

occurred, nor do scientists agree on a mechanism to cause a decrease, although uplift of the Himalayas or other mountain ranges has been suggested.

Intermediate climatic changes lasting for a few thousand to a few hundred thousand years, such as the Pleistocene glacial–interglacial stages, have also proved difficult to explain, but the Milankovitch theory, proposed many years ago, is now widely accepted.

The Milankovitch Theory

A particularly interesting hypothesis for intermediate-term climatic events was put forth by the Yugoslavian astronomer Milutin Milankovitch during the 1920s. He proposed that minor irregularities in Earth's rotation and orbit are sufficient to alter the amount of solar radiation that Earth receives at any given latitude and hence can affect climatic changes.

WHAT CAUSED PLEISTOCENE GLACIATION? **363**

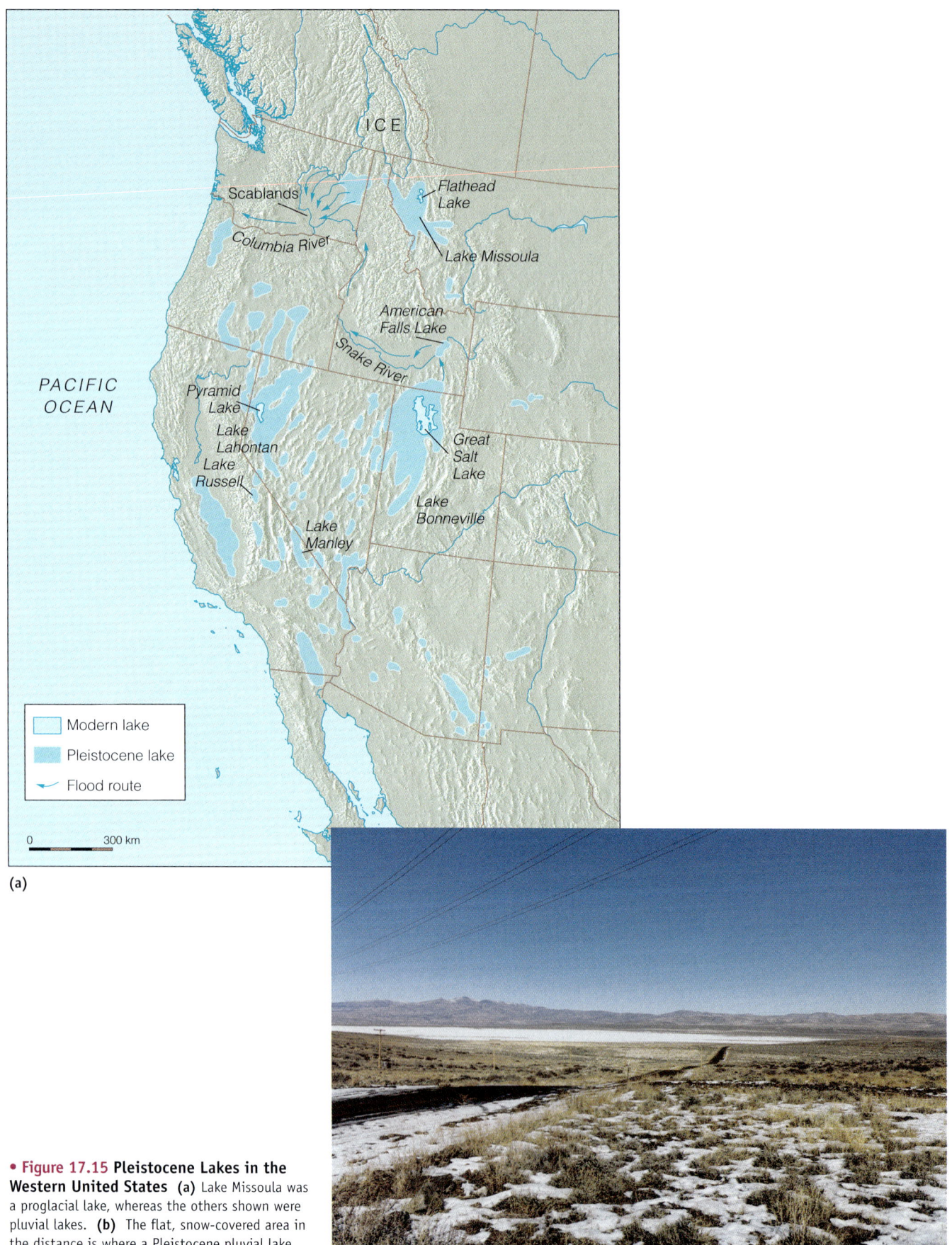

• **Figure 17.15 Pleistocene Lakes in the Western United States** (a) Lake Missoula was a proglacial lake, whereas the others shown were pluvial lakes. (b) The flat, snow-covered area in the distance is where a Pleistocene pluvial lake was present in northeastern California.

364 CHAPTER 17 CENOZOIC GEOLOGIC HISTORY: THE PLEISTOCENE AND HOLOCENE EPOCHS

(a)

(b)

• **Figure 17.16 Glacial Lake Missoula Emptied Rapidly Several Times When the Ice Dam Impounding the Lake Failed** (a) This rumpled surface is made up of gravel ridges, the so-called giant ripple marks that formed as glacial Lake Missoula drained across this area near Camas Prairie, Montana. (b) Palouse Falls in Washington is in a canyon eroded by the floodwaters from glacial Lake Missoula.

Now called the **Milankovitch theory,** it was initially ignored but has received renewed interest during the last 20 years.

Milankovitch attributed the onset of the Pleistocene Ice Age to variations in three parameters of Earth's orbit (• Figure 17.18). The first of these is orbital eccentricity, which is the degree to which the orbit departs from a perfect circle. Calculations indicate a roughly 100,000-year cycle between times of maximum eccentricity. This corresponds closely to 20 warm–cold climatic cycles that occurred during the Pleistocene. The second parameter is the angle between Earth's axis and a line perpendicular to the plane of its orbit around the Sun. This angle shifts about 1.5 degrees from its current

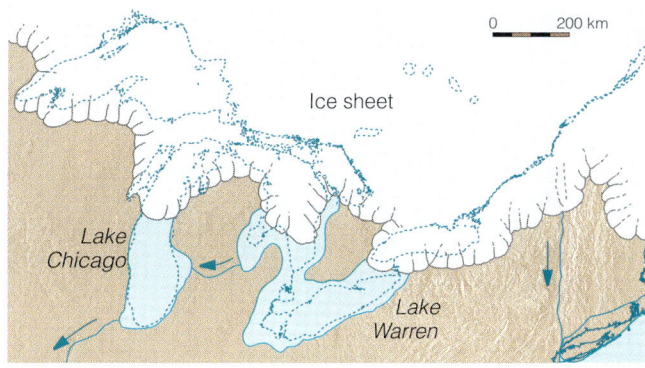

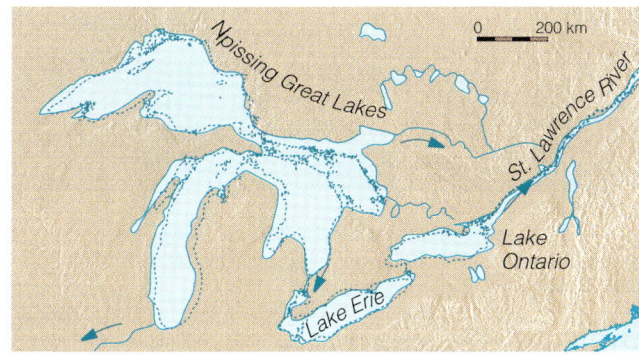

• **Figure 17.17 Four Stages in the Evolution of the Great Lakes** As the last continental glacier retreated northward, the lake basins began filling with meltwater. The dotted lines indicate the present-day shorelines of the lakes.

GLACIERS TODAY **365**

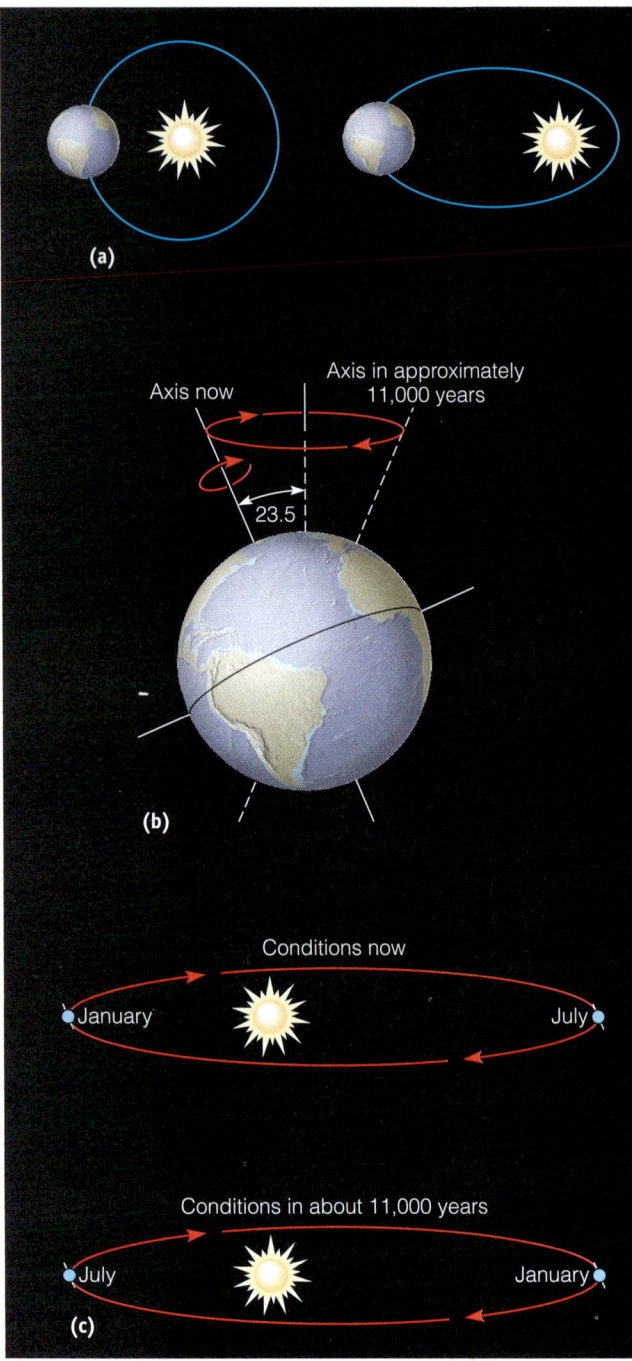

• **Figure 17.18 According to the Milankovitch Theory, Minor Irregularities in Earth's Rotation and Orbit May Affect Climatic Changes** (a) Earth's orbit varies from nearly a circle (left) to an ellipse (right) and back again in about 100,000 years. (b) Earth moves around its orbit while rotating on its axis, which is tilted to the plane of its orbit around the Sun at 23.5 degrees and points to the North Star. Earth's axis of rotation slowly moves and traces out a cone in space. (c) At present, Earth is closest to the Sun in January (top), when the Northern Hemisphere experiences winter. In about 11,000 years, however, as a result of precession, Earth will be closer to the Sun in July (bottom), when summer occurs in the Northern Hemisphere.

value of 23.5° during a 41,000-year cycle. The third parameter is the precession of the equinoxes, which causes the position of the equinoxes and solstices to shift slowly around Earth's elliptical orbit in a 23,000-year cycle.

Continuous changes in these three parameters cause the amount of solar heat received at any latitude to vary slightly over time. The total heat received by the planet, however, remains little changed. Milankovitch proposed, and now many scientists agree, that the interaction of these three parameters provides the triggering mechanism for the glacial–interglacial episodes of the Pleistocene.

Short-Term Climatic Changes

Climatic events having durations of several centuries, such as the Little Ice Age, are too short to be accounted for by plate tectonics or Milankovitch cycles. Several hypotheses have been proposed, including variations in solar energy and volcanism.

Variations in solar energy could result from changes within the Sun itself or from anything that would reduce the amount of energy Earth receives from the Sun. The latter could result from the solar system passing through clouds of interstellar dust and gas or from substances in the atmosphere reflecting solar radiation back into space. Records kept over the past 85 years, however, indicate that during this time the amount of solar radiation has varied only slightly. Thus, although variations in solar energy may influence short-term climatic events, such a correlation has not been demonstrated.

During large volcanic eruptions, tremendous amounts of ash and gases are spewed into the atmosphere, where they reflect incoming solar radiation and thus reduce atmospheric temperatures. Small droplets of sulfur gases remain in the atmosphere for years and can have a significant effect on the climate. Several such large-scale volcanic events have been recorded, such as the 1815 eruption of Tambora and the 1991 eruption of Mount Pinatubo, and are known to have had climatic effects. However, no relationship between periods of volcanic activity and periods of glaciation has yet been established.

Glaciers Today

Glaciers today are much more restricted in their distribution, but they nevertheless remain potent agents of erosion, sediment transport, and deposition. After all, even now they cover about 10% of Earth's land surface. Scientists monitor the behavior of glaciers to better understand the dynamics of moving bodies of ice, but they are also interested in glaciers as indicators of climatic change.

No doubt you have heard of *global warming*, a phenomenon of warming of Earth's atmosphere during the last 100 years or so. Many scientists are convinced that the

(a)

(b)

• **Figure 17.19 Glaciers in the Cascade Range** (a) Whitechuck Glacier on Glacier Peak in Washington State. The south branch of the glacier (foreground) has a small accumulation area, but the north branch no longer has one. (b) View of the lava dome and the newly formed glacier in the crater of Mount St. Helens on April 19, 2005. Notice the ash on the glacier's surface, which also has large crevasses.

cause of global warming is an increase of greenhouse gases, especially carbon dioxide, in the atmosphere as a result of burning fossil fuels. Others agree that surface temperatures have increased but attribute the increase to normal climatic variation. In either case, glaciers are good indicators of short-term climatic changes.

Any glacier's behavior depends on its budget—that is, its gains versus losses—which in turn is related to temperature and the amount of precipitation. So glaciers are very sensitive to changes in climate. According to one estimate there are about 160,000 valley glaciers and small ice caps outside Antarctica and Greenland, with Alaska alone having several tens of thousands. It is true that not many of these glaciers have been studied, but those that have show an alarming trend: Many are retreating, ceased moving entirely, or have disappeared.

For example, in 1850 there were about 150 glaciers in Glacier National Park in Montana, but now only a few remain, and nearly all the glaciers in the Cascade Range of the Pacific Northwest are retreating. Glacier Peak in Washington has more than a dozen glaciers, all of which are retreating, and Whitechuck Glacier will soon be inactive (• Figure 17.19a). When Mount St. Helens in Washington erupted in May 1980, all 12 of its glaciers were destroyed or considerably diminished. By 1982 the lava dome in the mountain's crater had cooled sufficiently for a new glacier to form; it is now about 190 m thick (Figure 17.19b). Remember, though, that Mount St. Helens already had the conditions for glaciers to exist, so the fact that one has become reestablished does counter the evidence from virtually all other glaciers in the range.

The ice sheet in Greenland has lost about 162 km^3 of ice during each of the years from 2003 through 2005, and many of the glaciers that flow into the sea from the ice sheet have speeded up markedly. For instance, the Kangerdlugssuag Glacier is moving at about 14 km per year (38.4 m/day), and its terminus retreated 5 km in 2005 alone. The termini of many glaciers in Alaska are also retreating, particularly valley glaciers that flow into the sea (the so-called tidewater glaciers). Two factors account for these phenomena. First, the glaciers are moving faster because more meltwater is present that percolates downward and facilitates basal slip. The other factor is that warmer ocean temperatures melt the glaciers where they flow into the sea.

Most of Antarctica shows no signs of a decreasing volume of ice because the continent is at such high latitudes and so cold that little melting takes place. The greatest concern here is that some of the ice shelves—the parts of vast glaciers that flow into the sea—will collapse and allow the glaciers inland to flow more rapidly. In fact, huge sections of ice shelves have broken off in recent years, allowing land-based glaciers to surge into the ocean. The ice shelves themselves are floating, so when they melt, that does not

What Would You Do?

Suppose that you live in western Nevada, and during one of your weekend excursions you notice a valley with a very flat floor and rocks that consist mostly of mud and sparse sand. Furthermore, you see several flat surfaces eroded into the hillsides around the valley. You have a geologic map that shows the deposits are of Pleistocene age, but it does not specify how the deposits formed or what the hillside erosion surfaces are. How would you interpret the geology of this area?

cause sea level to rise, but when the glacial ice on land flows into the ocean and melts, it does result in a rising sea level.

According to one prediction, glacial melting will cause a sea level rise of 21 cm by the 2050s, thus increasing the risk of coastal flooding in many areas. Global warming is a complex problem about which there remains considerable disagreement, particularly about its cause. Nearly everyone agrees, though, that the study of glaciers will help resolve some of the issues.

Pleistocene Mineral Resources

Many mineral deposits formed as a direct or indirect result of glacial activity during the Pleistocene and Holocene. We have already mentioned the vast salt deposits in Utah and the borax deposits in Death Valley, California, that originated when Pleistocene pluvial lakes evaporated. And some deposits of diatomite, rock composed of the shells of microscopic plants called *diatoms*, formed in the West Coast states during the Quaternary.

In many U.S. states as well as Canadian provinces, the most valuable mineral commodity is sand and gravel used in construction, much of which is recovered from glacial deposits, especially outwash. These same commodities are also recovered from deposits on the continental shelves and from stream deposits unrelated to glaciation. Silica sand is used in the manufacture of glass, and fine-grained glacial lake deposits are used to manufacture bricks and ceramics.

The California gold rush of the late 1840s and early 1850s was fueled by the discovery of Pleistocene and Holocene placer deposits of gold in the American River. Most of the $200 million in gold mined in California from 1848 to 1853 came from placer deposits. Discoveries of gold placer deposits in the Yukon Territory of Canada were primarily responsible for settlement of that area.

Peat consisting of semicarbonized plant material in bogs and swamps is an important resource that has been exploited in Canada and Ireland. It is burned as a fuel in some areas but also finds other uses, as in gardening.

SUMMARY

- The most recent part of geologic time is the Pleistocene (1.8 million to 10,000 years ago) and the Holocene or Recent epochs (10,000 years ago to the present).
- Although the Pleistocene is best known for widespread glaciers, it was also a time of volcanism and tectonism.
- Pleistocene glaciers covered about 30% of the land surface, and were most widespread on the Northern Hemisphere continents.
- At least four intervals of extensive Pleistocene glaciation took place in North America, each separated by interglacial stages. Fossils and oxygen isotope data indicate about 20 warm–cold cycles occurred during the Pleistocene.
- Areas far beyond the ice were affected by Pleistocene glaciers: Climate belts were compressed toward the equator, large pluvial lakes existed in what are now arid regions, and sea level was as much as 130 m lower than now.
- Moraines, striations, outwash, and various other glacial landforms are found throughout Canada, in the northern tier of states, and in many mountain ranges where valley glaciers were present.
- The tremendous weight of Pleistocene glaciers caused isostatic subsidence of Earth's crust. When the glaciers melted, isostatic rebound began and continues even now in some areas.
- Major glacial episodes separated by tens or hundreds of millions of years probably stem from changing positions of plates, which in turn profoundly affects oceanic and atmospheric circulation patterns.
- According to the Milankovitch theory, minor changes in Earth's rotation and orbit bring about climatic changes that produce glacial-interglacial intervals.
- The causes of short-term climatic changes such as occurred during the Little Ice Age are unknown; two proposed causes are variations in the amount of solar energy and volcanism.
- Pleistocene mineral resources include sand and gravel, placer deposits of gold, and some evaporite minerals such as borax.

IMPORTANT TERMS

cirque, p. 358
continental glacier, p. 348
glacial stage, p. 354
glacier, p. 348
ice cap, p. 348
ice-scoured plain, p. 357
interglacial stage, p. 354

isostasy, p. 358
isostatic rebound, p. 359
Little Ice Age, p. 356
Milankovitch theory, p. 365
moraine, p. 358
neoglaciation, p. 356
outwash, p. 358

pluvial lake, p. 359
pollen analysis, p. 356
proglacial lake, p. 359
U-shaped glacial trough, p. 358
valley glacier, p. 348
varve, p. 360

REVIEW QUESTIONS

1. The sediment deposited by streams that issue from melting glaciers is called
   a.____varve; b.____outwash; c.____pluvial; d.____isostatic; e.____cirque.
2. A formerly glaciated area with striated and polished bedrock, subdued topography, and poor drainage is a/an
   a.____glacial scabland; b.____recessional moraine; c.____ice-scoured plain; d.____neoglacial deposit; e.____pluvial lake.
3. Which one of the following statements is correct?
   a.____All of North America was ice covered during the Holocene; b.____The deposit that forms at the greatest extent of a glacier is a recessional moraine; c.____Glaciers advanced and retreated 40 times during the Pleistocene; d.____The theory that explains glacial-interglacial episodes is uniformitarianism; e.____The Pleistocene Epoch began 1.8 million years ago and ended 10,000 years ago.
4. The most recent episode of Pleistocene glaciation in North America was the
   a.____Wisconsinan; b.____Oklahoman; c.____Illinoian; d.____New Jerseyan; e.____Michiganian.
5. The phenomenon in which Earth's crust rises after unloading, as when vast glaciers melt, is known as
   a.____precession; b.____postglacial maxima; c.____isostatic rebound; d.____orogenic deformation; e.____neoglaciation.
6. Which one of the following statements is incorrect?
   a.____The Milankovitch theory relies on irregularities in Earth's orbit and rotation; b.____Glacial ice is enriched in oxygen 18 compared to oxygen 16; c.____The sediment deposited directly by glacial ice is a moraine; d.____The Little Ice Age took place from 1500 to the 1800s; e.____An ice cap is similar to but smaller than a continental glacier.
7. An important area of Pleistocene and Recent volcanism in North America is
   a.____the Cascade Range; b.____Cape Cod; c.____the interior lowlands; d.____glacial Lake Missoula; e.____the Atlantic Coastal Plain.
8. If a glacier has a balanced budget,
   a.____it flows into the sea; b.____it ceases flowing; c.____its terminus remains stationary; d.____it deposits varves and dropstones; e.____additions to the glacier exceed losses.
9. When continental glaciers were present during the Pleistocene, rivers and streams eroded more deeply as they adjusted to
   a.____more rainfall; b.____isostatic subsidence; c.____colder winters; d.____lower base level; e.____dryer summers.
10. A large Pleistocene pluvial lake and the forerunner of the Great Salt Lake was
    a.____Lake Missoula; b.____Lake Bonneville; c.____Lake Michigan; d.____Lake Niagara; e.____Lake Borax.
11. How do pluvial and proglacial lakes form?
12. Explain how the Milankovitch theory accounts for the onset of glacial ages.
13. Describe an end moraine, and explain what the difference is between a terminal moraine and a recessional moraine.
14. Give an account of the origin and subsequent history of Glacial Lake Missoula.
15. What was the Little Ice Age, when did it occur, and what impact did it have on humans?
16. What kinds of evidence indicate that isostatic rebound has taken place in Scandinavia and North America?
17. Where is the Cascade Range, what kinds of volcanoes are found there, and are they still active? Explain.
18. How does the landscape formed by erosion by continental glaciers differ from that eroded by valley glaciers?
19. Give a brief account for the origin of the Yellowstone caldera.
20. Why is it so difficult to correlate the sequence of warm–cold intervals recorded in seafloor sediment with the glacial record on land?

APPLY YOUR KNOWLEDGE

1. What kinds of evidence from glaciers indicate that global warming is taking place?
2. After carefully observing the same glacier for several years, you conclude (1) that the glacier's terminus has retreated 1 km, and yet (2) debris on the glacier's surface has moved several hundred meters toward the glacier's terminus. Can you think of an explanation for your observations?

FIELD QUESTION

1. You notice the following in a stream-cut gorge. In the lower part of the gorge you see a bedrock surface with linear scratches and polish that is overlain by a 12 m thick non-stratified mixture of mud, sand, and gravel; some of the gravel is up to 2 m in diameter. Next upward are moderately well-sorted layers of cross-bedded sand and horizontal layers of conglomerate. The upper part of the rock sequence is composed of thin (2–4 mm thick), alternating layers of light- and dark-colored clay, with a few boulders 10 to 15 cm across. Decipher the geologic history revealed by these rocks.

Erika Simons/Florida Museum of Natural History

CHAPTER 18
LIFE OF THE CENOZOIC ERA

Among the diverse Pliocene and Pleistocene mammals of Florida were 6 m long giant ground sloths and armored mammals known as glypotodonts that weighed more than 2 metric tons.

[OUTLINE]

INTRODUCTION

MARINE INVERTEBRATES AND PHYTOPLANKTON

CENOZOIC VEGETATION AND CLIMATE

CENOZOIC BIRDS

THE AGE OF MAMMALS BEGINS

DIVERSIFICATION OF PLACENTAL MAMMALS

PALEOGENE AND NEOGENE MAMMALS

Small Mammals—Insectivores, Rodents, Rabbits, and Bats

PERSPECTIVE 18.1 *A Miocene Catastrophe in Nebraska*

A Brief History of the Primates

The Meat Eaters—Carnivorous Mammals

The Ungulates or Hoofed Mammals

Giant Land-Dwelling Mammals—Elephants

Giant Aquatic Mammals—Whales

PLEISTOCENE FAUNAS

Ice Age Mammals

Pleistocene Extinctions

INTERCONTINENTAL MIGRATIONS

SUMMARY

ThomsonNOW™ Explore interactive tutorials, animations, or practice problems available on the ThomsonNow website at www.thomsonedu.com/login.

[CHAPTER OBJECTIVES]

At the end of this chapter, you will have learned that

- Survivors of the Mesozoic extinctions evolved and gave rise to the present-day invertebrate marine fauna.

- Angiosperms continued to diversify and to dominate land plant communities, but seedless vascular plants and gymnosperms are still common.

- Many of today's families and genera of birds evolved, and large flightless birds were important Early Cenozoic predators.

- If we could visit the Paleocene, we would not recognize many of the mammals, but more familiar ones evolved during the following epochs.

- Small mammals such as rodents, rabbits, insectivores, and bats adapted to the microhabitats unavailable to larger mammals.

- Carnivorous mammal fossils are not as common as those of herbivores, but there are enough to show their evolutionary trends and relationships to one another.

- The evolutionary histories of odd-toed and even-toed hoofed mammals are well documented by fossils.

- Today's giant land mammals (elephants) and giant marine mammals (whales) evolved from small Early Cenozoic ancestors.

- Extinctions at the end of the Pleistocene Epoch were most severe in the Americas and Australia, and the animals most affected were large land-dwelling mammals.

- As Pangaea continued to fragment during the Cenozoic, intercontinental migrations became increasingly difficult.

- A Late Cenozoic land connection formed between North and South America, resulting in migrations in both directions.

Introduction

When Earth first formed, it was hot, barren, and waterless; the atmosphere was noxious; it was bombarded by meteorites and comets; and no organisms existed. However, during the Precambrian and the following Paleozoic and Mesozoic eras, the planet and its biota evolved, and by Cenozoic time, Earth and its organisms were taking on their present-day appearance. Although the Cenozoic Era constitutes only 1.4% of all geologic time (see Figure 8.1), this comparatively brief 66 million years of Earth and life history was short only in the context of geologic time. From the human perspective it was far longer than we can even imagine, certainly long enough for many changes to take place in the biota.

Remember that mammals evolved during the Late Triassic Period, and some Mesozoic mammals retained characteristics of their ancestors, the cynodonts (see Chapter 15). In fact, the distinction between the earliest mammals and cynodonts is difficult to make, but by Cenozoic time, mammals had clearly differentiated from their ancestors. In this chapter we emphasize the evolution of mammals that really began to diversify following the end-of-Mesozoic extinctions. There were, however, other equally important biologic events taking place.

Angiosperms, or flowering plants, evolved during the Cretaceous and soon became the dominant land plants; now they constitute more than 90% of all land plant species. However, the geographic distribution of plants varied during the Cenozoic, depending on changing climates. Birds evolved during the Jurassic, but the families now common appeared during the Paleogene and Neogene periods, reached their maximum diversity during the Pleistocene Epoch, and have declined slightly since then. The marine invertebrates that survived the Mesozoic extinctions diversified and gave rise to the present-day marine fauna. Overall, we can think of the Cenozoic Era as the time during which the biota became increasingly familiar.

As we noted in Chapter 16, Cenozoic rocks are the most easily accessible at or near the surface, so we know more about Earth and life history for this time than for any of the previous eras. Cenozoic-aged rocks are especially widespread in western North America (see Figure 16.2), as well as along the Gulf and Atlantic coasts, and as a result, we have a particularly good fossil record for many organisms. Several of our national parks and monuments in the west feature displays of fossil mammals, including Agate Fossil Beds National Monument in Nebraska, Badlands National Park in South Dakota, and John Day Fossil Beds National Monument in Oregon (• Figure 18.1).

Deposits with land-dwelling fossil mammals are not nearly as common in the eastern part of the continent, but Florida is a notable exception (see the chapter opening photo). Furthermore, some eastern and southern states such as Maryland, South Carolina, and Alabama have deposits with the fossils of Cenozoic marine mammals as well as fossil invertebrates and sharks. Of course, mammal fossils are found on the other continents, too, but certainly one of the most remarkable fossil sites anywhere in the world is the Messel fossil beds in Germany. In fact, some of these fossils are truly remarkable because even hair, feathers, and color have been preserved.

• **Figure 18.1 Restoration of Fossils from the Eocene-aged Clarno Formation in John Day Fossil Beds National Monument, Oregon** The climate at this time was subtropical, and the lush forests of the region were occupied by **(1)** titanotheres standing 2.5 m high at the shoulder, **(2)** carnivores, **(3)** ancient horses, **(4)** tapirs, and **(5)** early rhinoceroses.

Marine Invertebrates and Phytoplankton

The Cenozoic marine ecosystem was populated mostly by plants, animals, and single-celled organisms that survived the terminal Mesozoic extinction. Gone were the ammonites, rudists, and most of the planktonic foraminifera. Especially prolific Cenozoic invertebrate groups were the foraminifera, radiolarians, corals, bryozoans, mollusks, and echinoids. The marine invertebrate community in general became more provincial during the Cenozoic because of changing ocean currents and latitudinal temperature gradients. In addition, the Cenozoic marine invertebrate faunas became more familiar in appearance.

Entire families of phytoplankton became extinct at the end of the Cretaceous. Only a few species in each major group survived into the Paleogene. These species diversified and expanded during the Cenozoic, perhaps because of decreased competitive pressures. Coccolithophores, diatoms, and dinoflagellates all recovered from their Late Cretaceous reduction in numbers to flourish during the Cenozoic. Diatoms were particularly abundant during the Miocene, probably because of increased volcanism during this time. Volcanic ash provided increased dissolved silica in seawater, which diatoms used to construct their skeletons. Massive Miocene diatomite rocks, made up of diatom shells, are present in several western states (• Figure 18.2)

The foraminifera were a major component of the Cenozoic marine invertebrate community. Although dominated by relatively small forms (• Figure 18.3), it included some exceptionally large forms that lived in the warm waters of the Cenozoic Tethys Sea. Shells of these larger forms accumulated to form thick limestones, some of which the ancient Egyptians used to construct the Sphinx and the Pyramids of Giza (Figure 18.3c).

Corals were perhaps the main beneficiary of the Mesozoic extinctions. Having relinquished their reef-building role to rudists, which are mollusks, during the mid-Cretaceous, corals again became the dominant reef builders. They formed extensive reefs in the warm waters of the Cenozoic oceans and were especially prolific in the Caribbean and Indo-Pacific regions (• Figure 18.4a).

Other suspension feeders such as bryozoans and crinoids were also abundant and successful during the Paleogene and Neogene. Bryozoans, in particular, were very abundant. Perhaps the least important of the Cenozoic marine invertebrates were brachiopods, with fewer than 60 genera surviving today. Brachiopods never recovered from their reduction in diversity at the end of the Paleozoic (see Chapter 12).

Just as during the Mesozoic, bivalves and gastropods were two of the major groups of marine invertebrates during the Cenozoic, and they had a markedly modern appearance.

After the extinction of ammonites and belemnites at the end of the Cretaceous, the Cenozoic cephalopod fauna consisted of nautiloids and shell-less cephalopods such as squids and octopuses.

Echinoids continued their expansion in the infaunal habitat and were very prolific during the Cenozoic. New forms such as sand dollars evolved during this time from biscuit-shaped ancestors (Figure 18.4b).

Cenozoic Vegetation and Climate

During the Cenozoic, angiosperms continued to diversify as more and more familiar varieties evolved, although seedless vascular plants and gymnosperms were also present in large

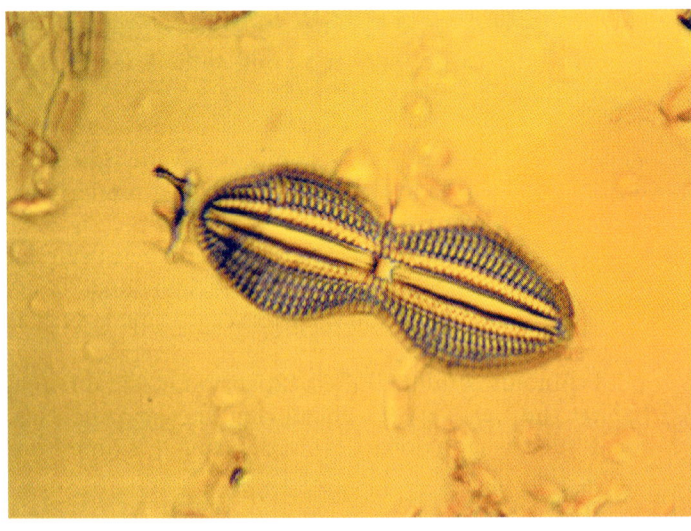

● Figure 18.2 **Miocene Diatomite** (a) Outcrop of the Monterey Formation at Newport Lagoon, California. (b) A pinnate and (c) a centric diatom from the Monterey Formation.

numbers. In fact, many Paleogene plants would be familiar to us today, but their geographic distribution was not what it is now, because changing climatic conditions along with shifting plant distributions were occurring.

The makeup of ancient floras and the types of leaves are good climatic indictors. Some plants today are confined to the tropics, whereas others have adapted to drier conditions, and we have every reason to think that climate was a strong control on plant distribution during the past. Furthermore, leaves with entire or smooth margins, many with pointed drip-tips, are dominant in areas with abundant rainfall and high annual temperatures. Smaller leaves with incised margins are more typical of cooler, drier areas (● Figure 18.5a). Accordingly, fossil floras with mostly smooth-margined leaves with drip-tips indicate wet, warm conditions, whereas a cool, dry climate is indicated by a predominance of small leaves with incised margins.

Paleocene rocks in the western interior of North America have fossil ferns and palms, both indicating a warm, subtropical climate. In a recently discovered

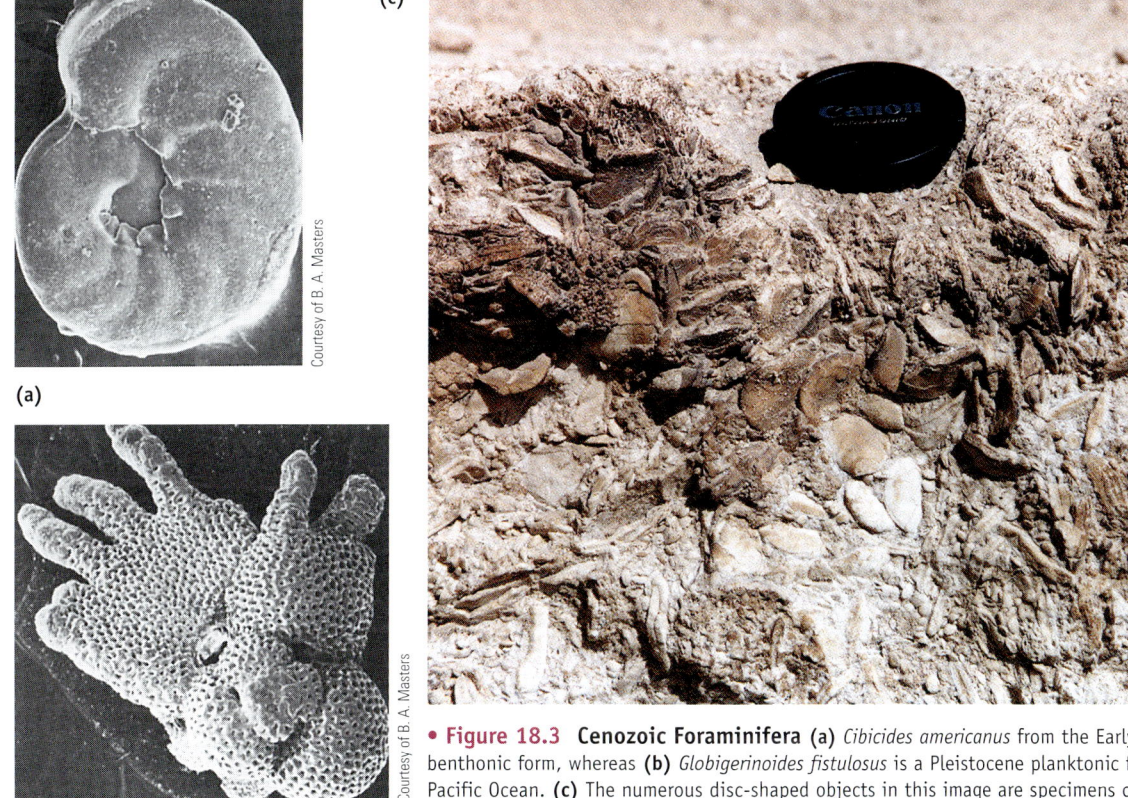

• **Figure 18.3 Cenozoic Foraminifera** (a) *Cibicides americanus* from the Early Miocene of California is a benthonic form, whereas (b) *Globigerinoides fistulosus* is a Pleistocene planktonic foraminifera from the South Pacific Ocean. (c) The numerous disc-shaped objects in this image are specimens of Eocene-aged *Nummulites*, a benthonic foraminifera in the limestone used to construct the pyramids on the Giza Plateau in Egypt.

Paleocene flora in Colorado with about 100 species of trees, nearly 70 percent of the leaves are smooth margined, and many have drip-tips. The nature of the leaves coupled with the diversity of plants is much like that found in today's rain forests. In fact, the Early Oligocene fossil plants at Florissant Fossil Beds National Monument indicate that a warm, wet climate persisted then.

Seafloor sediments and geochemical evidence indicate that about 55 million years ago an abrupt warming trend took place. During this time, known as the **Paleocene-Eocene Thermal Maximum**, large-scale oceanic circulation was disrupted so that heat transfer from equatorial regions to the poles diminished or ceased. As a result, deep oceanic water became warmer, resulting in extinctions of many deep-water foraminifera. Some scientists think that this deep, warm oceanic water released methane from seafloor methane hydrates, contributing a greenhouse gas to the atmosphere and either causing or contributing to the temperature increase at this time.

Subtropical conditions persisted into the Eocene in North America, probably the warmest of all the Cenozoic epochs. Fossil plants in the Eocene John Day Beds in Oregon include ferns, figs, and laurels, all of which today live in much more humid regions, as in parts of Mexico and Central America. Yellowstone National Park in Wyoming has a temperate climate now, with warm, dry summers and cold, snowy winters, certainly not an area where you would

• **Figure 18.4 Cenozoic Corals and Echinoids** (a) The dominant reef-building animals of the Cenozoic Era were corals such as this present-day colonial scleractinian. (b) Echinoids were especially abundant during the Cenozoic Era. New infaunal forms such as this sand dollar evolved from biscuit-shaped Mesozoic ancestors.

expect avocado, magnolia, and laurel trees to grow. Yet their presence there during the Eocene indicates the area then had a considerably warmer climate than it does now.

A major climatic change took place at the end of the Eocene, when mean annual temperatures dropped as much as 7°C in about 3 million years (Figure 18.5b). Since the Oligocene, mean annual temperatures have varied somewhat worldwide, but overall they have not changed much in the middle latitudes except during the Pleistocene Epoch.

A general decrease in precipitation during the last 25 million years took place in the midcontinent region of North America. As the climate became drier, the vast forests of the Oligocene gave way first to *savannah* conditions (grasslands with scattered trees) and finally to *steppe* environments (short-grass prairie of the desert margin). Many herbivorous mammals quickly adapted to these new conditions by developing chewing teeth suitable for a diet of grass.

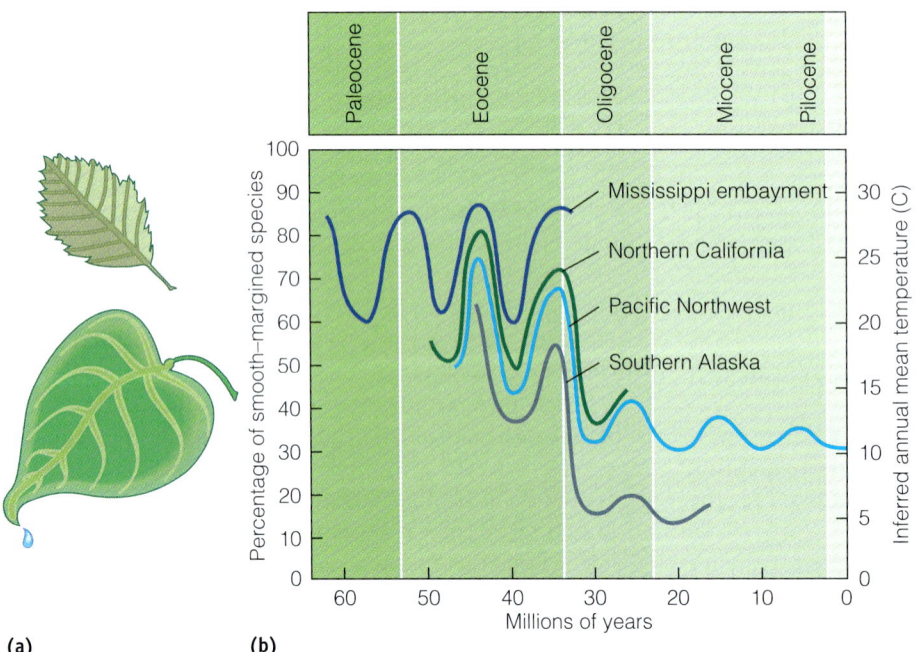

• **Figure 18.5 Vegetation and Climate** (a) Plants adapted to cool climates typically have small leaves with incised margins (top), whereas in humid, warm habitats they have larger, smooth-margined leaves, and many have drip-tips. (b) Climatic trends for four areas in North America based on the percentages of plant species with smooth-margined leaves.

Cenozoic Birds

Birds today are diverse and numerous, making them the most easily observed vertebrates. The first members of many of the living orders, including owls, hawks, ducks, penguins, and vultures, evolved during the Paleogene. Beginning during the Miocene, a marked increase in the variety of songbirds took place, and by 5 to 10 million years ago, many of the existing genera of birds were present. Birds adapted to numerous habitats and continued to diversify into the Pleistocene, but since then their diversity has decreased slightly.

Today, birds vary considerably in diet, habitat, adaptations, and size. Nevertheless, their basic skeletal structure has remained remarkably constant throughout the Cenozoic. Given that birds evolved from a creature very much like *Archaeopteryx* (see Figure 15.17), this uniformity is not surprising, because adaptations for flying limit variations in structure.

Penguins adapted to an aquatic environment, and in some large extinct and living flightless birds the skeleton became robust and the wings were reduced to vestiges. Indeed, one early adaptation in birds was the evolution of large, flightless predators such as *Diatryma* (• Figure 18.6). This remarkable bird stood more than 2 m tall, had a huge head and beak, toes with large claws, and small, vestigial wings. Its massive, short legs indicate that *Diatryma* was not very fast, but neither were the early mammals it preyed on. This extraordinary bird and related genera were widespread in North America and Europe during the Paleogene, and in South America they were the dominant predators until about 25 million years ago. Eventually, they died out, when they were replaced by carnivorous mammals.

Two of the most notable large flightless birds were the now extinct moas of New Zealand and elephant birds of Madagascar. Moas were up to 3 m tall; elephant birds were shorter but more massive, weighing up to 500 kg. They are known only from Pleistocene-age deposits, and both went extinct shortly after humans occupied their respective areas.

Large flightless birds are truly remarkable creatures, but the real success among birds belongs to the fliers. Even though few skeletal modifications occurred during the Cenozoic, a bewildering array of adaptive types arose. If number of species and habitats occupied is any measure of success, birds have certainly been at least as successful as mammals.

The Age of Mammals Begins

Mammals coexisted with dinosaurs for more than 140 million years, and yet, during this entire time they were not very diverse, and even the largest among them was only about 1 m long. Even at the end of the Cretaceous Period there were only a few families of mammals, a situation that was soon to change. With the demise of dinosaurs and their relatives, mammals quickly exploited the adaptive opportunities, beginning a diversification that continued throughout the Cenozoic Era. The Age of Mammals had begun.

• Figure 18.6 Restoration of *Diatryma* *Diatryma* was a flightless, predatory bird that stood more than 2 m tall. It lived during the Paleocene and Eocene in North America and Europe.

We have already mentioned that Cenozoic deposits are easily accessible at or near the surface, and overall they show fewer changes resulting from metamorphism and deformation when compared with older rocks. In addition, because mammals have teeth fully differentiated into various types (see Figure 15.19), they are easier to identify and classify than members of the other classes of vertebrates. In fact, mammal teeth not only differ from front to back of the mouth, but they also differ among various mammalian orders and even among genera and species. This is especially true of chewing teeth, the **premolars** and **molars**; a single chewing tooth is commonly enough to identify the genus from which it came.

All warm-blooded vertebrates with hair and mammary glands are members of the class Mammalia, which includes two fundamentally different kinds of mammals: the *prototheria* and the *theria* (or eutheria). The prototheria include some extinct animals, but the only living ones are the monotremes (order Monotremata) or egg-laying mammals—the platypus and spiny anteater of the Australian region. Therians, in contrast, include all mammals that give birth to live young such as marsupial mammals (order Marsupialia), commonly called "pouched mammals," and the placental mammals with about 18 living orders.

When the young of marsupial mammals are born, they are in an immature, almost embryonic condition and then undergo further development in the mother's pouch. Marsupials probably migrated to Australia, the only area in which they are common today, via Antarctica before Pangaea fragmented completely. However, they were also quite widespread in South America until only a few millions of years ago. Most South American marsupials died out when a land connection was established between the Americas and placental mammals migrated south. Now the only marsupials outside Australia and some nearby islands are species of opossums.

Like marsupials, placental mammals give birth to live young, but their reproductive method differs in important details. In placentals, the amnion of the amniote egg (see Figure 13.15) has fused with the walls of the uterus, forming a *placenta*. Nutrients and oxygen flow from mother to embryo through the placenta, permitting the young to develop much more fully before birth. Actually, marsupials also have a placenta, but it is less efficient, explaining why their newborn are not as fully developed.

A measure of the success of placental mammals is related in part to their method of reproduction—more than 90% of all mammals, fossil and extinct, are placentals.

Diversification of Placental Mammals

In our following discussion of placental mammals, we emphasize the origin and evolution of several of the 18 or so living orders (• Figure 18.7). Recall from Chapter 7 that in Linnaeus's classification scheme an order consists of one or more related families. For instance, the families Canidae (dogs), Felidae (cats), Ursidae (bears), and several others constitute the order Carnivora (see Figure 7.18).

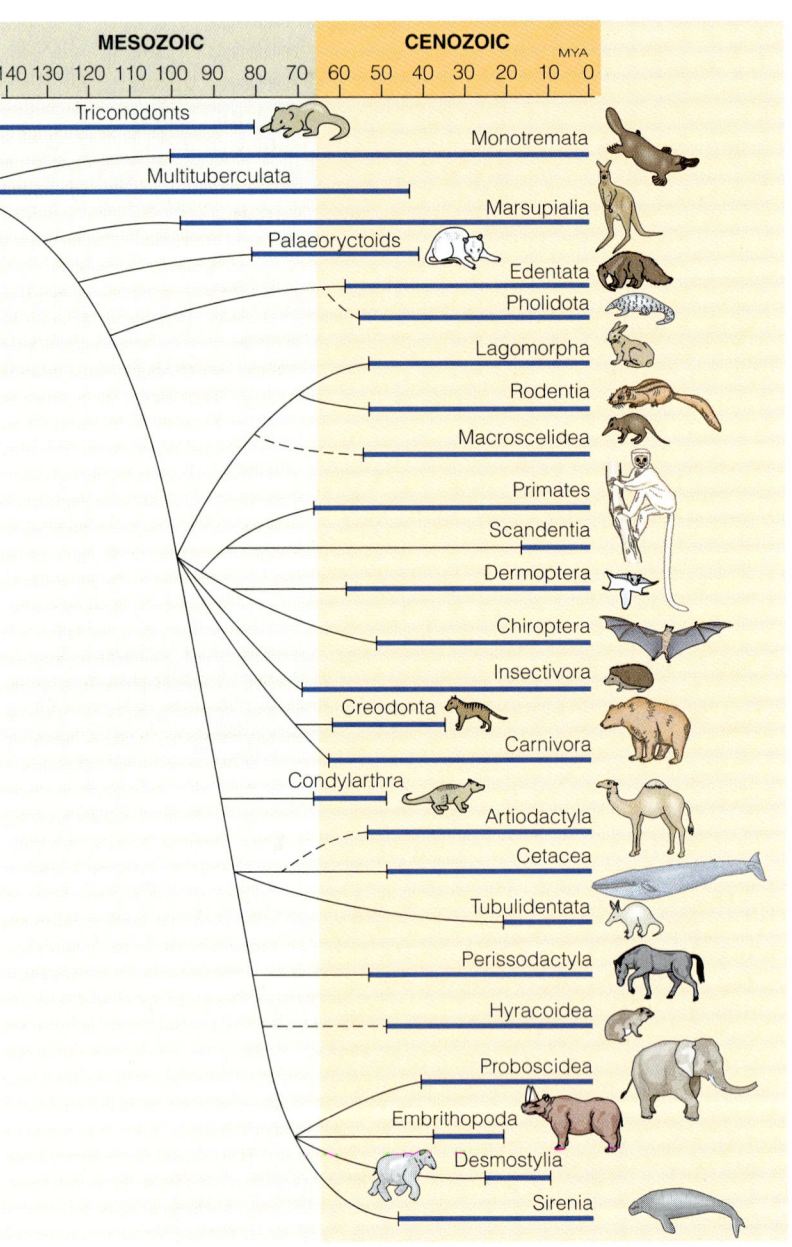

• **Figure 18.7 Several Orders of Mammals Existed During the Mesozoic, but Most Placental Mammals Diversified During the Paleocene and Eocene Epochs** Among the living orders of mammals, all are placentals except for the monotremes and marsupials. Several extinct orders are not shown. Bold lines indicate actual geologic ranges, whereas the thinner lines indicate the inferred branching of the various groups.

During the Paleocene Epoch several orders of mammals were present, but some were simply holdovers from the Mesozoic or belonged to new but short-lived groups that have no living descendants. These so-called *archaic* mammals, such as marsupials, insectivores, and the rodentlike multituberculates, occupied a world with several new mammalian orders (• Figure 18.8), including the first rodents, rabbits, primates, and carnivores, and the ancestors of hoofed mammals. Most of these Paleocene mammals, even those belonging to orders that still exist, had not yet become clearly differentiated from their ancestors, and the differences between herbivores and carnivores were slight.

The Paleocene mammalian fauna was also made up mostly of small creatures. By Late Paleocene time, though, some rather large mammals were around, although giant terrestrial mammals did not appear until the Eocene. With the evolution of a now extinct order known as the Dinocerata, better known as uintatheres, and the strange creature known as *Arsinoitherium*, giant mammals of one kind or another have been present ever since (• Figure 18.9).

Many mammalian orders that evolved during the Paleocene died out, but of the several that first appeared during the Eocene, only one has become extinct. Thus by Eocene time many of the mammalian orders existing now were present, yet if we could go back for a visit we would not recognize most of these animals. Surely we would know they were mammals and some would be at least vaguely familiar, but the ancestors of horses, camels, elephants, whales, and rhinoceroses bore little resemblance to their living descendants.

Warm, humid climates persisted throughout the Paleocene and Eocene of North America, but by Oligocene time drier and cooler conditions prevailed. Most of the archaic Paleocene mammals, as well as several groups that originated during the Eocene, had died out by this time. The large, rhinoceros-like titanotheres died out (Figure 18.1), and the uintatheres just mentioned also went extinct. In addition, some other groups of mammals suffered extinctions, including several types of herbivores loosely united as condylarths, carnivorous mammals known as *creodonts*, most of the remaining multituberculates, and some primates. All in all, this was a time of considerable biotic change.

By Oligocene time, most of the existing mammalian orders were present, but they continued to diversify as more and more familiar genera evolved. If we were to

• **Figure 18.8 Archaic Mammals of the Paleocene Epoch** The mammals include **(1)** *Protictus*, an early carnivore, **(2)** insectivores, **(3)** the 19 cm long, tree-dwelling, multituburculate *Ptilodus*, and **(4)** the pantodont known as *Pantolambda* that stood about 1 m high.

• **Figure 18.9 Some of the Earliest Large Mammals (a)** Skull of *Arsinoitherium*, a rhinoceros- to elephant-sized Early Oligocene animal. Its paired, hollow horns were more than 0.5 m long. **(b)** Scene from the Eocene showing the rhinoceros-sized mammal known as *Uintatherium*. It had three pairs of bony protuberances on the skull and saberlike upper canine teeth.

encounter some of these animals, we might think them a bit odd, but we would have little difficulty recognizing rhinoceroses (although some were hornless), elephants, horses, rodents, and many others. However, the large, horselike animals known as *chalicotheres*, with claws, and the large piglike entelodonts would be unfamiliar, and others would be found in areas where today we would not expect them—elephants in North America, for instance.

By Miocene and certainly Pliocene time, most mammals were quite similar to those existing now (• Figure 18.10 and Perspective 18.1). On close inspection, though, we would see horses with three toes, cats with huge canine teeth, deerlike

• **Figure 18.10 Pliocene Mammals of the Western North American Grasslands** The animals shown include **(1)** *Amebeledon*, a shovel-tusked mastodon, **(2)** *Teleoceras*, a short-legged rhinoceros, **(3)** *Cranioceras*, a horned, hoofed mammal, **(4)** a rodent, **(5)** a rabbit, **(6)** *Merycodus*, an extinct pronghorn, **(7)** *Synthetoceras*, a hoofed mammal with a horn on its snout, and **(8)** *Pliohippus*, a one-toed grazing horse.

animals with forked horns on their snouts, and very tall, slender camels. And we would still see a few rather odd mammals, but overall the fauna would be quite familiar.

Paleogene and Neogene Mammals

We know mammals evolved from mammal-like reptiles called *cynodonts* during the Late Triassic, and diversified during the Cenozoic, eventually giving rise to the present-day mammalian fauna. Now more than 4000 species exist, ranging from tiny shrews to giants such as whales and elephants.

When one mentions the term *mammal*, what immediately comes to mind are horses, pigs, cattle, deer, dogs, cats, and so on, but most often we do not think much about small mammals, rodents, shrews, rabbits, and bats. Yet most mammals are quite small, weighing less than 1 kg.

Small Mammals—Insectivores, Rodents, Rabbits, and Bats

Insectivores, rodents, rabbits, and bats are all placental mammals and accordingly share a common ancestor, but they have had separate evolutionary histories since they first evolved. With the exception of bats, the oldest of which is found in Eocene rocks, the others were present by the Late Mesozoic or Paleocene. The main reason we consider them together is that with few exceptions they are small and have adapted to the microhabitats unavailable to larger mammals. In addition to being small, bats are the only mammals capable of flight.

As you would expect from the name Insectivora, members of this group—today's shrews, moles, and hedgehogs—eat insects. Insectivores have probably not changed much since they appeared during the Late Cretaceous. In fact, an insectivore-like creature very likely lies at the base of the great diversification of placental mammals.

More than 40% of all living mammal species are members of the order Rodentia, most of which are very small animals. A few, though, including beavers and the capybara of South America, are sizable animals; the latter is more than 1 m long and weighs 45 kg. One Miocene beaver known as *Paleocastor* was not particularly large, but it constructed some remarkable burrows, and the Miocene rodent whimsically called "ratzilla" weighed an estimated 740 kg (• Figure 18.11). Rodents evolved during the Paleocene, diversified rapidly, and adapted to a wide range of habitats. One reason for their phenomenal success is that they can eat almost anything.

Rabbits (order Lagomorpha) superficially resemble rodents but differ from them in several anatomic details.

• **Figure 18.11 Fossil Rodents** (a) These spiral burrows in Miocene rocks in Nebraska were made by a land-dwelling beaver known as *Paleocastor*, which was about 30 cm long. (b) This restoration of *Phoberomys* from Miocene rocks in South America shows a huge rodent that weighed about 700 kg.

Perspective 18.1

A Miocene Catastrophe in Nebraska

Twelve million years ago, in what is now northeastern Nebraska, vast grasslands were inhabited by short-legged, aquatic rhinoceroses, horses, camels, saber-toothed deer, birds, land turtles, and many other animals (Figure 1a). This was a temperate savanna habitat with life as varied and abundant as it is now on the savannahs of East Africa. Many of these animals died when a huge cloud of volcanic ash rolled in, probably from a volcano in Idaho. The animals did not die immediately, but as they breathed in the abrasive particles of glass, their lungs were damaged and they started to die. Nothing disturbed the carcasses except scavengers, and they were buried as wind-blown ash covered the remains.

Paleontologists from the University of Nebraska have recovered or left exposed for viewing the skeletons of hundreds of victims of this prehistoric catastrophe (Figure 1b). The site is now Ashfall Fossil Beds State Historical Park. The magnitude of the ash-producing eruption is difficult to imagine, since the ash cloud probably came from at least 1000 km away. We know from historic eruptions that ash can indeed be carried this far. For example, during the April 11, 1815, eruption of Tambora in Indonesia, an ash layer 22 cm thick accumulated 400 km to the west, and some ash fell more than 1500 km from the volcano.

The initial ash fall was likely no more than 30 to 60 cm deep, but as it was redistributed by the wind, it accumulated in some areas to about 3 m thick. From the geologic evidence at the site, the area with most fossils was probably a low area with a water hole at which animals congregated. Many animals, including three-toed horses, camels, deer, turtles, and birds, perished here, and their skeletons show signs of partial decomposition, scavenging, and trampling before they were buried.

Soon after the initial ash fall, the depression was visited by herds of rhinoceroses and a few horses and camels. These animals also died as their lungs filled with ash, and then they were buried quickly, as indicated by the large number of complete skeletons. One of the most remarkable things about the fossils is the preserved detail. According to Michael Voorhies of the University of Nebraska, "Rarely found parts such as tongue bones, cartilages, tendons, and tiny bones in the middle ear all survive in exquisite detail and in their correct positions."* Only rarely are paleontologists fortunate enough to find and recover so many, well-preserved, associated vertebrate animals. A Late Miocene catastrophe turns out to be our good fortune because it provides us with a unique glimpse of what life was like in Nebraska 12 million years ago.

---

*"Ancient Ashfall Creates a Pompeii of Prehistoric Animals," *National Geographic*, 159, no. 1:69, 1981.

(a)

(b)

Figure 1 (a) This mural shows a restoration of some of the fossils recovered at Ashfall Fossil Beds Historical Park near Orchard, Nebraska; (1) one-toed horses, (2) small camels, (3) turtles, (4) rhinoceroses, (5) cranes, (6) giraffe-like camels, and (7) three-toed horses. In the distance you can see some carnivores and mastodons. (b) Paleontologists excavating fossil horses (foreground) and rhinoceroses from the 12-million-year-old ash.

Furthermore, since they arose from a common ancestor during the Paleocene, rabbits and rodents have had an independent evolutionary history. Like rodents, rabbits are gnawing animals, although details of their gnawing teeth differ. The development of long, powerful hind limbs for speed is the most obvious evolutionary trend in this group.

The oldest fossil bat (order Chiroptera) comes from the Eocene-age Green River Formation of Wyoming, but well-preserved specimens are known from several other areas, too. Apart from having forelimbs modified into wings, bats differ little from their immediate ancestors among the insectivores. Indeed, with the exception of wings they closely resemble living shrews. Unlike pterosaurs and birds, bats use a modification of the hand in which four long fingers support the wings (see Figure 15.14c).

A Brief History of the Primates

The order **Primates** includes the "lower primates" (tarsiers, lemurs, and lorises), and the monkeys, apes, and humans, collectively referred to as "higher primates." Much of the primate story is more fully told in Chapter 19, where we consider human evolution, so in this chapter we will be brief. Primates may have evolved by Late Cretaceous time, but by the Paleocene they were undoubtedly present.

Small Paleocene primates closely resembled their contemporaries, the shrewlike insectivores. By the Eocene, though, larger primates had evolved, and lemurs and tarsiers that resemble their present descendants lived in Asia and North America. And by Oligocene time primitive New World and Old World monkeys had developed in South America and Africa, respectively. The Hominoids, the group that includes apes and humans, evolved during the Miocene (see Chapter 19).

The Meat Eaters— Carnivorous Mammals

The order **Carnivora** is extremely varied, with animals as different as bears, seals, weasels, skunks, dogs, and cats. All are predators and therefore meat eaters, but their diets vary considerably. For example, cats rarely eat anything but meat, whereas bears, raccoons, and skunks have a varied diet and are thus *omnivorous*. Most carnivores have well-developed, sharp, pointed canine teeth and specialized shearing teeth known as **carnassials** for slicing meat (• Figure 18.12). Some land-dwelling carnivores depend on speed, agility, and intelligence to chase down prey, but others employ different tactics. Badgers, for instance, are not very fast—they dig prey from burrows, and some small cats depend on stealth and pouncing to catch their meals.

Fossils of carnivorous mammals are not nearly as common as those of many other mammals—but why should this be so? First, in populations of warm-blooded (endothermic) animals, carnivores constitute no more than 5% of the total population, usually less. Second, many, but not all, carnivores are solitary animals, so the chance of large numbers of them being preserved together is remote. Nevertheless, fossil carnivores are common enough for us to piece together their overall evolutionary relationships with some confidence.

The order Carnivora began to diversify when two distinct lines evolved from *creodonts* and *miacids* during the

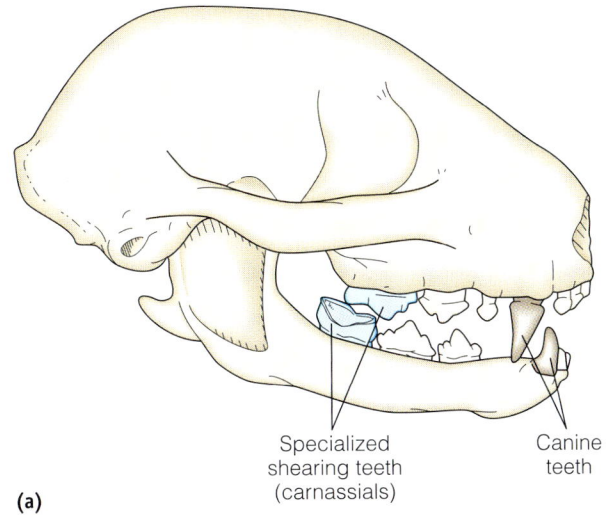
(a)

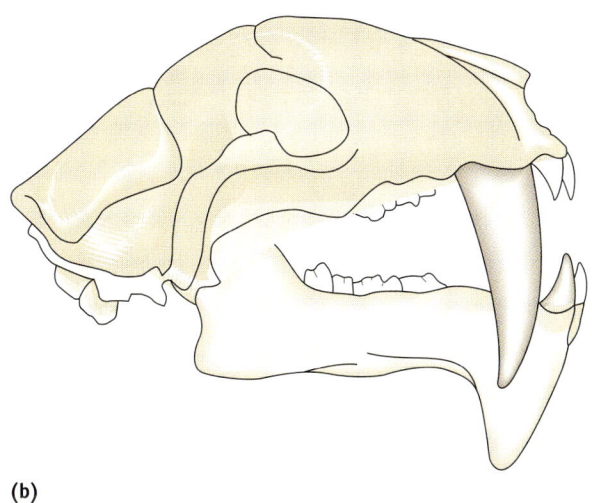
(b)

• **Figure 18.12 Teeth of Carnivorous Mammals** (a) This present-day skull and jaw of a large cat show the specialized sharp-crested shearing teeth of carnivorous mammals. The canine teeth are also large, but several Cenozoic saber-tooth cats had huge canine teeth. (b) The one shown here is the Oligocene saber-tooth *Eusmilus*.

Paleocene. Both had well-developed canines and carnassials, but they were rather short-limbed and flat-footed. Certainly they were not very fast, but neither was their prey. The creodont branch became extinct by Miocene time, so need not concern us further, but the other branch, evolving from miacids, led to all existing carnivorous mammals (• Figure 18.13).

Notice from Figure 18.13 that cats, hyenas, and viverrids (civits and mongooses) share a common ancestor but that dogs are rather distantly related to the somewhat similar appearing hyenas. In fact, dogs (family Canidae) and hyenas (family Hyenadae) not only are similar in appearance but also, with few exceptions, are pack hunters. Nevertheless, the fossil record and studies of living animals clearly indicate hyenas are more closely related to cats and mongooses; their similarity to dogs is another example of convergent evolution. One of the most remarkable developments in cats was the evolution of huge canines in the saber-tooth cats (Figure 18.12b). Saber-tooth cats existed throughout most of the Cenozoic Era and are particularly well known from Pleistocene-aged deposits.

The aquatic carnivores, seals, sea lions, and walruses, are most closely related to bears, but unfortunately, their ancestry is less well known than for other families of carnivores (Figure 18.13). Aquatic adaptations include a somewhat streamlined body, a layer of blubber for insulation, and limbs modified into paddles. Most are fish eaters and have rather simple, single-cusped teeth, except walruses, which have flattened teeth for crushing shells.

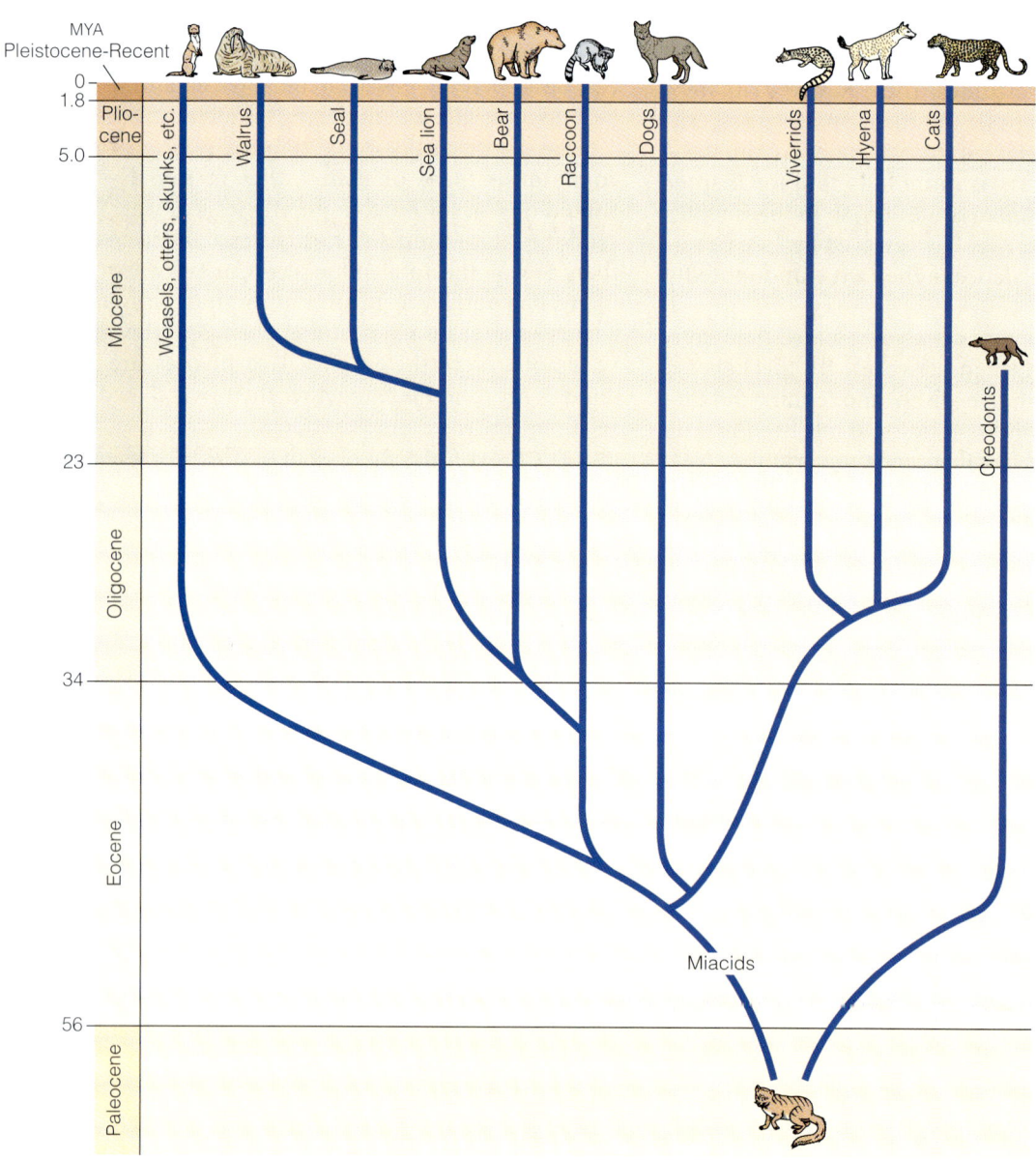

• Figure 18.13 Relationships Among Members of the Order Carnivora All living carnivores and the extinct creodonts evolved from ancient miacids. Notice that hyenas resemble dogs but are more closely related to cats—a good example of convergent evolution.

The Ungulates or Hoofed Mammals

The term **ungulate** is an informal one referring to several groups of living and extinct mammals, particularly the hoofed mammals that belong to the orders **Artiodactyla** and **Perissodactyla**. The artiodactyls, commonly called *even-toed hoofed mammals*, are the most diverse and numerous, with about 170 living species of cattle, goats, sheep, swine, antelope, deer, giraffes, hippopotamuses, camels, and several others. In marked contrast, the perissodactyls, or *odd-toed hoofed mammals*, have only 16 existing species of horses, rhinoceroses, and tapirs. During the Early Cenozoic, though, perissodactyls were more abundant than artiodactyls.

Some defining characteristics of these groups of hoofed mammals are the number of toes and how the animal's weight is borne on the toes. (Their teeth are also distinctive.) Artiodactyls have either two or four toes, and their weight is borne along an axis that passes between the third and fourth digits (• Figure 18.14a). For those artiodactyls with two toes, such as today's swine and deer, the first, second, and fifth digits have been lost or remain only as vestiges. Perissodactyls have one or three toes, although a few fossil species retained four toes on their forefeet. Nevertheless, their weight is borne on an axis passing through the third toe (Figure 18.14a). Even today's horses have vestigial side toes, and rarely they are born with three toes.

Many hoofed mammals such as antelope and horses depend on speed to escape from predators in their open-grasslands habitat. As a result they have long, slender limbs, giving them a greater stride length. Notice from Figure 18.14b that the bones of the palm and sole have become very long. In addition, these speedy runners have fewer bony elements in the feet, mostly because they have fewer toes. Not all hoofed mammals are long-limbed, speedy runners. Some are small and dart into heavy vegetation or a hole in the ground when threatened by predators. And size alone is adequate protection in some very large species such as rhinoceroses. In contrast to the long, slender limbs of horses, antelope, deer, and so on, these animals have developed massive, weight-supporting legs.

All artiodactyls and perissodactyls are herbivorous animals, with their chewing teeth—premolars and molars—modified for a diet of vegetation. One evolutionary trend

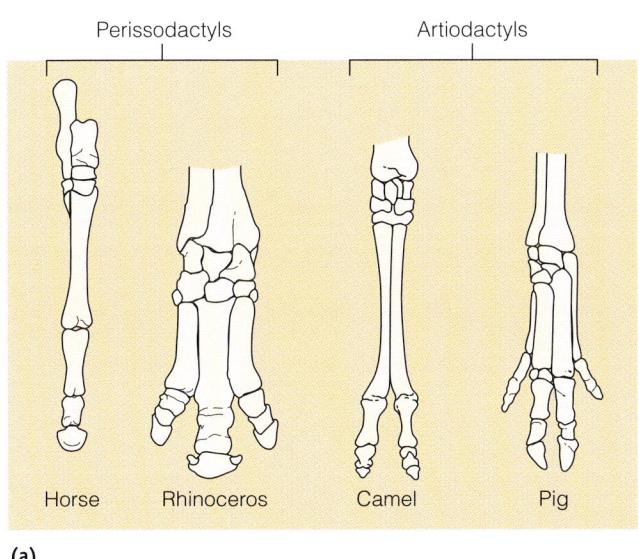

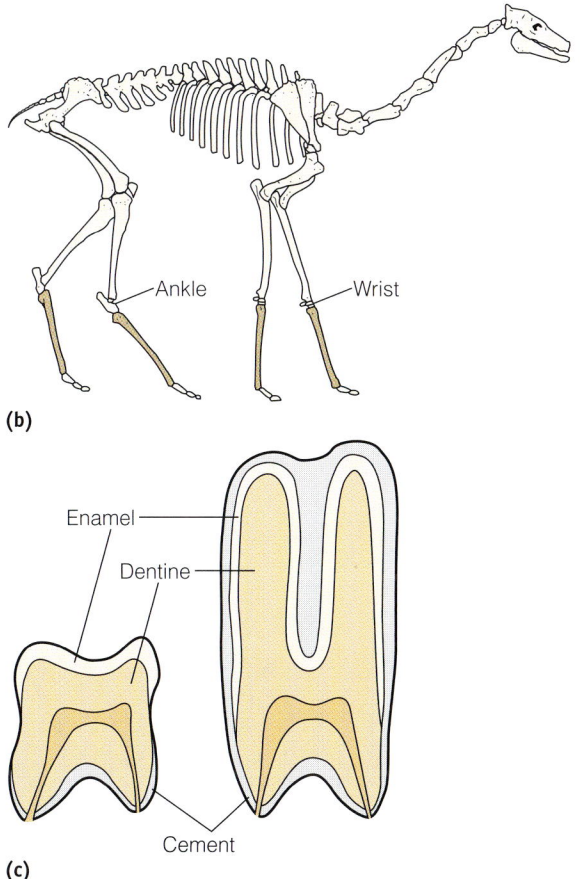

• **Figure 18.14 Evolutionary Trends in Hoofed Mammals (a)** Perissodactyls have one or three functional toes, whereas artiodactyls have one or three. Note that in perissodactyls the weight is borne on the third toe, but artiodactyls walk on toes two and four. **(b)** In many hoofed mammals, long, slender limbs evolved as bones between the wrist and toes and the ankle and toes became longer. **(c)** High-crowned, cement-covered chewing teeth (right) evolved in hoofed mammals that adapted to a diet of grass. Low-crowned teeth are found in many other mammals, including primates and pigs, both of which have a varied diet.

in these animals was **molarization,** a change in the premolars so that they are more like molars, thus providing a continuous row of grinding teeth. Some ungulates—horses, for example—are characterized as **grazers** because they eat grass, as opposed to **browsers,** which eat the tender leaves, twigs, and shoots of trees and shrubs. Grasses are very abrasive, because as they grow through soil they pick up tiny particles of silt and sand that quickly wear teeth down. As a result, once grasses had evolved many hoofed mammals became grazers and developed high-crowned, abrasion-resistant chewing teeth (Figure 18.14c).

Artiodactyls—Even-Toed Hoofed Mammals The oldest known artiodactyls were Early Eocene rabbit-sized animals that differed little from their ancestors. Yet these small creatures were ancestral to the myriad living and several extinct families of even-toed hoofed mammals (• Figure 18.15). Among the extinct families are the rather piglike oreodonts that were so common in North America until their extinction during the Pliocene and the peculiar genus *Synthetoceras* with forked horns on their snouts (Figure 18.10).

During much of the Cenozoic Era, especially in North America, camels of one kind or another were common. The earliest were small four-toed animals, but by Oligocene time all had two toes. Among the various types were very tall giraffe-like camels; slender, gazelle-like camels; and giants standing 3.5 m high at the shoulder. Most camel evolution took place in North America, but during the Pliocene they migrated to Asia and South America, where the only living species exist now. North American camels went extinct near the end of the Pleistocene Epoch.

Artiodactyls are numerous, and among them the family Bovidae is by far the most diverse, with dozens of species of cattle, bison, sheep, goats, and antelope. This family did not appear until the Miocene, but most of its diversification took place during Pliocene time on the northern continents. Bovids are now most numerous in Africa and southern Asia. North America still has its share of bovids such as bighorn sheep and mountain goats, but the most common ones during the Cenozoic were bison (which migrated from Asia), the pronghorn, and oreodonts, all of which roamed the western interior in vast herds.

Notice from Figure 18.15 that most living artiodactyls are **ruminants,** cud-chewing animals with complex three- or four-chambered stomachs in which food is processed to extract more nutrients. Perissodactyls lack such a complex digestive system. Perhaps the fact that artiodactyls use the same resources more effectively than do perissodactyls explains why artiodactyls have flourished and mostly replaced perissodactyls in the hoofed mammal fauna.

Perissodactyls—Odd-Toed Hoofed Mammals In Table 7.1, we said, "If we examine the fossil record of related organisms such as horses and rhinoceroses, we should find that they were quite similar when they diverged from a

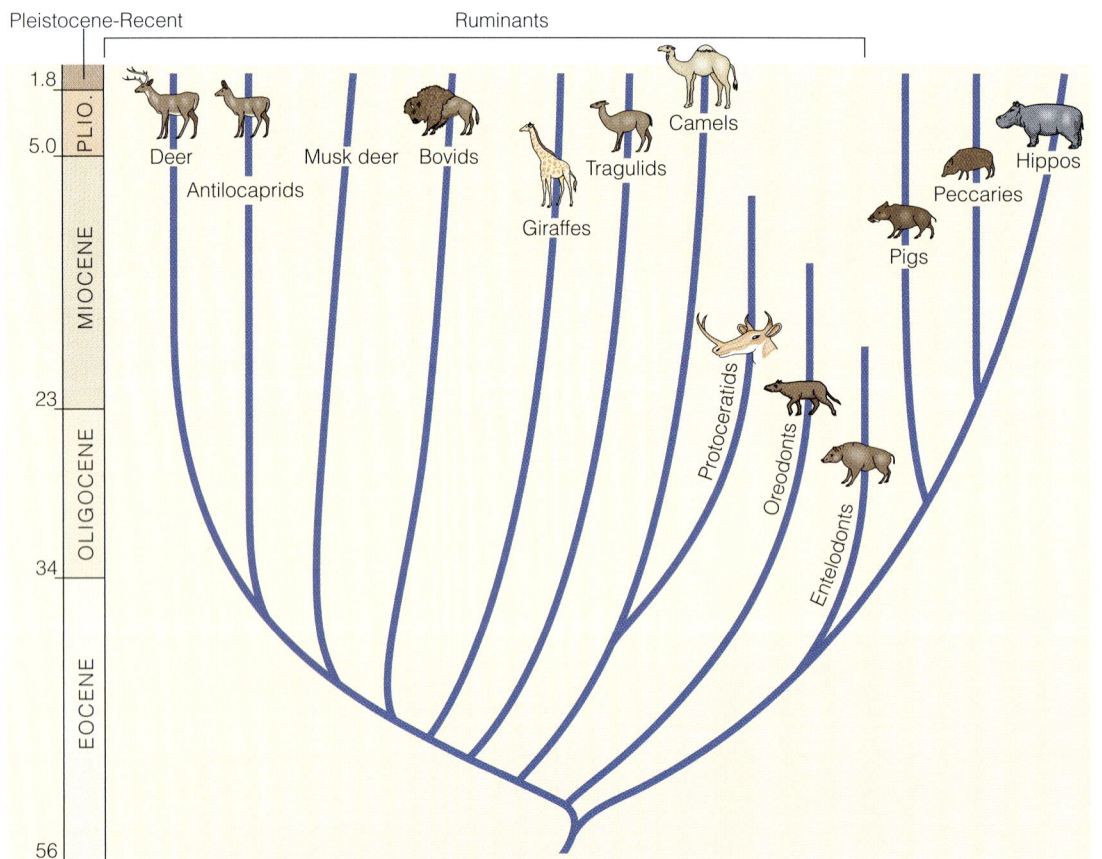

• **Figure 18.15 Relationships Among the Artiodactyls (Even-Toed Hoofed Mammals)** Most artiodactyls are ruminants—that is, cud-chewing animals with complex three- or four-chambered stomachs. The bovids, consisting of dozens of species of sheep, goats, bison, and antelope, are by far the most diverse and numerous artiodactyls.

TABLE 18.1	Trends in the Cenozoic Evolution of the Present-Day Horse *Equus*. A number of horse genera existed during the Cenozoic that evolved differently. For instance, some horses were browsers rather than grazers and never developed high-crowned chewing teeth, and retained three toes.

Trend

1. Size increase.
2. Legs and feet become longer, an adaptation for running.
3. Lateral toes reduced to vestiges. Only the third toe remains functional in *Equus*.
4. Straightening and stiffening of the back.
5. Incisor teeth become wider.
6. Molarization of premolars yielded a continuous row of teeth for grinding vegetation.
7. The chewing teeth, molars and premolars, become high-crowned and cement-covered for grinding abrasive grasses.
8. Chewing surfaces of premolars and molars become more complex—also an adaptation for grinding abrasive grasses.
9. Front part of skull and lower jaw become deeper to accommodate high-crowned premolars and molars.
10. Face in front of eye becomes longer to accommodate high-crowned teeth.
11. Larger, more complex brain.

common ancestor but became increasingly different as divergence continued." We also discussed how fossil records for horses, rhinoceroses, tapirs, and their extinct relatives, the chalicotheres and titanotheres, provide precisely this kind of evidence. In short, when these animals first appeared in the fossil record, they differed slightly in size and the structure of their teeth, but as they evolved, differences between them became more apparent. Perissodactyls evolved from a common ancestor during the Paleocene, reached their greatest diversity during the Oligocene, and have declined markedly since then.

With the possible exception of camels, probably no group of mammals has a better fossil record than do horses. Indeed, horse fossils are so common, especially in North America where most of their evolution took place, that their overall history and evolutionary trends are well known. The earliest member of the horse family (family Equidae) is the fox-sized animal known as **Hyracotherium** (• Figure 18.16). This small, forest-dwelling animal had four-toed forefeet and three-toed hind feet, but each toe was covered by a small hoof. Otherwise it possessed few of the features of present-day horses. So how can we be sure it belongs to the family Equidae at all?

Horse evolution was a complex, branching affair, with numerous genera and species existing at various times during the Cenozoic (Figure 18.16). Nevertheless, their exceptional fossil record clearly shows *Hyracotherium* is linked to the present-day horse, *Equus*, by a series of animals possessing intermediate characteristics. That is, Late Eocene and Early Oligocene horses, followed by more recent ones, show a progressive development of the characteristics found in present-day *Equus* (Table 18.1).

Figure 18.16 shows that horse evolution proceeded along two distinct branches. One led to three-toed browsing horses, all now extinct, and the other led to three-toed grazing horses and finally to one-toed grazers. The appearance of grazing horses, with high-crowned chewing teeth (Figure 18.14c), coincided with the evolution and spread of grasses during the Miocene. Speed was essential in this habitat, and horses' legs became longer and the number of toes was reduced finally to one (Figure 18.16). Pony-sized *Merychippus* is a good example of the early grazing horses; it had three toes, but its teeth were high-crowned and covered by abrasion-resistant cement.

The other living perissodactyls, rhinoceroses and tapirs, increased in size from Early Cenozoic ancestors, and both became more diverse and widespread than they are now. Most rhinoceroses evolved in the Old World, but North American rhinoceroses were common until they became extinct at the end of the Pleistocene. At more than 5 m high at the shoulder and weighing perhaps 13 or 14 metric tons, a hornless Oligocene-Miocene rhinoceros in Asia was the largest land-dwelling mammal ever.

For the remaining perissodactyls, chalicotheres and titanotheres, only the latter has a good fossil record. Chalicotheres, although never particularly abundant, are interesting, because the later members of this family, which were the size of large horses, had claws on their feet, rather than hooves. The prevailing opinion is that these claws were used to hook and pull down branches. Titanotheres existed only during the Eocene, giving them the distinction of being the shortest-lived perissodactyl family. They evolved from small ancestors to giants standing 2.5 m high at the shoulder (see Figures 7.16 and 18.1).

Giant Land-Dwelling Mammals—Elephants

A distinctive characteristic of elephants (order **Proboscidea**) is their long snout, or proboscis. During much of the Cenozoic, proboscideans of one kind or another were widespread on the northern continents, but now only two species exist, one in southeast Asia and one in Africa. The earliest member

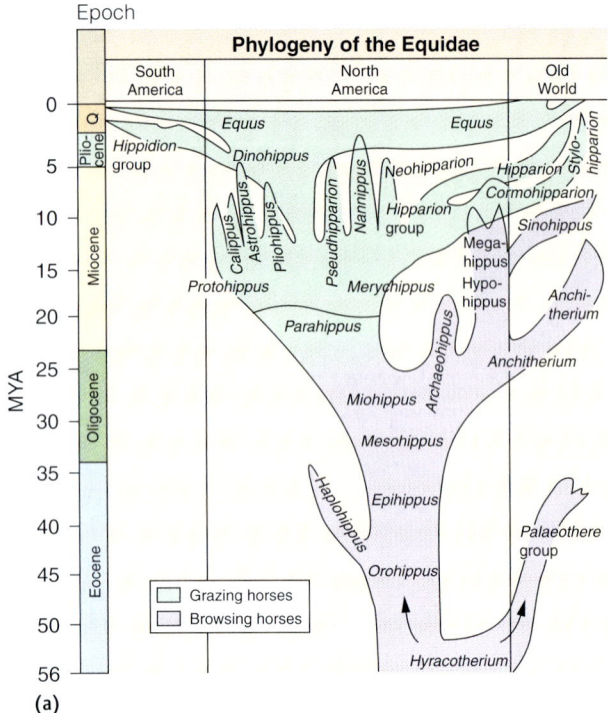
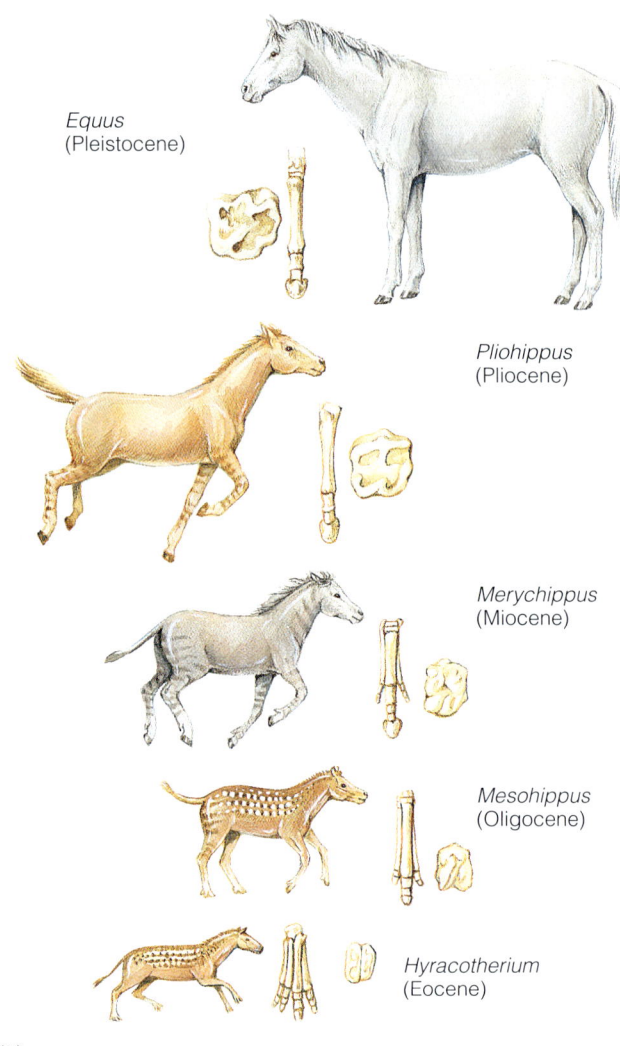

• **Figure 18.16 Evolution of Horses** (a) Summary chart showing the various genera of horses and their relationships. Note that during the Oligocene two separate lines emerged, one leading to three-toed browsers and the other to one-toed grazers, including the present-day horses. (b) Simplified diagram showing some of the evolutionary trends in horses. Important trends shown include an increase in size, lengthening of the limbs and reduction in the number of toes, and the development of high-crowned teeth with complex chewing surfaces.

of the order was a 100- to 200-kg creature called *Moeritherium* from the Eocene that possessed few elephant characteristics. It was probably semiaquatic.

By Oligocene time, elephants showed the trends toward large size and had developed a long proboscis and large tusks, which are enlarged incisors. Most elephants developed tusks in the upper jaw only, but a few had them in both jaws, and one, the deinotheres, had only lower tusks (• Figure 18.17).

The most familiar elephants, other than living ones, are the extinct *mastodons* and *mammoths*. Mastodons evolved in Africa, but from Miocene to Pleistocene time they spread over the Northern Hemisphere continents and one genus even reached South America. These large browsing animals died out only a few thousands of years ago. During the Pliocene and Pleistocene, mammoths and living elephants diverged (Figure 18.17). Most mammoths were no larger than elephants today, but they had the largest tusks of any elephant. In fact, mammoth tusks are common enough in Siberia that they have been and continue to be a source of ivory. Until their extinction near the end of the Pleistocene, mammoths lived on all Northern Hemisphere continents as well as in India and Africa.

Giant Aquatic Mammals—Whales

Our fascination with huge dinosaurs should not overshadow the fact that by far the largest animal ever is alive

What Would You Do?

You are a science teacher who, through remarkably good fortune, receives from a generous benefactor numerous unlabeled mammal and plant fossils. You're not too concerned with identifying genera and species, but you do want to show your students various mammalian adaptations for speed and diet. What features of the skulls, teeth, and bones would allow you to infer which animals were herbivores (grazers versus browsers) and carnivores, and which ones were speedy runners? Also, could you use the fossil leaves to make any inferences about ancient climates?

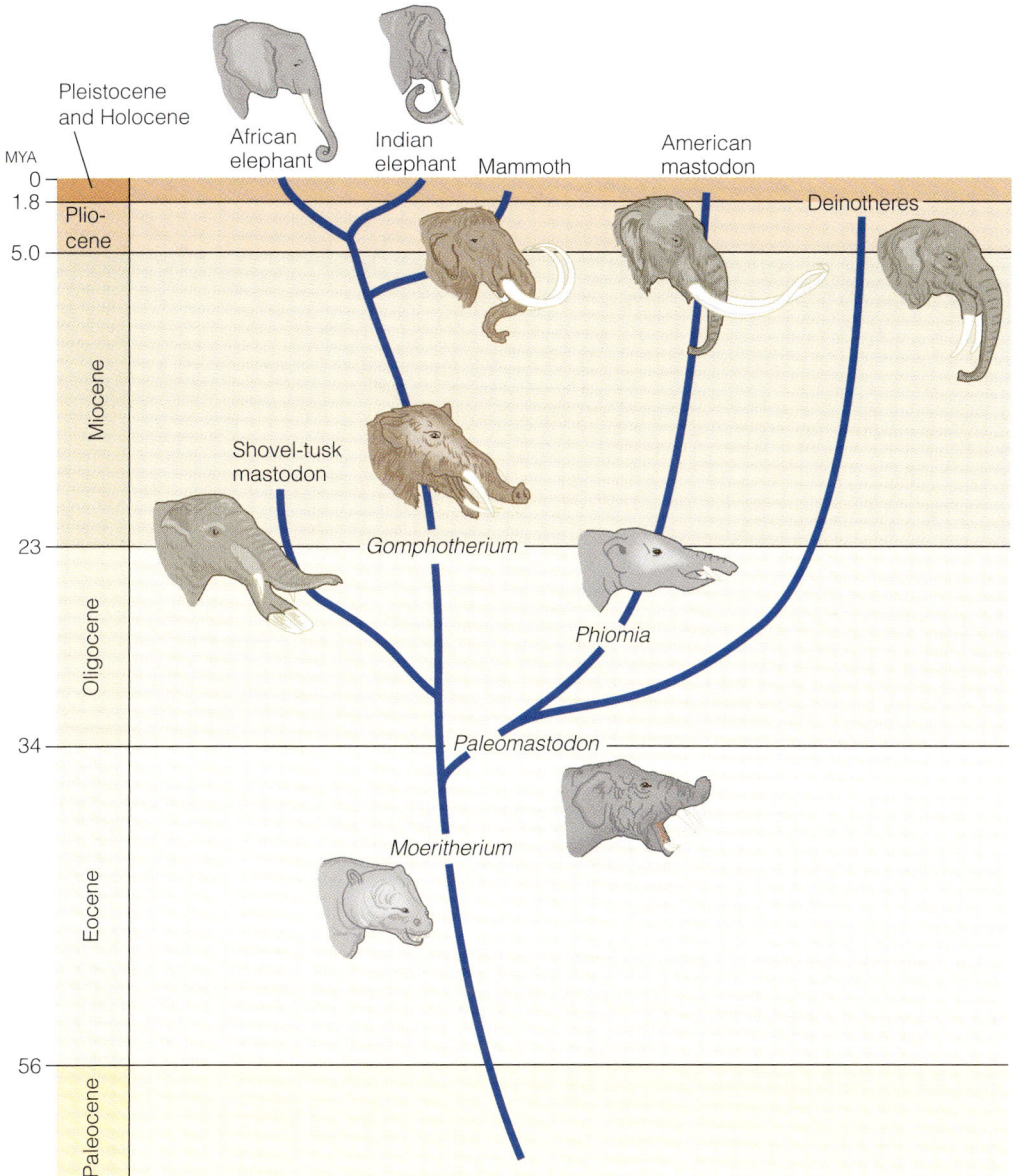

• **Figure 18.17 Phylogeny of Present-day Elephants and Some of Their Relatives.** Increased size and development of large tusks and a long proboscis were some of the evolutionary trends in this group. Several fossil elephants are not shown here, so they were actually more diverse than indicated in this illustration.

today. At more than 30 m long and weighing an estimated 130 metric tons, blue whales greatly exceed the size of any other living thing, except some plants such as redwood trees. But not all whales are large. Consider, for instance, dolphins and porpoises—both are sizable but hardly giants. Nevertheless, an important trend in whale evolution has been increase in body size.

Several kinds of mammals are aquatic or semiaquatic, but only sea cows (see Perspective 7.2) and whales, order **Cetacea**, are so thoroughly aquatic that they cannot come out onto land. Fossils discovered in Middle Eocene rocks in Pakistan indicate the land-dwelling ancestors of whales were among the artiodactyls, but some paleontologists think the ancestors were wolf-sized, meat-eating mammals (• Figure 18.18). During the transition from land-dwelling animals to aquatic whales, the front limbs modified into paddlelike flippers, the rear limbs were lost, and the nostrils migrated to the top of the head. In addition, whales have a large, horizontal tail fluke used for propulsion.

For many years, paleontologists had little fossil evidence that bridged the gap between land-dwelling animals and fully aquatic whales. As we mentioned in Chapter 7, though, this important transition took place in a part of the world where the fossil record was poorly known. Beginning about 20 years ago, paleontologists have made some remarkable finds that resolved this evolutionary enigma. For instance, the Early Eocene whale *Ambulocetus* still had limbs capable of support on land, whereas *Basilosaurus*, a 15 m long Late Eocene whale, had only tiny, vestigial rear limbs (Figure 18.18). The latter had teeth similar to those of their ancestors, and its nostrils were on the snout, but it was truly a whale, although differently proportioned from those living now. By Oligocene time, both presently existing whale groups—baleen whales and toothed whales—had evolved.

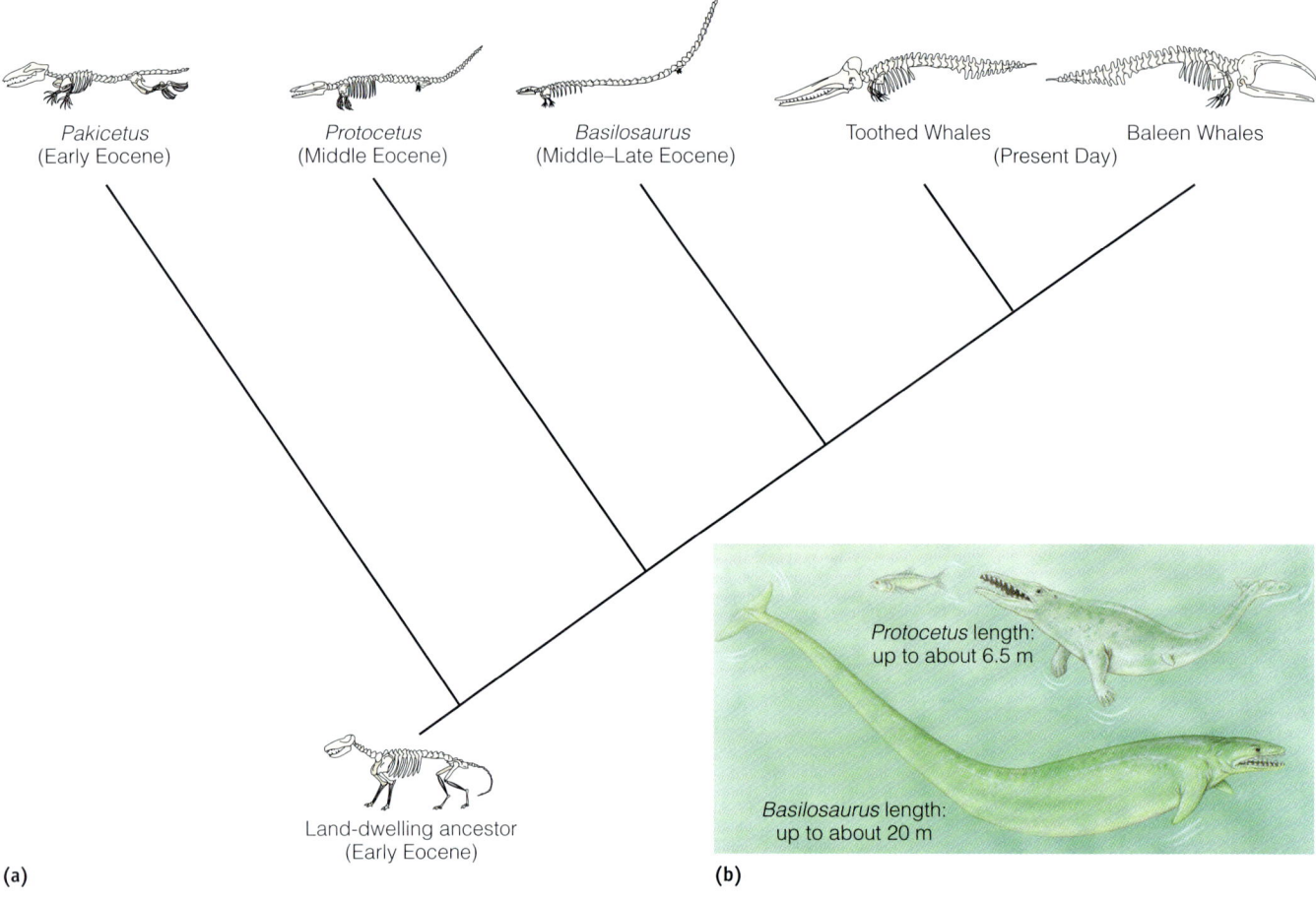

• **Figure 18.18 Cladogram Showing the Relationships Among Whales and Their Land-Dwelling Ancestors** (a) Note that *Pakicetus* had well-developed hind limbs, but only vestiges remain in *Protocetus* and *Basilosaurus*. (b) Restoration of *Protocetus* and *Basilosaurus*. Although *Basilosaurus* was a fully aquatic whale, it differed considerably from today's whales.

An interesting note on fossil whales is that during the 1840s, Albert Koch found what he claimed was a fossil sea serpent in Eocene rocks in Alabama, which was actually a nearly complete skeleton of an extinct whale. In his restoration Koch used the vertebrae of five different animals to render a creature nearly 35 m long. No one in the scientific community was fooled, but Koch took his creation on tour for viewing—for a fee, of course.

Pleistocene Faunas

As opposed to the fauna of the Paleocene Epoch with its archaic mammals, unfamiliar ancestors of living mammalian orders, and large, predatory birds, the fauna of the Pleistocene consists mostly of familiar animals. Even so, their geographic distribution might surprise us, because rhinoceroses, elephants, and camels still lived in North America, and a few unusual mammals, such as chalicotheres and the heavily armored glypotodonts, were present. In the avian fauna the giant moas and elephant birds were in New Zealand and in Madagascar, respectively.

Ice Age Mammals

The most remarkable aspect of the Pleistocene mammalian fauna is that so many very large species existed. Mastodons, mammoths, giant bison, huge ground sloths, immense camels, and beavers 2 m tall at the shoulder were present in North America. South America had its share of giants, too, especially sloths and glyptodonts. Elephants, cave bears, and giant deer known as Irish Elk lived in Europe and Asia (• Figure 18.19), and Australia had 3-m-tall kangaroos and wombats the size of rhinoceroses.

Of course, many smaller mammal species also existed, but one obvious trend among Pleistocene mammals was large body size. Perhaps this was an adaptation to the cooler conditions that prevailed during that time. Large animals have less surface area compared to their volume and thus retain heat more effectively than do smaller animals.

Some of the world's best-known fossils come from Pleistocene deposits. You have probably heard of the frozen mammals found in Siberia and Alaska, such as mammoths, bison, and a few others. These extraordinary fossils, although very rare, provide much more information than most fossils do (see Figure 5.14c). Contrary to what you

• **Figure 18.19 The Irish Elk** Restoration of the giant deer *Megaloceros giganteus*, commonly called the Irish Elk. It lived in Europe and Asia during the Pleistocene. Large males probably weighed 700 kg, about the same as a present-day moose, and had an antler spread of nearly 4 m.

Pleistocene Extinctions

During the Pleistocene, the continental interior of North America was teeming with horses, rhinoceroses, camels, mammoths, mastodons, bison, giant ground sloths, glyptodonts, saber-toothed cats, dire wolves, rodents, and rabbits. Beginning about 14,000 years ago, however, many of these animals became extinct, especially the larger ones. These Pleistocene extinctions were modest compared to those at the end of the Paleozoic and Mesozoic eras, but they were unusual in that they had a profound impact on large, land-dwelling mammals (those weighing more than 44 kg). Particularly hard hit were the mammalian faunas of Australia and the Americas.

In Australia 15 of the continent's 16 genera of large mammals died out. North America lost 33 of 45 genera of large mammals, and in South America 46 of 58 large mammal genera went extinct. In contrast, Europe had only 7 of 23 large genera die out, whereas Africa south of the Sahara lost only 2 of 44 genera. These data bring up three questions, none of which has been answered completely: (1) What caused Pleistocene extinctions? (2) Why did these extinctions eliminate mostly large mammals? (3) Why were extinctions most severe in Australia and the Americas? Scientists are currently debating two competing hypotheses for these extinctions. One holds that rapid changes in climate at the end of the Pleistocene caused extinctions, whereas the other, called *prehistoric overkill*, holds that human hunters were responsible.

Rapid changes in climate and vegetation occurred over much of Earth's surface during the Late Pleistocene, as glaciers began retreating. The North American and northern Eurasian open-steppe tundras were replaced by conifer and broadleaf forests as warmer and wetter conditions prevailed. The Arctic region flora changed from a productive herbaceous one that supported a variety of large mammals, to a relatively barren water-logged tundra that supported a far sparser fauna. The southwestern U.S. region also changed from a moist area with numerous lakes, where saber-tooth cats, giant ground sloths, and mammoths roamed, to a semiarid environment unable to support a diverse fauna of large mammals.

might hear in the popular press, all these frozen animals were partly decomposed, none were fresh enough to eat, and none were found in blocks of ice or icebergs. All were recovered from permanently frozen ground known as *permafrost*.

Paleontologists have recovered Pleistocene animals from many places in North America; two noteworthy areas are Florida and the La Brea tar pits at Rancho La Brea in southern California. In fact, Florida is one of the few places in the eastern United States where fossils of Cenozoic land-dwelling animals are common (see the chapter opening photo). At the La Brea tar pits,* at least 230 kinds of vertebrate animals have been found trapped in the sticky residue where liquid petroleum seeped out at the surface and then evaporated. Many of the fossils are carnivores, especially dire wolves and saber-tooth cats, that gathered to dine on mammals that became mired in the tar (• Figure 18.20).

What Would You Do?

While working in an area with well-exposed Pliocene, organic-rich mudstones and claystones you make a remarkable discovery: several fossil horses, camels, mastodons, and deerlike animals and dozens of skeletons of saber-toothed cats, dogs, and vultures. Why is this association of fossils anomalous, and what features of the rocks might help you resolve this apparent dilemma?

---

*The "tar" is really naturally formed asphalt, whereas tar is a product manufactured from peat or coal.

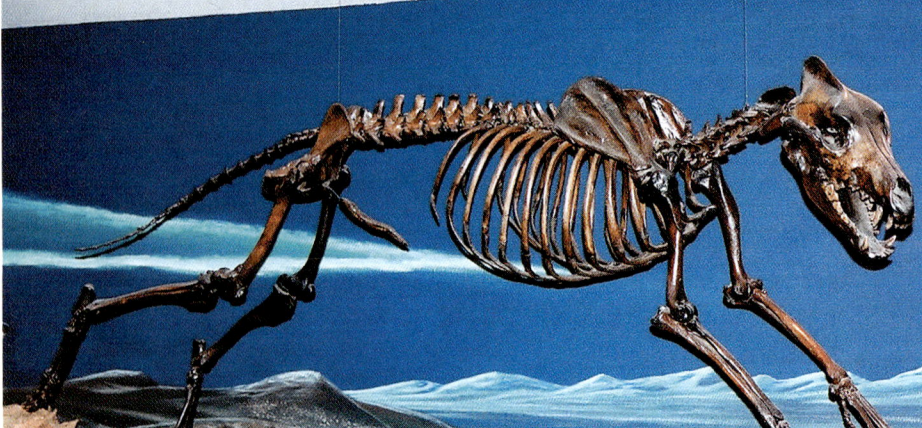

• Figure 18.20 The LaBrea Tar Pits in Los Angeles, California (a) Restoration of a mammoth trapped in the sticky tar (ashphalt) at a present-day oil seep. (b) Skeleton of a dire wolf (*Canis dirus*) in the George C. Page Museum in Los Angeles. Several thousand dire wolves have been recovered, making them the most common mammal in the deposits.

Proponents of the prehistoric overkill hypothesis argue that the mass extinctions in North and South America and Australia coincided closely with the arrival of humans. Perhaps hunters had a tremendous impact on the faunas of North and South America about 11,000 years ago because the animals had no previous experience with humans. The same thing happened much earlier in Australia soon after people arrived about 40,000 years ago. No large-scale extinctions occurred in Africa and most of Europe, because animals in those regions had long been familiar with humans.

One problem with the prehistoric overkill hypothesis is that archaeological evidence indicates the early human inhabitants of North and South America, as well as Australia, probably lived in small, scattered communities, gathering food and hunting. How could a few hunters decimate so many species of large mammals? However, it is true that humans have caused major extinctions on oceanic islands. For example, in a period of about 600 years after arriving in New Zealand, humans exterminated several species of the large, flightless birds called moas.

A second problem is that present-day hunters concentrate on smaller, abundant, and less dangerous animals. The remains of horses, reindeer, and other small animals are found in many prehistoric sites in Europe, whereas mammoth and woolly rhinoceros remains are scarce.

Finally, few human artifacts are found among the remains of extinct animals in North and South America, and there is usually little evidence that the animals were hunted.

Rapid changes in climate and vegetation can certainly affect animal populations, but there are several problems with the climate hypothesis. First, why didn't the large mammals migrate to more suitable habitats as the climate and vegetation changed? After all, many animal species did. For example, reindeer and the Arctic fox lived in southern France during the last glaciation and migrated to the Arctic when the climate became warmer. The second argument against the climate hypothesis is the apparent lack of correlation between extinctions and the earlier glacial advances and retreats throughout the Pleistocene Epoch. Previous changes in climate were not marked by episodes of mass extinctions.

Countering this argument is the assertion that the impact on the previously unhunted fauna was so swift as to leave little evidence.

The reason for the extinctions of large Pleistocene mammals is unresolved and probably will be for some time. It may turn out that the extinctions resulted from a combination of different circumstances. Populations that were already under stress from climate changes were perhaps more vulnerable to hunting, especially if small females and young animals were the preferred targets.

Intercontinental Migrations

The mammalian faunas of North America, Europe, and northern Asia exhibited many similarities throughout the Cenozoic. Even today, Asia and North America are only narrowly separated at the Bering Strait, which at several times during the Cenozoic formed a land corridor across which mammals migrated (see Figure 17.13). During the Early Cenozoic, a land connection between Europe and North America allowed mammals to roam across all the northern continents. Many did; camels and horses are only two examples.

However, the southern continents were largely separate island continents during much of the Cenozoic. Africa remained fairly close to Eurasia, and at times faunal interchange between those two continents was possible. For example, elephants first evolved in Africa, but they migrated to all the northern continents.

South America was isolated from all other landmasses from Late Cretaceous until a land connection with North America formed about 5 million years ago. Before the connection was established, the South American fauna was made up of marsupials and several orders of placental mammals that lived nowhere else. These animals thrived in isolation and showed remarkable convergence with North American placental mammals (see Figure 7.13a). When the Isthmus of Panama formed, migrants from North America soon replaced many of the indigenous South American mammals, whereas fewer migrants from the south were successful in North America (• Figure 18.21). As a result of this *great American interchange*, today about 50% of South America's mammalian fauna came from the north, but in

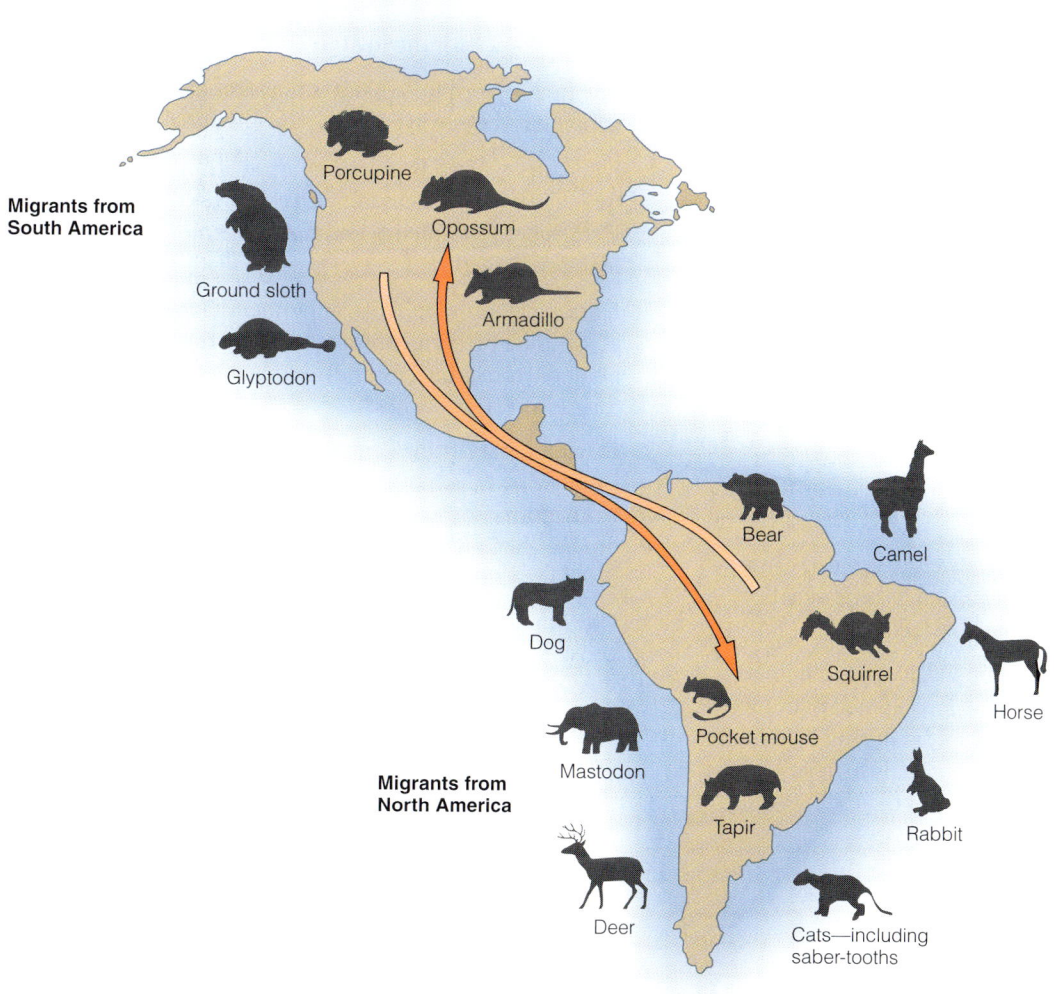

• **Figure 18.21 The Great American Interchange** South America was isolated during much of the Cenozoic, and its mammal fauna consisted of marsupials and placentals that lived nowhere else. When the Isthmus of Panama formed during the Late Pliocene, many placental mammals migrated south, and many South American marsupials became extinct. A few South American mammals migrated north and successfully occupied North America.

North America only 20% of its mammals came from the south. Even today, the coyote (*Canis latrans*) is extending its range from the north through Central America.

Most living species of marsupials are restricted to the Australian region. Recall from Chapter 15 that marsupials occupied Australia before its separation from Gondwana, but apparently placentals, other than bats and a few rodents, never got there until they were introduced by humans. So, unlike South America, which now has a connection with another continent, Australia has remained isolated, and its fauna is unique.

SUMMARY

- The marine invertebrate groups that survived the Mesozoic extinctions diversified throughout the Cenozoic. Bivalves, gastropods, corals, and several kinds of phytoplankton such as foraminifera proliferated.

- During much of the Early Cenozoic, North America was covered by subtropical and tropical forests, but the climate became drier by Oligocene and Miocene time, especially in the midcontinent region.

- Birds belonging to the living orders and families evolved during the Paleogene Period. Large, flightless predatory birds of the Paleogene were eventually replaced by mammalian predators.

- Evolutionary history is better known for mammals than for other classes of vertebrates, because mammals have a good fossil record and their teeth are so distinctive.

- Egg-laying mammals (monotremes) and marsupials exist mostly in the Australian region. The placental mammals—by far the most common mammals—owe their success to their method of reproduction.

- All placental and marsupial mammals descended from shrewlike ancestors that existed from Late Cretaceous to Paleogene time.

- Small mammals such as insectivores, rodents, and rabbits occupy the microhabitats unavailable to larger mammals. Bats, the only flying mammals, have forelimbs modified into wings but otherwise differ little from their ancestors.

- Most carnivorous mammals have well-developed canines and specialized shearing teeth, although some aquatic carnivores such as seals have peglike teeth.

- The most common ungulates are the even-toed hoofed mammals (artiodactyls) and odd-toed hoofed mammals (perissodactyls), both of which evolved during the Eocene. Many ungulates show evolutionary trends such as molarization of the premolars as well as lengthening of the legs for speed.

- During the Paleogene, perissodactyls were more common than artiodactyls but now their 16 living species constitute only about 10% of the world's hoofed mammal fauna.

- Although present-day *Equus* differs considerably from the oldest known member of the horse family, *Hyracotherium*, an excellent fossil record shows a continuous series of animals linking the two.

- Even though horses, rhinoceroses, and tapirs as well as the extinct titanotheres and chalicotheres do not closely resemble one another, fossils show they diverged from a common ancestor during the Eocene.

- The fossil record for whales is now complete enough to verify that they evolved from land-dwelling ancestors.

- Elephants evolved from rather small ancestors, became quite diverse and abundant, especially on the Northern Hemisphere continents, and then dwindled to only two living species.

- Horses, camels, elephants, and other mammals spread across the northern continents during the Cenozoic because land connections existed between those landmasses at various times.

- During most of the Cenozoic, South America was isolated, and its mammal fauna was unique. A land connection was established between the Americas during the Late Cenozoic, and migrations in both directions took place.

- One important evolutionary trend in Pleistocene mammals and some birds was toward giantism. Many of these large species died out, beginning about 40,000 years ago.

- Changes in habitat and prehistoric overkill are the two hypotheses explaining Pleistocene extinctions.

IMPORTANT TERMS

Artiodactyla, p. 385
browser, p. 386
carnassials, p. 383
Carnivora, p. 383
Cetacea, p. 389
grazer, p. 386
Hyracotherium, p. 387
molar, p. 378
molarization, p. 386
Paleocene-Eocene Thermal Maximum, p. 375
Perissodactyla, p. 385
premolar, p. 378
Primates, p. 383
Proboscidea, p. 387
ruminant, p. 386
ungulate, p. 385

REVIEW QUESTIONS

1. Mammals characterized as grazers eat
   a.____leaves; b.____meat; c.____grass; d.____plankton; e.____eggs.
2. The largest land-dwelling mammal to ever exist was a(n)
   a.____rhinoceros; b.____mammoth; c.____mastodon; d.____titanothere; e.____camel.
3. Which one of the following is correct?
   a.____Africa had the most extinctions at the end of the Pleistocene; b.____The Cenozoic Era began 1.8 million years ago; c.____The oldest known member of the horse family is *Megaloceros*; d.____The Paleocene mammal fauna included the giant uintatheres; e.____Ungulates are even- and odd-toed hoofed mammals.
4. One feature of Eocene whales that indicate they had land-dwelling ancestors is
   a.____high-crowned chewing teeth; b.____a long body and a small head; c.____vestigial rear limbs; d.____enlarged eyes; e.____teeth modified for a diet of seaweed.
5. Which one of the following was *not* a trend in the evolution of present-day horses from *Hyracotherium*?
   a.____lengthening of the legs; b.____size increase; c.____evolution of high-crowned chewing teeth; d.____lateral toes reduced to vestiges; e.____development of carnassials.
6. The only living egg-laying mammals are
   a.____multituberculates; b.____monotremes; c.____marsupials; d.____megadonts; e.____moerotheres.
7. One notable trend in Pleistocene mammals was
   a.____large body size; b.____reduced covering of hair; c.____shorter, stouter limbs; d.____increase in the number of teeth; e.____premolarization.
8. During the Cenozoic Era, the temperatures were probably highest during the
   a.____Pleistocene; b.____Miocene; c.____Eocene; d.____Oligocene; e.____Quaternary.
9. One indication of cool climates is
   a.____mammals with nostrils on top of the head; b.____increase in diversity of hoofed mammals; c.____evolution of large predatory birds; d.____small fossil leaves with incised margins; e.____extinctions of coccolithophores.
10. Cenozoic bryozoans were particularly abundant and successful
    a.____suspension feeders; b.____carnivorous mammals; c.____dinoflagellates; d.____predatory birds; e.____marine reptiles.
11. Why are so many fossil mammals found in the western United States but so few in the east?
12. Describe the changes that took place in the teeth and limbs of hoofed mammals that adapted to a diet of grass.
13. What do fossils reveal about the change that took place from *Hyracotherium* to *Equus*?
14. What is the hypothesis of prehistoric overkill, and what evidence is there for and against this hypothesis?
15. How do scientists use leaf structure in fossil plants to make inferences about ancient climates?
16. What were the dominant land-dwelling predators during the earliest Cenozoic, and what happened to them?
17. When did bats first evolve, and what sets them apart from their immediate ancestors?
18. Discuss three evolutionary trends seen in whales.
19. Explain how and why the South American fauna differed from faunas elsewhere and how it has changed.
20. Why do dogs and hyenas resemble one another? Are they closely related? Explain.

APPLY YOUR KNOWLEDGE

1. How does the fossil record for perissodactyls support the theory of evolution? See point 10 in Table 7.1.
2. How would you characterize the mammal faunas of the Paleocene, Miocene, and Pleistocene?

FIELD QUESTION

1. In your wanderings in South Dakota you encounter sandstone beds that have fossils of three-toed horses with high-crowned teeth, small but long-legged camels, and leaves with smooth margins and no drip-tips. What can you say about this ancient habitat?

CHAPTER 19

PRIMATE AND HUMAN EVOLUTION

Courtesy of Persisi Sturges

Olduvai Gorge on the eastern Serengeti Plain, Northern Tanzania, is often referred to as "The Cradle of Mankind" because of the many important hominid discoveries made there. The gorge, part of the East African Rift Valley, is 48 km long and 90 m deep and formed as a result of the tectonic forces shaping East Africa.

[OUTLINE]

INTRODUCTION
WHAT ARE PRIMATES?
PROSIMIANS
ANTHROPOIDS
HOMINIDS
Australopithecines

PERSPECTIVE 19.1 *Footprints at Laetoli*
The Human Lineage
Neanderthals
Cro-Magnons
SUMMARY

ThomsonNOW™ Explore interactive tutorials, animations, or practice problems available on the ThomsonNow website at www.thomsonedu.com/login.

[CHAPTER OBJECTIVES]

At the end of this chapter, you will have learned that

- Primates are difficult to characterize as an order because they lack strong specializations found in most other mammalian orders.

- Primates are divided into two suborders: the prosimians, which include lemurs and tarsiers, and the anthropoids, which include monkeys, apes, and humans.

- The hominids include present-day humans and their extinct ancestors.

- Human evolution is very complex and in a constant state of flux owing to new fossil and scientific discoveries.

- The most famous of all fossil humans are the Neanderthals, which were succeeded by the Cro-Magnons, about 30,000 years ago.

Introduction

Who are we? Where did we come from? What is the human genealogy? These are basic questions that probably everyone at some time or another has asked themselves. Just as many people enjoy tracing their own family history as far back as they can, paleoanthropologists are discovering, based on recent fossil finds, that the human family tree goes back much farther than we thought. In fact, a skull found in the African nation of Chad in 2002 and named *Sahelanthropus tchadensis* (but nicknamed *Tourmaï*, which means "hope of life" in the local Goran language) has pushed back the origins of humans to nearly 7 million years ago. Another discovery reported in 2006 provides strong evidence for an ancestor-descendant relationship between two early hominid lines, one of which leads to our own human lineage.

So where does this leave us, evolutionarily speaking? It leaves us at a very exciting time, as we seek to unravel the history of our species. Our understanding of our genealogy is presently in flux, and each new fossil hominid find sheds more light on our ancestry. Although some may find it frustrating, human evolution is just like that of other groups in that we have followed an uncertain evolutionary path. As new species evolved, they filled ecological niches and either gave rise to descendants better adapted to the changing environment or became extinct. So it should not surprise us that our own evolutionary history has many "dead-end" side branches.

In this chapter we examine the various primate groups, in particular the origin and evolution of the hominids, the group that includes our ancestors. However, we must point out that new discoveries of fossil hominids, as well as new techniques for scientific analysis, are leading to new hypotheses about our ancestry. In the Introduction of the third edition of this book, we stated that the earliest fossil evidence of hominids was from 4.4-million-year-old rocks in eastern Africa. In the fourth edition, we reported that new discoveries pushed that age back to almost 7 million years! In this edition, we can report that new finds in northeastern Ethiopia indicate a direct link between two early hominid groups that were previously thought to be closely related, but because of a sparse fossil record, that link could only be considered tenuous.

By the time you read this chapter, it is possible that new discoveries may change some of the conclusions stated here. Such is the nature of paleoanthropology—and one reason why the study of hominids is so exciting.

What Are Primates?

Primates are difficult to characterize as an order because they lack the strong specializations found in most other mammalian orders. We can, however, point to several trends in their evolution that help define primates and are related to their *arboreal*, or tree-dwelling, ancestry. These include changes in the skeleton and mode of locomotion; an increase in brain size; a shift toward smaller, fewer, and less specialized teeth; and the evolution of stereoscopic vision and a grasping hand with an opposable thumb. Not all these trends took place in every primate group, nor did they evolve at the same rate in each group. In fact, some primates have retained certain primitive features, whereas others show all or most of these trends.

The primate order is divided into two suborders (• Figure 19.1, Table 19.1). The *prosimians*, or lower primates, include the lemurs, lorises, tarsiers, and tree shrews, whereas the *anthropoids*, or higher primates, include monkeys, apes, and humans.

• Figure 19.1 **Primates** Primates are divided into two suborders: the prosimians, such as tarsiers **(a)** and ring-tailed lemurs **(b)**, and **(c)** through **(f)** the anthropoids. The anthropoids are further subdivided into three superfamilies: **(c)** New World monkeys, **(d)** Old World monkeys, and the great apes, such as gorillas **(e)** and chimpanzees **(f)**.

TABLE 19.1	Classification of the Primates

Order Primates: Lemurs, lorises, tarsiers, monkeys, apes, humans
   Suborder Prosimii: Lemurs, lorises, tarsiers, tree shrews (lower primates)
   Suborder Anthropoidea: Monkeys, apes, humans (higher primates)
      Superfamily Cercopithecoidea: Macaque, baboon, proboscis monkey (Old World monkeys)
      Superfamily Ceboidea: Howler, spider, and squirrel monkeys (New World monkeys)
      Superfamily Hominoidea: Apes, humans
         Family Pongidae: Chimpanzees, orangutans, gorillas
         Family Hylobatidae: Gibbons, siamangs
         Family Hominidae: Humans

Prosimians

Prosimians are generally small, ranging from species the size of a mouse up to those as large as a house cat. They are arboreal, have five digits on each hand and foot with either claws or nails, and are typically omnivorous. They have large, forwardly directed eyes specialized for night vision—hence, most are nocturnal (Figure 19.1a and b). As their name implies (*pro* means "before," and *simian* means "ape"), prosimians are the oldest primate lineage, and their fossil record extends back to the Paleocene.

During the Eocene, prosimians were abundant, diversified, and widespread in North America, Europe, and Asia (• Figure 19.2). As the continents moved northward during the Cenozoic and the climate changed from warm tropical to cooler midlatitude conditions, the prosimian population decreased in both abundance and diversity.

By the Oligocene, hardly any prosimians were left in the northern continents as the once widespread Eocene populations migrated south to the warmer latitudes of Africa, Asia, and Southeast Asia. Presently, prosimians are found only in the tropical regions of Asia, India, Africa, and Madagascar.

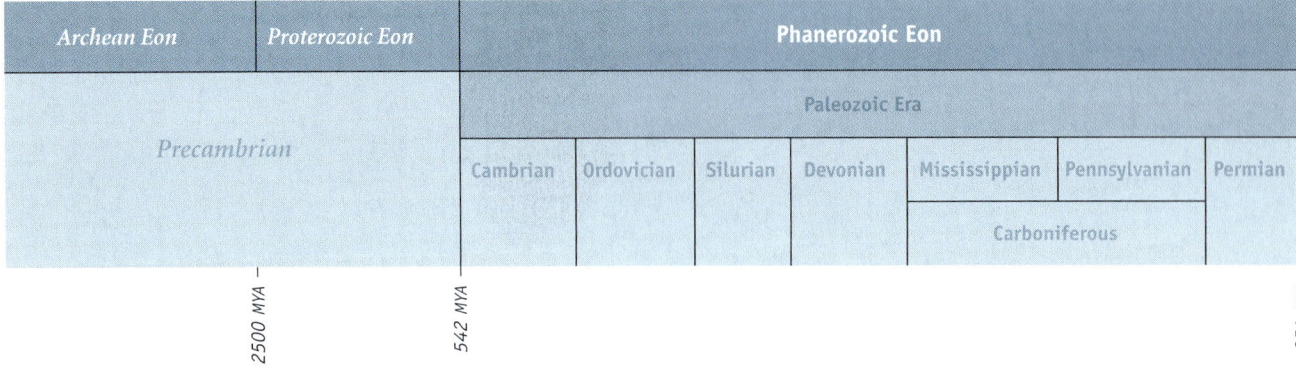

• **Figure 19.2 Eocene Prosimian** *Notharctus*, a primitive Eocene prosimian from North America.

• **Figure 19.3** *Aegyptopithecus zeuxis* Skull of *Aegyptopithecus zeuxis*, one of the earliest known anthropoids.

Anthropoids

Anthropoids evolved from a prosimian lineage sometime during the Late Eocene, and by the Oligocene they were well established. Much of our knowledge about the early evolutionary history of anthropoids comes from fossils found in the Fayum district, a small desert area southwest of Cairo, Egypt. During the Late Eocene and Oligocene, this region of Africa was a lush, tropical rain forest that supported a diverse and abundant fauna and flora. Within this forest lived many different arboreal anthropoids as well as various prosimians. In fact, several thousand fossil specimens representing more than 20 primate species have been recovered from rocks of this region. One of the earliest anthropoids was *Aegyptopithecus*, a small, Late Eocene, fruit-eating, arboreal primate that weighed about 5 kg (• Figure 19.3). *Aegyptopithecus* had not only monkey characteristics but features that were more like apes as well. As such, it is presently the closest link we have to the Old World primates.

Anthropoids are divided into three superfamilies: Old World monkeys, New World monkeys, and hominoids (Table 19.1).

Old World monkeys (superfamily Cercopithecoidea) are characterized by close-set, downward-directed nostrils (like those of apes and humans), grasping hands, and a nonprehensile tail. They include the macaque, baboon, and proboscis monkey (Figure 19.1d). Present-day Old World monkeys are distributed in the tropical regions of Africa and Asia and are thought to have evolved from a primitive anthropoid ancestor, like *Aegyptopithecus*, sometime during the Oligocene.

New World monkeys (superfamily Ceboidea) are found only in Central and South America. They probably evolved from African monkeys that migrated across the widening Atlantic sometime during the Early Oligocene, and they have continued evolving in isolation to this day. No evidence exists of any prosimian or other primitive primates in Central or South America nor of any contact with Old World monkeys after the initial immigration from Africa. New World monkeys are characterized by a prehensile tail, flattish face, and widely separated nostrils, and include the howler, spider, and squirrel monkeys (Figure 19.1c).

Hominoids (superfamily Hominoidea) consist of three families: the *great apes* (family Pongidae), which include chimpanzees, orangutans, and gorillas (Figure 19.1e and f); the *lesser apes* (family Hylobatidae), which are gibbons and siamangs; and the *hominids* (family Hominidae), which are humans and their extinct ancestors.

The hominoid lineage diverged from Old World monkeys sometime before the Miocene, but exactly when is still being debated. It is generally accepted, however, that

Phanerozoic Eon										
Mesozoic Era			Cenozoic Era							
Triassic	Jurassic	Cretaceous	Paleogene			Neogene		Quaternary		
			Paleocene	Eocene	Oligocene	Miocene	Pliocene	Pleistocene	Holocene	

251 MYA · 66 MYA

hominoids evolved in Africa, probably from the ancestral group that included *Aegyptopithecus*.

Recall that beginning in the Late Eocene the northward movement of the continents resulted in pronounced climatic shifts. In Africa, Europe, Asia, and elsewhere, a major cooling trend began, and the tropical and subtropical rain forests slowly began to change to a variety of mixed forests separated by savannas and open grasslands as temperatures and rainfall decreased. As the climate changed, the primate populations also changed. Prosimians and monkeys became rare, whereas hominoids diversified in the newly forming environments and became abundant. Ape populations became reproductively isolated from each other within the various forests, leading to adaptive radiation and increased diversity among the hominoids. During the Miocene, Africa collided with Eurasia, producing additional changes in the climate, as well as providing opportunities for migration of animals between the two landmasses.

Two apelike groups evolved during the Miocene that ultimately gave rise to present-day hominoids. Although there is still not agreement on the early evolutionary relationships among the hominoids, fossil evidence and molecular DNA similarities between modern hominoid families is providing a clearer picture of the evolutionary pathways and relationships among the hominoids.

The first group, the **dryopithecines,** evolved in Africa during the Miocene and subsequently spread to Eurasia following the collision between the two continents. The dryopithecines were a group of hominoids that varied in size, skeletal features, and lifestyle. The best known of all later hominoids is *Proconsul,* an apelike, fruit-eating animal that led a quadrupedal arboreal existence with limited activity on the ground (• Figure 19.4). The dryopithecines were very abundant and diverse during the Miocene and Pliocene, particularly in Africa.

The second group, the **sivapithecids,** evolved in Africa during the Miocene and then spread throughout Eurasia. The fossil remains of sivapithecids are plentiful and consist mostly of skulls, jaws, and isolated teeth. Body or limb bones are rare, limiting our knowledge about what they looked like and how they moved around. We do know that sivapithecids had powerful jaws and thick-enameled teeth with flat chewing surfaces, suggesting a diet of harder and coarser foods, including nuts.

• **Figure 19.4** *Proconsul* Probable appearance of *Proconsul*, a dryopithecine.

It is clear from the fossil evidence that sivapithecids were not involved in the evolutionary branch leading to humans, but they were probably the ancestral stock from which present-day orangutans evolved. In fact, one genus, *Gigantopithecus,* was a contemporary of early *Homo* in Eastern Asia.

Although many pieces are still missing, particularly during critical intervals in the African hominoid fossil record, molecular DNA as well as fossil evidence indicates the dryopithecines, African apes, and hominids form a closely related lineage. The sivapithecids and orangutans, as just discussed, form a different lineage that did not lead to humans.

Hominids

The **hominids** (family Hominidae), the primate family that includes present-day humans and their extinct ancestors (Table 19.1), have a fossil record extending back to almost 7 million years. Several features distinguish them from other hominoids. Hominids are bipedal; that is, they have an upright posture, which is indicated by several modifications in their skeleton (• Figure 19.5a and b). In addition, they show a trend toward a large and internally reorganized brain (Figure 19.5c–e). Other features include a reduced face and

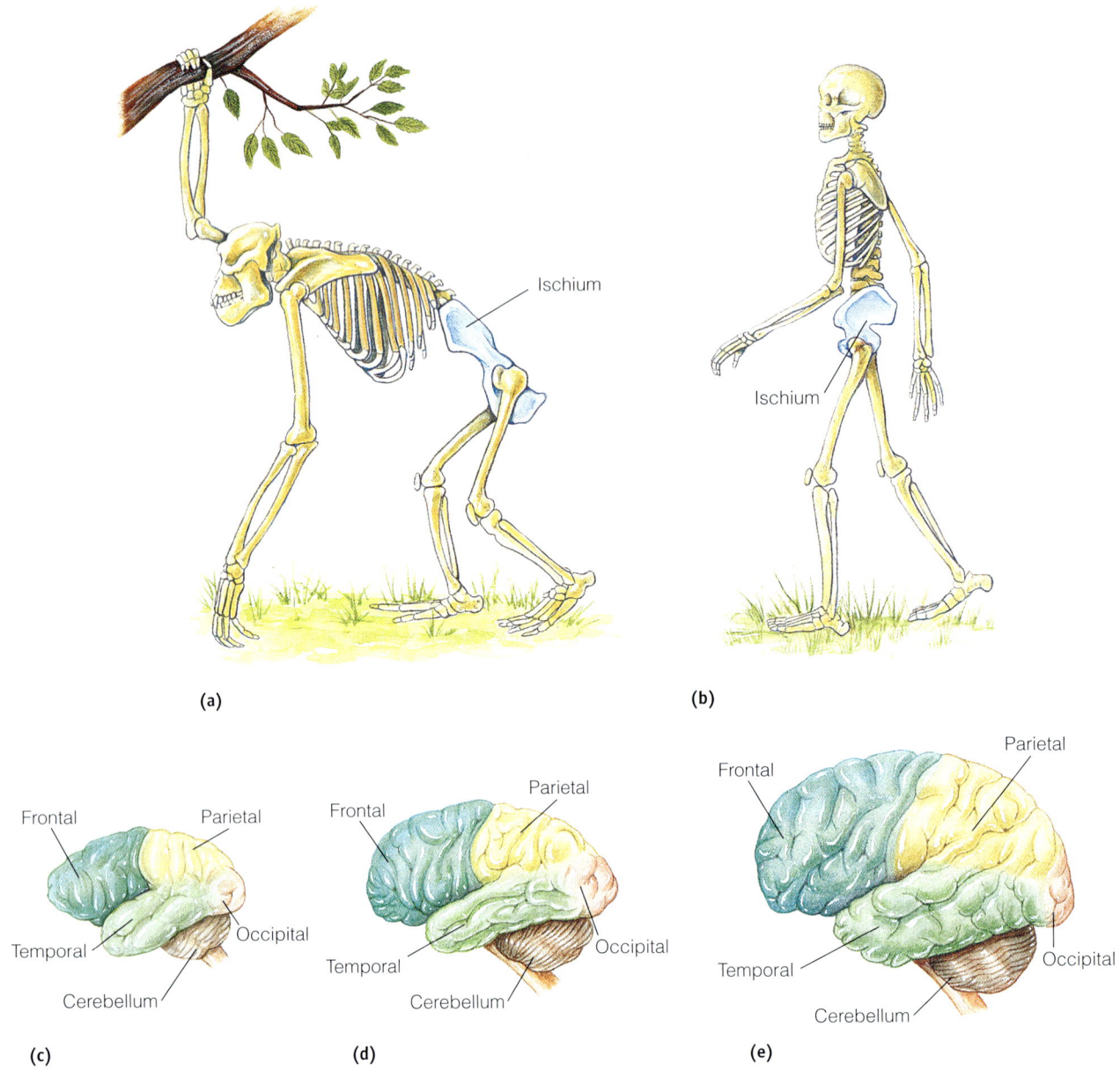

• **Figure 19.5 Comparison of Gorilla and Human Locomotion and Hominid Brain Size** Comparison between quadrupedal and bipedal locomotion in gorillas and humans. **(a)** In gorillas the ischium bone is long, and the entire pelvis is tilted toward the horizontal. **(b)** In humans the ischium bone is much shorter, and the pelvis is vertical. **(c–e)** An increase in brain size and organization is apparent in comparing the brains of **(c)** a New World monkey, **(d)** a great ape, and **(e)** a present-day human.

reduced canine teeth, omnivorous feeding, increased manual dexterity, and the use of sophisticated tools.

Many anthropologists think these hominid features evolved in response to major climatic changes that began during the Miocene and continued into the Pliocene. During this time, vast savannas replaced the African tropical rain forests where the lower primates and Old World monkeys had been so abundant. As the savannas and grasslands continued to expand, the hominids made the transition from true forest dwelling to life in an environment of mixed forests and grasslands.

At present, there is no clear consensus on the evolutionary history of the hominid lineage. This is due, in part, to the incomplete fossil record of hominids as well as new discoveries and also because some species are known only from partial specimens or fragments of bone.

Because of this, there is even disagreement on the total number of hominid species. A complete discussion of all the proposed hominid species and the various competing schemes of hominid evolution is beyond the scope of this chapter. However, we will discuss the generally accepted taxa (• Figure 19.6) and present some of the current theories of hominid evolution.

Remember that although the fossil record of hominid evolution is not complete, what exists is well documented. Furthermore, the interpretation of that fossil record

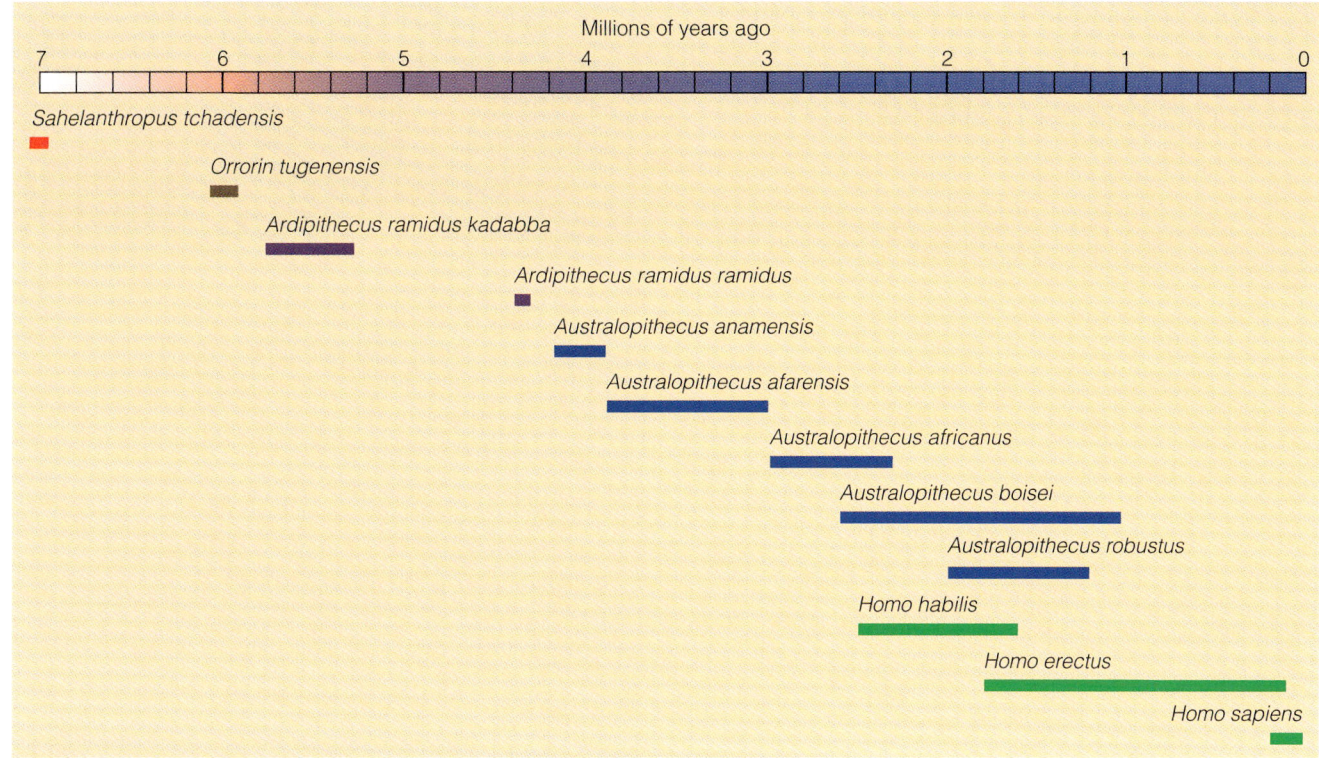

• **Figure 19.6 The Stratigraphic Record of Hominids** The geologic ranges for the commonly accepted species of hominids.

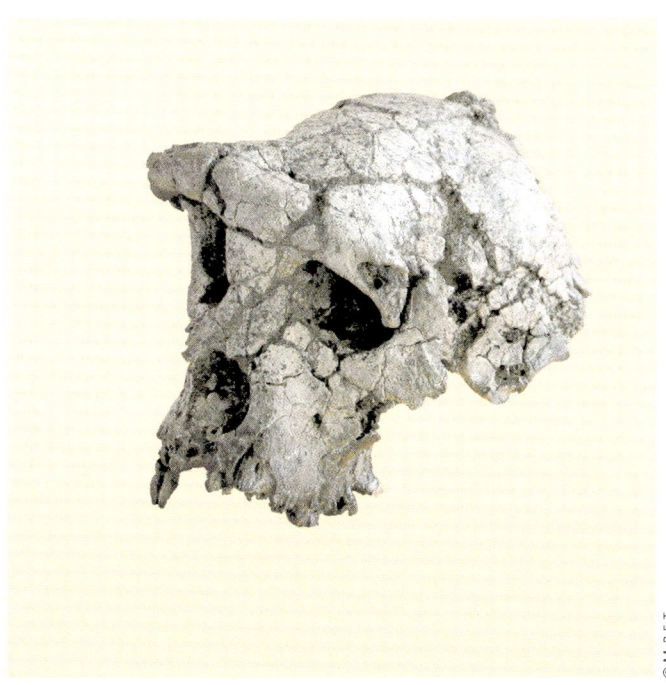

• **Figure 19.7** *Sahelanthropus tchadensis* Discovered in Chad in 2002, and dated at nearly 7 million years, this skull of *Sahelanthropus tchadensis* is presently the oldest known hominid.

precipitates the often vigorous and sometimes acrimonious debates concerning our evolutionary history.

Discovered in northern Chad's Djurab Desert in July 2002, the nearly 7-million-year-old skull and dental remains of *Sahelanthropus tchadensis* (• Figure 19.7) make it the oldest known hominid yet unearthed and at or very near to the time when humans diverged from our closest-living relative, the chimpanzee. Currently, most paleoanthropologists accept that the human-chimpanzee stock separated from gorillas about 8 million years ago and humans separated from chimpanzees about 5 million years ago.

Besides being the oldest hominid, *Sahelanthropus tchadensis* shows a mosaic of primitive and advanced features that has excited and puzzled paleoanthropologists. Its small brain case and most of its teeth (except the canines) are chimplike. However, the nose, which is fairly flat, and the prominent brow ridges are features only seen, until now, in the human genus *Homo*. It is hypothesized that *Sahelanthropus tchadensis* was probably bipedal in its walking habits, but until bones from its legs and feet are found, that supposition remains conjecture.

The next oldest hominid is *Orrorin tugenensis*, whose fossils have been dated at 6 million years old and consist of bits of jaw, isolated teeth, and finger, arm, and partial upper leg bones (Figure 19.6). At this time, there is still debate as to exactly where *Orrorin tugenensis* fits in the hominid lineage.

Sometime between 5.8 and 5.2 million years ago, another hominid was present in eastern Africa. *Ardipithecus ramidus kadabba* is older than its 4.4-million-year-old relative *Ardipithecus ramidus ramidus* (Figure 19.6). *Ardipithecus ramidus kadabba* is very similar in most features to *Ardipithecus ramidus ramidus* but in certain features of its teeth is more apelike than its younger relative.

Although many paleoanthropologists think both *Orrorin tugenensis* and *Ardipithecus ramidus kadabba* were habitual bipedal walkers and thus on a direct evolutionary line to humans, others are not as impressed with the fossil evidence and are reserving judgment. Until more fossil evidence is found and analyzed, any single evolutionary scheme of hominid evolution presented here would be premature.

Australopithecines

Australopithecine is a collective term for all members of the genus *Australopithecus*. Currently, five species are recognized: *A. anamensis, A. afarensis, A. africanus, A. robustus,* and *A. boisei*. Many paleontologists accept the evolutionary scheme in which *A. anamensis,* the oldest known australopithecine, is ancestral to *A. afarensis,* who in turn is ancestral to *A. africanus* and the genus *Homo,* as well as the side branch of australopithecines represented by *A. robustus* and *A. boisei.*

The oldest known australopithecine is *Australopithecus anamensis*. Discovered at Kanapoi, a site near Lake Turkana, Kenya, by Meave Leakey of the National Museums of Kenya and her colleagues, this 4.2-million-year-old bipedal species has many features in common with its younger relative, *A. afarensis,* yet is more primitive in other characteristics, such as its teeth and skull. *A. anamensis* is estimated to have been between 1.3 and 1.5 m tall and weighed between 33 and 50 kg.

A discovery, reported in 2006, of fossils of *Australopithecus anamensis* from the Middle Awash area in northeastern Ethiopia has shed new light on the transition between *Ardipithecus* and *Australopithecus*. Prior to this discovery, the origin of *Australopithecus* has been hampered by a sparse fossil record. The discovery of *Ardipithecus* in the same region of Africa and at the same time as the earliest *Australopithecus* provides strong evidence that *Ardipithecus* evolved into *Australopithecus* and links these two genera in the evolutionary lineage leading to humans.

Australopithecus afarensis (• Figure 19.8), who lived 3.9–3.0 million years ago, was fully bipedal (see Perspective 19.1) and exhibited great variability in size and weight. Members of this species ranged from just over 1 m to about 1.5 m tall and weighed between 29 and 45 kg. They had a brain size of 380–450 cubic centimeters (cc), larger than the 300–400 cc of a chimpanzee but much smaller than that of present-day humans (1350 cc average).

The skull of *A. afarensis* retained many apelike features, including massive brow ridges and a forward-jutting jaw, but its teeth were intermediate between those of apes and humans. The heavily enameled molars were probably an adaptation to chewing fruits, seeds, and roots (• Figure 19.9).

A. afarensis was stratigraphically succeeded by *Australopithecus africanus*, who lived 3.0–2.3 million years ago (• Figure 19.10). The differences between the two species are relatively minor. They were both about the same size

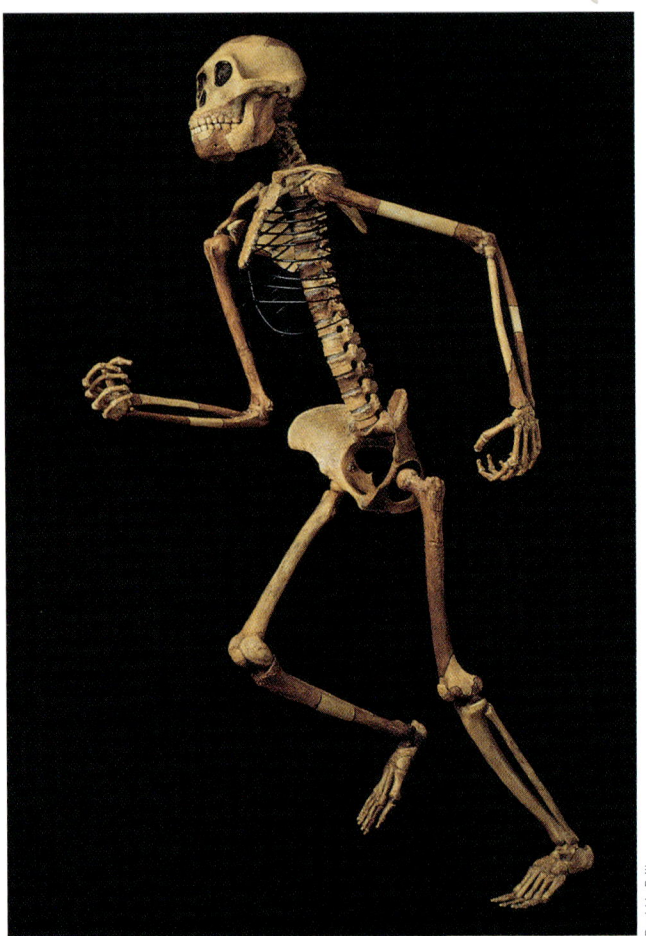

• **Figure 19.8** Skeleton of Lucy *(Australopithecus afarensis)* A reconstruction of Lucy's skeleton by Owen Lovejoy and his students at Kent State University, Ohio. Lucy, whose fossil remains were discovered by Donald Johanson, is an approximately 3.5-million-year-old *Australopithecus afarensis* individual. This reconstruction illustrates how adaptations in Lucy's hip, leg, and foot allowed a fully bipedal means of locomotion.

What Would You Do?

Because of the recent controversy concerning the teaching of evolution in the public schools, your local school board has asked you to make a 30-minute presentation on the evolutionary history of humans and how the fossil record of humans and their ancestors is evidence that evolution is a valid scientific theory. With only 30 minutes to make your case, what evidence in the fossil record would you emphasize, and how would you go about convincing the school board that humans have indeed evolved from earlier hominids?

Darwen and Vally Hennings

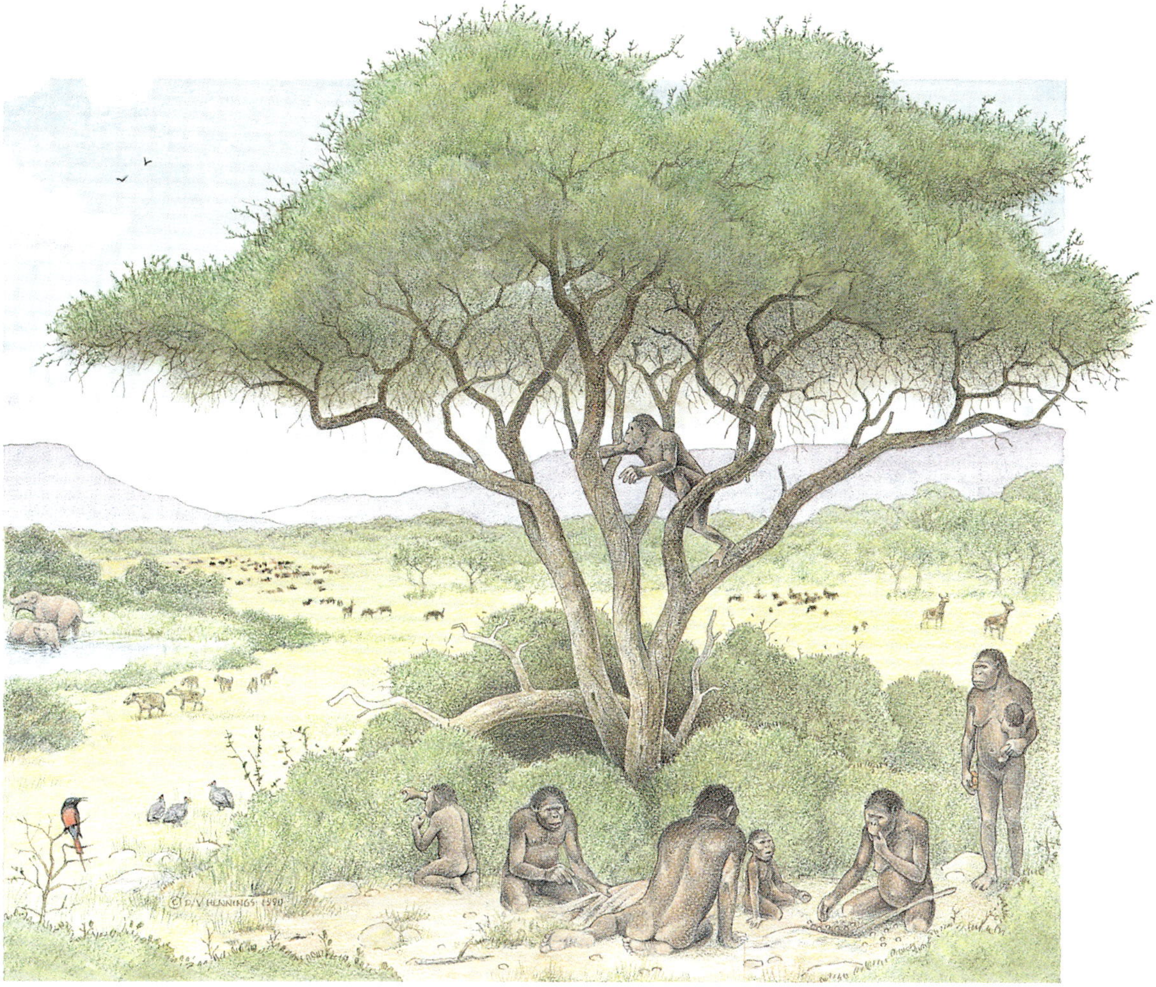

• **Figure 19.9 African Pliocene Landscape** Recreation of a Pliocene landscape showing members of *Australopithecus afarensis* gathering and eating various fruits and seeds.

Replica courtesy of Carolina Biological Supply, photo by Sue Monroe

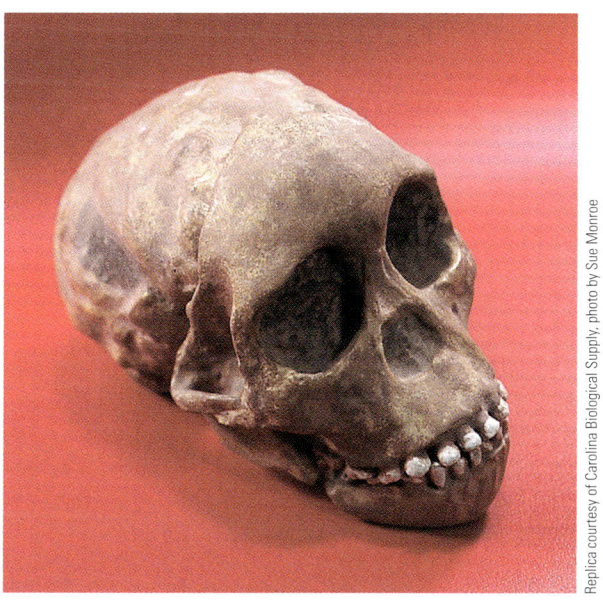

• **Figure 19.10 *Australopithecus africanus*** A reconstruction of the skull of *Australopithecus africanus*. This skull, known as that of the Taung child, was discovered by Raymond Dart in South Africa in 1924 and marks the beginning of modern paleoanthropology.

and weight, but *A. africanus* had a flatter face and somewhat larger brain. Furthermore, it appears the limbs of *A. africanus* may not have been as well adapted for bipedalism as those of *A. afarensis.*

Both *A. afarensis* and *A. africanus* differ markedly from the so-called robust species *A. boisei* (2.6–1.0 million years ago) and *A. robustus* (2.0–1.2 million years ago).

A. boisei was 1.2–1.4 m tall and weighed between 34 and 49 kg. It had a powerful upper body, a distinctive bony crest on the top of its skull, a flat face, and the largest molars of any hominid. *A. robustus,* in contrast, was somewhat smaller (1.1–1.3 m tall) and lighter (32–40 kg). It had a flat face, and the crown of its skull had an elevated bony crest that provided additional area for the attachment of strong jaw muscles (• Figure 19.11). Its broad, flat molars indicated *A. robustus* was a vegetarian.

Most scientists accept the idea that the robust australopithecines form a separate lineage from the other australopithecines that went extinct 1 million years ago.

19.1 Perspective

Footprints at Laetoli

During the summer of 1976, fossil footprints of such animals as giraffes, elephants, rhinoceroses, and several extinct mammals were found preserved in volcanic ash at Laetoli in northern Tanzania. Two years later, a member of Mary Leakey's archaeological team, which was searching for early hominid remains, found what appeared to be a human footprint in the same volcanic ash layer.

Dubbed the Footprint Tuff, a portion of this volcanic ash layer was excavated during the summers of 1978 and 1979, revealing two parallel trails of hominid footprints. This trackway stretched for 27 meters and consisted of 54 individual footprints (Figure 1). Radiometric dating of the ash indicates it was deposited between 3.8 and 3.4 million years ago (Pliocene Epoch), making it the oldest known hominid trackway, during one of several eruptions of ash from the Sadiman volcano, located approximately 20 km east of Laetoli.

In addition to the hominid footprints, there are approximately 18,000 other footprints, representing 17 families of mammals, found in the Laetoli area. Laetoli is part of the eastern branch of the Great Rift Valley of East Africa, a tectonically active area, which is separating from the rest of Africa along a divergent boundary (see Figure 3.16). During the Pliocene Epoch, Sadiman volcano erupted several times, spewing out tremendous quantities of ash that settled over the surrounding savannah. When light rains in the area moistened the ash, any animals walking over it left their footprints. As the ash dried, it hardened like cement, preserving whatever footprints had been made. Subsequent eruptions buried the footprint-bearing ash layer, thus further preserving the footprints.

What makes this find so exciting and scientifically valuable is that the footprints prove early hominids were fully bipedal and had an erect posture long before the advent of stone toolmaking or an increase in the size of the brain. Furthermore, the footprints showed that early hominids walked like modern humans by placing the full weight of the body on the ball of the heel. By examining how deep the impression in the ash is for various parts of the footprint, researchers can infer information about the soft tissue of the feet, something that can't be determined from fossil bones alone.

The question of who made the footprints and how many individuals were walking at the time the footprints were made has been debated since they were initially discovered. Most scientists think the footprints were made by *Australopithecus afarensis,* one of the earlier known hominids, whose fossil bones and teeth are found at Laetoli (Figure 19.9). *A. afarensis* lived from about 3.9 to 3.0 million years ago and exhibited great variability in size and weight. It is estimated that the largest of the hominids making the footprints, a male, was approximately 1.5 m tall, and the smallest hominid, either a female or a child, was approximately 1.2 m tall.

How many people made these parallel trails? There seems to be no argument that the people making these footprints were walking together, with one walking slightly behind the other. It was originally thought the footprints represented a male and female. However, closer examination of the footprints indicates there were probably three people. The larger footprints, which were probably made by a male, have features suggesting they are double prints. In this scenario, a second individual (possibly another male) followed the first one by deliberately stepping in its tracks, thus producing a double print. The smaller footprints are well defined and were probably made by a female or possibly a child. That the trackway was made by three individuals rather than two is now widely accepted.

To ensure these footprints are not destroyed and will be available for future generations to study, the trackway has been reburied. The Antiquities Department of the Tanzanian government, in cooperation with the Getty Conservation Institute, completely reburied the site under five layers of sand, soil, and erosion-control matting. Some of the layers were treated with root inhibitors to prevent roots from destroying the footprints. The site is currently topped with a bed of lava boulders to provide additional protection against erosion and to mark its location. A sacred ceremony was held in 1996 in which the site was included as a place revered by the Masai people.

Figure 1 Hominid footprints preserved in volcanic ash at the Laetoli site, Tanzania. Discovered in 1978 by Mary Leakey, these footprints proved hominids were bipedal walkers at least 3.5 million years ago. The footprints of two adults and possibly those of a child are clearly visible in this photograph.

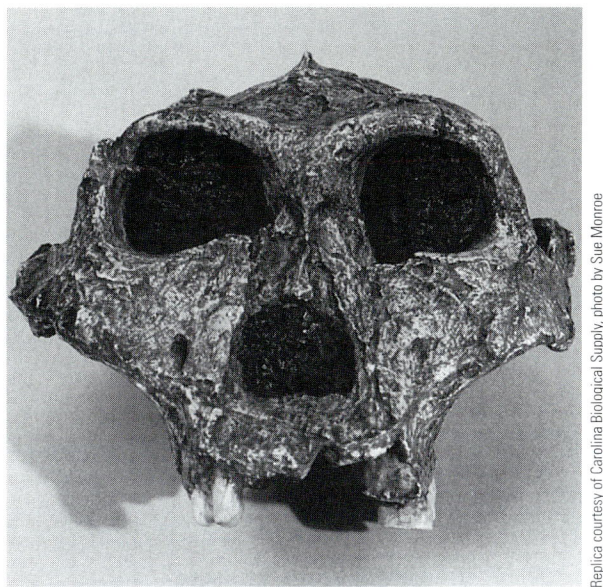

• **Figure 19.11** *Australopithecus robustus* The skull of *Australopithecus robustus*. This species had a massive jaw, powerful chewing muscles, and large, broad, flat chewing teeth apparently used for grinding up coarse plant food.

The Human Lineage

Homo habilis The earliest member of our own genus, **Homo**, is *Homo habilis*, who lived 2.5–1.6 million years ago. Its remains were first found at Olduvai Gorge in Tanzania (see chapter opening photo), but it is also known from Kenya, Ethiopia, and South Africa. *H. habilis* evolved from the *A. afarensis* and *A. africanus* lineage and coexisted with *A. africanus* for about 200,000 years (Figure 19.6). *H. habilis* had a larger brain (700 cc average) than its australopithecine ancestors but smaller teeth. It was about 1.2–1.3 m tall and only weighed 32–37 kg.

Homo erectus In contrast to the australopithecines and *H. habilis*, which are unknown outside Africa, *Homo erectus* was a widely distributed species, having migrated from Africa during the Pleistocene. Specimens have been found not only in Africa but also in Europe, India, China ("Peking Man"), and Indonesia ("Java Man"). *H. erectus* evolved in Africa 1.8 million years ago and by 1 million years ago was present in southeastern and eastern Asia, where it survived until about 100,000 years ago.

Although *H. erectus* developed regional variations in form, the species differed from modern humans in several ways. Its brain size of 800–1300 cc, although much larger than that of *H. habilis*, was still less than the average for *Homo sapiens* (1350 cc). *H. erectus*'s skull was thick-walled, its face was massive, it had prominent brow ridges, and its teeth were slightly larger than those of present-day humans (• Figure 19.12). *H. erectus* was comparable in size to modern humans, standing between 1.6 and 1.8 m tall and weighing between 53 and 63 kg.

The archaeological record indicates that *H. erectus* was a tool maker. Furthermore, some sites show evidence that

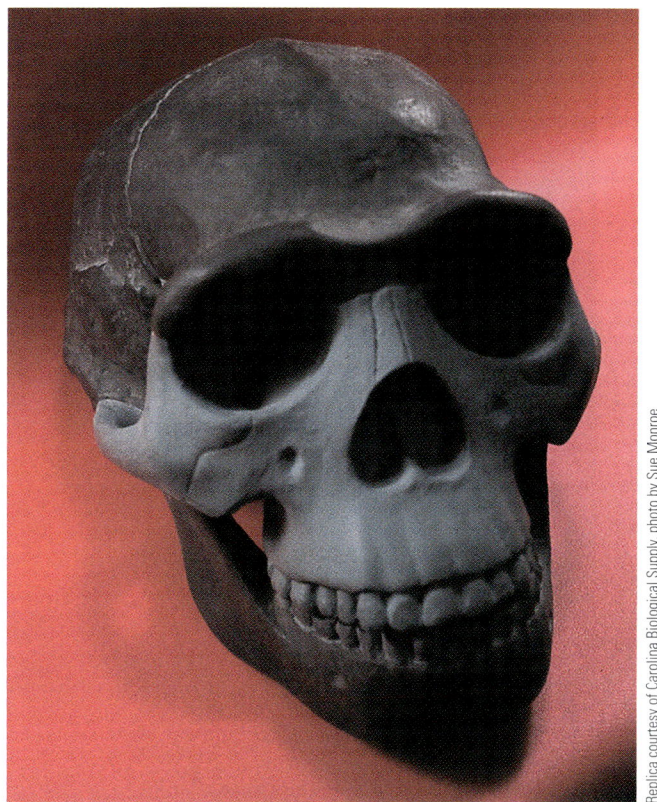

• **Figure 19.12** *Homo erectus* A reconstruction of the skull of *Homo erectus*, a widely distributed species whose remains have been found in Africa, Europe, India, China, and Indonesia.

its members used fire and lived in caves, an advantage for those living in more northerly climates (• Figure 19.13).

Debate still surrounds the transition from *H. erectus* to our own species, *Homo sapiens*. Paleoanthropologists are split into two camps. On the one side are those who support the "out of Africa" view. According to this camp, early modern humans evolved from a single woman in Africa, whose offspring then migrated from Africa, perhaps as recently as 100,000 years ago, and populated Europe and Asia, driving the earlier hominid populations to extinction. On the other side are those supporting the "multiregional" view. According to this hypothesis, early modern humans did not have an isolated origin in Africa but, rather, established separate populations throughout Eurasia. Occasional contact and interbreeding between these populations enabled our species to maintain its overall cohesiveness while still preserving the regional differences in people we see today. Regardless of which theory turns out to be correct, our species, *H. sapiens*, most certainly evolved from *H. erectus*.

Neanderthals

Perhaps the most famous of all fossil humans are the **Neanderthals**, who inhabited Europe and the Near East from about 200,000 to 30,000 years ago. Some paleoanthropologists regard the Neanderthals as a variety or subspecies (*Homo sapiens neanderthalensis*), whereas others regard

• **Figure 19.13 European Pleistocene Landscape with *Homo erectus*** Recreation of a Pleistocene setting in Europe in which members of *Homo erectus* are using fire and stone tools.

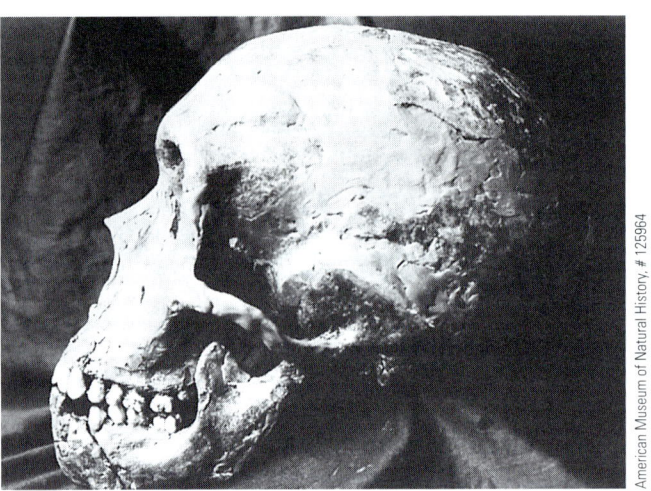

• **Figure 19.14 Reconstructed Neanderthal Skull** The Neanderthals were characterized by prominent heavy brow ridges and a weak chin. Their brain was also slightly larger on average than that of modern humans.

them as a separate species (*Homo neanderthalensis*). In any case, their name comes from the first specimens found in 1856 in the Neander Valley near Düsseldorf, Germany.

The most notable difference between Neanderthals and present-day humans is in the skull. Neanderthal skulls were long and low with heavy brow ridges, a projecting mouth, and a weak, receding chin (• Figure 19.14). Their brain was slightly larger on average than our own and somewhat differently shaped. The Neanderthal body was more massive and heavily muscled than ours, with rather short lower limbs, much like those of other cold-adapted people of today.

Given the specimens from more than 100 sites, we now know Neanderthals were not much different from us, only more robust. Europe's Neanderthals were the first humans to move into truly cold climates, enduring miserably long winters and short summers as they pushed north into tundra country (• Figure 19.15).

The remains of Neanderthals are found chiefly in caves and hutlike rock shelters, which also contain a variety of specialized stone tools and weapons. Furthermore, archaeological evidence indicates that Neanderthals commonly took care of their injured and buried their dead, frequently with such grave items as tools, food, and perhaps even flowers.

Cro-Magnons

About 30,000 years ago, humans closely resembling modern Europeans moved into the region inhabited by the Neanderthals and completely replaced them. **Cro-Magnons,** the name given to the successors of the Neanderthals in France, lived from about 35,000 to 10,000 years ago. During this period the development of art and technology far exceeded anything the world had seen before.

Highly skilled nomadic hunters, Cro-Magnons followed the herds in their seasonal migrations. They used a variety of specialized tools in their hunts, including perhaps the bow and arrow (• Figure 19.16). They sought refuge in caves and rock shelters and formed living groups of various sizes.

Cro-Magnons were also cave painters. Using paints made from manganese and iron oxides, Cro-Magnon

• **Figure 19.15 Pleistocene Cave Setting with Neanderthals** Archaeological evidence indicates that Neanderthals lived in caves and participated in ritual burials, as depicted in this painting of a burial ceremony such as occurred approximately 60,000 years ago at Shanidar Cave, Iraq.

• **Figure 19.16 Pleistocene Cro-Magnon Camp in Europe** Cro-Magnons were highly skilled hunters who formed living groups of various sizes.

people painted hundreds of scenes on the ceilings and walls of caves in France and Spain, where many of them are still preserved today (• Figure 19.17).

With the appearance of Cro-Magnons, human evolution has become almost entirely cultural rather than biological. Humans have spread throughout the world by devising means to deal with a broad range of environmental conditions.

Since the evolution of the Neanderthals about 200,000 years ago, humans have gone from a stone culture to a technology that has allowed us to visit other planets with space probes and land astronauts on the Moon. It remains to be seen how we will use this technology in the future and whether we will continue as a species, evolve into another species, or become extinct as many groups have before us.

• **Figure 19.17 Cro-Magnon Cave Painting** Cro-Magnons were very skilled cave painters. Shown is a painting of a horse from the cave of Niaux, France.

SUMMARY

- The primates evolved during the Paleocene. Several trends help characterize primates and differentiate them from other mammalian orders, including a change in overall skeletal structure and mode of locomotion, an increase in brain size, stereoscopic vision, and evolution of a grasping hand with opposable thumb.

- The primates are divided into two suborders: the prosimians and anthropoids. The prosimians are the oldest primate lineage and include lemurs, lorises, tarsiers, and tree shrews. The anthropoids include the New and Old World monkeys, apes, and hominids, which are humans and their extinct ancestors.

- The oldest known hominid is *Sahelanthropus tchadensis*, dated at nearly 7 million years. It was followed by *Orrorin tugenensis* at 6 million years, then two subspecies of *Ardipithecus* at 5.8 and 4.4 million years, respectively. These early hominids were succeeded by the australopithecines, a fully bipedal group that evolved in Africa 4.2 million years ago. Recent discoveries indicate *Ardipithecus* evolved into *Australopithecus*. Currently, five australopithecine species are known: *Australopithecus anamensis, A. afarensis, A. africanus, A. robustus,* and *A. boisei*.

- The human lineage began about 2.5 million years ago in Africa with the evolution of *Homo habilis,* which survived as a species until about 1.6 million years ago.

- *Homo erectus* evolved from *H. habilis* about 1.8 million years ago and was the first hominid to migrate out of Africa. Between 1.8 and 1 million years ago, *H. erectus* had spread to Europe, India, China, and Indonesia. *H. erectus* used fire, made tools, and lived in caves.

- Sometime between 200,000 and 100,000 years ago, *Homo sapiens* evolved from *H. erectus*. These early humans may be ancestors of Neanderthals.

- Neanderthals were not much different from present-day humans, only more robust and with differently shaped skulls. They made specialized tools and weapons, apparently took care of their injured, and buried their dead.

- The Cro-Magnons were the successors of the Neanderthals and lived from about 35,000 to 10,000 years ago. They were highly skilled nomadic hunters, formed living groups of various sizes, and were also skilled cave painters.

- Modern humans succeeded the Cro-Magnons about 10,000 years ago and have spread throughout the world.

IMPORTANT TERMS

anthropoid, p. 400
australopithecine, p. 404
Cro-Magnon, p. 408
dryopithecine, p. 401
hominid, p. 401

hominoid, p. 400
Homo, p. 407
Neanderthal, p. 407
New World monkey, p. 400
Old World monkey, p. 400

primate, p. 398
prosimian, p. 399
sivapithecid, p. 401

REVIEW QUESTIONS

1. The oldest currently known hominid is
   a. _____ *Sahelanthropus tchadensis*; b. _____ *Orrorin tugenensis*; c. _____ *Ardipithecus ramidus ramidus*; d. _____ *Australopithecus anamensis*; e. _____ *Homo erectus*.
2. Which extinct lineage of humans were skilled hunters and cave painters?
   a. _____ Acheuleans; b. _____ archaic; c. _____ Neanderthals; d. _____ Cro-Magnons; e. _____ none of the previous answers.
3. Which of the following features distinguish hominids from other hominoids?
   a. _____ a large and internally reorganized brain; b. _____ a reduced face and reduced canine teeth; c. _____ bipedalism; d. _____ use of sophisticated tools; e. _____ all of the previous answers.
4. The human lineage began with the evolution of which species?
   a. _____ *Orrorin tugenensis*; b. _____ *Ardipithecus ramidus*; c. _____ *Sahelanthropus tchadensis*; d. _____ *Homo habilis*; e. _____ *Australopithecus boisei*.
5. The first hominids to migrate out of Africa and from which we evolved were
   a. _____ *Australopithecus robustus*; b. _____ *Homo erectus*; c. _____ *Homo sapiens*; d. _____ *Ardipithecus ramidus ramidus*; e. _____ *Homo habilis*.
6. To which of the following species do Java Man and Peking Man belong?
   a. _____ *Homo sapiens*; b. _____ *Australopithecus robustus*; c. _____ *Homo erectus*; d. _____ *Australopithecus boisei*; e. _____ *Homo habilis*.
7. Which is the oldest primate lineage?
   a. _____ anthropoids; b. _____ prosimians; c. _____ insectivores; d. _____ omnivores; e. _____ hominids.
8. Which of the following evolutionary trends characterize primates?
   a. _____ change in overall skeletal structure; b. _____ grasping hand with opposable thumb; c. _____ increase in brain size; d. _____ stereoscopic vision; e. _____ all of the previous answers.
9. The oldest known australopithecine is *Australopithecus* _____.
   a. _____ *robustus*; b. _____ *afarensis*; c. _____ *anamensis*; d. _____ *boisei*; e. _____ *africanus*.
10. Which of the following is a hominid?
    a. _____ chimpanzee; b. _____ gibbon; c. _____ prosimian; d. _____ australopithecine; e. _____ gorilla.
11. When did primates evolve?
    a. _____ Paleocene; b. _____ Eocene; c. _____ Oligocene; d. _____ Miocene; e. _____ Pliocene.
12. According to archaeological evidence, which were the first hominids to use fire?
    a. _____ *Australopithecus robustus*; b. _____ *Homo sapiens*; c. _____ *Homo erectus*; d. _____ *Homo habilis*; e. _____ *Australopithecus boisei*.
13. Discuss the importance of the discovery that *Ardipithecus* evolved into *Australopithecus* in terms of hominid evolution.
14. What major evolutionary trends characterize the primates and set them apart from the other orders of mammals?
15. What are the main differences between the Neanderthals and Cro-Magnons?
16. Discuss the evolutionary history of the genus *Homo*.
17. Discuss the evolutionary history of the anthropoids.
18. Discuss the evolutionary history of hominids.
19. Discuss the differences between the prosimians and anthropoids.
20. Discuss the merits of the "out of Africa" and "multiregional" views concerning the transition between *Homo erectus* and *Homo sapiens*.

APPLY YOUR KNOWLEDGE

1. Based on what you now know about human evolution, as well as what you've read and witnessed about the rapid technological advances in science and their impact on society and the environment, speculate on the next 5000 years of human evolution.

Epilogue

The theme of this book is that Earth is a complex, dynamic planet that has changed continuously since its origin some 4.6 billion years ago. These changes and the present-day features we observe are the result of interactions between the various interrelated internal and external Earth systems, subsystems, and cycles. Furthermore, these interactions have also influenced the evolution of the biosphere.

The rock cycle (see Figure 2.7), with its recycling of Earth materials to form the three major rock groups, illustrates the interrelationships between Earth's internal and external processes. The hydrologic cycle is the continuous recycling of water from the oceans, to the atmosphere, to the land, and eventually back to the oceans again. Changes within this cycle can have profound effects on Earth's topography as well as its biota. For example, a rise in global temperature will cause the ice caps to melt, contributing to rising sea level, which will greatly affect coastal areas where many of the world's large population centers are presently located. We have seen the effect of changing sea level on continents in the past. Such rises and drops in sea level resulted in large-scale transgressions and regressions. Some of these were caused by growing and shrinking continental ice caps when land masses moved over the South Pole as a result of plate movements (see Chapter 11).

On a larger scale, the movement of plates has had a profound effect on the formation of landscapes, the distribution of mineral resources, and atmospheric and oceanic circulation patterns, as well as the evolution and diversification of life.

The launching in 1957 of *Sputnik 1*, the world's first artificial satellite, ushered in a new global consciousness in terms of how we view Earth and our place in the global ecosystem. Satellites have provided us with the ability to view not only the beauty of our planet, but also the fragility of Earth's biosphere and the role humans play in shaping and modifying the environment. The pollution of the atmosphere, oceans, and many of our lakes and streams; the denudation of huge areas of tropical forests; the scars from strip mining; the depletion of the ozone layer—all are visible in the satellite images beamed back from space and attest to the impact humans have had on the ecosystem.

Accordingly, we must understand that changes we make in the global ecosystem can have wide-ranging effects of which we may not even be aware. For this reason, an understanding of geology, and science in general, is of paramount importance so that disruption to the ecosystem is minimal. On the other hand, we must also remember that humans are part of the global ecosystem, and, like all other life-forms, our presence alone affects the ecosystem. We must therefore act in a responsible manner, based on sound scientific knowledge, so future generations will inherit a habitable environment.

When such environmental issues as acid rain, the greenhouse effect and global warming, and the depletion of the ozone layer are discussed and debated, it is important to remember that they are not isolated topics but are part of a larger system that involves the entire Earth. Furthermore, it is important to remember that Earth goes through cycles of much longer duration than the human perspective of time. Although they may have disastrous effects on the human species, global warming and cooling are part of a larger cycle that has resulted in numerous glacial advances and retreats during the past 1.8 million years (see Chapter 17). In fact, geologists can make important contributions to the debate on global warming because of their geologic perspective. Long-term trends can be studied by analyzing deep-sea sediments, ice cores, changes in sea level during the geologic past, and the distribution of plants and animals through time. As we have seen throughout this book, such studies have been done, and the results and synthesis of that information can be used to make intelligent decisions about how humans can better manage the environment and the effect we are having in altering the environment.

An example of an environmental imbalance is global warming caused by the greenhouse effect. Carbon dioxide is produced as a by-product of respiration and the burning of organic material. As such, it is a component of the global ecosystem and is constantly being recycled as part of the carbon cycle. The concern in recent years over the increase in atmospheric carbon dioxide has to do with its role in the greenhouse effect. The recycling of carbon dioxide between the crust and atmosphere is an important climatic regulator because carbon dioxide, as well as other gases such as methane, nitrous oxide, chlorofluorocarbons, and water vapor, allow sunlight to pass through them but trap the heat reflected back from Earth's surface. Heat is thus retained, causing the temperature of Earth's surface and, more importantly, the atmosphere to increase, producing the greenhouse effect.

Because of the increase in human-produced greenhouse gases during the last 200 years, many scientists are concerned that a global warming trend has already begun and will result in severe global climatic shifts. Most computer models based on the current rate of increase in greenhouse gases show Earth warming as a whole by as much as 5°C during the next 100 years. Such a temperature change will be uneven, however, with the greatest warming occurring in the higher latitudes. As a consequence of this warming, rainfall patterns will shift dramatically. This will have a major effect on the largest grain-producing areas of the world, such as the American Midwest. Drier and hotter conditions will intensify the severity and frequency of droughts, leading to more crop failures and higher food prices. With such shifts in climate, Earth may experience an increase in the expansion of deserts, which will remove valuable crop and grazing lands.

We cannot leave the subject of global warming without pointing out that many scientists are not convinced that the global warming trend is the direct result of increased human activity related to industrialization. They point out that whereas there has been an increase in greenhouse gases, there is still uncertainty about their rate of generation and rate of removal and about whether the 0.5°C rise in global temperature during the past century is the result of normal climatic variations through time or the result of human activity. Furthermore, they point out that even if there is a general global warming during the next 100 years, it is not certain that the dire predictions made by proponents of global warming will come true. Earth, as we know, is a remarkably complex system, with many feedback mechanisms and interconnections throughout its various subsystems. It is very difficult to predict all of the consequences that global warming would have for atmospheric and oceanic circulation patterns.

In conclusion, the most important lesson to be learned from the study of historical geology is that Earth is an extremely complex planet in which interactions between its various systems and subsystems has resulted in changes in the atmosphere, lithosphere, and biosphere through time. By studying how Earth has evolved in the past, we can apply the lessons learned from this study to better understanding how the different Earth systems and subsystems work and interact with each other, and more importantly, how our actions affect the delicate balance between these systems and subsystems. Historical geology is not a static science, but one that, like the dynamic Earth that it seeks to understand, is constantly evolving as new information becomes available.

Appendix A
English-Metric Conversion Chart

English Unit	Conversion Factor	Metric Unit	Conversion Factor	English Unit
Length				
Inches (in)	2.54	Centimeters (cm)	0.39	Inches (in)
Feet (ft)	0.305	Meters (m)	3.28	Feet (ft)
Miles (mi)	1.61	Kilometers (km)	0.62	Miles (mi)
Area				
Square inches (in^2)	6.45	Square centimeters (cm^2)	0.16	Square inches (in^2)
Square feet (ft^2)	0.093	Square meters (m^2)	10.8	Square feet (ft^2)
Square miles (mi^2)	2.59	Square kilometers (km^2)	0.39	Square miles (mi^2)
Volume				
Cubic inches (in^3)	16.4	Cubic centimeters (cm^3)	0.061	Cubic inches (in^3)
Cubic feet (ft^3)	0.028	Cubic meters (m^3)	35.3	Cubic feet (ft^3)
Cubic miles (mi^3)	4.17	Cubic kilometers (km^3)	0.24	Cubic miles (mi^3)
Weight				
Ounces (oz)	28.3	Grams (g)	0.035	Ounces (oz)
Pounds (lb)	0.45	Kilograms (kg)	2.20	Pounds (lb)
Short tons (st)	0.91	Metric tons (t)	1.10	Short tons (st)
Temperature				
Degrees Fahrenheit (°F)	$-32° \times 0.56$	Degrees Celsius (Celsius) (°C)	$\times 1.80 + 32°$	Degrees Fahrenheit (°F)

Examples: 10 inches = 25.4 centimeters; 10 centimeters = 3.9 inches
100 square feet = 9.3 square meters; 100 square meters = 1080 square feet
50°F = 10.08°C; 50°C = 122°F

Appendix B
Classification of Organisms

Any classification is an attempt to make order out of disorder and to group similar items into the same categories. All classifications are schemes that attempt to relate items to each other based on current knowledge and therefore are progress reports on the current state of knowledge for the items classified. Because classifications are to some extent subjective, classification of organisms may vary among different texts.

The classification that follows is based on the five-kingdom system of classification of Margulis and Schwartz.* We have not attempted to include all known life forms, but rather major categories of both living and fossil groups.

Kingdom Monera
Prokaryotes
- Phylum Archaebacteria—(Archean–Recent)
- Phylum Cyanobacteria—Blue-green algae or blue-green bacteria (Archean–Recent)

Kingdom Protoctista
Solitary or colonial unicellular eukaryotes
- Phylum Acritarcha—Organic-walled unicellular algae of unknown affinity (Proterozoic–Recent)
- Phylum Bacillariophyta—Diatoms (Jurassic–Recent)
- Phylum Charophyta—Stoneworts (Silurian–Recent)
- Phylum Chlorophyta—Green algae (Proterozoic–Recent)
- Phylum Chrysophyta—Golden-brown algae, silicoflagellates and coccolithophorids (Jurassic–Recent)
- Phylum Euglenophyta—Euglenids (Cretaceous–Recent)
- Phylum Myxomycophyta—Slime molds (Proterozoic–Recent)
- Phylum Phaeophyta—Brown algae, multicellular, kelp, seaweed (Proterozoic–Recent)
- Phylum Protozoa—Unicellular heterotrophs (Cambrian–Recent)
  - Class Sarcodina—Forms with pseudopodia for locomotion (Cambrian–Recent)
    - Order Foraminifera—Benthonic and planktonic sarcodinids most commonly with calcareous tests (Cambrian–Recent)
    - Order Radiolaria—Planktonic sarcodinids with siliceous tests (Cambrian–Recent)
- Phylum Pyrrophyta—Dinoflagellates (Silurian?, Permian–Recent)
- Phylum Rhodophyta—Red algae (Proterozoic–Recent)
- Phylum Xanthophyta—Yellow-green algae (Miocene–Recent)

Kingdom Fungi
- Phylum Zygomycota—Fungi that lack cross walls (Proterozoic–Recent)
- Phylum Basidiomycota—Mushrooms (Pennsylvanian–Recent)
- Phylum Ascomycota—Yeasts, bread molds, morels (Mississippian–Recent)

---

*Margulis, L., and K.V.S. Schwartz, 1982. *Five Kingdoms.* New York: W.H. Freeman and Co.

Kingdom Plantae

Photosynthetic eukaryotes

    Division* Bryophyta—Liverworts, mosses, hornworts (Devonian–Recent)

    Division Psilophyta—Small, primitive vascular plants with no true roots or leaves (Silurian–Recent)

    Division Lycopodophyta—Club mosses, simple vascular systems, true roots and small leaves, including scale trees of Paleozoic Era (lycopsids) (Devonian–Recent)

    Division Sphenophyta—Horsetails (scouring rushes), and sphenopsids such as the Carboniferous *Calamites* (Devonian–Recent)

    Division Pteridophyta—Ferns (Devonian–Recent)

    Division Pteridospermophyta—Seed ferns (Devonian–Jurassic)

    Division Coniferophyta—Conifers or cone-bearing gymnosperms (Carboniferous–Recent)

    Division Cycadophyta—Cycads (Triassic–Recent)

    Division Ginkgophyta—Maidenhair tree (Triassic–Recent)

    Division Angiospermophyta—Flowering plants and trees (Cretaceous–Recent)

Kingdom Animalia

Nonphotosynthetic multicellular eukaryotes (Proterozoic–Recent)

    Phylum Porifera—Sponges (Cambrian–Recent)

        Order Stromatoporoida—Extinct group of reef-building organisms (Cambrian–Oligocene)

    Phylum Archaeocyatha—Extinct spongelike organisms (Cambrian)

    Phylum Cnidaria—Hydrozoans, jellyfish, sea anemones, corals (Cambrian–Recent)

        Class Hydrozoa—Hydrozoans (Cambrian–Recent)

        Class Scyphozoa—Jellyfish (Proterozoic–Recent)

        Class Anthozoa—Sea anemones and corals (Cambrian–Recent)

            Order Tabulata—Exclusively colonial corals with reduced to nonexistent septa (Ordovician–Permian)

            Order Rugosa—Solitary and colonial corals with fourfold symmetry (Ordovician–Permian)

            Order Scleractinia—Solitary and colonial corals with sixfold symmetry. Most colonial forms have symbiotic dinoflagellates in their tissues. Important reef builders today (Triassic–Recent)

    Phylum Bryozoa—Exclusively colonial suspension feeding marine animals that are useful for correlation and ecological interpretations (Ordovician–Recent)

    Phylum Brachiopoda—Marine suspension feeding animals with two unequal sized valves. Each valve is bilaterally symmetrical (Cambrian–Recent)

        Class Inarticulata—Primitive chitinophosphatic or calcareous brachiopods that lack a hinging structure. They open and close their valves by means of complex muscles (Cambrian–Recent)

        Class Articulata—Advanced brachiopods with calcareous valves that are hinged (Cambrian–Recent)

    Phylum Mollusca—A highly diverse group of invertebrates (Cambrian–Recent)

        Class Monoplacophora—Segmented, bilaterally symmetrical crawling animals with cap-shaped shells (Cambrian–Recent)

        Class Amphineura—Chitons. Marine crawling forms, typically with 8 separate calcareous plates (Cambrian–Recent)

        Class Scaphopoda—Curved, tusk-shaped shells that are open at both ends (Ordovician–Recent)

        Class Gastropoda—Single shelled generally coiled crawling forms. Found in marine, brackish, and fresh water as well as terrestrial environments (Cambrian–Recent)

        Class Bivalvia—Mollusks with two valves that are mirror images of each other. Typically known as clams and oysters (Cambrian–Recent)

        Class Cephalopoda—Highly evolved swimming animals. Includes shelled sutured forms as well as non-shelled types such as octopus and squid (Cambrian–Recent)

            Order Nautiloidea—Forms in which the chamber partitions are connected to the wall along simple, slightly curved lines (Cambrian–Recent)

            Order Ammonoidea—Forms in which the chamber partitions are connected to the wall along wavy lines (Devonian–Cretaceous)

            Order Coleoidea—Forms in which the shell is reduced or lacking. Includes octopus, squid, and the extinct belemnoids (Mississippian–Recent)

    Phylum Annelida—Segmented worms. Responsible for many of the Phanerozoic burrow and trail trace fossils (Proterozoic–Recent)

    Phylum Arthropoda—The largest invertebrate group comprising about 80% of all known animals. Characterized by a segmented body and jointed appendages (Cambrian–Recent)

---

*In botany, division is the equivalent to phylum.

Class Trilobita—Earliest appearing arthropod class. Trilobites had a head, body, and tail and were bilaterally symmetrical (Cambrian–Permian)

Class Crustacea—Diverse class characterized by a fused head and body and an abdomen. Included are barnacles, copepods, crabs, ostracodes, and shrimp (Cambrian–Recent)

Class Insecta—Most diverse and common of all living invertebrates, but rare as fossils (Silurian–Recent)

Class Merostomata—Characterized by four pairs of appendages and a more flexible exoskeleton than crustaceans. Includes the extinct eurypterids, horseshoe crabs, scorpions, and spiders (Cambrian–Recent)

Phylum Echinodermata—Exclusively marine animals with fivefold radial symmetry and a unique water vascular system (Cambrian–Recent)

Subphylum Crinozoa—Forms attached by a calcareous jointed stem (Cambrian–Recent)

Class Crinoidea—Most important class of Paleozoic echinoderms. Suspension feeding forms that are either free-living or attach to sea floor by a stem (Cambrian–Recent)

Class Blastoidea—Small class of Paleozoic suspension feeding sessile forms with short stems (Ordovician–Permian)

Class Cystoidea—Globular to pear-shaped suspension feeding benthonic sessile forms with very short stems (Ordovician–Devonian)

Subphylum Homalozoa—A small group with flattened, asymmetrical bodies with no stems. Also called carpoids (Cambrian–Devonian)

Subphylum Echinozoa—Globose, predominantly benthonic mobile echinoderms (Ordovician–Recent)

Class Helioplacophora—Benthonic, mobile forms, shaped like a top with plates arranged in a helical spiral (Early Cambrian)

Class Edrioasteroidea—Benthonic, sessile or mobile, discoidal, globular- or cylindrical-shaped forms with five straight or curved feeding areas shaped like a starfish (Cambrian–Pennsylvanian)

Class Holothuroidea—Sea cucumbers. Sediment feeders having calcareous spicules embedded in a tough skin (Ordovician–Recent)

Class Echinoidea—Largest group of echinoderms. Globe- or disk-shaped with movable spines. Predominantly grazers or sediment feeders. Epifaunal and infaunal (Ordovician–Recent)

Subphylum Asterozoa—Stemless, benthonic mobile forms (Ordovician–Recent)

Class Asteroidea—Starfish. Arms merge into body (Ordovician–Recent)

Class Ophiuroidea—Brittle star. Distinct central body (Ordovician–Recent)

Phylum Hemichordata—Characterized by a notochord sometime during their life history. Modern acorn worms and extinct graptolites (Cambrian–Recent)

Class Graptolithina—Colonial marine hemichordates having a chitinous exoskeleton. Predominantly planktonic (Cambrian–Mississippian)

Phylum Chordata—Animals with notochord, hollow dorsal nerve cord, and gill slits during at least part of their life cycle (Cambrian–Recent)

Subphylum Urochordata—Sea squirts, tunicates. Larval forms with notochord in tail region.

Subphylum Cephalochordata—Small marine animals with notochords and small fish-like bodies (Cambrian–Recent)

Subphylum Vertebrata—Animals with a backbone of vertebrae (Cambrian–Recent)

Class Agnatha—Jawless fish. Includes the living lampreys and hagfish as well as extinct armored ostracoderms (Cambrian–Recent)

Class Acanthodii—Primitive jawed fish with numerous spiny fins (Silurian–Permian)

Class Placodermii—Primitive armored jawed fish (Silurian–Permian)

Class Chondrichthyes—Cartilaginous fish such as sharks and rays (Devonian–Recent)

Class Osteichthyes—Bony fish (Devonian–Recent)

Subclass Actinopterygii—Ray-finned fish (Devonian–Recent)

Subclass Sarcopterygii—Lobe-finned, air-breathing fish (Devonian–Recent)

Order Coelacanthimorpha—Lobe finned fish *Latimeria* (Devonian–Recent)

Order Crossoptergii—Lobe-finned fish that were ancestral to amphibians (Devonian–Permian)

Order Dipnoi—Lungfish (Devonian–Recent)

Class Amphibia—Amphibians. The first terrestrial vertebrates (Devonian–Recent)

Subclass Labyrinthodontia—Earliest amphibians. Solid skulls and complex tooth pattern (Devonian–Triassic)

Subclass Salientia—Frogs, toads, and their relatives (Triassic–Recent)

Subclass Condata—Salamanders and their relatives (Triassic–Recent)

Class Reptilia—Reptiles. A large and varied vertebrate group characterized by

having scales and laying an amniote egg (Mississippian–Recent)
- Subclass Anapsida—Reptiles whose skull has a solid roof with no openings (Mississippian–Recent)
  - Order Cotylosauria—One of the earliest reptile groups (Pennsylvanian–Triassic)
  - Order Chelonia—Turtles (Triassic–Recent)
- Subclass Euryapsida—Reptiles with one opening high on the side of the skull behind the eye. Mostly marine (Permian–Cretaceous)
  - Order Protorosauria—Land living ancestral euryapsids (Permian–Cretaceous)
  - Order Placodontia—Placodonts. Bulky, paddle-limbed marine reptiles with rounded teeth for crushing mollusks (Triassic)
  - Order Ichthyosauria—Ichthyosaurs. Dolphin-shaped swimming reptiles (Triassic–Cretaceous)
- Subclass Diapsida—Most diverse reptile class. Characterized by two openings in the skull behind the eye. Includes lizards, snakes, crocodiles, thecodonts, dinosaurs, and pterosaurs (Permian–Recent)

Infraclass Lepidosauria—Primitive diapsids including snakes, lizards, and the mosasaurs, a large Cretaceous marine reptile group (Permian–Recent)
- Order Mosasauria—Mosasaurs (Cretaceous)
- Order Plesiosauria—Plesiosaurs (Triassic–Cretaceous)
- Order Squamata—Lizards and snakes (Triassic–Recent)
- Order Rhynchocephalia—The living tuatara *Sphenodon* and its extinct relatives (Jurassic–Recent)

Infraclass Archosauria—Advanced diapsids (Triassic–Recent)
- Order Thecodontia—Thecodontians were a diverse group that was ancestral to the crocodilians, pterosaurs, and dinosaurs (Permian–Triassic)
- Order Crocodilia—Crocodiles, alligators, and gavials (Triassic–Recent)
- Order Pterosauria—Flying and gliding reptiles called pterosaurs (Triassic–Cretaceous)

Infraclass Dinosauria—Dinosaurs (Triassic–Cretaceous)
- Order Saurischia—Lizard-hipped dinosaurs (Triassic–Cretaceous)
  - Suborder Theropoda—Bipedal carnivores (Triassic–Cretaceous)
  - Suborder Sauropoda—Quadrupedal herbivores, including the largest known land animals (Jurassic–Cretaceous)
- Order Ornithischia—Bird-hipped dinosaurs (Triassic–Cretaceous)
  - Suborder Ornithopoda—Bipedal herbivores, including the duck-billed dinosaurs (Triassic–Cretaceous)
  - Suborder Stegosauria—Quadrupedal herbivores with bony spikes on their tails and bony plates on their backs (Jurassic–Cretaceous)
  - Suborder Pachycephalosauria—Bipedal herbivores with thickened bones of the skull roof (Cretaceous)
  - Suborder Ceratopsia—Quadrupedal herbivores typically with horns or a bony frill over the top of the neck (Cretaceous)
  - Suborder Ankylosauria—Heavily armored quadrupedal herbivores (Cretaceous)

- Subclass Synapsida—Mammal-like reptiles with one opening low on the side of the skull behind the eye (Pennsylvanian–Triassic)
  - Order Pelycosauria—Early mammal-like reptiles including those forms in which the vertebral spines were extended to support a "sail" (Pennsylvanian–Permian)
  - Order Therapsida—Advanced mammal-like reptiles with legs positioned beneath the body and the lower jaw formed largely of a single bone. Many therapsids may have been endothermic (Permian–Triassic)

Class Aves—Birds. Endothermic and feathered (Jurassic–Recent)

Class Mammalia—Mammals. Endothermic animals with hair (Triassic–Recent)
- Subclass Prototheria—Egg-laying mammals (Triassic–Recent)
  - Order Docodonta—Small, primitive mammals (Triassic)

Order Triconodonta—Small, primitive mammals with specialized teeth (Triassic–Cretaceous)
Order Monotremata—Duck-billed platypus, spiny anteater (Cretaceous–Recent)

Subclass Allotheria—Small, extinct early mammals with complex teeth (Jurassic–Eocene)
Order Multituberculata—The first mammalian herbivores and the most diverse of Mesozoic mammals (Jurassic–Eocene)

Subclass Theria—Mammals that give birth to live young (Jurassic–Recent)
Order Symmetrodonta—Small, primitive Mesozoic therian mammals (Jurassic–Cretaceous)
Order Upantotheria—Trituberculates (Jurassic–Cretaceous)
Order Creodonta—Extinct ancient carnivores (Cretaceous–Paleocene)
Order Condylartha—Extinct ancestral hoofed placentals (ungulates) (Cretaceous–Oligocene)
Order Marsupialia—Pouched mammals. Opossum, kangaroo, koala (Cretaceous–Recent)
Order Insectivora—Primitive insect-eating mammals. Shrew, mole, hedgehog (Cretaceous–Recent)
Order Xenungulata—Large South American mammals that broadly resemble pantodonts and uintatheres (Paleocene)
Order Taeniodonta—Includes some of the most highly specialized terrestrial placentals of the Late Paleocene and Early Eocene (Paleocene–Eocene)
Order Tillodontia—Large, massive placentals with clawed, five-toed feet (Paleocene–Eocene)
Order Dinocerata—Uintatheres. Large herbivores with bony protuberances on the skull and greatly elongated canine teeth (Paleocene–Eocene)
Order Pantodonta—North American forms are large sheep to rhinoceros-sized. Asian forms are as small as a rat (Paleocene–Eocene)
Order Astropotheria—Large placental mammals with slender rear legs, stout forelimbs, and elongate canine teeth (Paleocene–Miocene)
Order Notoungulata—Largest assemblage of South American ungulates with a wide range of body forms (Paleocene–Pleistocene)
Order Liptoterna—Extinct South American hoofed mammals (Paleocene–Pleistocene)
Order Rodentia—Squirrel, mouse, rat, beaver, porcupine, gopher (Paleocene–Recent)
Order Lagomorpha—Hare, rabbit, pika (Paleocene–Recent)
Order Primates—Lemur, tarsier, loris, monkey, human (Paleocene–Recent)
Order Edentata—Anteater, sloth, armadillo, glyptodont (Paleocene–Recent)
Order Carnivora—Modern carnivorous placentals. Dog, cat, bear, skunk, seal, weasel, hyena, raccoon, panda, sea lion, walrus (Paleocene–Recent)
Order Pyrotheria—Large mammals with long bodies and short columnar limbs (Eocene–Oligocene)
Order Chiroptera—Bats (Eocene–Recent)
Order Dermoptera—Flying lemur (Eocene–Recent)
Order Cetacea—Whale, dolphin, porpoise (Eocene–Recent)
Order Tubulidentata—Aardvark (Eocene–Recent)
Order Perissodactyla—Odd-toed ungulates (hoofed placentals). Horse, rhinoceros, tapir, titanothere, chalicothere (Eocene–Recent)
Order Artiodactyla—Even-toed ungulates. Pig, hippo, camel, deer, elk, bison, cattle, sheep, antelope, entelodont, oreodont (Eocene–Recent)
Order Proboscidea—Elephant, mammoth, mastodon (Eocene–Recent)
Order Sirenia—Sea cow, manatee, dugong (Eocene–Recent)
Order Embrithopodoa—Known primarily from a single locality in Egypt. Large mammals with two gigantic bony processes arising from the nose area (Oligocene)
Order Desmostyla—Amphibious or seal-like in habit. Front and hind limbs well developed, but hands and feet somewhat specialized as paddles (Oligocene–Miocene)
Order Hyracoidea—Hyrax (Oligocene–Recent)
Order Pholidota—Scaly anteater (Oligocene–Recent)

Appendix C
Mineral Identification

Geologists use various physical properties such as color, luster, crystal form, hardness, cleavage, specific gravity, and several others to identify most common minerals (Tables C1 and C2). Notice that the Mineral Identification Table (C3) is arranged with minerals having a metallic luster grouped separately from those with a nonmetallic luster. After determining luster, ascertain hardness and note that each part of the table is arranged with minerals in order of increasing hardness. Thus, if you have a nonmetallic mineral with a hardness of 6, it must be augite, hornblende, plagioclase, or one of the two potassium feldspars (orthoclase or microcline). If this hypothetical mineral is dark green or black, it must be augite or hornblende. Use other properties to make a final determination.

Table C1
Physical Properties Used to Identify Minerals

Mineral Property	Comment
Luster	Appearance in reflected light; if has appearance of a metal luster is metallic; those with nonmetallic luster do not look like metals
Color	Rather constant in minerals with metallic luster; varies in minerals with nonmetallic luster
Streak	Powdered mineral on an unglazed porcelain plate (streak plate) is more typical of a mineral's true color
Crystal form	Useful if crystals visible (see Figure 2.5)
Cleavage	Minerals with cleavage tend to break along a smooth plane or planes of weakness
Hardness	A mineral's resistance to abrasion (see Table C2)
Specific gravity	Ratio of a mineral's weight to an equal volume of water
Reaction with HCl (hydrochloric acid)	Calcite reacts vigorously, but dolomite reacts only when powdered
Other properties	Talc has a soapy feel; graphite writes on paper; magnetite is magnetic; closely spaced, parallel lines visible on plagioclase; halite tastes salty

Table C2
Moh's Hardness Scale

Austrian geologist Frederick Mohs devised this relative hardness scale for ten minerals. He assigned a value of 10 to diamond, the hardest mineral known, and lesser values to the other minerals. You can determine relative hardness of minerals by scratching one mineral with another or by using objects of known hardness.

Hardness	Mineral	Hardness of Some Common Objects
10	Diamond	
9	Corundum	
8	Topaz	
7	Quartz	
6	Orthoclase	Steel file (6½)
		Glass (5½—6)
5	Apatite	
4	Fluorite	
3	Calcite	Copper penny (3)
		Fingernail (2½)
2	Gypsum	
1	Talc	

Table C3
Mineral Identification Tables

Metallic Luster

Mineral	Chemical Composition	Color	Hardness / Specific Gravity	Other Features	Comments
Graphite	C	Black	1–2 / 2.09–2.33	Greasy feel; writes on paper; 1 direction of cleavage	Used for pencil "leads." Mostly in metamorphic rocks.
Galena	PbS	Lead gray	2½ / 7.6	Cubic crystals; 3 cleavages at right angles	The ore of lead. Mostly in hydrothermal rocks.
Chalcopyrite	$CuFeS_2$	Brassy yellow	3½–4 / 4.1–4.3	Usually massive; greenish black streak; iridescent tarnish	The most common copper mineral. Mostly in hydrothermal rocks.
Magnetite	Fe_3O_4	Black	5½–6½ / 5.2	Strong magnetism	An ore of iron. An accessory mineral in many rocks.
Hematite	Fe_2O_3	Red brown	6 / 4.8–5.3	Usually granular or massive; reddish brown streak	Important iron ore. An accessory mineral in many rocks.
Pyrite	FeS_2	Brassy yellow	6½ / 5.0	Cubic and octahedral crystals	Found in some igneous and hydrothermal rocks and in sedimentary rocks associated with coal.

Nonmetallic Luster

Mineral	Chemical Composition	Color	Hardness / Specific Gravity	Other Features	Comments
Talc	$Mg_3Si_4O_{10}(OH)_2$	White, green	1 / 2.82	1 cleavage direction; usually in compact masses	Formed by the alteration of magnesium silicates. Mostly in metamorphic rocks.
Clay minerals	Varies	Gray, buff, white	1–2 / 2.5–2.9	Earthy masses; particles too small to observe properties	Found in soils, mudrocks, slate, phyllite.
Chlorite	$(Mg,Fe)_3(Si,Al)_4O_{10}(Mg,Fe)_3(OH)_6$	Green	2 / 2.6–3.4	1 cleavage; occurs in scaly masses	Common in low-grade metamorphic rocks such as slate.
Gypsum	$CaSO_4 \cdot 2H_2O$	Colorless, white	2 / 2.32	Elongate crystals; fibrous and earthy masses	The most common sulfate mineral. Found mostly in evaporite deposits.
Muscovite (Mica)	$KAl_2Si_3O_{10}(OH)_2$	Colorless	2–2½ / 2.7–2.9	1 direction of cleavage; cleaves into thin sheets	Common in felsic igneous rocks, metamorphic rocks, and some sedimentary rocks.
Biotite (Mica)	$K(Mg,Fe)_3AlSi_3O_{10}(OH)_2$	Black, brown	2½ / 2.9–3.4	1 cleavage direction; cleaves into thin sheets	Occurs in both felsic and mafic igneous rocks, in metamorphic rocks, and in some sedimentary rocks.
Calcite	$CaCO_3$	Colorless, white	3 / 2.71	3 cleavages at oblique angles; cleaves into rhombs; reacts with dilute HCl	The most common carbonate mineral. Main component of limestone and marble.
Anhydrite	$CaSO_4$	White, gray	3½ / 2.9–3.0	Crystals with 2 cleavages; usually in granular masses	Found in limestones, evaporite deposits, and the cap rock of salt domes.

Table C3 (continued)
Mineral Identification Tables

Nonmetallic Luster

Mineral	Chemical Composition	Color	Hardness / Specific Gravity	Other Features	Comments
Halite	$NaCl$	Colorless, white	3–4 / 2.2	3 cleavages at right angles; cleaves into cubes; cubic crystals; salty taste	Occurs in evaporite deposits.
Dolomite	$CaMg(CO_3)_2$	White, yellow, gray, pink	3½–4 / 2.85	Cleavage as in calcite; reacts with dilute hydrochloric acid when powdered	The main constituent of dolostone. Also found associated with calcite in some limestones and marble.
Fluorite	CaF_2	Colorless, purple, green, brown	4 / 3.18	4 cleavage directions; cubic and octahedral crystals	Occurs mostly in hydrothermal rocks and in some limestones and dolostones.
Augite	$Ca(Mg,Fe,Al)(Al,Si)_2O_6$	Black, dark green	6 / 3.25–3.55	Short 8-sided crystals; 2 cleavages; cleavages nearly at right angles	The most common pyroxene mineral. Found mostly in mafic igneous rocks.
Hornblende	$NaCa_2(Mg,Fe,Al)_5(Si,Al)_8O_{22}(OH)_2$	Green, black	6 / 3.0–3.4	Elongate, 6-sided crystals; 2 cleavages intersecting at 56° and 124°	A common rock-forming amphibole mineral in igneous and metamorphic rocks.
Plagioclase feldspars	Varies from $CaAl_2Si_2O_8$ to $NaAlSi_3O_8$	White, gray, brown	6 / 2.56	2 cleavages at right angles	Common in igneous rocks and a variety of metamorphic rocks. Also in some arkoses.
Potassium Feldspars — Microcline	$KAlSi_3O_8$	White, pink, green	6 / 2.56	2 cleavages at right angles	Common in felsic igneous rocks, some metamorphic rocks, and arkoses.
Potassium Feldspars — Orthoclase	$KAlSi_3O_8$	White, pink	6 / 2.56	2 cleavages at right angles	
Olivine	$(Fe,Mg)_2SiO_4$	Olive green	6½ / 3.3–3.6	Small mineral grains in granular masses; conchoidal fracture	Common in mafic igneous rocks.
Quartz	SiO_2	Colorless, white, gray, pink, green	7 / 2.67	6-sided crystals; no cleavage; conchoidal fracture	A common rock-forming mineral in all rock groups and hydrothermal rocks. Also occurs in varieties known as chert, flint, agate, and chalcedony.
Garnet	$Fe_3Al_2(SiO_4)_3$	Dark red, green	7–7½ / 4.32	12-sided crystals common; uneven fracture	Found mostly in gneiss and schist.
Zircon	Zr_2SiO_4	Brown, gray	7½ / 3.9–4.7	4-sided, elongate crystals	Most common as an accessory in granitic rocks.
Topaz	$Al_2SiO_4(OH,F)$	Colorless, white, yellow, blue	8 / 3.5–3.6	High specific gravity; 1 cleavage direction	Found in pegmatites, granites, and hydrothermal rocks.
Corundum	Al_2O_3	Gray, blue, pink, brown	9 / 4.0	6-sided crystals and great hardness are distinctive	An accessory mineral in some igneous and metamorphic rocks.

Glossary

abiogenesis The origin of life from nonliving matter.

Absaroka Sequence A widespread succession of Pennsylvanian and Permian sedimentary rocks bounded above and below by unconformities; deposited during a transgressive–regressive cycle of the Absaroka Sea.

absolute dating Assigning an age in years before the present to geologic events; absolute dates are determined by various radioactive decay dating techniques.

Acadian orogeny A Devonian episode of mountain building in the northern Appalachian mobile belt resulting from a collision of Baltica with Laurentia.

acanthodian Any Early Silurian to Permian member of the class Acanthodii, the first fish with jaws or a jawlike structure.

acritarch Organic-walled microfossil that probably represents the cyst of a planktonic alga.

Alleghenian orogeny Pennsylvanian to Permian mountain building in the Appalachian mobile belt from New York to Alabama.

allele A variant form of a single gene.

allopatric speciation Model for the origin of a new species from a small population that became isolated from its parent population.

alluvial fan A cone-shaped accumulation of mostly sand and gravel where a stream flows from a mountain valley onto an adjacent lowland.

alpha decay A type of radioactive decay during which a particle made up of two protons and two neutrons is emitted from an atom's nucleus; decreases the atomic number by 2 and the atomic mass number by 4.

Alpine–Himalayan orogenic belt A linear zone of deformation extending from the Atlantic eastward across southern Europe and north Africa, through the Middle East and into Southeast Asia.

Alpine orogeny A Late Mesozoic–Early Cenozoic episode of mountain building affecting southern Europe and north Africa.

amniote egg An egg in which an embryo develops in a liquid-filled cavity (the amnion), a waste sac is present as well as a yolk sac for nourishment.

anaerobic Refers to organisms that do not depend on oxygen for respiration.

analogous structure Body part, such as wings of insects and birds, that serves the same function but differs in structure and development. (See *homologous structure*.)

Ancestral Rockies Late Paleozoic uplift in the southwestern part of the North American craton.

angiosperm Vascular plants having flowers and seeds; the flowering plants.

angular unconformity An unconformity below which strata generally dip at a steeper angle than those above. (See *disconformity and nonconformity*.)

anthropoid Any member of the primate suborder Anthropoidea; includes New World and Old World monkeys, apes, and humans.

Antler orogeny A Late Devonian to Mississippian episode of mountain building that affected the Cordilleran mobile belt from Nevada to Alberta, Canada.

Appalachian mobile belt A long narrow region of tectonic activity along the eastern margin of the North American craton extending from Newfoundland to Georgia; probably continuous to the southwest with the Ouachita mobile belt.

archaeocyathid A Cambrian-aged, benthonic, sessile suspension feeder that built reeflike structures.

Archaeopteryx The oldest positively identified fossil bird; it had feathers but retained many reptile characteristics; from Jurassic rocks in Germany.

archosaur A term referring to the ruling reptiles—dinosaurs, pterosaurs, crocodiles, and birds.

artificial selection The practice of selectively breeding plants and animals with desirable traits.

Artiodactyla The mammalian order whose members have two or four toes; the eventoed hoofed mammals such as deer, goats, sheep, antelope, bison, swine, and camels.

asthenosphere Part of the upper mantle over which the lithosphere moves; it behaves as a plastic and flows.

Atlantic Coastal Plain The broad, low relief area of eastern North America extending from the Appalachian Mountains to the Atlantic shoreline.

atom The smallest unit of matter that retains the characteristics of an element.

atomic mass number The total number of protons and neutrons in an atom's nucleus.

atomic number The number of protons in an atom's nucleus.

australopithecine A collective term for all species of the extinct genus *Australopithecus* that existed in South Africa during the Pliocene and Pleistocene.

autotrophic Describes organisms that synthesize their organic nutrients from inorganic raw materials; photosynthesizing bacteria and plants are autotrophs. (See *heterotrophic*.)

back-arc marginal basin A marine basin, such as the Sea of Japan, between a volcanic island arc and a continent; probably forms by back-arc spreading.

Baltica One of six major Paleozoic continents; composed of Russia west of the Ural Mountains, Scandinavia, Poland, and northern Germany.

banded iron formation (BIF) Sedimentary rocks made up of alternating thin layers of chert and iron minerals, mostly the iron oxides hematite and magnetite.

barrier island A long sand body more or less parallel with a shoreline but separated from it by a lagoon.

Basin and Range Province An area of Cenozoic block-faulting centered on Nevada but extending into adjacent states and northern Mexico.

bedding (stratification) The layering in sedimentary rocks; layers less than 1 cm thick are laminations, whereas beds are thicker.

benthos All bottom-dwelling marine organisms; may live on the seafloor or within seafloor sediments.

beta decay Radioactive decay during which a fast-moving electron emitted from a neutron is converted to a proton; results in an increase of 1 atomic number, but no change in atomic mass number.

Big Bang A theory for the evolution of the universe from a dense, hot state followed by expansion, cooling, and a less dense state.

biogenic sedimentary structure Any feature such as tracks, trails, and burrows in sedimentary rocks produced by the activities of organisms. (See *trace fossil.*)

biostratigraphic unit A unit of sedimentary rock defined solely by its fossil content.

bioturbation The churning of sediment by organisms that burrow through it.

biozone A general term referring to all biostratigraphic units such as range zones and concurrent range zones.

bipedal Walking on two legs as a means of locomotion as in birds and humans.

black smoker A submarine hydrothermal vent that emits a plume of black water colored by dissolved minerals. (See *submarine hydrothermal vent.*)

body fossil The shells, teeth, bones, or (rarely) the soft parts of organisms preserved in the fossil record.

bonding The processes whereby atoms join with other atoms.

bony fish Members of the class Osteichthyes that evolved during the Devonian; characterized by a bony internal skeleton; includes the ray-finned fishes and the lobe-finned fishes.

brachiopod Any member of the phylum Brachiopoda, a group of bivalved, suspension-feeding, marine invertebrates.

braided stream A stream with an intricate network of dividing and rejoining channels.

browser An animal that eats tender shoots, twigs, and leaves. (See *grazer.*)

Caledonian orogeny A Silurian–Devonian episode of mountain building that took place along the northwestern margin of Baltica, resulting from the collision of Baltica with Laurentia.

Canadian shield The Precambrian shield in North America; mostly in Canada but also exposed in Minnesota, Wisconsin, Michigan, and New York.

carbon 14 dating technique An absolute dating technique relying on the ratio of C14 to C12 in organic substances; useful back to about 70,000 years ago.

carbonate mineral Any mineral with the negatively charged carbonate ion $(CO_3)^{-2}$ (e.g., calcite $[CaCO_3]$ and dolomite $[CaMg(CO_3)_2]$).

carbonate rock Any rock composed mostly of carbonate minerals (such as limestone and dolostone).

carnassials A pair of specialized shearing teeth in members of the mammal order Carnivora.

Carnivora An order of mammals consisting of meat eaters such as dogs, cats, bears, weasels, and seals.

carnivore-scavenger Any animal that eats other animals, living or dead, as a source of nutrients.

cartilaginous fish Fish such as living sharks and their living and extinct relatives that have an internal skeleton of cartilage.

Cascade Range A mountain range made up of volcanoes stretching from northern California through Oregon and Washington and into British Columbia, Canada.

cast A replica of an object such as a shell or bone formed when a mold of that object is filled by sediment or minerals. (See *mold.*)

catastrophism A concept proposed by Baron Georges Cuvier explaining Earth's physical and biologic history by sudden, worldwide catastrophes; also holds that geologic processes acted with much greater intensity during the past.

Catskill Delta A Devonian clastic wedge deposited adjacent to the highlands that formed during the Acadian orogeny.

Cetacea The mammal order that includes whales, porpoises, and dolphins.

chemical sedimentary rock Rock formed of minerals derived from materials dissolved during weathering.

China One of six major Paleozoic continents; composed of all Southeast Asia, including China, Indochina, part of Thailand, and the Malay Peninsula.

chordate Any member of the phylum Chordata, all of which have a notochord, dorsal hollow nerve cord, and gill slits at some time during their life cycle.

chromosome Complex, double-stranded, helical molecule of deoxyribonucleic acid (DNA); specific segments of chromosome are genes.

circum-Pacific orogenic belt One of two major Mesozoic-Cenozoic areas of largescale deformation and the origin of mountains; includes orogens in South and Central America, the North American Cordillera, and the Aleutian, Japan, and Philippine arcs. (See *Alpine–Himalayan orogenic belt.*)

cirque A steep-walled, bowl-shaped depression formed on a mountainside by glacial erosion.

cladistics A type of analysis of organisms in which they are grouped together on the basis of derived as opposed to primitive characteristics.

cladogram A diagram showing the relationships among members of a clade, including their most recent common ancestor.

clastic wedge An extensive accumulation of mostly detrital sedimentary rocks eroded from and deposited adjacent to an area of uplift, as in the Catskill Delta or Queenston Delta.

Colorado Plateau A vast upland area in Colorado, Utah, Arizona, and New Mexico with only slightly deformed Phanerozoic rocks, deep canyons, and volcanic mountains.

compound A substance made up of different atoms bonded together (such as water $[H_2O]$ and quartz $[SiO_2]$).

concurrent range zone A biozone established by plotting the overlapping geologic ranges of fossils.

conformable Refers to a sequence of sedimentary rocks deposited one after the other with no or only minor discontinuities resulting from nondeposition or erosion.

conodont A small, toothlike fossil composed of calcium phosphate; located behind the mouth of an elongate animal where they probably functioned in feeding.

contact metamorphism Metamorphism taking place adjacent to a body of magma (a pluton) or beneath a lava flow from heat and chemically active fluids.

continental accretion The process whereby continents grow by additions of Earth materials along their margins.

continental–continental plate boundary A convergent plate boundary along which two continental lithospheric plates collide, such as the collision of India with Asia.

continental drift The theory proposed by Alfred Wegener that all continents were once joined into a single landmass that broke apart with the various fragments (continents) moving with respect to one another.

continental glacier A glacier covering at least 50,000 km^2 and unconfined by topography. Also called an *ice sheet.*

continental red bed Red-colored rock, especially mudrock and sandstone, on the continents. Ferric oxides account for their color.

continental rise The gently sloping part of the seafloor lying between the base of the continental slope and the deep seafloor.

continental shelf The area where the seafloor slopes gently seaward between a shoreline and the continental slope.

continental slope The relatively steep part of the seafloor between the continental shelf and continental rise or an oceanic trench.

convergent evolution The origin of similar features in distantly related organisms as they adapt in comparable ways, such as ichthyosaurs and porpoises. (See *parallel evolution.*)

convergent plate boundary The boundary between two plates that move toward one another. (See *continental–continental plate boundary, oceanic–continental plate boundary,* and *oceanic–oceanic plate boundary.*)

Cordilleran mobile belt An area of extensive deformation in western North America bounded by the Pacific Ocean and the Great Plains; it extends north–south from Alaska into central Mexico.

Cordilleran orogeny A period of deformation affecting the western part of North America from Jurassic to Early Cenozoic time; divided into three phases known as the Nevadan, Sevier, and Laramide orogenies.

core The inner part of Earth from a depth of about 2900 km consisting of a liquid outer part and a solid inner part; probably composed mostly of iron and nickel.

correlation Demonstration of the physical continuity of stratigraphic units over an area; also matching up time-equivalent events in different areas.

craton Name applied to a stable nucleus of a continent consisting of a Precambrian shield and a platform of buried ancient rocks.

cratonic sequence A widespread association of sedimentary rocks bounded above and below by unconformities that were deposited during a transgressive–regressive cycle of an epeiric sea, such as the Sauk Sequence.

Cretaceous Interior Seaway A Late Cretaceous arm of the sea that effectively divided North America into two large landmasses.

Cro-Magnon A race of *Homo sapiens* that lived mostly in Europe from 35,000 to 10,000 years ago.

cross-bedding A type of bedding in which individual layers are deposited at an angle to the surface on which they accumulate, as in sand dunes.

crossopterygian A specific type of lobe-finned fish that had lungs.

crust The upper part of Earth's lithosphere, which is separated from the mantle by the Moho; consists of continental crust with an overall granitic composition and thinner, denser oceanic crust made up of basalt and gabbro.

crystalline solid A solid with its atoms arranged in a regular three-dimensional framework.

Curie point The temperature at which iron-bearing minerals in a cooling magma attain their magnetism.

cyclothem A sequence of cyclically repeated sedimentary rocks resulting from alternating periods of marine and nonmarine deposition; commonly contain a coal bed.

cynodont A type of therapsid (advanced mammal-like reptile); ancestors of mammals are among the cynodonts.

daughter element An element formed by radioactive decay of another element, for example, argon 40 is the daughter element of potassium 40. (See *parent element.*)

delta A deposit of sediment where a stream or river enters a lake or the ocean.

deoxyribonucleic acid (**DNA**) The chemical substance of which chromosomes are composed.

depositional environment Any area where sediment is deposited; a depositional site where physical, chemical, and biological processes operate to yield a distinctive kind of deposit.

detrital sedimentary rock Rock made up of the solid particles derived from preexisting rocks as in sandstone.

dinosaur Any of the Mesozoic reptiles belonging to the orders Saurischia and Ornithischia.

disconformity A type of unconformity above and below which the strata are parallel. (See *angular unconformity* and *nonconformity.*)

divergent evolution The diversification of a species into two or more descendant species.

divergent plate boundary The boundary between two plates that move apart; characterized by seismicity, volcanism, and the origin of new oceanic lithosphere.

drift A collective term for all sediment deposited by glacial activity; includes till deposited directly by ice, and outwash deposited by streams discharging from glaciers.

dryopithecine Any of the members of a Miocene family of apelike primates; possible ancestors of apes and humans.

ectotherm Any of the cold-blooded vertebrates such as amphibians and reptiles; animals that depend on external heat to regulate body temperature. (See *endotherm.*)

Ediacaran fauna Name for all Late Proterozoic faunas with animal fossils similar to those of the Ediacara fauna of Australia.

electron capture decay Radioactive decay involving capture of an electron by a proton and its conversion to a neutron; results in a loss of 1 atomic number, but no change in atomic mass number.

element A substance composed of only one kind of atom (such as calcium [Ca] or silicon [Si]).

Ellesmere orogeny Mississippian episode of mountain building affecting the northern margin of Laurentia.

endosymbiosis A type of mutually beneficial symbiosis in which one symbiont lives within the other.

endotherm Any of the warm-blooded vertebrates such as birds and mammals who maintain their body temperature within narrow limits by internal processes. (See *ectotherm.*)

epeiric sea A broad shallow sea that covers part of a continent; six epeiric seas were present in North America during the Phanerozoic Eon, such as the Sauk Sea.

eukaryotic cell A cell with an internal membrane-bounded nucleus containing chromosomes and other internal structures such as mitochondria that are not present in prokaryotic cells. (See *prokaryotic cell.*)

evaporite Sedimentary rock formed by inorganic chemical precipitation from evaporating water (for example, rock salt and rock gypsum).

extrusive igneous (**volcanic**) **rock** An igneous rock that forms as lava cools and crystallizes or when pyroclastic materials are consolidated.

Farallon plate A Late Mesozoic–Cenozoic oceanic plate that was largely subducted beneath North America; the Cocos and Juan de Fuca plates are remnants.

fission-track dating The dating process in which small linear tracks (fission tracks) resulting from alpha decay are counted in mineral crystals.

fluvial Relating to streams and rivers and their deposits.

formation The basic lithostratigraphic unit; a mappable unit of strata with distinctive upper and lower boundaries.

fossil Remains or traces of prehistoric organisms preserved in rocks. (See *body fossil* and *trace fossil.*)

Franklin mobile belt The most northerly mobile belt in North America; extends from northwestern Greenland westward across the Canadian Arctic islands.

gene A specific segment of a chromosome constituting the basic unit of heredity. (See *allele.*)

geologic column A diagram showing a composite column of rocks arranged with the oldest at the bottom followed upward by progressively younger rocks.

geologic record The record of prehistoric physical and biologic events preserved in rocks.

geologic time scale A chart arranged so that the designation for the earliest part of geologic time appears at the bottom followed upward by progressively younger time designations. (See *geologic column.*)

geology The science concerned with the study of Earth; includes studies of Earth materials, internal and surface processes, and Earth and life history.

glacial stage A time of extensive glaciation that occurred several times in North America during the Pleistocene.

glacier A mass of ice on land that moves by plastic flow and basal slip.

Glossopteris **flora** A Late Paleozoic association of plants found only on the Southern Hemisphere continents and India; named after its best known genus, *Glossopteris.*

Gondwana One of six major Paleozoic continents; composed of South America, Africa, Australia, India, and parts of Southern Europe, Arabia, and Florida.

graded bedding A sediment layer in which grain size decreases from bottom up.

granite-gneiss complex One of the two main rock associations found in areas of Archean rocks.

graptolite Small, planktonic animal belonging to the phylum Hemichordata.

grazer An animal that eats low-growing vegetation, especially grasses. (See *browser.*)

greenstone belt A linear or podlike association of rocks particularly common in Archaen terranes; typically synclinal and consists of lower and middle volcanic units and an upper sedimentary unit.

Grenville orogeny An episode of deformation that took place in the eastern United States and Canada during the Neoproterozoic.

guide fossil Any easily identified fossil with a wide geographic distribution and short geologic range; useful for determining relative ages of strata in different areas.

Gulf Coastal Plain The broad low-relief area along the Gulf Coast of the United States.

gymnosperm A flowerless, seed-bearing plant.

half-life The time necessary for one-half of the original number of radioactive atoms of an element to decay to a stable daughter product; for example, the half-life of potassium 40 is 1.3 billion years.

herbivore An animal dependent on vegetation as a source of nutrients.

Hercynian orogeny Pennsylvanian to Permian deformation in the Hercynian mobile belt of southern Europe.

heterotrophic Organism such as an animal that depends on preformed organic molecules from its environment for nutrients. (See *autotrophic.*)

hominid Abbreviated term for Hominidae, the family that includes bipedal primates such as *Australopithecus* and *Homo*. (See *hominoid.*)

hominoid Abbreviated term for Hominoidea, the superfamily that includes apes and humans (See *hominid.*)

Homo The genus of hominids consisting of *Homo sapiens* and their ancestors *Homo erectus* and *Homo habilis*.

homologous structure Body part in different organisms with a similar structure, similar relationships to other organs, and similar development but does not necessarily serve the same function; such as forelimbs in whales, bats, and dogs. (See *analogous organ.*)

hot spot Localized zone of melting below the lithosphere; detected by volcanism at the surface.

hypothesis A provisional explanation for observations that is subject to continual testing and modification if necessary. If well supported by evidence, hypotheses may become theories.

Hyracotherium A small Early Eocene mammal that was ancestral to today's horses.

Iapetus Ocean A Paleozoic ocean between North America and Europe; it eventually closed as North America and Europe moved toward one another and collided during the Late Paleozoic.

ice cap A dome-shaped mass of glacial ice covering less than 50,000 km^2.

ice-scoured plain An area eroded by glaciers resulting in low-relief, extensive bedrock exposures with glacial polish and striations, and little soil.

ichthyosaur Any of the porpoiselike, Mesozoic marine reptiles.

igneous rock Rock formed when magma or lava cools and crystallizes and when pyroclastic materials become consolidated.

inheritance of acquired characteristics Jean-Baptiste de Lamarck's mechanism for evolution; holds that characteristics acquired during an individual's lifetime can be inherited by descendants.

interglacial stage A time of warmer temperatures between episodes of widespread glaciation.

Interior Lowlands An area in North America made up of the Great Plains and the Central Lowlands, bounded by the Rocky Mountains, the Canadian shield, the Appalachian Mountains, and parts of the Gulf Coastal Plain.

intrusive igneous (plutonic) rock Igneous rock that cools and crystallizes from magma intruded into or formed within the crust.

iridium anomaly The occurence of a higher-than-usual concentration of the element iridium at the Cretaceous-Tertiary boundary.

isostasy the concept of Earth's crust "floating" on the more dense underlying mantle. As a result of isostasy, thicker, less dense continental crust stands higher than oceanic crust.

isostatic rebound The phenomenon in which unloading of the crust causes it to rise, as when extensive glaciers melt, until it attains equilibrium.

Jovian planets Any of the planets with a low mean density that resembles Jupiter (Jupiter, Saturn, Uranus, and Neptune); the gas giants are composed largely of hydrogen, helium, and frozen compounds such as methane and ammonia.

Kaskaskia Sequence A widespread association of Devonian and Mississippian sedimentary rocks bounded above and below by uncomformities; deposited during a transgressive–regressive cycle of the Kaskaskia Sea.

Kazakhstania One of six major Paleozoic continents; a triangular-shaped continent centered on Kazakhstan.

labyrinthodont Any of the Devonian to Triassic amphibians characterized by complex folding in the enamel of their teeth.

Laramide orogeny Late Cretaceous to Early Paleogene phase of the Cordilleran orogeny; responsible for many of the structural features in the present-day Rocky Mountains.

Laurasia A Late Paleozoic, Northern Hemisphere continent made up of North America, Greenland, Europe, and Asia.

Laurentia A Proterozoic continent composed mostly of North America and Greenland, parts of Scotland, and perhaps parts of the Baltic shield of Scandinavia.

lava Magma that reaches the surface.

lithification The process of converting sediment into sedimentary rock.

lithosphere The outer, rigid part of Earth consisting of the upper mantle, oceanic crust, and continental crust; lies above the asthenosphere.

lithostratigraphic unit A body of sedimentary rock, such as a formation, defined solely by its physical attributes.

Little Ice Age An interval from about 1500 to the mid- to late-1800s during which glaciers expanded to their greatest historic extent.

living fossil An existing organism that has descended from ancient ancestors with little apparent change.

lobe-finned fish Fish with limbs containing a fleshy shaft and a series of articulating bones; one of the two main groups of bony fish.

macroevolution Evolutionary changes that account for the origin of new species, genera, orders, and so on.

magma Molten rock material below the surface.

magnetic anomaly Any change, such as the average strength, in Earth's magnetic field.

magnetic reversal The phenomenon involving the complete reversal of the north and south magnetic poles.

mantle The inner part of Earth surrounding the core, accounting for about 85% of the planet's volume; probably composed of peridotite.

marine regression Withdrawal of the sea from a continent or coastal area resulting from emergence of the land with a resulting seaward migration of the shoreline.

marine transgression Invasion of a coastal area or much of a continent by the sea as sea level rises resulting in a landward migration of the shoreline.

marsupial mammal The pouched mammals such as wombats and kangaroos that give birth to young in a very immature state.

mass extinction Greatly accelerated extinction rates resulting in marked decrease in biodiversity, such as the mass extinction at the end of the Cretaceous.

meandering stream A stream with a single, sinuous channel with broadly looping curves.

meiosis Cell division yielding sex cells, sperm and eggs in animals, and pollen and ovules in plants, in which the number of chromosomes is reduced by half. (See *mitosis.*)

metamorphic rock Any rock altered in the solid state from preexisting rocks by any combination of heat, pressure, and chemically active fluids.

micrite Microcrystalline calcium carbonate; commonly found as matrix in carbonate rocks.

microevolution Evolutionary changes within a species.

Midcontinent rift A Mesoproterozoic intracontinental rift in Lauentia in which volcanic and sedimentary rocks accumulated.

Milankovitch theory A theory that explains cyclic variations in climate and the onset of glacial episodes triggered by irregularities in Earth's rotation and orbit.

mineral Naturally occurring, inorganic, crystalline solid, having characteristic physical properties and a narrowly defined chemical composition.

mitosis Call division resulting in two cells with the same number of chromosomes as the parent cell; takes place in all cells except sex cells. (See *meiosis.*)

mobile belt Elongated area of deformation generally at the margins of a craton, such as the Appalachian mobile belt.

modern synthesis A combination of ideas of various scientists yielding a view of evolution that includes the chromosome theory of inheritance, mutations as a source of variation, and gradualism. It also rejects inheritance of acquired characteristics.

molar Any of a mammal's teeth that are used for grinding and chewing.

molarization An evolutionary trend in hoofed mammals in which the premolars become more like molars, giving the animals a continuous series of grinding teeth.

mold A cavity or impression of some kind of organic remains such as a bone or shell in sediment or sedimentary rock. (See *cast.*)

monomer A comparatively simple organic molecule, such as an amino acid, that can link with other monomers to form more complex polymers such as proteins. (See *polymer.*)

monotreme The egg-laying mammals; includes only the platypus and spiny anteater of the Australian region.

moraine A ridge or mound of unsorted, unstratified debris deposited by a glacier.

mosaic evolution The concept holding that not all parts of an organsim evolve at the same rate, thus yielding organisms with features retained from the ancestral condition as well as more recently evolved features.

mosasaur A term referring to a group of Mesozoic marine lizards.

mud crack A crack in clay-rich sediment that forms in response to drying and shrinkage.

multicelled organism Organism made up of many cells as opposed to a single cell; possesses cells specialized to perform specific functions.

mutation Any change in the genes of organisms; yields some of the variation on which natural selection acts.

Natural selection A mechanism accounting for differential survival and reproduction among members of a species; the mechanism proposed by Charles Darwin and Alfred Wallace to account for evolution.

Neanderthal A type of human that inhabited the Near East and Europe from 200,000 to 30,000 years ago; may be a subspecies (*Homo sapiens neanderthalensis*) of *Homo* or a separate species (*Homo neanderthalensis*).

nekton Actively swimming organisms, such as fish, whales, and squid.

neoglaciation An episode in Earth history from about 6000 years ago until the mid to late 1800s during which three periods of glacial expansion took place.

neptunism The discarded concept held by Abraham Gottlob Werner and others that all rocks formed in a specific order by precipitation from a worldwide ocean.

Nevadan orogeny Late Jurassic to Cretaceous phase of the Cordilleran orogeny; most strongly affected the western part of the Cordilleran mobile belt.

New World monkey Any of the monkeys native to Central and South America; a superclass of the anthropods.

nonconformity An unconformity in which stratified sedimentary rocks overlie an erosion surface cut into igneous or metamorphic rocks. (See *angular unconformity* and *disconformity.*)

nonvascular plant Plant lacking specialized tissues for transporting fluids.

North American Cordillera A complex mountainous region in western North America extending from Alaska into central Mexico.

oceanic-continental plate boundary A convergent plate boundary along which oceanic and continental lithosphere collide; characterized by subduction of the oceanic plate, seismicity, and volcanism.

oceanic-oceanic plate boundary A convergent plate boundary along which oceanic lithosphere collides with oceanic lithosphere; charcaterized by subduction of one of the oceanic plates, seismicity, and volcanism.

Old World monkey Any of the monkeys native to Africa, Asia, and southern Europe; a superfamily of the anthropoids.

ophiolite A sequence of igneous rocks consisting of peridotite from the upper mantle overlain successively by gabbro, basalt dikes, and pillow lava that represents oceanic crust; a fragment of oceanic lithosphere now on land.

organic evolution See *theory of evolution.*

organic reef A wave-resistant limestone structure with a framework of animal skeletons, such as a coral reef or stromatoporoid reef.

Ornithischia One of the two orders of dinosaurs; characterized by a birdlike pelvis; includes ornithopods, stegosaurs, ankylosaurs, pachycephalosaurs, and ceratopsians. (See *Saurischia.*)

orogen A linear part of Earth's crust that was or is being deformed during an orogeny; part of an orogenic belt.

orogeny An episode of mountain building involving deformation, usually accompanied by igneous activity, metamorphism, and crustal thickening.

ostracoderm The "bony-skinned" fish characterized by bony armor but no jaws or teeth; appeared during the Late Cambrian, making them the oldest known vertebrates.

Ouachita mobile belt An area of deformation along the southern margin of the North American craton; probably continuous to the northeast with the Appalachian mobile belt.

Ouachita orogeny A period of mountain building that took place in the Ouachita mobile belt during the Pennsylvanian Period.

outgassing The process whereby gases released from Earth's interior by volcanism formed an atmosphere.

outwash All sediment deposited by streams that issue from glaciers. (See *drift.*)

Pacific-Farallon ridge A spreading ridge that was located off the coast of western North America during part of the Cenozoic Era.

Paleocene-Eocene Thermal Maximum A warming trend that began abruptly about 55 million years ago.

paleogeography The study of Earth's ancient geography on a regional as well as a local scale.

paleomagnetism The study of the direction and strength of Earth's past magnetic field from remnent magnetism in rocks.

paleontology The use of fossils to study life history and relationships among organisms.

Pangaea Alfred Wegener's name for a Late Paleozoic supercontinent made up of most of Earth's landmasses.

Pannotia A supercontinent that existed during the Neoproterozoic.

Panthalassa A Late Paleozoic ocean that surrounded Pangaea.

parallel evolution Evolution of similar features in two separate but closely related lines of descent as a result of comparable adaptations. (See *convergent evolution.*)

parent element An unstable element that changes by radioactive decay into a stable daughter element. (See *daughter element.*)

pelycosaur Pennsylvanian to Permian reptile, many species with large fins on the back, that possessed some mammal characteristics.

period The fundamental unit in the hierarchy of time units; part of geologic time during which the rocks of a system were deposited.

Perissodactyla The order of odd-toed hoofed mammals; consists of present-day horses, tapirs, and rhinoceroses.

photochemical dissociation A process whereby water molecules in the upper atmosphere are disrupted by ultraviolet radiation, yielding oxygen (O_2) and hydrogen (H).

photosynthesis The metobolic process in which organic molecules are synthesized from water and carbon dioxide (CO_2), using the radiant energy of the Sun captured by chlorophyll-containing cells.

phyletic gradualism The concept that a species evolves gradually and continuously as it gives rise to new species. (See *puncuated equilibrium.*)

placental mammal All mammals with a placenta to nourish the developing embryo, as opposed to egg-laying mammals (monotremes) and pouched mammals (marsupials).

placoderm Late Silurian through Permian "plate-skinned" fish with jaws and bony armor, especially in the head-shoulder region.

plankton Animals and plants that float passively, such as phytoplankton and zooplankton. (See *nekton.*)

plate A segment of Earth's crust and upper mantle (lithosphere) varying from 50 to 250 km thick.

plate tectonic theory Theory holding that lithospheric plates move with respect to one another at divergent, convergent, and transform plate boundaries.

platform The buried extension of a Precambrian shield, which together with a shield makes up a craton.

playa lake A temporary lake in an arid region.

plesiosaur A type of Mesozoic marine reptile; short-necked and long-necked plesiosaurs existed.

pluvial lake Any lake that formed in nonglaciated areas during the Pleistocene as a result of increased precipitation and reduced evaporation rates during that time.

pollen analysis Identification and statistical analysis of pollen from sedimentary rocks provides information about ancient floras and climates.

polymer A comparatively complex organic molecule, such as nucleic acids and proteins, formed by monomers linking together. (See *monomer.*)

Precambrian shield An area in which a continent's ancient craton is exposed, as in the Canadian shield.

premolar Any of a mammal's teeth between the canines and the molars; premolars and molars together are a mammal's chewing teeth.

primary producer Organism in a food chain, such as bacteria and green plants, that manufacture their own organic molecules, and on which all other members of the food chain depend for sustenance. (See *autotrophic*.)

Primates The order of mammals that includes prosimians (lemurs and tarsiers), monkeys, apes, and humans.

principle of cross-cutting relationships A principle holding that an igneous intrusion or fault must be younger than the rocks it intrudes or cuts across.

principle of fossil succession A principle holding that fossils, especially groups or assemblages of fossils, succeed one another through time in a regular and determinable order.

principle of inclusions A principle holding that inclusions or fragments in a rock unit are older than the rock itself, such as granite inclusions in sandstone are older than the sandstone.

principle of lateral continuity A principle holding that rock layers extend outward in all directions until they terminate.

principle of original horizontality According to this principle, sediments are deposited in horizontal or nearly horizontal layers.

principle of superposition A principle holding that sedimentary rocks in a vertical sequence formed one on top of the other so that the oldest layer is at the bottom of the sequence whereas the youngest is at the top.

principle of uniformitarianism A principle holding that we can interpret past events by understanding present-day processes, based on the idea that natural processes have always operated as they do now.

Proboscidea The order of mammals that includes elephants and their extinct relatives.

proglacial lake A lake formed of meltwater accumulating along the margin of a glacier.

progradation The seaward (or lakeward) migration of a shoreline as a result of nearshore sedimentation.

prokaryotic cell A cell lacking a nucleus and organelles such as mitochondria and plastids; the cells of bacteria and archaea. (See *eukaryotic cell*.)

prosimian Any of the so-called lower primates, such as tree shrews, lemurs, lorises, and tarsiers.

protorothyrid A loosely grouped category of small, lizardlike reptiles.

pterosaur Any of the Mesozoic flying reptiles.

puncuated equilibrium A concept holding that new species evolve rapidly, in perhaps a few thousands of years, then remains much the same during its several million years of existence. (See *phyletic gradualism*.)

pyroclastic materials Fragmental materials such as ash explosively erupted from volcanoes.

quadrupedal A term referring to locomotion on all four limbs as in dogs and horses. (See *bipedal*.)

Queenston Delta A clastic wedge resulting from erosion of the highland formed during the Taconic orogeny.

radioactive decay The spontaneous change in an atom by emission of a particle from its nucleus (alpha and beta decay) or by electron capture, thus changing the atom to a different element.

range zone A biostratigraphic unit defined by the occurance of a single type of organism such as a species or a genus.

ray-finned fish A subclass (Actinopterygii) of the bony fish (class Osteichthyes) in which the fins are supported by thin bones that project from the body.

regional metamorphism Metamorphism taking place over a large but usually elongate area resulting from heat, pressure, and chemically active fluids.

relative dating The process of placing geologic events in their proper chronological order with no regard to when the events took place in number of years ago. (See *absolute dating*.)

relative geological time scale When it was first established the geologic time scale as deduced from the geologic column showed only relative time; that is, Silurian rocks are younger than those of the Ordovician but older than those designated Devonian.

Rio Grande rift A linear depression made up of several interconnected basins extending from Colorado into Mexico.

ripple mark Wavelike structure on a bedding plane, especially in sand, formed by unidirectional flow of air or water currents, or by oscillating currents as in waves.

rock An aggregate of one or more minerals as in granite (feldspars and quartz) and limestone (calcite), but also includes rocklike materials such as natural glass (obsidian) and consolidated organic material (coal).

rock cycle A sequence of processes through which Earth materials may pass as they are transformed from one rock type to another.

rock-forming mineral Any of about two dozen minerals common enough in rocks to be important for their identification and classification.

Rodinia The name of a Neoproterozoic supercontinent.

rounding The process involving abrasion of sedimentary particles during transport so that their sharp edges and corners are smoothed off.

ruminant Any cud-chewing placental mammal with a complex three- or four chambered stomach, such as deer, cattle, antelope, and camels.

San Andreas transform fault A major transform fault extending from the Gulf of Mexico through part of California to its termination in the Pacific Ocean off the north coast of California.

sand dune A ridge or mound of wind-deposited sand.

sandstone-carbonate-shale assemblage An association of sedimentary rocks typically found on passive continental margins.

Sauk Sequence A widespread association of sedimentary rocks bounded above and below by unconformities that was deposited during a Neoproterozoic to Early Ordovician transgressive–regressive cycle of the Saulk Sea.

Saurischia An order of dinosaurs; characterized by a lizardlike pelvis; includes theropods, prosauropods, and sauropods. (See *Ornithischia*.)

scientific method A logical, orderly approach involving data gathering, formulating and testing hypotheses, and proposing theories.

seafloor spreading The phenomenon involving the origin of new oceanic crust at spreading ridges that then moves away from ridges and is eventually consumed at subduction zones.

sedimentary facies Any aspect of sediment or sedimentary rocks that make them recognizably different from adjacent rocks of about the same age, such as a sandstone facies.

sedimentary rock Any rock composed of (1) particles of preexisting rocks, (2) or made up of minerals derived from solution by inorganic chemical processes or by the activities of organisms, and (3) masses of consolidated organic matter as in coal.

sedimentary structure All features in sedimentary rocks such as ripple marks, cross-beds, and burrows that formed as a result of physical or biological processes that operated in a depositional environment.

sediment-desposit feeder Animal that ingests sediment and extracts nutrients from it.

seedless vascular plant Plant with specialized tissues for transporting fluids and nutrients and that reproduces by spores rather than seeds, such as ferns and horsetail rushes.

sequence stratigraphy The study of rock relationships within a time-stratigraphic framework of related facies bounded by widespread unconformities.

Sevier orogeny Cretaceous phase of the Cordilleran orogeny that affected the continental shelf and slope areas of the Cordilleran mobile belt.

Siberia One of six major Paleozoic continents; composed of Russia east of the Ural Mountains, and Asia north of Kazakhstan and south of Mongolia.

silicate A mineral containing silica, a combination of silicon and oxygen, and usually one or more other elements.

sivapithecid Apelike hominoid that evolved in Africa during the Miocene; probable ancestor of present-day orangutans.

solar nebula theory An explanation for the origin and evolution of the solar system from a rotating cloud of gases.

Sonoma orogeny A Permian-Triassic orogeny caused by the collision of an island arc with the southwestern margin of North America.

sorting The process whereby sedimentary particles are selected by size during transport; deposits are poorly sorted to well sorted depending on the range of particle sizes present.

species A population of similar individuals that in nature can interbreed and produce fertile offspring.

spreading ridge A long elevated part of the crust, mostly beneath the oceans, where plates diverge, such as the Mid-Atlantic Ridge.

stratigraphy The branch of geology concerned with the composition, origin, and areal and age relationships of stratified (layered) rocks; concerned with all rock types but especially sediments and sedimentary rocks.

stromatolite A biogenic sedimentary structure, especially in limestone, produced by entrapment of sediment grains on sticky mats of photosynthesizing bacteria.

submarine hydrothermal vent A crack or fissure in the seafloor through which superheated water issues. (See *black smoker.*)

Sundance Sea A wide seaway that existed in western North America during the Middle Jurassic Period.

supercontinent A landmass consisting of most of Earth's continents (such as Pangaea).

suspension feeder Animal that consumes microscopic plants and animals or dissolved nutrients from water.

system The fundamental unit in the hierarchy of time-stratigraphic units, such as the Devonian System. A system is also a combination of related parts that interact in an organized manner. Earth's systems include the atmosphere, hydrosphere, biosphere, as well as Earth's lithosphere, mantle, and core.

Taconic orogeny An Ordovician episode of mountain building resulting in deformation of the Appalachian mobile belt.

Tejas epeiric sea A Cenozoic sea largely restricted to the Gulf and Atlantic Coastal Plains, coastal California, and the Mississippi Valley.

terrane A small lithospheric block with characteristics quite different from those of surrounding rocks. Terranes probably consist of seamounts, oceanic rises, and other seafloor features accreted to ontinents during orogenies.

terrestrial planet Any of the four, small inner planets (Mercury, Venus, Earth, and Mars) similar to Earth (Terra); all have high mean densities, indicating they are composed of rock. (See *Jovian planet.*)

theory An explanation for some natural phenomenon with a large body of supporting evidence; theories must be testable by experiments and/or observations, such as plate tectonic theory.

theory of evolution The theory holding that all living things are related and that they descended with modification from organisms living during the past.

therapsid Permian to Triassic mammal-like reptiles; the ancestors of mammals are among one group of therapsids known as cynodonts.

thermal convection cell A type of circulation of material in the asthenosphere during which hot material rises, moves laterally, cools and sinks, and is reheated and continues the cycle.

tidal flat A broad, extensive area along a coastline that is alternately water-covered at high tide and exposed at low tide.

till Sediment deposited directly by glacial ice, as in an end moraine.

time-stratigraphic unit A body of strata that was deposited during a specifc interval of geological time; for example, the Devonian System, a time-stratigraphic unit, was deposited during that part of geological time designated the Devonian Period.

time unit Any of the units such as eon, era, period, epoch, and age referring to specific intervals of geologic time.

Tippecanoe sequence A widespread body of sedimentary rocks bounded above and below by unconformities; deposited during an Ordovician to Early Devonian transgressive-regressive cycle of the Tippecanoe Sea.

trace fossil Any indication of prehistoric organic activity such as tracks, trails, burrows, and nests. (See *body fossil.*)

Transcontinental Arch Area extending from Minnesota to New Mexico that stood above sea level as several large islands during the Cambrian transgressions of the Sauk Sea.

transform fault A type of fault that changes one kind of motion between plates into another type of motion; recognized on land as a strike-slip fault. (See *San Andreas transform fault.*)

transform plate boundary Plate boundary along which adjacent plates slide past one another and crust is neither produced nor destroyed.

tree-ring dating The process of determining the age of a tree or wood in a structure by counting the number of annual growth rings.

trilobite A Paleozoic benthonic, detritus-feeding, marine invertabrate of the phylum Arthropoda.

trophic level The complex interrelationships among producers, consumers, and decomposers in a community of organisms in which several trophic levels exist through which energy flows in a feeding hierarchy.

unconformity A break or gap in the geologic record resulting from erosion or nondeposition or both. Also the surface separating younger from older rocks where a break in the geologic record is present. (See *angular unconformity, nonconformity,* and *disconformity.*)

ungulate An informal term referring to a variety of mammals but especially the hoofed mammals of the orders Artiodactyla and Perissodactyla.

U-shaped glacial trough A valley with steep or nearly vertical walls and a broad, concave, or rather flat floor; formed by movement of a glacier through a stream valley.

valley glacier A glacier confined to a mountain valley.

varve A dark-light couplet of sedimentary laminations representing an annual deposit in a glacial lake.

vascular plant A plant with specialized tissues for transporting fluids in land plants.

vertebrate Any animal possessing a segmented vertebral column as in fish, amphibians, reptiles, birds, and mammals; members of the subphylum Vertebrata.

vestigial structure In an organism, any structure that no longer serves any or only a limited function, such as dewclaws in dogs and wisdom teeth in humans.

volcanic island arc A curved chain of volcanic islands parallel with a deep sea trench where oceanic lithosphere is subducted, causing volcanism and the origin of volcanic islands.

Walther's law A concept holding that the facies in a conformable vertical sequence will be found laterally to one another.

Wilson cycle The relationship between mountain building (orogeny) and the opening and closing of ocean basins.

Zuni epeiric sea A widespread sea present in North America mostly during the Cretaceous, but it persisted into the Paleogene.

Zuni Sequence An Early Jurassic to Early Paleocene sequence of sedimentary rocks bounded above and below by unconformities; deposited during a transgressive–regressive cycle of the Zuni Sea.

Answers to Multiple Choice Review Questions

Chapter 1
1 c; 2 d; 3 b; 4 c; 5 a; 6 b; 7 e; 8 b; 9 d; 10 d; 11 d; 12 e

Chapter 2
1 a; 2 c; 3 d; 4 e; 5 a; 6 b; 7 d; 8 a; 9 c; 10 d

Chapter 3
1 d; 2 a; 3 b; 4 c; 5 a; 6 c; 7 b; 8 e; 9 e; 10 c; 11 a; 12 a; 13 a

Chapter 4
1 e; 2 a; 3 d; 4 a; 5 c; 6 d; 7 a; 8 e; 9 a; 10 b; 11 b

Chapter 5
1 a; 2 c; 3 b; 4 e; 5 d; 6 b; 7 d; 8 a; 9 e; 10 c

Chapter 6
1 b; 2 d; 3 a; 4 c; 5 d; 6 e; 7 c; 8 a; 9 b; 10 d

Chapter 7
1 c; 2 b; 3 b; 4 e; 5 c; 6 a; 7 d; 8 b; 9 c; 10 c

Chapter 8
1 b; 2 e; 3 d; 4 a; 5 c; 6 a; 7 d; 8 b; 9 a; 10 e

Chapter 9
1 b; 2 c; 3 e; 4 c; 5 a; 6 d; 7 b; 8 c; 9 d; 10 c

Chapter 10
1 b; 2 d; 3 b; 4 b; 5 b; 6 e; 7 a; 8 c; 9 d; 10 b; 11 d; 12 a; 13 d

Chapter 11
1 b; 2 d; 3 e; 4 b; 5 d; 6 a; 7 b; 8 c; 9 a; 10 b; 11 c; 12 e

Chapter 12
1 a; 2 a; 3 d; 4 e; 5 b; 6 c; 7 e; 8 e; 9 c; 10 c; 11 a; 12 d; 13 a; 14 c

Chapter 13
1 c; 2 b; 3 b; 4 c; 5 c; 6 a; 7 e; 8 b; 9 e; 10 c; 11 d; 12 c; 13 e; 14 b

Chapter 14
1 c; 2 a; 3 c; 4 a; 5 e; 6 b; 7 e; 8 c; 9 e; 10 c; 11 c; 12 b; 13 e

Chapter 15
1 a; 2 b; 3 d; 4 c; 5 e; 6 b; 7 a; 8 c; 9 c; 10 b

Chapter 16
1 b; 2 e; 3 a; 4 c; 5 d; 6 a; 7 c; 8 b; 9 c; 10 d

Chapter 17
1 b; 2 c; 3 e; 4 a; 5 c; 6 b; 7 a; 8 c; 9 d; 10 b

Chapter 18
1 c; 2 a; 3 e; 4 c; 5 e; 6 b; 7 a; 8 c; 9 d; 10 a

Chapter 19
1 a; 2 d; 3 e; 4 d; 5 b; 6 c; 7 b; 8 e; 9 c; 10 d; 11 a; 12 c

Index

A

Abiogenesis, 159
Absaroka Sea, 222–223
Absaroka sequence, 218–223
Absolute dating
    defined, 62
    fission-track dating, 72–73
    methods, 68–74
    radioactive decay and, 69–72
    radiocarbon dating, 73–74
    radiometric dating, 62, 69–72
    relative geologic time scale and, 97–98
    tree-ring dating, 74
Acadian orogeny, 211, 216, 225–226
Acanthodians, 252, 253–254
Acanthostega, 250, 255, 257
Acasta Gneiss, 149
Accessory minerals, 22
Accretion, 290
Acid rain, 317
Acquired characteristics, inheritance of, 125–126, 127
Acritarchs, 179, 240
Actualism, 67
Adenosine triphosphate (ATP), 162
Advanced bony fishes, 299
Aegyptopithecus, 400, 401
Afonin, S. A., 262
Africa
    continental drift and, 36, 38, 274
    fish, 255
    Great Rift Valley, 48
    hominids, 398, 403–404, 407
    in Jurassic, 314
    mammals, 386, 392
    plate tectonics and, 323, 325
Aftonian interglacial stage, 354
Agassiz, Louis, 348, 354
Ages, 93
Agnatha, 252
Alaska, 367
Aleutian Islands, 349
Algae
    blue-green, 174
    green, 262
    phytoplankton, 235
Alleghenian orogeny, 211, 279
Alleles, 128
Allopatric speciation, 132
Alluvial fans, 110
Alpha decay, 69
Alpine glaciers, 348
Alpine-Himalayan orogenic belt, 325, 326–328
Alpine orogeny, 326–327
Alps, 51
Altered remains, 88, 90, 91
Aluminum, 17
Amebeledon, 380
American River, 273
Amino acids, 130
Ammonoids, 241, 243, 297
Amnion, 258
Amniote eggs, 258, 259
Amphibians
    fish ancestors of, 255–256
    Mesozoic, 299–300
    Paleozoic, 257–258
    seedless vascular plants and, 264
Anaerobic organisms, 162
Analogous structures, 140–141
Anatolepis, 251–252
Ancestral Rockies, 221–222
Ancient history, 11–12
Ancient rifting, 47, 49
Andes Mountains, 50, 51, 56, 211, 329, 331, 349
Angiosperms, 300, 372, 373–374
Angular unconformity, 80–81
Animalia, 179
Animals
    See also Organisms; *specific types*
    emergence of, 250–251
    feeding systems, 235–237
    Neoproterozoic, 182–184
    Pleistocene, 390–393
Ankylosaurs, 305
Anning, Mary, 307, 309
Antarctica
    continental drift and, 274
    glaciers in, 367
Antelope, 386
Anthracite, 31, 228
Anthropoids, 398, 400–401
Antler orogeny, 211, 224
Apatosaurus, 304
Apes, 400–401
Aphanitic texture, 23, 24
Appalachian mobile belt, 191, 199, 204–205, 224–226
Appalachian Mountains, 51, 211, 279, 339
Aquatic carnivores, 384
Aquatic mammals, 388–390
Archaeocyathids, 201, 238, 240
Archaeopteryx, 144, 309–310, 313, 377
Archaic mammals, 379
Archean Eon
    atmosphere during, 157–158, 176–177
    cratons and, 150–152, 156–157
    mineral resources from, 163–164
    origin of life during, 159–163
    plate tectonics, 154, 156–157
Archean rocks, 152, 167–168
Archosaurs, 301
Ardipithecus ramidus kadabba, 403–404
Ardipithecus ramidus ramidus, 404
Argillite, 154
Aristotle, 124
Arkose, 27
Arsinoitherium, 379, 380
Articulate brachiopods, 237, 240
Artificial selection, 126
Artiodactyla, 385–387
Aseismic ridges, 52
Ashfall Fossil Beds State Historical Park, 382
Ash flows, determining absolute age from, 97
Asia
    continental-continental plate boundaries, 50
    hominids, 407
    mammals, 383, 386, 393
    plate tectonics and, 323, 325
Asthenosphere, 10, 45
Atlantic Coastal Plain, 340, 341–343
Atlantic Ocean, 280, 323, 325
Atlantic Ocean basin, topography of, 43

Atmosphere
    Archean, 157–158, 176–177
    formation and evolution of, 157–158
    plate tectonics and, 11
    Proterozoic, 176–177
    as subsystem, 2–4
Atomic mass number, 18, 69
Atomic number, 18, 68–69
Atoms, 18–19, 68–69
Augustine, St., 62
Australia, 274, 391, 394
Australopithecines, 404–405
Autotrophic organisms, 162, 236
Avalonia, 192, 194, 226–227

B

Back-arc marginal basins, 49, 155, 328
Background extinction, 137
Background radiation, 6
Bacteria
    classification of, 179
    cyanobacteria, 174
    fossil, 179
    reproduction, 131–132
    symbiotic, 109
Baltica, 192, 193, 194
Bam, Iran, 36
Banded iron formations (BIFs), 154, 170, 176–177, 178, 185
Barrier islands, 112, 115
Barrier reefs, 200–201
Baryonyx, 295
Basal rocks, 214
Basal slip, 348
Base level, 358
Basin and Range Province, 333–334
Bats, 381
Beartooth Plateau, 147
Bedding, 104–105
Bedding planes, 79, 105–106
Belemnoids, 297
Benthos organisms, 235
Bering Strait, 358, 393
Beta decay, 69
Big Bang, 6
Biochemical sedimentary rocks, 27, 29
Biogenic sedimentary structures, 108
Biological evidence, supporting evolution, 140–141
Biosphere, 2–4, 11
Biostratigraphic unit, 93
Biotic provinces, 56
Bioturbation, 108
Biozones, 93
    interval zones, 95
    range zone, 94–95
Birds
    See also specific types
    Cenozoic, 372, 377, 390
    evolution of, 296, 309–310
    Mesozoic, 309–310
Bitter Springs Formation, 179
Bituminous coal, 228
Bivalves, 297, 299, 373
Black Hills, 338
Black shales, 216–217
Black smokers, 161
Blanket geometry, 109
Blue-green algae, 174
Body fossils, 86, 88–89, 91
Bonding, 18–19, 20
Bony fish, 254, 299–300

Bovids, 386
Brachiopods, 237, 240, 241, 243, 373
Brachiosaurus, 304
Braided streams, 109–110
Bretz, Harlan, 360
Brongniart, Alexander, 89
Browsers, 386
Bryozoans, 243, 373
Buffon, Georges Louis de, 64
Burgess Shale, 231, 232, 238–239
Burrowing organisms, 299
Bushveld Complex, 184
Byrophytes, 262

C

Calamites, 267
Calcareous ooze, 115, 116
Calcite, 20, 21
Calcium, 20
Caledonian orogeny, 205, 211, 224–225
California gold rush, 272–273, 368
Cambrian, 193
    explosion, 232–233
    geologic time and, 91–92, 192
    of Grand Canyon, 197–198
    marine life, 237–239
    paleogeography of, 195–197
Cambrian rocks, 197–198
Camels, 386
Canada
    See also North America
    natural resources of, 185
    reef development in, 214–216
Canadian shield, 150–152, 198, 214
Canning Basin, 216
Cape Cod, Massachusetts, 358, 362
Capital Reef National Park, 94
Carbon 12, 73–74
Carbon 14, 73–74
Carbonate depositional environments, 115–117, 217
Carbonate minerals, 21
Carbonate rocks, 27–28
Carbon dating, 73–74
Carbon dioxide, 362–363
Carboniferous Period
    flora of, 266–268
    geologic time scale and, 92–93
    marine communities, 243–244
    paleogeography, 211
Carnassials, 383
Carnivore-scavengers, 235
Carnivorous mammals, 383–384
Carnotite, 291
Cartilaginous fish, 254
Cascade Range, 333, 334, 352, 353, 367
Casts, 88, 90, 91
Catastrophism, 66
Cats, 384
Catskill Delta, 225–226
Cave paintings, 408–409
Cell cleavage, 251
Cells
    eukaryotic, 162–163, 178–181
    evolution of living, 2
    prokaryotic, 162–163, 178, 180
Cellulose, 263
Cementation, 26
Cenozoic Era
    birds, 372, 377, 390
    climate during, 354, 373–376

    geologic time and, 322, 372, 374–375
    Holocene Epoch, 348
    introduction to, 322–323
    life during, 371–394
    mammals, 377–394
    marine invertebrates, 373
    mass extinctions of, 242
    mineral resources from, 343–344
    North American Cordillera, 328–337
    orogenic belts, 325–328
    paleogeography during, 324
    Pangaean breakup during, 274
    phytoplankton, 373
    plants, 372, 373–376
    plate tectonics, 323–325, 349–352
    Pleistocene Epoch, 348, 354–368
    stratigraphy, 352–355
    volcanism, 349–352, 373
Cenozoic rocks, 323, 372
Central Lowlands, 337–339
Cephalopods, 297, 373
Cetacea, 389
Chaleuria cirrosa, 266
Chalicotheres, 380
Chemical bonding, 18–19, 20
Chemical sedimentary rocks, 26–29
Chert, 28–29
Chicxulub structure, 317
Chimpanzees, 403
China, 192, 193, 310
Chinle Formation, 286, 287
Chlorine, 19
Chondrichthyes, 252
Chordates, 251
Christian theology, 62–63, 66–67, 78, 124–125
Chrome, 163
Chromosomes, 129–130
Chronostratigraphic units, 93
Circques, 358
Circum-Pacific orogenic belt, 325, 328
Clack, Jennifer, 250
Clade, 135
Cladistics, 135
Cladograms, 135, 136
Classification, of organisms, 138–140
Clastic wedge, 205
Clasts, 27
Claystone, 27
Climate
    biogeographic distribution and, 314
    Cenozoic, 354, 373–376
    Cretaceous, 314
    effect of Pangaean breakup on, 276–277
    Eocene, 375–376
    Mesozoic, 314
    Paleocene, 374–375
    Pleistocene, 355–356
Climate change
    Cenozoic, 354
    determining past, 62
    Earth's rotation and, 363–366
    Eocene, 376
    extinctions and, 262–263, 391–392
    geologic time and, 63
    glaciation and, 362–368
    humans' effect on, 5
    during Pleistocene, 354–356
    short-term, 366
    since the Pleistocene, 348
Closed systems, 71
Coal deposits, 29, 102, 214, 227–228, 276, 291, 343–344
Coal-forming swamps, 266–267
Coal transport, 189

Cobalt, 17
Coccolithospores, 299, 373
Coelacanths, 255
Coelophysis, 304
Coleoids, 297
Colorado Plateau, 286, 287, 334–335
Columbia Plateau, 332
Columnar sections, 85
Comets, 7–8
Comfortable strata, 80
Compaction, 26
Compounds, 18–19
Concurrent range zone, 95
Conglomerate, 27, 103
Conifers, 300
Conodonts, 240–241
Contact metamorphism, 30
Continental accretion, 156–157, 290
Continental-continental plate boundaries, 50
Continental crust, 10, 150
    Archean, 152
    origin of, 151
Continental drift
    See also Plate tectonics
    driving mechanism of, 53–55
    early ideas on, 36–41
    evidence for, 38–41
Continental environments, 109–111
Continental fit, 39
Continental glaciers, 348
Continental red beds, 170, 177
Continental rise, 114–115, 116
Continental shelf, 114–115, 116
Continental slope, 114–115, 116
Continents
    See also Plate tectonics; *specific continents*
    breakup of, during Mesozoic Era, 273–277
    cratons, 191
    effect pf proximity of, on animal evolution, 314
    evolution of Proterozoic, 168–172
    foundations of, 150–156
    mobile belts, 191
    Paleozoic, 190–194, 210–213
Convergent evolution, 133–134
Convergent plate boundaries, 46, 48–51
Cooksonia, 263–264
Copper deposits, 57, 291
Coprolite, 86
Coral reefs, 199–202, 297–298, 373
Corals, 199, 373, 376
Cordaite forest, 268
Cordiates, 268
Cordilleran batholiths, 285
Cordilleran mobile belt, 191, 224, 282–284, 285
Cordilleran orogeny, 278, 283–284, 285, 286, 328–332
Core, 2–4, 9
Correlation, 94–97
Covalent bonding, 19, 20
Cranioceras, 380
Cratonic sequences
    Absaroka Sequence, 218–223
    defined, 194–195
    Kaskaskia Sequence, 214–217
    of North America, 195
    Sauk Sequence, 195–198
    Tippecanoe Sequence, 198–204
    Zuni Sequence, 278–279
Cratonic uplift, 221–222
Cratons
    Archean, 150–152
    components of, 191
    origins of, 156–157
Creationism, 124–125

Creodonts, 379, 383–384
Cretaceous Interior Seaway, 289
Cretaceous Period, 274, 281–282
    climate during, 314
    invertebrates, 297–299
    paleogeography of North America during, 280
    sedimentation, 288–289
Crinoids, 243, 373
Crocodiles, 306, 307, 308
Cro-Magnons, 408–409
Cross-bedding, 104
Cross-cutting relationships, principle of, 64–65, 97, 105
Cross-dating, 74
Crossopterygians, 255, 256, 257, 299
Crust, 66
Crustal evolution, 167–168
Crust (Earth), 9, 10
    See also Continental crust; Oceanic crust
Crystalline solids, 20
Crystalline texture, 27
Curie, Marie, 68
Curie, Pierre, 68
Curie point, 41
Current ripple marks, 104, 106
Cutin, 263
Cuvier, George, 66, 125
Cyanobacteria, 174
Cycads, 300
Cyclothems, 218–220
Cynodonts, 310–312, 372, 381
Cynognathus, 40
Cyprus, 57

D

Darwin, Charles, 11, 123–124, 126, 128, 137, 141, 233
Dating techniques, 61–63
    absolute dating methods, 68–74
    fission-track dating, 72–73
    fossils and, 88–89
    radiocarbon dating, 73–74
    relative dating, 64–65
    tree-ring dating, 74
Daughter elements, 69–72
Da Vinci, Leonardo, 141
Dead Horse State Park, 101
Death Valley, 359
Deep-seafloor sediments, 115–116
Deep-sea stratigraphy, 354–355
Deep time, 2
Deinonychus, 304
Deltas, 111–114
Depositional environments, 102, 109–118
    continental environments, 109–111
    interpretation of, 117–118
    marine, 110, 114–117
    transitional, 110, 111–114
Descartes, René, 6
Desert environments, 110, 112
Detrital marine environments, 114–115
Detrital sedimentary rocks, 26–28
Devonian Great Barrier Reef, 216–217
Devonian Period, 211, 225–226
    amphibians, 257–258
    fish, 253–254
    geologic time and, 92–94
    marine communities, 241–243
    paleogeography of North America during, 214–216
    plants of, 261–266

Devonian Ridgeley Formation, 228
Diamonds, 17, 20, 291
Diatomite, 343, 368, 374
Diatoms, 299, 368, 373
Diatryma, 377
Differential pressure, 30
Dimetrodon, 260
Dinocerata, 379
Dinoflagellates, 299
Dinosaurs, 279, 288, 295, 301–307
    See also specific types
    archosaurs and, 301
    classification of, 303
    extinction of, 316–317
    Mesozoic, 296
    ornithischian, 304–305, 306
    relationships among, 302
    saurischian, 303–304
    warm-blooded, 305–307
Diplodocus, 304
Disconformity, 80–81
Divergent boundaries, 46–47
Divergent evolution, 133–134
DNA (deoxyribonucleic acid), 129–130
Dogs, 384
Dolomite, 21
Dolostone, 27–28, 103, 115–117
Drift, 111
Dropstones, 360
Dryopithecines, 401
Duck-billed dinosaurs, 305, 306
Dunkleosteus, 254
Dunton pegamite, 184
Du Toit, Alexander, 38

E

Early Paleozoic. *See* Paleozoic Era
Earth
    age of, 62–64, 78
    dynamic and evolving nature of, 9–11
    evolution of, 2
    formation of, 8
    layers of, 9–10
    magnetic field of, 41–42
    Moon and, 8–9, 149
    place of, in solar system, 8
    subsystems of, 2–4
    as a system, 2–5
Earth history, interpretation of, 5
Earthquakes, 36, 50, 51–52, 336–337
East African Rift, 46, 48, 397
Echinoids, 298, 299, 373, 376
Ectotherms, 305
Edaphosaurus, 260
Ediacaran fauna, 182–183, 233
Einstein, Albert, 6
Electromagnetic force, 6
Electron capture decay, 69
Electrons, 18, 68–69
Electron shells, 18
Elements, 18, 69–72
Elephants, 387–388, 389
Elk Point Basin, 228
Ellesmere orogeny, 211
Emperor Seamount-Hawaiian Island chain, 53
End moraines, 358, 360
Endosymbiosis, 179–180, 181
Endotherms, 260, 305–307
Energy resources, 227–228
Environment, humans' effect on, 5
Eoarchean, 149–150

INDEX **433**

Eocene
   See also Cenozoic Era
   climate during, 375–376
   mammals, 376, 379, 388, 389
Eocene Green River Formation, 343
Eomaia, 315
Eonothem, 93
Eons, 93
Epeiric seas, 191, 193, 239
   See also specific seas
Epifauna, 235, 236
Epiflora, 235, 236
Epochs, 93
Equisetum, 267
Equus, 387
Erathem, 93
Eukaryotic cells, 162–163, 178–181
Europe
   hominids, 407–409
   mammals, 391–392, 393
Eurypterids, 241
Eusthenopteron, 255, 257
Evaporite environments, 117
Evaporites, 28, 201–202, 204, 223
Even-toed hoofed mammals, 385–386
Evolution
   bird, 309–310
   cladistics, 135
   convergent, 133–134
   divergent, 133–134
   extinctions and, 137
   fossil evidence of, 122
   of greenstone belts, 154–155
   horse, 387–388
   human, 398, 401–409
   Lamarck's ideas on, 125–126
   macroevolution, 134–135
   of mammals, 310–312, 372
   microevolution, 134–135, 141
   mosaic, 135–137, 264
   organic, 11
   parallel, 133–134
   plant, 260–268
   rate of, 131–133
   vertebrate, 251
Evolutionary biology, 11
Evolutionary theory
   Darwin and, 123–124, 126, 128
   evidence supporting, 137–144
   genetics and, 128–129
   misconceptions about, 123–124, 128, 143
   missing links and, 143
   modern synthesis view of, 130–137
   natural selection and, 126, 128
   overview of, 124–129
   plate tectonic theory and, 55–56
   predictions from, 138
   speciation, 131–133
   variation, 130–131
Evolutionary trends, 135–137
Exoskeletons, 234
Extinctions, 137, 379
   See also Mass extinctions
Extraterrestrial, 11
Extrusive rocks, 23

F

Face of the Earth (Suess), 38
Facies, 81, 83
Farallon plate, 330–331, 335–336, 352
Fault-block basins, 279–281

Fauna. *See* Animals
Faunal succession, principle of, 89, 91
Felsic magma, 23
Fermentation, 162–163
Ferromagnesian silicates, 21
Fish
   See also specific types
   classification of, 254
   geologic ranges of, 252
   Mesozoic, 299–300
   Paleozoic, 251–257
Fission-track dating, 72–73
Five-kingdom classification, 179
Fixity of species, 123
Florida, fossils in, 391
Fluid activity, 30
Fluvial deposits, 109–111
Fluvial systems, 109–110
Flying reptiles, 307
Foliated metamorphic rocks, 30–31, 32
Foliated texture, 30–31
Footprint Tuff, 406
Foraminifera, 298, 373, 375
Fossil record, 88, 143
Fossils
   See also specific organisms, types of fossils
   Archean, 163
   body, 86, 88–89
   Burgess Shale, 232, 238–239
   Cenozoic, 372–373
   continental drift and, 40, 41
   dating techniques using, 88–89
   defined, 11, 85
   evidence for evolution in, 11, 122, 141–144
   formation of, 86–88
   guide, 94–95
   human, 398
   living, 136–137
   mammal, 372–373, 390–391
   Mazon Creek fossils, 210
   micofossils, 109
   Ordovician, 240
   plant, 374–375
   Proterozoic, 178, 182–184
   in sedimentary rocks, 109
   shelly, 233–234
   spores, 262
   tetrapod, 250
   trace, 86, 91, 104, 108
   uniformitarianism and, 87–88
Fossil succession, principle of, 89, 91, 94–97
Foster, C. B., 262
Four Corners region, 338–339
Fragmental texture, 23, 24
Franciscan complex, 283, 284
Franklin mobile belt, 191
Fungi, 179, 262
Fusulinids, 243–244, 245

G

Galápagos Islands, 123, 124
Garden of the Gods, 222
Gastropods, 373
Gemstones, 17, 184
Genes
   chromosomes and, 129–130
   discovery of, 126
   Mendel's experiments on, 128–129
Genesis, 124–125
Genetic drift, 131
Genetics, 128–130

Geologic columns, 92, 93
Geologic maps, 190
Geologic ranges, 96
Geologic record
   correlation and, 94–97
   fossils and, 85–91
   introduction to, 78
   relative geologic time scale and, 91–93, 97–98
   stratigraphy, 79–85, 93–94
Geologic systems, 91–93
Geologic time, 2, 11–12
   absolute dating methods, 68–74
   catastrophism and, 66
   changing concept of, 62–63
   climate change and, 63
   measurement of, 61–62
   neptunism and, 66
   relative dating and, 64–65
   uniformitarianism and, 66–67, 68
Geologic time scale, 12
   Cambrian, 91–92, 192
   Cenozoic Era, 322, 372, 374–375
   Precambrian, 147–148
   relative, 62, 91–93, 97–98
Geology
   debate in, 5
   defined, 4
   establishment of science of, 66–67
   historical. *See* Historical geology
   physical, 4
Ginkgos, 300
Glacial deposits, 113
Glacial environments, 111
Glacial evidence, for continental drift, 40
Glacial features, during Ice Age, 354
Glacial landforms, 357–358
Glacial melting, 367–368
Glacial stages, 354
Glaciation
   causes of, 362–366
   effects of, 357–361
Glacier National Park, 367
Glacier Peak, 367
Glaciers
   areas covered by Pleistocene, 354
   Carboniferous, 214
   continental, 348
   defined, 348
   formation of, 356–357
   isostasy and, 358–359
   of Late Paleozoic, 210–211
   Neoproterozoic, 176
   Paleoproterozoic, 175–176
   present-day, 366–368
   types of, 348, 349
   valley, 348
Glassy texture, 23, 24
Glauconite, 97
Global warming, 366–367
   See also Climate change
   extinctions and, 262–263
   geologic time and, 63
   problem of, 12
Globorotalia menardii, 355
Glossopteris flora, 38, 268
Glyptodonts, 88
Gneiss, 30
Gold, 20, 163, 291, 344, 368
Gold rush, 272–273
Gondwana, 38, 39, 192, 193, 210, 211
Gorillas, 402, 403
Gosse, Philip Henry, 78
Graded bedding, 104, 105
Grand Canyon, 60, 61, 81–83, 197–198

Grand Teton National Park, 153
Granite, 17
Granite-gneiss complexes, 152
Graptolites, 240, 241
Gravel, 368
Gravity, 6
Gravity-driven mechanisms, 54–55
Graywacke, 154
Grazers, 386
Great Lakes, 361
Great Plains, 337–339
Great Rift Valley of Africa, 48
Green algae, 262
Greenland, 274
Greenland Ice Sheet, 358, 367
Green River Formation, 383
Greenstone, 31
Greenstone belts, 152, 154–156, 168
Grenville orogeny, 171–172
Gressly, Armanz, 81
Guadalupe Mountains, 223
Guide fossils, 94–95
Gulf Coastal Plain, 340–341
Gulf of California, 46
Gulf of Mexico, 277, 281
Gymnosperms, 264–266, 268, 300, 373–374
Gypsum, 228

H

Half-lives, 69–72
Halite, 19
Heat, metamorphism and, 30
Helicoplacus, 238
Hemicyclaspis, 252
Herbivores, 235
Hercynian orogeny, 211
Hess, Henry, 42–43
Heterospory, 266
Heterotrophic organisms, 162
Himalayas, 50, 51, 326–328, 349, 363
Historical geology
 benefit of studying, 13
 defined, 4
 importance of sedimentary rocks in, 102
 interpretation of, 5
 rocks and, 33
 theory formulation and, 5–6
Holocene Epoch, 348, 349–352
Hominids, 400, 401–409
 Australopithecines, 404–405
 Cro-Magnons, 408–409
 human lineage, 407
 Neanderthals, 407–408
 stratigraphic record of, 403
Hominoids, 400–401
Homo erectus, 407
Homo habilis, 407
Homologous structures, 140–141
Homo sapiens, 407
Hoofed mammals, 385–387
Hornfels, 31
Horses, 387, 388
Hot spots, 52, 53
Hox genes, 233
Human evolution, 398, 401–409
Human history, geologic time and, 62–63
Human lineage, 407
Humans, effect of, on environment, 5
Hutton, James, 64–65, 66–67
Hydrocarbons, 291, 344
Hydrosphere, 2–4, 11, 158–159
Hyenas, 384
Hylonomus, 259
Hypotheses, 5
Hyracotherium, 387

I

Iapetus Ocean, 194, 204
Iberia-Armorica, 227
Ice, 17
Ice Age, 348, 354
 See also Pleistocene Epoch
 causes of glaciation in, 362–366
 mammals, 390–391
 onset of, 355–357
 sea level changes during, 358
Ice caps, 348
Ice-scoured plains, 357–358
Ice sheets. *See* Continental glaciers
Ice shelves, 367
Ichthyosaurs, 307, 310
Ichthyostega, 257, 258
Igneous activity
 in continental interior, 338
 Paleo- and Mesoproterozoic, 170
Igneous rocks, 23–26
 classification of, 23–26
 defined, 23
 texture and composition, 23–24
Illinoian glacial stage, 354
Inarticulate brachiopods, 237
Inclusions, principle of, 79–80
India, 274
Infauna, 235, 236, 299
Inheritance of acquired characteristics, 125–127
Inorganic matter, 20
Insectivores, 381
Intercontinental migrations, 393–394
Interglacial stages, 354
Interior Lowlands, 337–339
Intermediate bony fishes, 299
Intermediate magma, 23
Intertonguing, 81
Interval zones, 95
Intracontinental rift, formation of greenstone belt in, 156
Intrusive rocks, 23
Invertebrates
 See also specific types
 Cambrian, 232–233, 237–239
 Carboniferous, 243–244
 Cenozoic, 373
 changes in marine, during Paleozoic, 234–245
 Devonian, 241–243
 emergence of, 233–234
 feeding systems, 235–237
 major groups, 235
 mass extinctions of, 240–242, 244–245
 Mesozoic, 297–299, 297–299
 Ordovician, 239–240, 239–241
 Paleozoic, 231–246
 Permian, 243–244
 Silurian, 241–243
Ionic bonding, 19, 20
Ions, 19
Iran-Iraq War, 37
Iridium anomaly, 316–317
Iron ore, 17, 20, 164, 184, 185, 291
Isostasy, 358–359, 363
Isostatic rebound, 359, 361
Isotopes, 18, 68–69
Isthmus of Panama, 56

J

Jawed fish, 253–254
Jaws, mammal, 311–313
Joly, John, 64
Jovian planets, 7
Jupiter, 7
Jurassic Period, 274, 278, 280–281
 climate during, 314
 fish, 300
 mammals, 312
 paleogeography of North America during, 279
 sedimentation, 286–289

K

Kansan glacial stage, 354
Kaskaskia Sequence, 214–217
Kayenta Formation, 286
Kazakhstania, 192, 193, 211, 214
Kelvin, Lord, 68
Kerogen, 343
Kimberella, 184
Koch, Albert, 390
Komatiites, 154

L

La Brea Tar Pits, 88, 391, 392
Labyrinthodonts, 256, 258, 300
Laetoli, 406
Lake Agassiz, 359
Lake Bonneville, 359
Lake Manly, 359
Lake Missoula, 360–361, 364, 365
Lamarck, Jean-Baptise de, 125–126, 127
Laminations, 104
Lapilli, 24
Lapworth, Charles, 92
Laramide orogeny, 284, 329–332
Lassen Peak, 333
Late Paleozoic Era, 210–228
 Absaroka Sequence, 218–223
 evolution of North America during, 214–227
 Kaskaskia Sequence, 214–217
 mineral resources from, 227–228
 mobile belts during, 224–226
 paleogeography of, 210–213
Lateral continuity, principle of, 64–65, 81, 83
Lateral gradation, 81
Laurasia, 38, 226, 314
Laurentia, 192, 194, 211
 evolution of, 168–172
 Paleoproterozoic history of, 168–170
Lava, 23, 80, 97
 See also Volcanism
Lead, 206, 228
Leaves, 263
Lepidodendron, 267
Life
 See also specific life forms
 metabolism/reproduction criteria for, 159
 origin of, 159–163
 origins of human, 398
 plate tectonics and distribution of, 55–56

Life history, 11
　See also specific time periods
Lignin, 263
Limestone, 27–28, 103, 115–117, 228
Linnacus, Carolus, 138–140
Lithification, 26, 27
Lithosphere, 2–4, 10, 45
Lithostatic pressure, 30
Lithostratigraphic correlation, 94–95
Lithostratigraphic unit, 93
Little Ice Age, 347, 348, 356, 362, 366
Living cells, evolution of, 2
Living fossils, 136–137
Lizards, 308
Lobe-finned fish, 252, 254–255
Louann Salt, 291
Lovejoy, Owen, 404
Lower Silurian strata, 117, 118
Lucerne, Switzerland, 107
Lungfish, 255, 299
Lycopsids, 266–267
Lyell, Charles, 66, 67
Lysenko, Trofim Denisovich, 127
Lystrosaurus, 40

M

Macroevolution, 134–135
Madagascar, 377, 390
Mafic magma, 23
Magma (molten material), 10, 23, 36
Magnesium, 20
Magnetic anomalies, 43, 44, 52–53, 54
Magnetic field, 41–42
Magnetic poles, 41–42
Magnetic reversals, seafloor spreading and, 42–44
Magnetosphere, 157–158
Malthus, Thomas, 126
Mammal fossils, 372–373
Mammals
　See also specific types
　age of, 377–378
　archaic, 379
　Carnivora, 383–384
　Cenozoic, 377–394
　classification of, 378
　cynodonts and, 310–312
　elephants, 387–388
　evolution of, 296, 372
　extinctions, 379, 391–393
　Ice Age, 390–391
　intercontinental migrations, 393–394
　Neogene, 381–390
　origin and evolution of, 310–314
　Paleogene, 381–390
　placental, 378–381
　Pleistocene, 390–393
　primates, 383, 398–406
　small, 381–383
　ungulates, 385–387
　whales, 388–390
Mammal teeth, 378, 383, 384, 385–386
Mammoths, 388
Manatees, 143
Manganese, 17
Mantle, 2–4, 9, 10
Mantle plumes, 52, 351
Marble, 31
Marine deltas, 112
Marine environments, 110, 114–117, 235–236
Marine food systems, 235–237
Marine invertebrates. *See* Invertebrates

Marine life
　See also Fish; Invertebrates
　Cambrian, 237–239
　Carboniferous, 243–244
　Cenozoic, 372
　Devonian, 241–243
　Ordovician, 239–240
　Permian, 243–244
　Silurian, 241–243
Marine regressions, 81–85
Marine reptiles, 307–308
Marine transgressions, 81–85
Mars, 7
Marshall, James, 273
Marsupial mammals, 314, 315, 378, 394
Mass extinctions, 137, 240
　climate change and, 262–263
　Devonian, 241–242
　Mesozoic, 294, 296, 316–317
　Ordovician, 240
　Paleozoic, 296
　Permian, 244–245
　Pleistocene, 391–393
　possible causes of, 242
Mass spectometer, 70
Mastodons, 388
Matter, composition of, 18–19
Mazon Creek fossils, 210
Meandering streams, 109–110
Meiosis, 129, 131
Mélange, 51
Mendel, Gregor, 128–129
Mercury, 7
Merychippus, 387
Merycodus, 380
Mesoproterozoic
　orogenic activity, 171–172
　sedimentation, 172
Mesosaurus, 40, 41
Mesozoic Era
　amphibians, 299–300
　birds, 309–310
　climate changes during, 276–277
　Earth history during, 272–292
　fish, 299–300
　geologic events summary, 293
　invertebrates, 297–299
　major evolutionary and geologic events of, 318
　mammals, 310–314
　mass extinctions of, 242, 296, 316–317
　mineral resources from, 291
　North America in, 277–291
　paleobiogeography, 314–315
　paleogeography of, 274–275
　Pangaean breakup during, 273–277
　plants, 296, 300
　reptiles, 296, 300–308
　sedimentation, 284–289
　tectonics, 282–284
Messel fossils, 372
Metamorphic rocks, 29–32
Metamorphism, 29–30
　about, 29–30
　effects of, on radiometric dating, 72
　types of, 30
Meteorite hypothesis, 317
Methane hydrate, 344
Miacids, 383–384
Michigan Basin, 201, 202–203
Micrite, 115
Microcontinents, 192
Microevolution, 134–135, 141
Microfossils, 109

Microplates, 226–227
Microspheres, 159
Mid-Atlantic Ridge, 38, 42, 47
Midcontinent rift, 172, 173
Migmatites, 31
Migrations, intercontinental, 393–394
Milankovitch, Milutin, 363
Milankovitch theory, 363–366
Milky Way Galaxy, 6
Miller, Stanley, 160
Minerals
　accessory, 22
　Archean, 163–164
　carbonate, 21
　Cenozoic, 343–344
　composition of, 20
　defined, 17, 20
　distribution of, 57–58
　groups of, 21
　Late Paleozoic, 227–228
　Mesozoic, 291
　Paleozoic, 206
　Pleistocene, 368
　Proterozoic, 184–185
　rock-forming, 22–23
　silicate, 21
　uses of, 17
Miocene
　birds, 377
　Hominoids, 383
　mammals, 379, 380–381, 386
Miocene Dove Springs Formation, 321
Missing links, 143
Mississippian period, 217, 300
Mitosis, 130, 131
Mnomers, 160
Moas, 377
Mobile belts, 191, 224–226
Mobile organisms, 235
Modern synthesis, on evolutionary theory, 130–137
Moenkopi Formation, 286
Moeritherium, 388
Mojave National Preserve, 353
Molarization, 386
Molars, 378
Molds, 88, 90, 91
Mollusks, 297
Monera, 179
Mongooses, 384
Monkeys, 400
Monotremes, 314, 378
Moon
　formation of, 8–9
　interaction between Earth and, 149
Moraines, 111, 358, 360
Morrison Formation, 288–289
Mosaic evolution, 135–137, 264
Mosasaurs, 307–308
Mountain building
　mobile belts, 191
　during Paleozoic Era, 190
　plate tectonics and, 55
Mountain glaciers, 348
Mountain ranges
　See also specific mountains
　intracratonic, 221–222
Mount Bachelor, 353
Mount Pinatubo, 36
Mount St. Helens, 333, 367
Mud cracks, 104, 108
Mudrocks, 27
Mudstone, 27
Multicelled organisms, first, 180–183

Murchison, Roderick Impey, 91, 92–93
Mutagens, 131
Mutations, 130–131

N

Nappes, 326–327
Native elements, 20
Natural gas, 227, 228, 291, 344
Natural resources
    See also Minerals
    plate tectonics and distribution of, 57–58
Natural selection, 11, 126, 128
    extinctions and, 137
    variation and, 130–131
Nautiloids, 297
Navajo Sandstone, 102, 117–118, 286, 288
Nazca plate, 49–50
Neanderthals, 407–408
Nebraskan glacial stage, 354
Negaunee Iron Formation, 179, 181
Nekton, 235, 236
Neogene Period, 322, 323
    mammals, 381–390
    mineral resources from, 343–344
    rifting, 325
Neoglaciation, 356
Neoproterozoic
    animals, 182–184
    fossils, 182–184
    glaciers, 176
    sedimentation, 172
Neptune, 7
Neptunism, 66
Neutrons, 18, 68–69
Nevadan orogeny, 283–284
Newark Group, 279–280, 281
New England, 193
New World monkeys, 400
New Zealand, 377, 390
Nickel, 184
Noble gases, 19
Nonconformity, 80–81
Nonferromagnesian silicates, 21
Nonfoliated metamorphic rocks, 31, 32
Nonfoliated texture, 30–31
Nonvascular plants, 262
North America
    continental interior, 278–279, 337–339
    cratonic sequences of, 195
    Cretaceous paleogeography of, 280
    easter coastal region, 279–280
    glacial distribution in, 354
    Gulf Coastal region, 280–282
    Jurassic paleogeography of, 279
    Late Paleozoic evolution, 214–227
    mammals, 379–380, 382, 386, 390–391, 393–394
    Mesozoic, 277–290
    Paleozoic evolution of, 194–205
    rifting of South America and, 274
    Sauk Sequence, 195–198
    southern and eastern margins, 339–343
    Tippecanoe Sequence, 198–204
    Triassic paleogeography of, 278
    western region, 282–290
North American Cordillera, 328–337
    Basin and Range Province, 333–334
    Colorado Plateau, 334–335
    igneous activity, 332–333
    Pacific Coast, 335–337
    Rio Grande rift, 335

Nuclear force, 6
Nucleaus, 68–69

O

Obsidian, 23, 24, 26
Occipital condyle, 312
Ocean basins, age of, 43, 44
Ocean circulation patterns, effect of Pangaean breakup on, 276–277
Oceanic-continental plate boundaries, 49–50
Oceanic crust, 10
    formation of, 46–47, 323, 325
    seafloor spreading and, 42–44
Oceanic-oceanic plate boundaries, 48–49
Ocean salinity, 64
Odd-toed hoofed mammals, 385, 386–387
Oil, 37
Oil shale, 343
Old Red Sandstone, 226
Olduvai Gorge, 397
Old World monkeys, 400
Oligocene
    climate, 379
    mammals, 379–380, 389
    plants, 376
    primates, 383
Omnivores, 383
On the Origin of the Species (Darwin), 11, 126, 128, 141, 233
Ooids, 116
Oolitic limestone, 116
Ooze, 115, 116
Ophiolites, 51, 55, 156, 173, 175
Orangutans, 401
Ordovician Period, 193
    marine community, 239–240
    paleogeography of North America during, 200
Ordovician strata, 118
Organelles, 179
Organic evolution, 2, 11
Organic reefs, 199–202
Organisms
    See also specific types
    anaerobic, 162
    Archean, 159–163
    autotrophic, 162, 236
    burrowing, 299
    classification of, 138–140, 179
    evolutionary history of, 135–136
    heterotrophic, 162
    multicelled, 180–183
    oldest known, 162–163
    plate tectonics and distribution of, 55–56
    variation in, 130–131
Original horizontality, principle of, 64–65
Origin of Continent and Oceans, The (Wegener), 38
Origin of life, 159–163
    experimental evidence and, 159–161
    submarine hydrothermal vents and, 161–162
Ornithischian dinosaurs, 304–305, 306
Ornithopods, 304–305
Orogenic activity, 168–670, 171–172
Orogenic belts, 325–328
Orogenies, 55, 290, 325
    See also specific orogeny
Orogens, 168, 325
Orrorin tugenensis, 403
Osteichthyes, 252
Ostracoderms, 252–253, 257
Ouachita mobile belt, 191, 224, 225

Ouachita Mountains, 214
Ouachiya orogeny, 211
Our Wandering Continents (du Toit), 38
Outcrops, 79
Outgassing, 158
Outwash, 111, 358, 360, 368
Owen, Richard, 296
Oxygen isotope ratios, 355–356
Ozone layer, 177

P

Pachycephalosaurs, 305
Pacific Coast, 335–337
Pacific-Farallon ridge, 337
Pacific Ocean, 323
Painted Desert, 287
Paleobiogeography, Mesozoic, 314–315
Paleocene
    climate, 374–375
    mammals, 378–379
    plants, 375
    primates, 383
Paleocene-Eocene Thermal Maximum, 375
Paleocene rocks, 374–375
Paleoclimatology, 355–356
Paleocurrents, 105
Paleogene Period, 322, 323
    mammals, 381–390
    mineral resources from, 343–344
Paleogeographic maps, 119, 191–192
Paleogeography, 119
    See also specific periods
Paleomagnetism, 41–42
Paleontology
    Anning's contribution to, 309
    defined, 123
Paleoproterozoic, 151, 175–176
Paleozoic Era
    Absaroka Sequence, 218–223
    amphibians, 257–258
    Cambriab explosion, 232–233
    Carboniferous Period, 211
    continents during, 190–194, 210–213
    Devonian Period, 211
    evolution of North America during, 194–205, 214–227
    extinctions in, 296
    fish, 251–257
    introduction to, 190
    invertebrates, 231–246
    Kaskaskia Sequence, 214–217
    Late, 210–228
    major evolutionary and geologic events of, 246, 269
    mineral resources from, 206, 227–228
    mobile belts during, 224–226
    paleogeography of, 191–194, 210–213
    Permian Period, 211
    plants, 260–268
    reptiles, 258–260, 261
    vertebrates, 250–260
Palouse Falls, 365
Palynology, 262–263
Palynomorphs, 262
Panderichthys, 257
Pangaea, 38, 172, 211, 226–227
    breakup of, 273–277
    during Cenozoic Era, 323
Pannotia, 174, 190
Panthalassa, 211
Parallel evolution, 133–134

INDEX   **437**

Parent-daughter isotope pairs, 69–72
Parent elements, 69–72
Peat, 368
Pegmatites, 164, 184
Pelagic clay, 115, 116
Pelagic organisms, 235
Pelycosaurs, 259–260
Penguins, 377
Pennsylvanian Period, 210, 221–222, 266, 266–267
Periods, 93
Perissodactyla, 385–387
Permafrost, 391
Permian Delaware Basin, 228
Permian Period, 211, 222–223
    flora of, 266–268
    marine communities, 243–244
    mass extinctions of, 244–245, 262–263
    paleogeography of North America during, 222
    Pangaea during, 276
Persian Gulf region, 37, 291
Peru-Chile Trench, 50
Perunica, 227
Petrified Forest National Park, 286, 287
Petroleum, 37, 57, 227, 291, 344
Phaneritic texture, 23, 24
Phanerozoic Eon, 166, 245
Phosphorus, 343
Photochemical dissociation, 158
Photosynthesis, 158
Phyletic gradualism, 132
Phyllite, 30
Phylogenetic trees, 135
Phylogeny, 136
Physical geology, defined, 4
Phytoplankton, 235, 236, 297–299, 373
Pictured Rocks National Lakeshore, 196
Pillow lavas, 46
Pinnacle reefs, 201
Placental mammals, 314, 378, 378–381
Placentas, 378
Placer gold deposits, 273, 291, 368
Placoderms, 252, 254, 257
Planetismals, 6–7, 8
Planets
    Cenozoic, 373–376
    formation of, 7–8
    Jovian, 7
    terrestrial, 7
Plankton, 235, 236
Planktonic foraminifera, 298
Plantae, 179
Plants
    Carboniferous, 266–268
    Cenozoic, 372
    Devonian, 263–266
    Mesozoic, 296, 300
    nonvascular, 262
    Paleozoic, 260–268
    Permian, 266–268
    Silurian, 263–266
    vascular, 261–263, 265, 300
Plastic flow, 348
Plate boundaries
    continental-continental, 50
    convergent, 46, 48–51
    divergent, 46–47
    hot spots, 52, 53
    oceanic-continental, 49–50
    oceanic-oceanic, 48–49
    transform, 46, 51–52
    types of, 46
Plate movements, 10–11
    calculating rate of, 52–53
    determination of, 52–53
    driving mechanism of, 53–55
    influence of, 36
    mechanics of, 45
Plates, defined, 10, 45
Plate tectonics, 2, 3
    Archean, 154, 156–157
    Cenozoic, 323–325, 349–352
    convergent, 48–51
    distribution of natural resources and, 57–58
    driving mechanism, 53–55
    distribution of life and, 55–56
    Earth systems and, 11
    Holocene, 349–352
    Mesozoic, 282–284
    mountain building and, 55
    organic evolution and, 11
    Pleistocene, 349–352
    rock cycle and, 31–32
    terranes, 55
    theory, 6
Plate tectonic theory, 10–11
    consequences of, 36
    continental drift and, 36–41
    evolutionary theory and, 55–56
    magnetic reversals and, 42–44
    paleomagnetism and, 41–42
    plate boundaries and, 45–52
Platforms, 150, 191
Platinum, 163–164, 184
Playa lakes, 110
Pleistocene Epoch, 348
    animals, 372, 390–393
    causes of glaciation in, 362–366
    extinctions during, 391–393
    mineral resources from, 368
    onset of, 355–357
    stratigraphy, 352–355
    tectonism and volcanism during, 349–352
    volcanism, 353
Plesiosaurs, 307, 310
Pliocene, 380–381, 386
Pliohippus, 380
Plutonic rocks, 23
Pluvial lakes, 359–361
Point bar deposits, 110
Polar wandering, 42
Politics, oil and, 37
Pollen, 262, 263, 357
Pollen analysis, 356
Polymerization, 160
Polymers, 160
Polyploidy, 133
Populations
    speciation, 131–133
    varriation in, 130–131
Porphyritic texture, 23, 24
Porphyry, 23
Powell, John Wesley, 60, 61
Prebiotic stages, 159
Precambrian
    Archean eon, 147–164
    continental foundations, 150–156
    Eoarchean, 149–150
    geologic time scale, 147–148
    overview of, 148–149
    Proterozoic, 166–186
    subdivisions, 147–148
Precambrian rocks, 151–152
Precambrian shield, 150
Precambrian System, 93
Precious metals, 17, 163–164
Prediction, 138
Prehistoric overkill, 391, 392–393
Premolars, 378
Primary producers, 236, 299
Primates, 383
    anthropoids, 400–401
    classification of, 399
    defined, 398–399
    hominoids, 400–401
    prosimians, 399
Primitive bony fishes, 299
Principle of cross-cutting relationships, 64–65, 97
Principle of fossil succession, 89, 91
Principle of Geology (Lyell), 67
Principle of inclusions, 79–80
Principle of lateral continuity, 64–65, 81
Principle of original horizontality, 64–65
Principle of superposition, 64–65, 79–80
Principle of uniformitarianism, 12–13
Proconsul, 401
Proglacial lakes, 359–361
Progrades, 112
Prokaryotic cells, 162–163, 178, 180
Prosimians, 398, 399
Proterozoic Eon, 166–186
    atmosphere during, 176–177
    continental evolution, 168–172
    glaciation during, 175–176
    introduction to, 167–168
    life history events during, 178–184
    mineral resources from, 184–185
    supercontinents, 172–175
Proterozoic rocks, 167–168, 172, 174
Protistans, 179
Protoavis, 310
Protobionts, 160
Protoceratops, 304
Protons, 18, 68–69
Protorothyrids, 259
Prototheria, 378
Pteranodon, 308
Pteraspis, 252–253
Pterodoctylus, 308
Pterosaurs, 296, 307
Ptroteinoids, 160
Pumice, 26
Punctuated equilibrium, 132–133
Pyramids of Giza, 373
Pyroclastic materials, 23
Pyroclastic texture, 23

Q

Quartz, 17, 20
Quartzite, 31
Quartz sand, 103
Quartz sandstone, 27
Quaternary Period, 322–323, 348
Queen Charlotte fault, 336–337
Queenston Delta, 205

R

Rabbits, 381
Radiation, background, 6
Radioactive decay, 69–72
Radioactive isotopes, 68–69, 70–72
Radioactivity, discovery of, 68
Radiocarbon dating, 73–74
Radiometric dating, 62, 69–72
Range zones, 94–95, 96
Ray-finned fish, 252, 254, 255

Recent Epoch, 348
Recessional moraines, 358
Recrystallization, 30
Red Sea, 46, 48, 325
Redwall Limestone, 102
Reef development, 199–202
    Carboniferous Period, 243
    Cenozoic Era, 373
    Cretaceous Period, 281–282
    Devonian Period, 241042
    Mesozoic Era, 297–298
    Ordovician Period, 240
    Permian Period, 223, 243–244
    Silurian period, 241
    in Western Canada, 214–215
Regional metamorphism, 30
Relative dating, 61–62
    fossils and, 89
    fundamental principles of, 64–65
    methods, 64–65
Relative geologic time scale, 62, 91–93, 97–98
Relativity, theory of, 6
Reptiles
    See also specific types
    archosaurs, 301
    crocodiles, 308
    dinosaurs, 301–307
    evolution of, to birds, 309–310
    evolution of mammals from, 310–312
    flying, 307
    lizards, 308
    marine, 307–308
    Mesozoic, 296, 300–308
    Paleozoic, 258–260, 261
    snakes, 308
    stem, 301
    in Triassic, 279, 281
    turtles, 308
Resurgent domes, 351
Rhipidistian crossopterygians, 256, 257
Rhipidistians, 255
Rhizomes, 264
Ridge-push mechanism, 54–55
Rifting
    ancient, 47, 49
    Cenozoic, 323, 325
    greenstone belts and, 156
Rio Grande rift, 335, 336
Ripple marks, 104, 106
Rock cycle, 22–23, 26, 31–32
Rock-forming minerals, 22–23
Rock gypsum, 28, 103, 206
Rocks
    See also Sedimentary rocks
    defined, 17
    igneous, 23–26
    metamorphic, 29–31
    relative ages of, 85
    uses of, 17
Rock salt, 28, 103, 206
Rocky Mountains, 331–332
Rodents, 381
Rodinia, 173–174, 175
Roots, 263
Rounding, 103
Rudists, 297–298, 373
Ruminants, 386

S

Sahara Desert, 193
Sahelanthropus tchadensis, 398, 403

Salt deposits, 368
Salt domes, 291
San Andreas fault, 51–52, 336–337, 349
Sand dunes, 110
Sandstone, 27, 102, 117–118, 206, 286
Sandstone-carbonate-shale assemblages, 170
Sangamon interglacial stage, 354
Saturn, 7
Sauk Sequence, 195–198, 204
Saurischia, 303–304
Sauropods, 304
Savannah conditions, 376
Scablands, 361
Schistose foliation, 30
Schistosity, 30
Scientific debate, 5
Scientific method, 5
Scientific theories. *See* Theories
Scleractinians, 298
Seafloor, 115
Seafloor spreading, 42–44
Sea level, 84–85
    changes in, 191, 358
    glaciers melting and, 367–368
    rise in, 274, 277
Sea lions, 384
Seals, 384
Sedgwick, Adam, 91
Sediment, 26
Sedimentary breccia, 27
Sedimentary facies, 81, 83
Sedimentary rocks, 26–29
    See also Stratigraphy
    composition and texture, 103
    depositional environments, 109–118
    determining absolute age of, 97
    determining age of deformed, 107–108
    determining relative age of, 80
    detrital, 26–28
    fossils in, 109
    geometry of, 108–109
    in greenstone belts, 152, 154
    importance of, in historical geology, 102
    lithification, 26, 27
    paleogeography and, 119
    properties, 102–109
    Proterozoic, 172, 174
    types of, 26–29
    vertical stratigraphic relationships, 79–81
Sedimentary structures, 104–108
Sedimentation, Mesozoic, 284–289
Sediment-deposit feeders, 235
Sediment transport, 26
Seedless vascular plants, 262, 263–264, 265, 300, 373–374
Seed plants, 264–266
Seeds, 264
Sequence stratigraphy, 195
Sessile organism, 235
Sevier orogeny, 284, 286
Sex cells, 129–130
Shale, 27
Sharks, 299
Sheet geometry, 109
Shelly fossils, 233–234
Shields, 191
Shinarump Comglomerate, 286
Shiprock, New Mexico, 338–339
Shorelines movements, during transgression and regressions, 84–85
Siberia, 192, 193, 211, 214
Sierra Madre, 332
Sierra Nevada, 283, 334
Sigillaria, 267

Silica, 20, 23
Silica sand, 206, 228, 368
Silicate minerals, 21
Silica tetrahedron, 21
Siliceous ooze, 115, 116
Silicon, 19
Sills, determining relative age of, 80
Siltstone, 27
Silurian Clinton Formation, 206
Silurian Period, 193, 194
    flora of, 263–266
    marine communities, 241–243
    paleogeography of North America during, 202
Silurian System, 91–92
Sinodelphys, 315
Sivapithecids, 401
Skeletonized animals, 233–234
Slab-pull mechanism, 54–55
Slate, 30
Sloss, Laurence, 194
Smith, William, 4, 89, 189, 190
Snake River Plain, 332
Snakes, 308
Sodium, 19, 20
Solar nebula theory, 6–8
Solar system
    origin of, 6–8
    place of Earth in, 8
Sonomo orogeny, 278, 282–283
Sorting, 103
South Africa, 162, 184
South America, 393–394
    animals, 314
    in Jurassic, 314
    mammals, 386, 390, 393
    rifting of North America and, 274
Space—time continuum, 6
Speciation, 131–133
Species, defined, 131–132
Sphenopsids, 267
Sphynx, 373
Spores, 262, 266
Spreading ridges. *See* Divergent boundaries
Sprigg, R. C., 182
Stalagmites, 63
Stars, life cycle of, 6
Steel making, 185
Stegosaurus, 305
Stem reptiles, 301
Steno, Nicolas, 64, 81
Stillwater Complex, 163–164
Stössel, Iwan, 250
St. Peter Sandstone, 198–199
Strata/stratum, 79
    comformable, 80
Stratification, 104–105
Stratigraphic traps, 344
Stratigraphic units, 93–94
Stratigraphy, 79–85
    deep-sea, 354–355
    facies, 81
    marine transgressions and regressions, 81–85
    Pleistocene, 352–355
    sequence, 195
    terminology, 93–94
    terrestrial, 354
    vertical stratigraphic relationships, 79–81
Stratotypes, 93
Stromatolites, 162, 176
Strong nuclear force, 6
Structural traps, 344
Subduction, 48–49
Subduction complex, 48–49
Submarine hydrothermal vents, 161–162

Subsystems, interaction among Earth's, 2–4
Sudbury Basin, 184
Suess, Edward, 38
Sundance Sea, 286, 288
Supercontinents, 172–175
Superposition, principle of, 64–65, 79–80
Supervolcanoes, 350–351
Suspension-feeding animals, 235
Sutter's Mill, 273
Symbiosis, 179–180
Synthetoceras, 380
Systems
   defined, 2
   Earth as, 2–5
   geologic, 91–93
   interaction among Earth's, 2–4

T

Taconic orogeny, 204–205
Tapeats Sandstone, 198
Taylor, Frank, 38
Tectonic activity. *See* Plate tectonics
Tejas epeiric sea, 340
Teleoceras, 380
Temperature gradient, between poles and tropics, 276
Tenaya Lake, 16
Terminal moraines, 358
Terranes, 55, 226–227, 290
Terrestrial planets, 7
Terrestrial stratigraphy, 354
Tertiary Period, 322, 348
Tethys Sea, 274, 325, 355, 373
Tetrapods, 249, 250
Texture, 103
Theories
   defined, 5
   formulation of, 5–6
   on origin of universe and solar system, 6–8
Theory of evolution
   Darwin and, 123–124, 126, 128
   evidence supporting, 137–144
   genetics and, 128–129
   misconceptions about, 123–124, 128, 143
   missing links and, 143
   modern synthesis view of, 130–137
   natural selection and, 126, 128
   overview of, 124–129
   predictions from, 138
   speciation, 131–133
   variation, 130–131
Therapods, 303–304
Therapsids, 260, 261, 310
Theria, 378
Thermal convection cells, 42–43, 53–55
Tidal flats, 112, 114
Tiktaalik roseae, 257–258
Till, 111
Time-stratigraphic correlation, 94–97
Time-stratigraphic units, 93–94
Time transgression, 82
Time units, 93–94
Tippecanoe Sequence, 198–204

Trace fossils, 86, 91, 104, 108
Transcontinental Arch, 196
Transform boundaries, 46
Transform faults, 51–52
Transform plate boundaries, 51–52
Transgression, 201
Transitional environments, 110, 111–114
Tree-ring dating, 73–74
Triassic fault-block basin, 47, 49
Triassic Period, 274, 276–277
   invertebrates, 297, 298–299
   mammals, 312
   paleogeography of North America during, 278
   sedimentation, 284–286
Triceratops, 305
Trilobites, 237, 257
Trona, 343
Trophic levels, 236
Tullimonstrum gregarium, 209, 210
Tully Monster, 209, 210
Turtles, 306, 307, 308

U

Uintatherium, 380
Ultramafic magma, 23, 150, 154
Unaltered remains, 88–89, 91
Unconformities, 67, 80–81, 82
Ungulates, 385–387
Uniform, linear change, 70
Uniformitarianism, 12–13, 66–67, 68, 78, 87–88
United States
   See also North America
   coal deposits in, 227–228
   natural resources of, 185
Universe
   changing composition of, 6
   expansion of, 6
   origin of, 6
Uranium, 70, 291
Uranus, 7
U-shaped glacial troughs, 358
Ussher, James, 63

V

Valley glaciers, 348
Variation, in populations, 130–131
Varves, 111, 359–360
Vascular plants, 261–262, 262–263
   seedless, 262–265, 300, 373–374
Velociraptor, 304
Venus, 7
Vertebrates
   See also specific types
   amphibians, 257–258
   evolution of, 251
   fish, 251–257
   Paleozoic Era, 250–260
   reptiles, 258–260, 261
Vesicles, 23
Vesicular rocks, 23, 24

Vestigial structures, 140–141
Viruses, 159
Volcanic arc, 49
Volcanic breccia, 26
Volcanic glass, 26
Volcanic rocks, 23
Volcanism
   See also Lava
   Cenozoic, 349–352, 373
   climate change and, 366
   in Cordillera, 332–333, 334
   gas emissions, 158
   Holocene, 349–352
   Pleistocene, 349–352, 353
   supervolcanoes, 350–351
Voorhies, Michael, 382

W

Walcott, Charles, 162, 232
Wallace, Alfred, 126
Walruses, 384
Walther, Johannes, 83
Walther's law, 83
Warm-blooded dinosaurs, 305–307
Water vapor, 158, 159
Wave-formed ripple marks, 104, 106
Weak nuclear force, 6
Wegener, Alfred, 38, 39
Welded tuff, 24
Well cuttings, 109
Werner, Abraham Gottlob, 66
Westlothiana, 258–259
Whales, 388–390
Whewell, William, 66
Whitechuck Glacier, 367
Wilson, J. Tuzo, 55
Wilson cycle, 55, 168, 170
Wingate Sandstone, 286
Wisconsinan glacial stage, 354, 359
Wopmay orogen, 170
Worms, 299

Y

Yarmouth interglacial stage, 354
Yellowstone Calbera, 350–351
Yellowstone National Park, 350–351, 352
Yellowstone Plateau, 333
Yunnanozoon lividum, 251

Z

Zechstein deposits, 228
Zinc, 206, 228
Zion National Park, 102, 288
Zooplankton, 235, 236
Zuni epeiric sea, 337
Zuni Sequence, 278–279